# Human Blood Plasma Proteins
# Structure and Function

# Human Blood Plasma Proteins Structure and Function

**JOHANN SCHALLER, SIMON GERBER, URS KÄMPFER, SOFIA LEJON, CHRISTIAN TRACHSEL**

*Departement für Chemie und Biochemie*
*Universität Bern, Switzerland*

John Wiley & Sons, Ltd

Telephone (+44) 1243 779777

Email (for orders and customer service enquiries): cs-books@wiley.co.uk
Visit our Home Page on www.wileyeurope.com or www.wiley.com



***Other Wiley Editorial Offices***

John Wiley & Sons Inc., 111 River Street, Hoboken, NJ 07030, USA

Jossey-Bass, 989 Market Street, San Francisco, CA 94103-1741, USA

Wiley-VCH Verlag GmbH, Boschstr. 12, D-69469 Weinheim, Germany

John Wiley & Sons Australia Ltd, 42 McDougall Street, Milton, Queensland 4064, Australia

John Wiley & Sons (Asia) Pte Ltd, 2 Clementi Loop #02-01, Jin Xing Distripark, Singapore 129809

John Wiley & Sons Ltd, 6045 Freemont Blvd, Mississauga, Ontario L5R 4J3, Canada

Wiley also publishes its books in a variety of electronic formats. Some content that appears in print may not be available in electronic books.

***Library of Congress Cataloging-in-Publication Data***

Human blood plasma proteins : structure and function / Johann Schaller ... [et al.].
p. ; cm.
Includes bibliographical references and index.
ISBN 978-0-470-01674-9 (cloth : alk. paper)
1. Blood proteins. I. Schaller, Johann.
[DNLM: 1. Blood Proteins. 2. Blood Coagulation–physiology. 3.
Enzymes–blood. 4. Hormones–blood. 5. Peptides–blood. WH 400 H918 2007]
QP99.3.P7H86 2007
612.1′2–dc22

2007046618

***British Library Cataloguing in Publication Data***

A catalogue record for this book is available from the British Library
ISBN 9780470016749

Typeset in insert 9/11 pt Times by Thomson Digital, India
Printed and bounded by Markono Print Media Pte Ltd, Singapore

# Contents

# Preface

## *In memoriam Egon E. Rickli*

The endeavour to write a scientific textbook is a task that one cannot achieve without the help of many people with the required commitment and appropriate knowledge. Therefore, I would like to thank those people who have encouraged me to write this book and who have helped me to bring it to a successful end.

First of all, I would like to thank my four co-authors, Simon Gerber, Urs Kämpfer, Sofia Lejon and Christian Trachsel, all from the University of Bern, for their commitment and their invaluable contributions to this book. They did an excellent job and they were always present when I needed their help, advice and encouragement. It is mainly due to their effort that this book came into existence. In addition, I would also like to thank Dr Michael Locher for his help in the early phase of this book.

This book was written in memoriam of the late Prof Dr Egon E. Rickli, my supervisor and mentor and long-time friend.

The idea to write a book on human blood plasma proteins dates back to the end of the 1990s during the period when I was preparing a new lecture on this topic. The lack of a modern comprehensive textbook on structure and function of human blood plasma proteins convinced me over the years of the necessity to write such a book. In 2005 the University of Bern, Switzerland, granted me a sabbatical leave of six months, which actually gave me the opportunity to start this endeavour. I would like to thank the University of Bern and the Department of Chemistry and Biochemistry for their generous support over the years.

During my sabbatical leave I had the opportunity for a stay of nearly four months as a visiting scientist at the Institute for Protein Research (IPR), Osaka University, Japan. I would like to thank the IPR and especially its former director, Prof Dr Hideo Akutsu, for his generous support of this project and his encouragement.

Many thanks go to my colleague Prof Dr André Häberli, University of Bern, for leaving me the electronic version of his famous book: *Human Protein Data*. The data were of invaluable help for the preparation of this book.

I would also like to thank my colleague and friend Dr Stefan Schürch, University of Bern, for his support and encouragement. He always had time and an open heart when I needed it.

In addition, I would like to thank the Uchimura family, Sumiyo Uchimura, Yoshihito Uchimura and especially Dr Michitaka Uchimura, for their great hospitality, support and encouragement. Many things took their final shape under their roof in Kagoshima, Japan. Last but not least, I would like to thank my wife Shinobu Uchimura for her great heart, understanding and support. It was certainly not always easy for her during this period, especially when stress was rising.

And finally, I would also like to thank the whole team at John Wiley & Sons, Ltd for a great job.

*Johann Schaller*
*Bern, 31 May 2007*

# 1

# Introduction

The endeavour to write a textbook on the *Structure and Function of Human Blood Plasma Proteins* bears an inherent risk regarding the selection of the proteins to be discussed, because this selection will always be ambiguous depending on the applied definition of the term *blood plasma protein*. Based on the classification proposed in the 1970s by Frank W. Putnam in his famous book series *The Plasma Proteins: Structure, Function and Genetic Control*, in 2002 N. Leigh Anderson and Norman G. Anderson elaborated on the classification of human blood plasma proteins in a publication in *Molecular and Cellular Proteomics*, 'The human plasma proteome: history, character and diagnostic prospects' (for details see Chapter 3). The classification proposed by Anderson and Anderson clearly shows that human plasma contains the most comprehensive version of the human proteome. The complexity of the 'plasma proteome' is quickly understood when all the various forms of blood plasma proteins present in plasma are considered: precursor and mature forms, splice variants, degradation products and, of course, all combinations of posttranslational modifications. Of course, the scope and the available space in this book require a rigid selection of the blood plasma proteins discussed in some detail. The selection is primarily based on the classification introduced by Anderson and Anderson and on personal considerations of the authors.

The main sources of information used in this book including the references therein were the following:

1. *Databases*
   UniProt Knowledgebase (Swiss-Prot and TrEMBL)
   PROSITE: database of protein domains, families and functional sites
   Protein Data Bank (PDB): an information portal to biological macromolecular structures
   MIM: Mendelian Inheritance in Man (at the NCBI)

2. *Books*
   *Human Protein Data* (Haeberli, 1998)
   *Biochemical Pathways* (Michel, 1998)
   *Introduction to Protein Structure* (Branden and Tooze, 1999)
   *Biochemistry* (Voet and Voet, 2004)
   *Proteins: Structure and Function* (Whitford, 2005)

3. *Articles*
   'The human plasma proteome' (Anderson and Anderson, 2002)
   'The human plasma proteome' (Anderson *et al.*, 2004)
   'The human serum proteome' (Pieper *et al.*, 2003)
   'Exploring the human plasma proteome' (Omenn, 2005)

This book is divided into three parts:

*Part I*
Blood Components
Blood Plasma Proteins

*Human Blood Plasma Proteins: Structure and Function* Johann Schaller, Simon Gerber, Urs Kämpfer, Sofia Lejon and Christian Trachsel

*Part II*
- Domains, Motifs and Repeats
- Protein Families
- Posttranslational Modifications

*Part III*
- Blood Coagulation and Fibrinolysis
- The Complement System
- The Immune System
- Enzymes
- Inhibitors
- Lipoproteins
- Hormones
- Cytokines and Growth Factors
- Transport and Storage
- Additional Proteins

If not otherwise stated, the proteins discussed in this book are of human origin. The used protein name is the main name in Swiss-Prot and in some cases common synonyms are also given. All shown three-dimensional (3D) structures are from the Protein Data Bank (PDB). No model structures were included and the structures were either determined by X-ray diffraction or by nuclear magnetic resonance (NMR) spectroscopy. In a few cases nonhuman 3D structures are presented if the corresponding human 3D structure has not yet been determined. For enzymes the common EC classification system is given. Many domains, motifs and repeats or a certain stretch of sequence in a protein are characterised by a typical signature. In this book the signatures of the PROSITE database are used. In many cases diseases related to a certain protein are briefly mentioned and references to the disease database MIM (Mendelian Inheritance in Man) are given.

A limited number of references is given in each chapter and in the Data Sheets. Special emphasis was put on the quality of the references and the journals and their worldwide availability.

The information of each protein discussed in this book in some detail is summarised at the end of each chapter in a **Data Sheet**, where the most important data of each protein can be found at a glance. Proteins mentioned in the text but not discussed are compiled in the Appendix (Table A.1, human and Table A.2, nonhuman, with the corresponding reference to the database entry).

Each Data Sheet is divided into four sections:

**Fact Sheet**
**Description**
**Structure**
**Biological Function**

**Table 1.1** Useful universal resource locators (URL).

| URL | Description |
|---|---|
| www.expasy.org | ExPASy (Expert Protein Analysis System): proteomics server of the Swiss Institute of Bioinformatics (SIB) dedicated to protein analysis<br>Databases: UniProt (SwissProt + TrEMBL), PROSITE, SWISS-MODEL<br>Proteomics and Sequence Analysis Tools |
| pir.georgetown.edu | PIR (Protein Information Resource): integrated protein information resource for genomic and proteomic research<br>Databases: PIRSF, iProClass, iProLink |
| www.rcsb.org/pdb | RCSB (Research Collaboratory of Structural Bioinformatics): information portal to biological macromolecular structures<br>Database: PDB |
| www.ncbi.nlm.nih.gov | NCBI (National Center for Biotechnology Information): national resource for molecular biology information, USA |
| www.ebi.ac.uk | EBI (European Bioinformatics Institute): freely available data and bioinformatic services |

The Fact Sheet contains the following data:

| | |
|---|---|
| Classification: | Swiss-Prot |
| Abbreviations: | Most common |
| Structures/motifs: | Swiss-Prot and PROSITE |
| DB/PDB entries: | Swiss-Prot/Protein Data Bank |
| MW/length: | Mature Protein (without PTMs) |
| Concentration: | Approximate or range (if available), adult |
| Half-life: | Approximate or range (if available) |
| PTMs: | Swiss-Prot (most common) |
| References: | Key reference(s) on sequence and 3D structure |

In the section Description each protein is briefly described (usually by only one sentence).

In the section Structure the main structural features of a protein are briefly summarised and, if available, a short description of the 3D structure is given. If available, a figure of the 3D structure is included.

Finally, in the section Biological Function the main, usually physiological, functions are briefly summarised.

A limited number of useful universal resource locators (URLs) containing reliable protein data from curated and annotated databases are tabulated in Table 1.1.

## REFERENCES


Anderson and Anderson, 2002, The human plasma proteome, history, character and diagnostic prospects, *Mol. Cell. Proteomics*, **1**, 845–867.

Anderson *et al.*, 2004, The human plasma proteome. A nonredundant list developed by combination of four separate sources, *Mol. Cell. Proteomics*, **3**, 311–326.

Branden and Tooze, 1999, *Introduction to Protein Structure*, Garland Publishing, New York.

Haeberli, 1998, *Human Protein Data: Fourth Installment*, John Wiley & Sons, Ltd, Chichester, West Sussex.

Michel, 1998, *Biochemical Pathways: An Atlas of Biochemistry and Molecular Biology*, Wiley–Spektrun, New York.

Omenn, 2005, Exploring the human plasma proteome. The HUPO plasma proteome project (HPPP). *Proteomics*, **5** (13), 3223–3550.

Pieper *et al.*, 2003, The human serum proteome: display of nearly 3700 chromatographically separated protein spots on two-dimensional electrophoresis gels and identification of 325 distinct proteins. *Proteomics*, **3**, 1345–1364.

Putnam, 1975–1987, *The Plasma Proteins: Structure Function and Genetic control*, Academic Press, New York.

Voet and Voet, 2004, *Biochemistry*, John Wiley & Sons, Ltd, Chichester West Sussex.

Whitford, 2005, *Proteins: Structure and Function*, John Wiley & Sons, Ltd, Chichester West Sussex.

# PART I

# 2

# Blood Components

## 2.1 INTRODUCTION

Blood can be divided into two main parts, the blood plasma and the blood cells, and both will be discussed later in this chapter. Many medical terms contain hemo- or hemato as a prefix, derived from the Greek word *haima* (blood). Blood is distributed throughout an organism by a circulatory system and there exist three types of circulatory systems:

- No circulatory system exists for instance in flatworms (*Platyhelminthes*).
- An open circulatory system is present in many invertebrates like molluscs and arthropods. The circulatory fluid is called hemolymph and there is no distinction between blood and the interstitial fluid.
- The closed circulatory system is present in all vertebrates. The blood never leaves the blood vessels system, or cardiovascular system, which is composed of arteries, veins and capillaries.

## 2.2 SHORT HISTORY

Already in ancient times people realised that blood was a special liquid with special but somehow obscure and unexplainable properties. The Greeks had already realised the importance of blood and Hippocrates of Kos (ca. 460 BC–ca. 370 BC), often referred to as 'The Father of Medicine', proposed that diseases are based on an imbalance of the four humors, blood, phlegm, yellow bile and black bile, implying a physical cause of diseases and not a divine origin. He also introduced the Hippocratic Oath, a prescription of practices for physicians, which remains in use today. The Greek physician Galen (129–ca. 200), working in Rome, knew that the blood vessels carry blood and he identified venous (dark red) and arterial (brighter and thinner) blood and assigned them distinct and separate functions. Galen's observations, hypotheses and ideas and his triadic plan of physiology had a great influence on medicine for more than a thousand years until the Renaissance. Ibn-el-Nafis (1208–1288), born near Damascus, is thought to have been the first to describe accurately the blood circulation in the human body. The famous Italian Leonardo da Vinci (1452–1519), a genius of extraordinary originality, despite of his own extensive and thoroughly executed studies and observations, was not in the position to develop a new comprehensive theory of the circulatory system and remained with the principles of the Galenic school. The English physician William Harvey (1578–1657) is credited to have been the first to describe correctly and in great detail blood being pumped by the heart through the blood circulatory system.

Already in ancient times first experiments dealing with blood transfusion have been described, but with little success. Blood coagulation and fibrinolysis were unknown processes and in consequence substances with anticoagulant properties were not available. The first well-documented and successful blood transfusion was performed by the French physician Jean-Baptiste Denys (1625–1704) in 1667 using sheep blood and administering it to a 15 year old boy. At the beginning of the nineteenth century initial blood transfusion experiments from men to men were performed and the English physician James Blundell (1791–1878) was the first to administer a successful human-to-human transfusion (in 1818) by extracting blood from the patient's husband and then transfusing it to his wife. It took until the end of the nineteenth century before the first successful blood transfusions from vena to vena were carried out.

*Human Blood Plasma Proteins: Structure and Function* Johann Schaller, Simon Gerber, Urs Kämpfer, Sofia Lejon and Christian Trachsel

The observation of blood group substances and the description of the ABO blood group system (described later in this chapter) at the beginning of the twentieth century by the Austrian biologist and physician Karl Landsteiner (1868–1943) was a great step in blood research and in blood transfusion. Landsteiner was rewarded with the Nobel Prise in Physiology and Medicine in 1930 for his achievements. In addition, together with the American physician Alexander S. Wiener (1907–1976), Landsteiner was also the first to identify the Rhesus factor in 1937. The description of the anticoagulant properties of sodium citrate in 1914 led in 1915 to the first blood transfusions with conserved blood, which eventually saved many lives during the First World War.

## 2.3 BLOOD COMPONENTS

Blood is a very complex mixture of many types of components with very diverse properties and is often referred to as 'liquid organ'. Its distribution via the vascular system throughout the whole body is essential for the existence of the organism. Blood exhibits a wide variety of functions and the description of one type of component, the blood plasma proteins, is the topic of this book. The main functions of blood are the following:

- Transport system of many types of components:

  Blood cells, e.g. erythrocytes to transport oxygen
  Low molecular weight compounds, e.g. salts and ions
  High molecular weight compounds, e.g. proteins

- Defence system against hostile pathogens such as bacteria, virus and fungi, thus maintaining a balance between the organism and the environment:

  Specialised cells, e.g. lymphocytes, monocytes and granulocytes
  Antibodies, e.g. IgG
  Components of the complement system

- The wound sealing and wound healing system, life-saving precautions in the case of injuries:

  Blood cells, e.g. blood platelets
  Blood coagulation and fibrinolysis

- The balance of heat distribution throughout the body, thus guaranteeing a constant body temperature.

Blood can be separated by sedimentation into two main parts:

1. The *blood cells*, which are primarily synthesised in the bone marrow, represent approximately 45 % (by vol.) of entire blood.
2. The *blood plasma* or liquid portion of the blood represents approximately 55 % of the entire blood.

### 2.3.1 Blood cells

In a process termed *haematopoiesis* the adult blood cells are slowly differentiated under the influence of growth factors such as the colony-stimulating factor (CSF) and erythropoietin (see Chapter 14) from a homogeneous progenitor cell population, the *pluripotent stem cells*, into the various adult blood cells. In the fetal stage of development this process takes place primarily in the placenta, the fetal liver and in the bone marrow; in the adult stage this process is limited to the bone marrow. In the first stage of differentiation two types of precursor cells are developed, which are the progenitor cells of all subsequent cell types:

1. *Lymphoid precursor cells*, which are differentiated into the various types of lymphocytes such as B lymphocytes and T lymphocytes.
2. *Myeloid precursor cells*, which are further differentiated into erythrocytes, leukocytes and thrombocytes.

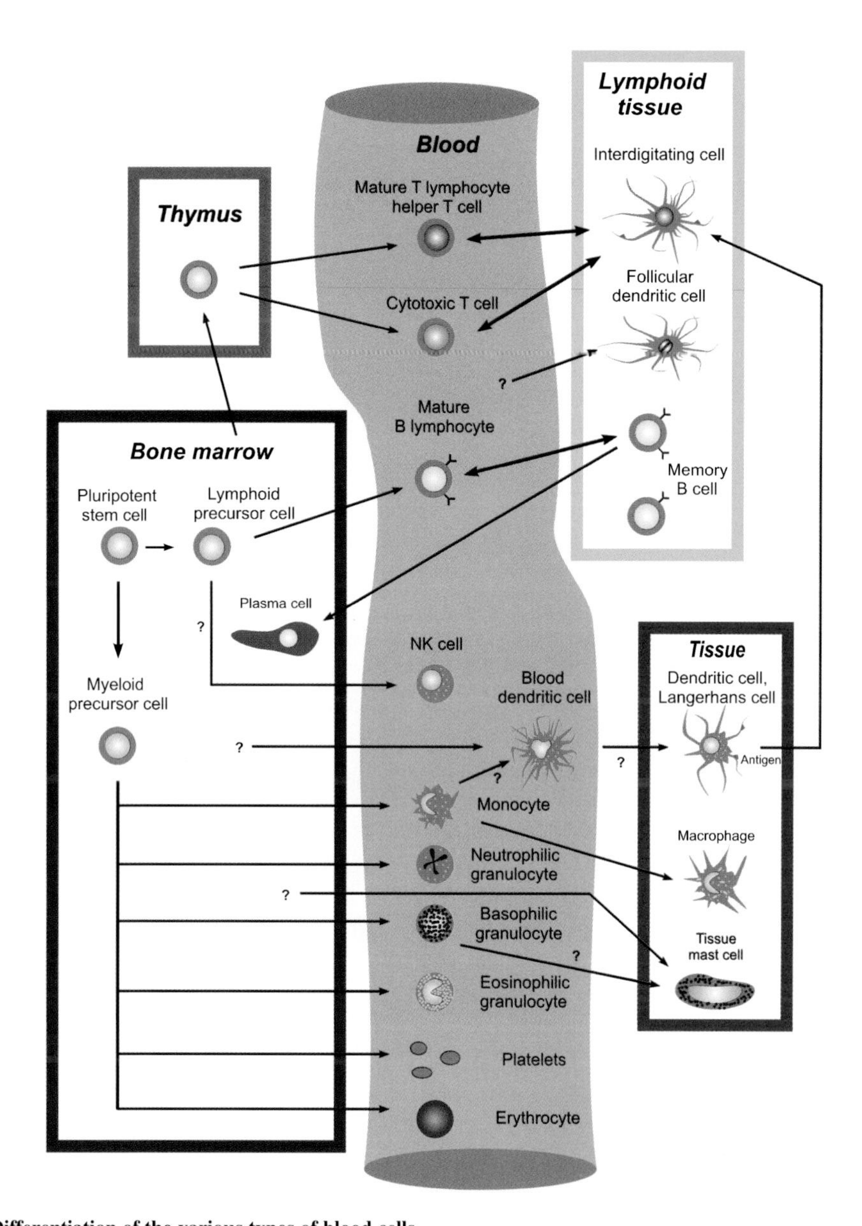


**Figure 2.1 Differentiation of the various types of blood cells**
The various types of blood cells are slowly differentiated from homogeneous pluripotent stem cells into lymphoid and myeloid precursor cells and finally into adult blood cells.

In Figure 2.1 the differentiation of the various blood cell types from precursor cells is depicted.
Blood contains three main types of cells:

1. **Erythrocytes** or **red blood cells.** The term is derived from two Greek words: *erythros* (red) and *kytos* (hollow). Erythrocytes are biconcave, flattened cells with a discoid shape containing no nucleus and have a diameter of approximately $7.5 \times 10^{-3}$ mm, a thickness of approximately $2 \times 10^{-3}$ mm and an expected lifetime of 100–120 days. Blood contains approximately $4\text{–}5 \times 10^{9}$ erythrocytes/ml of blood, representing approximately 96 % of all blood cells. In adults erythrocytes are produced in the bone marrow and are disposed in the spleen. Erythrocytes contain the protein

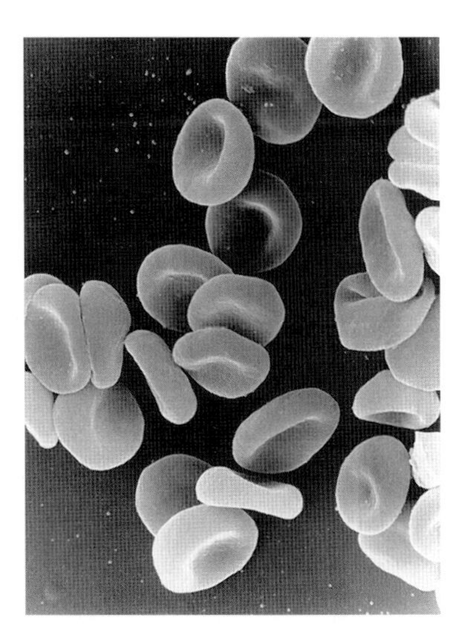

**Figure 2.2 Electron micrograph of erythrocytes**
Erythrocytes are biconcave, flattened cells with a discoid shape without a nucleus. Diameter: approx. $7.5 \times 10^{-3}$ mm; Thickness: approx. $2 \times 10^{-3}$ mm. Magnification: 7800. Zink *et al.* (1991) (with permission, Copyright available).

hemoglobin in vast amounts (see Chapter 15), responsible for the red colour of blood. Each erythrocyte carries approximately $3 \times 10^8$ hemoglobin molecules. Erythrocytes are the main carrier of oxygen from the lung to the peripheral tissues such as muscles, liver and intestines. Erythrocytes are quite easily deformed, thus enabling a safe passage through the narrow blood capillary system. An electron micrograph of erythrocytes is shown in Figure 2.2.

2. **Leukocytes** or **white blood cells.** The term is derived from two Greek words: *leukos* (white) and *kytos* (hollow). Blood contains approximately $4–8 \times 10^6$ leukocytes/ml of blood with an approximate size of $7–15 \times 10^{-3}$ mm, representing only approximately 3 % of all blood cells. Leukocytes are involved in the immune defence and produce antibodies on one side and are differentiated into memory cells on the other side. Three main subtypes exist:

   (a) **Granulocytes.** The expression is derived from the Latin word *granula* (granule). Granulocytes represent approximately 70 % of leukocytes and contain a nucleolus and their lifetime is only a few hours. Three types of granulocytes exist, termed according to their staining properties:

      (i) **Neutrophils** represent approximately 65 % of leukocytes and are usually the first to respond to bacterial infection and are involved in smaller inflammatory processes.
      (ii) **Eosinophils** represent approximately 4 % of leukocytes and primarily react against infections by parasites.
      (iii) **Basophils** represent <1 % of leukocytes and are mainly responsible for antigenic and allergic response by histamine release, thus causing inflammation.

   They are all polymorphic nuclear cells.

   (b) **Monocytes** or **macrophages** (once they have entered tissue). Monocytes are the largest leukocytes with a lifetime of one to two days and represent approximately 6 % of leukocytes. The term macrophage is derived from two Greek words: *macros* (large) and *phagein* (eat). They are large devouring cells, which ingest foreign cells, dead cells and cell debris by endocytosis.

   (c) **Lymphocytes.** Lymphocytes represent approximately 25 % of leukocytes and produce the antibodies, approximately 2000 molecules/second, and lymphocyte. There are three types of lymphocytes in blood:

      (i) **B cells** (B stands for the origin from the bone marrow) produce the antibodies and have a memory for previously ingested pathogens in the form of memory B cells and therefore their lifetime might reach several tenths of years.
      (ii) **T cells** (T stands for the origin from the thymus gland) play a central role in cell-mediated immunity. Several subtypes of T cells exist with distinct functions:

—**Cytotoxic T cells** ($T_c$ cells) or **CD8$^+$T cells** are involved in the destruction of virus-infected cells and tumour cells.
—**Helper T cells** ($T_h$ cells) or **CD4$^+$T cells** are mediators in the immune response.
—**Memory T cells** are the memory cells of the immune response and remain over a very long period after an infection and are either CD4$^+$ or CD8$^+$ cells.
—**Regulatory T cells** ($T_{reg}$ cells) or **suppressor T cells** are essential for keeping immune tolerance.
—**Natural killer T cells** (NKT cells) are cells that bridge the adaptive and the innate immune system.
—**$\gamma\delta$ T cells** are T cells containing a special T cell receptor (TCR).

(iii) **Natural killer (NK) cells** are able to kill cells carrying no signal for destruction, such as undetected virus-infected cells or cells that have turned cancerous.

Monocytes and lymphocytes are so-called *agranulocytes*, characterised by the absence of granules in the cytoplasm. The various types of leukocytes are shown in Figure 2.3. Neutrophils represent approximately 65 % of all leukocytes and exhibit a rather spherical shape (Figure 2.3(a)), with a diameter of $12–15 \times 10^{-3}$ mm. Macrophages are the largest leukocytes and represent approximately 6 % of all leukocytes (Figure 2.3(b)). Lymphocytes represent approximately 25 % of all leukocytes. B lymphocytes are shown as an example in Figure 2.3(c).

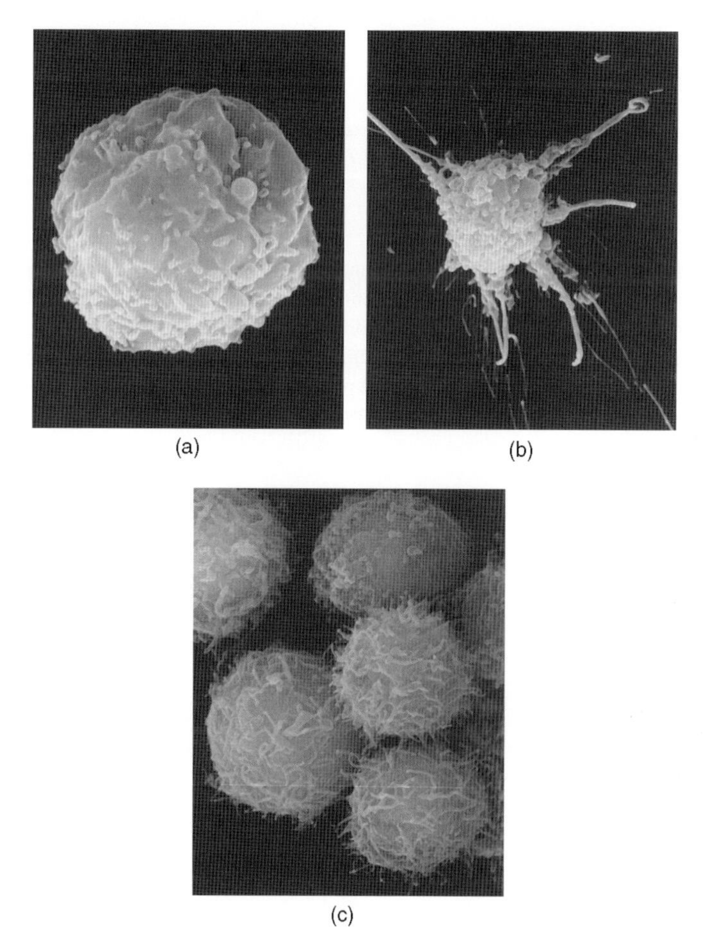


**Figure 2.3 Electron micrographs of various types of leukocytes**
(a) Neutrophil granulocyte: Neutrophils represent approx. 65% of all leukocytes and exhibit a rather spherical shape. Diameter: $12 - 15 \times 10^{-3}$ mm. Magnification: 19 000 ×. (b) Macrophage: Macrophages are the largest leukocytes and represent approx 6% of all leukocytes. Magnification: 10 000 ×. (c) B Lymphocytes: Lymphocytes represent approx. 25% of all leukocytes. Magnification: 13 000 ×. Zink *et al.* (1991) (with permission, Copyright available).

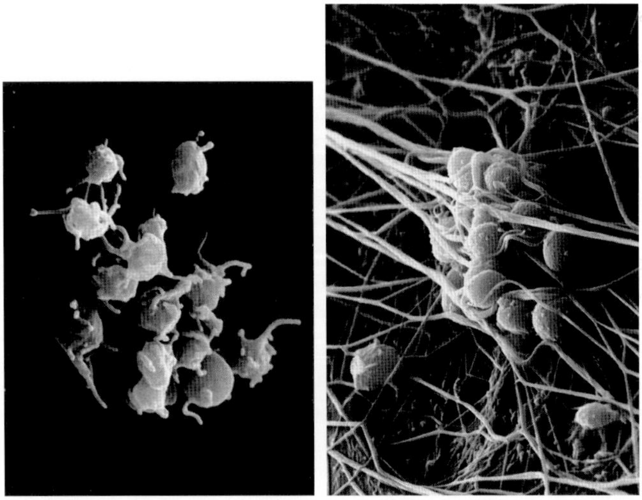

(a) (b)

**Figure 2.4 Electron micrographs of platelets**
(a) Activated platelets exhibit a nearly spherical shape exposing spines (pseudopodia). Magnification: 13 000 ×. (b) Spherical platelets are embedded in the fibrin network at the beginning of clot formation. Zink *et al.* (1991) (with permission, Copyright available).

3. **Thrombocytes** or **platelets.** Platelets contain no nucleus and exhibit a discoid shape with a diameter of approximately 1–$3 \times 10^{-3}$ mm and their life time is 8–10 days. Blood contains approximately 2–3 × $10^8$ platelets/ml of blood. Blood platelets are essential in the healing process of vascular injuries by closing the site of injury via intercalation with the fibrin network during the coagulation cascade (see Chapter 7). During this process activated platelets undergo a dramatic shape change from discoid to spherical, exposing spines termed pseudopodia (shown in Figure 2.4(a)), which become sticky, thus annealing the injured blood vessel. Figure 2.4(b) shows spherical platelets embedded in the fibrin network at the beginning of clot formation.

### 2.3.2 Blood plasma

The liquid part of blood, the blood plasma, represents approximately 55 % (by volume) of entire blood. There are many types of components dissolved in blood plasma, which exhibit very different functions. The components of blood plasma can be divided into several groups:

1. **Water.** Water is by far the main component of blood plasma, approximately 90 %.
2. **Mineral salts and ions.** There are many salts and ions dissolved in blood plasma, e.g. sodium chloride (at physiological concentration), buffer salts such as bicarbonate to guarantee a constant pH or metal ions such as calcium, copper or iron, which are essential in many biological processes and are contained in many blood plasma proteins, as will be seen later.
3. **Low molecular weight components.** Blood plasma contains many types of low molecular weight compounds: carbohydrates such as glucose and fructose, the whole set of amino acids, nucleotides such as ATP and cAMP, many vitamins, hormones, fatty acids, lipids and triglycerides, bile acids, urea and ammonia and many more components.
4. **High molecular weight components.** Peptides and (glyco)-proteins, oligosaccharides and polysaccharides, oligonucleotides and polynucleotides, e.g. DNA and RNA.
5. **Gases in soluble form.** Many gases such as oxygen, carbon dioxide and nitric oxide are dissolved in blood.
6. **Metabolites.** Blood plasma serves not only as the medium to transport the above-mentioned components but also the products of the metabolism.

### 2.3.3 The blood group system

At the beginning of the twentieth century Karl Landsteiner described blood group substances that evolved into the ABO blood group system. The cell surface of cells such as erythrocytes or endothelial cells of the blood vessels are composed of a *glycocalyx* containing glycolipids and glycoproteins with complex carbohydrate structures. The glycocalyx exhibits many functions such as protection from physical and chemical damage, reducing friction in blood flow and preventing loss of fluid through the blood vessel wall. The membrane of erythrocytes contains more than 100 blood group determinants, of which at least 15 are genetically different. However, there are mainly three systems in clinical use:

1. **The ABO blood group system.** The ABO blood group system contains three antigens: A, B and O(H). The blood group genes code for glycosyltransferases (EC 2.4):
   (a) A: *N*-acetylgalactoseamine transferase.
   (b) B: galactosyltransferase (this sequence differs from A at only four positions).
   (c) O(H): neither enzyme A or B, fucose is the terminal sugar moiety and is not antigenic. The O phenotype results from a single base deletion in the N-terminal region of the gene, leading to a frameshift.

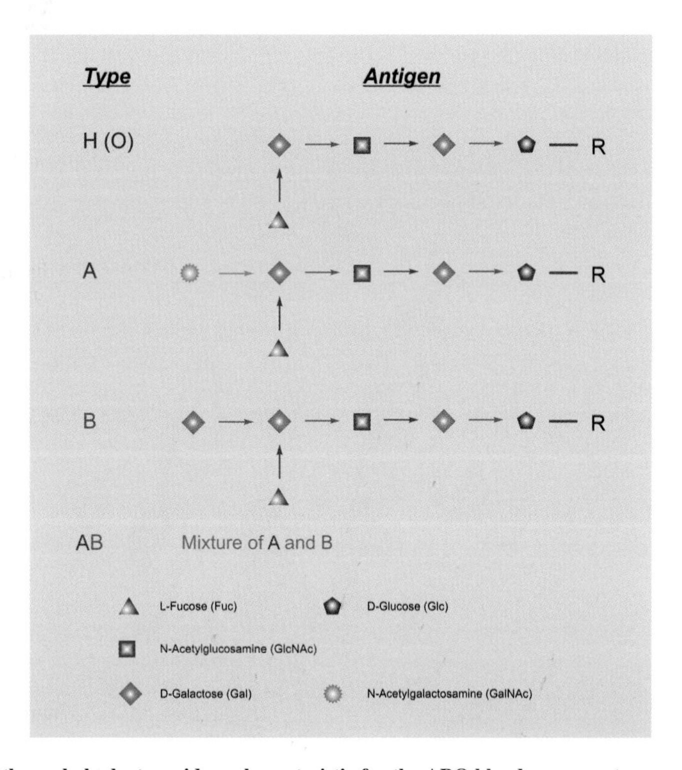


**Figure 2.5 Diagram of the carbohydrate residues characteristic for the ABO blood group system**
The O(H), A and B antigens are characterised by their terminal carbohydrate structures. Glc: Glucose; Gal: Galactose; Fuc: Fucose; NAcGlc: N-acetyl glucosamine; NAcGal: N-acetyl galactosamine.

The A and B antigens are derived from a common precursor, the O(H) antigen, and are converted into the A or B antigens by glycosyltransferases A and B. The ABO blood group antigens are characterised and distinguishable by the terminal carbohydrate residues depicted in Figure 2.5. The O(H) antigen has the basic CHO–structure:

–Glc–Gal–NAcGlc–Gal–Fuc

where Glc is glucose, Gal is galactose, NAcGlc is *N*-acetyl glucosamine and Fuc is fucose. In the A antigen an additional NAcGal (*N*-acetyl galactosamine) residue is attached to the penultimate Gal residue; in the B antigen it is an additional Gal residue. The AB blood group is a mixture of the A and B antigens.

The histo-blood group ABO system transferase (NAGAT) (see Data Sheet) contains two enzymes, the glycoprotein-fucosylgalactoside alpha-*N*-acetylgalactosaminyltransferase or A transferase (EC 2.4.1.40) and the glycoprotein-fucosylgalactoside alpha-galactosyltransferase or B transferase (EC 2.4.1.37), representing the basis of the ABO blood group system. The two glycosyltransferases transfer NAcGal and Gal from the corresponding sugar nucleotides UDP–NAcGal and UDP–Gal to the corresponding glycoprotein. NAGAT catalyses the two reactions:

$$\text{UDP–NAcGal} + \text{glycoprotein–Gal–Fuc} \leftrightarrow \text{UDP} + \text{glycoprotein–Gal–Fuc/NAcGal}$$

and

$$\text{UDP–Gal} + \text{glycoprotein–Gal–Fuc} \leftrightarrow \text{UDP} + \text{glycoprotein–Gal–Fuc/Gal}$$

The B transferase differs from the A transferase in four positions: aa 176: Arg → Gly; aa 235: Gly → Ser; aa 266: Leu → Met; aa 268: Gly → Ala. The last two variants are important for the specificity.

**Histo-blood group ABO system transferase** (NAGAT, EC 2.4.1.40) exists in two forms, a membrane-type form (354 aa: Met 1–Pro 354) and a soluble form (301 aa: Ala 54–Pro 354). The membrane form is a type II membrane glycoprotein containing a potential N-terminal 21 residue transmembrane segment (Gly 33–Met 53). The soluble form is derived from the membrane form by proteolytic processing and is devoid of a 53 residue N-terminal peptide and the conserved sequence DXD is involved in manganese binding as a cofactor (Asp 211 and 213).

The 3D structure of an N-terminally truncated soluble form of the glycosyltransferase A (292 aa: Val 64–Pro 354) was determined by X-ray diffraction (PDB: 1LZI) and is shown in Figure 2.6. The enzyme consists of an N-terminal (in green)

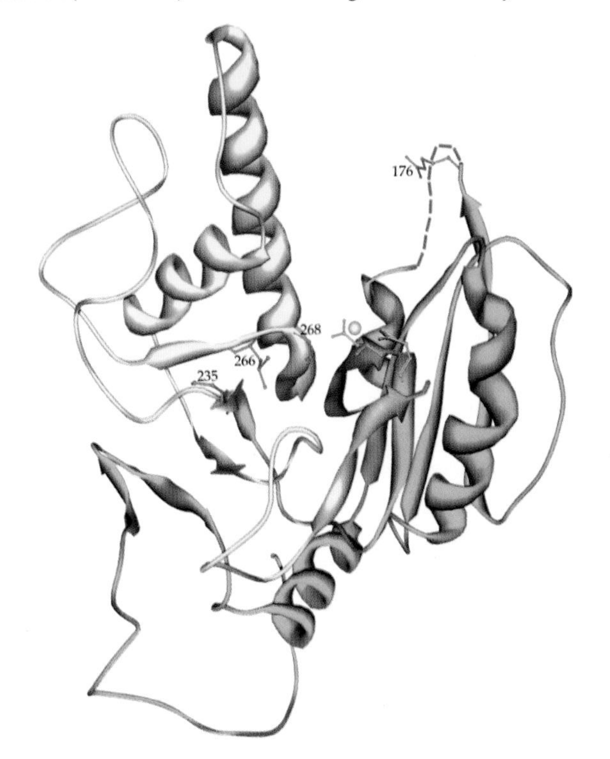


**Figure 2.6 3D structure of glycosyltransferase A (PDB: 1LZI)**
Glycosyltransferase A consists of an N-terminal (in green) and a C-terminal (in yellow) domain separated by a wide central cleft containing the active site residues Asp 211 and Asp 213 (in pink) coordinated to a $Mn^{2+}$ ion (orange ball). The four residues making the difference between the A and the B transferase are indicated in blue.

**Table 2.1** Distribution of blood groups in the ABO blood group system (%).

| Population | O | A | B | AB |
|---|---|---|---|---|
| Germans | 43 | 42 | 11 | 4 |
| French | 43 | 45 | 9 | 3 |
| Russians | 33 | 36 | 23 | 8 |
| Japanese | 30 | 40 | 20 | 10 |
| Kenyan | 60 | 19 | 20 | 1 |
| Native Australians | 44 | 56 | | |
| Native South Americans | 100 | | | |

and a C-terminal (in yellow) domain separated by a wide central cleft containing the active site residues Asp 211 and Asp 213 (in pink) coordinated to an $Mn^{2+}$ ion (orange ball). All four critical residues making the difference between the A and the B transferase are indicated in blue: Arg/Gly 176, Gly/Ser 235, Leu/Met 266 and Gly/Ala 268 (reconstructed region is shown in grey dashed lines).

2. **The Rhesus factor (Rh-factor).** In 1937 Karl Landsteiner and Alexander S. Wiener discovered the Rhesus factor system in Rhesus monkeys. The Rhesus C/E antigens or blood group Rh(CE) polypeptide (416 aa, RHCE_HUMAN/P18577) and the Rhesus D antigen or blood group Rh(D) polypeptide (416 aa, RHD_HUMAN/Q02161) are integral membrane proteins containing 11 potential 21 residues long transmembrane segments sharing a sequence identity of 92 %. Single amino acid variants exist: at position 102 antigen C (Rh2) carries a Ser and variant c (Rh4) a Pro residue; at position 225 antigen E (Rh3) carries a Pro and variant e (Rh5) an Ala residue. Polymorphism at position 109 in the antigen D results in a Tar (Rh40) variant. The largest part of the population is Rh+: Japanese and Chinese as well as people of African descent have over 99 %, Caucasians 84 % and by far the lowest are European Basque with 65 %. The rest of the population possesses no antigen and is therefore Rh−.

   In the case of blood transfusions Rh- individuals produce antibodies if they obtain Rh+ blood. There is a well-known hemolytic disease of newly born babies. During the delivery of her first child the mother produces antibodies that will lead to complications during subsequent pregnancies. Therefore, the mother is given antibodies against the Rhesus factor after the first birth.

   The largest part of the population has blood groups O or A and Table 2.1 shows the distribution of some selected populations. In Europe, the distribution is quite similar throughout the different countries. In populations of native origin the distributions are quite unique like in Native South Americans with 100 % O and Australian Aborigines with 44 % O and 56 % A.

3. **The human leukocyte antigen (HLA) system.** The human leukocyte antigen (HLA) system represents the major histocompatibility complex (MHC). Everybody has an individual, genetically determined HLA constellation, primarily present in leukocytes and tissues. The chance of two individuals having identical HLA proteins is very low, except for siblings. Over 1000 variants exist of MHC class I major antigens (HLA A, B and C) and over 5700 potential combinations in the MHC class II antigens. In the case of recurrent blood transfusions the body starts to produce antibodies against the HLA antigens. Therefore, in such cases the blood donations are said to be fine-tuned.

## REFERENCES

Cianchi, 1998, *Leonardo. The Anatomy*, Giunti, Prato, Italy.
Lodish *et al.*, 2003, *Molecular Cell Biology*, Freeman, New York.
Lutz, 2002, *The Rise of Experimental Biology. An Illustrated History*, Humana Press, Totowa, New Jersey.
Marsh *et al.*, 2005, Nomenclature for factors of the HLA system, 2004, *Tissue Antigens*, **65**, 301–369.
Michel, 1998, *Biochemical Pathways: An Atlas of Biochemistry and Molecular Biology*, Wiley–Spektrun, New York.
Zink *et al.*, 1991, *The Living Microcosm. Electron-microscopic Pictures*, Editiones Roche, Basel.

DATA SHEET

## *Histo-blood group ABO system transferase NAGAT: glycoprotein-fucosylgalactoside alpha-N-acetylgalactosaminyltransferase (EC 2.4.1.40) + glycoprotein-fucosylgalactoside alpha-galactosyltransferase (EC 2.4.1.37)*

### Fact Sheet

| | | | |
|---|---|---|---|
| ***Classification*** | Glycosyltransferase | ***Abbreviations*** | NAGAT |
| ***Structures/motifs*** | | ***DB entries*** | BGAT_HUMAN/P16442 |
| ***MW/length*** | 40 934 Da/354 aa | ***PDB entries*** | 1LZI |
| ***Concentration*** | | ***Half-life*** | |
| ***PTMs*** | 1 N-CHO (potential) | | |
| ***References*** | Yamamoto *et al.*, 1990, *Nature* **345**, 229–233.<br>Patenaude *et al.*, 2002, *Nat. Struct. Biol.* **9**, 685–690. | | |

### Description

**Histo-blood group ABO system transferase** (NAGAT) is the basis of the ABO blood group system.

### Structure

NAGAT is a type II membrane glycoprotein containing a potential 21 residue transmembrane segment (Gly 33–Met 53). The soluble form of NAGAT (301 aa) is derived from the membrane form by proteolytic processing and is devoid of a 53 residue N-terminal peptide. The conserved sequence **DXD** is involved in manganese binding as a cofactor (Asp 211 and 213). The A transferase and the B transferase differ in only four residues. The 3D structure of the soluble form consists of a N-terminal (in green) and a C-terminal (in yellow) domain separated by a wide cleft containing the active site residues Asp 211 and Asp 213 (in pink) coordinated to a $Mn^{2+}$ ion (orange ball) (**1LZI**).

### Biological Functions

NAGAT catalyses the reactions UDP-NAcGal + glycoprotein-Gal-Fuc ↔ UDP + glycoprotein-Gal-Fuc/NAcGal and UDP-Gal + glycoprotein-Gal-Fuc ↔ UDP + glycoprotein-Gal-Fuc/Gal.

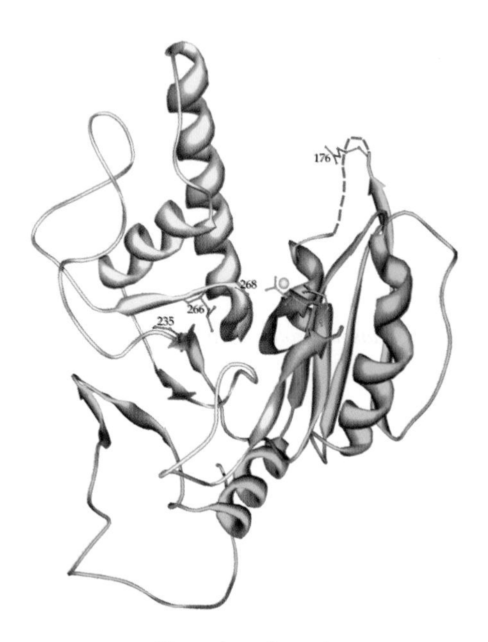


Glycosyltransferase A

# 3

# Blood Plasma Proteins

## 3.1 INTRODUCTION

As discussed in the previous chapter, blood is a very complex system containing various types of blood cells and blood plasma. Blood plasma contains many different water-soluble components, among them the blood plasma proteins. In principle, any protein present in the body can become at least temporarily a blood plasma protein depending on the actual state of the body, e.g. especially in the case of a pathological situation. A protein that is usually not present in blood plasma may accumulate in blood plasma and function as a characteristic diagnostic marker for a certain disease. Most plasma proteins carry posttranslational modifications (PTMs), either permanently or temporarily, and the vast majority are glycoproteins, thus ensuring a reasonable solubility (see Chapter 6). Blood plasma proteins exhibit a wide variety of functions and have different structural properties.

## 3.2 SHORT HISTORY

The Greeks had realised the importance of blood, as seen in the previous chapter. In 1777 the French chemist Pierre Macquer (1718–1784) coined the term 'albumin' to represent substances that coagulate upon applying heat. In 1838 the Swedish chemist Jöns Berzelius (1779–1848) introduced the term 'protein' (from the Greek word *proteuein*, meaning to be the first). In the nineteenth century blood plasma became more and more a subject of interest and in the 1830s the German chemists Justus von Liebig (1803–1873) and Gerardus J. Mulder (1802–1880) described a substance termed 'albumin'. In 1862 the German chemist Carl Schmidt (1822–1894) introduced the term 'globulin' for proteins insoluble in pure water and in 1894 horse serum albumin was crystallised for the first time by the German physician August Gürber (1864–1937).

The introduction of the ABO blood group system by Karl Landsteiner (1868–1943) in the early twentieth century, already described in the previous chapter, was also a big leap forward for blood plasma proteins. The introduction of blood plasma fractionation by the American protein chemists Edwin J. Cohn (1892–1953) and John T. Edsall (1902–2002) resulted in the production of large amounts of albumin and $\gamma$-globulin for therapeutic treatment and saved thousands of lives during the Second World War. The use of an unusual precipitation technique with rivanol introduced by the Behring Institute led to the preparation and distribution of many blood plasma proteins.

## 3.3 CLASSIFICATION OF BLOOD PLASMA PROTEINS

The classification of blood plasma proteins has always been under debate and an unambiguous classification is still quite difficult today. In his famous book series *The Plasma Proteins: Structure, Function and Genetic Control*, F.W. Putnam has defined in the 1970s the *true plasma proteins* as those proteins carrying out their function in circulation. This definition excludes proteins present in blood plasma exhibiting a function as messenger between tissues such as peptide hormones or cytokines, proteins leaking into the blood plasma as a result of tissue damage such as myoglobin in the case of a myocardial infarction or proteins from aberrant secretions released from tumours or other diseased tissues.

*Human Blood Plasma Proteins: Structure and Function* Johann Schaller, Simon Gerber, Urs Kämpfer, Sofia Lejon and Christian Trachsel

The reasons why a protein is permanently present in blood plasma or is residing there only temporarily can be manifold. The introduction of several new and sufficiently sensitive methods in the field of protein chemistry and protein analysis during the last 20 years has led to a new concept based on these new and sophisticated analytical tools. This 'proteome' concept suggests a general and comprehensive identification and characterisation of every protein present in plasma by state-of-the-art protein analytical means, resulting in the so-called 'plasma proteome', which is distinct from the plasma proteins.

However, based on the classification of plasma proteins by their function introduced by Putnam, and taking into account the developments in protein analysis N.L. Anderson and N.G. Anderson introduced in 2002 in their famous article 'The human plasma proteome: history, character and diagnostic prospects', an alternative and more comprehensive classification of the protein content in plasma. As an introductory remark it should be noted that the lifetime of a blood plasma protein is strongly dependent on its molecular mass regulated by the filtration cut-off of the kidney (approximately 45 kDa). The classification comprises the following categories:

1. **Proteins secreted by solid tissues that act in plasma.** This group mainly comprises the *classical plasma proteins*, which are primarily secreted by the liver and intestines. Assumptions suggest an approximate number of 500 'true plasma proteins', of which most are glycosylated, resulting in an average of 20 differently glycosylated forms for each protein. In addition, for each protein there are on average five different masses resulting from the precursor and mature forms, splice variants and degradation products. This results in an approximate number of 50 000 different molecular forms of classical plasma proteins.
2. **Immunoglobulins.** The immunoglobulins are typical plasma proteins, but due to their extraordinary complexity they represent a special class of proteins. An estimated number of at least 10 Mio different sequences of immunoglobulins exist in an adult human being.
3. **'Long-distance' receptor ligands.** This group comprises the classical peptide and protein hormones such as insulin or erythropoietin. The large range of their size regulates the timescale of their presence in plasma.
4. **'Local' receptor ligands.** This group contains cytokines and short-distance mediators, which are usually present in plasma only during a short period due to their rather small size below the filtration cut-off of the kidney.
5. **Temporary passengers.** This group includes nonhormone proteins on the passage through plasma from the site of synthesis and secretion to the site of action, such as lysosomal proteins.
6. **Tissue leakage products.** This group contains proteins that usually act within cells but are released into plasma upon cell death or cell damage. These proteins represent a class of important diagnostic markers, e.g. creatine kinase and myoglobin in the case of myocardial infarction. In principle, this group includes the entire human proteome, probably 50 000 gene products, on average, with 10 different forms such as splice variants, posttranslational modifications or cleavage products, resulting in approximately 500 000 different molecular forms of proteins.
7. **Aberrant secretion.** This group of proteins is released from tumours or other disease tissues, which are usually not present in plasma, and therefore represent cancer markers.
8. **Foreign proteins.** These proteins originate from infectious organisms and pathogens and are released into plasma.

The selection of the human blood plasma proteins described in this book is based on this classification. Two articles, both called 'The human plasma proteome' by N.L. Anderson and N.G. Anderson in 2002 and by N.L. Anderson *et al.* in 2004, and an article 'The human serum proteome' by R. Pieper *et al.* in 2003 and personal considerations were used as criteria for the final selection of the plasma proteins described in this book. The selection is arbitrary and is of course limited, and is therefore in the end a personal selection.

The selected proteins are grouped into the following topics (see Part III):

1. **Blood coagulation and fibrinolysis.** Chapter 7 describes proteins from primary haemostasis, the intrinsic, extrinsic and common pathways, the fibrinolytic system, regulatory proteins (the serpins involved in blood coagulation and fibrinolysis are described in Chapter 11) and the kinin and angiotensin/renin system (compiled in the Appendix, Table A.3).
2. **The complement system.** Chapter 8 includes proteins from the classical pathway, the alternative pathway, the membrane attack complex, the complement-activating components and the complement regulatory proteins (compiled in the Appendix, Table A.4).
3. **The immune system.** Chapter 9 covers the classical immunoglobulins, the HLA histocompatibility antigens, proteins of the innate immune system and additional immune reactive proteins (compiled in the Appendix, Table A.5).

4. **Enzymes.** Chapter 10 discusses enzymes not described in Chapters 7 and 8. It includes peroxidases, transferases, esterases, glycosylases, metallocarboxypeptidases, serine proteases, cysteine proteases, aspartic proteases and oxidoreductases (compiled in the Appendix, Table A.6).
5. **Inhibitors.** Chapter 11 describes various types of inhibitors: serine protease inhibitors (serpins), general protease inhibitors, Kunitz-type and Kazal-type inhibitors and thiol protease inhibitors (compiled in the Appendix, Table A.7).
6. **Lipoproteins.** Chapter 12 covers enzymes involved in lipoprotein metabolism, the various apolipoproteins, low-density and very low-density lipoprotein receptors and serum amyloid A protein (compiled in the Appendix, Table A.8).
7. **Hormones.** Chapter 13 describes the various types of hormones: pancreatic hormones, gastrointestinal hormones, hormones involved in calcium metabolism, hormone releasing factors, their trophic hormones and released hormones. In addition, somatostatin, choriogonadotropin and prolactin, vasopressin and oxytocin, opioid peptides, vasoactive peptides that are not described elsewhere, erythropoietin and thrombopoietin and adiponectin (compiled in the Appendix, Table A.9).
8. **Cytokines and growth factors.** Chapter 14 covers the various types of cytokines and growth factors (compiled in the Appendix, Table A.10).
9. **Transport and storage.** Chapter 15 describes proteins involved in transport and storage: the albumin and globin families, iron transport and storage proteins, proteins involved in the transport of hormones, steroids and vitamins and selenoprotein P (compiled in the Appendix, Table A.11).
10. **Additional proteins.** Chapter 16 is a compendium of plasma proteins that have not been discussed in previous chapters (compiled in the Appendix, Table A.12).

Although blood plasma contains a large number of proteins, only a handful constitute over 90 % of the total amount of all blood plasma proteins, with serum albumin representing approximately 55 %. Plasma proteins cover a very wide concentration range of at least 10 orders of magnitude. This is an extraordinary challenge for the isolation, identification and characterisation of low-abundance blood plasma proteins.

The abundance of a selection of blood plasma proteins is plotted in Figure 3.1. High-abundance serum albumin is in the range of 35–50 mg/ml (35–50 × $10^9$ pg/ml). With a half-life of about 21 days this results in an astonishing

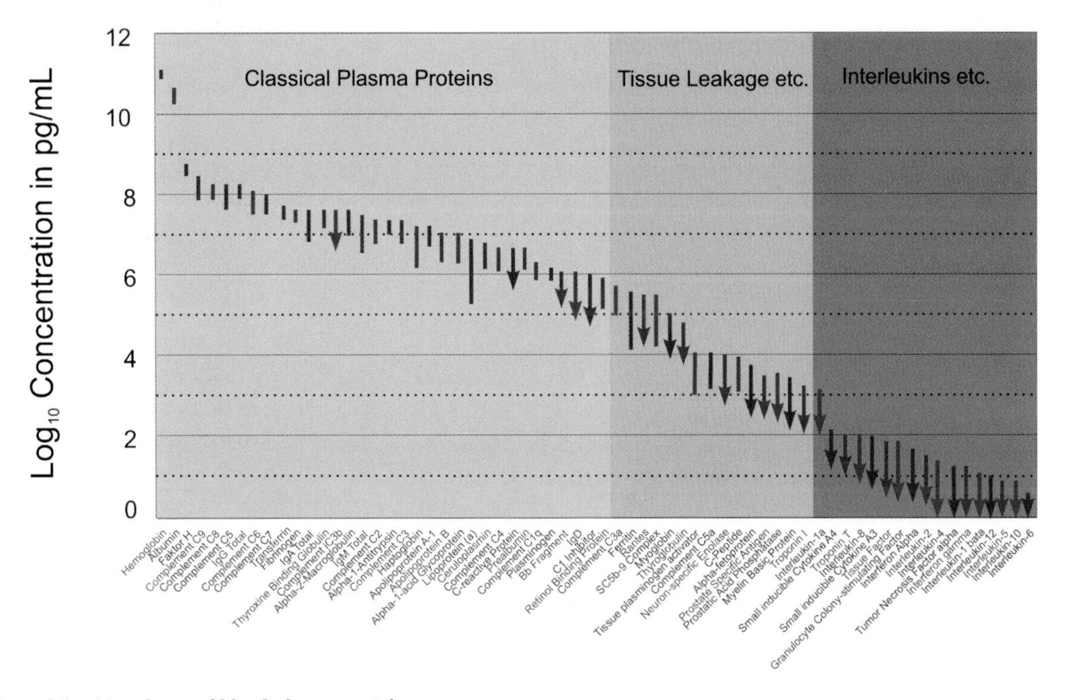


**Figure 3.1 Abundance of blood plasma proteins**
The abundance of a selection of blood plasma proteins is plotted on a log scale. An arrowhead indicates that only an upper limit is quoted. The classical plasma proteins are in the range of $10^{10}$– $10^6$ pg/ml, tissue leakage proteins in the range of $10^6$ – $10^3$ pg/ml and cytokines such as interleukins in the range $10^2$ – a few pg/ml. (data from Anderson and Anderson, 2002).

synthesis rate of approximately 12 g/day in the liver. Classical blood plasma proteins such as complement components or coagulation factors are in the range of 1 mg–1 μg/ml ($10^9$–$10^6$ pg/ml), proteins from tissue leakage are in the range of 1 μg–1 ng/ml ($10^6$–$10^3$ pg/ml) and cytokines such as interleukins in the range of 100–1 pg/ml. Low abundant interleukin-6, an indicator for inflammation and infection, has a concentration of only a few pg/ml.

## REFERENCES

Anderson and Anderson, 2002, The human plasma proteome: history, character, and diagnostic prospects, *Mol. Cell. Proteomics*, **1**, 845–867.

Anderson *et al.*, 2004, The human plasma proteome: a nonredundant list developed by combination of four separate sources, *Mol. Cell. Proteomics*, **3**, 311–326.

Cohn, 1948, The history of blood plasma fractionation, *Adv. Mil. Med.*, **1**, 364–443.

Pieper *et al.*, 2003, The human serum proteome: display of nearly 3700 chromatographically separated protein spots on two-dimensional electrophoresis gels and identification of 325 distinct proteins, *Proteomics*, **3**, 1345–1364.

Putnam, 1975–1987, *The Plasma Proteins: Structure, Function and Genetic Control*, Academic Press, New York.

Schultze and Heremans, 1966, *Molecular Biology of Human Proteins with Special Reference to Plasma Proteins*, Elsevier, New York.

# PART II

# 4

# Domains, Motifs and Repeats

## 4.1 INTRODUCTION

Human blood contains a wide variety of peptides and proteins with very different structural and functional properties. They differ in size from small peptides like the vasoactive nonapeptide bradykinin to large proteins like apolipoprotein B-100 (4536 aa) that is involved in the transport of cholesterol. They exhibit very diverse biological functions like oxygen transport in the case of hemoglobin, they act as hormones like insulin or they are involved in regulatory and inhibitory processes like the general protease inhibitor $\alpha_2$-macroglobulin. Plasma proteins have very different physicochemical parameters like a wide range of isoelectric points pI (theoretical pIs: pepsin A: 3.36, lysozyme: 9.28, sperm protamine P1: 12.08 and salmon protamine: 13.30) and they have very different half-life times in plasma, e.g. hemoglobin approximately 120 days and tissue plasminogen activator approximately 6 min. The abundance of plasma proteins spans at least 12 orders of magnitude with hemoglobin (about 150 g/l) and albumin (approximately 42 g/l) on the high end and cytokines with a few pg/l on the low end. Over 90 % of the blood plasma proteins contain at least one posttranslational modification and the vast majority are glycoproteins.

Many plasma proteins are multidomain proteins composed of a wide variety of domains, motifs and repeats, which differ in size, structure and function, such as the epidermal growth factor (EGF)-like domain originally identified in the epidermal growth factor and primarily found in proteins of the blood coagulation cascade and in the complement system (see Section 4.2) or the low-density lipoprotein (LDL) receptor class A (LDLRA) and class B (LDLRB) domains first identified in low-density lipoprotein receptors and present in many proteins of the complement system (see Section 4.12).

The type of domain, their number and their arrangement in the parent protein are responsible for the specific structure and function of a multidomain protein. Many of these domains are well characterised by a domain signature, their disulfide bridge pattern, their binding properties, their biological function and also their 3D structure, which is often known either from X-ray crystallographic or high-resolution NMR data. In order to facilitate the understanding of the structure–function relationship in the following chapters a detailed enumeration, description and characterisation of the major domains, motifs and repeats present in plasma proteins is given in this chapter. Many properties described for the individual domains, usually generated by limited proteolysis or by recombinant means, are again found as such or at least in part in the parent multidomain protein. In a multidomain protein many characteristics of the individual domains are present in an additive way. Therefore, a detailed description of the individual domains seems to be an adequate approach for a comprehensive understanding of the characteristics and functions of multidomain proteins described in later chapters.

Each domain is described by the domain signature and the enumeration of the plasma proteins in which the corresponding domains are found and, if available, the disulfide bridge pattern, the 3D structure, the binding properties and the biological function are given. The domains usually have a characteristic disulfide bridge pattern. The disulfide bridge pattern of the most common domains present in plasma proteins together with the corresponding signature are given in Table 4.1.

Many domains are present in multiple copies in their parent protein, e.g. 38 kringle structures in apolipoprotein(a) or 30 sushi/CCP/SCR domains in complement receptor type 1, or they are present in combination with other domains, e.g. coagulation factor XII contains two EGF-like, one kringle, one fibronectin type I and type II and a serine protease domain or complement component C6 contains one LDL-receptor class A, two sushi/CCP/SCR, one EGF-like and three TSP1 domains.

*Human Blood Plasma Proteins: Structure and Function* Johann Schaller, Simon Gerber, Urs Kämpfer, Sofia Lejon and Christian Trachsel

**Table 4.1** Disulfide bridge pattern of domains in human blood plasma proteins.

| Domain[a] | Number of bridges | Pattern | Signature |
|---|---|---|---|
| EGF-like | 3 | 1–3, 2–4, 5–6 | PS00022/PS01186 |
| Kringle | 3 | 1–6, 2–4, 3–5 | PS00021 |
| PAN/Apple | 3 | 1–6, 2–5, 3–4 | PS00495 |
| Sushi/CCP/SCR | 2 | 1–3, 2–4 | PS50923 |
| FN1 + FN2 | 2 | 1–3, 2–4 | PS00023/PS01253 |
| TSP1 | 3 | 1– 5, 2–6, 3–4 | PS50092 |
| LDLRA | 3 | 1–3, 2–5, 4–6 | PS01209 |
| CTCK | 5 | 1–6, 2–7, 3–8, 4–9, 5–10 | PS01185 |
| Anaphylatoxin | 3 | 1–4, 2–5, 3–6 | PS01177 |
| NTR | 3 | 1–3, 2–6, 4–5 | PS50189 |
| CTL | 2 | 1–2, 3–6, 4–5 | PS00615 |

[a]EGF-like: epidermal growth factor-like; sushi/CCP/SCR: sushi/complement control protein/short consensus repeat; FN1 + FN2: fibronectin type I + type II collagen-binding; TSP1: thrombospondin type I repeat; LDLRA: low-density lipoprotein receptor class A; CTCK: C-terminal cystine knot structure; NTR: netrin; CTL: C-type lectin.

## 4.2 THE EPIDERMAL GROWTH FACTOR (EGF)-LIKE DOMAIN

The ***EGF-like domain*** was originally found in the sequence of the *epidermal growth factor (EGF)* and was later identified in many vertebrate proteins in a more or less conserved form. EGF-like domains are found in the extracellular region of membrane-bound proteins or in secreted proteins. Blood plasma proteins containing EGF-like domains are primarily found in coagulation and fibrinolysis and in the complement system. A compilation of the EGF-like domain containing proteins in human blood plasma is given in Table 4.2.

The EGF-like domain is a polypeptide of about 50 residues with three intradomain disulfide bridges that are linked in the following pattern:

**Cys 1–Cys 3, Cys 2–Cys 4, Cys 5–Cys 6**

The EGF-like domain is characterised by two signatures. The three Cys residues in both signatures are involved in disulfide bridges:

EGF_1 (PS00022): **C**–x–**C**–x(5)–G–x(2)–**C**
EGF_2 (PS01186): **C**–x–**C**–x(2)–[GP]–[FYW]–x(4, 8)–**C**

Many of the proteins containing EGF-like domains require calcium for their biological function. A calcium-binding site has been identified in the N-terminal portion of the EGF-like domain and is characterised by the following consensus sequence, where the Cys residues are involved in disulfide bridges:

EGF_CA (PS01187): [DENQ]–x–[DENQ](2)–**C**–x(3, 14)–**C**–x(3, 7)–**C**–x–[DN]–x(4)–[FY]–x–**C**

The 3D structure of the calcium-binding EGF-like domains is very similar to the noncalcium-binding EGF-like domains with the same disulfide bridge pattern and the structural changes upon calcium binding are strictly local.

The 3D structure of several EGF-like domains has been solved and as an example the structure of the first EGF-like calcium-binding domain in coagulation factor VII (PDB: 1BF9; 41 aa: Ser 45–Lys 85) is given in Figure 4.1, where the domain is stabilised by three intrachain disulfide bridges (in pink). Factor VII is a multidomain protein containing one Gla, two EGF-like and a serine protease domain.

**Table 4.2** Plasma proteins containing EGF-like domains (in parenthesis is the number of Ca-binding EGF-like domains).

| Protein | Data bank entry | Number of copies (Ca-binding) |
|---|---|---|
| ***Coagulation and fibrinolysis*** | | |
| Factor VII | FA7_HUMAN/P08709 | 2 (1) |
| Factor IX | FA9_HUMAN/P00740 | 2 (1) |
| Factor X | FA10_HUMAN/P00742 | 2 (1) |
| Factor XII | FA12_HUMAN/P00748 | 2 |
| Protein C | PROC_HUMAN/04070 | 2 |
| Protein S | PROS_HUMAN/P07225 | 4 (3) |
| Protein Z | PROZ_HUMAN/P22891 | 2 |
| Thrombospondin-1 | TSP1_HUMAN/P07996 | 3 (1) |
| Thrombomodulin | TRBM_HUMAN/P07204 | 6 (2) |
| Tissue-type plasminogen activator | TPA_HUMAN/P00750 | 1 |
| Urokinase-type plasminogen activator | UROK_HUMAN/P00749 | 1 |
| ***Complement System*** | | |
| C1r subcomponent | C1R_HUMAN/P00736 | 1 (1) |
| C1s subcomponent | C1S_HUMAN/P09871 | 1 (1) |
| Component C6 | CO6_HUMAN/P13671 | 1 |
| Component C7 | CO7_HUMAN/P10643 | 1 |
| Component C8 alpha | CO8A_HUMAN/P07357 | 1 |
| Component C8 beta | CO8B_HUMAN/P07358 | 1 |
| Component C9 | CO9_HUMAN/P02748 | 1 |
| Perforin | PERF_HUMAN/P14222 | 1 |
| Complement-activating component | MASP1_HUMAN/P48740 | 1 |
| ***Others*** | | |
| LDL receptor | LDLR_HUMAN/P01130 | 3 (1) |
| VLDL receptor | VLDLR_HUMAN/P98155 | 3 |
| Epidermal growth factor | EGF_HUMAN/P01133 | 9 (3) |

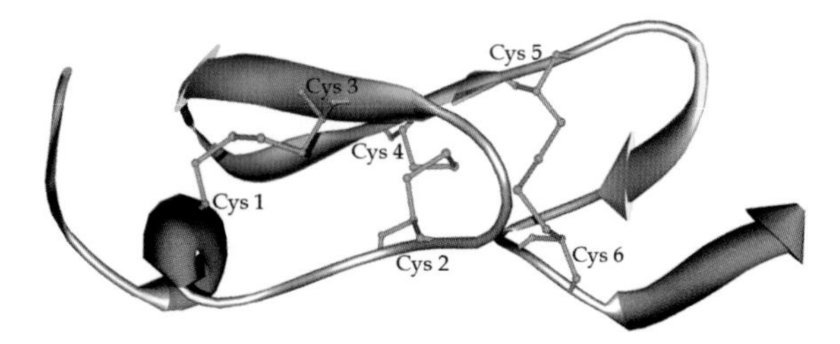


**Figure 4.1 3D structure of the first EGF-like domain in coagulation factor VII (PDB: 1BF9)**
3D structure of the first EGF-like calcium binding domain in coagulation factor VII (41 aa: Ser 45 – Lys 85) stabilised by three intrachain disulfide bridges (in pink).

**Table 4.3** Examples of 3D protein structures containing EGF-like domains.

| Protein | EGF-like (number of copies) | Other domains | PDB entry |
|---|---|---|---|
| Factor VII | 2 | Catalytic chain | 1QFK |
| Factor IX | 1 | Catalytic chain | 1RFN |
| Factor X | 2 | 1 Gla/catalytic chain | 1KSN |
| Protein C | 2 | Catalytic chain | 1AUT |
| Protein S | 2 | | 1Z6C |
| Thrombomodulin | 3 | α-Thrombin (complex) | 1DX5 |

The structure of the EGF-like domain consists of a two-stranded β-sheet followed by a loop connected to a C-terminal short two-stranded β-sheet. The 3D structure is stabilised by three intrachain disulfide bridges (in pink). Several examples of 3D protein structures containing EGF-like domains are compiled in Table 4.3. The 3D structures are given in the Data Sheets of the corresponding proteins.

EGF-like domains mediate protein–protein and protein–cell interactions. The EGF-like domains of various proteins are involved in receptor–ligand interactions. They bind with high affinity to specific cell-surface receptors, inducing their dimerisation, which leads to the activation of tyrosine kinases and signal transduction resulting in DNA synthesis and cell proliferation.

## 4.3 THE KRINGLE DOMAIN

***Kringle domains*** are mainly found in proteins of blood coagulation and fibrinolysis. The kringle-containing plasma proteins are listed in Table 4.4.

The kringle domain is a triple-loop structure of approximately 80 residues with three intradomain disulfide bridges linked in the following pattern:

**Cys 1–Cys 6, Cys 2–Cys 4, Cys 3–Cys 5**

The kringle domain is characterised by the following signature with the Cys residues involved in intrakringle disulfide bridges:

KRINGLE_1 (PS00021): [FY]–**C**–[RH]–[NS]–x(7, 8)–[WY]–**C**

The 3D structure of several kringle domains has been solved and as an example the 3D structure of recombinant mutated plasminogen kringle 2 (PDB: 1B2I; 83 aa: Cys162Thr–Thr 244) determined by NMR spectroscopy is shown in Figure 4.2. The kringle structure is characterised by a Cys cluster consisting of the two inner disulfide bridges (Cys 2–Cys 4 and

**Table 4.4** Plasma proteins with Kringle domains.

| Protein | Data bank entry | Number of copies |
|---|---|---|
| ***Coagulation and fibrinolysis*** | | |
| Prothrombin | THRB_HUMAN/P00734 | 2 |
| Factor XII | FA12_HUMAN/P00748 | 1 |
| Plasminogen | PLMN_HUMAN/P00747 | 5 |
| Tissue-type plasminogen activator | TPA_HUMAN/P00750 | 2 |
| Urokinase-type plasminogen activator | UROK_HUMAN/P00749 | 1 |
| Hepatocyte growth factor | HGF_HUMAN/P14210 | 4 |
| Hepatocyte growth factor-like | HGFL_HUMAN/P26927 | 4 |
| Hepatocyte growth factor activator | HGFA_HUMAN/Q04756 | 1 |
| Hepatocyte growth factor activator-like | Q14520_HUMAN/Q14520 | 1 |
| ***Lipoproteins*** | | |
| Apolipoprotein(a) | APOA_HUMAN/P08519 | 38 |

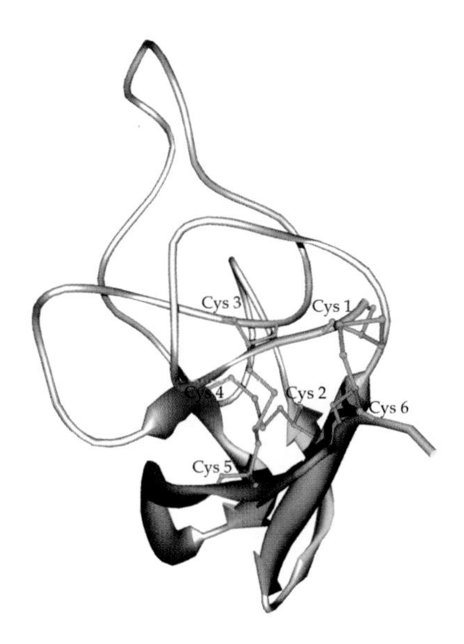


**Figure 4.2 3D structure of plasminogen kringle 2 (PDB: 1B2I)**
In the 3D structure of plasminogen kringle 2 (83 aa: Cys162Thr/Glu163Ser – Thr 244) the inner disulphide bridges Cys 2 – Cys 4 and Cys 3 – Cys 5 are almost perpendicular to each other and Cys 1 – Cys 6 forms the terminal disulfide bridge.

Cys 3–Cys 5) that are almost perpendicular to each other and a terminal disulfide bridge Cys 1–Cys 6. The binding of the ligand induces little conformational change in kringle 2 but stabilises its conformation.

Kringle domains are thought to mediate binding to proteins and phospholipids in membranes and are likely to play a role in the regulation of proteolytic activity. Some kringles contain lysine-binding sites (LBS) involved in the binding process (for details see Section 7.4 in Chapter 7).

## 4.4 THE γ-CARBOXYGLUTAMIC ACID-RICH (Gla) DOMAIN

***Gla domains*** are mainly found in the N-terminal region of proteins involved in blood coagulation and fibrinolysis. They are compiled in Table 4.5. The Gla domain is a membrane-binding motif that is required for the binding of coagulation factors to the phospholipid membrane in the presence of calcium ions. The posttranslational modification of specific Glu residues in the N-terminal region of these proteins is a vitamin K-dependent enzymatic reaction catalysed by a γ-carboxylase, as shown in Figure 4.3. PIVKA is the protein induced by vitamin K absence or vitamin K antagonists (non-γ-carboxylated proteins).

**Table 4.5** Plasma proteins containing Gla domains.

| Protein | Data bank entry | Number of γ-Gla residues |
|---|---|---|
| ***Coagulation and fibrinolysis*** | | |
| Prothrombin | THRB_HUMAN/P00734 | 10 |
| Factor VII | FA7_HUMAN/P08709 | 10 |
| Factor IX | FA9_HUMAN/P00740 | 12 |
| Factor X | FA10_HUMAN/P00742 | 11 |
| Protein C | PROC_HUMAN/P04070 | 9 |
| Protein S | PROS_HUMAN/P07225 | 11 |
| Protein Z | PROZ_HUMAN/P22891 | 13 |

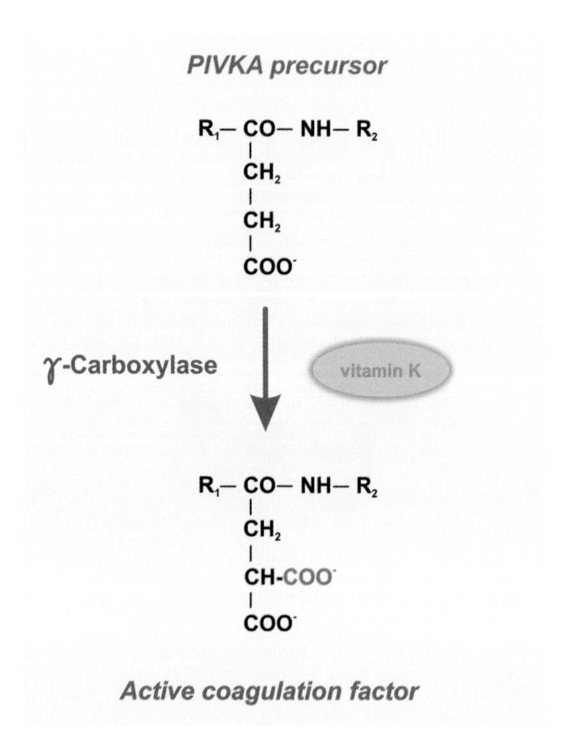


**Figure 4.3 γ-Carboxylation of Glu residues**
Vitamin K-dependent γ-carboxylation of Glu residues in the Gla domain of proteins involved in blood coagulation and fibrinolysis. PIVKA: Protein Induced by Vitamin K Absence or Vitamin K Antagonists (non γ-carboxylated proteins).

The Gla domain is characterised by the following signature, where the two Glu residues are carboxylation sites:

GLA_1 (PS00011): **E**–X(2)–[ERK]–**E**–X–C–X(6)–[EDR]–x(10, 11)–[FYA]–[YW]

The 3D structure of the Gla domain of coagulation factor IX (PDB: 1CFH; 47 aa: Tyr 1–Asp 47) determined by NMR spectroscopy is shown in Figure 4.4 (a). The calcium-bound structure is significantly more ordered with a higher helical content than the apo-Gla domain, as can be seen in Figure 4.4 (b). The nine N-terminal Gla residues are oriented towards the interior of the domain, forming an internal calcium-binding pocket. Five calcium ions are located in this pocket at the core of the Gla domain coordinated to the carboxyl groups of the Gla residues, as shown in Figure 4.4 (c). The 3D structure of factor Xa comprising the entire catalytic chain, the Gla domain and the two EGF-like domains has been solved (PDB: 1KSN; for details see Chapter 7).

Calcium ions are required for the correct folding of the Gla domain. A characteristic structural feature of the Gla domain is the clustering of N-terminal hydrophobic residues into a hydrophobic patch that mediates interaction with phospholipid membranes.

## 4.5 THE PAN/APPLE DOMAIN

Plasma kallikrein and coagulation factor XI contain four tandem repeats of approximately 90 residues each, which can be drawn in the shape of an apple and contain three intradomain disulfide bridges arranged in the following pattern:

**Cys 1– Cys 6, Cys 2– Cys 5, Cys 3– Cys 4**

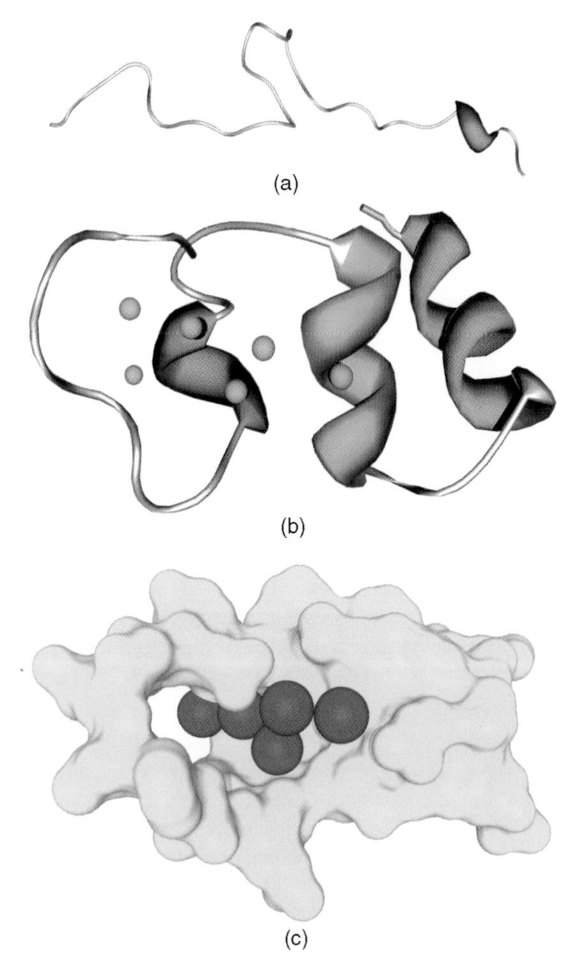


**Figure 4.4 The 3D structure of the Gla domain of coagulation factor IX (PDB: 1CFH)**
(a) 3D structure of the Gla domain of coagulation factor IX (47 aa: Tyr 1 – Asp 47) without calcium is rather unstructured. (b) The Gla domain with six calcium ions (pink balls) is more structured with a higher helical content than without calcium (PDB: 1NL0). (c) Surface representation of the calcium ions located in a pocket at the core of the Gla domain (PDB: 1NL0).

The **apple domain** is characterised by the following signature with all six Cys involved in intradomain disulfide bridges:

APPLE (PS00495): **C**–x(3) – [LIVMFY]–x(5)–[LIVMFY]–x(3)–[DENQ]–[LIVMFY]–x(10)–**C**–x(3)–**C**–T
–x(4)–**C**–x – [LIVMFY]–F–x–[FY]–x(13, 14)–**C**–x–[LIVMFY]–[RK]–x–[ST]–x(14, 15)
–S–G–x–[ST]–[LIVMFY]–x(2)–**C**

The apple domains mediate binding, in prekallikrein, to HMW kininogen and in factor XI to factors IX and XIIa, kininogen, heparin and platelets.

The apple domain displays similarity to the N-terminal regions of plasminogen, to the plasminogen-related protein B (Q02325), to the hepatocyte growth factor and to the hepatocyte growth factor-like protein (see Table 4.4). All these domains belong to a superfamily termed 'PAN'. The structure of the PAN *domain* (in green) linked to the kringle 1 domain (in yellow) in the hepatocyte growth factor (PDB: 1BHT; 176 aa: Arg 4–Glu 179) has been solved as shown in Figure 4.5. The PAN domain consists of a five-stranded antiparallel β-sheet flanked by a α-helix and a two-stranded β-ribbon, and the structure is stabilised by

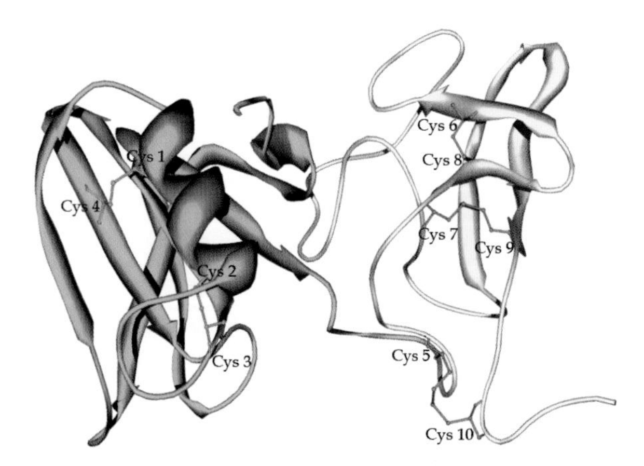


**Figure 4.5 3D structure of the PAN domain linked to the kringle 1 domain of the hepatocyte growth factor (PDB: 1BHT)** 3D structure of the PAN domain (in green) containing two intrachain disulfide bridges (in pink) linked to the kringle 1 domain (in yellow) containing three intrachain disulfide bridges (in blue) of the hepatocyte growth factor (176 aa: Arg 4 – Glu 179).

two intrachain disulfide bridges (in pink). The kringle structure containing three intrachain disulfide bridges (in blue) of the hepatocyte growth factor is the known structure for individual kringles (for details see Section 4.3).

The PAN module is approximately 60 residues in size and contains a characteristic hairpin-loop structure. It is stabilised by two intrachain disulfide bridges arranged in the pattern **Cys 1–Cys 4, Cys 2–Cys 3**. Cys 1 and Cys 6 in the apple domain are not conserved in the PAN module.

## 4.6 THE SUSHI/CCP/SCR DOMAIN

The ***sushi domains*** are mainly found in the complement system and in factor XIIIb and $\beta_2$-glycoprotein of the blood coagulation cascade. The sushi domain is also known as the complement control protein (CCP) module or the short consensus repeat (SCR). The sushi-containing proteins are summarised in Table 4.6.

**Table 4.6** Plasma proteins containing sushi/CCP/SCR domains.

| Protein | Data bank entry | Number of copies |
|---|---|---|
| ***Complement system*** | | |
| C1r subcomponent | C1R_HUMAN/P00736 | 2 |
| C1s subcomponent | C1S_HUMAN/P09871 | 2 |
| Component C2 | CO2_HUMAN/P06681 | 3 |
| Component C6 | CO6_HUMAN/P13671 | 2 |
| Component C7 | CO7_HUMAN/P10643 | 2 |
| Factor B | CFAB_HUMAN/P00751 | 3 |
| Factor H | CFAH_HUMAN/P08603 | 20 |
| C4b-binding protein | | |
| α-Chain | C4BP_HUMAN/P04003 | 8 |
| β-Chain | C4BB_HUMAN/P20851 | 3 |
| Decay accelerating factor (DAF) | DAF_HUMAN/P08174 | 4 |
| Membrane cofactor protein (MCP) | MCP_HUMAN/P15529 | 4 |
| Complement receptor type 1 | CR1_HUMAN/P17927 | 30 |
| Complement-activating component | MASP1_HUMAN/P48740 | 2 |
| ***Coagulation and fibrinolysis*** | | |
| Factor XIII B chain | F13B_HUMAN/P05160 | 10 |
| $\beta_2$-Glycoprotein | APOH_HUMAN/P02749 | 4 |
| ***Storage and transport*** | | |
| Haptoglobin | HPT_HUMAN/P00738 | 2 |

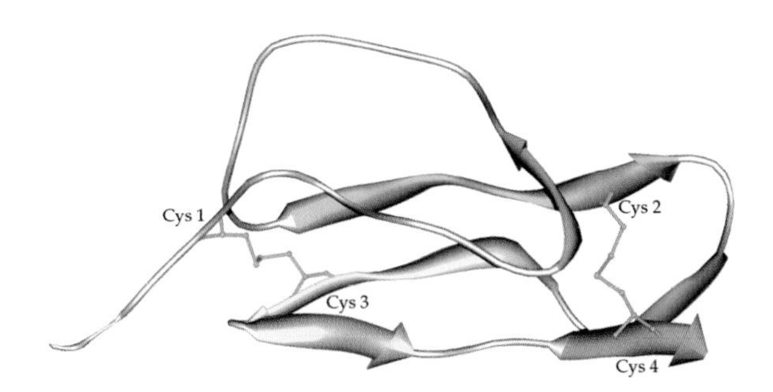


**Figure 4.6 3D structure of the sushi domain 16 in complement factor H (PDB: 1HCC)**
3D structure of the single sushi domain 16 in complement factor H (59 aa: Glu 909 – Ile 967) containing two intrachain disulfide bridges (in pink).

The sushi domain is characterised by a consensus sequence of approximately 60 residues with two conserved disulfide bridges arranged in the pattern **Cys 1–Cys 3, Cys 2–Cys 4**, a highly conserved Trp and conserved Gly, Pro and hydrophobic residues. The structure of several sushi domains has been solved and the 3D structure of the single sushi domain 16 in complement factor H (PDB: 1HCC; 59 aa: Glu 909–Ile 967) determined in solution by NMR spectroscopy is shown in Figure 4.6. The structure is stabilised by two intrachain disulphide bridges (in pink). In addition, the 3D structure of the entire $\beta_2$-glycoprotein containing four sushi and one sushi-like domain has been solved by X-ray diffraction (PDB: 1C1Z; 326 aa: Gly 1–Cys 326). The five distinct domains are arranged like beads on a string forming a J-shaped protein (for details see Chapter 7). The 3D structures of parts of complement C1r containing the two sushi domains linked to the catalytic domain (PDB: 1GPZ) and parts of complement C1s containing the second sushi domain linked to the catalytic domain (PDB: 1ELV) have been determined (for details see Chapter 8). Moreover, the 3D structures of the four sushi domains in complement decay-accelerating factor (DAF) (PDB: 1OK1) and of the two-sushi domains linked to the catalytic chain in mannan-binding lectin serine protease 2 (MASP-2) (PDB: 1ZJK) have also been determined (for details see Chapter 8).

The sushi fold contains a compact hydrophobic core surrounded by six β-strands that are stabilised by the two intradomain disulfide bridges. The sushi domains are known to be involved in many recognition processes like the binding of several complement factors to complement fragments C3b and C4b.

## 4.7 THE FIBRONECTIN TYPE I, THE FIBRONECTIN TYPE II COLLAGEN-BINDING AND THE FIBRONECTIN TYPE III DOMAINS

The ***fibronectin type I*** *(FN1)*, the ***fibronectin type II collagen binding*** *(FN2)* and the ***fibronectin type III*** *(FN3)* ***domains*** are mainly found in proteins involved in blood coagulation and fibrinolysis. The fibronectin-containing plasma proteins are compiled in Table 4.7.

**Table 4.7** Plasma proteins containing fibronectin type I, fibronectin type II collagen-binding and fibronectin type III domains.

| Protein | Data bank entry | Number of Copies | | |
|---|---|---|---|---|
| | | Type I | Type II | Type III |
| ***Coagulation and fibrinolysis*** | | | | |
| Tissue factor | TF_HUMAN/P13726 | | | 2 |
| Factor XII | FA12_HUMAN/P00748 | 1 | 1 | |
| Tissue-type plasminogen activator | TPA_HUMAN/P00750 | 1 | | |
| Hepatocyte growth factor activator | HGFA_HUMAN/Q04756 | 1 | 1 | |
| Fibronectin | FINC_HUMAN/P02751 | 12 | 2 | 16 |
| ***Others*** | | | | |
| Interleukin-12 beta | IL12B_HUMAN/P29460 | | | 1 |
| Tenascin | TENA_HUMAN/P24821 | | | 15 |

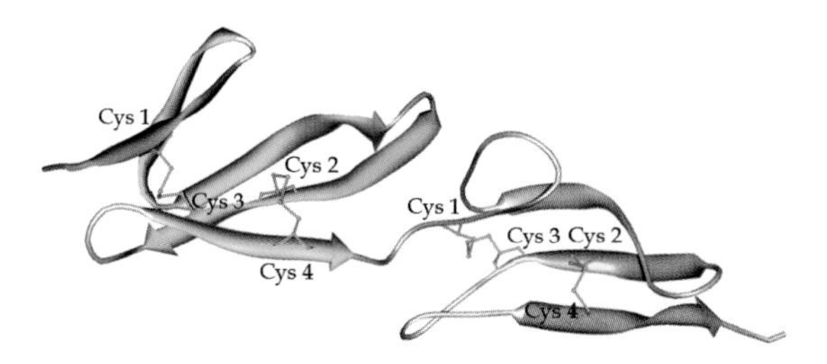


**Figure 4.7 3D structure of the fibronectin type I (FN1) domains 4 + 5 in fibronectin (PDB:1FBR)**
3D structure of the fibronectin type I (FN1) domains 4 + 5 in fibronectin (93 aa: Ala 152 – Ser 244) containing two disulfide bridges (in pink) in each domain.

The FN1 and FN2 domains are approximately 40 residues in length with two conserved disulfide bridges arranged in the pattern **Cys 1–Cys 3, Cys 2–Cys 4**. The FN1 and FN2 domains are characterised by the following signatures, where all four Cys are involved in intradomain disulfide bridges:

FN1_1 (PS01253): **C**–x(6, 8)–[LFY]–x(5)–[FYW]–x–[RK]–x(8, 10)–**C**–x–**C**–x(6, 9)–**C**

FN2_1 (PS00023): **C**–x(2)–P–F–x–[FYWIV]–x(7)–**C**–x(8, 10)–W–**C**–x(4)–[DNSR]–[FYW]–x(3, 5)–[FYW]–x–[FYWI]–**C**

The 3D structure of the FN1 domain has been solved and the structure of the fibronectin type I (FN1) domains 4 + 5 in fibronectin (PDB: 1FBR; 93 aa: Ala 152–Ser 244) determined in solution by NMR spectroscopy is shown as an example in Figure 4.7. Each FN1 domain is stabilised by two disulfide bridges (in pink).

FN1 consists of two antiparallel β-sheets, a double-stranded and a triple-stranded sheet linked by a disulfide bridge. The second disulfide bridge links the C-terminal adjacent strands of the domain.

The 3D structure of the FN2 domain has been solved and the structure of the second fibronectin type II (FN2) domain in fibronectin (PDB: 2FN2; 59 aa: Val 375–Ala 433) determined in solution by NMR spectroscopy is given as an example in Figure 4.8. The FN2 structure consists of two double-stranded antiparallel β-sheets oriented nearly perpendicular to each other. The two β-sheets are connected by an irregular loop and the two strands in the second β-sheet are separated by another loop. The structure is stabilised by two intrachain disulphide bridges (in pink).

The minimal collagen-binding region FN1–FN2–FN2 in fibronectin adopts a hairpin structure, where the conserved aromatic residues of FN2 form a hydrophobic pocket thought to provide a binding site for nonpolar residues in collagen.

The FN3 domain is approximately 90 residues in length and in contrast to FN1 and FN2 lacks a conserved disulfide bridge pattern. The 3D structure has been solved and the structure of the first fibronectin type III (FN3) domain in fibronectin (PDB: 1OWW; 98 aa: Ser 578–Ser 670 + N-terminal pentapeptide) determined by NMR spectroscopy is shown as an example in Figure 4.9.

The FN3 structure is characterised by a conserved β-sandwich composed of two β-sheets with four and three strands, respectively.

The various FN domains in fibronectin exhibit binding to collagen, fibrin, actin, heparin and DNA. FN3 domain number 10 in fibronectin contains an **RGD** sequence that can modulate various types of cell adhesion events in thrombosis, inflammation and tumour metastasis.

## 4.8 THE COAGULATION FACTORS 5/8 TYPE C DOMAIN (FA58C)

Blood **coagulation factors V** and **VIII** are large plasma glycoproteins (>2000 aa) each containing in their C-terminal region three F5/8 type A domains (FA58A) of 300–340 amino acids in length with each domain composed of two plastocyanin-like domains and two F5/8 type C domains (FA58C). In addition, the storage and transport protein ceruloplasmin contains three FA58A domains.

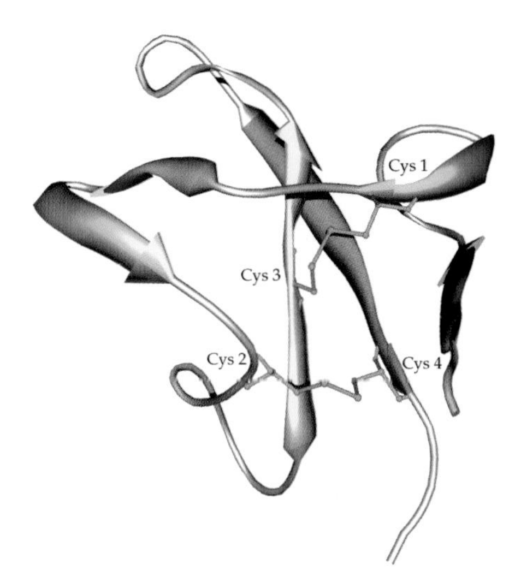


**Figure 4.8 3D structure of the second fibronectin type II (FN2) domain in fibronectin (PDB: 2FN2)**
3D structure of the second fibronectin type II (FN2) domain in fibronectin (59 aa: Val 375 – Ala 433) containing two intrachain disulfide bridges (in pink).

The FA58C domain contains approximately 150 residues and is characterised by two signatures:

FA58C_1 (PS01285): [GASP]–W–x(7, 15)–[FYW]–[LIV]–x–[LIVFA]–[GSTDEN]–x(6)–[LIVF]–x(2)
–[IV]–x–[LIVT]–[QKMT]–G

FA58C_2 (PS01286): P–x(8, 10)–[LM]–R–x–[GE]–[LIVP]–x–G–C

The 3D structure of the FA58C domain has been solved and the structure of the second FA58C domain in the C-terminal region of coagulation factor V (PDB: 1CZS; 160 aa: Gly 2037–Tyr 2196) is shown in Figure 4.10 containing a single intrachain disulfide bridge (in pink). The 3D structure of the second FA58C domain in coagulation factor VIII (159 aa: Leu 2171–Gln 2329) (PDB: 1D7P) is very similar in structure and function.

The FA58C domain exhibits a distorted β-barrel motif consisting of eight antiparallel strands arranged in two β-sheets. The lower part of the β-barrel is characterised by several basic residues and three protruding loops that play a key role in lipid binding. The ends of the FA58C domain are linked by a conserved disulfide bridge (in pink). The domains promote binding to anionic phospholipids on the surface of platelets and endothelial cells.

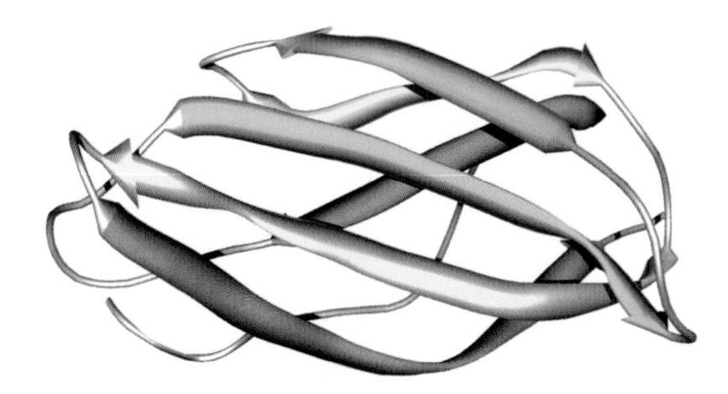

**Figure 4.9 3D structure of the first fibronectin type III (FN3) domain in fibronectin (PDB: 1OWW)**
3D structure of the first fibronectin type III (FN3) domain in fibronectin (98 aa: Ser 578 – Ser 670 + N-terminal pentapeptide) lacks a conserved disulfide bridge pattern.

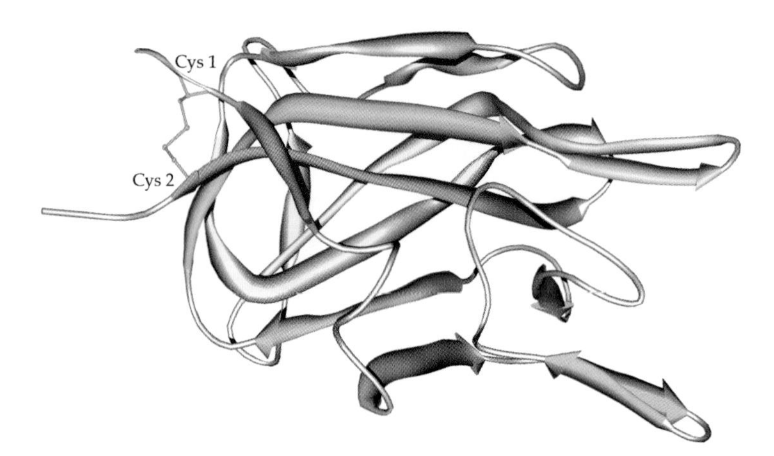


**Figure 4.10 3D structure of the second FA58C domain of coagulation factor V (PDB: 1CZS)**
3D structure of the second FA58C domain of coagulation factor V (160 aa: Gly 2037 – Tyr 2196) containing a single intrachain disulfide bridge (in pink).

## 4.9 THE THROMBOSPONDIN TYPE I REPEAT (TSP1)

The ***thrombospondin type I repeat*** (TSP1) was first identified in thrombospondin-1 and -2. In addition, TSP1 domains are contained in several proteins of the complement system. Table 4.8 shows their enumeration.

The TSP1 domain is approximately 55 residues in length and contains three conserved intradomain disulfide bridges arranged in the following pattern:

**Cys 1–Cys 5, Cys 2–Cys 6, Cys 3–Cys 4**

Many TSP1 modules contain the conserved sequence motifs **WSXW** and **CSVTCG**. The 3D structure of the TSP1 domains 2 + 3 in thrombospondin-1 (PDB: 1LSL; 113 aa: Gln 416–Cys 528) is given in Figure 4.11. Each TSP1 domain is stabilised by three intrachain disulfide bridges (in pink). The structure is characterised by an antiparallel, three-stranded fold consisting of alternating stacked layers of Trp and Arg residues from respective strands stabilised by disulfide bonds on each strand. In addition, thrombospondin-1 and –2 contain seven TSP type 3 domains (approximately 25–35 residues), one TSP N-terminal (TSPN) and one TSP C-terminal (TSPC) domain, each with approximately 200 residues.

The 3D structure of the TSPN domain in TSP1 (PDB: 1ZA4) and of a C-terminal fragment of TSP1 comprising the TSP type 3 repeats 5–7 and TSPC (PDB: 1UX6) was determined (for details see Chapter 7). The various TSP domains in thrombospondin mediate cell-to-cell and cell-to-matrix interactions.

**Table 4.8** Plasma proteins containing the thrombospondin type I repeat (TSP1).

| Protein | Data bank entry | Number of copies |
|---|---|---|
| ***Coagulation and fibrinolysis*** | | |
| Thrombospondin-1 | TSP1_HUMAN/P07996 | 3 |
| Thrombospondin-2 | TSP2_HUMAN/P35442 | 3 |
| ***Complement system*** | | |
| Component C6 | CO6_HUMAN/P13671 | 3 |
| Component C7 | CO7_HUMAN/P10643 | 2 |
| Component C8 alpha | CO8A_HUMAN/P07357 | 2 |
| Component C8 beta | CO8B_HUMAN/P07358 | 2 |
| Component C9 | CO9_HUMAN/P02748 | 1 |
| Properdin | PROP_HUMAN/P27918 | 6 |

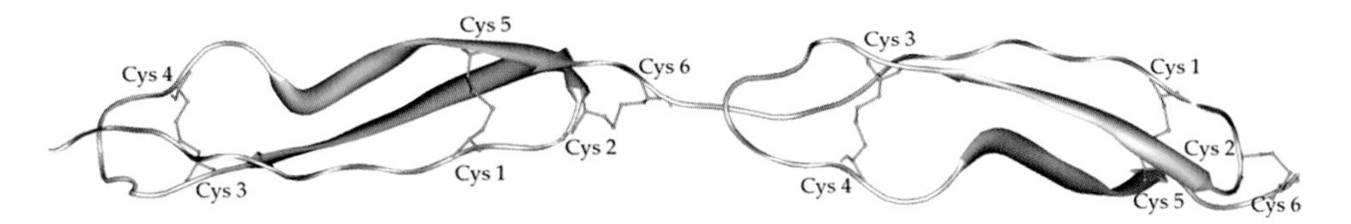


**Figure 4.11 3D structure of TSP1 domains 2 + 3 in thrombospondin-1 (PDB: 1LSL)**
3D structure of TSP1 domains 2 + 3 in thrombospondin-1 (113 aa: Gln 416 – Cys 528) each containing three intrachain disulphide bridges (in pink).

## 4.10 THE VWFA, VWFC AND VWFD DOMAINS

The ***VWFA, VWFC*** and ***VWFD*** *domains* are named after the von Willebrand factor (vWF). Plasma proteins containing these domains are given in Table 4.9.

The VWFA domain is approximately 170–200 residues long and the 3D structure of the VWFA domain 1 in the von Willebrand factor (PDB: 1AUQ; 208 aa: Asp 498–Thr 705) was determined by X-ray diffraction and is shown in Figure 4.12. The 3D structure of the VWFA domain linked to the catalytic chain in complement factor B was determined by X-ray diffraction (PDB: 1RRK) (for details see Chapter 8).

The domain adopts the dinucleotide binding or α/β 'Rossmann' fold characterised by a central β-sheet surrounded by α-helices. The VWFA domain 1 is the principal binding site for glycoprotein Ib (GPIb) and the VWFA domains bind to collagen and heparin. The inter-alpha-trypsin inhibitor heavy chains H1, H2, H3 and H4 each contain one VWFA domain.

The VWFC domain is approximately 70 residues in length and contains 10 well-conserved Cys residues. The VWFC domain is characterised by the following signature:

VWFC_1 (PS01208): **C**–x(2, 3)–**C**–{CG}–**C**–x(6, 14)–**C**–x(3, 4)–**C**–x(2, 10)–**C**–x(9, 16)–**C**–**C**–x(2, 4)–**C**

It seems that the VWFC domain is involved in the formation of larger protein complexes, e.g. oligomerisation of vWF. Little is known of the VWFD domain with a length of approximately 350 residues.

**Table 4.9** Plasma proteins containing the VWFA, VWFC and VWFD domains.

| Protein | Data bank entry | Number of copies | | |
|---|---|---|---|---|
| | | VWFA | VWFC | WWFD |
| ***Coagulation and fibrinolysis*** | | | | |
| Von Willebrand factor | VWF_HUMAN/P04275 | 3 | 3 | 2 |
| Thrombospondin-1 | TSP1_HUMAN/P07996 | | 1 | |
| ***Complement system*** | | | | |
| Complement C2 | CO2_HUMAN/P06681 | 1 | | |
| Factor B | CFAB_HUMAN/P00751 | 1 | | |
| ***Platelet membrane glycoproteins*** | | | | |
| Integrin alpha-2 | ITA2_HUMAN/P17301 | 1 | | |
| Integrin beta-3 | ITB3_HUMAN/P05106 | 1 | | |
| ***Inhibitors*** | | | | |
| Inter-alpha-trypsin inhibitor | | | | |
| Heavy chain H1 | ITIH1_HUMAN/P19827 | 1 | | |
| Heavy chain H2 | ITIH2_HUMAN/P19823 | 1 | | |
| Heavy chain H3 | ITIH3_HUMAN/Q06033 | 1 | | |
| Heavy chain H4 | ITIH4_HUMAN/Q14624 | 1 | | |

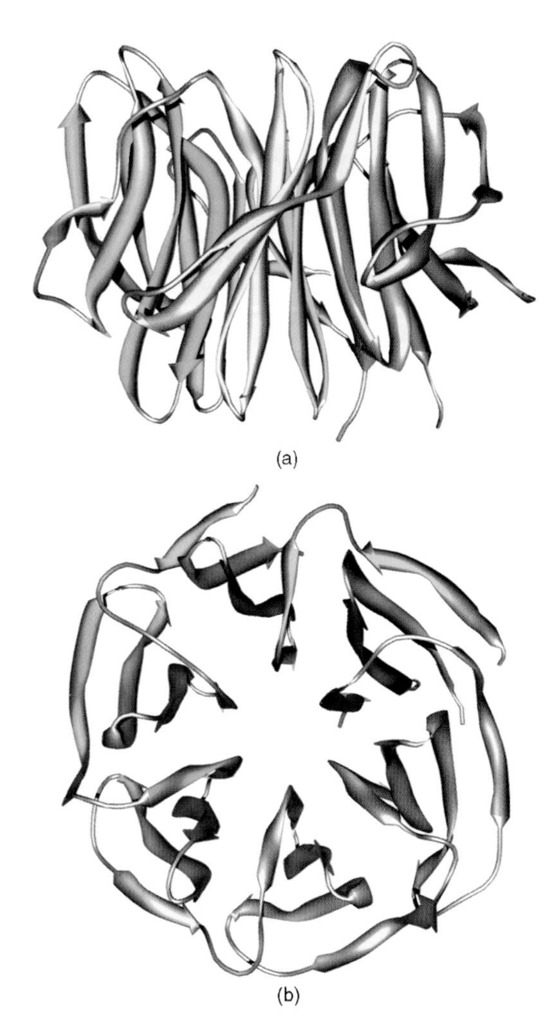


**Figure 4.14 3D structure of the six LDLRB domains in the low-density lipoprotein receptor (PDB: 1IJQ)**
(a) 3D structure of the six LDLRB domains linked to one EGF-like domain in the low-density lipoprotein receptor (316 aa: Ile 377 – Thr 692). The structure is characterised by a six-bladed β-propeller.
(b) Top view.

receptor (PDB:1IJQ; 316 aa: Ile 377–Thr 692) determined by X-ray diffraction is shown in Figure 4.14(a), with the top view shown in figure 4.14(b). The YWTD repeats form a six-bladed β-propeller tightly packed against the C-terminal EGF-like domain. The structure of the extracellular domain of the low-density lipoprotein receptor was determined by X-ray diffraction (for details see Chapter 12).

## 4.13 THE C-TERMINAL CYSTINE KNOT (CTCK) STRUCTURE

Several growth factors like the transforming growth factor TGF-β, the nerve growth factor NGF and the platelet-derived growth factor PDGF as well as the gonadotropins follitropin β-chain and lutropin β-chain contain a so-called ***cystine knot structure*** (for details see Chapters 13 and 14). Functionally diverse modular proteins like the von Willebrand factor contain in their C-terminal cysteine-rich region a structure thought to be related to the cystine knot family and is therefore called the

C-terminal cystine knot (CTCK) structure. The CTCK domain is approximately 90 residues in length and contains five disulfide bridges arranged in the following pattern:

$$\mathbf{Cys\,1-Cys\,6,\ Cys\,2-Cys\,7,\ Cys\,3-Cys\,8,\ Cys\,4-Cys\,9,\ Cys\,5-Cys\,10}$$

The CTCK domain is characterised by the following signature, where all six Cys are involved in intradomain disulfide bridges:

$$\text{CTCK_1 (PS01185)} : \mathbf{C}-\mathbf{C}-\text{x}(13)-\mathbf{C}-\text{x}(2)-[\text{GN}]-\text{x}(12)-\mathbf{C}-\text{x}-\mathbf{C}-\text{x}(2,4)-\mathbf{C}$$

The disulfide bonds 3–8 and 4–9 form a cystine ring and bond 1–6 passes through this ring. The 3D structure of the entire follitropin β-chain (PDB: 1XWD; 111 aa: Asn 1–Glu 111) is given in Figure 4.15(a). The disulfide bridge Cys 1–Cys 6 passes through the cystine ring formed by disulfide bridges Cys 4–Cys 8 and Cys 5–Cys 9. The other three disulfide bridges are detached; Cys 2–Cys 7 and Cys 3–Cys 12 are partially conserved in CTCK domains, while Cys 10–Cys 11 is additional in the follitropin β-chain. The core disulfide bridge pattern is shown in detail in Figure 4.15(b).

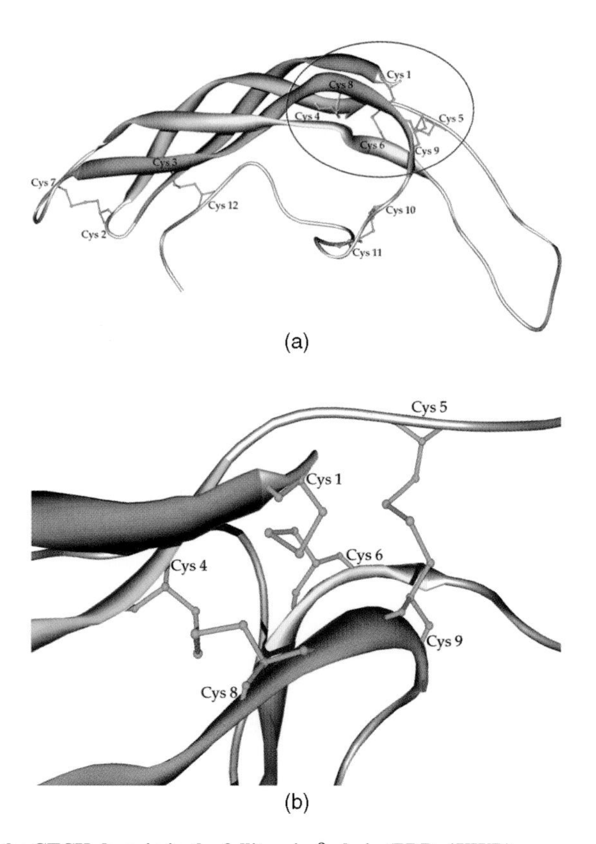


**Figure 4.15 3D structure of the CTCK domain in the follitropin β-chain (PDB: 1XWD)**
(a) In the 3D structure of the CTCK domain in the follitropin β-chain (111 aa: Asn 1 – Glu 111) the disulfide bridge Cys 1 – Cys 6 passes through the cystine ring formed by disulfide bridges Cys 4 – Cys 8 and Cys 5 – Cys 9. The other three disulfide bridges Cys 2 – Cys 7, Cys 3 – Cys 12 and Cys 10 – Cys 11 are detached.
(b) The structure of the core disulfide bridge pattern.

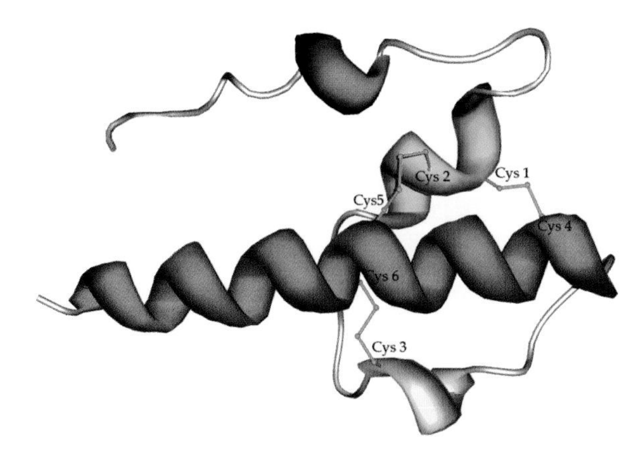


**Figure 4.16 3D structure of anaphylatoxin C3a from the complement C3 (PDB: 2A73)**
3D structure of anaphylatoxin C3a (77 aa: Ser 650 – Arg 726) is derived from the crystal structure of entire complement C3 containing three intrachain disulfide bridges (in pink).

## 4.14 THE ANAPHYLATOXIN DOMAIN

***Anaphylatoxins*** are mediators of local inflammatory processes. The complement components C3, C4 and C5 each contain one anaphylatoxin domain: anaphylatoxins C3a, C4a and C5a. The cysteine-rich region shares similarities with three repeats in fibulin-1 (674 aa: FBLN1_HUMAN/P23142), which is involved in haemostasis and thrombosis. The anaphylatoxins are characterised by three conserved disulfide bridges arranged in the following pattern:

**Cys 1–Cys 4, Cys 2–Cys 5, Cys 3–Cys 6**

The anaphylatoxin domain is characterised by the following signature and all six Cys are involved in intradomain disulfide bridges:

ANAPHYLATOXIN_1 (PS01177): [CSH]–**C**–x(2)–[GAP]–x(7, 8)–[GASTDEQR]–**C**–[GASTDEQL]
–x(3, 9)–[GASTDEQN]–x(2)–[**C**E]–x(6, 7)–**C**–**C**

The 3D structure of anaphylatoxin C3a (PDB: 2A73; 77 aa: Ser 650–Arg 726) derived from the crystal structure of the entire complement C3 is shown in Figure 4.16 (for the crystal structure of the entire complement C3 see Chapter 8). The α-helices in C3a are stabilised by three disulfide bridges (in pink).

## 4.15 THE CUB DOMAIN

The ***CUB domain*** is approximately 115 residues in length and is found in functionally diverse, mainly developmentally regulated, proteins. In blood plasma the complement components C1r and C1s, the complement-activating component and the mannan-binding lectin serine protease 2 each contain two CUB domains. Many CUB domains contain two conserved disulfide bridges arranged in the pattern **Cys 1–Cys 2, Cys 3–Cys 4**. The 3D structure of the CUB2 domain (PDB: 1NT0; 114 aa: Cys 169–Ala 281) in the mannan-binding lectin serine protease 2 (MASP-2), which is part of the larger CUB1–EGF-like–CUB2 region, was determined by X-ray diffraction (278 aa: Pro5Ser–Ala 282 + His-tag) and is shown in Figure 4.17. The structure is characterised by seven β-strands stabilised by two intrachain disulfide bridges (in pink). Part of the loop containing Cys 3–Cys 4 is modelled (grey dashed line). The CUB1-EGF-CUB2 triple domain fragment reveals an elongated structure with a prominent concave surface proposed to contain the binding site for the mannose-binding proteins.

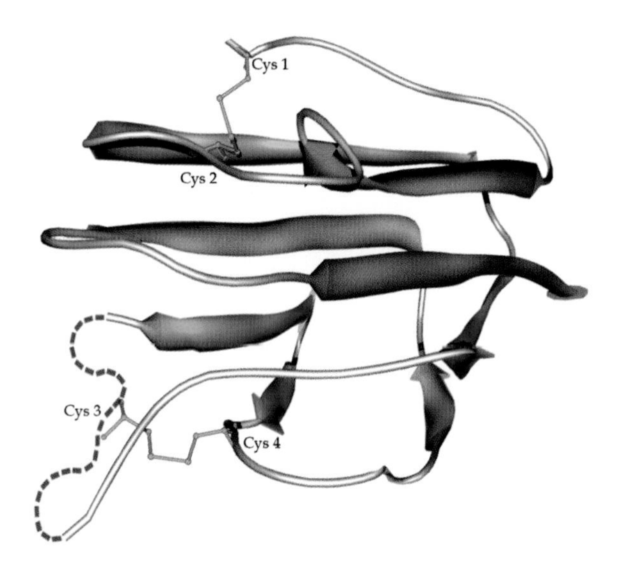


**Figure 4.17 3D structure of the CUB2 domain in the mannan-binding lectin serine protease 2 (PDB: 1NT0)**
3D structure of the CUB2 (114 aa: Cys 169 – Ala 281) domain in the mannan-binding lectin serine protease 2 (MASP-2) contains two intrachain disulfide bridges (in pink). Part of the loop containing Cys 3 – Cys 4 was modelled (dashed grey line).

## 4.16 THE NTR DOMAIN

The ***netrin*** (NTR) ***domain*** is found in proteins of different origin, among them the C-terminal region of netrins and complement components C3, C4 and C5. The NTR module is a basic domain of about 140 amino acids containing six conserved Cys residues most likely arranged in the disulfide bridge pattern

**Cys 1–Cys 3, Cys 2–Cys 6, Cys 4–Cys 5**

and several conserved hydrophobic sequences including a **YLLLG**-like motif. The 3D structure of the C-terminal NTR domain of complement C5 (PDB: 1XWE; 151 aa: Ala 1512–Cys 1658 + N-terminal tetrapeptide) determined by NMR spectroscopy is shown in Figure 4.18. The NTR domain adopts the known oligosaccharide/oligonucleotide binding fold consisting of a five-stranded $\beta$-barrel with two terminal $\alpha$-helices, stabilised by three intrachain disulfide bridges (in pink).

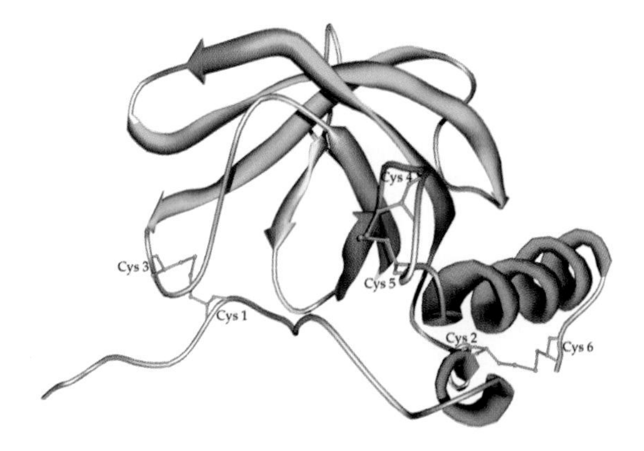


**Figure 4.18 3D structure of the C-terminal NTR domain of complement C5 (PDB: 1XWE)**
3D structure of the C-terminal NTR domain of complement C5 (151 aa: Ala 1512 – Cys 1658 + N-terminal tetrapeptide) containing three intrachain disulfide bridges (in pink).

The NTR domain in complement C5 interacts with the C5 convertase (C4b2b3b) and binds with high affinity the complement components C6 and C7, important events in the initial phase of the formation of the membrane attack complex (MAC).

## 4.17 THE C-TYPE LECTIN (CTL) DOMAIN

The **C-type lectin** (CTL) ***domain*** was first characterised in lectins and is thought to function as a calcium-dependent carbohydrate-recognition domain (CRD). The CTL domain is present in many different protein families and is also contained in some plasma proteins like thrombomodulin, tetranectin and mannose-binding protein C. The CTL domain is 110–140 residues long and contains three disulfide bridges arranged in the following pattern (bridges two and three are strictly conserved):

**Cys 1–Cys 2, Cys 3–Cys 6, Cys 4–Cys 5**

The CTL domain is characterised by the following signature and the three Cys are involved in disulfide bridges:

C_TYPE_LECTIN_1 (PS00615): **C**–[LIVMFYATG]–x(5, 12)–[WL]–x–[DNSR]–{C}–{LI}–**C**
–x(5, 6)–[FYWLIVSTA]–[LIVMSTA]–**C**

The 3D structure of the CTL domain of tetranectin (PDB: 1TN3; 137 aa: Ala 45–Val 181) determined by X-ray diffraction is shown in Figure 4.19. The CTL structure consists of two distinct regions: region 1 (in green) contains six β-strands and two α-helices stabilised by two disulfide bridges (Cys 1–Cys 2 and Cys 3–Cys 6) and region 2 (in yellow) is composed of four loops with one disulfide bridge (Cys 4–Cys 5) and binds two calcium ions (red balls). The overall fold is similar to other CRD structures. In addition, the 3D structure of the CTL domain in mannose-binding protein C (MBP-C) was also determined (PDB: 1HUP) (for details see Chapter 8).

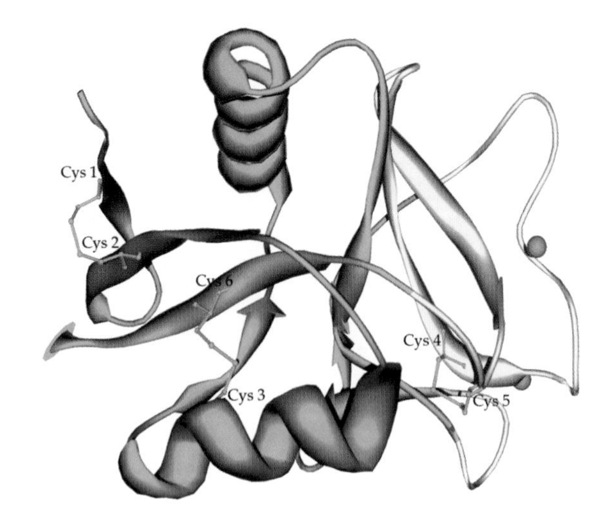


**Figure 4.19 3D structure of the CTL domain of tetranectin (PDB: 1TN3)**
3D structure of the CTL domain of tetranectin (137 aa: Ala 45 – Val 181) consists of two regions: Region 1 (in green) contains two disulfide bridges: Cys 1 – Cys 2 and Cys 3 – Cys 6 and region 2 (in yellow) one disulfide bridge (Cys 4 – Cys 5) with two calcium ions (orange balls).

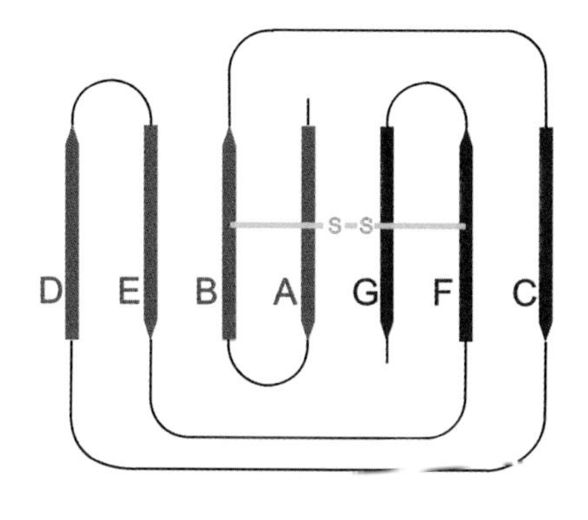


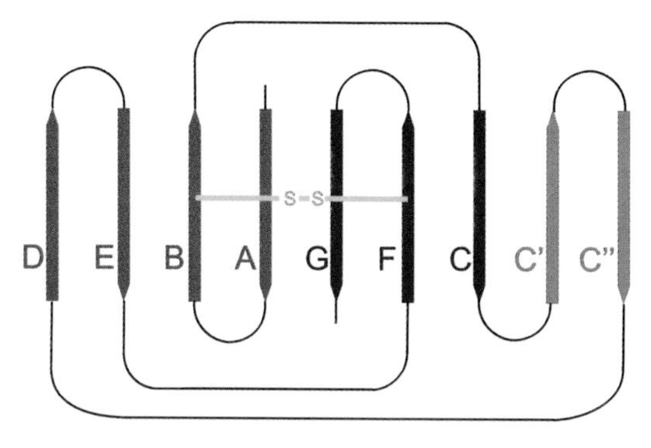


**Figure 4.20 Schematic representation of Ig-like domains**
Top: C1-type domain contains two sheets: sheet I (in grey) with ABED strands and sheet II (in blue) with CFG strands. The single disulfide bridge (in yellow) connects strands B and F. Bottom: V-type domain contains two sheets: sheet I (in grey) with ABED strands and sheet II (in blue) with C″C′CFG strands. Extra strands C′ and C″ (in light blue) are inserted between strand C and D. The single disulfide bridge (in yellow) connects strands B and F.

## 4.18 THE IG-LIKE DOMAIN

The ***Ig-like domain*** is one of the most widespread domains in the animal kingdom and is present in immunoglobulins and the major histocompatibility complex (MHC) of the immune system and in certain receptors. The Ig-like domains are primarily involved in binding functions and the ligands have a wide range: antigens, hormones and large-scale molecules. The Ig-like domains contain 7–10 β-strands arranged into two β-sheets with a typical topology in the form of a Greek key β-barrel. The schematic representation of typical Ig-like domains is shown in Figure 4.20.

The Ig-like domains are classified according to the numbers of β-strands:

*C1-type*. The classical Ig-like domain is described in Figure 4.20 and contains two sheets: Sheet I with ABED strands and Sheet II with CFG strands. The C1-type is primarily present in immunoglobulins (Ig), the major histocompatibility complex (MHC) and T-cell receptors.

*C2-type*. Strand D is deleted and is replaced by strand C′, which is connected to strand E. The C2-type is present in various types of receptors and adhesion molecules.

*V-type*. Extra strands C′ and C″ are inserted between strand C and D. There are two sheets: Sheet I with ABED strands and Sheet II with C″C′CFG strands. The V-type is present in the variable region of immunoglobulin heavy chains and other proteins.

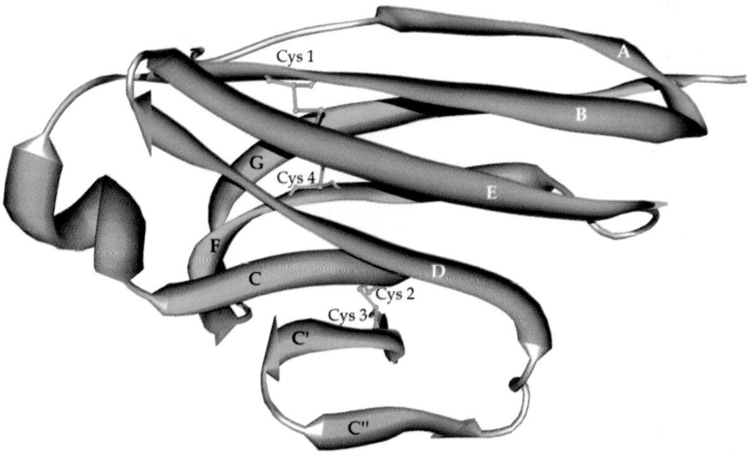

**Figure 4.21 3D structure of the first Ig-like V-type domain of the polymeric immunoglobulin receptor (PDB: 1XED)**
The 3D structure is characterised by two β-sheets: Sheet I (β-strands A, B, E and D) and sheet II (β-strands C, C′, C″, F and G). The two intrachain disulfide bridges are in pink.

The 3D structure of the first Ig-like V-type domain of the polymeric immunoglobulin receptor (PDB: 1XED; 109 aa + His-tag: Lys 1–Val 109 + His-tag) was determined by X-ray diffraction and is shown in Figure 4.21. The structure represents the V-type of the Ig-like domain containing two intrachain disulfide bridges Cys 1–Cys 4 and Cys 2–Cys 3 (in pink) and is characterised by two β-sheets: sheet I (in blue) contains β-strands A, B, E and D and sheet II (in green) strands C, C′, C″, F and G and β-strands B and C are connected with a short helix (in red). The two extra strands C′ and C″ are inserted between strands C and D, compared with the classical C1-type Ig-like domain.

## REFERENCES


Amoresano, *et al.*, 2001, Assignment of the complete disulfide bridge pattern in the human recombinant follitropin beta-chain, *Biol. Chem.*, **382**, 961–968.

An, *et al.*, 1998, Structural/functional properties of the Glu 1–HSer 57 N-terminal fragment of human plasminogen: conformational characterization and interaction with kringle domains, *Protein Sci.*, **7**, 1947–1959.

Appella, *et al.*, 1988, Structure and function of epidermal growth factor-like regions in proteins, *FEBS Lett.*, **231**, 1–4.

Banner, *et al.*, 1996, The crystal structure of the complex of blood coagulation factor VIIa with soluble tissue factor, *Nature*, **380**, 41–46.

Banyai and Patthy, 1999, The NTR module: domains of netrin, secreted frizzled related protein and type I procollagen C-proteinase enhancer protein are homologous with tissue inhibitors of metalloproteinases, *Protein Sci.*, **8**, 1636–1642.

Baron, *et al.*, 1990, Structure of the fibronectin type 1 module, *Nature*, **345**, 642–646.

Bork, 1991, Shuffled domains in extracellular proteins, *FEBS Lett.*, **286**, 47–54.

Bork and Beckmann, 1993, The CUB domain. A widespread module in developmentally regulated proteins, *J. Mol. Biol.*, **231**, 539–545.

Bramham, *et al.*, 2005, Functional insights from the structure of the multifunctional C345C domain of C5 of complement, *J. Biol. Chem.*, **280**, 10636–10645.

Daly, *et al.*, 1995, Three-dimensional structure of a cysteine-rich repeat from the low-density lipoprotein receptor, *Proc. Natl. Acad. Sci. USA*, **92**, 6334–6338.

Dolmer, *et al.*, 1993, Disulfide bridges in human complement component C3b, *FEBS Lett.*, **315**, 85–90.

Drickamer, 1988, Two distinct classes of carbohydrate-recognition domains in animal lectins, *J. Biol. Chem.*, **263**, 9557–9560.

Emsley, *et al.*, 1998, Crystal structure of the von Willebrand factor A1 domain and implications for collagen binding, *J. Biol. Chem.*, **273**, 10396–10401.

Fan, *et al.*, 2005, Structure of human follicle-stimulating hormone in complex with its receptor, *Nature*, **433**, 269–277.

Feinberg, *et al.*, 2003, Crystal structure of the CUB1-EGF-CUB2 region of mannose-binding protein associated serine protease 2, *EMBO J.*, **22**, 2348–2359.

Freedman, *et al.*, 1995, Structure of the calcium-ion bound gamma-carboxyglutamic acid-rich domain of factor IX, *Biochemistry*, **34**, 12126–12137.
Gao, *et al.*, 2003, Structure and functional significance of mechanically unfolded fibronectin type III 1 intermediates, *Proc. Natl. Acad. Sci. USA*, **100**, 14784–14789.
Guertin *et al.*, 2002, Optimization of the β-aminoester class factor Xa inhibitors. Part 2: identification of FXV673 as a potent and selective inhibitor with excellent in vivo anticoagulant activity, *Bioorg. Med. Chem. Lett.*, **12**, 1671–1674.
Halaby, *et al.*, 1999, The immunoglobulin fold family: sequence analysis and 3D structure comparisons, *Protein Eng.*, **12**, 563–571.
Hamburger, *et al.*, 2004, Crystal structure of a polymeric immunoglobulin binding fragment of the human polymeric immunoglobulin receptor, *Structure*, **12**, 1925–1935.
Huang, *et al.*, 2004, Crystal structure of the calcium-stabilized human factor IX Gla domain bound to a conformation-specific anti-factor IX antibody, *J. Biol. Chem.*, **279**, 14338–14346.
Hugli, 1986, Biochemistry and biology of anaphylatoxins, *Complement*, **3**, 111–127.
Huizinga, *et al.*, 1997, Crystal structure of the A3 domain of the human von Willebrand factor: implications for collagen binding, *Structure*, **5**, 1147–1156.
Janssen, *et al.*, 2005, Structures of the complement C3 provide insights into the function and evolution of immunity, *Nature*, **437**, 505–511.
Jeon, *et al.*, 2001, Implications for familial hypercholesterolemia from the structure of the LDL receptor YWTD-EGF domain pair, *Nat. Struct. Biol.*, **8**, 499–504.
Kane and Davie, 1988, Blood coagulation factors V and VIII: structural and functional similarities and their relationship to hemorrhagic and thrombotic disorders, *Blood*, **71**, 539–555.
Kastrup, *et al.*, 1998, Structure of the C-type lectin carbohydrate recognition domain of human tetranectin, *Acta Crystallogr. D*, **54**, 757–766.
Kvansakul, *et al.*, 2004, Structure of the thrombospondin C-terminal fragment reveals a novel calcium core in the type 3 repeats, *EMBO J.*, **23**, 1223–1233.
Lawler and Hynes, 1986, The structure of human thrombospondin, an adhesive glycoprotein with multiple calcium-binding sites and homologies with several different proteins, *J. Cell. Biol.*, **103**, 1635–1648.
Leahy, *et al.*, 1992, Structure of a fibronectin type III domain from tenascin phased by MAD analysis of the selenomethionyl protein, *Science*, **258**, 987–991.
Marti, *et al.*, 1999, Solution structure and dynamics of the plasminogen kringle 2–AMCHA complex: 3(1)–helix in homologous domains, *Biochemistry*, **38**, 15741–15755.
Macedo-Ribeiro *et al.*, 1999, Crystal structure of the membrane-binding C2 domain of human coagulation factor V, *Nature*, **402**, 434–439.
Muranyi, *et al.*, 1998, Solution structure of the N-terminal EGF-like domain from human factor VII, *Biochemistry*, **37**, 10605–10615.
Norman, *et al.*, 1991, Three-dimensional structure of the complement control protein module in solution, *J. Biol. Chem.*, **219**, 717–725.
Patthy, *et al.*, 1985, Evolution of the proteases of blood coagulation and fibrinolysis by assembly from modules, *Cell*, **41**, 657–663.
Pickford, *et al.*, 2001, The hairpin structure of the (6)F1(1)F2(2)F2 fragment from human fibronectin enhances gelatin binding, *EMBO J.*, **20**, 1519–1529.
Pratte *et al.*, 1999, Structure of the C2 domain of human factor VIII at 1.5 Angstrom resolution, *Nature*, **402**, 439–442.
Reid and Day, 1989, Structure–function relationships of the complement components, *Immunol. Today*, **10**, 177–180.
Rudenko, *et al.*, 2002, Structure of the LDL receptor extracellular domain at endosomal pH, *Science*, **298**, 2353–2358.
Selander-Sunnerhagen, *et al.*, 1992, How an epidermal growth factor (EGF)-like domain binds calcium. High resolution NMR structure of the calcium form of the, $NH_2$-terminal EGF-like domain in coagulation factor X, *J. Biol. Chem.*, **267**, 19642–19649.
Schwarzenbacher, *et al.*, 1999, Crystal structure of human $\beta_2$-glycoprotein I: implications for phospholipid binding and the antiphospholipid syndrome, *EMBO J.*, **18**, 6228–6239.
Springer 1998, An extracellular beta-propeller module predicted in lipoprotein and scavenger receptors, tyrosine kinases, epidermal growth factor precursor, and extracellular matrix components, *J. Mol. Biol.*, **283**, 837–862.
Sticht, *et al.*, 1998, Solution structure of the glycosylated second type 2 module of fibronectin, *J. Mol. Biol.*, **276**, 177–187.
Sun and Davies, 1995, The cystine-knot growth-factor superfamily, *Annu. Rev. Biophys. Biomol. Struct.*, **24**, 269–291.
Tan, *et al.*, 2002, Crystal structure of the TSP-1 type 1 repeat: a novel layered fold and its biological implication, *J. Cell. Biol.*, **159**, 373–382.
Tan, *et al.*, 2006, The structures of the thrombospondin-1 N-terminal domain and its complex with a synthetic pentameric heparin, *Structure*, **14**, 33–42.
Tordai, *et al.*, 1999, The PAN module: the N-terminal domains of plasminogen and hepatocyte growth factor are homologous with the apple domains of the prekallikrein family and with a novel domain found in numerous nematode proteins, *FEBS Lett.*, **461**, 63–67.
Turk and Bode, 1991, The cystatins: protein inhibitors of cysteine proteinases, *FEBS Lett.*, **285**, 213–219.
Ultsch, *et al.*, 1998, Crystal structure of the NK1 fragment of human hepatocyte growth factor at 2.0 Å resolution, *Structure*, **6**, 1383–1393.
Vermeer, 1990, Gamma-carboxyglutamate-containing proteins and the vitamin-K dependent carboxylase, *Biochem. J.*, **266**, 625–636.
Williams, *et al.*, 1994, Solution structure of a pair of fibronectin type I modules with fibrin binding activity, *J. Mol. Biol.*, **235**, 1302–1311.

# 5

# Protein Families

## 5.1 INTRODUCTION

Many plasma proteins are multidomain proteins composed of various types of domains in single or multiple copies. The vast majority of the plasma proteins are glycoproteins and over 90 % carry at least one posttranslational modification. Therefore, many of them share similar structural properties and exhibit similar biological functions. Many plasma proteins are assigned to distinct protein families, such as the serine protease inhibitor (serpin) family (see Section 20035.3) or the serum albumin family (see Section 20035.5). The various protein families cover a wide range of the proteins present in plasma and a selection of the main plasma protein families is described in this chapter.

## 5.2 SERINE PROTEASES: THE TRYPSIN FAMILY

Many plasma proteins are synthesised as zymogens of peptidases or contain in their multidomain architecture a peptidase domain. As an example the serine protease trypsin family belonging to the family S1 in the classification of peptidases is described in some detail. A selection of plasma proteins belonging to the peptidase S1 family is given in Table 5.1.

The protease domain of the serine protease trypsin family is about 25 kDa in size and contains a partially conserved disulfide bridge pattern. Many serine protease domains contain four conserved disulfide bridges arranged in the following pattern:

**Cys 1–Cys 2, Cys 3–Cys 7, Cys 4–Cys 5, Cys 6–Cys 8**

The trypsin family is characterised by two signatures, covering the sequences in the vicinity of the His and Ser residues in the catalytic triad, respectively:

TRYPSIN_HIS (PS00134): [LIVM]–[ST]–A–[STAG]–**H**–C
TRYPSIN_SER (PS00135): [DNSTAGC]–[GSTAPIMVQH]–x(2)–G–[DE]–**S**–G–[GS]–[SAPHV]–[LIVMFYWH]–[LIVMFYSTANQH]

The catalytic activity of the serine proteases from the trypsin family is usually generated from an inactive zymogen form by limited proteolysis with a concomitant large conformational change that leads to a charge relay system with the catalytic triad composed of an Asp, His and Ser residue (for details see Chapter 10).

The 3D structure of the serine protease domain is usually characterised by twelve β-strands grouped into two β-sheets surrounded by α-helical and $3_{10}$-helical segments. Of the many available 3D structures of the peptidase S1 family the 3D structure of trypsin I (PDB: 1TRN; 224 aa: Ile 1–Ser 224) determined by X-ray diffraction is shown as an example in Figure 5.1. The fold of trypsin is very similar to other serine protease domains and the residues of the catalytic triad His 40, Asp 84 and Ser 177 are shown in pink. Trypsin I contains five disulfide bridges (in light green) arranged in the pattern Cys 1–Cys 5, Cys 2–Cys 3, Cys 4–Cys 9, Cys 6–Cys 7, Cys 8–Cys 10.

*Human Blood Plasma Proteins: Structure and Function* Johann Schaller, Simon Gerber, Urs Kämpfer, Sofia Lejon and Christian Trachsel

**Table 5.1** Plasma proteins belonging to the peptidase S1 family.

| Protein | Data bank entry | Nomenclature |
|---|---|---|
| *Coagulation and fibrinolysis* | | |
| Thrombin (factor II) | THRB_HUMAN/P00734 | EC 3.4.21.5 |
| Factor VIIa (proconvertin) | FA7_HUMAN/P08709 | EC 3.4.21.21 |
| Factor IXa (Christmas factor) | FA9_HUMAN/P00740 | EC 3.4.21.22 |
| Factor Xa (Stuart factor) | FA10_HUMAN/P00742 | EC 3.4.21.6 |
| Factor XIa | FA11_HUMAN/P03951 | EC 3.4.21.27 |
| Factor XIIa (Hagemann factor) | FA12_HUMAN/P00748 | EC 3.4.21.38 |
| Protein C (factor XIV) | PROC_HUMAN/P04070 | EC 3.4.21.69 |
| Protein Z | PROZ_HUMAN/P22891 | No activity |
| Plasmin | PLMN_HUMAN/P00747 | EC 3.4.21.7 |
| Plasma kallikrein | KLKB1_HUMAN/P03952 | EC 3.4.21.34 |
| Tissue-type plasminogen activator | TPA_HUMAN/P00750 | EC 3.4.21.68 |
| Urokinase-type plasminogen activator | UROK_HUMAN/P00749 | EC 3.4.21.73 |
| Hepatocyte growth factor activator | HGFA_HUMAN/Q04756 | EC 3.4.21.- |
| Hepatocyte growth factor activator-like | HABP2_HUMAN/Q14520 | No activity |
| Hepatocyte growth factor (scatter factor) | HGF_HUMAN/P14210 | No activity |
| Hepatocyte growth factor-like | HGFL_HUMAN/P26927 | No activity |
| *Complement System* | | |
| C1r subcomponent | C1R_HUMAN/P00736 | EC 3.4.21.41 |
| C1s subcomponent | C1S_HUMAN/P09871 | EC 3.4.21.42 |
| C3/C5 convertase (complement C2) | CO2_HUMAN/P06681 | EC 3.4.21.43 |
| C3/C5 convertase (factor B) | CFAB_HUMAN/P00751 | EC 3.4.21.47 |
| Factor D (C3 convertase activator) | CFAD_HUMAN/P00746 | EC 3.4.21.46 |
| Factor I (C3b/C4b inactivator) | CFAI_HUMAN/P05156 | EC 3.4.21.45 |
| Complement-activating component | MASP1_HUMAN/P48740 | EC 3.4.21.- |
| *Enzymes* | | |
| Chymotrypsin B | CTRB1_HUMAN/P17538 | EC 3.4.21.1 |
| Trypsin I (cationic) | TRY1_HUMAN/P07477 | EC 3.4.21.4 |
| Trypsin II (anionic) | TRY2_HUMAN/P07478 | EC 3.4.21.4 |
| Chymase | MCTP1_HUMAN/P23946 | EC 3.4.21.39 |
| Leukocyte elastase | ELNE_HUMAN/P08246 | EC 3.4.21.37 |
| Prostate specific antigen | KLK3_HUMAN/P07288 | EC 3.4.21.77 |
| Tryptase beta-2 | TRYB2_HUMAN/P20231 | EC 3.4.21.59 |
| Cathepsin G | CATG_HUMAN/P08311 | EC 3.4.21.20 |
| *Lipoproteins* | | |
| Apolipoprotein(a) | APOA_HUMAN/P08519 | EC 3.4.21.- |

## 5.3 SERINE PROTEASE INHIBITORS (SERPINS)

Serine protease inhibitors (serpins) are the most important inhibitor family of serine proteases. The enumeration of important plasma serpins is given in Table 5.2.

Serpins are high molecular weight serine protease inhibitors of approximately 400–500 amino acids in length. They are termed suicide inhibitors because in an irreversible reaction the reactive site (P1–P1′) in the reactive loop peptide forms a covalent bond with the Ser side chain of the catalytic triad in the serine protease and the subsequent cleavage of the reactive site in the serpin (for details see Chapter 11). The serpins are characterised by the following signature:

SERPIN (PS00284): [LIVMFY]–{G}–[LIVMFYAC]–[DNQ]–[RKHQS]–[PST]–F–[LIVMFY]
–[LIVMFYC]–x–[LIVMFAH]

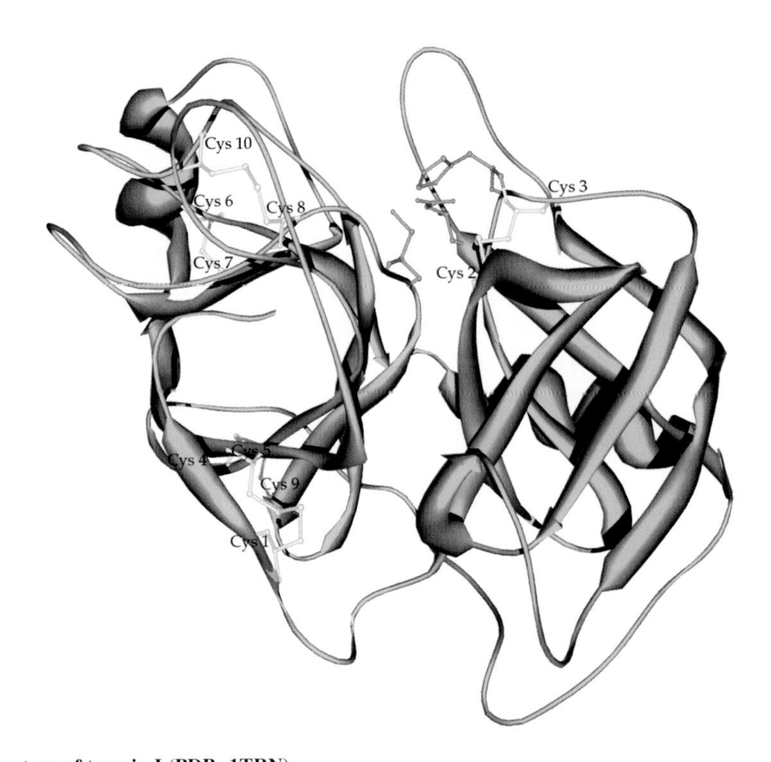


**Figure 5.1 3D structure of trypsin I (PDB: 1TRN)**
3D structure of trypsin 1 (224 aa: Ile 1 – Ser 224) with the residues of the catalytic triad His 40, Asp 84 and Ser 177 and the five disulfide bridges (in pink).

**Table 5.2** Blood plasma serpins.

| Protein | Data bank entry | Specificity |
|---|---|---|
| $\alpha_1$-Antitrypsin | A1AT_HUMAN/P01009 | Elastase, thrombin, FXa, FXIa, protein C |
| $\alpha_1$-Antichymotrypsin | AACT_HUMAN/P01011 | Chymotrypsin, cathepsin G, chymase, prostate specific antigen |
| Antithrombin-III | ANT3_HUMAN/P01008 | Thrombin, FIXa, FXa, FXIa |
| $\alpha_2$-Antiplasmin | A2AP_HUMAN/P08697 | Plasmin, FXIIa |
| Plasminogen activator inhibitor-1 | PAI1_HUMAN/P05121 | tPA, uPA, thrombin, plasmin, protein C |
| Plasminogen activator inhibitor-2 | PAI2_HUMAN/P05120 | tPA, uPA |
| Plasma serine protease inhibitor | IPSP_HUMAN/P5154 | Protein C, plasma kallikrein, thrombin, FXa, FXIa, uPA |
| Plasma protease C1 inhibitor | IC1_HUMAN/P05155 | C1r, C1s, FXIIa, kallikrein |
| Heparin cofactor II | HEP2_HUMAN/P05546 | Thrombin, chymotrypsin, cathepsin G |
| Kallistatin | KAIN_HUMAN/P29622 | Tissue kallikrein |
| Protein Z-dependent protease inhibitor | ZPI_HUMAN/Q9UK55 | FXa |
| Pigment epithelium-derived factor | PEDF_HUMAN/P36955 | Inhibitor angiogenesis |
| Thyroxine-binding globulin | THBG_HUMAN/P05543 | No inhibitory activity |
| Corticosteroid-binding globulin | CBG_HUMAN/P08185 | No inhibitory activity |
| Angiotensinogen | ANGT_HUMAN/P01019 | No inhibitory activity |
| Leukocyte elastase inhibitor | ILEU_HUMAN/P30740 | Neutrophile elastase, cathepsin G |
| Neuroserpin | NEUS_HUMAN/Q99574 | Protects neurons (tPA) |
| Ovalbumin | OVAL_CHICK/P01012 | No inhibitory activity |

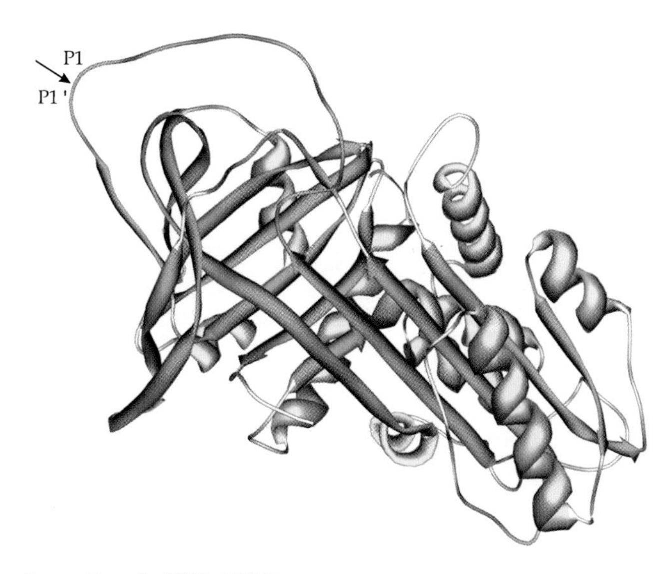


**Figure 5.2 3D structure of $\alpha_1$-antitrypsin (PDB: 1QLP)**
3D structure of intact $\alpha_1$-antitrypsin (394 aa: Glu1Met – Lys 394) is characterised by nine $\alpha$-helices, three $\beta$-sheets and the reactive loop peptide (in pink) with the reactive bond P1 – P1′ (Met 358 – Ser 359).

The 3D structure of serpins is usually characterised by nine $\alpha$-helices, three $\beta$-sheets and the reactive loop peptide with the reactive site located on the surface of the serpin. Disulfide bridges in serpins are not conserved and existing disulfide bridges are often not essential for their biological activity. Of the many available 3D structures of serpins the 3D structure of intact $\alpha_1$-antitrypsin (PDB: 1QLP; 394 aa: Glu1Met–Lys 394) determined by X-ray diffraction is given in Figure 5.2. The structure is characterised by the typical structural folds of serpins containing nine $\alpha$-helices, three $\beta$-sheets and the reactive loop peptide (in pink) with the reactive bond P1–P1′ (Met 358–Ser 359).

## 5.4 THE $\alpha_2$-MACROGLOBULIN FAMILY

The members of the $\alpha_2$-macroglobulin family are very important protease inhibitors in blood plasma and belong to the protease inhibitor I39 family. The members are compiled in Table 5.3.

The $\alpha_2$-macroglobulin-like proteins are able to inhibit all four classes of proteases: serine proteases, cysteine proteases, aspartyl proteases and metalloproteases, by a 'trapping' mechanism. They have a peptide stretch termed 'bait region' which contains specific cleavage sites for all proteases. Upon cleavage of the bait region by the protease a conformational change is induced in the inhibitor, thus trapping the protease. Subsequently, the thiol ester bond formed between the side chains of a Cys and a Gln residue in the inhibitor is cleaved and mediates the covalent binding of the inhibitor to the protease. The members of

**Table 5.3** Members of the $\alpha_2$-macroglobulin family (protease inhibitor I39 family).

| Protein | Data bank entry | Comments |
|---|---|---|
| $\alpha_2$-Macroglobulin | A2MG_HUMAN/P01023 | |
| Pregnancy zone protein | PZP_HUMAN/P20742 | |
| Complement C3 | CO3_HUMAN/P01024 | No inhibitory activity |
| Complement C4 | COA_HUMAN/P0C0L4 | No inhibitory activity |
| | COB_HUMAN/P0C0L5 | |
| Complement C5 | CO5_HUMAN/P01031 | No inhibitory activity |

the protease inhibitor I39 ($\alpha_2$-macroglobulin) family are characterised by the following signature, where Cys and Glu/Gln are involved in the thiol ester bond:

ALPHA_2_MACROGLOBULIN (PS00477): [PG]–x–[GS]–**C**–[GA]–E–[**EQ**]–x–[LIVM]

The complement components C3, C4 and C5 exhibit no inhibitory activity.

As an example the 3D structure of two-chain complement C3 consisting of the β-chain (643 aa: Ser 1–Pro 643) and the α-chain (991 aa: Val 651–Asn 1641) linked by a single disulfide bridge (indicated by a pink circle) was determined by X-ray diffraction and is shown in Figure 5.3 (PDB: 2A73). The structure reveals 13 domains, among them eight macroglobulin domains (MG1–MG8), exhibiting a fibronectin type 3-like core fold, of which 5.5 are in the β-chain and 2.5 in the α-chain. A small hydrophobic core (LNK) formed by a α-helix and a β-sheet is wedged between MG1, MG4 and MG5. The N-terminal region of the α chain contains the C3a anaphylatoxin (ANA) domain followed by an extended loop structure (α′NT), a CUB domain and the region (TED) containing the reactive isoglutamyl cysteine thioester bond (**Cys–Gly–Glu–Gln**) and finally an anchor region linked to the C-terminal netrin domain (C345C). The various domains are shown in different colours. Nine of the 13 domains are unpredicted, suggesting that the members of the $\alpha_2$-macroglobulin family evolved from a core of eight homologous domains.

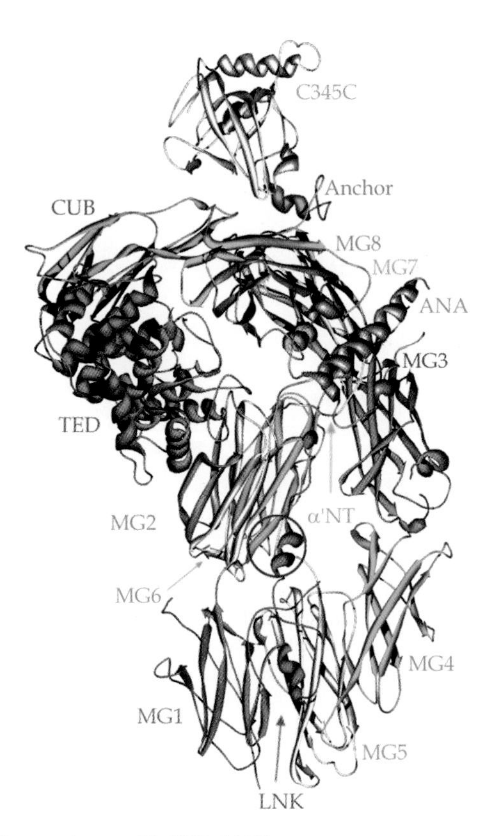


**Figure 5.3 3D structure of two-chain complement C3 (PDB: 2A73)**
Two-chain complement C3 consists of the β-chain (643 aa: Ser 1 – Pro 643) and the α-chain (991 aa: Val 651 – Asn 1641) linked by a single disulfide bridge (indicated by a pink circle). The multidomain protein contains eight macroglobulin domains (MG1 – MG8), the C3a anaphylatoxin (ANA) domain and an extended loop structure (α′NT), one CUB domain and the region containing the thioester bond (TED), and the C-terminal netrin domain (C345C). In addition, there exists a hydrophobic linker domain (LNK) and an anchor region. The various domains are shown in different colours.

## 5.5 THE SERUM ALBUMIN FAMILY

The transport proteins in plasma serum albumin, afamin, α-fetoprotein and vitamin D-binding protein are evolutionary related and belong to the serum albumin family. They all contain three homologous domains of about 190 residues. Each domain contains five to six disulfide bridges arranged in the following pattern:

**Cys 1–Cys 3, Cys 2–Cys 4, Cys 5–Cys 7, Cys 6–Cys 8, Cys 9–Cys 11, Cys 10–Cys 12**

The serum albumin domain is characterised by the following signature, where the three Cys are involved in disulphide bridges:

ALBUMIN (PS00212): [FY]–x(6)–**C**–**C**–x(2)–{C}–x(4)–**C**–[LFY]–x(6)–[LIVMFYW]

As an example the 3D structure of dimeric serum albumin (PDB: 1AO6; 585 aa: Asp 1–Leu 585) determined by X-ray diffraction is shown in Figure 5.4. The structure is characterised by a high content of α-helices and no β-strands and each monomer consists of three homologous domains: domain 1 (in yellow) (195 aa: Asp 1–Lys 195), domain 2 (in red) (188 aa: Gln 196–Glu 383) and domain 3 (in green) (202 aa: Pro 384–Leu 585). The tentative binding sites for long-chain fatty acids are located on the surface of each domain and domains 1 and 3 each contain a pocket lined out primarily with hydrophobic and positively charged residues enabling the binding of various compounds (for details see Chapter 15).

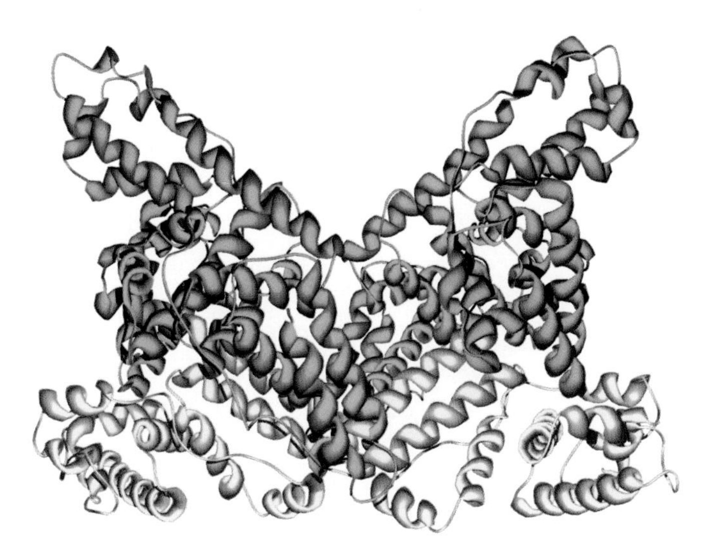

**Figure 5.4 3D structure of serum albumin (PDB: 1AO6)**
3D structure of dimeric serum albumin 585 aa: Asp 1 – Leu 585) is characterised by a high content of α-helices and no β-strands and each monomer consists of three homologous domains: Domain 1 (in yellow), domain 2 (in red) and domain 3 (in green).

## 5.6 THE PANCREATIC TRYPSIN INHIBITOR (KUNITZ) FAMILY SIGNATURE

The pancreatic trypsin inhibitor (Kunitz) family is one of the numerous families of serine protease inhibitors and its best-characterised member is the bovine pancreatic trypsin inhibitor (BPTI) (BPT1_BOVIN/P00974). The inter-α-trypsin inhibitor and the tissue factor pathway inhibitor contain two and three domains, respectively. The domain is approximately 50 residues long and contains three conserved disulfide bridges arranged in the following pattern (for details see Chapter 11):

**Cys 1–Cys 6, Cys 2–Cys 4, Cys 3–Cys 5**

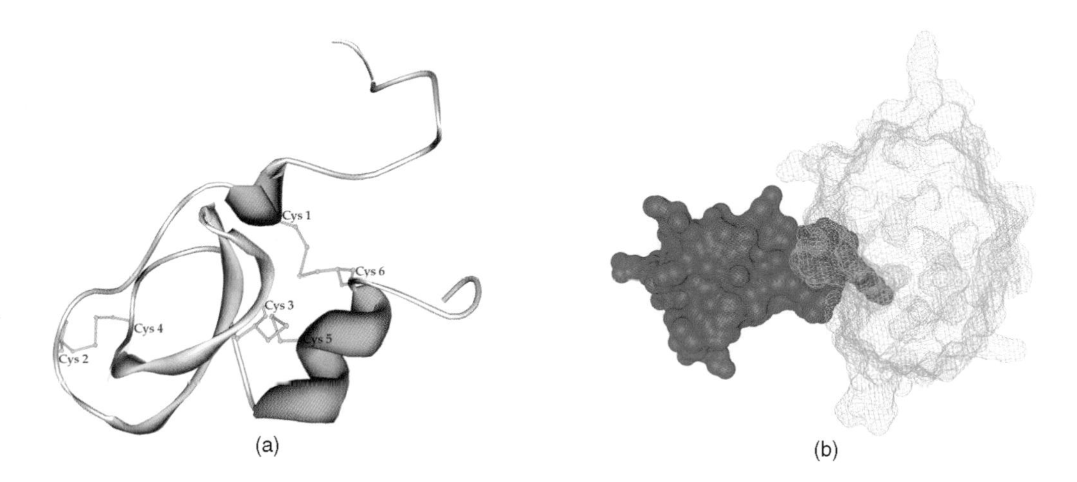


**Figure 5.5 3D structures of (a) the second Kunitz domain of tissue factor pathway inhibitor (PDB: 1ADZ) and (b) the complex of anionic salmon trypsin with bovine pancreatic trypsin inhibitor (PDB: 1BZX)**
(a) The second Kunitz domain (71 aa: Lys 93 – Phe 154 + nonapeptide) in TFPI containing three intrachain disulfide bridges (in pink). (b) The surface representation clearly shows the interaction of BPTI (in red) in the active site of salmon trypsin (in blue, mesh representation).

The Kunitz domain is characterised by the following signature where the two Cys residues are involved in disulfide bridges:

BPTI_KUNITZ_1 (PS00280): F–x(3)–G–**C**–x(6)–[FY]–x(5)–**C**

The 3D structure of the second Kunitz domain of tissue factor pathway inhibitor (71 aa: Lys 93–Phe 154 in TFPI + nonapeptide) determined by NMR spectroscopy is shown in Figure 5.5(a) (PDB: 1ADZ). The three disulfide bridges (in pink) are arranged in the typical pattern of the Kunitz domain. Because of the lack of human examples, the 3D structure of the complex of anionic salmon trypsin (222 aa: Ile 1–Tyr 222; P35031) with bovine pancreatic trypsin inhibitor (58 aa: Arg 1–Ala 58) determined by X-ray diffraction is shown in Figure 5.5(b) (PDB: 1BZX). The surface representation of the complex clearly demonstrates the interaction of the inhibitor BPTI (in red) with the active site of the enzyme trypsin (in blue, mesh representation). The structure may serve as a typical model for a Michaelis–Menten complex.

## 5.7 THE KAZAL SERINE PROTEASE INHIBITORS FAMILY SIGNATURE

The Kazal inhibitors family is one of the many serine protease inhibitor families and its best-characterised member is the pancreatic secretory trypsin inhibitor (PSTI). PSTI prevents the trypsin-catalysed premature activation of zymogens in the pancreas. The domain is approximately 55 residues long and contains three conserved disulfide bridges arranged in the following pattern (for details see Chapter 11):

**Cys 1–Cys 5, Cys 2–Cys 4, Cys 3–Cys 6**

The Kazal domain is characterised by the following signature where the four Cys are involved in disulfide bridges:

KAZAL (PS00282): **C**–x(4)–{C}–x(2)–**C**–x–{A}–x(4)–Y–x(3)–**C**–x(2, 3)–**C**

The 3D structure of a recombinant mutant pancreatic secretory trypsin inhibitor (PDB: 1HPT; 56 aa: Asp 1–Cys 56) was determined by X-ray diffraction and is shown in Figure 5.6. The structure is characterised by a β-sheet and a short α-helix. The three disulfide bridges (in pink) are arranged in the typical pattern of the Kazal domain. Differences with other known PSTI structures mainly occur within the flexible N-terminal region.

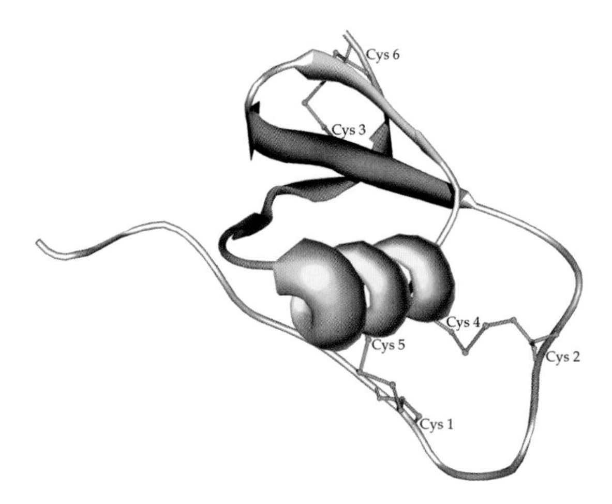


**Figure 5.6 3D structure of pancreatic secretory trypsin inhibitor (PDB: 1HPT)**
3D structure of a recombinant mutant pancreatic secretory trypsin inhibitor (56 aa: Asp 1 – Cys 56) is characterised by a β-sheet and a short α-helix and three intrachain disulfide bridges (in pink).

## 5.8 THE MULTICOPPER OXIDASE FAMILY

Multicopper oxidases are enzymes that possess three different copper centres called type 1 (blue), type 2 (normal) and type 3 (coupled binuclear), with plasma ceruloplasmin (ferroxidase) as a typical member. Based on structural similarities, coagulation factors FV and FVIII also belong to this family. Multicopper oxidases are characterised by two signatures. Signature 1 is not related to copper binding and thus recognises proteins like FV and FVIII that have lost their ability to bind copper and signature 2 is specific for copper-binding proteins. The first two His in signature 2 are part of the type 3 copper-binding site:

MULTICOPPER_OXIDASE1 (PS00079): G–x–[FYW]–x–[LIVMFYW]–x–[CST]–x–{PR}–x(5)–{LFH}–G–[LM]–x(3)–[LIVMFYW]

MULTICOPPER_OXIDASE2 (PS00080): **H**–C–**H**–x(3)–H–x(3)–[AG]–[LM]

The 3D structure of entire ceruloplasmin (PDB: 1KCW; 1046 aa: Lys 1–Gly 1046) determined by X-ray diffraction is given in Figure 5.7. The structure comprises six plastocyanin-like domains which are arranged in a triangular array and are numbered. Three of the six copper atoms (pink balls) with bicarbonate (grey balls) form a trinuclear cluster at the interface of domains 1 and 6. Two of the copper atoms coordinate to three His residues (type 3 coppers) and the remaining only to two His residues (type 2 copper). The other three copper atoms form mononuclear sites in domains 2, 4 and 6 (type 1 coppers). The mononuclear copper atoms coordinate to two His and a Cys residue and those in domains 4 and 6 also to a Met residue. In domain 2 the Met is replaced by a Leu residue.

## 5.9 THE LIPOCALIN FAMILY

The members of the lipocalin protein family transport small hydrophobic molecules like steroids, bilins, retinoids and lipids. They share regions with sequence similarity and a common architecture of the tertiary structure, which is characterised by an eight-stranded antiparallel β-barrel that encloses an internal ligand-binding site. The members of the lipocalin family are compiled in Table 5.4.

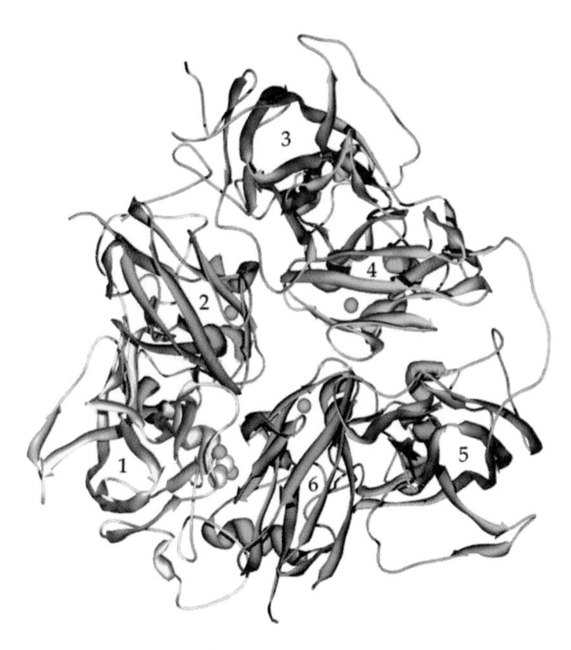


**Figure 5.7 3D structure of ceruloplasmin (PDB: 1KCW)**
3D structure of entire ceruloplasmin (1046 aa: Lys 1 – Gly 1046) contains six plastocyanin-type domains, which are arranged in a triangular array and are numbered. Three of the six copper atoms (pink balls) with bicarbonate (grey balls) form a trinuclear cluster at the interface of domains 1 and 6, the other three copper atoms form mononuclear sites in domains 2, 4 and 6.

The lipocalin family is characterised by a signature located near the first β-strand:

$$\text{LIPOCALIN (PS00213): [DENG]}-\{\text{A}\}-\text{[DENQGSTARK]}-\text{x}\{0,2\}-\text{[DENQARK]}-\text{[LIVFY]}$$
$$-\{\text{CP}\}-\text{G}-\{\text{C}\}-\text{W}-\text{[FYWLRH]}-\{\text{D}\}-\text{LIVMTA]}$$

The 3D structure of the entire γ-chain of complement component C8 (PDB: 1LF7; 182 aa: Gln 1–Arg 182) determined by X-ray diffraction is shown in Figure 5.8(a). The structure displays the typical lipocalin fold with an eight-stranded antiparallel β-barrel and a single α-helix and the ligand citric acid is indicated (in pink). The surface representation shown in Figure 5.8(b) demonstrates the binding pocket for citric acid.

## 5.10 THE GLOBIN FAMILY

Globins are heme-containing proteins that are involved in the binding and transport of oxygen. In vertebrates, there are two globin groups: hemoglobins (Hb) and myoglobins (Mg).

**Table 5.4** Members of the lipocalin family.

| Protein | Data bank entry | Comments |
|---|---|---|
| Complement component C8 gamma chain | CO8G_HUMAN/P07360 | Complement system |
| Plasma retinol-binding protein | RETBP_HUMAN/P02753 | Storage/transport |
| Apolipoprotein D | APOD_HUMAN/P05090 | Lipoprotein |
| $\alpha_1$-Microglobulin | AMBP_HUMAN/P02760 | Inhibitor |
| $\alpha_1$-Acid glycoprotein 1 | A1AG1_HUMAN/P02763 | Immune system |
| $\alpha_1$-Acid glycoprotein 2 | A1AG2_HUMAN/P19652 | Immune system |

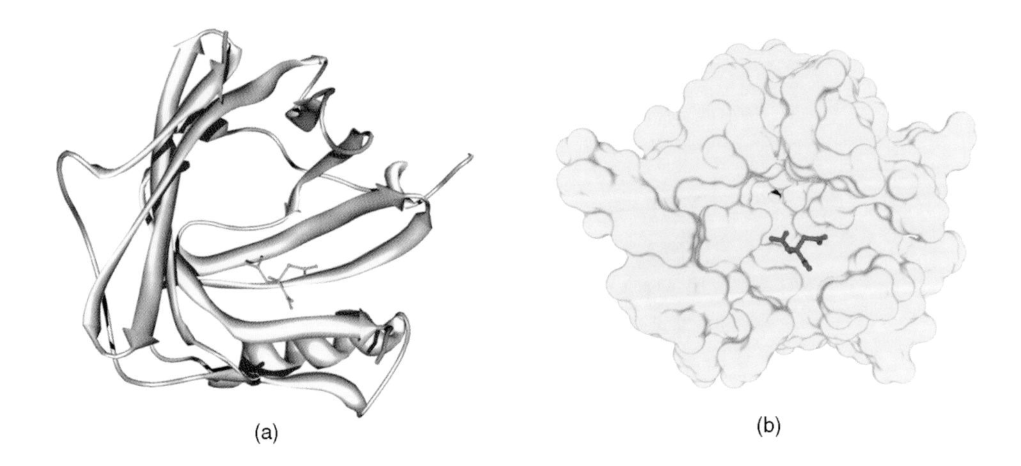


**Figure 5.8 3D structure of the γ-chain of complement component C8 (PDB: 1LF7)**
(a) The 3D structure of the entire γ-chain of complement component C8 (182 aa: Gln 1 – Arg 182) consists of an eight-stranded β-barrel and a single α-helix and the binding pocket contains citric acid as ligand (in pink).
(b) Surface representation of the binding pocket in the complement C8 γ-chain for citric acid (in pink).

Hb transports the oxygen from the lung to the tissues and is a tetramer of two α- and two β-chains. Several forms of hemoglobins exist with different types of chains:

- heterotetramer of two α- and two δ-chains in hemoglobin HbA2, representing less than 3.5 % of adult hemoglobin;
- heterotetramer of two ζ- and two γ-chains in hemoglobin Portland-1.

In addition, several specific fetal forms of hemoglobin exist where the α- or β-chains are replaced by chains with a higher affinity for oxygen:

- heterotetramer of two α- and either two $\gamma_1$- or $\gamma_2$- chains;
- heterotetramer of two ζ- and two ε-chains in early embryonic hemoglobin Gower-1;
- heterotetramer of two α- and two ε-chains in early embryonic hemoglobin Gower-2.

Mg is a monomeric protein responsible for the storage of oxygen in muscles.

Globins exist in almost all organisms. A wide variety of globins are found in invertebrates, leghemoglobins in leguminous plants and flavohemoproteins in bacteria and fungi. It is likely that all these globins have evolved from a common ancestral gene (for details see Chapter 15).

The 3D structure of a mutant hemoglobin (PDB: 1A00) (α-chain, 141 aa: Val 1–Arg 141; β-chain, 146 aa: Val1Met–His 146) determined by X-ray diffraction is shown in Figure 5.9(a). The tetramer contains two α-chains (yellow and red) and two β-chains (green and blue) and each chain contains a heme group (pink), where the iron ions are coordinated to His 58 (distal) and His 87 (proximal) in the α-chain and to His 63 (distal) and His 92 (proximal) in the β-chain. The coordination of the heme group (pink) to the two His residues (blue) in the monomer (ribbon) is shown in Figure 5.9(b) and the surface representation of the monomer demonstrates the heme binding in the binding pocket (Figure 5.9(c)). Both chains are characterised by a high content of α-helices but no β-strands.

## 5.11 THE GLUCAGON/GASTRIC INHIBITORY POLYPEPTIDE (GIP)/SECRETIN/VASOACTIVE INTESTINAL PEPTIDE (VIP) FAMILY

Several polypeptide hormones with a length of 25–45 aa that are mainly expressed in the intestine or in the pancreas belong to the structurally related peptide family of glucagon/GIP/secretin/VIP. The main members of this peptide family are compiled in Table 5.5.

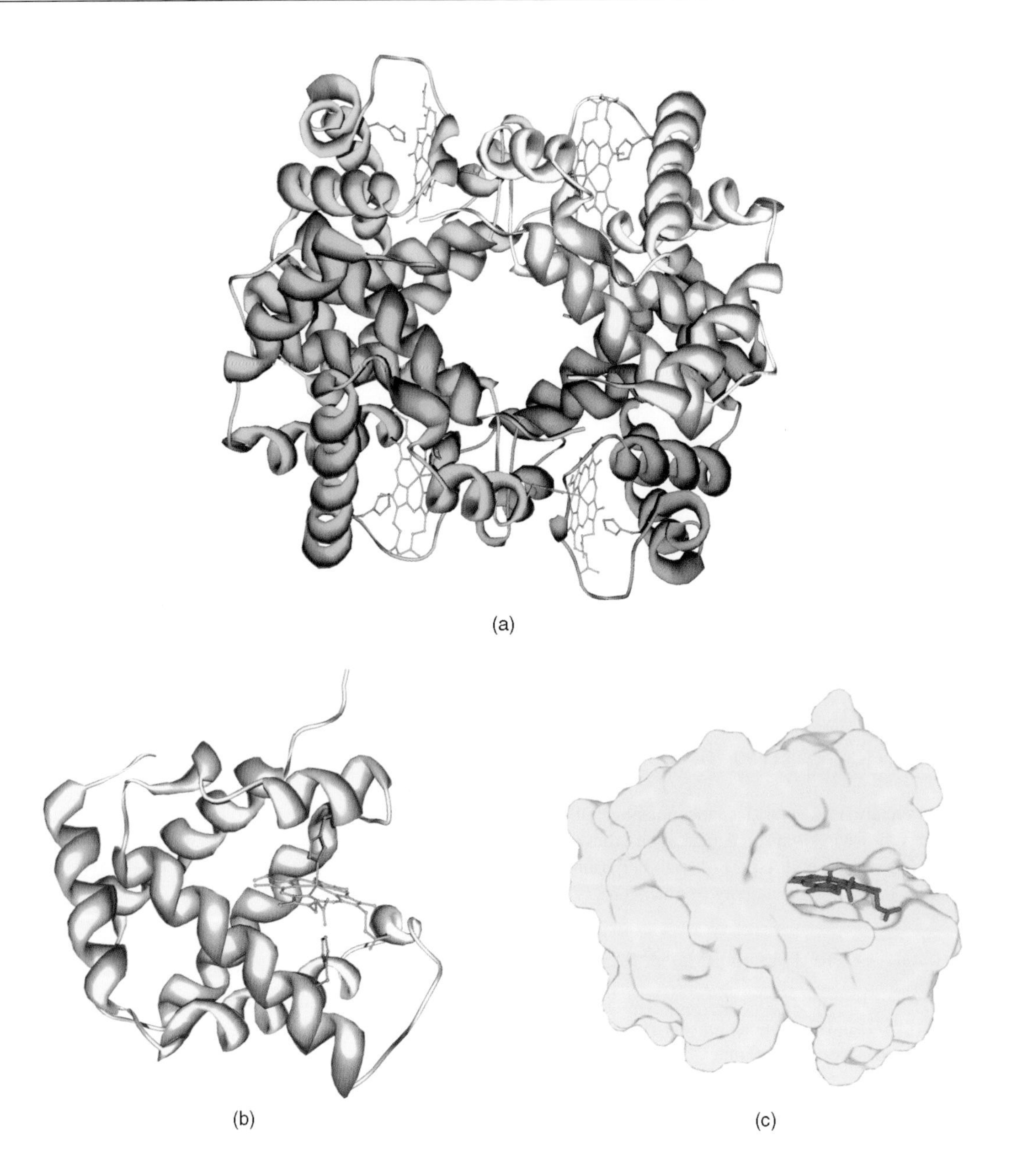


**Figure 5.9 3D structure of hemoglobin (PDB: 1A00)**

(a) The tetramer contains two α-chains (141 aa: Val 1 – Arg 141, in yellow and red) and two β-chains (146 aa: Val1Met – His 146, in green and blue) and each chain contains a heme group (in pink), where the iron ions are coordinated to two His residues.
(b) Coordination of the heme group (in pink) with the two His residues (in blue) in the monomer.
(c) Surface representation of the heme-binding in the binding pocket in the monomer of hemoglobin.

**Table 5.5** Members of the glucagon/GIP/secretin/VIP family.

| Protein | Length | Data bank entry | Comments |
|---|---|---|---|
| Glucagon | 29 aa | GLUC_HUMAN/P01275 | Proglucagon |
| Glucagon-like peptide 1 | 37 aa | GLUC_HUMAN/P01275 | Proglucagon |
| Glucagon-like peptide 2 | 33 aa | GLUC_HUMAN/P01275 | Proglucagon |
| Gastric inhibitory peptide (GIP) | 42 aa | GIP_HUMAN/P09681 | |
| Secretin | 27 aa | SECR_HUMAN/P09683 | |
| Vasoactive intestinal peptide (VIP) | 28 aa | VIP_HUMAN/P01282 | VIP precursor |
| Intestinal peptide PHV-42 | 42 aa | VIP_HUMAN/P01282 | VIP precursor |
| Intestinal peptide PHM-27 | 27 aa | VIP_HUMAN/P01282 | VIP precursor |
| Somatoliberin | 44 aa | SLIB_HUMAN/P1286 | |

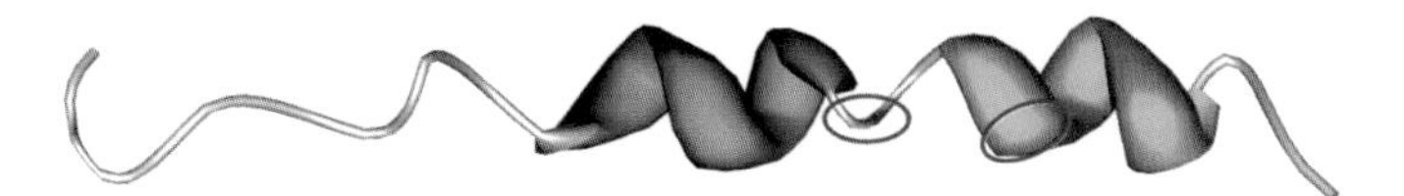

**Figure 5.10 3D structure of glucagon (PDB: 1BH0)**
3D structure of a glucagon mutant (29 aa: His 1 – Thr 29) exhibits an extended fold and the charged residues 17 and 18 (Arg → Lys) and 21 (Asp → Glu), which are important for the biological activity are indicated by pink ellipsoids.

The glucagon/GIP/secretin/VIP peptide hormone family is characterised by a consensus pattern of the more or less conserved first 10 residues and a conserved hydrophobic residue at position 23:

GLUCAGON (PS00260): [YH]–[STAIVGD]–[DEQ]–[AGF]–[LIVMSTE]–[FYLR]–x–[DENSTAK]
–[DENSTA]–[LIVMFYG]–x(8)–{K}–[KREQL]–[KRDENQL]–[LVFYWG]–[LIVQ]

The 3D structures of several glucagon mutants (PDB: 1BH0; 29 aa: His 1–Thr 29) were determined by X-ray diffraction and an example is shown in Figure 5.10. Glucagon exhibits an extended structure with a α-helical segment. Charged residues Arg 17 and 18 (Arg → Lys) and Asp 21 (Asp → Glu) (indicated by pink ellipsoids) as well as the ability to form salt bridges are important for the biological activity of glucagon.

## 5.12 THE GLYCOPROTEIN HORMONE FAMILY

The glycoprotein hormones (gonadotropins) are a family of protein hormones that include follitropin (FSH), lutropin (LSH), thyrotropin (TSH) and choriogonadotropin (CG). They all share a common (identical) α-chain and the β-chains are different but similar in structure. The members are compiled in Table 5.6.

The common α-chain is 92 residues long and contains five disulfide bridges that are arranged in the following pattern:

**Cys 1–Cys 4, Cys 2–Cys 7, Cys 3–Cys 8, Cys 5–Cys 9, Cys 6–Cys 10**

**Table 5.6** Members of the glycoprotein hormone family.

| Protein | Length | Data bank entry | Comments |
|---|---|---|---|
| Glycoprotein hormones alpha chain | 92 aa | GLHA_HUMAN/P01215 | Common α-chain |
| Follitropin (FSH) beta-chain | 111 aa | FSHB_HUMAN/P01225 | β-Chain |
| Lutropin (LSH) beta-chain | 121 aa | LSHB_HUMAN/P01229 | β-Chain |
| Thyrotropin (TSH) beta-chain | 112 aa | TSHB_HUMAN/P01222 | β-Chain |
| Choriogonadotropin (CG) beta-chain | 145 aa | CGHB_HUMAN/P01233 | β-Chain |

Two patterns exist, each containing three conserved Cys residues. In both patterns the three conserved Cys residues are involved in intrachain disulfide bridges:

GLYCO_HORMONE_ALPA_1 (PS00779): **C**–x–G–**C**–**C**–[FY]–S–[RQS]–A–[FY]–P–T–P

GLYCO_HORMONE_ALPA_2 (PS00780): N–H–T–x–**C**–x–**C**–x–T–**C**–x(2)–H–K

The different but structurally similar β-chains are 110 to 150 residues long and share six common disulfide bridges that are arranged in the following pattern:

**Cys 1–Cys 6, Cys 2–Cys 7, Cys 3–Cys 12, Cys 4–Cys 8, Cys 5–Cys 9, Cys 10–Cys 11**

Two consensus patterns exist: the first in the N-terminal region is thought to be involved in the association of the two chains and the second in the C-terminal region with a cluster of five conserved Cys residues:

GLYCO_HORMONE_BETA_1 (PS00261): **C**–[STAGM]–G–[HFYL]–**C**–x–[ST]

GLYCO_HORMONE_BETA_2 (PS00689): [PA]–V–A–x(2)–**C**–x–**C**–x(2)–**C**–x(4)–[STDA]
–[DEY]–**C**–x(6, 8)–[PGSTAVMI]–x(2)–**C**

In both patterns the three conserved Cys residues are involved in intrachain disulfide bridges.

The 3D structure of the heterodimer of a mutant follitropin consisting of the α-chain (92 aa: Ala 1–Ser 92) and the β-chain (111 aa: Asn 1–Glu 111) determined by X-ray diffraction is shown in Figure 5.11 (PDB: 1FL7). The α-chain (in yellow) with five intrachain disulfide bridges (in green) and the β-chain (in blue) with six intrachain disulfide bridges (in pink) have similar folds consisting of central cystine knot motifs from which three β-hairpins extend. The two chains are associated in a head-to-tail arrangement resulting in an elongated, slightly curved structure.

## 5.13 MEMBRANE ATTACK COMPLEX COMPONENTS/PERFORIN SIGNATURE (THE COMPLEMENT C6/C7/C8/C9 FAMILY)

The membrane attack complex (MAC) of the complement system that forms the transmembrane channels and thus disrupts the phospholipid bilayer of target cells leading to cell lysis and cell death is composed of the complement components C5b, C6, C7, C8 and C9. Complement components C6, C7, C8α, C8β, C9 and perforin belong to the complement C6/C7/C8/C9 family. All these proteins share several regions of sequence similarity (for details see Chapter 8, Figure 8). The best conserved sequence stretch is located in the tentative membrane-spanning segment and is characterised by the following

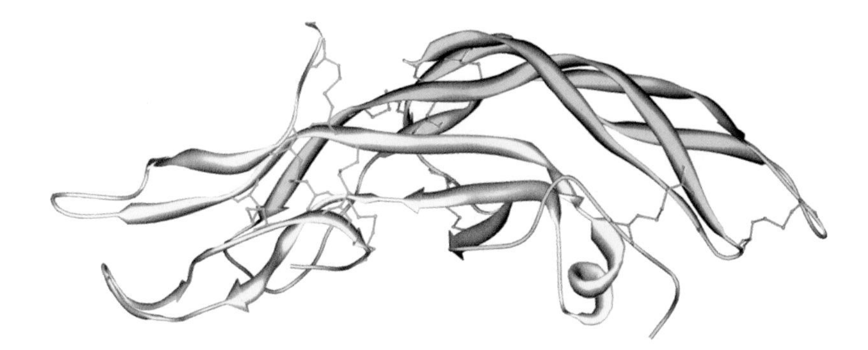

**Figure 5.11 3D structure of follitropin (PDB: 1FL7)**
3D structure of the heterodimer of a mutant follitropin consists of the α-chain (92 aa: Ala 1 – Ser 92) and the β-chain (111 aa: Asn 1 - Glu 111). The α-chain (in yellow) contains five intrachain disulfide bridges (in green), the β-chain (in blue) six (in red).

consensus pattern:

MAC_PERFORIN (PS00279): Y–x(6)–[FY]–G–T–H–[FY]

No entire 3D structure of a member of the complement C6/C7/C8/C9 family is available. However, the 3D structures of the individual domains in these multidomain proteins have been determined (for details see Chapter 4):

- LDL-receptor class A (LDLRA) domain
- Sushi/complement control protein (CCP)/short consensus repeat (SCR) domain
- Epidermal growth factor-like (EGF-like) domain
- Thrombospondin type I (TSP1) repeat

## 5.14 THE LIPASE FAMILY

Lipolytic enzymes like hepatic triacylglycerol lipase (EC 3.1.1.3) and lipoprotein lipase (EC 3.1.1.34) hydrolyse triglycerides and are members of the lipase family. The lecithin-cholesterol acyltransferase (LCAT) (EC 2.3.1.43) that catalyses the fatty acid transfer between phosphatidylcholine and cholesterol also belongs to this family. The most conserved region of these enzymes is located around the Ser residue that together with a His and an Asp residue participates in the charge relay system. The Ser active site region of lipases is characterised by the following consensus pattern, where the Ser residue is part of the active site:

LIPASE_SER (PS00120): [LIV]–{KG}–[LIVFY]–[LIVMST]–G– [HYWV]–**S**–{YAG}–G–[GSTAC]

The 3D structure of a human lipase is not available. Therefore, the 3D structure of the dog pancreatic lipase-related protein 1 (PDB: 1RP1; 450 aa: Lys 1–Cys 450; P06857) determined by X-ray diffraction is shown in Figure 5.12. The dog protein exhibits a sequence similarity of 56 % with human hepatic triacylglycerol lipase. The structure belongs to the α/β-hydrolase fold consisting of two main globular domains. The core of the N-terminal domain is characterised by a tightly packed β-sheet structure (in green) surrounded by five helices (in orange) and the C-terminal domain (in yellow) exhibits a β-sandwich structure. Although the structure is similar to classical pancreatic lipases, no significant catalytic activity was observed because of a closed conformation of the domain that controls the access to the active site.

**Figure 5.12 3D structure of dog pancreatic lipase related protein 1 (PDB: 1RP1)**
Dog pancreatic lipase related protein 1 (450 aa: Lys 1 – Cys 450) consists of two main domains: The N-terminal domain is characterised by a tightly packed β-sheet structure (in green) surrounded by five helices and the C-terminal domain (in yellow) exhibits a β-sandwich structure.

## 5.15 HORMONE FAMILIES AND SIGNATURES

Many hormones belong to a family and/or are characterised by a signature with a specific pattern. A selection of these hormones is summarised in this section of the chapter.

### 5.15.1 The erythropoietin/thrombopoietin signature

Erythropoietin (EPO) involved in the regulation of erythrocyte differentiation and thrombopoietin (TPO) responsible for the regulation of the circulating platelet numbers share a common consensus pattern located at the N-terminal end of EPO and TPO. The two Cys residues are involved in disulfide bonds:

EPO_TPO (PS00817): P–x(4)–**C**–D–x–R–[LIVM](2)–x–[KR]–x(14)–**C**

The 3D structure of the entire mutant erythropoietin (PDB: 1BUY; 166 aa: Ala 1–Arg 166) determined by NMR spectroscopy is shown in Figure 5.13. The structure is characterised by an antiparallel four-helix bundle connected by two long and one short loop and by three short β-strands. The ellipsoidal fold is stabilised by two intrachain disulfide bridges (in pink) on either side of the fold (Cys 7–Cys 161, Cys 29–Cys 33).

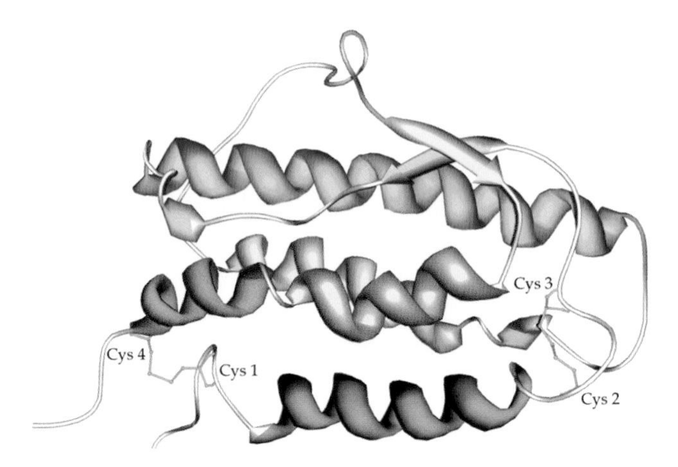


**Figure 5.13 3D structure of erythropoietin (PDB: 1BUY)**
3D structure of the entire mutant erythropoietin (166 aa: Ala 1 – Arg 166) is characterised by an antiparallel four-helix bundle and three short β-strands and two intrachain disulfide bridges (in pink).

### 5.15.2 The corticotropin-releasing factor family

The corticotropin-releasing factor (CRF, corticoliberin) that regulates the release of corticotropin shares a consensus pattern with other peptide hormones of approximately 40 residues, which are C-terminally amidated and processed from larger precursor proteins:

CRF (PS00511): [PQA]–x–[LIVM]–S–[LIVM]–x(2)–[PST]–[LIVMF]–x–[LIVM]–L–R–x(2)–[LIVMW]

The 3D solution structure of corticoliberin (PDB: 1GO9; 41 aa: Ser 1–Ile 41) determined by NMR spectroscopy is shown in Figure 5.14. Corticoliberin adopts an almost liner helical structure with a helical content of over 80 % and is C-terminally amidated (Ile-$NH_2$, pink circle).

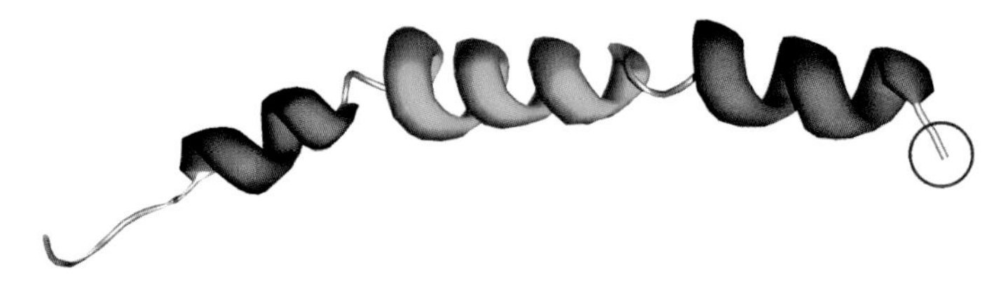

**Figure 5.14 3D structure of corticoliberin (PDB: 1GO9)**
Corticoliberin (41 aa: Ser 1 – Ile 41) exhibits an almost linear helical structure with an amidated C-terminus (Ile-NH2, pink circle).

### 5.15.3 The gonadotropin-releasing hormone signature

The gonadotropin-releasing hormones (GnRH, gonadoliberin) are a peptide family that plays a key role in reproduction like the stimulation of the synthesis and secretion of luteinizing and follicle-stimulating hormones. GnRH is a decapeptide that is C-terminally amidated (Gly-$NH_2$) and processed from a larger precursor protein and is characterised by the following consensus pattern:

GNRH (PS00473): Q–[HY]–[FYW]–S–x(4)–P–**G**

### 5.15.4 The calcitonin/CGRP/IAPP family

Calcitonin is a 32 residue C-terminally amidated peptide hormone that causes the rapid but short-lived decrease of calcium and phosphate in blood by promoting their incorporation in the bones. Calcitonin is structurally related to other 37 residue peptide hormones like calcitonin gene-related peptide I and II (P06881, P10092) and islet amyloid polypeptide (P10997) and they share a common consensus pattern in the N-terminal region characterised by a single intrachain disulfide bridge and the two Cys residues are part of the pattern:

CALCITONIN (PS00258): **C**–[SAGDN]–[STN]–x(0, 1)–[SA]–T–**C**–[VMA]–x(3)–[LYF]–x(3)–[LYF]

The 3D structure exhibits an extended fold with a α-helical segment in the N-terminal region and a C-terminal Pro-$NH_2$ residue.

### 5.15.5 The gastrin/cholecystokinin family

Gastrin and cholecystokinin (CCK) are structurally related peptide hormones involved in the regulation of various digestive processes and feeding behaviours. The biological activity of gastrin and CCK is associated with the very C-terminal sequence containing a conserved sulfated Tyr residue that is characterised by the following consensus pattern:

GASTRIN (PS00259): **Y**–x(0, 1)–[GD]–[WH]–M–[DR]–F

The Tyr residue is sulfated and sulfation increases their biological activity.

### 5.15.6 The insulin family

Insulin together with other evolutionary-related active peptides like the relaxins H1, H2 and H3 (P04808, P04090, Q8WXF3) or insulin-like peptides INSL5 and INSL6 (Q9Y5Q6, Q9Y581) form a family of structurally related peptides consisting of two peptide chains A and B linked by two disulfide bridges. They share a conserved arrangement of four Cys residues in the A chain with Cys 1 and Cys 3 forming an intrachain disulfide bridge and Cys 2 and Cys 4 two interchain disulfide bridges to two Cys in chain B. The four conserved Cys residues in chain A are characteristic for the consensus pattern. All four Cys residues are involved in disulfide bridges:

INSULIN (PS00262): **C**–**C**–{P}–{P}–x–**C**–[STDNEKPI]–x(3)–[LIVMFS]–x(3)–**C**

The 3D structure of a hexameric form of insulin (A-chain, 21 aa: Gly 1–Asn 21; B-chain, 30 aa: Phe 1–Thr 30) in solution determined by NMR spectroscopy is shown in Figure 5.15 (PDB: 3AIY) (a). The surface representation of the hexamer clearly depicts a highly symmetrical triangular array. The monomer is composed of an A-chain with two short α-helical segments and a B-chain with a longer α-helix and a short β-strand. Figure 5.15(b) depicts the monomeric structure with two interchain disulfide bridges (Cys 2A–Cys 1B, Cys 4A–Cys 2B) (in blue) and one intrachain disulfide

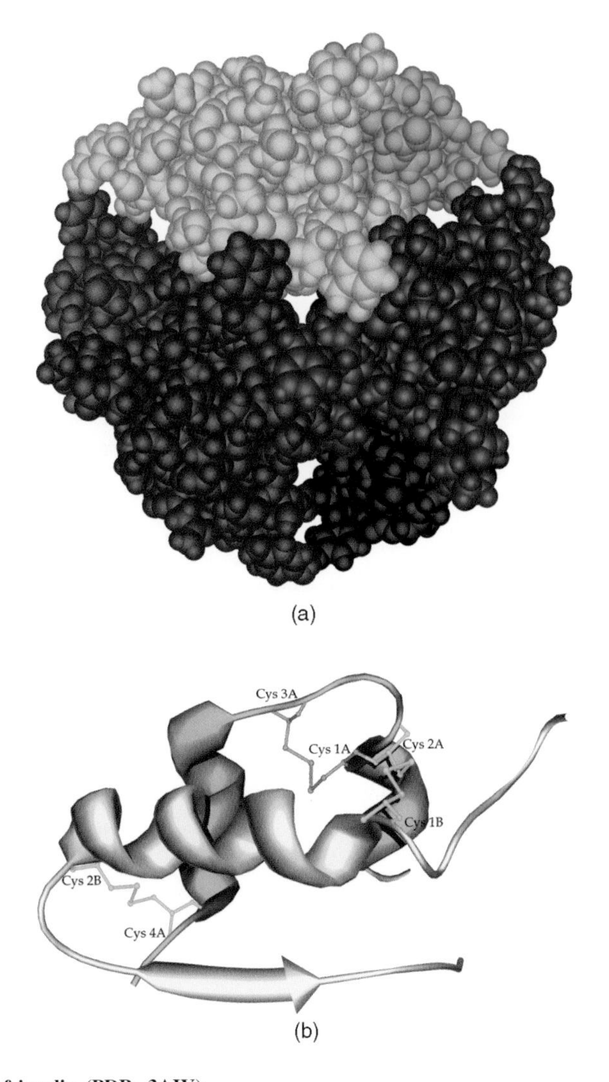


**Figure 5.15 3D structure of insulin (PDB: 3AIY)**
(a) The surface representation of hexameric insulin clearly depicts a highly symmetrical triangular array.
(b) Monomeric insulin consisting of the A chain (21 aa: Gly 1 – Asn 21) and the B chain (30 aa: Phe 1 – Thr 30) are linked by two interchain disulfide bridges (in blue) and chain A contains a single intrachain disulfide bridge (in pink).

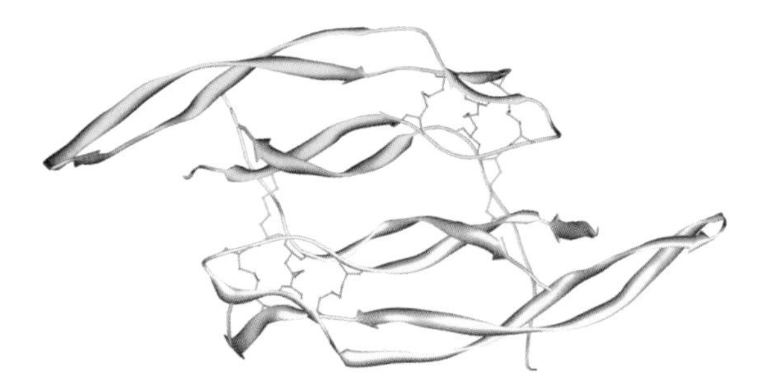

**Figure 5.16 3D structure of the platelet-derived growth factor B chain (PDB: 1PDG)**
The homodimeric platelet-derived growth factor B chain (109 aa: Ser 1 – Thr 109) contains two highly twisted antiparallel pairs of β-strands with three intrachain disulfide bridges (in pink) arranged in a tight disulfide knot and the dimer is linked by two interchain disulfide bridges (in green).

bridge in chain A (Cys 1A–Cys 3A) (in pink). Both NMR and crystallographic studies have revealed a highly flexible insulin molecule.

## 5.16 GROWTH FACTOR FAMILIES

### 5.16.1 The platelet-derived growth factor family

Platelet-derived growth factor (PDGF) is a potent mitogen for cells of mesenchymal origin such as smooth muscle cells and glial cells. PDGF is structurally related to other growth factors like the vascular endothelial growth factors VEGF-A, VEGF-B, VEGF-C and VEGF-D (for details see Chapter 14). PDGF consists of two different but closely related A and B chains which form disulfide-linked homo- and heterodimers (A–A, B–B, A–B). The PDGF family is characterised by a consensus pattern containing four out of eight conserved Cys residues. The four Cys residues are involved in intra- and interchain disulfide bridges:

PDGF_1 (PS00249): P–[PSR]–**C**–V–x(3)–R–**C**–[GSTA]–G–**C**–**C**

The 3D structure of the homodimeric (BB) platelet-derived growth factor B chain (109 aa: Ser 1–Thr 109) determined by X-ray diffraction is shown in Figure 5.16 (PDB:1PDG). PDGF is folded into two highly twisted antiparallel pairs of β-strands with an unusual tight knot of three intrachain disulfide bonds (in pink). The homodimer is linked by two interchain disulfide bridges (in green).

### 5.16.2 The transforming growth factor-β family

Transforming growth factor-β (TGF-β) is a multifunctional protein controlling the proliferation, differentiation and other functions in many cell types. Three main TGF-β forms exist, TGF-β-1, TGF-β-2 and TGF-β-3 (for details see Chapter 14). Other members of the TGF-β family are the bone morphogenetic proteins, the growth/differentiation factors and inhibins. TGF-β proteins are only active as homo- or heterodimers linked by a single disulfide bond. The structure is stabilised by four intrachain disulfide bridges in the following arrangement:

**Cys 1–Cys 3, Cys 2–Cys 6, Cys 4–Cys 7, Cys 5–Cys 8**

Two of the conserved Cys residues are included in the consensus pattern of TGF-β and are involved in disulfide bridges:

TGF_BETA_1 (PS00250): [LIVM]–x(2)–P–x(2)–[FY]–x(4)–**C**–x–G–x–**C**

The 3D structure of homodimeric transforming growth factor-β-2 (112 aa: Ala 1–Ser 112) determined by X-ray diffraction is shown in Figure 5.17 (PDB:1TFG). The monomer consists of two antiparallel β-strands forming a flat curved surface and a

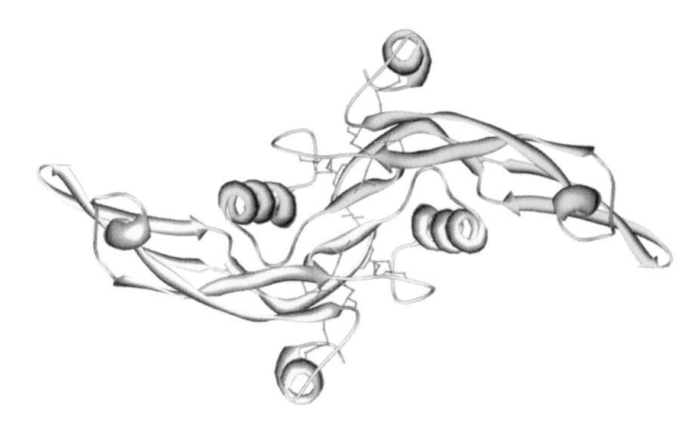

**Figure 5.17 3D structure of the transforming growth factor-β-2 (PDB: 1TFG)**
Each monomer (in blue and yellow) of the homodimeric transforming growth factor-β-2 (112 aa: Ala 1 – Ser 112) contains a disulfide-rich core structure with four intrachain disulfide bridges (in pink) and the monomers are connected by a single interchain disulfide bridge (in green) drawn schematically.

separate long α-helix. Each monomer (in blue and yellow) is characterised by a disulfide-rich core with four intrachain disulfide bridges (in pink). The monomers are connected by a single interchain disulfide bridge (in green) drawn schematically and the helix of one monomer interacts with the concave β-sheet surface of the other.

## 5.17 CYTOKINE FAMILIES

### 5.17.1 The tumor necrosis factor family

Tumor necrosis factor (TNF, cachectin or TNF-α) is structurally and functionally related to a series of other cytokines like lymphotoxin-α and lymphotoxin-β. The TNF family is characterised by a consensus pattern located in a β-strand in the central region of the proteins:

TNF_1 (PS00251): [LV]–x–[LIVM]–{V}–x–{L}–G–[LIVMF]–Y–[LIVMFY](2)–x(2)–[QEKHL]–[LIVMGT]
–x–[LIVMFY]

The 3D structure of a mutant soluble form of tumor necrosis factor α (PDB: 4TSV; 150 aa: Pro 8–Leu 157 in the soluble form) determined by X-ray diffraction is shown in Figure 5.18. The structure consists of an antiparallel β-sandwich similar to the 'jelly-roll' structural motif and is stabilised by one intrachain disulfide bridge (in pink).

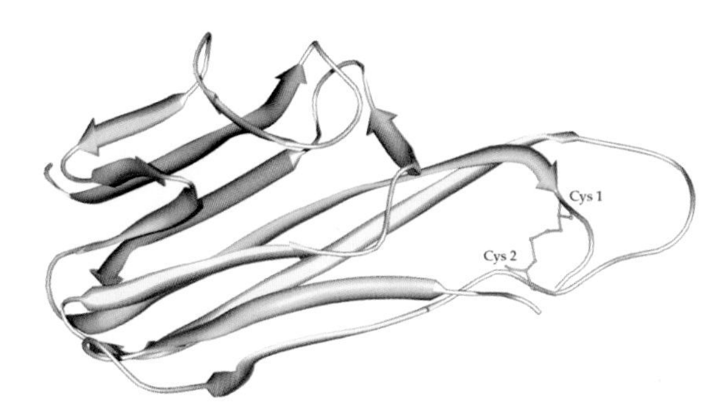


**Figure 5.18 3D structure of tumor necrosis factor-α (PDB: 4TSV)**
A mutant soluble form of tumor necrosis factor-α (150 aa: Pro 8 – Leu 157 in the soluble form) consists of an antiparallel β-sandwich stabilised by a single disulfide bridge (in pink).

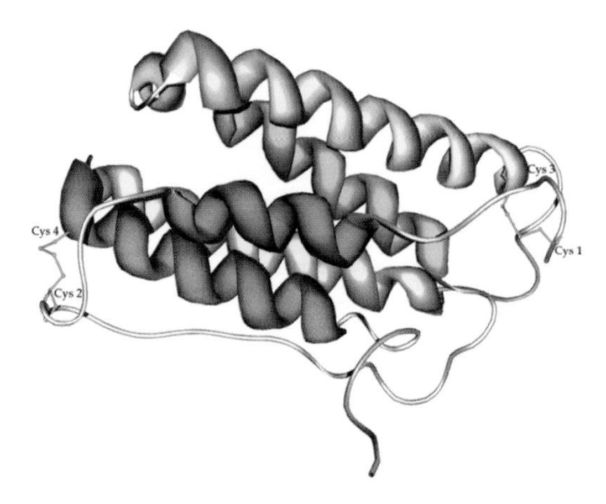


**Figure 5.19 3D structure of interferon α-2a (PDB: 1ITF)**
Interferon α-2a (165 aa: Cys 1 – Glu 165) is characterised by a four helix bundle structure and two intrachain disulfide bridges (in pink).

### 5.17.2 The interferon α, β and δ family

Interferons are cytokines that are involved in antiviral and antiproliferative responses in the cells. Interferon α, α-II (ω), β and δ (P37290, trophoblast) are structurally related (160–170 residues) and form a family that is best characterised by the following consensus pattern, where the second Cys residue is involved in a disulfide bridge:

INTERFERON_A_B_D (PS00252): [FYH]–[FY]–x–[GNRCDS]–[LIVM]–x(2)–[FYL]–L–x(7)–[**C**Y]–[AT]–W

The 3D solution structure of interferon α-2a (PDB: 1ITF; 165 aa: Cys 1–Glu 165) determined by NMR spectroscopy is shown in Figure 5.19. The structure is characterised by a cluster of five α-helices, of which four form a left-handed helix bundle arranged in an up–up down–down topology structurally related to other four-helix bundle cytokines. The structure is stabilised by two intrachain disulfide bridges (in pink) located on either side of the fold.

## 5.18 INTERLEUKIN FAMILIES AND SIGNATURES

### 5.18.1 The interleukin-1 signature

The two forms of interleukin-1 (IL-1), IL-1α (hematopoietin) and IL-1β (catabolin), are involved in many biological processes, such as T lymphocyte and together with interleukin-2 (IL-2) B lymphocyte activation and proliferation, induction of prostaglandin synthesis; both are pyrogens. The interleukin-1 signature is located in a conserved C-terminal region and is characterised by the following consensus pattern:

INTERLEUKIN_1 (PS00253): [FC]–x–S–[ASLV]–x(2)–P–x(2)–[FYLIV]–[LI]–[SCA]–T–x(7)–[LIVM]

The 3D structure of entire recombinant interleukin-1β (PDB: 1I1B; 153 aa: Ala 1–Ser 153) determined by X-ray diffraction is shown in Figure 5.20. The structure is characterised by 12 antiparallel β-strands folded into a six-stranded β-barrel similar in architecture to the Kunitz-type trypsin inhibitor.

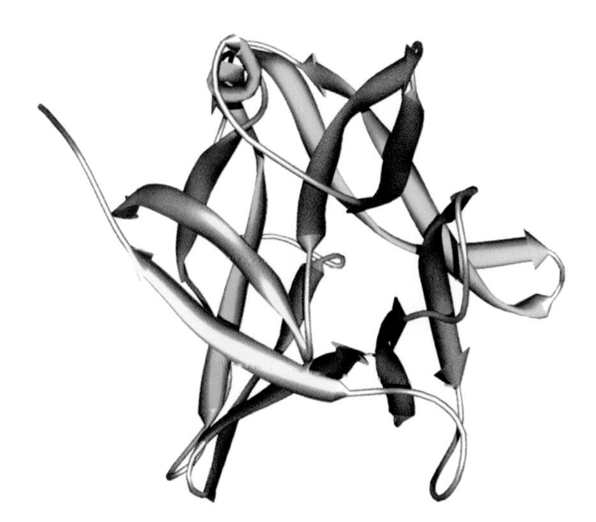

**Figure 5.20 3D structure of interleukin-1β (PDB: 1I1B)**
3D structure of entire recombinant interleukin-1β (153 aa: Ala 1 – Ser 153) is characterised by 12 antiparallel β-strands folded into a six-stranded β-barrel.

### 5.18.2 The interleukin-2 signature

Interleukin-2 (IL-2) is required for T cell proliferation and processes that are crucial for regulation of the immune response. The interleukin-2 signature is positioned around the first Cys of the single intrachain disulfide bond and is characterised by the following consensus pattern:

INTERLEUKIN_2 (PS00424): T–E–[LF]–x(2)–L–x–**C**–L–x(2)–E–L

The 3D structure of entire interleukin-2 (PDB: 1M47; 133 aa: Ala–Thr 133) determined by X-ray diffraction is shown in Figure 5.21. The structure is characterised by a four-helix bundle structure and two short β-strands and contains a single intrachain disulfide bridge (in pink).

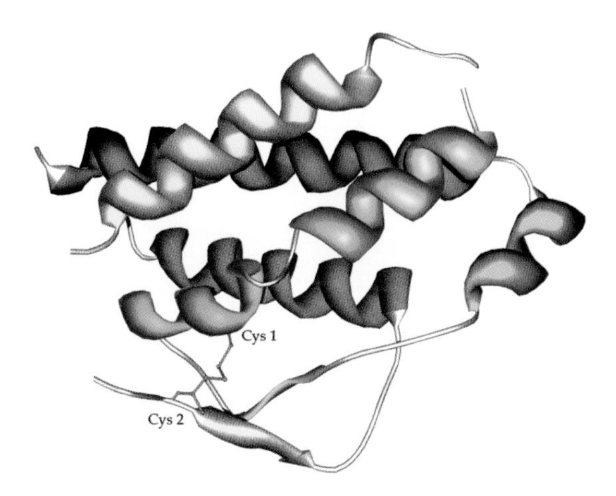


**Figure 5.21 3D structure of interleukin-2 (PDB: 1M47)**
Interleukin-2 (133 aa: Ala 1 – Thr 133) contains a four-helix bundle structure and two short β-strands and a single intrachain disulfide bridge (in pink).

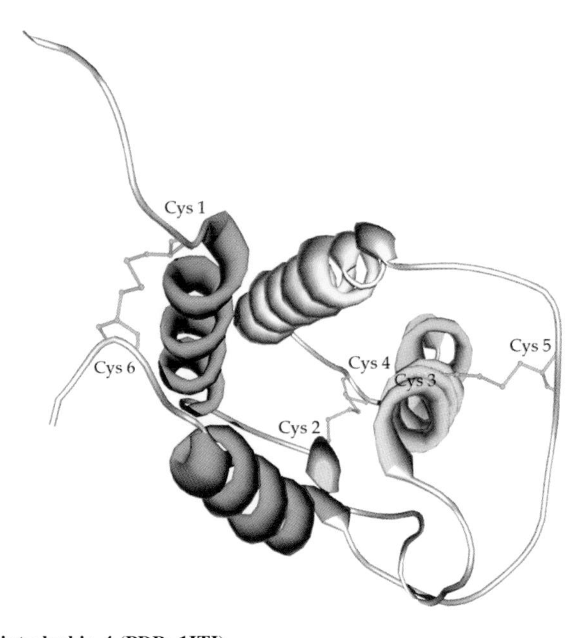


**Figure 5.22 3D structure of interleukin-4 (PDB: 1ITI)**
Interleukin-4 (129 aa: His 1 – Ser 129 + N-terminal tetrapeptide) is characterised by a left-handed four-helix bundle containing three intrachain disulfide bridges (in pink).

### 5.18.3 The interleukin-4/-13 signature

Interleukin-4 (IL-4) plays an important role in the control and regulation of the immune and inflammatory system. Interleukin-13 (IL-13) is a lymphokine regulating inflammatory and immune responses. IL-4 and IL-13 are distantly related and share a consensus pattern in the best conserved N-terminal region located around a Cys residue involved in the sole conserved disulfide bond:

INTERLEUKIN_4_13 (PS00838): [LI]–x–E–[LIVM](2)–x(4,5)–[LIVM]–[TL]–x(5,7)–**C**–x(4)–[IVA]–x–[DNS]–[LIVMA]

The 3D solution structure of recombinant interleukin-4 (PDB: 1ITI; 129 aa: His 1–Ser 129 + N-terminal tetrapeptide) determined by NMR spectroscopy is shown in Figure 5.22. The structure is characterised by a left-handed four-helix bundle linked by either long loops, small helical turns or short strands. IL-4 contains three intrachain disulfide bridges (in pink) arranged in the following pattern:

**Cys 1–Cys 6, Cys 2–Cys 4, Cys 3–Cys 5**

### 5.18.4 The interleukin-6/granulocyte colony-stimulating factor/myelomonocytic growth factor family

Interleukin-6 (IL-6) or interferon β-2, involved in a wide variety of biological functions, granulocyte colony-stimulating factor (G-CSF), regulating hematopoietic cell proliferation and differentiation, and avian myelomonocytic growth factor (P13854) belong to a single family containing four conserved Cys residues involved in two disulfide bonds. The consensus pattern is located around the second disulfide bridge:

INTERLEUKIN_6 (PS00254): **C**–x(9)–**C**–x(6)–G–L–x(2)–[FY]–x(3)–L

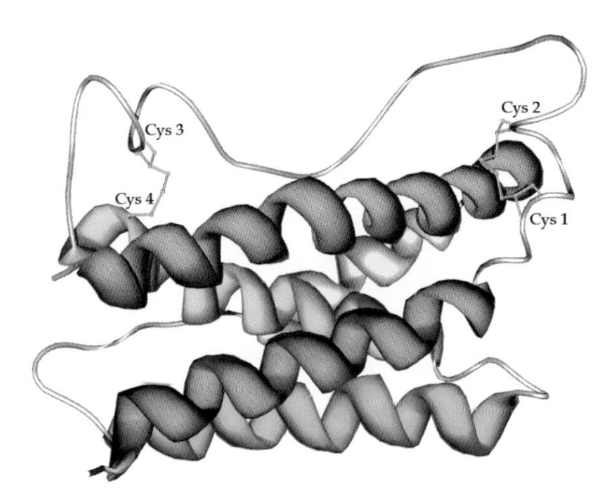


**Figure 5.23 3D structure of interleukin-6 (PDB: 1IL6)**
Recombinant interleukin-6 (185 aa: Ala 1 – Met 185) exhibits a classical four helix bundle structure stabilised by two intrachain disulfide bridges (in pink).

The 3D solution structure of recombinant interleukin-6 (PDB: 1IL6; 185 aa: Ala 1–Met 185; the UniProt databank entry contains only 183 aa, lacking the N-terminal dipeptide Ala–Pro) determined by NMR spectroscopy is shown in Figure 5.23. The structure is characterised by a classical four-helix bundle linked with loops of variable length and the fifth helix (in yellow) located in a connecting loop. The structure is stabilised by two intrachain disulfide bridges (in pink) located on either side of the fold and are arranged in the following pattern:

**Cys 1–Cys 2, Cys 3–Cys 4**

### 5.18.5 The interleukin-7 and -9 signature

Interleukin-7 (IL-7) is a growth factor involved in the proliferation of lymphoid progenitors of both B and T cells. Interleukin-9 (IL-9) supports the interleukin-2 and interleukin-4 independent growth of helper T cells. IL-7 and IL-9 are evolutionary related and are characterised by the following consensus pattern:

INTERLEUKIN_7_9 (PS00255): N–x–[LAP]–[SCT]–F–L–K–x–L–L

IL-7 contains three intrachain disulfide bridges arranged in the following pattern:

**Cys 1–Cys 6, Cys 2–Cys 5, Cys 3–Cys 4**

### 5.18.6 Interleukin-10 family

Interleukin-10 (IL-10) inhibits the synthesis of various cytokines such as interferon γ, interleukin-2 and interleukin-3, tumor necrosis factor and granulocyte-macrophage colony-stimulating factor. Together with other interleukins (IL-19: Q9UHD0; IL-20: Q9NYY1; IL-22: Q9GZX6 and IL-24: Q13007) it belongs to the interleukin-10 family that contains four conserved Cys residues involved in two disulfide bridges. The consensus pattern is located around the second Cys residue:

INTERLEUKIN_10 (PS00520): [KQS]–x(4)–**C**–[QYC]–x(4)–[LIVM](2)–x–[FL]–[FYT]–[LMV]–x
–[DERT]–[IV]–[LMF]

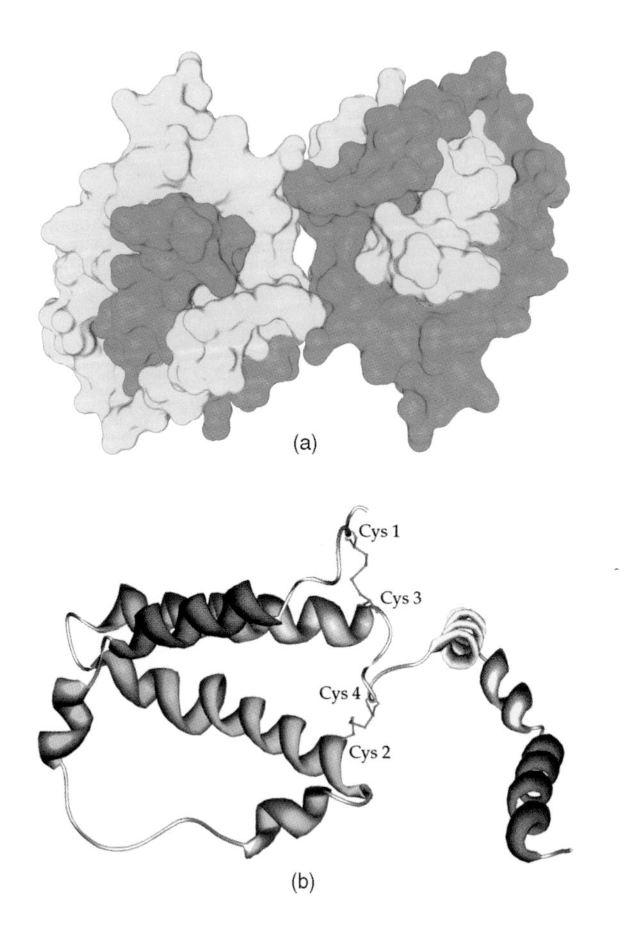


**Figure 5.24 3D structure of interleukin-10 (PDB: 1ILK)**
(a) Surface representation of the tight interaction in the dimer of recombinant interleukin-10 (151 aa: Asn 10 – Asn 160 in native IL-10). (b) Monomeric interleukin-10 contains six α-helices, of which four form a classical helix bundle stabilised by two disulfide bridges (in pink).

The 3D structure of dimeric recombinant interleukin-10 (PDB: 1ILK; 151 aa: Asn 10–Asn 160 in native IL-10) determined by X-ray diffraction is shown in Figure 5.24(a). The surface representation clearly shows the tight interaction in the dimer, which forms a V-shaped structure presumed to participate in receptor binding. The monomer contains six α-helices, of which four form a classical helix bundle in the up–up down–down orientation as can be seen in Figure 5.24(b). The structure is stabilised by two intrachain disulfide bridges (in pink) arranged in the following pattern:

**Cys 1–Cys 3, Cys 2–Cys 4**

## 5.19 THE SMALL CYTOKINE (INTERCRINE/CHEMOKINE) FAMILY AND SIGNATURE

The small cytokines also termed intercrines or chemokines exhibit mitogenic, chemotactic or inflammatory activities. They are small cationic proteins of approximately 70–100 residues in length and are characterised by two conserved disulfide bridges arranged in the following pattern:

**Cys 1–Cys 3, Cys 2–Cys 4**

The small cytokines are sorted into two groups: in the first group Cys 1 and Cys 2 are separated by a single residue (**Cys–Xaa–Cys**); in group two they are adjacent (**Cys–Cys**). Among others platelet factor 4, platelet basic protein and interleukin-8 belong to the Cys–Xaa–Cys group and small inducible cytokine A2 and A5 to the Cys–Cys group. The two types of small cytokines are characterised by the following consensus patterns:

SMALL_CYTOKINES_CXC (PS00471): **C**–x–**C**–[LIVM]–x(5, 6)–[LIVMFY]–x(2)–[RKSEQ]–x –[LIVM]–x(2)–[LIVM]–x(5)–[STAG]–x(2)–**C**–x(3)–[EQ]–[LIVM](2)–x(9, 10)–**C**–L–[DN]

SMALL_CYTOKINES_CC (PS00472): **C**–**C**–[LIFYT]–x(5, 6)–[LI]–x(4)–[LIVMF]–x(2)–[FYW] – x(6, 8)–**C**–x(3, 4)–[SAG]–[LIVM](2)–[FL]–x(8)–**C**–[STA]

In both cases all four Cys residues are involved in two intrachain disulfide bridges.The 3D structure of tetrameric platelet factor 4 (PDB:1RHP; 70 aa: Glu 1–Ser 70) determined by X-ray diffraction is shown in Figure 5.25(a). Under physiological conditions platelet factor 4 assembles to a homotetramer, which is thought to be its active form. The tetrameric structure is

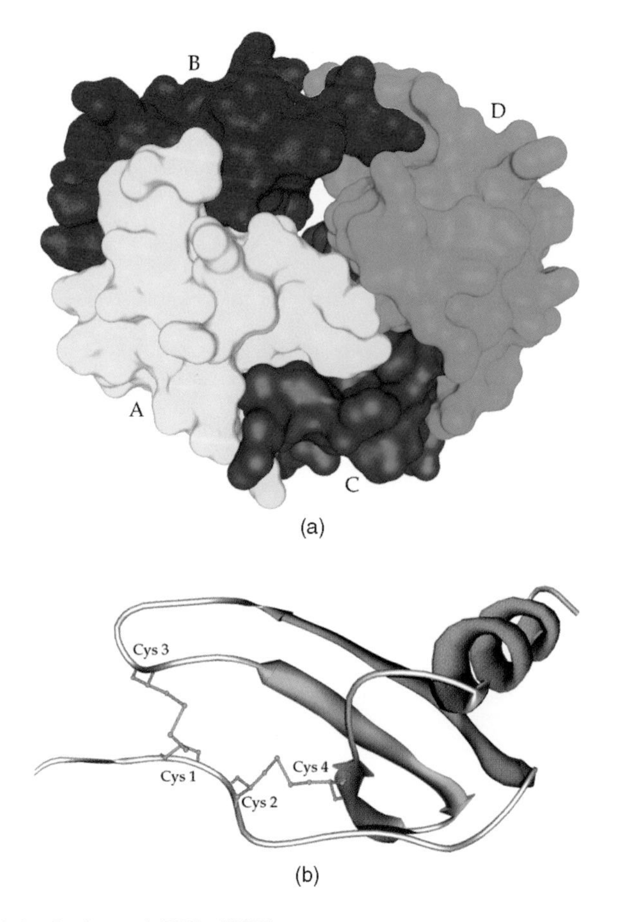


**Figure 5.25 3D structure of platelet factor 4 (PDB: 1RHP)**
(a) Tetrameric platelet factor 4 (70 aa: Glu 1 – Ser 70) consists of two A/B (in yellow/red) and C/D (in green/blue) dimers.
(b) The monomer consists of a N-terminal antiparallel β-sheet-like structure linked to a C-terminal α-helix stabilised by two intrachain disulfide bridges (in pink).

stabilised by electrostatic interactions and hydrogen bonding at the A/B (in yellow/red) and C/D (in green/blue) dimeric interface. In its N-terminal region the monomer forms an antiparallel β-sheet-like structure connected to a C-terminal α-helix stabilised by two intrachain disulfide bridges (in pink), as can be seen in Figure 5.25b.

## REFERENCES


Arkin *et al.*, 2003, Binding of small molecules to an adaptive protein-protein interface, *Proc. Natl. Acad. Sci. USA*, **100**, 1603–1608.

Bazan, 1993, Emerging families of cytokines and receptors, *Curr. Biol.*, **3**, 603–606.

Blundel *et al.*, 1980, Hormone families: pancreatic hormones and homologous growth factors, *Nature*, **287**, 781–787.

Bouley *et al.*, 1993, Hematopoietin sub-family classification based on size, gene organisation and sequence homology, *Curr. Biol.*, **3**, 573–581.

Breimer *et al.*, 1988, Peptides from the calcitonin genes: molecular genetics, structure and function, *Biochem. J.*, **255**, 377–390.

Cha *et al.*, 1998, High resolution crystal structure of a human tumor necrosis factor-alpha mutant with low systemic toxicity, *J. Biol. Chem.*, **273**, 2153–2160.

Cheetham *et al.*, 1998, NMR structure of human erythropoietin and a comparison with its receptor bound conformation, *Nat. Struct. Biol.*, **5**, 861–866.

De Maeyer *et al.*, (eds.) 1988, *Interferons and Other Regulated Cytokines*, John Wiley & Sons, Inc. New York.

Delain *et al.*, 1992, Ultrastructure of alpha 2-macroglobulin, *Electron Microsc. Rev.*, **5**, 231–281.

Dinarello, 1988, Biology of interleukin 1, *FASEB J.*, **2**, 108–115.

Elliott *et al.*, 2000, Topology of a 2.0 A structure of alpha 1-antitrypsin reveals targets for rational drug design to prevent conformational disease, *Protein Sci.*, **9**, 1274–1281.

Finzel *et al.*, 1989, Crystal structure of recombinant human interleukin-1 beta at 2.0 A resolution, *J. Mol. Biol.*, **209**, 779–791.

Finkelman *et al.*, 1990, Lymphokine control of in *vivo* immunoglobulin isotype selection, *Annu. Rev. Immunol.*, **8**, 303–333.

Flower *et al.*, 1993, Structure and sequence relationships in the lipocalins and related proteins, *Protein Sci.*, **2**, 753–761.

Fox *et al.*, 2001, Three-dimensional structure of human follicle-stimulating hormone, *Mol. Endocrinol.*, **15**, 378–389.

Friedrich *et al.*, 1993, A Kazal-type inhibitor with thrombin specificity from *Rhodnius proxilus*, *J. Biol. Chem.*, **268**, 16216–16222.

Gaboriaud *et al.*, 1996, Crystal structure of human trypsin 1: unexpected phosphorylation of Tyr151, *J. Mol. Biol.*, **259**, 995–1010.

Goodman *et al.*, 1988, An evolutionary tree for invertebrate globin sequences, *J. Mol. Evol.*, **27**, 236–249.

Haefliger *et al.*, 1989, Amphibian albumins as members of the albumin, alpha-fetoprotein, vitamin D-binding protein multigene family, *J. Mol. Evol.*, **29**, 344–354.

Hect *et al.*, 1992, Three-dimensional structure of a recombinant variant of human pancreatic secretory trypsin inhibitor (Kazal type), *J. Mol. Biol.*, **225**, 1095–1103.

Heldin, 1992, Structural and functional studies on platelet-derived growth factor, *EMBO J.*, **11**, 4251–4259.

Helland *et al.*, 1998, The crystal structure of anionic salmon trypsin in complex with bovine pancreatic trypsin inhibitor, *Eur. J. Biochem.*, **256**, 317–324.

Hinck *et al.*, 1996, Transforming growth factor beta 1: three dimensional structure in solution and comparison with X-ray structure of transforming growth factor beta 2, *Biochemistry*, **35**, 8517–8534.

Janssen *et al.*, 2005, Structures of complement component C3 provide insights into the function and evolution of immunity, *Nature*, **437**, 505–511.

Kavanaugh *et al.*, 1998, High-resolution crystal structures of human hemoglobin with mutations at tryptophan 37beta: structural basis for a high-affinity T-state, *Biochemistry*, **37**, 4358–4373.

Kishimoto *et al.*, 1988, A new interleukin with pleiotropic activities, *BioEssays*, **9**, 11–15.

Klaus *et al.*, 1997, The three-dimensional high resolution structure of human interferon alpha-2a determined by heteronuclear NMR spectroscopy in solution, *J. Mol. Biol.*, **274**, 661–675.

Koury *et al.*, 1992, The molecular mechanism of erythropoietin action, *Eur. J. Biochem.*, **210**, 649–663.

Lapthorn *et al.*, 1994, Crystal structure of human chorionic gonadotropin, *Nature*, **369**, 455–461.

Lederis *et al.*, 1990, Evolutionary aspects of corticotropin releasing hormones, *Prog. Clin. Biol. Res.*, **342**, 467–472.

Mann *et al.*, 1988, Cofactor proteins in the assembly and expression of blood clotting enzyme complexes, *Annu. Rev. Biochem.*, **57**, 915–956.

Maurits *et al.*, 1997, The second Kunitz domain of human tissue factor pathway inhibitor: cloning, structure determination and interaction with factor Xa, *J. Mol. Biol.*, **269**, 395–407.

Metcalf, 1985, The granulocyte-macrophage colony-stimulating factors, *Science*, **229**, 16–22.

Minty *et al.*, 1993, Interleukin-13 is a new lymphokine regulating inflammatory and immune responses, *Nature*, **362**, 248–250.

Mizel, 1989, The interleukins, *FASEB J.*, **3**, 2379–2388.

Moore *et al.*, 1993, Interleukin-10, *Annu. Rev. Immunol.*, **11**, 165–190.

Mutt, 1988, Vasoactive intestinal polypeptide and related peptides. Isolation and chemistry, *Ann. NY Acad. Sci.*, **527**, 1–19.

O'Donoghue *et al.*, 2000, Unravelling the symmetry ambiguity in a hexamer: calculation of the R6 human insulin structure, *J. Biomol. NMR*, **16**, 93–108.
Oefner *et al.*, 1992, Crystal structure of human platelet-derived growth factor BB, *EMBO J.*, **11**, 3921–3926.
Oppenheim *et al.*, 1991, Properties of the novel proinflammatory supergene 'intercrine' cytokine family, *Annu. Rev. Immunol.*, **9**, 617–648.
Ortlund *et al.*, 2002, Crystal structure of human complement protein C8gamma at 1.2 A resolution reveals a lipocalin fold and a distinct ligand-binding site, *Biochemistry*, **41**, 7030–7037.
Ouzounis *et al.*, 1991, A structure-derived sequence pattern for the detection of type I copper binding domains in distantly related proteins, *FEBS Lett.*, **279**, 73–78.
Peitsch, *et al.*, 1991, Assembly of macromolecular pores by immune defence systems, *Curr. Opin. Cell. Biol.*, **3**, 710–716.
Persson *et al.*, 1989, Structural features of lipoprotein lipases. Lipase family relationships, binding interactions, non-equivalence of lipase cofactors, vitellogenin similarities and functional subdivision of lipoprotein lipase, *Eur. J. Biochem.*, **179**, 39–45.
Pierce *et al.*, 1991, Glycoprotein hormones: structure and function, *Annu. Rev. Biochem.*, **50**, 465–495.
Powers *et al.*, 1993, The high-resolution, three-dimensional solution structure of human interleukin-4 determined by multidimensional heteronuclear magnetic resonance spectroscopy, *Biochemistry*, **32**, 6744–6762.
Rawlings and Barrett, 1994, Families of serine peptidases, *Meth. Enzymol.*, **244**, 19–61.
Reid, 1988, in *Molecular Immunology* (eds. Hames *et al.*), IRL-Press, Oxford, pp. 189–241.
Roussel *et al.*, 1998, Reactivation of the totally inactive pancreatic lipase RP1 by structure-predicted point mutations, *Proteins*, **32**, 523–531.
Salier, 1990, Inter-alpha-trypsin inhibitor. Emergence of a family within the Kunitz-type protease inhibitor superfamily, *Trends Biochem. Sci.*, **15**, 435–439.
Schlunegger *et al.*, 1992, An unusual feature revealed by the crystal structure at 2.2 A resolution of human transforming growth factor-beta 2, *Nature*, **358**, 430–434.
Sherwood, 1987, The GnRH family of peptides, *Trends in Neurosci.*, **10**, 129–132.
Smith, 1988, Interleukin-2: Inception, impact and implications, *Science*, **240**, 1169–1176.
Sottrup-Jensen, 1989, Alpha-macroglobulins: structure, shape, and mechanism of proteinase complex formation, *J. Biol. Chem.*, **264**, 11539–11542.
Spyroulias *et al.*, 2002, Monitoring the structural consequences of Phe12 → D-Phe and Leu15 (Aib substitution in human/rat corticotropin releasing hormone, *Eur. J. Biochem.*, **269**, 6009–6019.
Sturm *et al.*, 1998, Structure-function studies on positions 17,18 and 21 replacement analogues of glucagon: the importance of charged residues and salt bridges in glucagon biological activity, *J. Med. Chem.*, **41**, 2693–2700.
Sugio *et al.*, 1999, Crystal structure of human serum albumin at 2.5 A resolution, *Protein Eng.*, **12**, 439–446.
Whisstock *et al.*, 1998, An atlas of serpin conformations, *Trends Biochem. Sci.*, **23**, 63–67.
Wiborg *et al.*, 1984, Structure of a human gastrin gene, *Proc. Natl. Acad. Sci. USA*, **81**, 1067–1069.
Xu *et al.*, 1997, Solution structure of recombinant human interleukin-6, *J. Mol. Biol.*, **268**, 468–481.
Zaitseva *et al.*, 1996, The X-ray structure of human serum ceruloplasmin at 3.1 A: nature of the copper centres, *J. Biol. Inorg. Chem.*, **1**, 15–23.
Zdanov *et al.*, 1995, Crystal structure of interleukin-10 reveals the functional dimer with an unexpected topological similarity to interferon gamma, *Structure*, **3**, 591–601.
Zhang *et al.*, 1994, Crystal structure of recombinant human platelet factor 4, *Biochemistry*, **33**, 8361–8366.

# 6

# Posttranslational Modifications

## 6.1 INTRODUCTION

Almost all blood plasma proteins carry *posttranslational modifications* (PTMs). Many hundred PTMs have been characterised in some detail in proteins of diverse origin and with very different structural and functional properties. The PTMs usually have a more or less pronounced influence on the characteristics of proteins like their overall or local charge, their hydrophobicity/hydrophilicity patterns, their conformation, their 3D structure, their stability and their correct biological function.

In eukaryotes over 95 % of the proteins are modified to a certain extent. Many of the modifications take place in special compartments: intracellular, e.g. in the nucleus, in mitochondria, in chloroplasts, in the Golgi apparatus or in the cytosol, extracellular, e.g. in the extracellular matrix (ECM) or in the body fluid, and on the cell surface, e.g. on the plasma membrane.

The identification and characterisation of PTMs is often difficult and very cumbersome. Usually, mass spectrometry and NMR spectroscopy are the methods of choice for their structural elucidation. The identification of the biological function of PTMs or their relevance for a specific protein is in many cases very difficult or even impossible. The PTMs can lead to dozens of isoforms in proteins, thus complicating the situation quite dramatically. Many isoforms can be separated by two-dimensional (2D) gel electrophoresis or sometimes also by chromatographic methods.

In this chapter some of the most common and relevant PTMs are described: Disulfide bridges, *N*-linked (*N*-CHO) and *O*-linked (*O*-CHO) glycosylations, phosphorylations and sulfations, hydroxylations and carboxylations, acylations and alkylations. The most common PTMs in plasma proteins are summarised in Table 6.1.

## 6.2 DATABASES FOR POSTTRANSLATIONAL MODIFICATIONS (PTMs)

A series of valuable databases specific for PTMs exists (see Table 6.2). The exact mass differences of the PTMs are compiled in the Delta Mass databank. The RESID database of protein modifications is a 'comprehensive collection of annotations and structures for protein modifications including amino-terminal, carboxyl-terminal and peptide crosslinks, pre-, co- and post-translational modifications'. The DSDBASE database contains proteins that have been used to determine experimentally or model the disulfide bridge pattern. The two carbohydrate databases, GlycoSuite and O-GLYCBASE, contain valuable information regarding the glycosylation of proteins. Phospho.ELM is a database containing eukaryotic proteins with experimentally verified phosphorylation sites.

Programs (servers) exist that predict several types of PTMs: NetNGlyc and NetOGlyc for glycosylations, NetAcet for N-terminal acetylations, NetPhos for phosphorylations and Sulfinator for sulfation. In addition, signal peptides can be predicted by SignalP.

## 6.3 DISULFIDE BRIDGES

Almost all plasma proteins contain disulfide bridges ranging in number from one bridge, e.g. in $\alpha_2$-antiplasmin, to dozens of bridges, e.g. in the von Willebrand factor and apolipoprotein (a). The correct arrangement of the disulfide bridges is essential for the native 3D structure of a protein and in consequence for its biological function. Many blood plasma proteins are multidomain proteins with different types of domains or repeats, quite often with more than one copy, e.g. coagulation factor XII with two

*Human Blood Plasma Proteins: Structure and Function* Johann Schaller, Simon Gerber, Urs Kämpfer, Sofia Lejon and Christian Trachsel

**Table 6.1** Common posttranslational modifications (PTMs) in plasma proteins.

| Modification | aa/Consensus sequence | Comments |
|---|---|---|
| Disulfide bridges | **Cys–Cys** | Most plasma proteins |
| Glycosylation | | Most plasma proteins |
| *N*-glycosylation | **Asn**–Xaa–Thr/Ser/Cys | Xaa ≠ Pro |
| *O*-glycosylation | Xaa–Xaa–**Ser/Thr**–Xaa-Xaa | Xaa = Pro/Ser/Thr/Ala |
| Phosphorylation | **Ser, Thr, Tyr**, (His, Cys) | Often: nonpermanent, low level |
| Acylation/alkylation | **α-/ε-amino, thiol** | Acetyl, formyl, methyl, pyro-Glu |
| Sulfation | **Tyr** | |
| Hydroxylation | **Lys, Asp, Asn** | |
| | Gly–Xaa–**Pro**–Gly–Xaa | |
| Carboxylation | Glu → **γ-Gla** | γ-Carboxylation |
| | Asp → **β-Asp** | β-Carboxylation |
| Proteolytic processing | N- or C-terminal truncation | |

**Table 6.2** Descriptions of databases specific for PTMs.

| Database | URL | Description |
|---|---|---|
| Delta Mass | http://www.abrf.org/index.cfm/dm.home | Mass of PTMs |
| RESID | http://www.ebi.ac.uk/RESID/ | Annotation of PTMs in Swiss-Prot |
| DSDBASE | http://www.ncbs.res.in/~faculty/mini/dsdbase/ | Disulfide bridge pattern |
| GlycoSuite | http://tmat.proteomesystems.com/glycosuite/ | Carbohydrates in proteins |
| O-GLYCBASE | http://www.cbs.dtu.dk/databases/OGLYCBASE/ | Mammalian *O*-glycosylation |
| Phospho.ELM | http://phospho.elm.eu.org/ | Phosphorylation sites (experimental) |
| NetNGlyc | http://www.cbs.dtu.dk/services/NetNGlyc | Prediction *N*-glycosylation |
| NetOGlyc | http://www.cbs.dtu.dk/services/NetOGlyc | Prediction *O*-glycosylation |
| NetAcet | http://www.cbs.dtu.dk/services/NetAcet | Prediction N-terminal acetylation |
| NetPhos | http://www.cbs.dtu.dk/services/NetPhos | Prediction phosphorylation |
| Sulfinator | http://www.expasy.org/tools/sulfinator | Prediction sulfation |
| SignalP | http://www.cbs.dtu.dk/services/SignalP | Prediction of signal peptides |

EGF-like, one kringle and one fibronectin type I and type II domains or complement component C6 with one LDLRA, two sushi/CCP/SCR, one EGF-like and two TSP1 domains. The individual domains usually represent independent folding units, very often stabilised by intradomain disulfide bridges (for more details see Chapter 4).

Protein folding and the formation of the correct disulfide bonds to achieve the native 3D structure of a protein is a complex process. Classical experiments by Anfinsen *et al.* (1961) provided evidence that the formation of disulfide bonds is a spontaneous process and the polypeptide chain as such is sufficient for achieving the correct disulfide bridge pattern in a native protein. The endoplasmic reticulum (ER) provides an optimized environment for oxidative protein folding. According to recent experiments in yeast the disulfide bond formation is driven by a protein relay involving the FAD-dependent enzyme Ero1p that oxidises the protein disulfide isomerase (PDI), which subsequently oxidises the folding protein directly. Ero1p plays an analogous role as the periplasmic DsbB in oxidative folding in bacteria. A schematic representation of oxidative protein folding in the ER of yeast is shown in Figure 6.1.

FAD-bound Ero1p transfers electrons to molecular oxygen, probably resulting in reactive oxygen species (ROS). In addition to PDI, some of the four homologues of PDI, Eugp1, Mpd1p, Mpd2p and Eps1, seem to be involved in disulfide isomerisation and reduction. Reduced glutathione (GSH) seems to assist disulfide reduction, resulting in oxidised glutathione (GSSG).

Folding studies of small disulfide-rich proteins such as bovine pancreatic trypsin inhibitor (BPTI), ribonuclease, lysozyme, α-lactalbumin and hirudin showed a high level of diversity in folding mechanisms, differing in the heterogeneity and the number of correctly formed disulfide bonds in the intermediates.

In the databank entries of the individual proteins in UniProt (SwissProt, TrEMBL, PIR) experimentally determined disulfide bridges are listed, and assignments by similarity or probable or potential disulfide bonds are also given and interchain disulfide

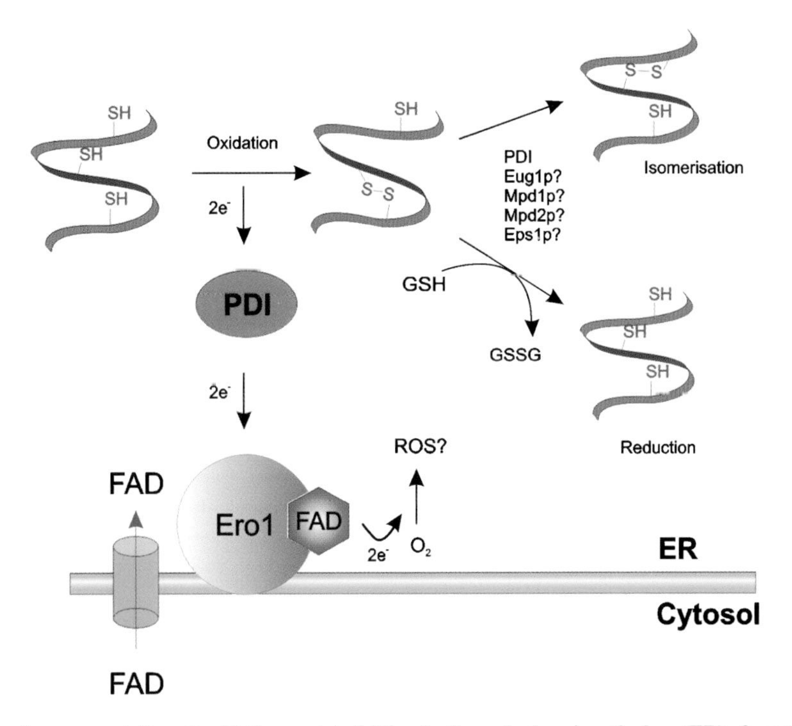


**Figure 6.1 Schematic representation of oxidative protein folding in the endoplasmic reticulum (ER) of yeast**
The formation of disulfide bonds in the ER is driven by Ero1p. The FAD-bound Ero1p oxidises the protein disulfide isomerase (PDI) which in turn oxidises the folding proteins directly. FAD-bound Ero1p transfers electrons to molecular oxygen, probably resulting in reactive oxygen species (ROS). In addition to PDI, some of the four homologues of PDI, Eugp1, Mpd1p, Mpd2p and Eps1 seem to be involved in disulfide isomerisation and reduction. Reduced glutathione (GSH) seems to assist disulfide reduction resulting in oxidised glutathione (GSSG). (Adapted from Tu *et al.*, J Cell Biol 164: 341–346. Copyright (2004), with permission from Rockefeller University Press.)

bridges are indicated. Experimentally determined disulfide bridges of proteins are compiled in the DSDBASE database (http://www.ncbs.res.in/~faculty/mini/dsdbase).

## 6.4 GLYCOSYLATION

Many plasma proteins are glycoproteins and glycosylation is the most common posttranslational modification in proteins. There are two main glycosylation sites: the *N*-glycosidic site (*N*-CHO) linked to the side chain of Asn and the *O*-glycosidic site (*O*-CHO) linked to the side chains of Ser and Thr. The *N*-CHO is characterised by the following consensus sequence:

**Asn**–Xaa–**Thr/Ser**/(Cys)

The *N*-glycosylation consensus pattern has a high probability of occurrence:

ASN_GLYCOSYLATIN (PS00001): **N**–{P}–[ST]–{P}

The presence of this consensus tripeptide is not sufficient in all cases for *N*-glycosylation at Asn as the folding of the protein plays an important role in the regulation of the glycosylation reaction. If the Xaa between Asn and Thr/Ser is a Pro, *N*-glycosylation will be impossible. Furthermore, *N*-glycosylation is in quite a few cases only partial, leading to further heterogeneities of the proteins. In addition, a few cases of *N*-glycosylation at the **Asn**–Xaa–**Cys** pattern are known, e.g. in plasma protein C (Asn(CHO) 329–Glu–Cys).

## N-Glycosylation

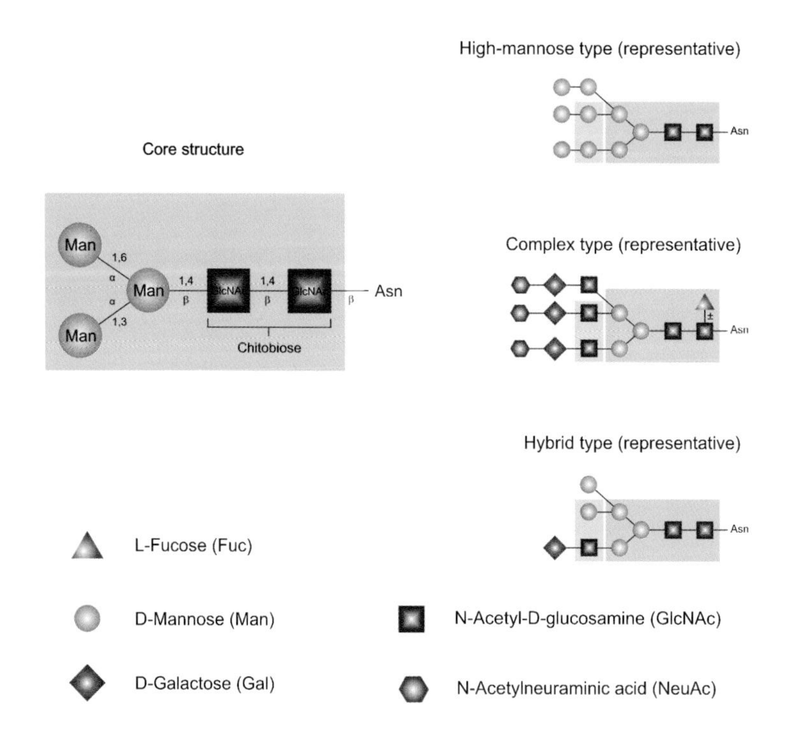


## O-Glycosylation

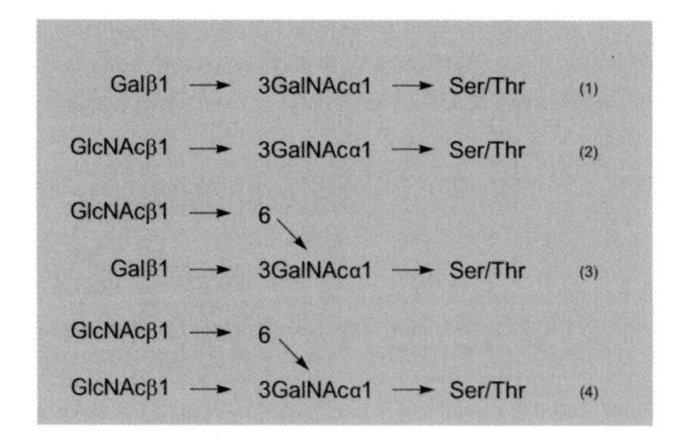


**Figure 6.2 Main types of glycosylation: *N*-glycosylation and *O*-glycosylation**
The core structure and representative forms of N-glycosylation are shown: high-mannose type, complex type and the hybrid type. The main forms of O-glycosylation with di- and tri-saccharides are indicated.

No predominant consensus sequence exists for *O*-CHO. However, there is a certain preference for the following sequence:

Xaa–Xaa–**Ser/Thr**–Xaa–Xaa with Xaa ( Pro/Ser/Thr/Ala)

Also in the case of *O*-glycosylation the glycosylation reaction is not always complete, thus leading to additional heterogeneity.

A few cases of C-linked glycosylations have been reported, e.g. in thrombospondin-1 where the first Trp within the recognition pattern **WXXW** carries a mannose at the C-2 atom (Trp 367, 420, 423 and 480) and in complement components C6 6 Trp and C7 4 Trp are mannosylated. In the interleukin-12 β-chain Trp 297 carries a mannose. A rare case of S-linked glycosylation has been reported in the inter-α-trypsin inhibitor heavy chain H1 at Cys 26.

The most common glycosylation patterns in *N*-glycosylation and *O*-glycosylation are shown in Figure 6.2. *N*-glycosylation is characterised by a core structure composed of two GlcNAc and three mannose molecules in branched form. Attached to mannose are a variable number and types of sugar molecules, resulting in three main types of *N*-glycosylation:

- high-mannose type;
- complex type;
- hybrid type.

The main forms of *O*-glycosylation with di- and tri-saccharides are indicated.

Valuable information regarding the glycosylation of proteins are compiled in the databases GlycoSuite (http://tmat.proteomesystems.com/glycosuite) and O-GLYCBASE (http://www.cbs.dtu.dk/databases/OGLYCBASE). *N*-glycosylations and *O*-glycosylations are predictable with NetNGlyc (http://www.cbs.dtu.dk/services/NetNGlyc) and NetOGlyc (http://www.cbs.dtu.dk/services/NetOGlyc).

No general rules of the function of glycosylations in proteins exist. However, the frequency of glycosylations are suggestive of the importance of carbohydrates in proteins. Depending on the type of protein, glycosylations influence various properties of glycoproteins:

- protein folding;
- physical stability;
- solubility;
- specific activity;
- resistance against proteolytic degradation by proteases;
- degradation rate of glycoproteins in the bloodstream which is dependent on the integrity of the terminal sialic acid residues in the carbohydrate structures;
- cell-to-cell recognition via receptors on the cell surfaces for cell invasion in the case of cancer.

## 6.5 PHOSPHORYLATION

Phosphorylation reactions catalysed by kinases and dephosphorylation reactions catalysed by phosphatases are important reactions in many biological processes. They usually occur as regulation of the biological activity of a protein and are as such transient, e.g. as on/off signals in biochemical pathways. The main phosphorylation sites are the hydroxyl groups of Ser, Thr and Tyr, resulting in **P-Ser, P-Thr** and **P-Tyr**, and to a much lower extent His and Cys leading to **P-His** and **P-Cys**. The extent of phosphorylation is often only partial at a rather low level and is often present only during a limited time period. Phosphorylated amino acids are usually rather labile and their detection is therefore often difficult, especially also at their low level of existence. A selection of plasma proteins with documented phosphorylation sites is compiled in Table 6.3.

Experimentally verified phosphorylation sites in eukaryotic proteins are compiled in the Phospho.ELM database (http://phospho.elm.eu.org). Phosphorylation sites are predictable with NetPhos (http://www.cbs.dtu.dk/services/NetPhos).

**Table 6.3** Plasma proteins with documented phosphorylation sites.

| Protein | Data bank entry | Site of phosphorylation |
|---|---|---|
| ***Coagulation and fibrinolysis*** | | |
| Fibrinogen α-chain | FIBA_HUMAN/P02671 | Ser 3 |
| Factor IX | FA9_HUMAN/P00740 | Ser 68, 158 |
| Plasminogen | PLMN_HUMAN/P00747 | Ser 578 |
| Urokinase-type plasminogen activator | UROK_HUMAN/P00749 | Ser 138, 303 |
| Fibronectin | FINC_HUMAN/P02751 | Ser 2353 |
| Vitronectin | VTNC_HUMAN/P04004 | Ser 378 |
| ***Complement system*** | | |
| C1r subcomponent | C1R_HUMAN/P00736 | Ser 189 |
| ***Inhibitors*** | | |
| Tissue factor pathway inhibitor | TFPI1_HUMAN/P10646 | Ser 2 |
| ***Storage and transport*** | | |
| Ferritin heavy chain | FRIH_HUMAN/P02794 | Ser 178 |
| ***Enzymes*** | | |
| Trypsin I | TRY1_HUMAN/P07477 | Tyr 131 |
| Pepsin A | PEPA_HUMAN/P00790 | Ser 68 (potential) |
| ***Hormones*** | | |
| Somatotropin | SOMA_HUMAN/P01241 | Ser 106, 150 |
| ***Cytokines and growth factors*** | | |
| Insulin-like growth factor binding protein 1 | IBP1_HUMAN/P08833 | Ser 101, 119, 169 |
| ***Additional proteins*** | | |
| $\alpha_2$-HS-glycoprotein | FETUA_HUMAN/P02765 | Ser 120 (partial) |
| Chromogranin A | CMGA_HUMAN/P10645 | Ser 200, 252, 304, 315 |
| Secretogranin-1 | SCG1_HUMAN/P05060 | Ser 129, 385 |

### 6.5.1 Ser/Thr phosphorylation sites

*In vivo*, protein kinase C exhibits a preference for the phosphorylation of Ser/Thr found in close proximity to a C-terminal basic residue with the following consensus pattern:

PKC-PHOSPHO_SITE (PS00005): [**ST**]–x–[RK]

cAMP- and cGMP-dependent protein kinase phosphorylations exhibit a preference for phosphorylation of Ser/Thr in close proximity to at least two N-terminal basic residues with the following consensus pattern:

CAMP_PHOSPHO_SITE (PS00004): [RK](2)–x–[**ST**]

Casein kinase II exhibits a preference for the phosphorylation of Ser/Thr in close proximity to at least one acidic C-terminal residue with the following consensus pattern:

CK2_PHOSPHO_SITE (PS00006): [**ST**]–x(2)–[DE]

The removal of the phosphate groups attached to the Ser/Thr residues is catalysed by specific enzymes, the Ser/Thr-specific protein phosphatases (EC 3.1.3.16). These enzymes are very important in controlling intracellular events in eukaryotic cells. Several types of phosphatases exist that are evolutionary related. The Ser/Thr-specific phosphatases are characterised by the following consensus pattern:

SER_THR_PHOSPHATASE (PS00125): [LIVMN]–[KR]–G–N–H–E

### 6.5.2 Tyr phosphorylation site

For tyrosine kinase phosphorylation two consensus patterns exist that are often characterised by a Lys/Arg seven residues N-terminal to Tyr and a Asp/Glu three or four residues N-terminal to Tyr:

TYR_PHOSPHO_SITE (PS00007): [RK]–x(2)–[DE]–x(3)–**Y** or
[RK]–x(3)–[DE]–x(2)–**Y**

As in the case of Ser/Thr dephosphorylation specific enzymes exist for the removal of the phosphate group attached to Tyr residues, the Tyr-specific protein phosphatases (EC 3.1.3.48). These enzymes are very important in controlling cell growth, proliferation, differentiation and transformation. Multiple forms of these phosphatases exist that are classified into two categories, soluble and transmembrane forms. A consensus pattern describes the active site of Tyr-specific protein phosphatases. The Cys residue is the active site:

TYR_PHOSPHATASE_1 (PS00383): [LIVMF]–H–**C**–x(2)–G–x(3)–[STC]–[STAGP]–x–[LIVMFY]

## 6.6 HYDROXYLATION

Hydroxylations are quite common in several types of proteins. In collagens and related proteins, e.g. in the A, B and C chains of the complement C1q subcomponent, the side chains of Pro and Lys are quite often hydroxylated, resulting in 4-hydroxyproline (**OH-Pro**) and 5-hydroxylysine (**OH-Lys**), respectively. The generation of OH-Pro is favoured by the following consensus sequence and the hydroxylation of Pro is influenced by Xaa:

**Gly**–Xaa–**Pro**–**Gly**–Xaa

Lysyl hydroxylases (EC 1.14.11.4) catalyse the hydroxylation of Lys in the sequence **Xaa–Lys–Gly** in collagens, resulting in the attachment of carbohydrate to OH-Lys that are essential for the stability of the intermolecular crosslinks in collagen. A consensus pattern exists for the hydroxylation of Lys residues and the first His and Asp are ligands for iron binding:

LYS_HYDROXYLASE (PS01325): P–**H**–H–**D**–[SA]–S–T–F

In blood coagulation and in the complement system several proteins at specific sites contain hydroxylated side chains of Asp and Asn, leading to β-hydroxyaspartic acid (**OH-Asp**) and 3-hydroxyasparagine (**OH-Asn**) (for details see Section 6.6.1).

### 6.6.1 Aspartic acid and asparagine hydroxylation site

Several proteins in blood coagulation and in the complement system as well as the LDL receptor and the epidermal growth factor contain a single OH-Asp or OH-Asn at specific sites summarised in Table 6.4.

A consensus pattern of the Asx hydroxylation site is located in the N-terminal part of the EGF-like domain where Asp/Asn are hydroxylated:

ASX_HYDROXYL (PS00010): C–x–[**DN**]–x(4)–[FY]–x–C–x–C

**Table 6.4** Plasma proteins with an Asp or an Asn hydroxylation site.

| Protein | Data bank entry | Hydroxylation site |
|---|---|---|
| ***Coagulation and fibrinolysis*** | | |
| Factor VII | FA7_HUMAN/P08709 | Asp 63 |
| Factor IX | FA9_HUMAN/P00740 | Asp 64 |
| Factor X | FA10_HUMAN/P00742 | Asp 63 |
| Protein C | PROC_HUMAN/P04070 | Asp 71 |
| Protein S | PROS_HUMAN/P07225 | Asp 95 |
| Protein Z | PROZ_HUMAN/P22891 | Asp 64 |
| Thrombomodulin | TRBM_HUMAN/P07204 | Asn 324 |
| ***Complement system*** | | |
| C1r subcomponent | C1R_HUMAN/P00736 | Asn 150 |
| C1s subcomponent | C1S_HUMAN/P09871 | Asn 134 |
| Complement-activating component | MASP1_HUMAN/P48740 | Asn 140 |
| Mannan-binding lectin serine protease 2 | MASP2_HUMAN/O00187 | Asn 143 |
| ***Others*** | | |
| LDL receptor | LDLR_HUMAN/P01130 | Asx? |
| Epidermal growth factor | EGF_HUMAN/P01133 | Asx? |

## 6.7 SULFATION

In several plasma proteins the hydroxyl group in the side chain of Tyr is sulfated, resulting in sulfotyrosine (**S-Tyr**), summarised in Table 6.5.

Typically, at least three acidic and no more than one basic and three hydrophobic residues are located within the −5 to +5 region of S-Tyr. In addition, the vicinity of the S-Tyr site is characterised by the absence of disulfide bridges and *N*-glycans.

**Table 6.5** Plasma proteins with Tyr sulfations.

| Protein | Data bank entry | Number of sulfation |
|---|---|---|
| ***Coagulation and fibrinolysis*** | | |
| Fibrinogen gamma-chain | FIBG_HUMAN/P02679 | 2 (418, 422) |
| Factor V | FA5_HUMAN/P12259 | 7 (potential) |
| Factor VIII | FA8_HUMAN/P00451 | 8 (2 probable) |
| Factor IX | FA9_HUMAN/P00740 | 1 (155) |
| Fibronectin | FINC_HUMAN/P02751 | 2 (potential) |
| Vitronectin | VTNC_HUMAN/P04004 | 5 (potential) |
| ***Complement system*** | | |
| Complement C4 | CO4A_HUMAN/P0C0L4<br>CO4B_HUMAN/P0C0L5 | 3 (1398, 1401, 1403) |
| ***Inhibitors*** | | |
| Alpha-2-antiplasmin | A2AP_HUMAN/P08697 | 1 (445) |
| Heparin cofactor II | HEP2_HUMAN/P05546 | 2 (60, 73) |
| ***Hormones*** | | |
| Thyroglobulin | THYG_HUMAN/P01266 | 1 (5) |
| Gastrin | GAST_HUMAN/P01350 | 1 (12) |
| Cholecystokinin 33 | CCKN_HUMAN/P06307 | 1 (27) |
| ***Additional proteins*** | | |
| Chromogranin A | CMGA_HUMAN/P10645 | ? |
| Secretogranin-1 | SCG1_HUMAN/P05060 | 2 (153(partial), 321) |

**Table 6.6** Plasma proteins with acylations and alkylations.

| Protein | Data bank entry | Modification |
|---|---|---|
| ***Coagulation and fibrinolysis*** | | |
| Fibrinogen beta-chain | FIBB_HUMAN/P02675 | Pyro-Glu |
| Tissue factor | TF_HUMAN/P13726 | Palmitoyl (Cys 245) |
| Factor XIII A-chain | F13A_HUMAN/P00488 | *N*-acetyl (Gly) |
| Fibronectin | FINC_HUMAN/P02751 | Pyro-Glu |
| Kininogen | KNG1_HUMAN/P01042 | Pyro-Glu |
| Annexin 5 | ANXA5_HUMAN/P08758 | *N*-acetyl (Ala) |
| ***Complement system*** | | |
| Subcomponent C1q B-chain | C1QB_HUMAN/P02746 | Pyro-Glu |
| Component C8 gamma-chain | CO8G_HUMAN/P07360 | Pyro-Glu |
| Complement receptor type 1 | CR1_HUMAN/P17927 | Pyro-Glu (potential) |
| C-reactive protein | CRP_HUMAN/P02741 | Pyro-Glu |
| ***Inhibitors*** | | |
| Pigment epithelium-derived factor | PEDF_HUMAN/P36955 | Pyro-Glu |
| Heparin cofactor II | HEP2_HUMAN/P05546 | Pyro-Glu |
| Leukocyte elastase inhibitor | ILEU_HUMAN/P30740 | *N*-acetyl (Met) |
| Cystatin B | CYTB_HUMAN/P04080 | *N*-acetyl (Met, probable) |
| ***Lipoproteins*** | | |
| Apolipoprotein A-I | APOA1_HUMAN/P02647 | Palmitoyl (site unknown) |
| Apolipoprotein A-II | APOA2_HUMAN/P02652 | Pyro-Glu |
| Apolipoprotein B-100 | APOB_HUMAN/P04114 | Palmitoyl (Cys 1085) |
| Apolipoprotein D | APOD_HUMAN/P05090 | Pyro-Glu |
| Serum amyloid A protein | SAA_HUMAN/P02735 | *N*,*N*-dimethyl-Asn (83) |
| ***Storage and transport*** | | |
| Ferritin light chain | FRIL_HUMAN/P02792 | *N*-acetyl (Ser) |
| ***Enzymes*** | | |
| Angiogenin | ANGI_HUMAN/P03950 | Pyro-Glu |
| Glutathione peroxidase | GPX3_HUMAN/P22352 | Pyro-Glu |
| ***Hormones*** | | |
| (Pro)-gonadoliberin 1 | GON1_HUMAN/P01148 | Pyro-Glu |
| (Pro)-gonadoliberin 2 | GON2_HUMAN/O43555 | Pyro-Glu |
| Thyroliberin | TRH_HUMAN/P20396 | Pyro-Glu |
| Gastrin | GAST_HUMAN/P01350 | Pyro-Glu |
| ***Cytokines and growth factors*** | | |
| Interferon $\gamma$ | IFNG_HUMAN/P01579 | Pyro-Glu |
| Small inducible cytokine A2 | CCL2_HUMAN/P13500 | Pyro-Glu |
| ***Immune system*** | | |
| Zn-$\alpha_2$-Glycoprotein | ZA2G_HUMAN/P25311 | Pyro-Glu |
| Alpha-1-acid glycoprotein 1 | A1AG1_HUMAN/P02763 | Pyro-Glu |
| Alpha-1-acid glycoprotein 2 | A1AG2_HUMAN/P19652 | Pyro-Glu |
| Beta-2-microglobulin | B2MG_HUMAN/P61769 | Pyro-Glu |
| ***Additional proteins*** | | |
| Ficolin-3 | FCN3_HUMAN/O75636 | Pyro-Glu |
| Platelet epithelium-derived factor | PEDF_HUMAN/P36955 | Pyro-Glu |

Furthermore, sulfation seems to be favoured by the presence of turn-inducing amino acids Pro, Gly, Asp, Ser and Asn in the vicinity of Tyr.

The *O*-sulfation of Tyr is catalysed by the enzyme tyrosylprotein sulfotransferase (TPST) (EC 2.8.2.20) that transfers the sulfate from the universal sulfate donor adenosine 3′-phosphate 5′-phosphosulfate (PAPS) to the hydroxyl group of Tyr. Sulfations are predicable with the Sulfinator (http://www.expasy.org/tools/sulfinator).

## 6.8 ACYLATION AND ALKYLATION

The α- and ε-amino groups and thiols can be acylated or alkylated, resulting among others in acetylation, formylation, methylation and pyroglutamylation. A selection of acylated and alkylated plasma proteins with the corresponding modified amino acid are compiled in Table 6.6.

Many proteins are modified at their N-terminus, mainly by acetylation, formylation and pyroglutamylation.

**Acetylation** takes place at the ε-amino group of Lys residues or at the N-terminus of proteins. In histones, Lys residues are acetylated and deacetylated involving enzymes with 'histone acetyltransferase' and 'histone deacetyltransferase' activity and acetyl-CoA is the source of the acetyl group. Prior to N-terminal acetylation, an initiator Met residue is cleaved and the newly generated N-terminus is acetylated with acetyl-CoA as the donor.

**Methylation** takes place at Arg or Lys residues, resulting in mono- or dimethyl-Arg or mono-, di- or trimethyl-Lys with the transfer of the methyl group from *S*-adenosyl-Met catalysed by methyltransferases (Arg or Lys).

**Pyroglutamylation** is a cyclisation reaction of an N-terminal Gln resulting in a pyrrolidone carboxylic acid residue (**Pyro-Glu**). N-terminally modified proteins cannot be identified by Edman degradation as the required free amino group is cyclised. The enzyme pyrrolidone-carboxylate peptidase (pyroglutamyl peptidase (EC 3.4.19.3): 209 aa, PGPI_HUMAN/Q9NXJ5) removes pyroglutamate (Pyro-Glu) selectively from the N-terminus of peptides and proteins. They belong to the Cys proteases with the catalytic triad Cys, His and Glu. The active site of the enzyme is characterised by two consensus patterns and the Glu and Cys residues are the active site residues, respectively:

PYRASE_GLU (PS01333): G–x(2)–[GAP]–x(4)–[LIV]–[ST]–x–**E**–[KR]–[LIVC]–[AG]–x–[NG]

PYRASE_CYS (PS01334): [LIVF]–x–[GSAVC]–x–[LIVM]–S–x–[STAD]–A–G–x–[FY]–[LIVN]–**C**–[DNS]

Based on a yeast dataset N-terminal acetylation can be predicted by NetAcet (http://www.cbs.dtu.dk/services/NetAcet). Using an independent dataset of mammalian *N*-acetylated proteins, 74 % of acetylated Ser residues in mammalian proteins were correctly predicted. This prediction rate is similar to the case for yeast proteins.

## 6.9 AMIDATION

Precursors of hormones and other biologically active peptides are quite often C-terminally amidated, which is in most cases important for their biological activity and their biostability. The peptide amidation reaction is an enzyme-catalysed process where the C-terminal amino acid is directly followed by a Gly residue that provides the required amide group. A consensus pattern exists where the required Gly residue is followed by two consecutive basic residues (Arg/Lys). In principal, all C-terminal amino acids can be amidated, but there is a preference for neutral, hydrophobic residues (Val, Phe), while charged amino acids like Asp and Arg are much less reactive.

AMIDATION (PS00009): **x**–G–[RK]–[RK]

## 6.10 CARBOXYLATION

The introduction of a carboxyl group at the γ position of the side chain of Glu is quite common in several blood plasma proteins (for details see Section 4.4), resulting in γ-carboxyglutamic acid (**γ-Gla**). The vitamin K-dependent carboxylation reaction is catalysed by γ-carboxylase (vitamin K-dependent γ-carboxylase: 758 aa, VKGC_HUMAN/P38435) and γ-Gla residues are found in the Gla domain in the N-terminal region of several proteins in blood coagulation (for details see Chapter 4). In addition, in some proteins carboxylation at the β position of the side chain of Asp results in the generation of β-carboxyaspartic acid (**β-Asp**).

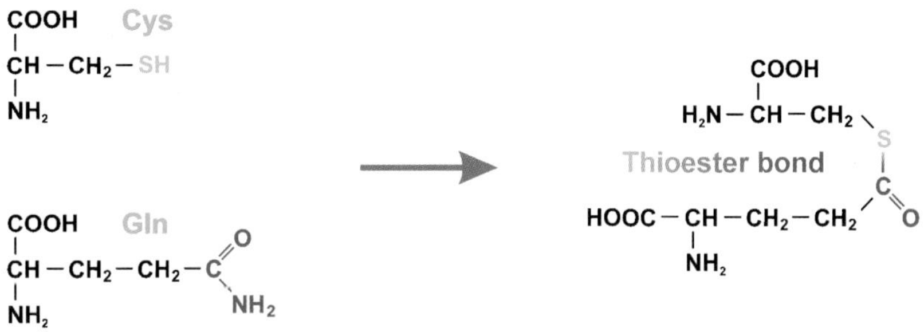

**Figure 6.3 Formation of a thioester bond**
The thiol group of Cys and the amide group of Gln form a thioester bond.

## 6.11 CROSSLINKS

Intrachain as well as interchain crosslinking reactions are important for obtaining structural stability or for the generation of stable structures in proteins. The most important and most widely spread crosslinking reaction is the formation of disulfide bridges, which are essential for obtaining and maintaining structural stability in most proteins (for details see Section 6.3). However, there are other crosslinking reactions that are important for biological function, e.g. the formation of an intrachain isoglutamyl cysteine thioester bond (**Cys–Gln**) and the formation of an interchain isopeptide bond between the side chains of Lys and Gln (**Lys–Gln**).

### 6.11.1 The isoglutamyl cysteine thioester bond

The most prominent Cys–Gln bonds are present in complement components C3 and C4 and in $\alpha_2$-macroglobulin ($\alpha_2$M): The sequences **Cys988–Gly–Glu–Gln991** in C3, **Cys991–Gly–Glu–Gln994** in C4 and **Cys949–Gly–Glu–Gln952** in $\alpha_2$M are the sites for the intrachain crosslinks. The formation of the reactive thioester bond from the thiol group of Cys and the amide group of Gln is shown in Figure 6.3.

In the complement cascade the reactive thioester bonds of C3 and C4 bind covalently to amino or hydroxyl groups on the surfaces of pathogens. In $\alpha_2$M the cleavage of the thioester bond is part of the inhibitory reaction. The active thioesters are rapidly inactivated by hydrolysis with a water molecule (for details see chapter 8).

### 6.11.2 The isopeptide bond

The formation of isopeptide bonds is an important crosslinking reaction to stabilise protein structures. In the coagulation cascade the fibrin network is stabilised by the formation of isopeptide bonds, catalysed by coagulation factor XIIIa, a

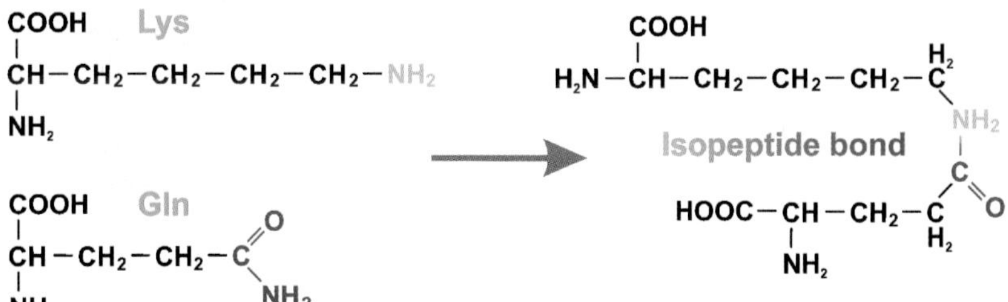

**Figure 6.4 Formation of an isopeptide bond**
The amine group of the Lys side chain and the amide group of Gln form an isopeptide bond.

transglutaminase. In fibrinogen α–α chain crosslinks involve Gln 328 and 366 and Lys 508, 539, 556, 562 and 580 and γ–γ chain crosslinks Gln 398 and Lys 406. The formation of the isopeptide bond from the amine group of the Lys side chain and the amide group of Gln is depicted in Figure 6.4.

The formation of an isopeptide bond between Lys 303 in the α-chain of fibrinogen and Gln 2 in the $\alpha_2$-plasmin inhibitor prevents the fibrin clot from a premature degradation by plasmin. Gln 3, 32 and 1412 in fibronectin seem to form interchain isopeptide bonds with so far unidentified Lys residues.

## REFERENCES

Ahvazi *et al.*, 2003, A model of the reaction mechanisms of the transglutaminase 3 enzyme, *Exp. Mol. Med.*, **35**, 228–242.

Anfinsen *et al.*, 1961, The kinetics of formation of native ribonuclease during oxidation of the reduced polypeptide chain, *Proc. Natl. Acad. Sci. USA*, **47**, 1309–1314.

Apweiler *et al.*, 2004, UniProt: the universal protein knowledgebase, *Nucleic Acids Res.*, **32**, (database use), D115–D119.

Arolas *et al.*, 2006, Folding of small disulfide-rich proteins: clarifying the puzzle, *Trends Biochem. Sci.*, **31**, 292–301.

Blom *et al.*, 1999, Sequence- and structure-based prediction of eukaryotic protein phosphorylation sites, *J. Mol. Biol.*, **294**, 1351–1362.

Bradbury *et al.*, 1991, Peptide amidation, *Trends Biochem. Sci.*, **16**, 112–115.

Cohen, 1989, The structure and regulation of protein phosphatases, *Annu. Rev. Biochem.*, **58**, 453–508.

Diella *et al.*, 2004, Phospho.ELM: a database of experimentally verified phosphorylation sites in eukaryotic proteins, *BMC Bioinformatics*, **5**, 79.

Driessen *et al.*, 1985, The mechanism of N-terminal acetylation of proteins, *CRC Crit. Rev. Biochem.*, **18**, 281–325.

Dyrlov *et al.*, 2004, Improved prediction of signal peptides: SignalP 3.0, *J. Mol. Biol.*, **340**, 783–795.

Farriol-Mathis *et al.*, 2004, Annotation of post-translational modifications in the Swiss-Prot knowledge base, *Proteomics*, **4**, 1537–1550.

Fischer *et al.*, 1991, Protein tyrosine phosphatases: a diverse family of intracellular and transmembrane enzymes, *Science*, **253**, 401–406.

Fremisco *et al.*, 1980, Optimal spatial requirements for the location of basic residues in peptide substrates for the cyclic AMP-dependent protein kinase, *J. Biol. Chem.*, **255**, 4240–4245.

Gavel and von Heijne, 1990, Sequence differences between glycosylated and non-glycosylated Asn-X-Thr/Ser acceptor sites: implications for protein engineering, *Prot. Eng.*, **3**, 433–442.

Gupta *et al.*, Prediction of *N*-glycosylation sites in human proteins (in preparation).

Hansen *et al.*, 1995, prediction of *O*-glycosylation of mammalian proteins: specificity patterns of UDP-GalNAc:polypeptide *N*-acetylgalactosaminyltransferase, *Biochem. J.*, **388**, 801–813.

Hauschka *et al.*, 1980, Quantitative analysis and comparative decarboxylation of aminomalonic acid, beta-carboxyaspartic acid and gamma-carboxyglutamic acid, *Anal. Biochem.*, **108**, 57–63.

Horwitz *et al.*, 1984, Localization of a fibrin gamma-chain polymerisation site within segment Thr-374 to Glu 396 in human fibrinogen, *Proc. Natl. Acad. Sci. USA*, **81**, 5980–5984.

Julenius *et al.*, 2004, Prediction, conservation analysis, and structural characterization of mammalian mucin-type *O*-glycosylation sites, *Glycobiology*, **15**, 153–164.

Kiemer *et al.*, 2005, NetAcet: prediction of N-terminal acetylation sites, *Bioinformatics*, **21**, 1269–1270.

Kivirikko *et al.*, 1991, in *Post-translational modifications of proteins*, (eds. Crabbe and Harding), CRC Press, Bocca Raton, pp. 1–51.

Kreil, 1984, Occurrence, detection, and biosynthesis of carboxy-terminal amides, *Methods Enzymol.*, **106**, 218–223.

Monigatti *et al.*, 2002, The sulfinator: predicting tyrosine sulfation sites in protein sequences, *Bioinformatics*, **18**, 769–770.

Moore, 2003, The biology and enzymology of protein tyrosine *O*-sulfation, *J. Biol. Chem.*, **278**, 24243–24246.

Nicholas *et al.*, 1999, Reevaluation of the determinants of tyrosine sulfation, *Endocrine*, **11**, 285–292.

Patschinsky *et al.*, 1982, Analysis of the sequence of amino acids surrounding sites of tyrosine phosphorylation, *Proc. Natl. Acad. Sci. USA*, **79**, 973–977.

Pinna, 1990, Casein kinase 2: an 'eminence grise' in cellular regulation, *Biochim. Biophys. Acta*, **1054**, 267–284.

Pless and Lennarz, 1977, Enzymatic conversion of proteins to glycoproteins, *Proc. Natl. Acad. Sci. USA*, **74**, 134–138.

Rawlings and Barrett, 1994, Families of cysteine peptidases, *Methods Enzymol.*, **244**, 461–486.

Simon *et al.*, 1988, The glutamine residues reactive in transglutaminase-catalyzed cross-linking of involucrin, *J. Biol. Chem.*, **263**, 18093–18098.

Sowdhamini *et al.*, 1989, Stereochemical modelling of disulfide bridges: criteria for introduction into proteins by site-directed mutagenesis, *Prot. Engng.*, **3**, 95–103.

Srinivasan *et al.*, 1990, Conformations of disulfide bridges in proteins and peptides, *Int. J. Pept. Prot. Res.*, **36**, 147–155.

Stenflo *et al.*, 1988, Beta-hydroxyaspartic acid or beta-hydroxyasparagine in bovine low density lipoprotein receptor and in bovine thrombomodulin, *J. Biol. Chem.*, **263**, 21–24.

Thomas *et al.*, 1983, Identification and alignment of a thiol ester site in the third component of guinea pig complement, *Biochemistry*, **22**, 942–947.

Tu *et al.*, 2004, Oxidative protein folding in eukaryotes: mechanisms and consequences, *J. Cell. Biol.*, **164**, 341–346.

Vermeer, 1990, Gamma-carboxyglutamate-containing proteins and the vitamin-K dependent carboxylase, *Biochem. J.*, **266**, 625–636.

Wilkins *et al.*, 1997, *Proteome Research: New Frontiers in Functional Genomics*, Springer.

Woodget *et al.*, 1986, Substrate specificity of protein kinase C. Use of synthetic peptides corresponding to physiological sites as probes for substrate recognition requirements, *Eur. J. Biochem.*, **161**, 177–184.

# PART III

# 7

# Blood Coagulation and Fibrinolysis

## 7.1 INTRODUCTION

At rest, the blood flow in the vascular system of the body exhibits neither a tendency for bleeding nor for thrombosis. This balance between blood coagulation and fibrinolysis is termed *haemostasis*. Under normal conditions the vascular endothelium supports vasodilatation, inhibits the adhesion and activation of platelets, keeps coagulation at a low level and enhances the cleavage of eventually occurring fibrin clots and is anti-inflammatory in its nature. Haemostasis contributes in many different ways to life saving and life-supporting processes of the body and regulates many events that are absolutely essential for the well being of an organism. Haemostasis prevents the loss of blood and controls the normal blood flow and supports the repair of the injured vasculature and tissues. Already during haemostasis the formation of new connective tissue and revascularisation is initiated. In addition, the formation of a temporary platelet and fibrin plug helps the body to prevent attacks from microorganisms, at least to a certain extent. At rest, the interactions of the various components involved in haemostasis are ongoing and their consumption and synthesis is a continuing process, however at a low level.

### 7.1.1 Definition of haemostasis

Haemostasis comprises the processes of blood clotting and the dissolution of the blood clots with the subsequent repair of the injured tissues and is the well-balanced interaction of blood cells, the vascular system, plasma proteins and low molecular weight components.

**Perfect haemostasis**: neither bleeding nor thrombosis

The major components that are involved in haemostasis are compiled in Table 7.1.

Upon an injury of a blood vessel all processes involved in blood coagulation and fibrinolysis and the synthesis of the involved components is accelerated extremely rapidly. The onset of the activation process of coagulation and fibrinolysis may also be favoured under pathological conditions like autoimmune diseases, arteriosclerosis or cancer.

There are various types of proteins involved in haemostasis: enzymes, usually synthesised as zymogens and thus requiring an activation step, cofactors, inhibitors and adhesive proteins, cytokines and hormones, as well as membrane and intercellular proteins.

**Table 7.1** Components involved in haemostasis.

| |
|---|
| **Blood cells**: thrombocytes (platelets) and other blood cells |
| **The vascular cell wall**: endothelium and subendothelium |
| **Coagulation factors** |
| **Factors of fibrinolysis** |
| **Inhibitors**: serpins, $\alpha_2$-macroglobulin, other inhibitors |
| **Proteins**: associated with coagulation and fibrinolysis |
| **Calcium ions** |
| **Low molecular weight (LMW) components**: phospholipids, prostaglandins, others |

*Human Blood Plasma Proteins: Structure and Function* Johann Schaller, Simon Gerber, Urs Kämpfer, Sofia Lejon and Christian Trachsel

**Table 7.2** Major events in haemostasis.

| Primary haemostasis | Secondary haemostasis | Fibrinolysis |
|---|---|---|
| Vasoconstriction<br>⇒ **immediately**<br>Platelet adhesion<br>⇒ **seconds**<br>Platelet aggregation<br>⇒ **minutes** | Activation of coagulation factors<br>⇒ **seconds**<br>Formation of fibrin network<br>⇒ **minutes** | Activation of fibrinolysis<br>⇒ **minutes**<br>Lysis of the fibrin clot<br>⇒ **hours** |

The components of haemostasis usually interact on surfaces, e.g. on activated platelets or on the vasculature. The walls of intact blood vessels are composed of endothelial cells, which have a highly active surface that is strongly antithrombotic as an uncontrolled and unregulated thrombosis might be life threatening and the intact cell surface ensures biocompatibility. In addition, shear forces and blood viscosity are important factors that influence haemostasis. In principle, haemostasis can be divided schematically into three major events: primary haemostasis, secondary haemostasis and fibrinolysis (Table 7.2).

## 7.2 PRIMARY HAEMOSTASIS

Immediately upon the loss of the vascular integrity vascular constriction is onset, leading to a restricted flow of blood to the site of the injury in the initial phase of primary haemostasis. Vasoconstriction, the constriction of the blood vessels, is a very fast process and an efficient way to prevent blood loss, especially in the small blood vessels, the microvasculature. Vasoconstriction is mediated by a complex process of the autonomous nerve system, muscle cells and a series of mediators like serotonin, epinephrine, noradrenaline and others. The reverse process, vasodilatation, is mainly mediated by prostaglandins.

Platelet adhesion is a very complex process initiated within seconds of the injury upon contact of the platelets with nonphysiological surfaces such as an injured blood vessel or the extracellular matrix (ECM). In the resting state platelets are nonthrombogenic. The cytoskeleton of platelets contributes significantly to their pronounced shape change, which takes place after their activation and helps them to spread out over the injured area (see Chapter 2).

The main proteins involved in the cytoskeleton superstructure of platelets and their pronounced shape change are the α-chain (2418 aa: P02549) and the β-chain (2136 aa: P11277) of spectrin, β-actin (374 aa: P60709) and the band 4.1 protein (864 aa: P11171). The collagen-rich ECM and the subendothelium play a crucial role in platelet activation, platelet adhesion and later platelet aggregation. The interaction between thrombocytes and the exposed collagen is mainly mediated by glycoprotein receptors on the surface of platelets. Other activating components are thrombin and the LMWs ADP, serotonin, epinephrine, thromboxane A2 and arachidoncic acid. Furthermore, interactions are mediated by the plasma proteins von Willebrand factor (vWF) and fibrinogen (FI).

**Von Willebrand factor** (vWF) (see Data Sheet) synthesised in endothelial cells and megakaryocytes is a very large single-chain glycoprotein (19 % CHO) containing three VWFA, three VWFC and two VWFD domains, two trypsin inhibitory-like (TIL) domains and one C-terminal cystine knot (CTCK) structure. Mature vWF is generated from the proform (2791 aa: Ala 1–Lys 2791) by releasing a 741 residues propeptide, the von Willebrand antigen II. vWF circulates in blood plasma in multimeric forms, the smallest being the dimer (540 kDa) held together by disulfide bridges in the C-terminal region of the subunits. Multimers with up to 50 dimers have been observed (approximately up to 20 000 kDa) that are linked together by interdimeric disulfide bridges in the N-terminal regions of the dimers. The oligomerisation process of the dimers forms long linear filaments of up to 3 mm length, the largest known plasma protein. vWF has two main functions:

1. It serves as a carrier protein for **coagulation factor VIII** (FVIII) protecting it from proteolytic degradation and inactivation. The binding site for FVIII is located in the N-terminal region of vWF and the binding ratio is approximately 100:1 (w/w).
2. It is required for the interaction of platelets with the subendothelium of the injured blood vessel. The binding sites for collagen are located in the VWFA 1 (aa 514–690) and the VWFA 3 (aa 928–1108) domains, for the platelet glycoprotein GPIb in the regions between VWFD 3 and VWFA 1 (aa 474–488) and VWFA 1 and VWFA 2 (aa 694–708), respectively. The single **RGD** sequence (aa 1744–1746) binds to the platelet glycoprotein complex GPIIb/IIIa and the N-terminal

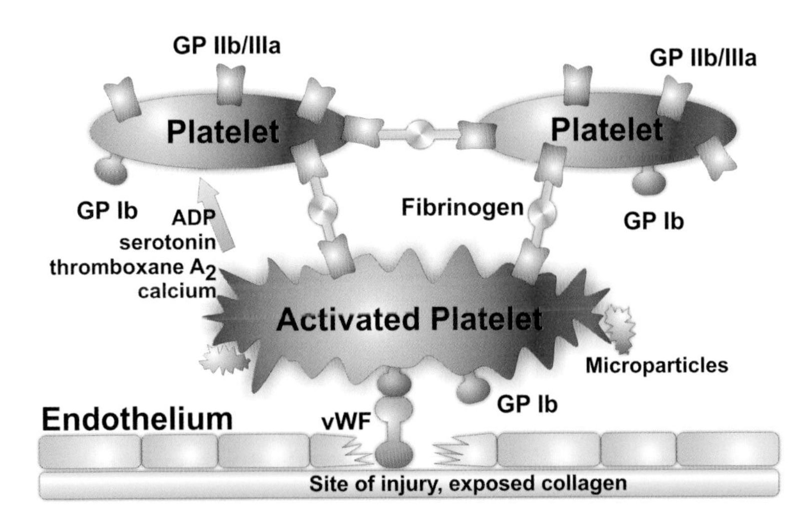


**Figure 7.1 Platelet adhesion and aggregation**
Schematic representation of platelet adhesion and aggregation at the site of an injured blood vessel with the major components involved in these processes. GP Ib, platelet glycoprotein Ib; GP IIb/IIIa, complex integrin α-IIb/integrin β-3; vWF, von Willebrand factor.

region of vWF is involved in binding to heparin. The 3D structure of the first VWFA domain (PDB: 1IJB; 202 aa: Ser 500–Ala 701 in vWF) has been elucidated by X-ray diffraction and is characterised by seven α-helices and six β-sheets (see Chapter 4).

Based on these multiple interactions the binding affinities for platelets and collagen are greatly increased in the multimeric forms of vWF and thus vWF works as a kind of 'molecular glue'. Under shear stress vWF can undergo a shape change from its usual globular form to a rod-like structure.

The von Willebrand disease is the most common bleeding disorder, with a prevalence of about 1 %, and is classified into three categories: in type 1 the vWF deficiency is partial quantitative, in type 2 qualitative and in type 3 virtually absent (for details see MIM: 193400 and 277480).

Platelet aggregation represents the final phase of primary haemostasis and is closely connected with the previously mentioned platelet adhesion. In platelet adhesion, platelets interact with the injured blood vessel wall and adhesive proteins interact with each other mediated by fibrinogen and vWF, as depicted in Figure 7.1.

Adhesive proteins play an important role in many physiological processes and are required for the interaction of cells with surfaces, e.g. with specific receptors, collagens, structural proteins or glycosaminoglycans. In haemostasis adhesive proteins like the vWF, fibrinogen, fibronectin (FN), thrombospondin (TSP), vitronectin (VN) and others are of special interest.

**Fibronectin** (FN) (see Data Sheet) or cold-insoluble protein synthesised in hepatocytes is a large single-chain glycoprotein (approximately 5 % CHO) that contains 12 fibronectin type I (FN1), two type II (FN2) and 16 type III (FN3) domains. The N-terminus is modified by pyroglutamylation, Tyr 845 and 850 are potential sulfation sites and Ser 2353 is phosphorylated. In addition, it contains a single **RGD** sequence (aa 1493–1495) as a cell-binding motif. FN contains three probable interchain crosslinking sites forming isopeptide bonds between Gln 3, 32 and 1412 and so far unidentified Lys residues (see Chapter 6). FNs are involved in cell adhesion, cell motility, tissue turnover and wound healing. They bind to cell surfaces and interact among others with fibrin, collagens, actin and heparin. A first fibrin- and heparin-binding region is located in the N-terminal area of FN comprising FN1 domains 1–5 followed by a collagen-binding region comprising FN1 domains 6–9 and FN2 domains 1 + 2. A second fibrin-binding region is located in the C-terminal area of FN comprising FN1 domains 10–12 and a second heparin-binding region comprising FN3 domains 13–15. Circulating fibronectin is thought to play a major role as a precursor for tissue fibronectin. There are no known inhibitors of fibronectin. At least 12 splice variants exist.

FN exhibits an elongated structure of approximately 120 nm length and 2.5 nm diameter. The modules seem to be arranged like 'beads on a string'. The 3D structure of the N-terminal part of the collagen-binding region in FN comprising FN1 domain 6 and FN2 domain 1 (PDB: 1QO6; 101 aa: Ser 274–Thr 374 in FN) was determined by NMR spectroscopy. The structure is characterised by nine β-strands and no α-helices and the structures conform with previously determined consensus folds of FN repeats. The structure of a four-domain segment of FN comprising the FN3 domains 7–10 (FN3 domain 10 containing the RGD sequence; PDB: 1FNF; 368 aa: Pro 1142–Thr 1509) was determined by X-ray diffraction and is characterised by 32 β-strands and no α-helices. The RGD loop is well-ordered and easily accessible (see Chapter 4).

**Thrombospondin-1** (TSP-1) (see Data Sheet) is a large glycoprotein that circulates in blood as a disulfide-linked homotrimer (Cys 252 and Cys 256 form interchain disulfide bridges) belonging to the thrombospondin family. TSP-1 exhibits with 54 % identical aa, a high structural similarity with TSP-2 (1154 aa: P35442) and TSP-1 and TSP-2 form disulfide-linked heterotrimers. TSP-1 is a multidomain protein with three EGF-like, one TSP C-terminal (TSPC) and one TSP N-terminal (TSPN), three TSP1 and seven TSP3 domains and one VWFC domain. In addition, it contains a single **RGD** sequence (aa 908–910) as a cell-binding motif. TSP-1 and TSP-2 are adhesive glycoproteins that mediate cell-to-cell and cell-to-matrix interactions. They can bind to many types of proteins like plasmin, fibrinogen, fibronectin, laminins (α-chain, 3058 aa: P25391; β-chain, 1765 aa: P07942; γ-chain, 1576 aa: P11047), collagen type V and various types of integrins. In addition, they also bind heparin and calcium ions. These interactions enable TSP-1/TSP-2 to modulate cell shape, cell migration and cell proliferation. There are no known inhibitors of TSP-1/TSP-2. Electron microscopic images show N-terminal and C-terminal globular regions that are connected by a region that appears to be rather thin and flexible. Besides TSP-1 and TSP-2 the TSP family consists of TSP-3 (934 aa: P49746) and TSP-4 (935 aa: P35443) which lack the TSP type 1 and the VWFC domains but are otherwise similar in structure and function.

The 3D structure of the N-terminal domain of TSP-1 (TSPN) which mediates the interactions during cellular adhesion was determined by X-ray diffraction in complex with a synthetic pentameric heparin homologue (PDB: 1ZA4; 240 aa: Asn 1–Ile 236 + N-terminal tetrapeptide). A positively charged patch containing Arg 29, Arg 42 and Arg 77 at the bottom of TSPN specifically interacts with the sulfate groups of heparin.

The 3D structure of a C-terminal fragment of TSP-1 comprising TSP-3 domains 5–7 and TSPC (PDB: 1UX6; 337 aa: Asp 816–Pro 1152 + N-terminal His-tag peptide) was determined by X-ray diffraction. The TSP-3 domains are arranged around a calcium core and each domain contains two Asp-rich motifs binding two calcium ions. The availability of the RGD sequence located in TSP-3 domain 7 is modulated by calcium loading (see Chapter 4).

**Vitronectin** (VN) (see Data Sheet) or serum spreading factor mainly synthesised in the liver and present in platelets and monocytes is a glycoprotein (10–15 % CHO) that circulates in blood as a monomer, a dimer and probably also as an oligomer. It contains two homopexin-like and one somatomedin B domain, it has five potential sulfation sites and is phosphorylated at Ser 378. Two forms exist of vitronectin: a single-chain form (V75) and a two-chain form obtained by cleavage of the Arg 379–Ala 380 peptide bond in V75. The two chains V65 and V10 are linked by a disulfide bond. It also contains a single **RGD** sequence (aa 45–47) as a cell-binding motif. Vitronectin is an important adhesive protein that mediates adhesion and binding to cell surfaces. It binds to many types of proteins like the vitronectin receptor (1018 aa: P06756), to certain integrins, to the urokinase plasminogen activator surface receptor uPAR (283 aa: Q03405), to the C1q-binding protein (209 aa: Q07021), to the plasminogen activator inhibitor-1 (PAI-1) and to plasminogen. It also interacts with glycosaminoglycans and proteoglycans. There are no known inhibitors of vitronectin.

**Somatomedin B** (44 aa: Asp 1–Thr 44) generated by limited proteolysis from the N-terminal region of its parent protein vitronectin is a growth hormone-dependent plasma protein containing four intrachain disulfide bridges (Cys 5–Cys 9, Cys 19–Cys 21, Cys 25–Cys 31, Cys 32–Cys 39) arranged in the following pattern:

**Cys 1–Cys 2, Cys 3–Cys 4, Cys 5–Cys 6, Cys 7–Cys 8**

The somatomedin B domain is characterised by a consensus sequence comprising six out of eight Cys residues all involved in intrachain disulfide bonds:

SMB_1 (PS00524): **C**–x–**C**–x(3)–**C**–x(5,6)–**C**–**C**–x–[DN]–[FY]–x(3)–**C**

The 3D structure of the somatomedin B domain of VN (51 aa: Asp 1–Met 51) in complex with the serpin plasminogen activator inhibitor-1 (379 aa) has been solved by X-ray diffraction (PDB: 1OC0). The binding of the somatomedin B domain of VN to PAI-1 stabilises the active conformation of PAI-1 and thus its lifetime.

Many platelet membrane glycoproteins are involved in platelet adhesion and aggregation and are receptors and/or binding partners for many proteins involved in blood coagulation and fibrinolysis: vWF, fibrinogen, collagen, fibronectin, thrombospondin, vitronectin, prothrombin, plasminogen and others. Table 7.3 lists the major platelet membrane glycoproteins.

**Table 7.3** Major platelet membrane glycoproteins.

| Name | Synonyms | Code | Size (Da/number aa) | Type |
|---|---|---|---|---|
| Integrin α-2 | GPIa | ITA2_HUMAN/P17301 | 126 388/1152 | Type I membrane |
| Platelet glycoprotein Ib | | | | |
| α-Chain | GPIb-α | GP1BA_HUMAN/P07359 | 67 193/610 | Type I membrane |
| β-Chain | GPIb-β | GP1BB_HUMAN/P13224 | 19 217/180 | Type I membrane |
| Integrin α-IIb | GPIIb | ITA2B_HUMAN/P08514 | 110 021/1008 | Type I membrane |
| Integrin β-3 | GPIIIa | ITB3_HUMAN/P05106 | 84 518/762 | Type I membrane |
| Platelet glycoprotein IV | GPIV/GPIIIb | CD36_HUMAN/P16671 | 52 922/471 | Integral membrane |
| Platelet glycoprotein V | GPV | GPV_HUMAN/P40197 | 59 277/544 | Type I membrane |
| Platelet glycoprotein IX | GPIX | GPIX_HUMAN/P14770 | 17 316/161 | Type I membrane |

All platelet membrane glycoproteins contain a potential transmembrane segment of 20–25 residues length; GPIV as an integral membrane protein has two such segments.

GPIa is a receptor for collagen, laminin and fibronectin and is responsible for the adhesion of platelets and other cells to exposed collagen of the subendothelium. The complex of GPIb/GPV/GPIX functions as a receptor for vWF and mediates the vWF-dependent adhesion of the platelets to the injured blood vessel wall, which is a critical step in the initial phase of haemostasis. The GPIb α- and β-chains are linked with a disulfide bridge and this heterodimer is in complex with GPIX. GPIIb/GPIIIa is a receptor for many proteins involved in blood coagulation and fibrinolysis: fibrinogen, plasminogen, prothrombin, fibronectin, thrombospondin and vitronectin. It recognises **RGD** sequences involved in cell attachment in many ligands and the sequence **HHLGGGAKQAGDV** in the γ-chain of fibrinogen leads to rapid platelet aggregation, thus plugging the ruptured endothelial cell surface. GPIV binds collagen and thrombospondin and seems to be involved in cell adhesion processes.

## 7.3 THE COAGULATION CASCADE

Secondary haemostasis is initiated with the activation of various coagulation factors. With the exception of the membrane-bound tissue factor (TF) all coagulation factors require an activation step. This activation process is based on limited proteolysis at specific sites. Limited proteolysis is also a very common process in many other systems in the body such as:

- the complement system;
- the regulation of blood pressure with the angiotensin/renin system and its main component, the octapeptide angiotensin II (see angiotensinogen in Section 7.6);
- the kinin system with the nonapeptide bradykinin (see kininogen in Section 7.6);
- the activation of matrix metalloproteases.

The activation process leads to a major conformational change upon cleavage of at least one peptide bond, often with the concomitant release of a (pre-)activation peptide and the subsequent exposure of the active site of the generated, active enzyme. All coagulation factors that exhibit protease activity belong to the class of serine proteases (for details consult Chapters 5 and 10). With the exception of fibrinogen (approximately 9 μM) and prothrombin (approximately 1.5 μM) and tissue factor that is usually not detectable in circulation, all coagulation factors are present in plasma at *nM* concentrations.

Blood coagulation is a process that is taking place continually in the body at a very low level. There are two major activation pathways, the extrinsic pathway and the intrinsic pathway that converge at the level of factor X and then lead into the common pathway. Although the extrinsic and the intrinsic pathways are interdependent, at least to some extent, for historical reasons they are usually described separately. A schematic representation in Figure 7.2 depicts the major components participating in the coagulation cascade.

The contact of the bloodstream with free subendothelial cells and thus the interaction with collagen leads to the immediate extrinsic activation of the coagulation cascade. This initial physiological activation process primarily takes place via the tissue factor (TF). The subendothelium is rich in TF and after an injury or cell stimulation, such as in inflammation, TF is exposed or synthesised. Biosynthesis of TF is induced by cytokines like interleukin-1 (IL-1) and tumor necrosis factor (TNF) (see Chapter 14), thrombin and anaphylatoxin C5a from the complement system (see Chapter 8). In the presence of phospholipids and $Ca^{2+}$ ions TF forms a complex with factor VIIa that leads to the activation of factor X to FXa. The major components participating in blood coagulation are shown in Table 7.4.

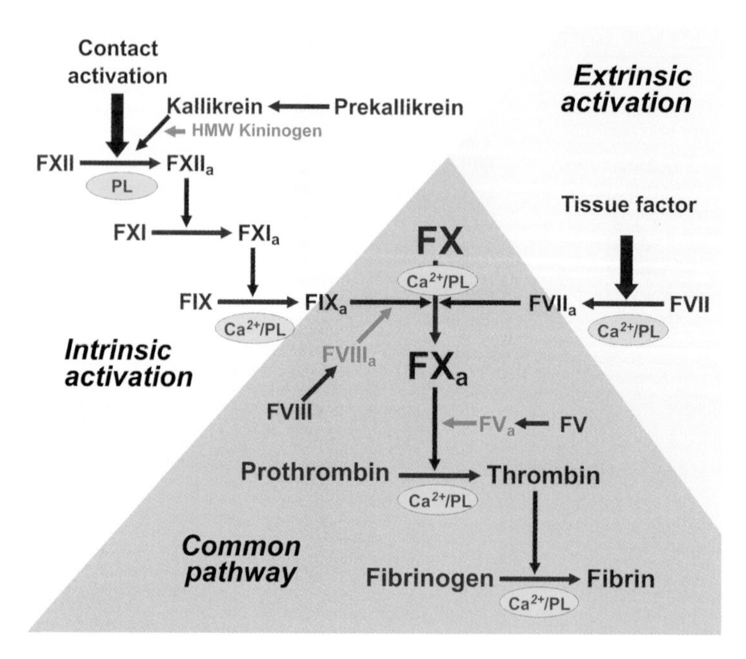


**Figure 7.2 Schematic representation of the coagulation cascade**
A general overview of the extrinsic, intrinsic and common pathways in blood coagulation. The major components involved in blood coagulation are depicted. For abbreviations: see Table 7.4.

**Tissue factor** (TF) (see Data Sheet) or coagulation factor III or thromboplastin is a single-chain type I membrane glycoprotein belonging to the tissue factor family and is usually not present in circulation. TF consists of an extracellular domain (approximately 220 aa) containing two FN3 domains and two disulfide bonds (Cys 49–Cys 57, Cys 186–Cys 209) and three **WKS** motifs, a potential transmembrane segment (Ile 220–Leu 242) and a short cytoplasmic C-terminal region,

**Table 7.4** Characteristics of the major components of blood coagulation.

| Factor | Synonyms | Mass (kDa)[a] | Concentration (nM)[b] | Function |
|---|---|---|---|---|
| Factor I | Fibrinogen | 340 | 7600 | Precursor |
| Factor II | Prothrombin | 72 | 1400 | Zymogen |
| Factor III | Tissue factor | 44 | Membrane protein | Cofactor |
| Factor IV | Calcium ions | 0.04 | 2 | Cofactor |
| Factor V | Proaccelerin | 330 | 20 | Cofactor |
| Factor VII | Proconvertin | 50 | 10 | Zymogen |
| Factor VIII | Antihemophilic factor | 285 | 0.7 | Cofactor |
| Factor IX | Christmas factor | 57 | 90 | Zymogen |
| Factor X | Stuart factor | 59 | 170 | Zymogen |
| Factor XI | Thromboplastin antecedent | 160 | 30 | Zymogen |
| Factor XII | Hagemann factor | 84 | 500 | Zymogen |
| Factor XIII | Transglutaminase | 320 | 90 | Zymogen |
| Factor XIV | Protein C | 62 | 60 | Zymogen |
| | Protein S | 69 | 300 | Cofactor |
| | Prekallikrein | 90 | 670 | Zymogen |
| | HMW kininogen | 110 | 430 | Cofactor |
| | von Willebrand factor | >540 | 20 | Cofactor |

[a]Values obtained by SDS-PAGE.
[b]Represent average values.

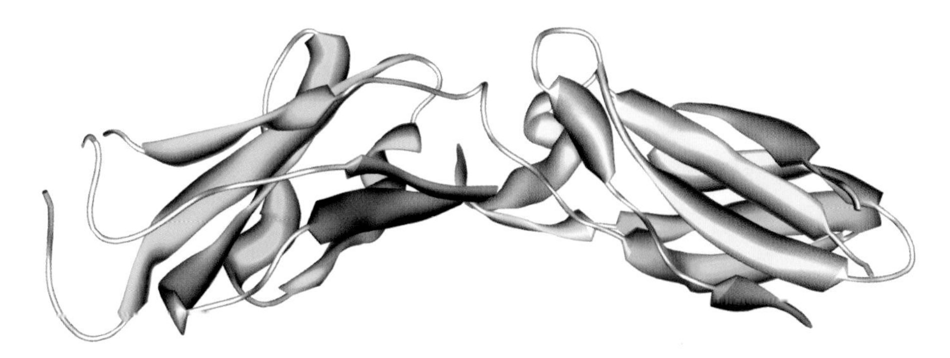

**Figure 7.3 3D structure of the extracellular domain of tissue factor (PDB: 2HFT)**
The extracellular domain of tissue factor consists of two FN3 domains containing mainly β-strands and their hydrophobic cores merge at the domain-domain interface.

which is palmitoylated at Cys 245. TF is the only coagulation factor containing a transmembrane segment and is characterised by the following signature with the two Cys forming a disulfide bridge:

TISSUE_FACTOR (PS00621): W–K–x–K–**C**–x(2)–T–x–[DEN]–T–E–**C**–D–[LIVM]–T–D-E

In complex with coagulation factor FVIIa, phospholipids and $Ca^{2+}$ ions TF initiates the extrinsic activation of the coagulation cascade via the activation of FX to FXa. The Kunitz-type tissue factor pathway inhibitor (TFPI) is a potent inhibitor of the TF-FVIIa complex, but it requires the presence of FXa to form a stable quaternary complex resulting in inhibited TF.

The 3D structure of the extracellular domain of TF (PDB: 2HFT; 218 aa: Ser 1–Gly 218) consisting of two FN3 domains (see also Chapter 4) has been solved by X-ray diffraction and is shown in Figure 7.3. The structure is characterised by three α-helices and 17 β-strands and the hydrophobic cores of the two modules merge in the domain–domain interface, suggesting that the extracellular region is a rather rigid template for the binding of factor FVIIa.

**Coagulation factor VII** (FVII, EC 3.4.21.21) (see Data Sheet) or proconvertin synthesised in the liver is a single-chain, vitamin K-dependent glycoprotein belonging to the peptidase S1 family and participates in the early phases of the extrinsic pathway of blood coagulation. FVII circulates in blood as zymogen and is activated to the active, two-chain FVIIa by coagulation factors FIXa, FXa, FXIIa and thrombin cleaving the Arg 152–Ile 153 peptide bond. The light chain of FVIIa contains a $Ca^{2+}$ ion-binding Gla domain, an OH-Asp residue at position 63 and two EGF-like domains, and the heavy chain comprises the serine protease part. FVII shares considerable structural similarity with other vitamin K-dependent proteins in blood coagulation, like prothrombin, FIX, FX, and proteins C, S and Z. FVIIa seems to be the only coagulation factor that is always present in plasma at a very low concentration in its activated form. Together with TF and $Ca^{2+}$ ions FVIIa activates FX to FXa and FIX to FIXa by limited proteolysis. In the presence of $Ca^{2+}$ ions TFPI neutralises FVIIa within the ternary complex TFPI-FVIIa-TF-FXa.

The 3D structure of an engineered construct (PDB: 1QFK) comprising the entire catalytic domain (254 aa: Ile 1–Pro 254, in blue) and the C-terminal part of the light chain containing the two EGF-like domains (104 aa: Gln 49–Arg 152 in the light chain, in red) but lacking the entire Gla domain of FVIIa in complex with the inhibitor D–Phe–Phe–Arg methyl ketone (not shown) has been solved by X-ray diffraction and is shown in Figure 7.4. The construct adopts an extended conformation similar to that observed for the full-length protein in complex with the first FN3 domain of tissue factor (PDB: 1DAN). The catalytic chain is characterised by 2 α-helices and 18 β-strands and is disulfide-linked to the light chain (Cys 110 in the catalytic chain and Cys 87 in the light chain) containing one α-helix and six β-strands (see Chapter 4). The residues of the active site are shown in pink and the interchain disulfide bridge is in green.

Like FVII, several other coagulation factors are vitamin K-dependent glycoproteins that contain an N-terminal γ-carboxyglutamic acid-rich (Gla) domain with 9–13 γ-Gla residues involved in the binding of $Ca^{2+}$ ions (see Chapter 4).

A schematic representation of the extrinsic activation of the coagulation pathway is shown in Figure 7.5. Involved are the glycoproteins TF, FVIIa, FX, phospholipids of the membrane lipid bilayer and $Ca^{2+}$ ions.

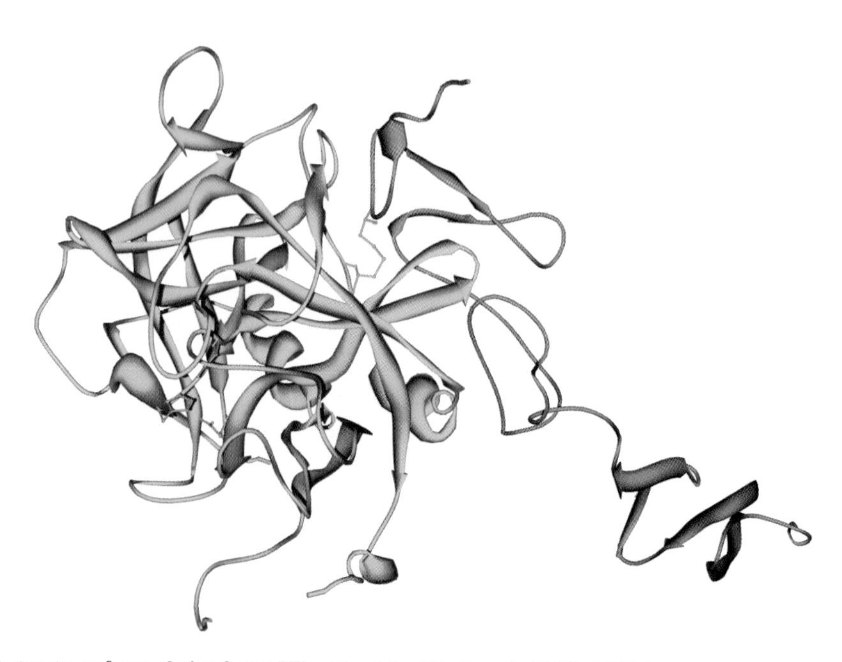

**Figure 7.4 3D structure of coagulation factor VII without the Gla domain (PDB: 1QFK)**
The construct consisting of two EGF-like domains (in red) and the catalytic chain (in blue) adopts an extended conformation. The active site residues (in pink) and the interchain disulfide bridge (in green) are indicated.

The interaction of the bloodstream with artificial materials like negatively charged glass or plastic surfaces or the exposure of negatively charged biological surfaces such as sulfatides upon an injury leads to the contact phase activation of blood coagulation via the intrinsic pathway system. HMW kininogen, prekallikrein, FXII and FXI are the major factors involved in the surface-mediated activation pathway. In the presence of HMW kininogen, prekallikrein is activated to kallikrein by activated FXIIa. In return, kallikrein activates FXII to FXIIa. The surface-associated ternary complex of HMW kininogen,

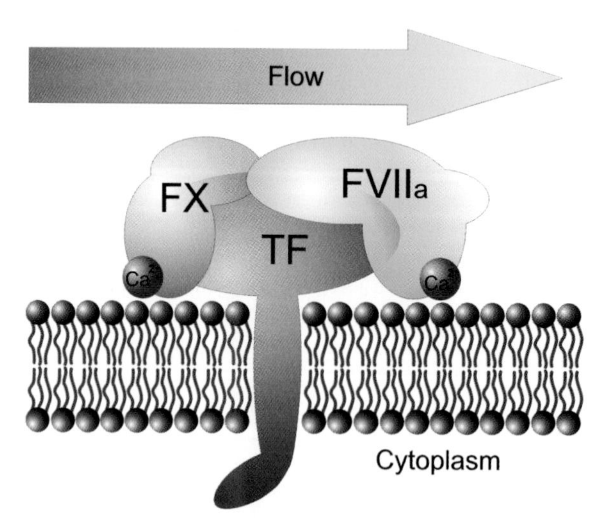


**Figure 7.5 Activation of the extrinsic pathway**
A schematic representation of the extrinsic activation of the coagulation pathway is shown with the components TF, FVIIa, FX, the lipid bilayer, phospholipids and calcium ions.

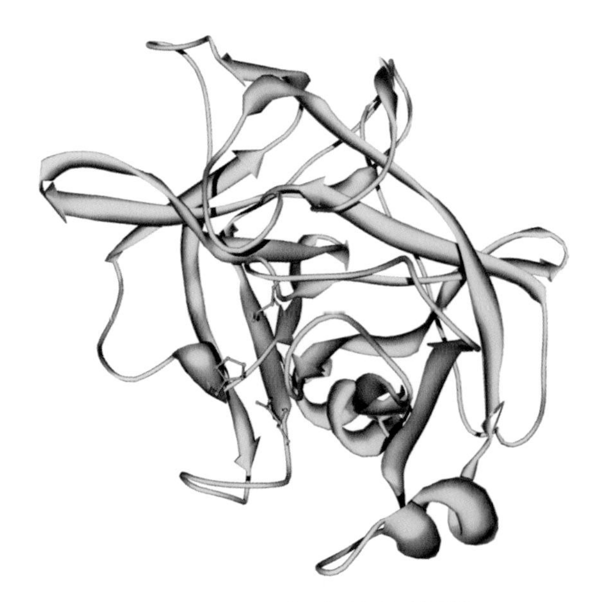

**Figure 7.6 3D structure of the catalytic chain of plasma kallikrein (PDB: 2ANW)**
Plasma kallikrein exhibits the classical features of a chymotrypsin-like serine protease. The residues of the active site are indicated (in pink).

FXIIa and FXI leads to the activation of FXI to FXIa. In the presence of HMW kininogen and $Zn^{2+}$ ions FXIa binds to activated platelets, resulting in the activation of FIX to FIXa. FIXa binds to a specific receptor on platelets or endothelial cells and forms a complex with the nonenzymatic cofactor FVIIIa, resulting in the activation of FX to FXa. This step is comparable to the activation by the FVIIa/TF complex in the extrinsic pathway. For details on kininogens and bradykinin see Section 7.6.

**Plasma (pre)kallikrein** (PK, EC 3.4.21.34) (see Data Sheet) or Fletcher factor or kininogenin synthesised in the liver is a single-chain glycoprotein (15 % CHO) belonging to the peptidase S1 family and circulates in blood in complex with HMW kininogen. FXIIa activates prekallikrein to two-chain kallikrein by cleaving the Arg 371–Ile 372 peptide bond. The heavy chain contains four apple domains (see Chapter 4) and the light chain comprises the serine protease part. Together with FXII and HMW kininogen, prekallikrein initiates the intrinsic pathway system. Plasma protease C1 inhibitor, kallistatin and $\alpha_2$-macroglobulin are the main physiological inhibitors of kallikrein.

The 3D structure of the catalytic domain of kallikrein (241 aa: Ile 1–Lys 241, lacking the C-terminal heptapeptide) has been solved by X-ray diffraction (PDB: 2ANW) and is shown in Figure 7.6. The structure exhibits the classical features of a chymotrypsin-like serine protease conformation and the residues of the catalytic triad are indicated in pink.

The Fletcher factor deficiency is caused by defects in the plasma kallikrein gene leading to a blood coagulation defect (for details see MIM: 229000).

**Coagulation factor XII** (FXII, EC 3.4.21.38) (see Data Sheet) or Hagemann factor is a single-chain glycoprotein (17 % CHO) belonging to the peptidase S1 family and is present in blood in its precursor form. In an autocatalytic reaction surface-bound FXII is able to generate catalytic amounts of two-chain $\alpha$-FXIIa by cleaving the Arg 353–Val 354 peptide bond. The heavy chain contains one FN1 and one FN3, two EGF-like and one kringle domain, and the light chain comprises the serine protease part. Kallikrein converts two-chain $\alpha$-FXIIa into two-chain $\beta$-FXIIa by cleaving the Arg 334–Asn 335 and the Arg 343–Leu 344 peptide bonds. The former heavy chain is reduced to a nine-residue minichain linked with a single disulfide bridge to the intact catalytic chain (Cys 6 in the minichain and Cys 114 in the catalytic chain). In complex with HMW kininogen, $\alpha$-FXIIa initiates the intrinsic coagulation cascade by activating FXI to FXIa. $\alpha$-FXIIa activates plasma prekallikrein to kallikrein and in consequence initiates the generation of vasoactive bradykinin. Plasma protease C1 inhibitor is the main inhibitor of $\alpha$-FXIIa and $\beta$-FXIIa involved in the formation of a 1:1 stoichiometric complex. In addition, $\alpha_2$-antiplasmin, antithrombin III and $\alpha_2$-macroglobulin are also known to inhibit FXIIa.

**Coagulation factor XI** (FXI, EC 3.4.21.27) (see Data Sheet) or plasma thromboplastin antecedent is a glycoprotein (5 % CHO) that belongs to the peptidase S1 family and circulates in blood in complex with HMW kininogen as a homodimer linked by a disulfide bridge between Cys 321 of the two monomers. FXI is activated by FXIIa to the active two-chain FXIa by

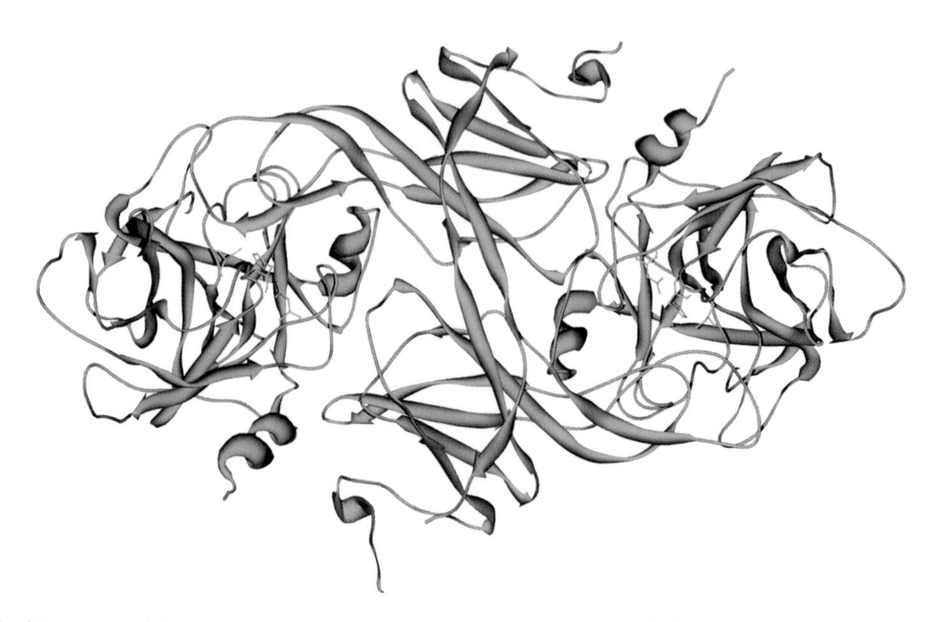

**Figure 7.7 3D structure of the catalytic chain of coagulation factor FXI in complex with the protease inhibitor ecotin (PDB: 1XX9)** Two molecules of the catalytic chain of factor XI (in blue) form a sandwich-like structure with the dimer of the inhibitor ecotin (in red) and the active site residues are in pink.

cleaving the Arg 369–Ile 370 peptide bond. The heavy chain of FXIa contains four apple domains (see Chapter 4) and the light chain comprises the serine protease part. FXI together with FXII, prekallikrein and HMW kininogen are the major factors in the surface-mediated activation pathway. The surface-associated ternary complex of FXI, HMW kininogen and FXIIa results in the proteolytic activation of FXI to FXIa. In the presence of HMW kininogen and $Zn^{2+}$ ions, FXIa binds to activated platelets, leading to the activation of FIX to FIXa. The most potent physiological inhibitors of FXIa are $\alpha_1$-antitrypsin and the antithrombin III–heparin complex. Plasma protease C1 inhibitor, $\alpha_2$-antiplasmin and $\alpha_2$-macroglobulin are less potent inhibitors of FXIa.

The 3D structure of the catalytic domain of FXI (238 aa: Ile 1–Val 238) in complex with ecotin mutants (a protease inhibitor from *Escherichia coli*; 142 aa) was determined by X-ray diffraction (PDB: 1XX9) and is shown in Figure 7.7. Two molecules of the catalytic chain of FXI (in blue) form a sandwich-like complex with a dimer of the inhibitor ecotin (in red), revealing substrate-like interactions between the serine protease portion of FXI and the inhibitor. The catalytic chain of FXI exhibits a chymotrypsin-like fold containing two $\beta$-barrels with a few helical segments and many loops and the classical catalytic triad of serine proteases (in pink).

The Rosenthal syndrome (thromboplastin antecedent deficiency) is caused by defects in the F11 gene leading to an excessive bleeding tendency after a haemostatic challenge. This coagulation disorder is highly frequent in Ashkenazi Jews (for details see MIM: 264900).

**Coagulation factor IX** (FIX, EC 3.4.21.22) (see Data Sheet) or Christmas factor synthesised in the liver is a single-chain, vitamin K-dependent glycoprotein (18 % CHO) belonging to the peptidase S1 family and is present in blood as a zymogen. In the presence of $Ca^{2+}$ ions FIX is activated either by FXIa or the FVIIa/TF complex to the active, two-chain FIXa by cleaving the Arg 145–Val 146 and the Arg 180–Val 181 peptide bonds and the concomitant release of the 35 residue activation peptide. The light chain of FIXa contains a $Ca^{2+}$ ion-binding Gla domain containing 12 $\gamma$-Gla residues, an OH-Asp residue (aa 64) and a partially phosphorylated P-Ser residue (aa 68), and two EGF-like domains, and the heavy chain comprises the serine protease part. FIX shares considerable structural similarity with other vitamin K-dependent proteins in blood coagulation, like prothrombin, FVII, FX and proteins C, S and Z. In complex with the cofactor FVIIIa it activates FX to FXa. The main physiological inhibitor of FIXa is antithrombin III, forming an equimolar complex with FIXa. In the presence of heparin the inactivation reaction rate is increased several hundred-fold.

**Figure 7.8 3D structure of the catalytic chain and the second EGF-like domain of coagulation factor IX (PDB: 1RFN)**
The EGF-like domain (in red) is linked by a single disulphide bridge (in green) to the catalytic chain (in blue) with the residues of the catalytic triad (in pink).

The 3D structure of an engineered construct (PDB: 1RFN) comprising the complete catalytic domain (235 aa: Val 1–Thr 235) and a mini light chain containing the second EGF-like domain (57 aa: Val86Met–Lys 142 in the light chain) of FIXa in complex with *p*-amino benzamidine (not shown) has been solved by X-ray diffraction and is shown in Figure 7.8. The catalytic chain (in blue) is characterised by 3 α-helices and 16 β-strands and is linked to the mini light chain (in red) containing four β-strands via a single disulfide bridge (in green, Cys 109 in the catalytic chain and Cys 47 in the mini light chain) (see Chapter 4). The residues of the catalytic triad are shown in pink.

Haemophilia B, a deficiency of or a defect in FIX, is a gender-linked recessive coagulation disorder located on the X chromosome with an incidence of approximately 1 in 50 000 and occurring almost exclusively in males. Several hundred genetic defects have been shown to cause haemophilia B affecting the catalytic activation, the EGF-like, the Gla and the propeptide regions of the protein (for details see MIM: 306900).

**Coagulation factor VIII** (FVIII) (see Data Sheet) or antihemophilic factor primarily synthesised in the liver is a large single-chain glycoprotein belonging to the multicopper oxidase family and circulates in blood in complex with vWF. FVIII contains three F5/8 type A, two F5/8 type C and six plastocyanin-like domains and up to eight S-Tyr residues. The cleavage of the Arg 372–Ser 373, Arg 740–Ser 741 and Arg 1689–Ser 1690 peptide bonds by thrombin is a prerequisite for the activation of FVIII to FVIIIa by cleavage of the Arg 1313–Ala 1314 and the Arg 1648–Glu 1649 peptide bonds. The two chains in active FVIIIa are noncovalently linked in a Ca-dependent manner. Factor VIII exhibits structural similarity with factor V. Complex formation of FVIII with vWF is essential for its biological function and its stability.

The 3D structure of the second F5/8 type C domain (PDB: 1D7P; 159 aa: Leu 2171–Gln 2329 in FVIII) in the very C-terminal region of FVIII has been solved by X-ray diffraction. The structure contains the binding sites to vWF and to negatively charged phospholipid surfaces and is characterised by 19 β-strands and no α-helices. The structure reveals a β-sandwich core from which two β-turns and a loop exhibit several solvent-exposed hydrophobic residues and underneath lies a cluster of positively charged residues. This structure implies a membrane-binding mechanism involving hydrophobic and electrostatic interactions (see Chapters 4 and 5).

Haemophilia A is a recessive X chromosome-linked coagulation disorder in males with an incidence of approximately 1–2 in 10 000. FVIII is essential for blood coagulation and concentrations below 20 % of the normal level lead to bleeding disorders. At concentrations below 1 % of the normal level severe bleeding disorders are observed with a spontaneous bleeding tendency. Genetic mutations are responsible for the abolishment of thrombin-cleavage sites and the introduction of new *N*-glycosylation sites in FVIII (for details see MIM: 306700).

At the FX level the extrinsic and the intrinsic pathways converge, leading to the common pathway. The TF/FVIIa complex from the extrinsic pathway and the FIXa/FVIIIa complex from the intrinsic pathway activate FX to FXa. On membrane surfaces in the presence of phospholipids and $Ca^{2+}$ ions FXa forms with activated FVa, an essential cofactor in blood coagulation, the prothrombinase complex (FXa/FVa complex) that leads to the rapid conversion of prothrombin to thrombin. FVa is required for the assembly of the prothrombinase enzyme complex and for efficient binding of the substrate prothrombin and to facilitate its rapid proteolysis by FXa.

**Coagulation factor X** (FX, EC 3.4.21.6) (see Data Sheet) or Stuart factor synthesised in the liver is a single-chain, vitamin K-dependent glycoprotein belonging to the peptidase S1 family and is present in blood as a precursor. FX is activated intrinsically by FIXa and extrinsically by FVIIa to the active two-chain FXa by limited proteolysis of the Arg 142–Ser 143 and Arg 194–Ile 195 peptide bonds releasing a 52-residue activation peptide. The light chain of FXa contains a Gla domain with 11 $\gamma$-Gla residues, an OH-Asp residue (aa 63) and two EGF-like domains, and the heavy chain consists of the protease part. FX shares considerable structural similarity with other vitamin K-dependent proteins in blood coagulation like prothrombin, FVII, FIX and proteins C, S and Z. In complex with FVa, phospholipids and $Ca^{2+}$ ions, FXa activates prothrombin to thrombin and in a positive feedback mechanism it can also activate FVII to FVIIa (see Figure 7.12). In addition, it can also convert the cofactors FV to FVa and FVIII to FVIIIa.

The main inhibitors of FXa (PDB: 1KSN) are antithrombin III, especially in complex with heparin, resulting in an increase of the inactivation rate by approximately a factor of 1000, $\alpha_1$-antitrypsin, tissue factor pathway inhibitor (TFPI), antistasin from the Mexican leech (119 aa: P15358) and the soybean trypsin inhibitor (181 aa: P01070).

The 3D structure of FXa comprising the entire catalytic domain (PDB: 1KSN 254 aa: Ile 1–Lys 254) and the light chain without the N-terminal pentapeptide (134 aa: Glu 6–Arg 139 in the light chain of FX) containing the Gla domain and the two EGF-like domains in complex with a $\beta$-aminester class of inhibitor has been solved by X-ray diffraction and is shown in Figure 7.9. The first EGF-like domain is flexibly disordered and the second EGF-like domain is positionally ordered, making contacts with the catalytic chain. The catalytic chain (in blue) containing the active site residues of the catalytic triad (in pink) exhibits a similar

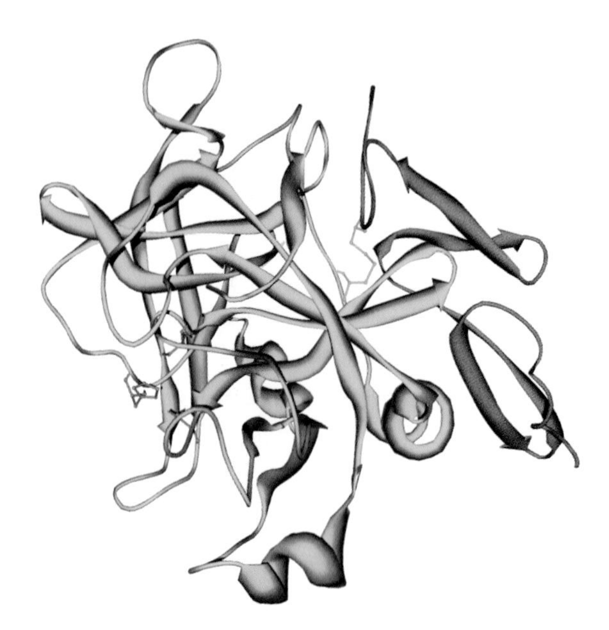

**Figure 7.9 3D structure of activated coagulation factor FXa (PDB: 1KSN)**
Coagulation factor FXa contains a Gla domain, two EGF-like domains and the catalytic chain. The second EGF-like domain (in red) is linked by a single disulphide bridge (in green) to the catalytic chain (in blue), which exhibits a similar fold as $\alpha$-thrombin. The active site residues are shown in pink.

folding as α-thrombin and is characterised by 3 α-helices and 17 β-strands and is linked to the light chain (in red) containing four β-strands by a single disulfide bridge (in green, Cys 108 in the catalytic chain and Cys 127 in the light chain) (see Chapter 4).

**Coagulation factor V** (FV) (see Data Sheet) or proaccelerin is a large single-chain glycoprotein that belongs to the multicopper oxidase family. FV contains three F5/8 type A, two F5/8 type C and six plastocyanin-like domains, seven potential Tyr-sulfation sites and two 17 residues long repeats and 35 repeats of 9 residues length with the conserved sequence **[TNP]LSPDLSQT**. The cleavage of the Arg 709–Ser 710, Arg 1018–Thr 1019 and Arg 1545–Ser 1546 peptide bonds and the concomitant release of two large activation peptides by thrombin or by FXa leads to the active, two-chain FVa. The two chains are noncovalently linked in a Ca-dependent manner. FV shares structural similarity with FVIII. FVa is required for the assembly of the prothrombinase enzyme complex and for efficient binding of the substrate prothrombin. Factor FVa is an essential cofactor in the prothrombinase complex, leading to the rapid activation of prothrombin to thrombin. FVa is proteolytically inactivated by activated vitamin K-dependent protein C by cleaving at the Arg 306–Asn 307, Arg 506–Gly 507 and Arg 679–Lys 680 peptide bonds in the heavy chain of FVa. Inactivated FVa is no longer able to mediate the assembly of the prothrombinase complex.

The 3D structure of the second F5/8 type C domain (PDB: 1CZS; 160 aa: Gly 2037–Tyr 2196 in FV) in the very C-terminal region of FV has been solved by X-ray diffraction. The structure is characterised by 16 β-strands and no α-helices. The structure is similar as the corresponding F5/8 type C domain in FVIII (see Chapter 4).

Owren parahemophilia is a hemorrhagic diathesis (bleeding tendency) caused by defects in the F5 gene (for details see MIM: 227400). Resistance to activated protein C (APCR) leading to thrombotic diathesis (thrombotic tendency) is also caused by defects in the F5 gene, and is found in about 5 % of the population (for details see MIM: 188055).

**Vitamin K-dependent protein C** (FXIV, EC 3.4.21.69) (see Data Sheet) or coagulation factor XIV predominantly synthesised in the liver is a single-chain, vitamin K-dependent glycoprotein belonging to the peptidase S1 family and is present in blood as a precursor. It is cleaved into a two-chain molecule by the concomitant release of an N-terminal 10 residue propeptide, resulting in a light chain with a Gla domain containing nine γ-Gla residues, an OH-Asp residue (aa 71) and two EGF-like domains, and a heavy chain comprising the serine protease part. Protein C is activated by thrombin in conjunction with the endothelial cell membrane protein thrombomodulin, releasing an N-terminal 12 residue activation peptide from the N-terminus of the heavy chain. Protein C shares considerable structural similarity with other vitamin K-dependent proteins in blood coagulation, like prothrombin, FVII, FIX, FX and proteins S and Z. Activated protein C is involved in the regulation of blood coagulation by the selective inactivation of the two important cofactors, FVa and FVIIa, in the presence of $Ca^{2+}$ ions and phospholipids. Protein S and FV function as syner-gistic cofactors in the protein C controlled regulation process. The main physiological inhibitors are the plasma serine protease inhibitor (PAI-3), which forms an equimolar complex with protein C, $\alpha_1$-antitrypsin and $\alpha_2$-macroglobulin.

The 3D structure of the activated protein C (PDB: 1AUT) comprising the complete catalytic chain (250 aa: Leu 1–Pro 250) and the C-terminal part of the light chain without the Gla domain (114 aa: Ser 42–Leu 155 in the light chain) containing two EGF-like domains in complex with a synthetic tripeptide (FP-deoxomethyl-R) (not shown) was solved by X-ray diffraction and is shown in Figure 7.10. The two chains are linked by a single disulfide bridge (in green, Cys 108 in the catalytic chain and Cys 100 in the partial light chain). The catalytic chain (in blue) containing the active site residues of the catalytic triad (in pink) is characterised by three α-helices and 16 β-strands, and the partial light chain (in red) is devoid of α-helices but contains eight β-strands. The catalytic chain exhibits an overall trypsin-like structure but contains a large insertion loop at the edge of the active site, a third helix, a cationic patch similar to the anion-binding exosite I in thrombin and a calcium-binding site like in trypsin.

**Thrombomodulin** (TM) (see Data Sheet) or fetomodulin is a single-chain type I membrane glycoprotein containing six EGF-like and one C-type lectin domain, an OH-Asn residue (aa 324) and an approximately 24 residue long presumed transmembrane segment located in the C-terminal region (Gly 498–Leu 521). TM is a natural anticoagulant that forms a 1:1 complex with thrombin, thus accelerating the activation of protein C by thrombin and in consequence regulating blood coagulation by inactivation of FVa and FVIIIa by activated protein C. Tumor necrosis factor (TNF), the cytokine interleukin-1 and endotoxins decrease the cell surface activity of thrombomodulin.

The 3D structure of the EGF-like 4–6 domains in TM (118 aa: Val 345–Cys 462 in TM) in complex with α-thrombin consisting of the complete catalytic chain (259 aa: Ile 1–Glu 259) and the entire light chain (36 aa: Thr 1–Arg 36) containing the synthetic thrombin inhibitor EG-deoxy-methyl-R was determined with X-ray diffraction (PDB: 1DX5) (see Chapter 4). The EGF-like 4–6 domains in TM are characterised by one α-helix and 12 β-strands and the structure of α-thrombin in the complex is comparable to the structure of α-thrombin alone (PDB: 1A3B). The Y-shaped TM fragment binds to anion-binding exosite I in thrombin, preventing the binding of procoagulant substrates.

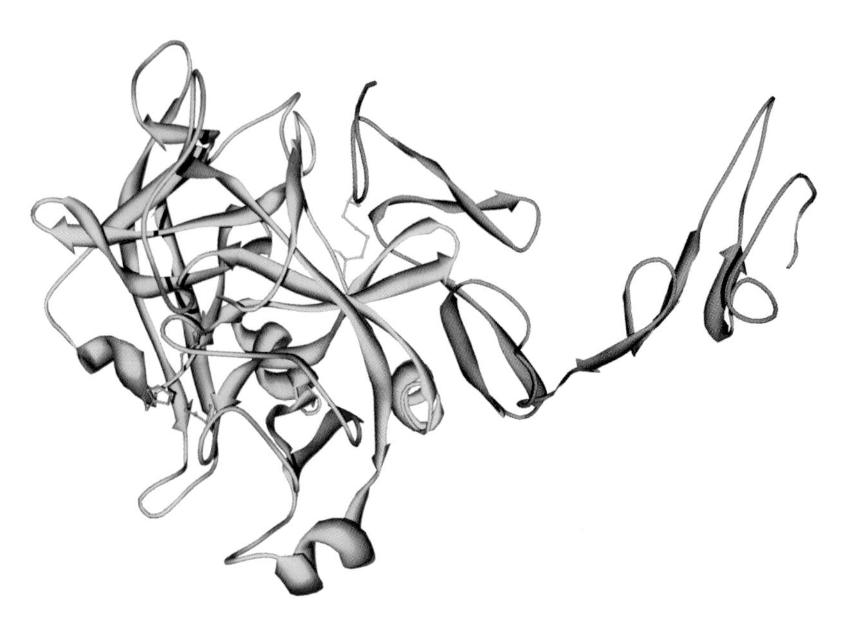

**Figure 7.10 3D structure of activated protein C (PDB: 1AUT)**
The two EGF-like domains (in red) are linked by a single disulphide bridge (in green) to the catalytic chain (in blue) of activated protein C. The active site residues are shown in pink.

**Vitamin K-dependent protein S** (see Data Sheet) synthesised in the liver and present in platelets is a single-chain glycoprotein containing an N–terminal Gla domain with 11 γ-Gla residues, an OH-Asp residue (aa 95), four EGF-like and two laminin G-like domains. Protein S shares considerable structural similarity with other vitamin K-dependent proteins in blood coagulation, like prothrombin, FVII, FIX, FX and proteins C and Z. The C-terminal region of protein C exhibits similarity with sex hormone binding globulin. Protein S is involved in the regulation of blood coagulation as a cofactor to activated protein C, which degrades FVa and FVIIIa. Protein S enhances the binding of activated protein C to phospholipid surfaces, thus abrogating FXa- and FIXa-dependent protection of the degradation of FVa and FVIIIa by activated protein C. Protein S is reported to inhibit the activation of prothrombin by direct interactions with FVa and FXa. Protein S helps to prevent coagulation and thus stimulates fibrinolysis.

The 3D structure of the EGF-like domains 3 + 4 of protein S carrying the calcium-binding sites have been solved by NMR spectroscopy (PDB: 1Z6C; 87 aa: Lys 159–Ser 245). The bent conformation of the EGF-like domain 3 is different from other calcium-binding EGF-like domains which usually show an extended conformation (see Chapter 4).

The activation of FX to FXa and the formation of the prothrombinase enzyme complex (FXa/FVa complex) initiates the common pathway. The conversion of prothrombin to thrombin is one of the key steps in the whole blood coagulation system.

**Prothrombin** (FII, EC 3.4.21.5) (see Data Sheet) or coagulation factor II synthesised in the liver is a single-chain, vitamin K-dependent glycoprotein belonging to the peptidase S1 family and is present in blood as a precursor. Prothrombin is activated by FXa to the active two-chain α-thrombin by cleaving the Arg 284–Thr 285 and the Arg 320–Ile 321 peptide bonds, releasing two activation peptides: fragment 1 (Ala 1–Arg 155) that contains a Gla domain with 10 γ-Gla residues and kringle 1 and fragment 2 (Ser 156–Arg 284) that comprise the kringle 2 domain. Two-chain α-thrombin contains a 36 residue light chain and the heavy chain that comprises the serine protease part linked by a single disulfide bond (Cys 9 in the light chain and Cys 119 in the catalytic chain). Prothrombin shares considerable structural similarity with other vitamin K-dependent proteins in blood coagulation, like FVII, FIX, FX and proteins C, S and Z. Thrombin is the key component of the whole blood coagulation system. Prothrombin binds via its Gla domain in fragment 1 to phospholipids of damaged cell membranes and blood platelets. Fragment 2 mediates the binding of prothrombin to cofactor FVa. Thrombin releases the fibrinopeptides A and B from the fibrinogen α- and β-chains, respectively, thus exposing the crosslinking sites in fibrinogen that leads to fibrin polymerisation. Protein Z appears to assist haemostasis by binding thrombin. Main inhibitors of thrombin are antithrombin III, especially in complex with heparin, resulting in an increase in the inactivation rate by approximately a factor of 10 000, heparin

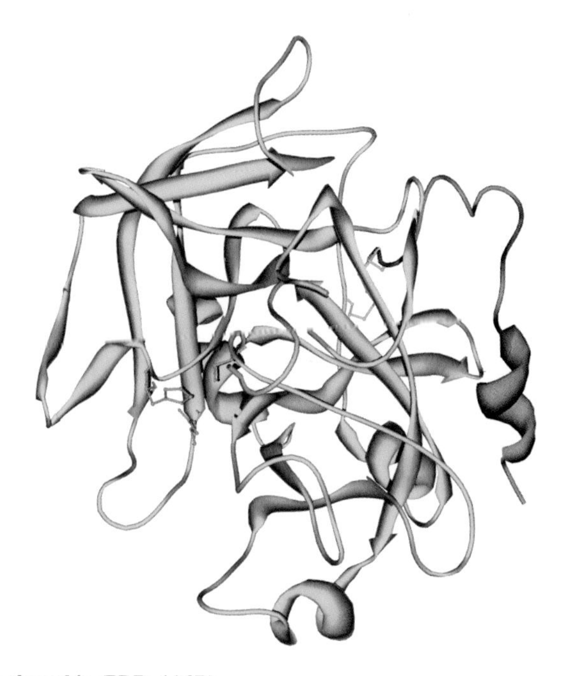

**Figure 7.11 3D structure of α-thrombin (PDB: 1A3B)**
The mini chain (in red) is linked by a single disulphide bridge (in green) to the catalytic chain (in blue) of α-thrombin, which exhibits the typical serine protease conformation. The residues of the active site are shown in pink.

cofactor II, especially in the presence of glycosaminoglycans, resulting in an increase of the inactivation rate up to a factor of 10 000, $\alpha_1$-antitrypsin, $\alpha_2$-macroglobulin and hirudin from the medicinal leech (65 aa: P01050).

The 3D structure of α-thrombin (PDB: 1A3B) in complex with the nonhydrolysable bifunctional inhibitors hirutonin-2 and hirutonin-6 (not shown) mimicking the scissile peptide bond has been determined by X-ray diffraction and is shown in Figure 7.11. The complete catalytic chain (259 aa: Ile 1–Glu 259, in blue) is characterised by 3 α-helices and 19 β-strands and the entire light chain (36 aa: Thr 1–Arg 36, in red) contains one α-helix. Both chains are linked by a single interchain disulfide bridge (in green, Cys 119 in the catalytic chain and Cys 9 in the light chain). The catalytic chain exhibits the typical fold of serine proteases and the residues of the active site are indicated in pink.

**Vitamin K-dependent protein Z** (see Data Sheet) is a single-chain glycoprotein belonging to the peptidase S1 family and contains an N-terminal Gla domain with 13 γ-Gla residues, one OH-Asp residue (aa 64), two EGF-like domains and a serine protease part. Protein Z shares considerable structural similarity with other vitamin K-dependent proteins in blood coagulation, like prothrombin, FVII, FIX, FX and proteins C and S. Protein Z exhibits no measurable proteolytic activity as His and Ser of the supposed catalytic triad are replaced by Lys and a nonassignable residue Xaa, respectively. Protein Z appears to assist haemostasis by binding thrombin and thus promoting its association with phospholipids. In addition, protein Z acts as a cofactor in the inhibition process of FXa by the protein Z-dependent protease inhibitor, a serpin (see Chapter 11).

The regulation of many cascade systems is governed by positive and negative feedback reactions. A positive feedback mechanism is characterised by participation of the reaction product in its own creation and thus resulting in a massive amplification. A negative feedback mechanism halts the creation of the reaction product by stopping its own formation. In haemostasis positive and negative feedback reactions as well as inhibition by different types of protease inhibitors are key events for its efficient regulation. Figure 7.12 depicts a schematic representation of the positive feedback reactions (dark brown) involved in blood coagulation, mainly by thrombin and FXa.

The activation of prothrombin to thrombin initiates the final steps in blood coagulation. The formation of an insoluble fibrin network from the soluble plasma protein fibrinogen is characterised by a series of proteolytic reactions in the fibrinogen

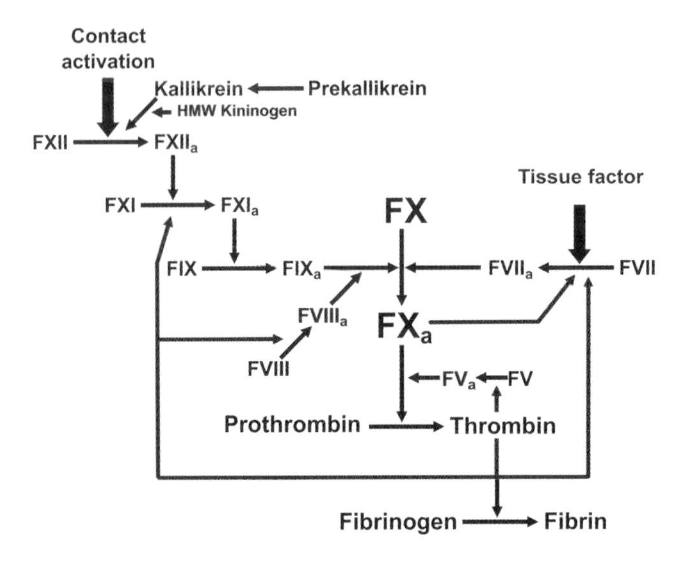


**Figure 7.12 Feedback mechanisms in blood coagulation**
A schematic representation of the positive feedback reactions (dark brown) in the blood coagulation cascade with the main components thrombin and FXa.

molecule carried out by thrombin and crosslinking reactions between the three chains of the fibrin monomers, especially the formation of isopeptide bonds between the side chains of Lys and Gln residues, catalysed by coagulation factor FXIIIa, which acts as a transglutaminase, in the presence of calcium ions. The formation of an isopeptide bond is depicted in Figure 7.13.

**Fibrinogen** (FI) (see Data Sheet) or coagulation factor I synthesised in the liver is a 340 kDa glycoprotein (3 % CHO) present in plasma as a dimer of three nonidentical Aα-, Bβ- and γγ-chains which are linked by disulfide bridges. The α-chain carries a P-Ser residue (aa 3), the β-chain an N-terminal pyro-Glu residue and the γ-chain two S-Tyr residues (aa 418, 422). The N-terminal regions of the three chains are evolutionary related and contain the Cys residues involved in the crosslinking of the chains. The β- and the γ-chains each contain one fibrinogen C-terminal domain of about 250 residues with two conserved disulfide bridges characterised by the following signature:

FIBRIN_AG_C_DOMAIN (PS00514): W–W–[LIVMFYW]–x(2)–C–x(2)–[GSA]–x(2)–N–G

The fibrinogen Aα-chains contain two **RGD** sequences (aa 95–97 and 572–574) that act as cell-binding motifs. The two heterotrimers of nonidentical Aα-(red), Bβ-(blue) and γγ-chains (green) are in head-to-head conformation with the N-termini of each chain in a small central domain from which fibrinopeptides A and B are released by thrombin. The α-, β- and γ-chains are linked by a series of interchain disulfide bridge (in yellow), as depicted in Figure 7.14.

The interchain disulfide bridges between the α-, β- and γ-chains are connected in the following way:

α-chain–α-chain: Cys 28–Cys 28
γ-chain–γ-chain: Cys 8–Cys 9, Cys 9–Cys 8
α-chain–β-chain: Cys 36–Cys 65, Cys 49–Cys 76, Cys 165–Cys 193
α-chain–γ-chain: Cys 45–Cys 23, Cys 161–Cys 139

Thrombin converts fibrinogen to the fibrin monomer by limited proteolysis in this central domain, releasing the 16 residue **fibrinopeptide A** (FPA) from the Aα chains by cleaving the Arg 16–Gly 17 peptide bond and the 14 residue **fibrinopeptide B** (FPB) from the Bβ chains by cleaving the Arg 14–Gly 15 peptide bond. The reaction product is called **des-AB-fibrin** or **fibrin monomer**. The removal of FPA and FPB uncovers the N-terminal polymerisation sites in the α- and β-chains of fibrinogen and leads to spontaneous binding to complementary sites in the C-terminal domains of neighbouring fibrin(ogen)

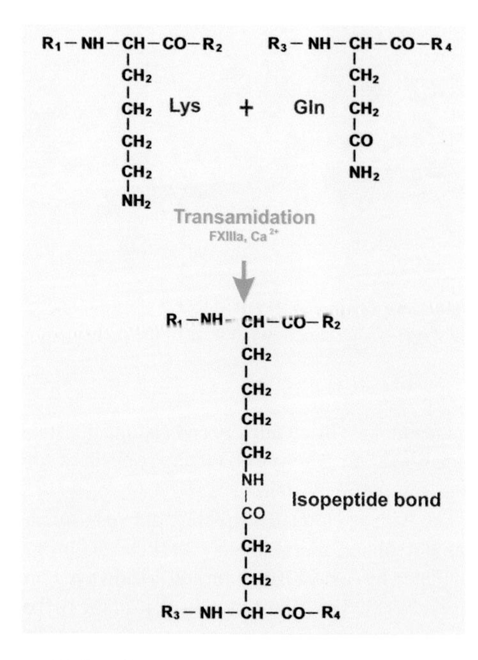


**Figure 7.13 The formation of an isopeptide bond**
The formation of an isopeptide bond between the side chains of a Lys and a Gln residue is catalysed by the transglutaminase FXIIIa in the presence of calcium ions.

molecules, resulting in protofibrils that interact laterally to multistranded fibres and leads to the formation of the soft clot. The soft clot is converted to the hard clot by various crosslinking reactions:

1. crosslinks between the α-chain region Gly 17–Arg 19 and the γ-chain region Thr 374–Glu 396;
2. crosslinks of the Gly 15–Arg 17 region in the β-chain to another fibrin monomer;
3. the formation of several interchain isopeptide bonds between α-chains (Gln 328 and 366 and Lys 508, 539, 556, 562 and 580 were identified as potential crosslinking sites) and of one interchain isopeptide bond between Gln 398 and Lys 406 of two γ-chains, catalysed by FXIIIa that acts as a transglutaminase.

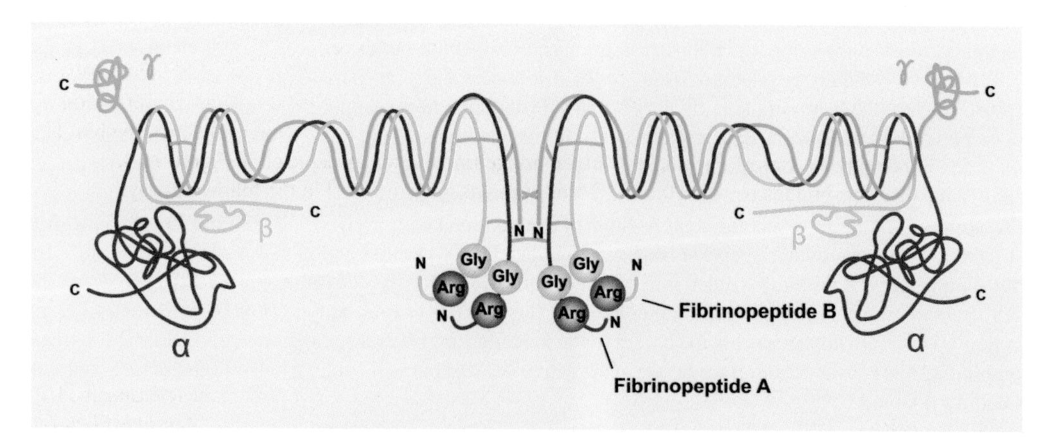


**Figure 7.14 Schematic representation of fibrinogen**
The non-identical Aα- (red), Bβ- (blue) and γγ-chains (green) are in head to head conformation forming a small central domain from which fibrinopeptides A and B are released by thrombin. The interchain disulfide bridges are shown in yellow. (Adapted from Doolittle RF. Annu Rev Biochem 53: 195–229. Copyright (1984), with permission from Annual Reviews.)

**Figure 7.15 3D structure of the D fragment of fibrinogen (PDB: 1FZG)**
Fragment D is a dimer comprising 87 aa of the $\alpha$-chain (in yellow), 328 aa of the $\beta$-chain (in green) and 319 aa of the $\gamma$-chain (in blue).

The crosslinking of fibrin by FXIIIa results in a mechanically and chemically stable fibrin clot. In addition, FXIIIa also catalyses the formation of an isopeptide bond between Lys 303 in the $\alpha$-chain of fibrin and Gln 2 in $\alpha_2$-antiplasmin, thus protecting fibrin from a premature degradation by plasmin.

Fibrinogen is a large protein (340 kDa) with an elongated structure of two terminal D and one central E domain. The 3D structure of the plasmin-derived D fragment (dimer, approximately 200 kDa) of fibrinogen (PDB:1FZG) in complex with the tetrapeptide GHRD-$NH_2$ has been elucidated by X-ray diffraction and is shown in Figure 7.15. Fragment D comprises 87 aa (Val 111–Arg 197) of the $\alpha$-chain (in yellow), 328 aa (Asp 134–Gln 461) of the $\beta$-chain (in green) and 319 aa (Lys 88–Lys 406) of the $\gamma$-chain (in blue). The $\alpha$-chain fragment is characterised by extended $\alpha$-helices and the $\beta$-chain and the $\gamma$-chain fragments have similar contents of $\alpha$-helices and $\beta$-strands.

Defects in the FGA, FGB and FGG genes are the cause of congenital afibrinogenemia, characterised by mild to severe bleeding disorders due to the lack or malfunctioning of fibrinogen (for details see MIM: 202400).

**Coagulation factor XIII** (FXIII, EC 2.3.2.13) (see Data Sheet) or transglutaminase is a tetramer of two A-chains and two B-chains that are noncovalently linked. Thrombin releases an N-terminal 37 residue activation peptide from the A-chain followed by a calcium-dependent dissociation of the B-chains. The noncatalytic B-chain contains 10 sushi (CCP/SCR) domains and one **RGD** sequence (aa 597–599) that acts as a cell-binding motif. The active site of transglutaminases is characterised by the following consensus pattern, where the first Cys is part of the active site:

TRANSGLUTAMINASES (PS00547): [GT]–Q–[**C**A]–W–V–x–[SA]–[GAS]–[IVT]–x(2)–T–x–[LMSC]–R–[CSAG]–[LV]–G

FXIIIa is primarily responsible for the crosslinking reactions of the three chains in the fibrin monomer. FXIIIa acts as transglutaminase that catalyses the formation of intermolecular isopeptide bonds between the side chains of a Lys and a Gln residue, leading to $\gamma$–$\gamma$ dimers, $\alpha$–$\gamma$ heterodimers, $\alpha$-chain polymers and $\gamma$-chain multimers of fibrin. As mentioned above, FXIIIa also crosslinks the $\alpha_2$-plasmin inhibitor to the $\alpha$-chain of fibrin, thus preventing a premature degradation of the fibrin network by plasmin. In addition, FXIIIa also crosslinks fibronectin to itself and to fibrin and collagen. These reactions are physiologically important and support the attachment of fibrin to the blood vessel wall and facilitate tissue repair and clot retraction. There are no known natural inhibitors of FXIIIa in plasma.

The 3D structure of the entire recombinant A-chain of coagulation factor FXIII (PDB: 1GGU) 731 aa: Ser 1–Met 731) including the 37 residue activation peptide has been elucidated by X-ray diffraction and is shown in Figure 7.16. Each chain of the homodimeric protein is folded into four sequential domains: the activation peptide AP (in red) exhibits a $\beta$-sandwich structure, the core domain (in blue) containing the residues of the catalytic triad (Cys 314, His 373 and Asp 396) (in pink), which are not accessible to solvent in the zymogen, and barrel 1 (in yellow) and barrel 2 (in green). The main ion-binding site is near the interface between the active site and barrel 1 (orange ball). The structure contains 11 $\alpha$-helices and 42 $\beta$-strands.

At the end of the coagulation cascade stands the resultant insoluble fibrin network generated from soluble fibrinogen by limited proteolysis with thrombin and crosslinking reactions catalysed by FXIIIa, as already seen. The fibrin network contains many types of blood cells, but mainly two types of plugs are formed: a platelet plug that is usually almost devoid of

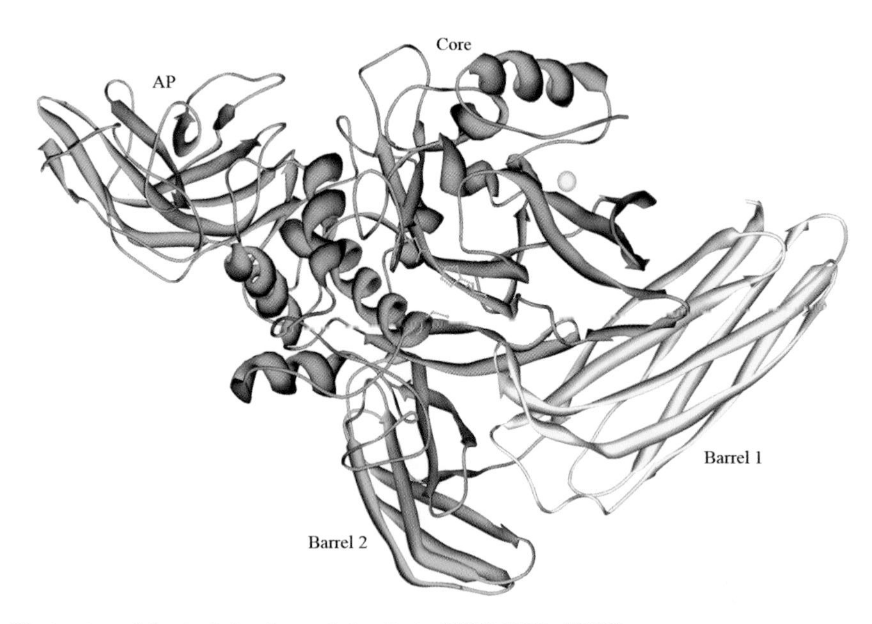


**Figure 7.16 3D structure of the A-chain of coagulation factor FXIII (PDB: 1GGU)**
The A-chain of coagulation factor FXIII is folded into four sequential domains: The activation peptide AP (in red), the core domain (in blue) with the residues of the catalytic triad (in pink) and the main ion binding site (orange ball) and barrel 1 (in yellow) and barrel 2 (in green) (Adapted from Mathews *et al.*, Biochemistry 35: 2567–2576. Copyright (1996), with permission from the American Chemical Society.)

red blood cells and therefore termed 'white clot' and a plug primarily containing erythrocytes and thus termed 'red clot'. The formed thrombus is a natural seal of the injured blood vessel wall but it only helps to solve an acute phase problem, e.g. to limit the blood loss to an acceptable level and to help to protect the body from the invasion of pathogens through open wounds. On longer terms the thrombus represents a potential risk at the site of injury by reducing the blood flow and thus leading to necrosis in the surrounding tissue. In addition, the clots still contain active thrombin and released parts of the clots or emboli represent a high thrombogenic risk if they are transported to other areas of the blood vessel system, and may thus be the cause of strokes or myocardial infarction. Therefore, this initial 'repair' is only temporarily and has to be replaced by the repair and re-generation of the injured tissues of the vascular system.

## 7.4 THE FIBRINOLYTIC SYSTEM

In general, the efficient control of chemical reactions in biological systems are key events to obtain, maintain and guarantee the integrity of a living organism. The various cascades and biochemical pathways have to be controlled and regulated, resulting in usually well-balanced systems under normal conditions. As seen in the previous section the clots generated by the coagulation cascade represent a potential risk factor and therefore another proteolytic cascade, the fibrinolytic system, works as a counterbalance of blood coagulation.

With the onset of blood coagulation the fibrinolytic system is also initiated, but at the beginning of the coagulation cascade fibrinolysis plays only a minor role, mainly because of the following events: the crosslinking of $\alpha_2$-antiplasmin to fibrin that inhibits generated plasmin to degrade the fibrin network, the release of plasminogen activator inhibitors-1, -2 and -3 that prevent the activation of plasminogen and the generation of the thrombin-activatable fibrinolysis inhibitor (TAFI), a carboxypeptidase that degrades the plasminogen-binding region in fibrin by cleavage of C-terminal Lys or Arg residues, resulting in a loss of plasminogen binding and activation capability. With the progress of the coagulation cascade, however, fibrinolysis becomes more and more dominant. Some of the key components of the fibrinolytic system are listed in Table 7.5.

The key component of the fibrinolytic system is plasminogen (Pg) that is activated to plasmin, mainly by two activators: tissue-type (tPA) and urokinase-type (uPA) plasminogen activators.

**Plasminogen** (Pg, EC 3.4.21.7) (see Data Sheet) synthesised in the liver and the kidney is a single-chain glycoprotein (2 % CHO) present in plasma as a zymogen belonging to the peptidase S1 family. In plasma, plasminogen is partially bound

**Table 7.5** Characteristics of the major components of the fibrinolytic system.

| Protein | Mass (kDa)[a] | Concentration (mg/l)[b] | Function |
|---|---|---|---|
| Plasminogen | 90 | 150 | Zymogen of plasmin |
| Tissue-type plasminogen activator (tPA) | 70 | 0.005 | Activator of plasmin |
| Urokinase-type plasminogen activator (uPA) | 55 | 0.01 | Activator of plasmin |
| Prekallikrein | 90 | 45 | Zymogen of kallikrein |
| $\alpha_2$-Antiplasmin | 67 | 70 | Inhibitor of plasmin |
| $\alpha_2$-Macroglobulin | 160 | 1200 | General protease inhibitor |
| Carboxypeptidase B2 (TAFI) | 60 | 5 | Carboxypeptidase |
| Plasminogen activator inhibitor-1 | 50 | 0.06 | Inhibitor of tPA |
| Plasminogen activator inhibitor-2 | 50 | 0.005 | Inhibitor of tPA and uPA |

[a]Values obtained by SDS-PAGE.
[b]Represent average values.

to the histidine-rich glycoprotein. Plasminogen is activated by tPA and uPA to the active two-chain plasmin by cleavage of the Arg 561–Val 562 peptide bond and the concomitant release of a 77 residue (pre-)activation peptide. Via a positive feedback mechanism plasmin can induce its own proteolytic activation. Upon activation the molecule undergoes a large conformational change from a compact and closed form in plasminogen to a more open form in plasmin. The heavy chain of plasmin contains five kringle domains that carry **lysine binding sites** (LBS) which are responsible for an efficient binding to the substrate fibrin and to the inhibitor $\alpha_2$-antiplasmin and the light chain comprises the serine protease part with a partial phosphorylation site at Ser 578. The structure of the well-characterised LBS is shown in Figure 7.17. The LBS consist of anionic and cationic centres interspaced by a hydrophobic groove that is outlined with a series of aromatic residues.

Plasmin cleaves the insoluble fibrin polymers at specific sites resulting in soluble fragments. A schematic representation of fibrinogen, the fibrin polymer and the cleavage pattern by plasmin is depicted in Figure 7.18.

Electron microscopy and small-angle neutron scattering of plasminogen in solution suggest a compact, spiral-like structure with the (pre-) activation peptide in close proximity to the kringle 5 domain. The fragment angiostatin comprising the first three or four kringles of plasminogen (Tyr 80–Val 338/354, Lys 78–Ala 440) is an angiogenesis inhibitor. The complex of streptokinase with human plasminogen can activate plasminogen to plasmin. Similar binding activation mechanisms also occur in some key steps of blood coagulation. Streptokinase (414 aa: P10520) from *Streptococcus pyogenes* is widely used in the treatment of blood-clotting disorders.

The 3D structure of the kringle 2 domain (PDB: 1B2I; 83 aa: Cys162Thr/Glu163Ser–Thr 244) in complex with *trans*-4-aminomethylcyclohexane carboxylic acid was determined by NMR spectroscopy. The 3D structure of angiostatin comprising the first three kringles of plasminogen (PDB: 1KI0; 253 aa: Leu 81–Cys 333) was determined by X-ray diffraction. Both structures are characterised by $\beta$-strands but no $\alpha$-helices (see Chapter 4).

The 3D structure of the plasmin light chain containing an N-terminal 20 residue extension from the heavy chain (250 aa: Ala 542–Asn 791) in complex with streptokinase from *Streptococcus equisimilis* (362 aa: Ser 12–Pro 373, P00779) was determined by X-ray diffraction (PDB: 1BML) and is shown in Figure 7.19. The C-terminal domain of streptokinase (in red) binds near the activation loop of the plasmin light chain (in green) and seems to be responsible for the contact activation of plasminogen in the complex. In the catalytic triad (in pink) Ser is mutated to Ala (in blue).

Plasmin also acts as a proteolytic factor in many other physiological processes like wound healing, tissue remodelling, angiogenesis, embryogenesis and pathogen and tumor cell invasion. The main physiological inhibitor of plasmin is $\alpha_2$-antiplasmin.

**Tissue-type plasminogen activator** (tPA, EC 3.4.21.68) (see Data Sheet) synthesised and secreted by endothelial cells is a single-chain glycoprotein belonging to the peptidase S1 family that exhibits almost full activity in its single-chain form. It is completely activated to two-chain tPA by plasmin, kallikrein or FXa. The A-chain contains one EGF-like, two kringle and one FN1 domain and the B-chain comprises the serine protease part. Both chains are linked by a single disulfide bridge (Cys 264 in the A-chain and Cys 120 in the catalytic chain). Together with uPA, tPA activates plasminogen to plasmin by cleaving a single peptide bond Arg 561–Val 562. By controlling plasmin-mediated proteolysis tPA plays an important role in many physiological processes, like tissue remodelling and degradation and cell migration. The main physiological inhibitor of tPA is plasminogen activator inhibitor-1 (PAI-1) with which it forms an irreversible inactive stoichiometric complex.

Genetically engineered constructs of tPA (activase and retravase) are used therapeutically to initiate fibrinolysis in acute myocardial infarction, in acute ischemic stroke and in pulmonary embolism.

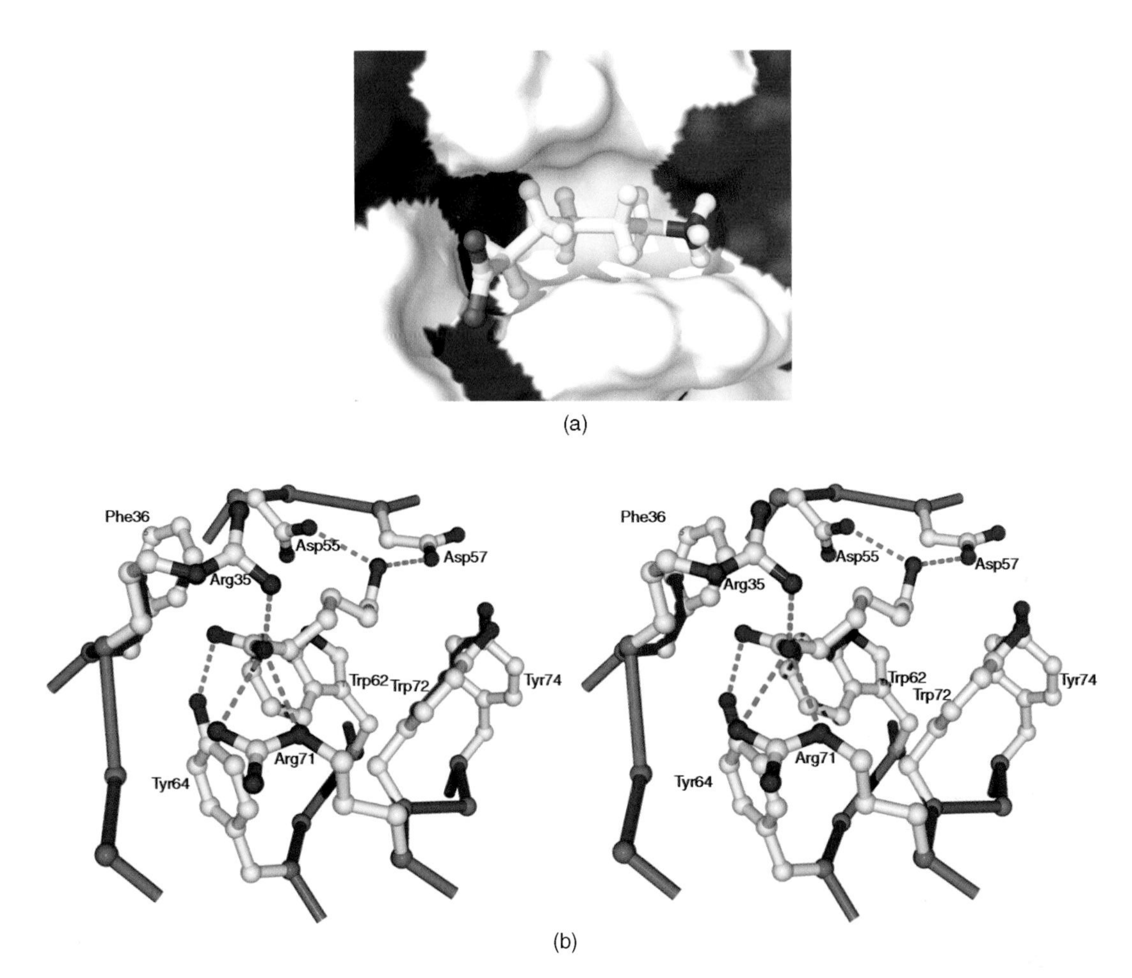


**Figure 7.17 Structure of the lysine binding site (LBS) in kringles**
The LBS consists of cationic and anionic centres interspaced by an aromatic hydrophobic groove. (a) View of the calculated electrostatic surface potential of the LBS in kringle 1 of plasminogen in complex with the ligand 6-aminohexanoic acid (red, negative charges; blue, positive charges). (b) Stereo view of the LBS in kringle 1 of plasminogen in complex with the ligand 6-aminohexanoic acid. The calculated hydrogen bonds are indicated by dashed green lines. (Both adapted from Mathews *et al.*,1996).

The 3D structure of the kringle 2 domain of tPA (PDB: 1PK2; 90 aa: Ser 174–Thr 263 in the A-chain of tPA) was determined by NMR spectroscopy and is characterised by one α-helix and three β-strands. The 3D structure of the single FN1 domain in tPA (PDB: 1TPM; 50 aa: Ser 1–Ser 50 in the A-chain of tPA) was determined by NMR spectroscopy and is characterised by four β-strands and no α-helices (see Chapter 4).

The 3D structure of the catalytic chain containing an N-terminal 13 residue extension from the A-chain (265 aa: Thr 263–Pro 527 in single-chain tPA) in complex with the inhibitor dansyl-EGR-chloromethyl ketone (in red) was determined by X-ray diffraction (PDB: 1BDA) and is shown in Figure 7.20(a). The inhibitor is covalently bound to Ser 578 and His 322 (single-chain numbering) of the active site residues of the catalytic triad (in pink). Lys 429 forms a salt bridge with Asp 477 (single-chain numbering of tPA) promoting an active conformation in single-chain tPA. The surface representation in Figure 7.20(b) clearly shows the interaction of the inhibitor with the active site.

**Urokinase-type plasminogen activator** (uPA, EC 3.4.21.73) (see Data Sheet) is a single-chain glycoprotein belonging to the peptidase S1 family. It is converted to the active two-chain uPA autocatalytically and by plasmin.Two active forms of uPA exist: a HMW form with a long A-chain of 157 amino acids containing an EGF-like and a kringle domain and a LMW form with a short A-chain of only 22 residues. Both forms contain an identical B-chain that comprises the serine protease part

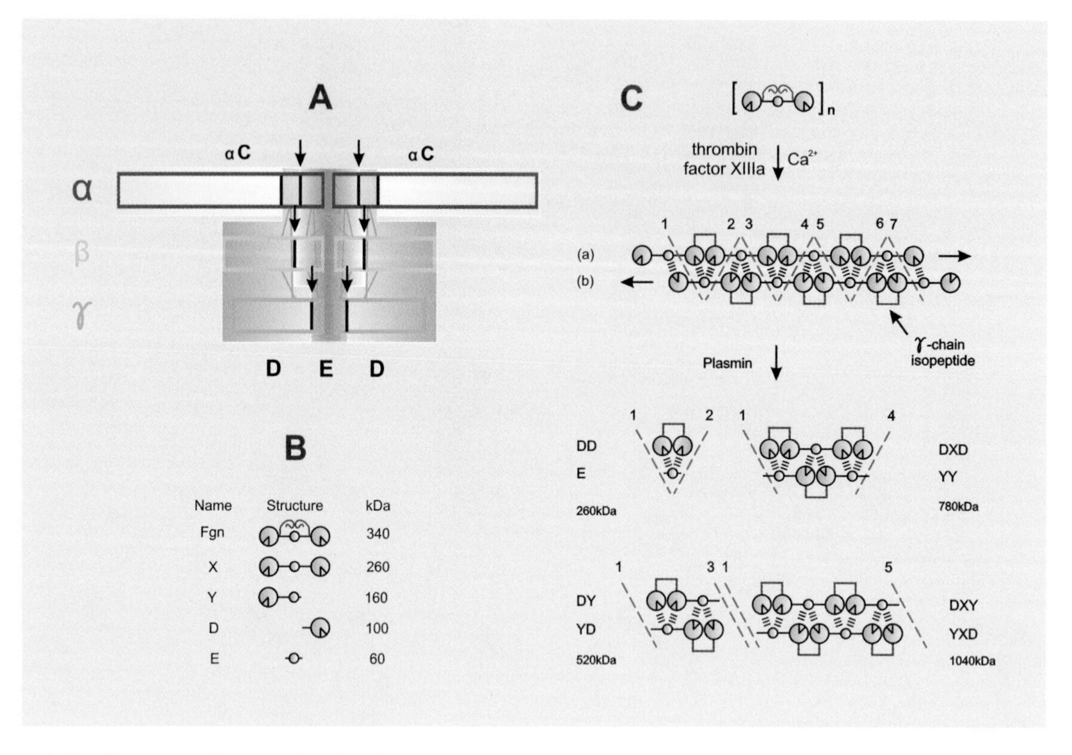


**Figure 7.18 Cleavage of fibrinogen by plasmin**
Schematic representation of fibrinogen, the fibrin polymer, the cleavage pattern by plasmin and the generated fragments. (A) Schematic representation of fibrinogen. The arrows indicate the plasmin cleavage sites. (B) Cleavage of fibrin by plasmin leads to the main fragments X, Y, D and E. (C) Schematic representation of the possible cleavage products of the fibrin polymer by plasmin. (Adapted from Walker and Nesheim. J Biol Chem 274: 5201–5212. Copyright (1999), The American Society for Biochemistry and Molecular Biology.)

and is linked to the A-chains by a single interchain disulfide bridge. uPA is phosphorylated at Ser 138 in the HMW A-chain and at Ser 303 in the B-chain. Together with tPA, uPA is the main activator of plasminogen and, as tPA, plays a similar role in the control of many physiological processes.

The 3D structure of the LMW form (PDB: 1GJA) comprising the catalytic chain (253 aa: Ile 1–Leu 253) and the mini chain (23 aa: Lys 1–Lys 23) linked by a single disulfide bridge (in blue, Cys 121 in the catalytic chain and Cys 13 in the mini chain) was solved by X-ray diffraction and is shown in Figure 7.21. The catalytic chain (in green) is characterised by three α-helices and 20 β-strands and the mini chain (in red) is not especially structured. The enzyme has the expected topology of a trypsin-like serine protease and the solvent-accessible S3 pocket is capable of binding a wide range of residues. The active site residues of the catalytic triad are shown in pink.

**Thrombin-activatable fibrinolysis inhibitor** (TAFI, EC 3.4.17.20) (see Data Sheet) or **carboxypeptidase B2** synthesised in the liver is a single-chain glycoprotein belonging to the peptidase M14 family. Active TAFI is generated by the release of a 92 residue activation peptide from the N-terminus of the proform. The active site of TAFI comprises Glu 271 and a binding site for zinc ions consists of His 67, Glu 70 and His 196. TAFI is an active carboxypeptidase that removes C-terminal Arg or Lys residues from biologically active peptides such as kinins or anaphylatoxins, thus regulating their activity. TAFI also cleaves C-terminal Lys and Arg residues in the plasminogen-binding region of fibrin, thus preventing an efficient binding of plasminogen via its LBS to fibrin and reducing its activation.

The 3D structure of pancreatic TAFI (307 aa: P15086) exhibiting 42 % sequence identity with the plasma form of TAFI was determined by X-ray diffraction (PDB: 1KWM; 402 aa: His 1–Tyr 402, containing the 95 residues activation peptide). The fold of the proenzyme is similar to other procarboxypeptidase structures.

**Histidine-rich glycoprotein** (HRG) (see Data Sheet) synthesised in the liver and present in platelets is a single-chain glycoprotein (14 % CHO) with an unusually high content of His and Pro (each 13 %). The N-terminal region contains two

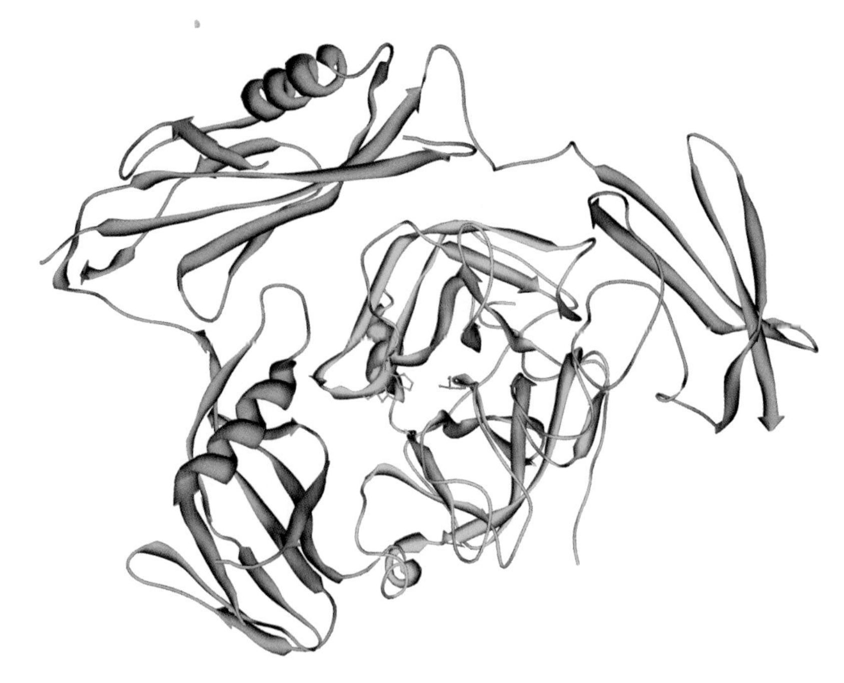

**Figure 7.19 3D structure of the plasmin light chain in complex with streptokinase from *Streptococcus equisimilis* (PDB: 1BML)**
The C-terminal domain of streptokinase (in red) binds near the activation loop of the plasmin light chain (in green), where in the catalytic triad (in pink) Ser is mutated to Ala (in blue).

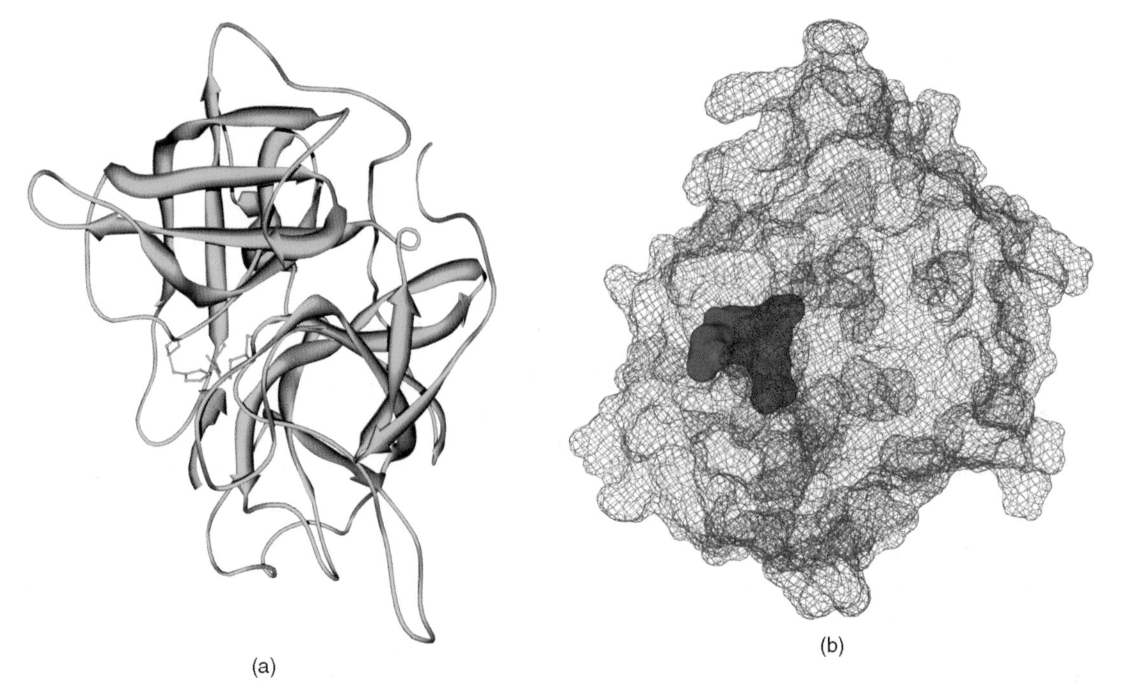


**Figure 7.20 3D structure of the catalytic chain of tissue-type plasminogen activator (PDB: 1BDA)**
a) The catalytic chain of tissue-type plasminogen activator in complex with the inhibitor dansyl-EGR-chloromethyl ketone (in red) exhibits the typical fold of serine proteases. The active site residues are shown in pink.
b) Surface representation of the interaction of the inhibitor with the active site.

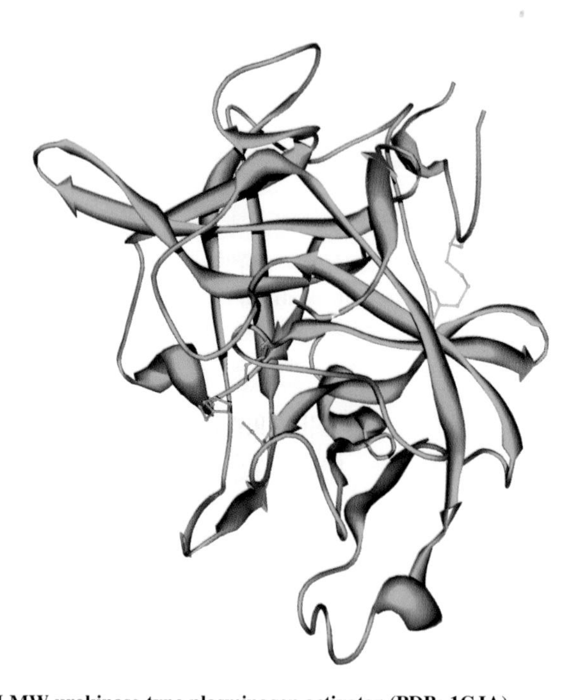

**Figure 7.21 3D structure of LMW urokinase-type plasminogen activator (PDB: 1GJA)**
The catalytic chain (in green) of LMW urokinase-type plasminogen activator is linked to a mini chain (in red) by a single disulfide bridge (in blue). The active site residues are shown in pink.

cystatin-like domains that exhibit similarity with antithrombin III. The His-rich central part shows similarity with HMW kininogen and contains 12 tandem repeats of five residues with the sequence **GHHPH**. The His-rich region is flanked by two Pro-rich regions of approximately 30 amino acids. Although the primary physiological role of HRG is not really clear, HRG is known to interact with heparin, thrombospondin and the LBS of plasminogen and seems to play a role in the contact phase activation of the intrinsic blood coagulation cascade. In addition, it binds heme, dyes and divalent ions. There are no known physiological inhibitors.

**Tetranectin** (TN) (see Data Sheet) or plasminogen-kringle 4 binding protein is a single-chain glycoprotein containing a C-type lectin domain and is present in plasma as a noncovalently linked homotrimer. Each monomer contains three intrachain disulfide bridges (Cys 50–Cys 60, Cys 77–Cys 176, Cys 152–Cys 168). TN exhibits binding to plasminogen kringle 4 and to kringle 4 in apolipoprotein(a). Its biological function is not very well understood, but TN seems to be involved in the packaging of molecules destined for exocytosis.

The 3D structure of engineered TN (PDB: 1HTN; 182 aa: Gly 1–Val 182 containing an additional N-terminal Gly) was determined by X-ray diffraction and is shown in Figure 7.22. Tetranectin is a homotrimer forming a triple α-helical coiled coil lectin structure. Each monomer consists of a **CHO-recognition domain** (CDR) connected to a long α-helix. The structure is characterised by four α-helices and seven β-sheets.

## 7.5 THE REGULATION OF BLOOD COAGULATION AND FIBRINOLYSIS

The various cascades and biochemical pathways have to be controlled and regulated in order to obtain and maintain a well-balanced system. Blood coagulation and fibrinolysis are heavily regulated and a whole set of inhibitors is present in the bloodstream for an efficient control and regulation. The description of the various types of inhibitors in blood is given in a separate chapter on inhibitors (see Chapter 11). However, the importance of regulation and inhibition of blood coagulation and fibrinolysis requires a short description in this chapter.

There are several regulatory mechanisms in blood coagulation and fibrinolysis: primarily, protease inhibitors directly inhibit activated factors. Many factors are present in blood as zymogens and are activated by specific activators, usually by limited proteolysis with a concomitant and large conformational change. Many factors in blood coagulation and fibrinolysis

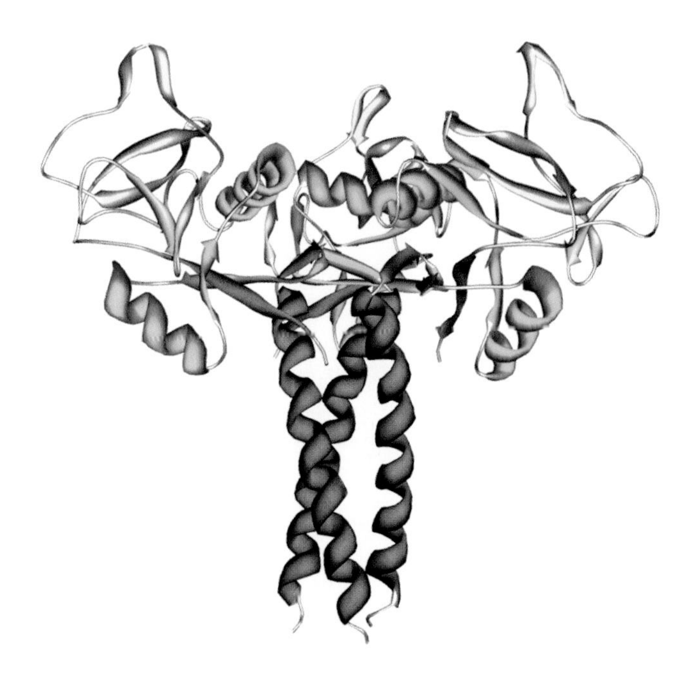

**Figure 7.22 3D structure of trimeric tetranectin (PDB: 1HTN)**
Trimeric tetranectin forms a triple α-helical coiled coil lectin structure. Each monomer consists of a CHO-recognition (CDR) domain connected to a long α-helix.

are serine proteases and for almost all of them specific serine protease inhibitors or serpins exist. Usually, the concentration of the serpin in plasma is similar as the corresponding serine protease. Some of the main inhibitors involved in blood coagulation and fibrinolysis are compiled in Table 7.6.

Additionally, there are several other mechanisms to regulate blood coagulation and fibrinolysis:

**Table 7.6** Characteristics of the major inhibitors of blood coagulation and fibrinolysis.

| Inhibitor | Mass (kDa)[a] | Concentration (mg/l)[b] | Specificity[c] |
|---|---|---|---|
| Antithrombin-III | 60 | 150 | Thrombin, FIXa, FXa, FXIa |
| $\alpha_1$-Antitrypsin | 51 | 1300 | Elastase, thrombin, FXa, FXIa, protein C |
| Plasma protease C1 inhibitor | 105 | 200 | FXIIa, plasma kallikrein, C1r, C1s |
| $\alpha_2$-Antiplasmin | 67 | 70 | Plasmin, FXIIa |
| $\alpha_2$-Macroglobulin | 720 | 1200 | All proteases |
| Tissue factor pathway inhibitor | 43 | 0.1 | Complex TF/FVIIa, FXa |
| Heparin cofactor II | 66 | 80 | Thrombin, chymotrypsin, cathepsin G |
| Kallistatin | 58 | 20 | Tissue kallikrein |
| Plasminogen activator inhibitor-1 | 50 | 0.06 | tPA, uPA, thrombin, plasmin, protein C |
| Plasminogen activator inhibitor-2 | 50 | 0.005 | uPA, tPA |
| Plasma serine protease inhibitor | 57 | — | Protein C, plasma kallikrein, thrombin, FXa, FXIa, uPA |
| Protein C | 60 | 4 | FVa |

[a] Values obtained by SDS-PAGE.
[b] Represent average values.
[c] Main specificity.

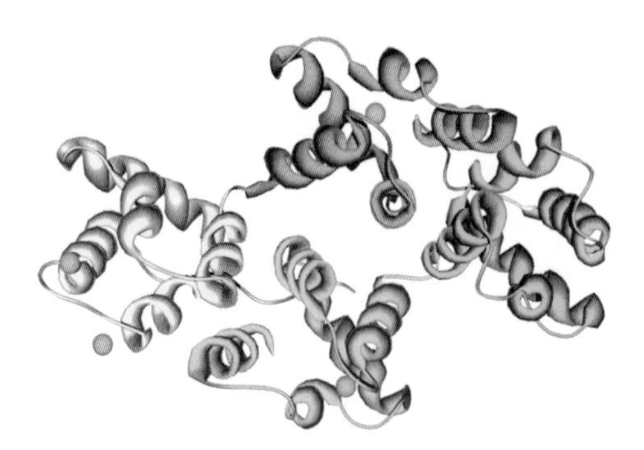

**Figure 7.23 3D structure of annexin A5 (PDB: 1HVD)**
Annexin A5 contains four repeats (in yellow, red, green and blue) exhibiting similar structures with five α-helical segments arranged into a right-handed compact superhelix. The calcium ions (pink balls) are located on the surface.

- the enzymatic inactivation of cofactors, e.g. the proteolytic inactivation of FVa and FVIIIa by activated protein C;
- the neutralisation of phospholipids by annexin V resulting in the prevention of the formation of active complexes required for the activation of coagulation, e.g. prevention of the complex formation of TF/FVIIa or FXa/FVa both requiring $Ca^{2+}$ ions and phospholipids;
- the binding to surfaces, thus maintaining a high local concentration but preventing a spreading to surrounding areas, e.g. the binding of thrombin to fibrin and the binding of vitamin K-dependent factors to phospholipids.

There are also other proteins involved in regulatory processes, e.g. $\beta_2$-glycoprotein I that binds negatively charged components such as heparin, dextran sulfate and phospholipids and may prevent the activation of the intrinsic coagulation pathway.

**Annexin A5** (see Data Sheet) or lipocortin is a single-chain plasma protein containing four annexin repeats and an N-terminal acetyl-Ala residue. Annexin A5 belongs to the annexin family and exhibits similarity with other annexins, especially with annexin A8 (vascular anticoagulant-beta, 327 aa: P13928). Annexin A5 is an anticoagulant protein involved in the neutralisation of phospholipids and thus prevents the formation of active complexes in blood coagulation and also acts as an indirect inhibitor of the thrombospondin-specific complex.

The 3D structure of an annexin A5 mutant (PDB: 1HVD; 319 aa: Ala 1–Asp 319) has been solved by X-ray diffraction and is shown in Figure 7.23. Annexin A5 is folded into a planar cyclic arrangement of four repeats (in yellow, red, green and blue) exhibiting similar structures with five α-helical segments arranged into a right-handed compact superhelix. Annexin A5 binds phospholipids in a Ca-dependent manner and exhibits calcium channel activity with the calcium ions (pink balls) located on the surface.

**$\beta_2$-Glycoprotein I** (see Data Sheet) or apolipoprotein H synthesised in the liver is a single-chain glycoprotein containing four sushi/CCP/SCR domains and a C-terminal sushi-like region. $\beta_2$-Glycoprotein exhibits a strong tendency to bind negatively charged substances like heparin, dextran sulfate and phospholipids. It may prevent the activation of the intrinsic coagulation cascade by binding to phospholipids on the surface of damaged cells.

The 3D structure of $\beta_2$-Glycoprotein (PDB: 1C1Z; 326 aa: Gly 1–Cys 326) has been solved by X-ray diffraction and is shown in Figure 7.24. The structure clearly shows the four distinct sushi/CCP/SCR domains and the special folding of the sushi-like region (in violet), which together form an elongated J-shaped molecule (like beads on a string).

A detailed description of the serpins involved in blood coagulation and fibrinolysis, the Kunitz-type inhibitors, the general protease inhibitor $\alpha_2$-macroglobulin and other proteins with inhibitory function is given in a separate chapter on inhibitors (Chapter 11).

## 7.6 THE KININ AND ANGIOTENSIN/RENIN SYSTEMS

Two important systems exist that regulate the vascular tonus and are involved in the regulation of the volume and the mineral balance of body fluids:

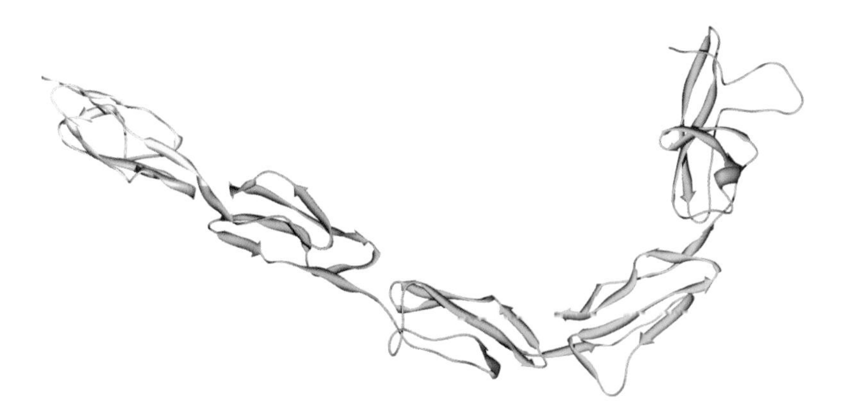

**Figure 7.24 3D structure of $\beta_2$-glycoprotein (PDB: 1C1Z)**
$\beta_2$-Glycoprotein contains four distinct sushi/CCP/SCR domains and the special folding of the sushi-like region (in violet) forming an elongated J-shaped molecule.

1. the **kinin system** with its main components HMW and LMW kininogens, from which the vasoactive peptides bradykinin and lysyl-bradykinin are generated;
2. The **angiotensin/renin system** with its main components angiotensinogen from which renin releases the decapeptide angiotensin I and the angiotensin-converting enzyme (ACE) that releases the C-terminal dipeptide His–Leu from angiotensin I, thus generating the vasoconstrictiv octapeptide angiotensin II.

**Kininogens** (see Data Sheet) or $\alpha_2$-thiol proteinase inhibitors are the major thiol protease inhibitors in human plasma, e.g. in inflammatory processes, and work as cofactors in blood coagulation. Kininogens exists in plasma in two single-chain glycoprotein forms: high molecular weight kininogen (**HMW kininogen**) and low molecular weight kininogen (**LMW kininogen**), comprising 409 aa of the N-terminal region of HMW kininogen. Kallikrein cleaves the Met 361–Lys 362 and the Arg 371–Ser 372 peptide bonds, releasing the vasoactive kinins, the nonapeptide **bradykinin** (see Data Sheet for kininogen) or kallidin I (**RP(OH)-PGFSPFR**) and the decapeptide **lysyl-bradykinin** or kallidin II, both carrying an OH-Pro residue. Eventually, aminopeptidases remove the N-terminal Lys residue from lysyl-bradykinin, thus generating bradykinin. The resulting two-chain kininogens consist of a common heavy chain (362 aa) with three cystatin-like domains and an N-terminal pyro-Glu residue and a short (LMW: 38 aa) and a long (HMW: 255 aa) light chain. Bradykinin is the major component responsible for vasodilatation, thus lowering and regulating blood pressure. In addition, bradykinin exhibits many other physiological effects like influence on the smooth muscle contraction, induction of hypotension, natriuresis and diuresis, decrease of the glucose level in blood, mediator of inflammation, increase of vascular permeability and release of mediators of inflammation-like prostaglandins.

The 3D structure of bradykinin in SDS micelles determined by NMR spectroscopy exhibits a $\beta$-turn-like structure in its C-terminal region (Lee *et al.*, 1990).

As a member of the cysteine protease inhibitors kininogens are characterised by a specific consensus pattern also found in cystatins:

CYSTATIN (PS00287): [GSTEQKRV]–Q–[LIVT]–[VAF]–[SAGQ]–G–{DG}–[LIVMNK]–{TK}–x–[LIVMFY]–x–[LIVMFYA]–[DENQKRHSIV]

**Angiotensinogen** (see Data Sheet) synthesised in the liver is a single-chain glycoprotein belonging to the serpin family and is the parent protein of the vasoactive peptides **angiotensin I (DRVYIHPFHL)**, **angiotensin II (DRVYIHPF)** and **angiotensin III (RVYIHPF)**. In response to a decreased blood pressure the peptidase renin releases the decapeptide angiotensin I from the N-terminus of the precursor angiotensinogen. The dipeptidyl carboxypeptidase **angiotensin-converting enzyme** (ACE) removes the C-terminal dipeptide His–Leu from angiotensin I to generate the octapeptide angiotensin II. The release of the N-terminal Asp from angiotensin II results in the heptapeptide angiotensin III. Angiotensin II is responsible for vasoconstriction and thus an increase in blood pressure. Angiotensin II and bradykinin involved in

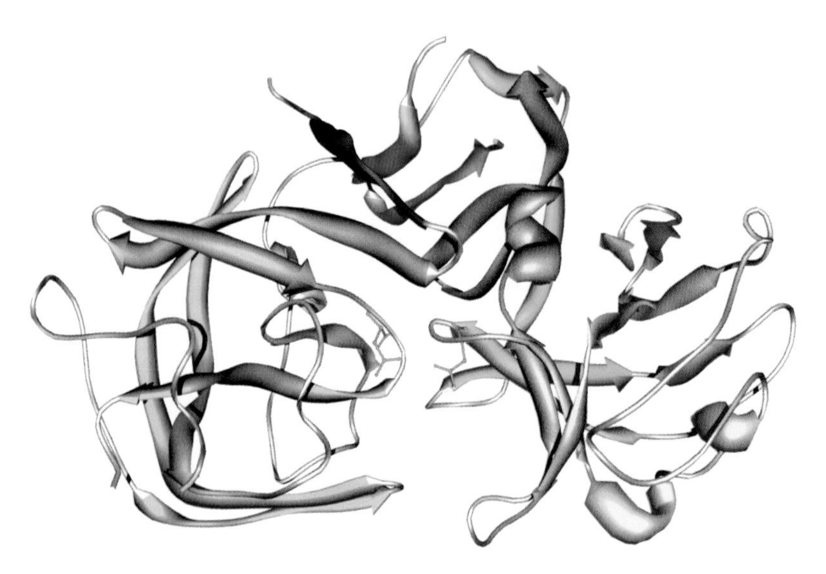

**Figure 7.25 3D structure of activated renin (PDB: 2REN)**
Renin exhibits the typical fold of aspartyl proteases predominantly characterised by β-sheet conformation. The substrate binding cleft containing the two Asp residues of the active site (in pink) is at the junction of two structurally similar domains.

vasodilatation are the main components responsible for the regulation of the vascular tonus. Angiotensin II helps to regulate the volume and the mineral content of body fluids. Angiotensin III stimulates the release of aldosterone. In phospholipid micelles angiotensin II adopts a well-defined hairpin structure with its C- and N-termini in close proximity (Carpenter *et al.*, 1998).

**Renin** (EC 3.4.23.15) (see Data Sheet) or angiotensinogenase is a single-chain glycoprotein belonging to the peptidase A1 family (aspartyl or acid proteases) and is generated from its proform by releasing a 43-residue activation peptide from the N-terminus. The active site centre of renin comprises Asp 38 and Asp 226. Aspartyl proteases are characterised by the following signature where Asp is part of the active site:

ASP_PROTEASE (PS00141): [LIVMFGAC]–[LIVMTADN]–[LIVFSA]–**D**–[ST]–G–[STAV]–[STAPDENQ]–{GQ}–[LIVMFSTNC]–{EGK}–[LIVMFGTA]

Renin is a highly specific endopeptidase with the only known function to generate angiotensin I from angiotensinogen by cleaving the Leu 10–Val 11 peptide bond.

The 3D structure of recombinant, activated renin (PDB: 2REN; 340 aa: Leu 1–Arg 340) has been solved by X-ray diffraction and is shown in Figure 7.25. Renin folds predominantly in a β-sheet conformation. This fold is characteristic for aspartyl proteases (see Chapter 10) and renin is closest related in structure to pepsinogen. The substrate binding cleft containing the two Asp residues of the active site (in pink) is at the junction of two structurally similar domains.

Somatic **angiotensin-converting enzyme** (ACE, EC 3.4.15.1) (see Data Sheet) or kininase II is a single-chain type I membrane glycoprotein (26 % CHO) belonging to the peptidase M2 family and contains a C-terminal transmembrane segment (Trp 1231–Leu 1247) and two binding sites for two zinc ions consisting of His 361 and 365 and His 959 and 963, respectively. The zinc-dependent metallopeptidase superfamily is characterised by the following signature, where the two His form a zinc-binding site and Glu is part of the active site:

ZINC_PROTEASE (PS00142): [GSTALIVN]–{PCHR}–{KND}–**H**–**E**–[LIVMFYW]–{DEHRKP}–**H**–x–[LIVMFYWGSPQ]

ACE is a dipeptidyl carboxypeptidase with two Glu residues (aa 362 and 960) in its active centre that releases dipeptides from the C-terminus of unmodified peptides. The plasma form is generated by limited proteolysis from endothelial ACE. Vascular ACE participates in blood pressure regulation by converting angiotensin I to angiotensin II through the release of

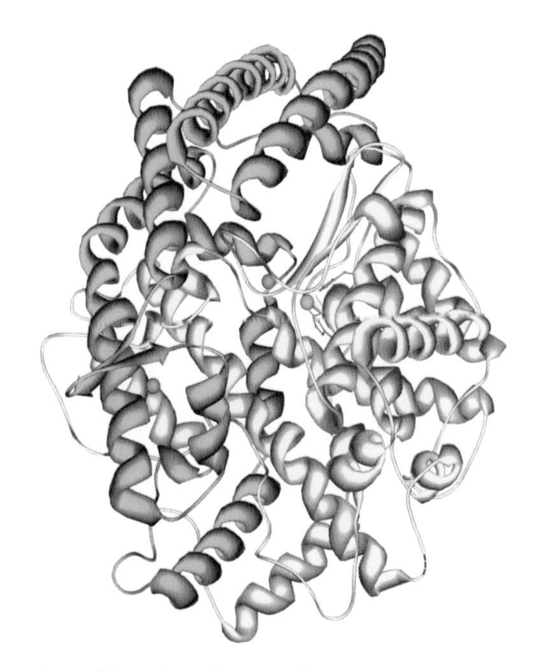

**Figure 7.26 3D structure of the testis-specific angiotensin-converting enzyme (PDB: 1O8A)**
Testis-specific angiotensin-converting enzyme is characterised by a high abundance of α-helices and the zinc ion (pink ball) in the active site is bound to His residues (in blue) in the zinc-binding motif and two chloride ions (red balls).

the C-terminal dipeptide His–Leu. ACE is able to inactivate the vasoactive bradykinin by releasing the C-terminal dipeptide Phe–Arg.

The 3D structure of the testis-specific engineered isoforms of ACE (701 aa: P22966) comprising the sequence Leu 37–Ser 625 (PDB: 1O8A; 589 aa without the 36 residue N-terminal peptide and without the 76 residue C-terminal region containing the potential 17 residue transmembrane segment) has been solved by X-ray diffraction and is shown in Figure 7.26. Testis-specific ACE consists of two subdomains (in yellow and green) containing in the active site a zinc ion (pink ball) coordinated to two His residues and a Glu residue (all in blue) in the conserved **HEXXH** zinc-binding motif and two chloride ions (red balls). The structure is characterised by an abundance of 30 α-helices and only six β-strands. In addition, the 3D structure of the N-terminal region of somatic ACE was determined recently by X-ray diffraction (PDB: 2C6N; 612 aa: Leu 1–Asp 612).

## REFERENCES

Adkins *et al.*, 2002, Toward a human blood serum proteome. Analysis by multidimensional separation coupled with mass spectrometry, *Mol. Cell. Proteomics*, **1**, 947–955.

Anderson and Anderson, 2002, The human plasma proteome. History, character and prospects, *Mol. Cell. Proteomics*, **1**, 845–867.

Anderson *et al.*, 2004, The human plasma proteome. A nonredundant list developed by combination of four separate resources, *Mol. Cell. Proteomics*, **3**, 311–326.

Campbell, 2003, The rennin–angiotensin and the kallikrein–kinin systems, *Int. J. Biochem.*, **35**, 784–791.

Carpenter *et al.*, 1998, The octapeptide angiotensin II adopts a well-defined structure in a phospholipid environment, *Eur. J. Biochem.*, **251**, 448–453.

Castellino *et al.*, 2005, Structure and function of the plasminogen/plasmin system, *Thromb. Haemost.*, **93**, 647–654.

Dahlbäck *et al.*, 2005, The anticoagulant protein C pathway, *FEBS Lett.*, **579**, 3310–3316.

Davie, 2003, A brief historical review of the waterfall/cascade of blood coagulation, *J. Biol. Chem.*, **278**, 50819–50832.

Doolittle, 1984, Fibrinogen and fibrin, *Annu. Rev. Biochem.*, **53**, 195–229.

Esmon, 2000, Regulation of blood coagulation, *Biochim. Biophys. Acta*, **1477**, 349–360.

Haeberli, 1998, *Human Protein Data: Fourth Installment*, John Wiley & Sons, Ltd, Chichester.

Kolde, 2001, *Haemostasis: Physiology, Pathology, Diagnostics*, Pentatharm Ltd, Basel.

Lee *et al.*, 1990, Three-dimensional structure of bradykinin in SDS miscelles. Study using nuclear magnetic resonance, distance geometry, and restrained molecular mechanics and dynamics, *Int. J. Pept. Protein Res.*, **35**, 367–377.

Longstaff *et al.*, 2005, Understanding the enzymology of fibrinolysis and improving thrombolytic therapy, *FEBS Lett.*, **579**, 3303–3309.

Mathews *et al.*, 1996, Crystal structures of the recombinant kringle 1 domain of human plasminogen in complices with the ligands ε-aminocaproic acid and *trans*-4-(amino methyl) cyclohexane-1-carboxylic acid, *Biochemistry*, **35**, 2567–2576.

Omenn *et al.*, 2005, Overview of the HUPO plasma proteome project: results from the pilot phase with 35 collaborating laboratories and multiple analytical groups, generating a core dataset of 3020 proteins and a publicly-available database, *Proteomics*, **5**, 3226–3245.

Pieper *et al.*, 2003, The human serum proteome: display of nearly 3700 chromatographically separated protein spots on two-dimensional electrophoresis gels and identification of 325 distinct proteins, *Proteomics*, **3**, 1345–1364.

Pike *et al.*, 2005, Control of the coagulation system by serpins. Getting by with a little help from glycosaminoglycans, *FEBS J.*, **272**, 4842–4851.

Qian *et al.*, 2005, Comparative proteome analyses of human plasma following *in vivo* lipopolysaccharide administration using multidimensional separations coupled with tandem mass spectrometry, *Proteomics*, **5**, 572–584.

Schenone *et al.*, 2004, The blood coagulation cascade, *Curr. Opin. Haematol.*, **11**, 272–277.

Shen *et al.*, 2004, Ultra-high efficiency strong cation exchange LC/RPLC/MS/MS for high dynamic range characterization of the human plasma proteome, *Anal. Chem.*, **76**, 1134–1144.

Walker and Nesheim, 1999, The molecular weights mass distribution, chain composition, and structure of soluble fibrin degradation products released from a fibrin clot perfused with plasmin, *J. Biol. Chem.*, **274**, 5201–5212.

DATA SHEETS

# *Angiotensin-converting enzyme (Kininase II; EC 3.4.15.1)*

## Fact Sheet

| | | | |
|---|---|---|---|
| ***Classification*** | Peptidase M2 | ***Abbreviations*** | ACE |
| ***Structures/motifs*** | | ***DB entries*** | ACE_HUMAN/P12821 |
| ***MW/length*** | 146 829 Da/1277 aa | ***PDB entries*** | 1O8A; 2C6N |
| ***Concentration*** | 300–500 mg/l | ***Half-life*** | |
| ***PTMs*** | 17 N-CHO (potential) | | |
| ***References*** | Ehlers *et al.*, 1989, *Proc. Natl. Acad. Sci. USA*, **86**, 7741–7745.<br>Jongeneel *et al.*, 1989, *FEBS Lett.*, **242**, 211–214.<br>Natesh *et al.*, 2003, *Nature*, **421**, 551–554.<br>Soubrier *et al.*, 1988, *Proc. Natl. Acad. Sci. USA*, **85**, 9386–9390. | | |

## Description

The plasma form is derived from the somatic form of the **angiotensin-converting enzyme** (ACE) by proteolysis. The main function of ACE is to generate vasoconstrictive angiotensin II from angiotensin I.

## Structure

Somatic ACE (1277 aa) is a single-chain type I membrane glycoprotein (26 % CHO) containing one C-terminal transmembrane segment (Trp 1231–Leu 1247), two binding sites for zinc ions consisting of two His residues (His 361 and 365, His 959 and 963) and the active centre with two Glu residues (362 and 960 aa). The plasma form of ACE (701 aa) is generated from vascular ACE by limited proteolysis. The 3D structure of the testis-specific engineered isoforms (701 aa) of ACE (589 aa: Leu 37–Ser 625) contains 30 $\alpha$-helices and six $\beta$-strands (**1O8A**). The zinc ion (in pink) in the active site is bound to two His residues and a Glu residue (all in blue) in the **HEXXH** zinc-binding motif.

## Biological Functions

Vascular ACE participates in blood pressure regulation by converting the decapeptide angiotensin I to angiotensin II (cleavage of the C-terminal dipeptide His–Leu). The dipeptidyl carboxypeptidase ACE is able to inactivate vasoconstrictive bradykinin by releasing the C-terminal dipeptide Phe–Arg.

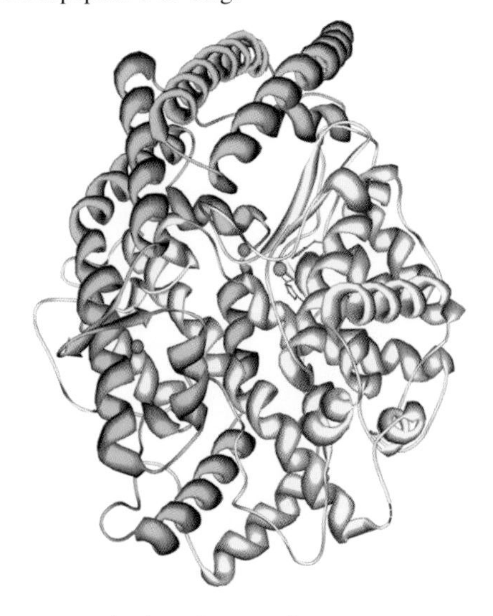

Angiotensin-converting enzyme

## *Angiotensinogen/Angiotensin I, II and III*

### Fact Sheet

| | | | |
|---|---|---|---|
| ***Classification*** | Serpin | ***Abbreviations*** | |
| ***Structures/motifs*** | | ***DB entries*** | ANGT_HUMAN/P01019 |
| ***MW/length*** | Angiotensinogen: 49 761 Da/452 aa<br>Angiotensin I: 1296 Da/10 aa<br>Angiotensin II: 1046 Da/10 aa<br>Angiotensin III: 931 Da/7 aa | ***PDB entries*** | |
| ***Concentration*** | approx. 45 mg/l | ***Half-life*** | |
| ***PTMs*** | 4 N-CHO | | |
| ***References*** | Kageyama *et al.*, 1984, *Biochemistry*, **23**, 3603–3609. | | |

### Description

**Angiotensinogen** is a glycoprotein synthesised in the liver. It is the parent protein of **angiotensin I, II** and **III**. Angiotensin II is responsible for vasoconstriction and in consequence for the increase in blood pressure.

### Structure

Angiotensinogen is a single-chain glycoprotein and the precursor of angiotensin I, II and III. The peptidase renin cleaves the decapeptide angiotensin I from angiotensinogen. The dipeptidyl carboxypeptidase **angiotensin converting enzyme** (ACE) removes the C-terminal dipeptide His–Leu from angiotensin I to generate the octapeptide angiotensin II. The heptapeptide angiotensin III is generated by removing the N-terminal Asp from angiotensin II. Angiotensinogen exhibits structural similarity with other serpins.

### Biological Functions

In response to a decreased blood pressure renin generates angiotensin I from angiotensinogen. Angiotensin II is responsible for vasoconstriction and in consequence for an increase in blood pressure. Angiotensin II helps to regulate the volume and the mineral content of body fluids. Angiotensin III stimulates the release of aldosterone.

## *Annexin A5 (Lipocortin)*

### Fact Sheet

| | | | |
|---|---|---|---|
| ***Classification*** | Annexin | ***Abbreviations*** | |
| ***Structures/motifs*** | 4 annexin | ***DB entries*** | ANXA5_HUMAN/P08758 |
| ***MW/length*** | 35 806 Da/319 aa | ***PDB entries*** | 1HVD |
| ***Concentration*** | | ***Half-life*** | |
| ***PTMs*** | 1 *N*-acetyl-Ala | | |
| ***References*** | Burger *et al.*, 1994, *J. Mol. Biol.*, **237**, 479–499.<br>Funakoshi *et al.*, 1987, *Biochemistry*, **26**, 8087–8092. | | |

### Description

**Annexin A5** is an anticoagulant protein involved in blood coagulation. It shows extensive similarity with other annexins, e.g. annexin A8.

### Structure

Annexin A5 is a single-chain plasma protein containing four annexin repeats and an N-terminal *N*-acetyl-Ala and belongs to the annexin family. The 3D structure of a mutant contains 21 α-helices and no β-strands (**1HVD**).

### Biological Functions

Annexin A5 is an anticoagulant protein that neutralises phospholipids and acts as an indirect inhibitor of the thrombospondin-specific complex which is involved in the coagulation cascade.

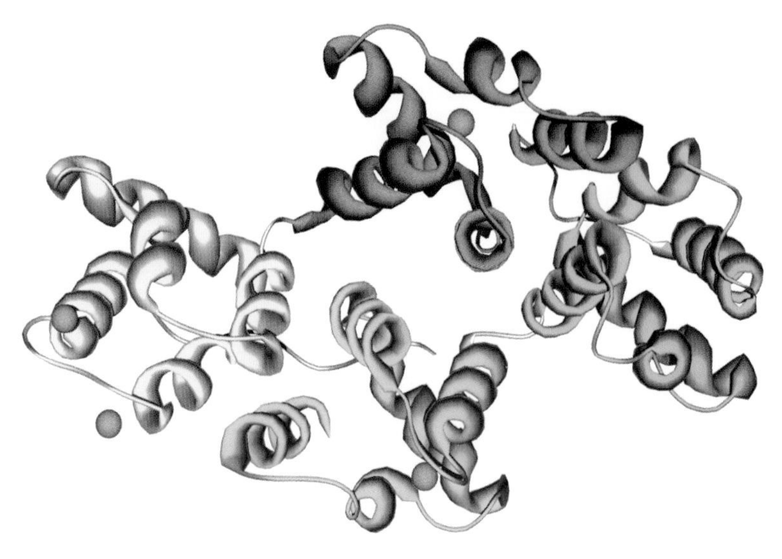

Annexin

## *Coagulation factor V (Proaccelerin; Protein C cofactor)*

### Fact Sheet

| | | | |
|---|---|---|---|
| ***Classification*** | Multicopper oxidase | ***Abbreviations*** | FV; FVa |
| ***Structures/motifs*** | 6 Plastocyanin-like; 3 F5/8 type A; 2 F5/8 type C | ***DB entries*** | FA5_HUMAN/P12259 |
| ***MW/length*** | 248 665 Da/2196 aa | ***PDB entries*** | 1CZS |
| ***Concentration*** | 4–10 mg/l | ***Half-life*** | approx. 12 h |
| ***PTMs*** | 26 N-CHO (potential); 7 S-Tyr (potential) | | |
| ***References*** | Cripe *et al.*, 1992, *Biochemistry*, **31**, 3777–3785. | | |

### Description

**Coagulation factor V** (FV) is a large procofactor glycoprotein that is converted by thrombin or FXa in a Ca-dependent manner to the active cofactor FVa. On membrane surfaces FVa forms together with the serine protease FXa the prothrombinase complex that catalyses the rapid conversion of the zymogen prothrombin to thrombin. FV shares structural similarities with FVIII.

### Structure

FV is a large single-chain glycoprotein that contains three F5/8 type A, two F5/8 type C and six plastocyanin-like domains. Selective proteolysis by either thrombin or FXa releases two large activation peptides (cleavages at Arg 709, 1018 and 1545) and leads to the active two-chain FVa which are noncovalently linked in a Ca-dependent manner. FV contains two 17 residues long repeats and 35 repeats of nine residues with the sequence **[TNP]LSPDLSQT** and seven potential Tyr-sulfation sites.

### Biological Functions

FVa is an essential cofactor of the prothrombinase enzyme complex and exhibits two main functions: (1) it serves as a receptor for FXa on membrane surfaces and (2) binds the substrate prothrombin in a manner to facilitate its rapid proteolysis by FXa. FVa is inactivated proteolytically by activated protein C.

# *Coagulation factor VII (Proconvertin; EC 3.4.21.21)*

## Fact Sheet

| | | | |
|---|---|---|---|
| ***Classification*** | Peptidase S1 | ***Abbreviations*** | FVII; FVIIa |
| ***Structures/motifs*** | 2 EGF-like; 1 Gla | ***DB entries*** | FA7_HUMAN/P08709 |
| ***MW/length*** | 45 079 Da/406 aa | ***PDB entries*** | 1QFK; 1DAN |
| ***Concentration*** | 0.4–0.6 mg/l | ***Half-life*** | 2–2.5 h |
| ***PTMs*** | 2 N-CHO; 2 O-CHO; 1 OH-Asp (aa 63); 10 γ-Gla | | |
| ***References*** | Hagen *et al.*, 1986, *Proc. Natl. Acad. Sci. USA*, **83**, 2412–2416.<br>Pike *et al.*, 1999, *Proc. Natl. Acad. Sci. USA*, **96**, 8925–8930. | | |

## Description

**Coagulation factor VII** (FVII) is a glycoprotein synthesised in the liver and belongs to the extrinsic pathway system. It shares considerable structural similarities with other vitamin K-dependent proteins in blood coagulation like FIX, FX, prothrombin, proteins C, S and Z, suggesting an evolution from a common ancestral gene.

## Structure

FVII is a single-chain glycoprotein that is activated by either FIXa, FXa, FXIIa or thrombin to the active, two-chain FVIIa (cleavage of Arg 152–Ile 153). The light chain contains a Gla domain, an OH-Asp residue (63 aa) and 2 EGF-like domains and the heavy chain comprises the serine protease part. The 3D structure shows the catalytic chain (254 aa, in blue) and the C-terminal part of the light chain (104 aa, in red) containing the two EGF-like domains in complex with a tripeptide inhibitor (**1QFK**).

## Biological Functions

FVIIa participates in the extrinsic pathway. In the presence of tissue factor and $Ca^{2+}$ ions FVIIa converts FX to FXa and FIX to FIXa by limited proteolysis. The Kunitz-type inhibitor **tissue factor pathway inhibitor** (TFPI) neutralises the ternary complex FVIIa-TF-FXa in the presence of $Ca^{2+}$ ions.

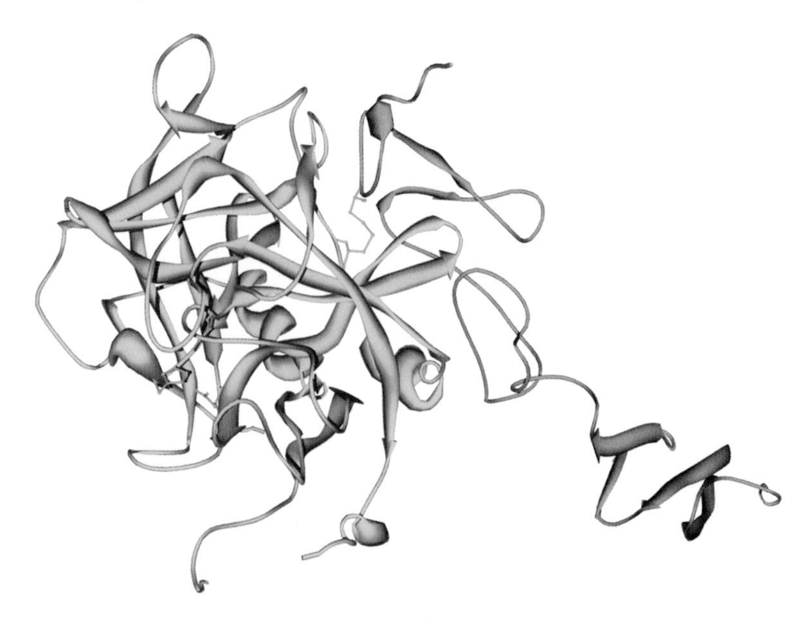

Coagulation factor VII

## *Coagulation factor VIII (Antihemophilic factor)*

### Fact Sheet

| | | | |
|---|---|---|---|
| ***Classification*** | Multicopper oxidase | ***Abbreviations*** | FVIII; AHF |
| ***Structures/motifs*** | 6 Plastocyanin-like; 3 F5/8 type A; 2 F5/8 type C | ***DB entries*** | FA8_HUMAN/P00451 |
| ***MW/length*** | 264 726 Da/2322 aa | ***PDB entries*** | 1D7P |
| ***Concentration*** | 0.1–0.4 mg/l | ***Half-life*** | 8–12 h |
| ***PTMs*** | 22 N-CHO (potential); 8 S-Tyr (2 probable) | | |
| ***References*** | Toole *et al.*, 1984, *Nature*, **312**, 342–347. | | |

### Description

**Coagulation factor** VIII (FVIII) is a single chain cofactor glycoprotein primarily synthesised in the liver and circulates in blood in complex with the von Willebrand factor. FVIII shares structural similarities with FV.

### Structure

FVIII is a large single-chain glycoprotein containing 3 F5/8 type A, 2 F5/8 type C and 6 plastocyanin-like domains. Prior to activation the cofactor function requires the cleavage by thrombin at Arg 372, 740 and 1689. Activation (cleavage at Arg 1313 and 1648) leads to the active, two-chain FVIIIa noncovalently linked in a Ca-dependent manner. FVIII contains up to eight sulfated Tyr.

### Biological Functions

FVIII is an essential cofactor in the activation process of FX to FXa. The complex formation of FVIII with the von Willebrand factor is a prerequisite for its release from cells, its stability and its effectiveness in haemostasis.

## *Coagulation factor IX (Christmas factor; EC 3.4.21.22)*

### Fact Sheet

| | | | |
|---|---|---|---|
| ***Classification*** | Peptidase S1 | ***Abbreviations*** | FIX; FIXa |
| ***Structures/motifs*** | 2 EGF-like; 1 Gla | ***DB entries*** | FA9_HUMAN/P00740 |
| ***MW/length*** | 46 548 Da/415 aa | ***PDB entries*** | 1RFN |
| ***Concentration*** | 3–5 mg/l | ***Half-life*** | 18–24 h |
| ***PTMs*** | 2 N-CHO; 6 O-CHO; 1 OH-Asp (64 aa); 2 P-Ser (68, 158 aa); 1 S-Tyr (155 aa); 12 γ-Gla | | |
| ***References*** | Hopfner *et al.*, 1999, *Structure*, **7**, 989–996. | | |
| | Yoshitake *et al.*, 1985, *Biochemistry*, **24**, 3736–3750. | | |

### Description

**Coagulation factor IX** (FIX) is a plasma glycoprotein synthesised in the liver and belongs to the intrinsic pathway system. It shares considerable structural similarities with other vitamin K-dependent proteins in blood coagulation like FVII, FX, prothrombin, proteins C, S and Z, suggesting an evolution from a common ancestral gene.

### Structure

FIX is a single-chain glycoprotein (18 % CHO). It is activated by FXIa or the FVIIa/TF complex in the presence of $Ca^{2+}$ ions to the active, two-chain FIXa, releasing the activation peptide (35 aa) from the zymogen (cleavage of Arg 145–Val 146 and Arg 180–Val 181). The light chain contains a Gla domain, an OH-Asp residue (64 aa) and a P-Ser residue (68 aa) and two EGF-like domains and the heavy chain comprises the serine protease part. The 3D structure shows the catalytic chain (235 aa, in blue) and the C-terminal part of the light chain (57 aa, in red) containing the second EGF-like domain in complex with *p*-amino benzamidine (**1RFN**).

### Biological Functions

FIXa binds to a high-affinity, specific receptor on platelets or endothelial cells and forms a complex with the nonenzymatic cofactor FVIIIa. This complex activates FX to FXa in the intrinsic pathway similar to that in the extrinsic pathway with the FVIIa/TF complex. The main physiological inhibitor antithrombin III forms an equimolar complex with FIXa.

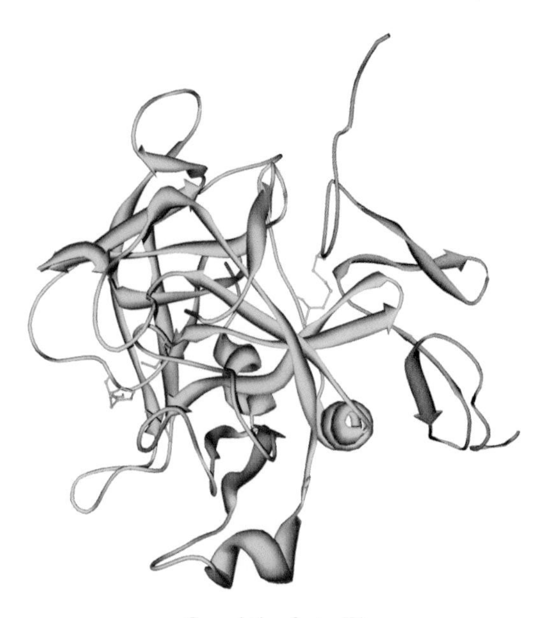

Coagulation factor IX

## *Coagulation factor X (Stuart factor; EC 3.4.21.6)*

### Fact Sheet

| | | | |
|---|---|---|---|
| ***Classification*** | Peptidase S1 | ***Abbreviations*** | FX; FXa |
| ***Structures/motifs*** | 2 EGF-like; 1 Gla | ***DB entries*** | FA10_HUMAN/P00742 |
| ***MW/length*** | 50 336 Da/448 aa | ***PDB entries*** | 1KSN |
| ***Concentration*** | 7–10 mg/l | ***Half-life*** | 40 h |
| ***PTMs*** | 2 N-CHO; 6 O-CHO; 1 OH-Asp (aa 63); 11 $\gamma$-Gla | | |
| ***References*** | Leytus *et al.*, 1986, *Biochemistry*, **25**, 5098–5102.<br>Guertin *et al.*, 2002, *Bioorg. Med. Chem. Lett.*, **12**, 1671–1674. | | |

### Description

**Coagulation factor X** (FX) is a plasma glycoprotein synthesised in the liver. It shares considerable structural similarities with other vitamin K-dependent proteins in blood coagulation like FVII, FIX, prothrombin, proteins C, S and Z, suggesting an evolution from a common ancestral gene.

### Structure

FX is a single-chain plasma glycoprotein. FX is activated intrinsically by FIXa or extrinsically by FVIIa to the active, two-chain FXa, releasing the activation peptide (52 aa) from the zymogen (cleavage of Arg 142–Ser 143 and Arg 194–Ile 195). The light chain contains a Gla-domain, an OH-Asp residue (63 aa) and 2 EGF-like domains and the heavy chain comprises the serine protease part. The 3D structure shows the catalytic chain (254 aa, in blue) and the light chain without the N-terminal pentapeptide (134 aa, in red) containing the Gla domain and the two EGF-like domains in complex with an inhibitor ($\beta$-aminester). The catalytic domain exhibits a similar folding as $\alpha$-thrombin and the second EGF-like domain makes contact with the catalytic chain (**1KSN**).

### Biological Functions

In complex with FVa, FXa binds to negatively charged membrane phospholipids in the presence of calcium ions, thus leading to the activation of prothrombin to thrombin. FXa can also activate FVII to FVIIa and can convert the cofactor proteins FV to FVa and FVIII to FVIIIa. The main inhibitors of FXa are antithrombin III, $\alpha_1$-antitrypsin and tissue factor pathway inhibitor (TFPI).

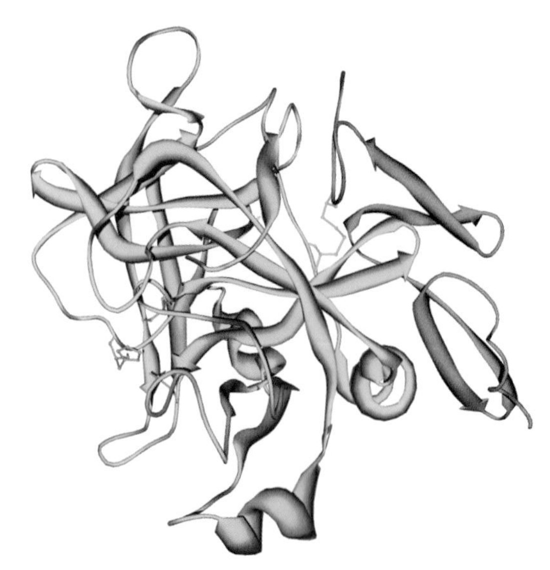

Coagulation factor X

## *Coagulation factor XI (Plasma thromboplastin antecedent; EC 3.4.21.27)*

### Fact Sheet

| | | | |
|---|---|---|---|
| ***Classification*** | Peptidase S1 | ***Abbreviations*** | FXI; FXIa |
| ***Structures/motifs*** | 4 Apple | ***DB entries*** | FA11_HUMAN/P03951 |
| ***MW/length*** | 68 026 Da/607 aa | ***PDB entries*** | 1XX9 |
| ***Concentration*** | 4–6 mg/l | ***Half-life*** | 60–80 h |
| ***PTMs*** | 5 N-CHO | | |
| ***References*** | Badellino and Walsh, 2001, *Biochemistry*, **40**, 7569–7580.<br>Fujikawa *et al.*, 1986, *Biochemistry*, **25**, 2417–2424. | | |

### Description

**Coagulation factor XI** (FXI) is a unique glycoprotein present in plasma as a homodimer. It circulates in complex with HMW kininogen. Together with FXII, prekallikrein and HMW kininogen, it is referred to as a contact-phase activation component. It shares considerable structural similarities with plasma prekallikrein, suggesting that they have evolved from a common ancestral gene.

### Structure

FXI is a disulfide-linked homodimer (disulfide bridge between Cys 321 of the two monomers) glycoprotein (5 % CHO). FXI is activated by FXIIa to the active two-chain FXIa (cleavage of Arg 369–Ile 370). The heavy chain contains four Apple domains and the light chain comprises the serine protease part. The 3D structure of two molecules of the light chain (in blue) in complex with the dimer of the protease inhibitor ecotin (from *E. coli*, in red) reveals a substrate-like interaction (**1XX9**).

### Biological Functions

FXI, FXII, prekallikrein and HMW kininogen are the major factors in the surface-mediated activation pathway. The ternary complex FXI, HMW kininogen and FXIIa activates FXI to FXIa. In the presence of HMW kininogen and $Zn^{2+}$ ions FXIa binds to activated platelets, resulting in the activation of FIX to FIXa. The most potent physiological inhibitors of FXIa are $\alpha_1$-antitrypsin and the complex of antithrombin–heparin.

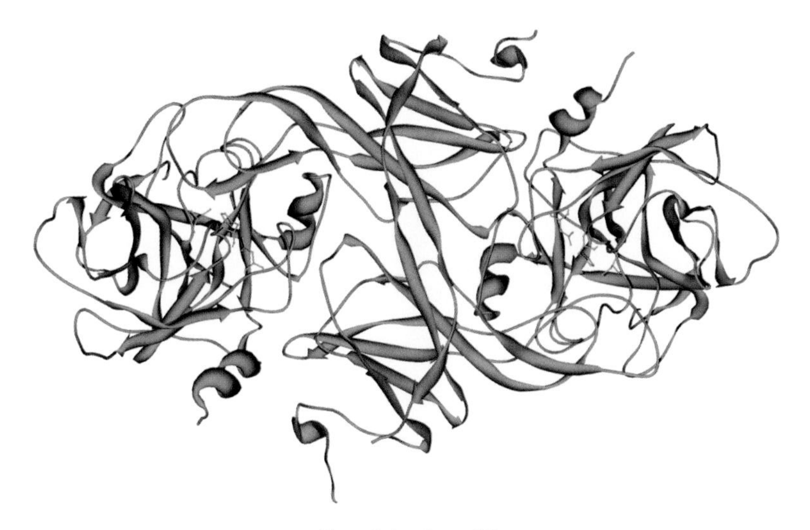

Coagulation factor XI

## *Coagulation factor XII* (*Hagemann factor*; *EC* 3.4.21.38)

### Fact Sheet

| | | | |
|---|---|---|---|
| ***Classification*** | Peptidase S1 | ***Abbreviations*** | FXII; FXIIa; HF |
| ***Structures/motifs*** | 2 EGF-like; 1 kringle; 1 FN1; 1 FN3 | ***DB entries*** | FA12_HUMAN/P00748 |
| ***MW/length*** | 65 760 Da/596 aa | ***PDB entries*** | |
| ***Concentration*** | approx. 24 mg/l | ***Half-life*** | 50–70 h |
| ***PTMs*** | 1 N-CHO; 7 O-CHO (possible) | | |
| ***References*** | Cool and McGillivray, 1987, *J. Biol. Chem.*, **262**, 13 662–13 673. | | |

### Description

**Coagulation factor XII** (FXII) is a multidomain glycoprotein and circulates in blood as a zymogen. Together with FXI, plasma kallikrein and HMW kininogen it is referred to as the contact-phase activation factors.

### Structure

FXII is a single-chain glycoprotein (17 % CHO). In an autocatalytic process surface-bound FXII may generate initial amounts of the two-chain $\alpha$-FXIIa (cleavage of Arg 353–Val 354). The heavy chain contains 1 FN1 and 1 FN3, 2 EGF-like and 1 kringle domain and the light chain comprises the serine protease part. Subsequently, two-chain $\alpha$-FXIIa is converted by kallikrein to two-chain $\beta$-FXIIa (cleavage of Arg 334–Asn 335 and Arg 343–Leu 344). From the former heavy chain (353 aa) in $\alpha$-FXIIa only 9 aa remain in $\beta$-FXIIa, whereas the catalytic chain remains intact. The two chains are linked by a single disulfide bridge (Cys 6 in the mini chain and Cys 114 in the catalytic domain).

### Biological Functions

$\alpha$-FXIIa initiates the intrinsic coagulation cascade by activating FXI to FXIa on surfaces in complex with HMW kininogen. $\alpha$-FXIIa activates plasma prekallikrein to kallikrein, thus initiating the generation of bradykinin and fibrinolysis. The main physiological inhibitors of $\alpha$-FXIIa and $\beta$-FXIIa are plasma protease C1 inhibitor, $\alpha_2$-antiplasmin, antithrombin III and $\alpha_2$-macroglobulin.

# *Coagulation factor XIII (Transglutaminase; EC 2.3.2.13)*

## Fact Sheet

| | | | |
|---|---|---|---|
| ***Classification*** | Transglutaminase | ***Abbreviations*** | FXIII; FXIIIa |
| ***Structures/motifs*** | B: 10 sushi/CCP/SCR | ***DB entries*** | F13A_HUMAN/P00488<br>F13B_HUMAN/P05160 |
| ***MW/length*** | A: 79 245 Da/694 aa<br>B: 73 188 Da/641 aa | ***PDB entries*** | 1GGU |
| ***Concentration*** | A2B2: 22 mg/l | ***Half-life*** | 10–14 days |
| ***PTMs*** | A-chain: N-acetyl-Gly | | |
| ***References*** | Ichinose *et al.*, 1986, *Biochemistry*, **25**, 4633–4638 and 6900–6906.<br>Ichinose *et al.*, 1990, *J. Biol. Chem.*, **265**, 13 411–13 414. | | |

## Description

**Coagulation factor XIII** (FXIII) is present in plasma as a tetrameric zymogen composed of two A-chains and two B-chains. FXIII is activated by thrombin to the active transglutaminase FXIIIa.

## Structure

FXIII is a tetramer of two A-chains and two B-chains that are noncovalently linked. The activation by thrombin releases an N-terminal activation peptide (37 aa) from the A-chain (694 aa) followed by the calcium-dependent dissociation of the B-chains. The B-chain contains 10 sushi/CCP/SCR domains and an **RGD** sequence (597–599 aa) as a cell-binding motif. The shown 3D structure comprises the whole A-chain folded in four sequential domains: The activation peptide AP (in red), the core domain (in blue), barrel 1 (in yellow) and barrel 2 (in green). The active site in the core domain (in pink) and the main calcium binding site (orange ball) are indicated (**1GGU**).

## Biological Functions

FXIIIa is primarily responsible for the crosslinking of fibrin. FXIIIa catalyses the formation of intermolecular isopeptide bonds between the side chains of Gln and Lys, resulting in $\gamma$–$\gamma$ dimmers, $\alpha$-chain polymers, $\alpha$–$\gamma$ heterodimers and $\gamma$-chain multimers of fibrin. FXIIIa also crosslinks the $\alpha_2$-plasmin inhibitor (Gln 2) to the $\alpha$-chain of fibrin (Lys 303), thus preventing the premature degradation of fibrin by plasmin.

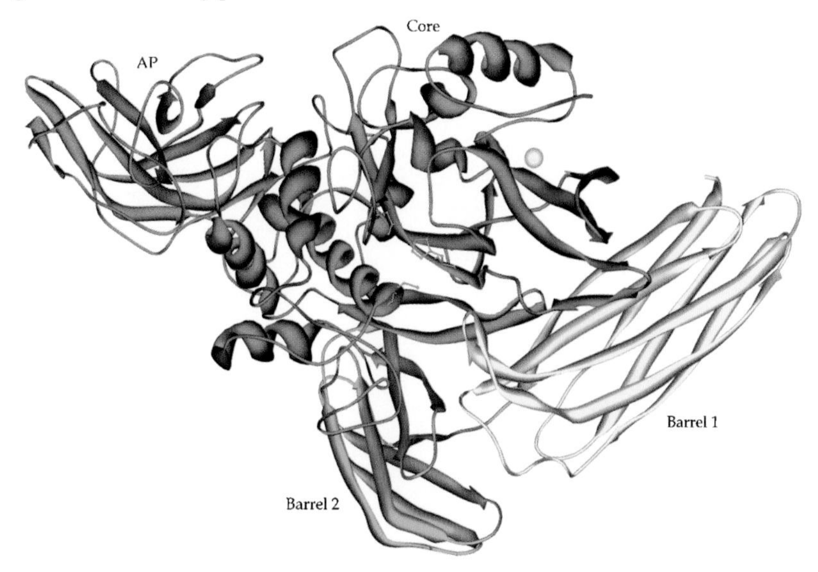


Coagulation factor XIII A-chain

# *Fibrinogen (Coagulation factor I)*

## Fact Sheet

***Classification***

***Structures/motifs*** β-chain: 1 fibrinogen C-terminal
γ-chain: 1 fibrinogen C-terminal

***MW/length*** α-chain: 92 877 Da/847 aa
β-chain: 52 315 Da/461 aa
γ-chain: 48 883 Da/427 aa

***Concentration*** 2–4 g/l

***PTMs*** α-chain: 1 P-Ser (aa 3)
β-chain: 1 Pyro-Glu
γ-chain: 2 S-Tyr (418, 422 aa); 1isopeptide bond (Gln 398 – Lys 406)

***References***
Chung *et al.*, 1990, *Adv. Exp. med. Biol.*, **281**, 39–48.
Doolittle, 1984, *Annu. Rev. Biochem.*, 53, 195–229.
Everse *et al.*, 1999, *Biochemistry*, **38**, 2941–2946.
Rixon *et al.*, 1985, *Biochemistry*, **24**, 2077–2086.

***Abbreviations*** FI; Fg

***DB entries*** FIBA_HUMAN/P02671
FIBB_HUMAN/P02675
FIBG_HUMAN/P02679

***PDB entries*** 1FZG

***Half-life*** 3.5–4.5 days

## Description

**Fibrinogen** (FI) is a large glycoprotein (3 % CHO) synthesised in the liver and is present in plasma as a dimer of three nonidentical chains. Fibrinogen is essential for blood coagulation and concentrations below 0.5 g/l may cause severe coagulation disorders.

## Structure

Fibrinogen (approx. 340 kDa) is a dimer of three nonidentical chains: Aα-, Bβ- and γ-chains. All chains are linked by disulfide bridges and the β-and γ-chains contain one fibrinogen C-terminal domain. The two heterotrimers are in head-to-head conformation with the N-termini in a small central domain. Thrombin converts fibrinogen to the fibrin monomer by releasing fibrinopeptide A (16 aa) from the Aα-chains (cleavage of Arg 16–Gly 17) and fibrinopeptide B (14 aa) from the Bβ-chains (cleavage of Arg 14–Gly 15). The Aα-chains contain two **RGD** sequences (95–97 aa and 572–574 aa) as a cell-binding motif. The shown 3D structure represents the dimer of the plasmin-derived fragment D (approx. 200 kDa) in complex with the tetrapeptide GHRP-$NH_2$. The chains have the following colours: $\alpha_1$, $\alpha_2$, yellow; $\beta_1$, $\beta_2$, green; $\gamma_1$, $\gamma_2$, blue (**1FZG**).

## Biological Functions

The removal of the fibrinopeptides A and B uncovers the N-terminal polymerisation sites in the α- and β-chains responsible for the formation of the soft clot. The soft clot is converted to the hard clot by various crosslinking reactions: (1) crosslinks between the α-chain (Gly 17–Arg 19) and the γ-chain (Thr 374–Glu 396); (2) crosslinks of β-chains (Gly 15–Arg 17) to another fibrin monomer; (3) the formation of several isopeptide bonds between different α-chains and one isopeptide bond between the γ-chains (Gln 398 and Lys 406), all catalysed by the transglutaminase FXIIIa.

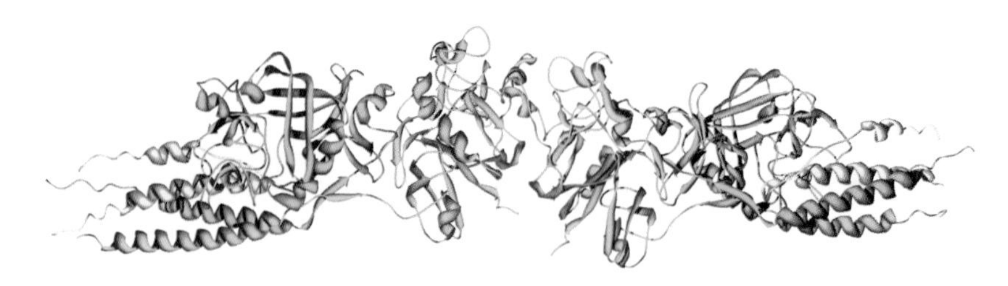

Fibrinogen

# *Fibronectin (Cold-insoluble globulin)*

## Fact Sheet

| | | | |
|---|---|---|---|
| ***Classification*** | | ***Abbreviations*** | FN |
| ***Structures/motifs*** | 12 FN1, 2 FN2, 16 FN3 | ***DB entries*** | FINC_HUMAN/P02751 |
| ***MW/length*** | 259 545 Da/2355 aa | ***PDB entries*** | 1QO6; 1FNF |
| ***Concentration*** | approx. 300 mg/l | ***Half-life*** | 1–2 days |
| ***PTMs*** | 7 N-CHO; 2 S-Tyr (potential); 1 P-Ser (aa 2353); 1 Pyro-Glu | | |
| ***References*** | Kornblihtt *et al.*, 1985, *EMBO J.*, **4**, 1755–1759. | | |

## Description

Plasma **fibronectin** (FN) is a large glycoprotein synthesised in hepatocytes and circulates in plasma as heterodimer or as multimers of alternatively spliced variants linked by two disulfide bridges in the C-terminal region.

## Structure

FN is a single-chain glycoprotein (approx. 5 % CHO) that contains 12 FN1, two FN2 and 16 FN3 domains. The N-terminus is modified (Pyro-Glu), it has two potential sulfation sites at Tyr 845 and 850 and it is phosphorylated at Ser 2353. It contains an **RGD** sequence (1493–1495 aa) as a cell-binding motif. There are three probable interchain isopeptide crosslinking sites.

## Biological Functions

FNs bind to cell surfaces and interact, among others, with fibrin, collagens, actin and heparin. FNs are involved in cell adhesion, cell motility, tissue turnover and wound healing.

# $\beta_2$-*Glycoprotein I (Apolipoprotein H)*

## Fact Sheet

| | | | |
|---|---|---|---|
| ***Classification*** | | ***Abbreviations*** | Apo-H |
| ***Structures/motifs*** | 4 sushi/CCP/SCR; 1 sushi-like | ***DB entries*** | APOH_HUMAN/P02749 |
| ***MW/length*** | 36 255 Da/326 aa | ***PDB entries*** | 1C1Z |
| ***Concentration*** | approx. 220 mg/l | ***Half-life*** | |
| ***PTMs*** | 4 N-CHO; 1 O-CHO | | |
| ***References*** | Lozier *et al.*, 1984, *Proc. Natl. Acad. Sci. USA*, **81**, 3640–3644.<br>Schwarzenbacher *et al.*, 1999, *EMBO J.*, **18**, 6228–6239. | | |

## Description

**$\beta_2$-Glycoprotein** I (Apo-H) is a plasma glycoprotein synthesised in the liver. It seems to be involved in the regulation of the intrinsic coagulation cascade.

## Structure

Apo-H is a single-chain glycoprotein containing four sushi (CCP/SCR) domains and a C-terminal sushi-like region. The 3D structure of Apo-H contains four sushi and one sushi-like domain (in violet), forming an elongated J-shaped molecule (**1C1Z**).

## Biological Functions

Apo-H binds to negatively charged components such as heparin, phospholipids and dextran sulfate. It seems to be involved in the regulation of the intrinsic coagulation cascade by binding to phospholipids on the surface of damaged cells.

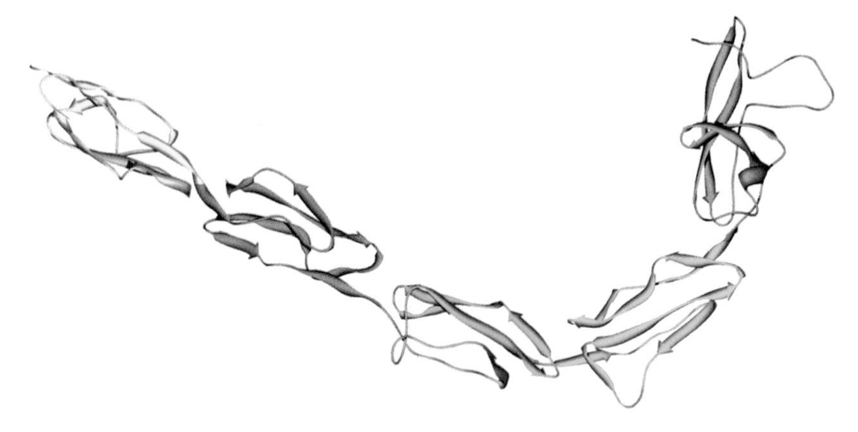

$\beta_2$-Glycoprotein I

# *Histidine-rich glycoprotein*

## Fact Sheet

| | | | |
|---|---|---|---|
| ***Classification*** | | ***Abbreviations*** | HRG |
| ***Structures/motifs*** | 2 cystatin-like; 12 repeats (GHHPH) | ***DB entries*** | HRG_HUMAN/P04196 |
| ***MW/length*** | 57 660 Da/507 aa | ***PDB entries*** | |
| ***Concentration*** | 60–140 mg/l | ***Half-life*** | 2.9 days |
| ***PTMs*** | 5 N-CHO (potential) | | |
| ***References*** | Koide *et al.*, 1986, *Biochemistry*, **25**, 2220–2225. | | |

## Description

**Histidine-rich glycoprotein** (HRG) is a plasma glycoprotein synthesised in the liver and present in platelets.

## Structure

HRG is a single-chain glycoprotein (14 % CHO) with an unusually high content of His and Pro (each 13 %). The N-terminal region contains two cystatin-like domains and the His-rich central part contains 12 tandem repeats of five residues (**GHHPH**). On either side of the His-rich region are two Pro-rich regions of approx. 30 residues.

## Biological Functions

Although the primary physiological role of HRG is still not yet clear it interacts with heparin, thrombospondin and the LBS of plasminogen and seems to play a role in the intrinsic pathway.

## *Kininogen ($\alpha_2$-Thiolproteinase inhibitor)/Bradykinin*

### Fact Sheet

| | | | |
|---|---|---|---|
| ***Classification*** | | ***Abbreviations*** | HMW; LMW |
| ***Structures/motifs*** | 3 cystatin-like | ***DB entries*** | KNG1_HUMAN/P01042 |
| ***MW/length*** | HMW: 69 885 Da/626 aa<br>LMW: 45 822 Da/409 aa<br>Bradykinin: 1060 Da/9 aa<br>Lysyl-bradykinin: 1188 Da/10 aa | ***PDB entries*** | |
| ***Concentration*** | HMW: 55–93 mg/l<br>LMW: 109–161 mg/l | ***Half-life*** | 2–3 days |
| ***PTMs*** | 3 N-CHO; 9 O-CHO; 1 OH-Pro (aa 365);<br>1 Pyro-Glu | | |
| ***References*** | Rawlings *et al.*, 1990, *J. Mol. Evol.*, **30**, 60–71.<br>Takagi *et al.*, 1985, *J. Biol. Chem.*, **260**, 8601–8609. | | |

### Description

There are two forms of **kininogen**, **HMW-kininogen** and **LMW-kininogen**, both synthesised in the liver. They are the parent proteins of bradykinin, a nonapeptide responsible for vasodilatation and in consequence the decrease of blood pressure.

### Structure

HMW-kininogen (626 aa) and LMW-kininogen (409 aa) are two forms of kininogen containing a heavy chain (362 aa) region with three cystatin-like domains, an N-terminal pyro-Glu, a kinin segment (9 aa) with an OH-Pro residue (aa 365) and two light chain forms, the LMW light chain (38 aa) and the HMW light chain (255 aa). Kallikrein releases the vasoactive nonapeptide bradykinin and the decapeptide lysyl-bradykinin from kininogen. Eventually, aminopeptidases release the N-terminal Lys in Lys-bradykinin generating bradykinin.

### Biological Functions

Kininogens are multifunctional proteins that exhibit several biological roles, e.g. as precursors of the vasoactive nonapeptide bradykinin, as cofactors of the blood coagulation pathway and as thiol protease inhibitors in inflammatory processes.

## *Plasma (pre)kallikrein (Fletcher factor; Kininogenin; EC 3.4.21.34)*

### Fact Sheet

| | | | |
|---|---|---|---|
| ***Classification*** | Peptidase S1 | ***Abbreviations*** | PK |
| ***Structures/motifs*** | 4 Apple | ***DB entries*** | KLKB1_HUMAN/P03952 |
| ***MW/length*** | 69 204 Da/619 aa | ***PDB entries*** | 2ANW |
| ***Concentration*** | 35–50 mg/l | ***Half-life*** | |
| ***PTMs*** | 5 N-CHO | | |
| ***References*** | Chung *et al.*, 1986, *Biochemistry*, **25**, 2410–2417<br>Tang *et al.*, 2005, *J. Biol. Chem.*, **280**, 41077–41089. | | |

### Description

**Plasma prekallikrein** (PK) is a glycoprotein synthesised in the liver as a zymogen and is distinct from tissue kallikrein. It circulates in complex with HMW kininogen and participates in the intrinsic activation pathway. PK shares considerable structural similarities with FXI.

### Structure

PK is a single-chain glycoprotein (15 % CHO) that is activated by FXIIa to two-chain kallikrein (cleavage of Arg 371–Ile 372). The heavy chain contains four Apple domains and the light chain comprises the serine protease part. The 3D structure of the serine protease domain has been solved by X-ray diffraction and exhibits the classical chymotrypsin-like conformation (**2ANW**).

### Biological Functions

On negatively charged surfaces PK together with FXII and HMW kininogen initiates the intrinsic pathway system and generates the vasoactive peptide bradykinin from HMW kininogen. FXIIa converts prekallikrein to kallikrein and reciprocally kallikrein activates FXII to FXIIa. The main inhibitors are plasma protease C1 inhibitor and $\alpha_2$-macroglobulin.

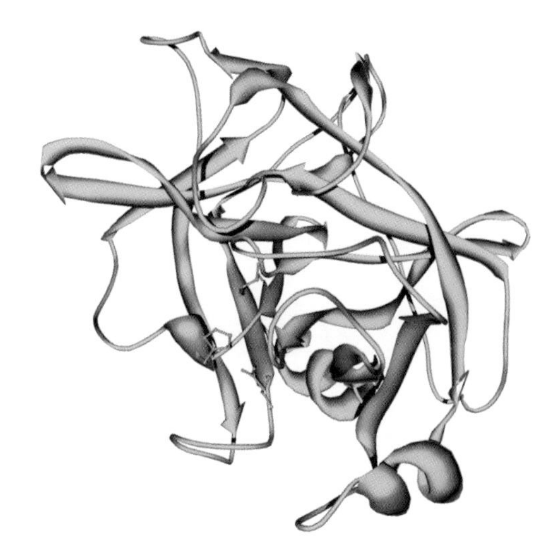

Plasma kallikrein

# *Plasminogen (EC 3.4.21.7)*

## Fact Sheet

| | | | |
|---|---|---|---|
| ***Classification*** | Peptidase S1 | ***Abbreviations*** | Pg |
| ***Structures/motifs*** | 1 PAN; 5 kringle | ***DB entries*** | PLMN_HUMAN/P00747 |
| ***MW/length*** | 88 432 Da/791 aa | ***PDB entries*** | 1B2I; 1KI0; 1BML |
| ***Concentration*** | 70–200 mg/l | ***Half-life*** | 2.2 h |
| ***PTMs*** | 1 N-CHO; 2 O-CHO; 1 P-Ser (partial, 578 aa) | | |
| ***References*** | Petersen *et al.*, 1990, *J. Biol. Chem.*, **265**, 6104–6111.<br>Ponting *et al.*, 1992, *Biochim. Biophys. Acta*, **1159**,155–161.<br>Wang *et al.*, 1998, *Science*, **281**, 1662–1665. | | |

## Description

**Plasminogen** (Pg) is a multidomain glycoprotein synthesised in the liver and in the kidney and is the key component of the fibrinolytic system. The active form plasmin is responsible for cleaving the fibrin polymer network in blood clots.

## Structure

Pg is a single-chain glycoprotein (2 % CHO) activated by plasminogen activators tPA and uPA to two-chain plasmin by cleavage of Arg 561–Val 562 and concomitant release of the N-terminal preactivation peptide (77 aa). The heavy chain contains five kringle domains and the light chain comprises the serine protease part with a partial phosphorylation site at Ser 578. Electron microscopy and small-angle neutron scattering are indicative for a compact, spiral-like structure. The 3D structure shows the catalytic domain of plasminogen (in green) containing an N-terminal 20 residue extension from the heavy chain (250 aa: Ala 542–Asn 791) in complex with streptokinase (362 aa: Ser 12–Pro 373; in red) (**1BML**). In the catalytic site of plasmin (in pink) Ser is mutated to Ala (in blue).

## Biological Functions

Plasmin cleaves the fibrin polymers in the blood clots, thus leading to fibrinolysis, but also acts as a proteolytic factor in a variety of other important physiological processes, like wound healing, tissue remodelling, angiogenesis, embryogenesis and pathogen and tumour cell invasion. The main physiological inhibitor of plasmin is $\alpha_2$-antiplasmin.

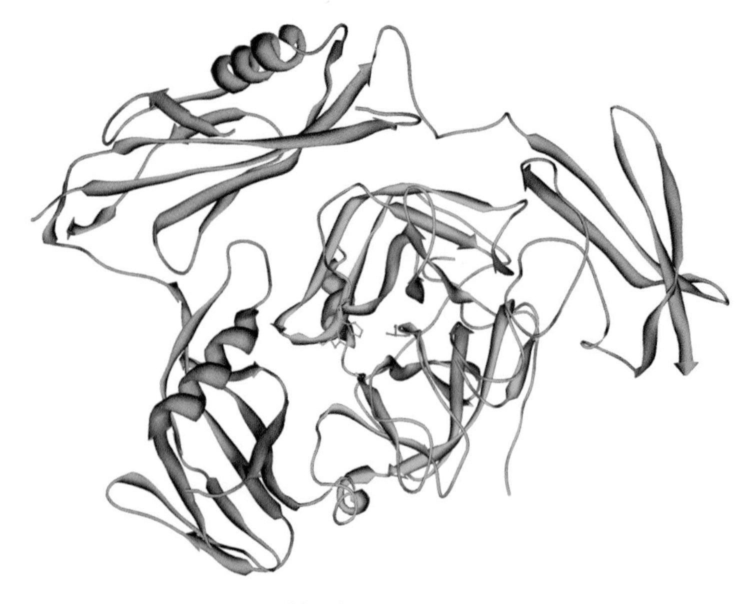

Plasmin streptokinase

## *Prothrombin (Coagulation factor II; EC 3.4.21.5)*

### Fact Sheet

| | | | |
|---|---|---|---|
| ***Classification*** | Peptidase S1 | ***Abbreviations*** | FII |
| ***Structures/motifs*** | 2 kringle; 1 Gla | ***DB entries*** | THRB_HUMAN/P00734 |
| ***MW/length*** | 65 308 Da/579 aa | ***PDB entries*** | 1A3B |
| ***Concentration*** | approx. 110 mg/l | ***Half-life*** | 60 h |
| ***PTMs*** | 3 N-CHO; 10 $\gamma$-Gla | | |
| ***References*** | Degen *et al.*, 1987, *Biochemistry*, **26**, 6165–6177.<br>Zdanov *et al.*, 1993, *Proteins*, **17**, 252–265. | | |

### Description

**Prothrombin** (FII) is the key component of the blood coagulation system. Prothrombin is synthesised in the liver in a vitamin K-dependent way and circulates in blood as a zymogen. It shares considerable structural similarities with other vitamin K-dependent proteins in blood coagulation, like FVII, FIX, FX and proteins C, S and Z, suggesting an evolution from a common ancestral gene.

### Structure

Prothrombin is a single-chain multidomain glycoprotein. It contains a Gla domain, two kringle structures and the protease domain. Prothrombin is activated by FXa to two-chain $\alpha$-thrombin (cleavage of Arg 284–Thr 285 and Arg 320–Ile 321). Fragment 1 (Ala 1–Arg 155) comprises the Gla-domain and kringle 1, fragment 2 (Ser 156–Arg 284) the kringle 2 domain. Two-chain $\alpha$-thrombin contains a light chain (36 aa) and the heavy chain with the serine protease part linked by a single disulfide bridge (Cys 9 in the light chain and Cys 119 in the catalytic domain in green). The 3D structure of $\alpha$-thrombin consists of the catalytic chain (259 aa, in blue) containing three $\alpha$-helices and 19 $\beta$-strands and the light chain (36 aa, in red) with one $\alpha$-helix (**1A3B**).

### Biological Functions

Thrombin releases the fibrinopeptides A and B from fibrinogen, thus exposing the crosslinking sites and resulting in fibrin polymerisation. Prothrombin binds via the Gla domain in fragment 1 to phospholipids of damaged cell membranes and platelets. Fragment 2 mediates binding of prothrombin to the cofactor FVa. Naturally occurring inhibitors of thrombin are antithrombin III, heparin cofactor II, $\alpha_1$-antitrypsin, $\alpha_2$-macroglobulin and hirudin.

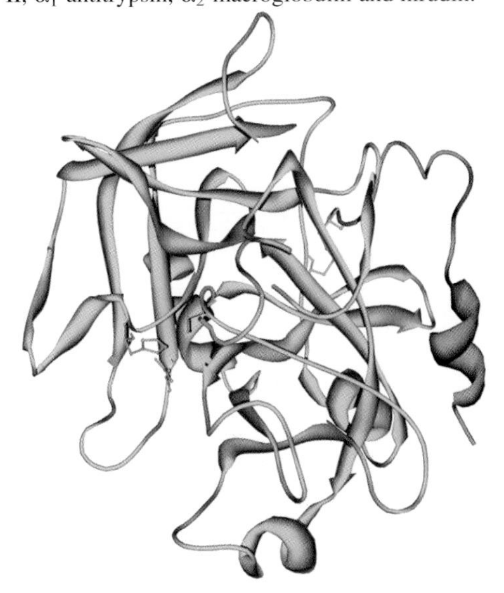

Alpha Thrombin

## *Renin (Angiotensinogenase; EC 3.4.23.15)*

### Fact Sheet

| | | | |
|---|---|---|---|
| ***Classification*** | Peptidase A1 | ***Abbreviations*** | |
| ***Structures/motifs*** | | ***DB entries*** | RENI_HUMAN/P00797 |
| ***MW/length*** | 37 236 Da/340 aa | ***PDB entries*** | 2REN |
| ***Concentration*** | | ***Half-life*** | |
| ***PTMs*** | 2 N-CHO | | |
| ***References*** | Imai *et al.*, 1983, *Proc. Natl. Acad. Sci. USA*, **80**, 7405–7409.<br>Rawlings *et al.*, 1995, *Meth. Enzymol.*, **248**, 105–120.<br>Sielecki *et al.*, 1989, *Science*, **243**, 1346–1351. | | |

### Description

**Renin** is a plasma glycoprotein responsible for the generation of the decapeptide angiotensin I from angiotensinogen.

### Structure

Renin is a single-chain glycoprotein generated from its proform by removing a 43 residue activation peptide. The active centre of renin is formed by Asp 38 and 226 (in pink). The 3D structure shows the activated form of renin typical for aspartyl proteases (**2REN**).

### Biological Functions

Renin is a highly specific endopeptidase with the only known function to generate the decapeptide angiotensin I from angiotensinogen (cleavage of Leu 10–Val 11).

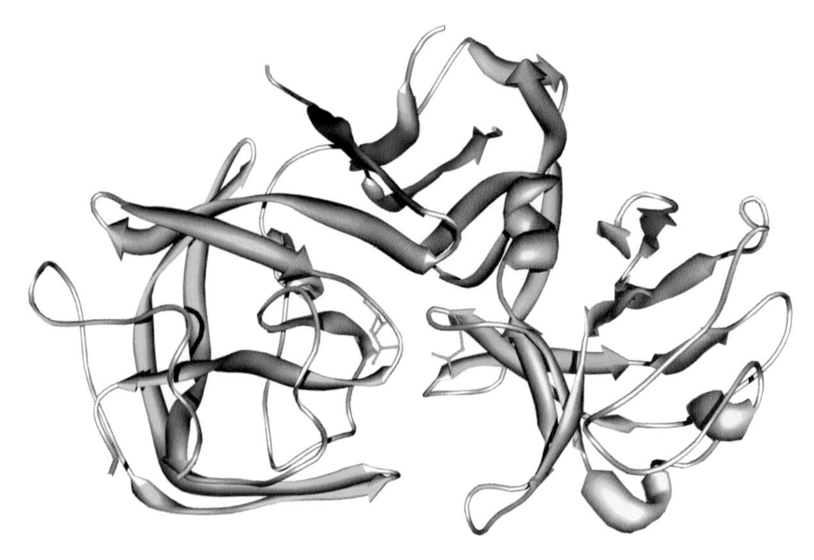

Renin

## *Tetranectin (Plasminogen-kringle 4 binding protein)*

### Fact Sheet

| | | | |
|---|---|---|---|
| ***Classification*** | | ***Abbreviations*** | TN |
| ***Structures/motifs*** | 1 C-type lectin | ***DB entries*** | TETN_HUMAN/P05452 |
| ***MW/length*** | 20 169 Da/181 aa | ***PDB entries*** | 1HTN |
| ***Concentration*** | 10–12 mg/l | ***Half-life*** | |
| ***PTMs*** | 1 O-CHO | | |
| ***References*** | Berglund and Petersen, 1992, *FEBS Lett.*, **309**, 15–19.<br>Nielsen *et al.*, 1997, *FEBS Lett.*, **412**, 388–396. | | |

### Description

**Tetranectin** (TN) is a glycoprotein circulating in blood as a noncovalently linked homotrimer. It binds plasminogen and apolipoprotein(a).

### Structure

Tetranectin is a single-chain glycoprotein containing a C-type lectin domain and three intrachain disulfide bridges (Cys 50–Cys 60, Cys 77–Cys 176, Cys 152–Cys 168). The 3D structure of TN shows an $\alpha$-helical coiled coil lectin structure (**1HTN**).

### Biological Functions

Tetranectin binds to kringle 4 in plasminogen and to kringle 4 in apolipoprotein (a). Its main biological function is still unknown.

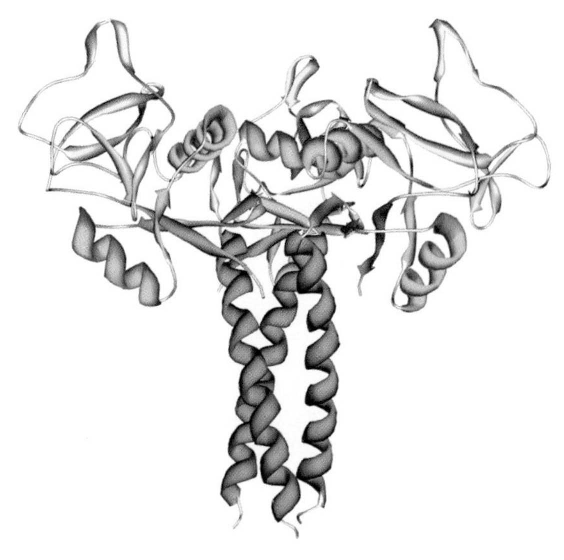

Tetranectin

## *Thrombin-activatable fibrinolysis inhibitor (Carboxypeptidase B2; EC 3.4.17.20)*

### Fact Sheet

| | | | |
|---|---|---|---|
| ***Classification*** | Peptidase M14 | ***Abbreviations*** | TAFI |
| ***Structures/motifs*** | | ***DB entries*** | CBPB2_HUMAN/Q96IY4 |
| ***MW/length*** | 35 788 Da/309 aa | ***PDB entries*** | 1KWM |
| ***Concentration*** | 2–5 mg/l | ***Half-life*** | minutes |
| ***PTMs*** | 5 N-CHO (potential) | | |
| ***References*** | Eaton *et al.*, 1991, *J. Biol. Chem.*, **266**, 21833–21838. | | |

### Description

Carboxypeptidase B2 is better known as **thrombin-activatable fibrinolysis inhibitor** (TAFI) and is synthesised in the liver. TAFI regulates biologically active peptides like kinins and anaphylatoxins.

### Structure

TAFI is a single-chain glycoprotein generated by removing a 92 residue activation peptide from the N-terminal region of the proform. TAFI carries a binding site for zinc ions and the active site consists of Glu 271.

### Biological Functions

TAFI removes C-terminal Arg or Lys residues from biologically active peptides such as kinins and anaphylatoxins, thus regulating their activity. It also cleaves C-terminal Lys and Arg in the plasminogen-binding region of fibrin.

## *Thrombomodulin (Fetomodulin)*

### Fact Sheet

| | | | |
|---|---|---|---|
| ***Classification*** | Type I membrane | ***Abbreviations*** | TM |
| ***Structures/motifs*** | 6 EGF-like; 1 C-type lectin | ***DB entries*** | TRBM_HUMAN/P07204 |
| ***MW/length*** | 58 635 Da/557 aa | ***PDB entries*** | 1DX5 |
| ***Concentration*** | approx. 20 mg/l | ***Half-life*** | |
| ***PTMs*** | 5 N-CHO (potential); 1 O-CHO; 1 OH-Asn (324 aa) | | |
| ***References*** | Jackman *et al.*, 1987, *Proc. Natl. Acad. Sci. USA*, **84**, 6425–6429. | | |

### Description

**Thrombomodulin** (TM) is a cell surface transmembrane glycoprotein but it also circulates in blood.

### Structure

TM is a single-chain type I membrane glycoprotein that contains six EGF-like and one C-type lectin domain and one OH-Asn (324 aa). The 24 residue long presumed transmembrane segment is located in the C-terminal region (Gly 498–Leu 521).

### Biological Functions

TM forms a 1:1 complex with thrombin which converts protein C to its activated form. Activated protein C inactivates FVa and FVIIIa. Thrombin bound to thrombomodulin is a less efficient procoagulant. Tumour necrosis factor, interleukin-1 and endotoxin decrease the cell surface activity of thrombomodulin.

# *Thrombospondin-1*

## Fact Sheet

| | | | |
|---|---|---|---|
| ***Classification*** | Thrombospondin | ***Abbreviations*** | TSP-1 |
| ***Structures/motifs*** | 3 EGF-like; 1 TSPN; 1 TSPC; 3 TSP1; 7 TSP3; 1 VWFC | ***DB entries*** | TSP1_HUMAN/P07996 |
| ***MW/length*** | 127 525 Da/1152 aa | ***PDB entries*** | 1ZA4; 1UX6 |
| ***Concentration*** | 100–160 mg/l | ***Half-life*** | |
| ***PTMs*** | 4 N-CHO (potential); 3 O-CHO; 4 C-CHO (Trp) | | |
| ***References*** | Hofsteenge *et al.*, 2001, *J. Biol. Chem.*, **276**, 6485–6498. Lawler and Hynes, 1986, *J. Cell. Biol.*, **103**, 1635–1648. | | |

## Description

**Thrombospondin-1** (TSP1) is a large multidomain glycoprotein and circulates in blood as disulfide-linked homotrimer (420 kDa). It can bind large amounts of calcium ions.

## Structure

TSP1 shows a high similarity with TSP2 and they form homo- and heterotrimers linked by disulfide bridges. TSP1/TSP2 contain many different types of domains: three EGF-like, one TSP C-terminal and TSP N-terminal, three TSP1, seven TSP3 and one VWFC. In addition, they contain an **RGD** sequence (908–910 aa) as a cell-binding motif. The closely related proteins TSP3 and TSP4 lack the TPSP type 1 and VWFC domains.

## Biological Functions

TSP1 is an adhesive glycoprotein that mediates cell-to-cell and cell-to-matrix interactions. It can bind many types of proteins like plasmin, fibrinogen, fibronectin, laminin, collagen type V and various types of integrins.

## *Tissue factor (Coagulation factor III; Thromboplastin)*

### Fact Sheet

| | | | |
|---|---|---|---|
| ***Classification*** | Tissue factor | ***Abbreviations*** | TF |
| ***Structures/motifs*** | 3 WKS motifs; 2 FN3 | ***DB entries*** | TF_HUMAN/P13726 |
| ***MW/length*** | 29 593 Da/263 aa | ***PDB entries*** | 2HFT |
| ***Concentration*** | | ***Half-life*** | |
| ***PTMs*** | Palmitoylated at Cys 245; 2 N-CHO | | |
| ***References*** | Andrews *et al.*, 1991, *Gene*, **98**, 265–269.<br>Mackman *et al.*, 1989, *Biochemistry*, **28**, 175–1762.<br>Muller *et al.*, 1996, *J. Mol. Biol.*, **256**, 144–159. | | |

### Description

**Tissue factor** (TF) is the only membrane-bound factor in blood coagulation and is usually not present in circulation. TF is an essential cofactor for the initiation of blood coagulation.

### Structure

TF is a type I membrane glycoprotein with one potential transmembrane segment (Ile 220–Leu 242). TF contains two FN3 domains with two disulfide bridges (Cys 49–Cys 57, Cys 186–Cys 209), three **WKS** motifs and is palmitoylated at Cys 245. The 3D structure shows the extracellular region of TF consisting of two FN3 domains characterised by three α-helices and 17 β-strands (**2HFT**).

### Biological Functions

In TF-initiated coagulation FXa is generated directly by the action of the TF–FVIIa complex and indirectly by activation of FIX to FIXa and subsequent activation of FX to FXa in the presence of FVIII. The tissue factor pathway inhibitor (TFPI) efficiently inhibits TF in the quaternary complex TF–FVIIa–FXa.

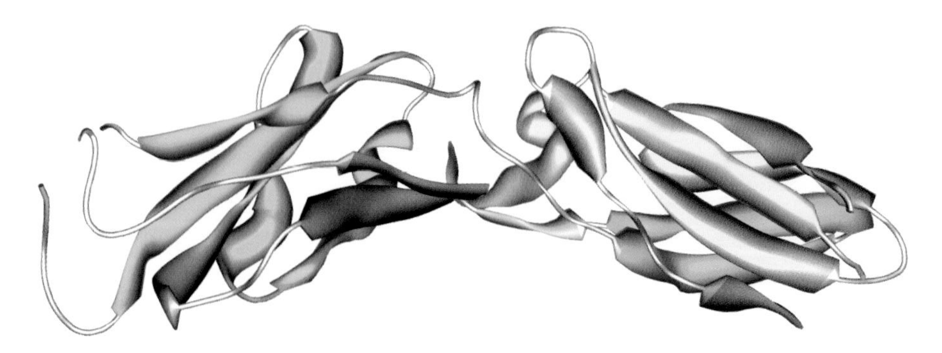

Tissue factor

## *Tissue-type plasminogen activator (EC 3.4.21.68)*

### Fact Sheet

| | | | |
|---|---|---|---|
| ***Classification*** | Peptidase S1 | ***Abbreviations*** | tPA |
| ***Structures/motifs*** | 1 EGF-like; 2 kringle; 1 FN1 | ***DB entries*** | TPA_HUMAN/P00750 |
| ***MW/length*** | 59 042 Da/527 aa | ***PDB entries*** | 1PK2; 1TPM; 1BDA |
| ***Concentration*** | 5–10 mg/l | ***Half-life*** | approx. 6 min |
| ***PTMs*** | 3 N-CHO (only in variants); 1 O-CHO | | |
| ***References*** | Friezner Degen *et al.*, 1986, *J. Biol. Chem.*, **261**, 6972–6985<br>Renatus *et al.*, 1997, *EMBO J.*, **16**, 4797–4805. | | |

### Description

**Tissue-type plasminogen activator** (tPA) is a multidomain glycoprotein synthesised and secreted by endothelial cells. Together with the urokinase-type plasminogen activator it is the main activator of the key component of fibrinolysis, plasminogen.

### Structure

The partially active single-chain tPA is converted to the fully active, two-chain tPA (cleavage of Arg 275–Ile 276) by plasmin, kallikrein or FXa. The A-chain contains one EGF-like, two kringle and one FN1 domain and the B-chain comprises the serine protease domain. The 3D structure of the catalytic domain with an N-terminal 13 residues extension from the A-chain (265 aa: Thr 263–Pro 527 in single-chain tPA) in complex with dansyl-EGR-chloromethyl ketone (in red) shows a salt bridge between Lys 429 and Asp 477 promoting an active conformation in single-chain tPA (**1BDA**).

### Biological Functions

Together with uPA, tPA activates plasminogen to plasmin (cleavage of Arg 561–Val 562), thus leading to the dissolution of fibrin clots. In controlling plasmin-mediated proteolysis it plays an important role in tissue remodelling and degradation, cell migration and other pathophysiological processes. The main physiological inhibitor is the plasminogen activator inhibitor-1 (PAI-1).

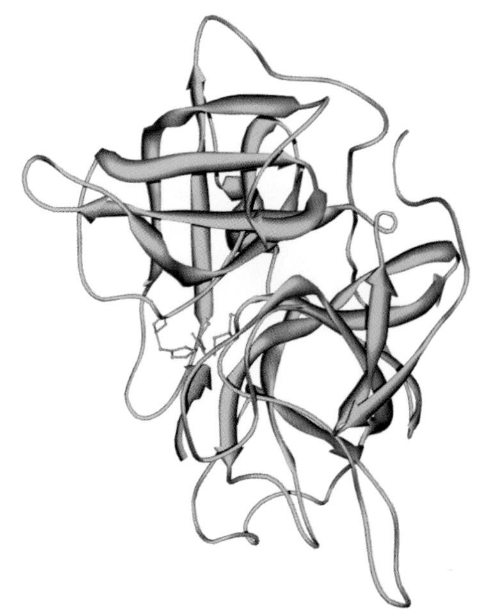

Tissue-type plasminogen activator

## *Urokinase-type plasminogen activator (EC 3.4.21.73)*

### Fact Sheet

| | | | |
|---|---|---|---|
| ***Classification*** | Peptidase S1 | ***Abbreviations*** | uPA |
| ***Structures/motifs*** | 1 EGF-like; 1 kringle | ***DB entries*** | UROK_HUMAN/P00749 |
| ***MW/length*** | 46 386 Da/411 aa | ***PDB entries*** | 1GJA |
| ***Concentration*** | approx. 12 mg/l | ***Half-life*** | |
| ***PTMs*** | 1 N-CHO; 2 P-Ser (aa 138 and 303) | | |
| ***References*** | Katz *et al.*, 2001, *Chem. Biol.*, **8**, 1107–1121.<br>Riccio *et al.*, 1985, *Nucleic Acids Res.*, **13**, 2759–2771. | | |

### Description

**Urokinase-type plasminogen activator** (uPA) is a multidomain glycoprotein. Together with tissue-type plasminogen activator it is the main activator of the key component of fibrinolysis, plasminogen.

### Structure

The single-chain uPA is converted to two-chain uPA by plasmin autocatalytically. The A-chain (long-chain form with 157 aa) contains one EGF-like and one kringle domain (a short-chain form with 22 aa is devoid of the two domains) and the B-chain comprises the serine protease domain. uPA is phosphorylated at Ser 138 in the A-chain and at Ser 303 in the B-chain. The 3D structure shows the catalytic domain (253 aa, in green) and the mini chain (23 aa, in red) with the expected topology of a trypsin-like serine protease (**1GJA**).

### Biological Functions

Together with tPA, uPA activates plasminogen to plasmin (cleavage of Arg 561–Val 562), thus leading to the dissolution of fibrin clots. In controlling plasmin-mediated proteolysis it plays an important role in tissue remodelling and degradation, cell migration and other pathophysiological processes.

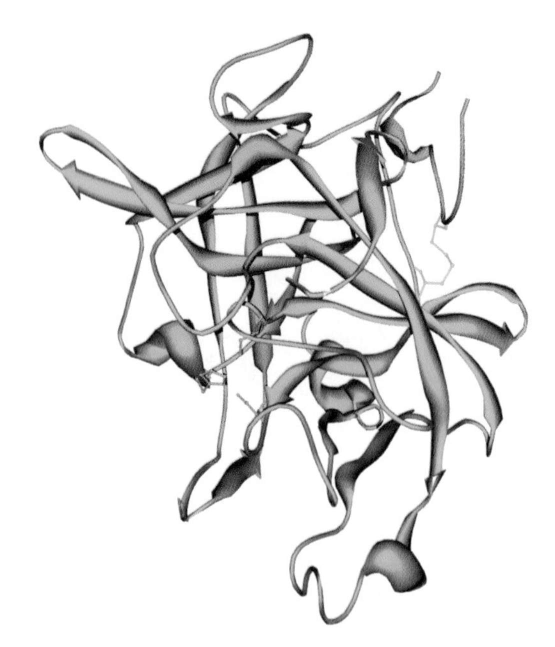

Urokinase-type plasminogen activator

# *Vitamin K-dependent protein C (Coagulation factor XIV; EC 3.4.21.69)*

## Fact Sheet

| | | | |
|---|---|---|---|
| ***Classification*** | Peptidase S1 | ***Abbreviations*** | |
| ***Structures/motifs*** | 2 EGF-like; 1 Gla | ***DB entries*** | PROC_HUMAN/P04070 |
| ***MW/length*** | 47 333 Da/419 aa | ***PDB entries*** | 1AUT |
| ***Concentration*** | 3–5 mg/l | ***Half-life*** | 6–8 h |
| ***PTMs*** | 4 N-CHO; 1 OH-Asp (aa 71); 9 $\gamma$-Gla | | |
| ***References*** | Foster *et al.*, 1985, *Proc. Natl. Acad. Sci. USA*, **82**, 4673–4677.<br>Mather *et al.*, 1996, *EMBO J.*, **15**, 6822–6831. | | |

## Description

**Protein C** is a multidomain plasma glycoprotein predominantly synthesised in the liver. It shares considerable structural similarities with other vitamin K-dependent proteins in blood coagulation like FVII, FIX, FX, prothrombin, proteins S and Z, suggesting an evolution from a common ancestral gene.

## Structure

Protein C is a single-chain glycoprotein that is cleaved into a two-chain molecule with the concomitant release of an N-terminal 10 residue propeptide resulting in a light chain with a Gla-domain, a OH-Asp residue (aa 71) and two EGF-like domains and a heavy chain comprising the serine protease part. Protein C is activated by thrombin releasing a 12 residue activation peptide from the N-terminus of the heavy chain (cleavage of Arg 157–Asp 158 and Arg 169–Leu 170). The shown 3D structure comprises the catalytic domain (250 aa, in blue) and the disulfide-linked C-terminal part of the light chain (114 aa, in red) containing two EGF-like domains in complex with a synthetic tripeptide (**1AUT**).

## Biological Functions

Activated protein C participates in the regulation of blood coagulation by selectively inactivating the two cofactors FVa and FVIIIa in the presence of $Ca^{2+}$ ions and phospholipids. Protein S and FV function as synergistic cofactors in this process. The main physiological inhibitors are the plasma serine protease inhibitor (PAI-3), $\alpha_1$-antitrypsin and $\alpha_2$-macroglobulin.

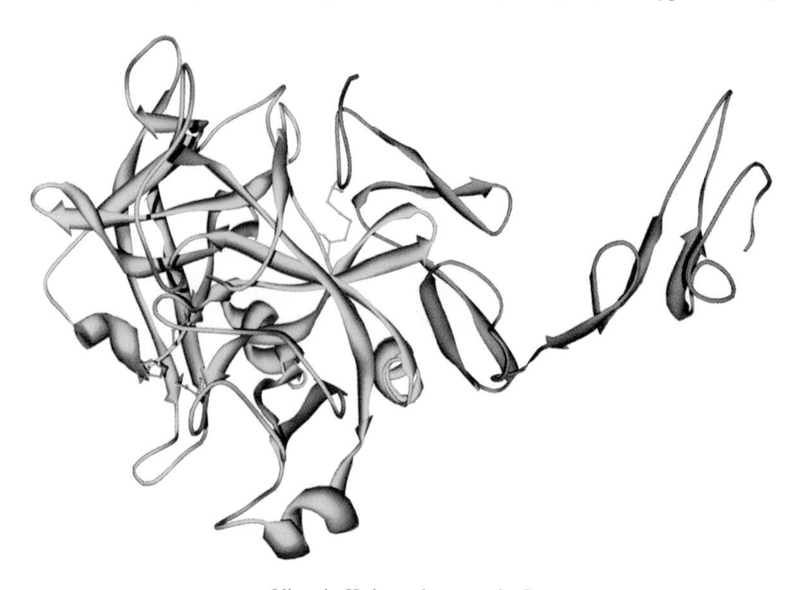

Vitamin K-dependent protein C

## *Vitamin K-dependent protein S*

### Fact Sheet

| | | | |
|---|---|---|---|
| ***Classification*** | | ***Abbreviations*** | |
| ***Structures/motifs*** | 4 EGF-like; 1 Gla; 2 laminin-G like | ***DB entries*** | PROS_HUMAN/P07225 |
| ***MW/length*** | 70 645 Da/635 aa | ***PDB entries*** | 1Z6C |
| ***Concentration*** | 17–33 mg/l | ***Half-life*** | |
| ***PTMs*** | 3 N-CHO (possible); 1 OH-Asp (95 aa); 11 $\gamma$-Gla | | |
| ***References*** | Schmidel *et al.*, 1990, *Biochemistry*, **29**, 7845–7852. | | |

### Description

**Protein S** is a multidomain plasma glycoprotein synthesised in the liver and present in platelets. It shares considerable structural similarities with other vitamin K-dependent proteins in blood coagulation like FVII, FIX, FX, prothrombin, proteins C and Z, suggesting an evolution from a common ancestral gene.

### Structure

Protein S is a single-chain glycoprotein containing an N-terminal Gla domain, one OH-Asp (95 aa), four EGF-like and two laminin G-like domains.

### Biological Functions

Protein S is involved in the regulation of blood coagulation as a cofactor to activated protein C involved in the degradation of FVa and FVIIIa.

## *Vitamin K-dependent protein Z*

### Fact Sheet

| | | | |
|---|---|---|---|
| ***Classification*** | Peptidase S1 | ***Abbreviations*** | |
| ***Structures/motifs*** | 2 EGF-like; 1 Gla | ***DB entries*** | PROZ_HUMAN/P22891 |
| ***MW/length*** | 40 365 Da/360 aa | ***PDB entries*** | |
| ***Concentration*** | | ***Half-life*** | |
| ***PTMs*** | 1 OH-Asp (aa 64); 5 N-CHO (potential); 5 O-CHO (potential); 13 $\gamma$-Gla | | |
| ***References*** | Ichinose *et al.*, 1990, *Biochem. Biophys. Res. Commun.*, **172**, 1139–1144. | | |

### Description

**Protein Z** is associated with haemostasis by binding thrombin. It shares considerable structural similarities with other vitamin K-dependent proteins in blood coagulation like FVII, FIX, FX, prothrombin and proteins C and S, suggesting an evolution from a common ancestral gene.

### Structure

Protein Z is a single-chain glycoprotein containing an N-terminal Gla domain, one OH-Asp residue (64 aa), two EGF-like domains and the serine protease part. Protein Z exhibits no proteolytic activity (His and Ser of the catalytic triad are replaced).

### Biological Functions

Protein Z binds to thrombin and thus promotes its association with phospholipids. It acts as a cofactor in the inhibition of FXa by the protein Z-dependent protease inhibitor.

# *Vitronectin (Serum spreading factor)*

## Fact Sheet

| | | | |
|---|---|---|---|
| ***Classification*** | | ***Abbreviations*** | VN |
| ***Structures/motifs*** | 2 homopexin-like; 1 somatomedin B | ***DB entries*** | VTNC_HUMAN/P04004 |
| ***MW/length*** | 52 278 Da/459 aa | ***PDB entries*** | 1OC0; 1S4G |
| ***Concentration*** | 250–500 mg/l | ***Half-life*** | |
| ***PTMs*** | 3 N-CHO; 5 S-Tyr (potential); 1 P-Ser (378 aa) | | |
| ***References*** | Kamikubo *et al.*, 2002, *J. Biol. Chem.*, **277**, 27109–27119.<br>Suzuki *et al.*, 1985, *EMBO J.*, **4**, 2519–2524.<br>Zhou *et al.*, 2003, *Nat. Struct. Biol.*, **10**, 541–544. | | |

## Description

**Vitronectin** (VN) is a plasma glycoprotein mainly synthesised in the liver and present in platelets and monocytes. It circulates in blood as monomer, dimer and probably oligomer and forms a binary complex with plasminogen activator inhibitor–1 and a ternary complex with thrombin–antithrombin III.

## Structure

VN is a glycoprotein (10–15 % CHO) with two homopexin-like and one somatomedin B domain. It has five potential sulfation sites and is phosphorylated at Ser 378. There are two forms of vitronectin: a single-chain form (V75) and a two-chain form obtained by cleavage of Arg 379–Ala 380 and the chains V65 and V10 are linked by a disulfide bond. It contains an **RGD** sequence (45–47 aa) as a cell-binding motif. Somatomedin B is generated from the N-terminal region of VN by limited proteolysis and contains four intrachain disulfide bridges (Cys 5–Cys 9, Cys 19–Cys 21, Cys 25–Cys 31, Cys 32–Cys 39).

## Biological Functions

VN is an important adhesive protein that mediates adhesion and binding to cell surfaces. It interacts with many types of proteins like the vitronectin receptor, certain integrins, the urokinase plasminogen activator receptor uPAR, the C1q-binding protein, the plasminogen activator inhibitor-1 and plasminogen.

## *Von Willebrand factor*

### Fact Sheet

| | | | |
|---|---|---|---|
| ***Classification*** | | ***Abbreviations*** | vWF |
| ***Structures/motifs*** | 3 VWFA; 3 VWFC; 2 VWFD; 1 CTCK; 2 TIL | ***DB entries*** | VWF_HUMAN/P04275 |
| ***MW/length*** | 225 717 Da/2050 aa | ***PDB entries*** | 1IJB |
| ***Concentration*** | approx. 10 mg/l | ***Half-life*** | 8–12 h |
| ***PTMs*** | 12 N-CHO; 10 O-CHO (potential) | | |
| ***References*** | Mancuso *et al.*, 1989, *J. Biol. Chem.*, **264**, 19514–19527.<br>Ruggeri, 2003, *J. Thromb. Haemost.*, **1**, 1335–1342.<br>Sadler, 1998, *Annu. Rev. Biochem.*, **67**, 395–424. | | |

### Description

**Von Willebrand factor** (vWF) is a large glycoprotein associated with blood coagulation. It is synthesised in endothelial cells and megakaryocytes and circulates in blood in complex with coagulation factor FVIII.

### Structure

vWF is a single-chain glycoprotein (19 % CHO) containing three VWFA, three VWFC and two VWFD domains, two TIL (trypsin inhibitory-like) domains and one CTCK structure. Mature vWF is generated from the proform (2791 aa) by releasing a 741 residues propeptide, the von Willebrand antigen II. vWF circulates in blood in multimeric forms, the smallest being dimers held together by disulfide bridges in the C-terminal region of two subunits. Interdimeric disulfide bridges linking the N-terminal domains of two dimers lead to linear multimers of up to 50 dimers. It contains an **RGD** sequence (1744–1746 aa) involved in binding to the platelet glycoprotein complex GPIIb/IIIa.

### Biological Functions

vWF has two functions: (1) it serves as a carrier for coagulation factor FVIII protecting it from proteolytic degradation and inactivation and (2) is required for interactions of platelets with the subendothelium of the injured vessel wall. It binds to platelet glycoproteins GPIb and GPIIb/IIIa and to collagens type I and III. vWF functions as a kind of 'molecular glue'.

# 8

# The Complement System

## 8.1 INTRODUCTION

Maintaining the integrity of a living organism is a vital process. In the never-ending race between host and pathogens various mechanisms have evolved to obtain and maintain an equilibrated balance within an organism on one hand and between the host and the pathogens on the other. In vertebrates the formation of antibodies is the key response to the hostile attack of pathogens. The humoral immune response to an infection is composed of the production of antibodies by B lymphocytes, the binding of the antibodies to the pathogen and its subsequent elimination by accessory cells and molecules of the humoral immune system. Almost any substance can trigger an immune response and the subsequent production of antibodies.

Many years ago ***complement*** was found as a heat-sensitive part of plasma that was able to increase the capacity of antibodies in the opsonisation process of bacteria and to facilitate their killing by some antibodies. This activity was thought to complement the antibacterial activity of antibodies and was therefore termed 'complement'.

The complement system consists of a whole set of plasma proteins with very different structural and functional properties. The activation process by antibodies of the first component of the complement system C1 triggers a complex reaction cascade similar to the cascades in blood coagulation and fibrinolysis. Some proteins of the complement system bind covalently to the surface of bacteria, leading to their opsonisation and thus their uptake and destruction by phagocytes. Small fragments released from complement proteins termed anaphylatoxins lead to a local inflammatory response and in consequence attract phagocytes that are required for the elimination of the pathogens. Finally, the terminal complement components damage the membranes of the pathogen via pore formation, resulting in the destruction of the pathogens by phagocytes.

There are three principal ways to activate the effector function of the complement system:

1. The ***classical pathway*** is part of the adaptive, humoral immune response and is initiated by antibodies and the formation of antigen–antibody complexes.
2. The ***alternative pathway*** is part of the innate immune response and is initiated if a spontaneously activated complement component binds to the surface of a pathogen.
3. The ***lectin pathway*** is initiated by serum proteins that bind to mannose-containing proteins or carbohydrates on the surface of bacteria or viruses.

The activation of the complement system leads to three main events:

1. the attraction of inflammatory cells;
2. the opsonisation of the pathogens;
3. the destruction of the pathogens.

A schematic overview of the events involved in the complement cascade with the main components is given in Figures 8.1 (a) and (b).

*Human Blood Plasma Proteins: Structure and Function* Johann Schaller, Simon Gerber, Urs Kämpfer, Sofia Lejon and Christian Trachsel

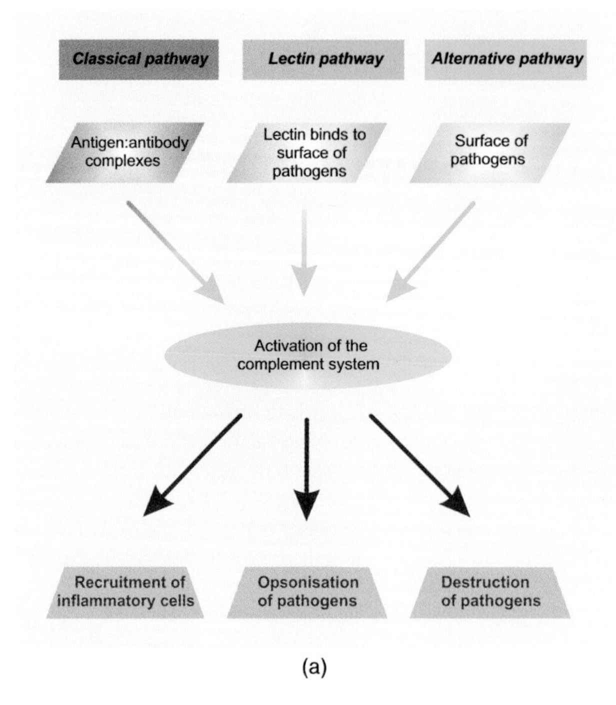


**Figure 8.1 The complement cascade**
(a) Schematic representation of the events involved in the complement cascade: the classical, the alternative and the lectin pathways.

### 8.1.1 Nomenclature

The nomenclature of the proteins of the complement system is quite difficult to understand. However, the following common nomenclature is used:

1. The components of the classical pathway are indicated with the capital letter C, followed by a number indicating the native components: C1, C2, C3, .... Unfortunately, the numbering is according to their historical detection and not according to their sequence in the complement cascade: C1, C4, C2, C3, C5, C6, C7, C8 and C9.
2. The cleavage products are indicated by additional small letters, the larger fragments with the letter b and the smaller fragments with the letter a. Thus C3 is cleaved into C3b and C3a, C4 into C4b and C4a, and so on.
3. The components of the alternative pathway are indicated by different capital letters: B, D, and so on. The cleavage products are indicated in the same way as in the classical pathway. Thus B is cleaved into Bb and Ba, and so on.
4. In the literature and in some textbooks activated components are often indicated with a horizontal bar above the corresponding component. However, this nomenclature will not be used in this book.

## 8.2 THE CLASSICAL PATHWAY

A prerequisite for the initiation of the activation of the classical pathway of the complement system is either the binding of pentameric IgM or IgG molecules to antigens on the surface of a pathogen. Due to special structural requirements for the binding of the C1q subcomponent of complement component C1 to antibodies no activation of the complement cascade can be achieved with antibodies in solution. The cascade is only triggered if the antibodies cover many sites on the surface of a pathogen.

The first component of the classical pathway of complement activation is the complement component C1 present as a trimolecular complex of three proteins, C1q, C1r and C1s, in the molar ratio of 1 : 2 : 2, forming the Ca-dependent $C1qC1r_2C1s_2$ complex with an approximate mass of 740 kDa. In solution, the pentameric IgM exhibits a planar, star-shaped conformation to which C1q is not able to bind. Upon binding to the surface of a pathogen a conformational change of the

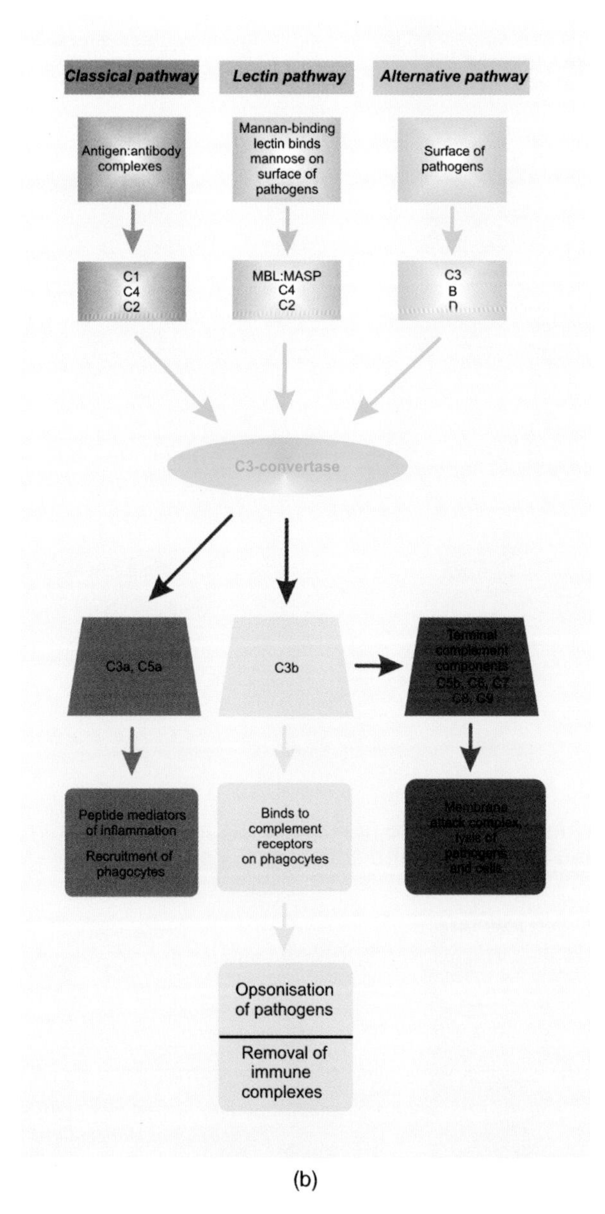


**Figure 8.1** (*Continued*) (b) Overview of the main components of the complement system and the involved events. The activation of the complement cascade results in the formation of the C3 convertase. Finally, the formation of the membrane attack complex leads to the destruction of the pathogens. (Adapted from Janeway *et al.*, Immunobiology. Copyright (2005), Garland.)

pentameric IgM molecule to a crab-like, 'staple' conformation exposes the binding sites for C1q. The globular heads of the C1q subcomponent bind to the Fc-fragment of pentameric IgM in its 'staple' conformation or to the Fc-fragment of two or more IgG molecules on the surface of a pathogen, as depicted in Figure 8.2. This binding leads to the activation of the complement C1r subcomponent, resulting in its active two-chain structure. As a consequence, activated C1r is able to activate the complement C1s subcomponent, resulting in the active two-chain C1s molecule and thus terminating the first step of the activation cascade of the classical pathway. A compilation of the components of the classical pathway of the complement system is given in Table 8.1.

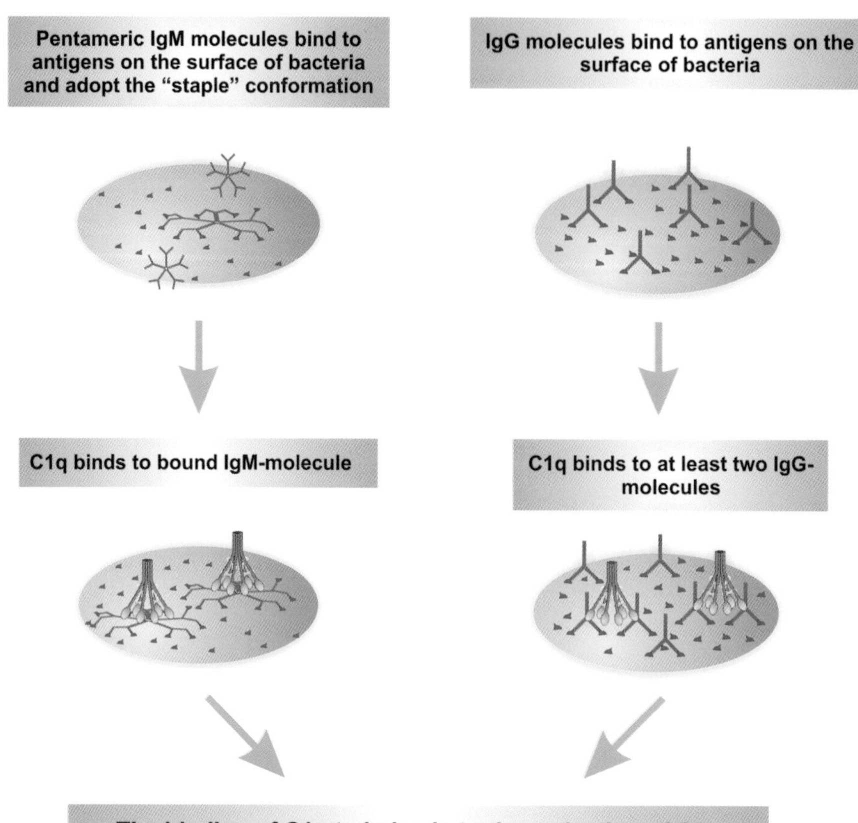


**Figure 8.2 Activation of the classical pathway**
C1q binds either to pentameric IgM in its 'staple' conformation or to at least two IgG molecules on the surface of pathogens, which leads to the activation of C1r and the subsequent activation of C1s.

**Table 8.1** Characteristics of the components of the classical pathway of the complement system.

| Component | Mass (kDa)[a] | Concentration (mg/l)[b] | Function |
|---|---|---|---|
| C1: $C1qC1r_2C1s_2$ | 750 | | Initiation of complement activation |
| C1q: 6 C1qA, B, C of each | 460 | 70–180 | Binding of IgM or IgG |
| C1q: A–B dimer | 52.8 | | C1q subcomponent |
| C1q: C–C dimer | 47.6 | | C1q subcomponent |
| $C1r_2$ | 173 | 100 | Zymogen |
| $C1s_2$ (C1 esterase) | 155 | 30 | Zymogen |
| C2 (C3/C5 convertase) | 85 | 15 | Zymogen |
| C2a | | | Precursor of vasoactive C2-kinins |
| C2b | | | Serine protease |
| C3 | 185 | 1000–2000 | Precursor; key component complement |
| C3a anaphylatoxin | | | Inflammatory mediator |
| C3b | | | Binding to pathogen surface |
| C4 | 202 | 200–600 | Precursor |
| C4a anaphylatoxin | | | Inflammatory mediator |
| C4b | | | Binding to pathogen surface |

[a]Values obtained by either SDS-PAGE or sedimentation equilibrium.
[b]Represent average values.

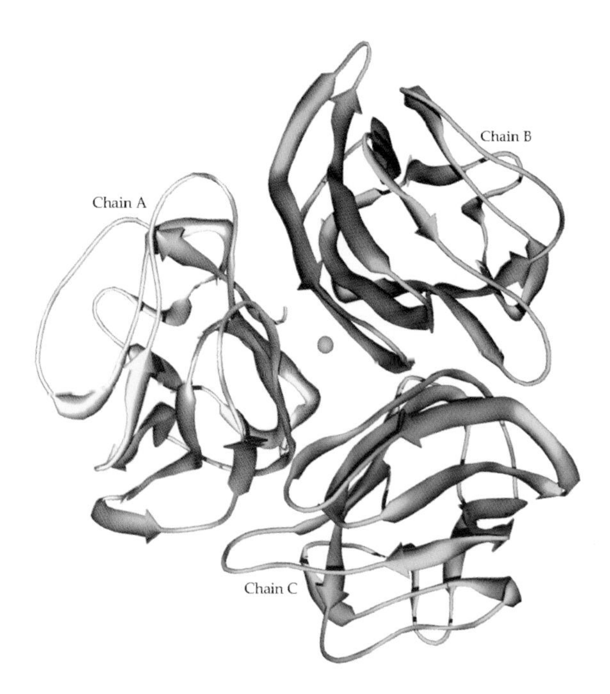


**Figure 8.3 3D structure of the globular C1q domain of C1q (PDB: 1PK6)**
The heterotrimer of the C1q globular domains of chains A (yellow), B (green) and C (blue) exhibits a compact, almost spherical structure and the $Ca^{2+}$ (pink ball) is located on top of the assembly.

**Complement C1q subcomponent** (C1q) (see Data Sheet) is one of the most cationic proteins in human plasma. C1q is a large glycoprotein (8 % CHO) with an approximate mass of 400 kDa containing six A-, six B- and six C-chains. The six A- and six B-chains form six heterodimers and the six C-chains form three homodimers all linked by disulfide bridges. The A-, B- and C-chains are very similar in structure, characterised by an N-terminal collagen-like domain of approximately 80 aa rich in Gly, OH-Lys and OH-Pro residues containing a series of **Gly–Xaa–Yaa** repeats characteristic for collagen-like structures followed by a globular C-terminal C1q domain of approximately 130 aa. The structure of C1q resembles a 'bunch of tulips' consisting of six globular heads (each can bind to one Fc domain either in IgM or in IgG) that are connected to six collagen-like stalks that form a central fibril-like stem.

The 3D structure of the globular domain of C1q (C1qA; 133 aa: Gln 90–Ser 222; C1qB; 132 aa: Thr 92–Asp 223; C1qC; 129 aa: Lys 89–Asp 217) was determined by X-ray diffraction (PDB: 1PK6) and is shown in Figure 8.3. The heterotrimeric assembly of the C1q globular domains of chain A (yellow), chain B (green) and chain C (blue) exhibits a compact, almost spherical structure with the $Ca^{2+}$-binding site ($Ca^{2+}$ shown as a pink ball) located at the top of the assembly and the homotrimer is similar in structure to the collagen X homotrimer. C1q is thought to be a key factor in binding physiological ligands such as IgG and C-reactive protein.

**Complement C1r subcomponent** (C1r, EC 3.4.21.41) (see Data Sheet) is a single-chain multidomain glycoprotein (9 % CHO) that is present in plasma as a zymogen and belongs to the peptidase S1 family. The binding of C1q to pentameric IgM in its 'staple' conformation or to at least two IgG molecules leads to the activation of C1r, resulting in the active two-chain C1r by cleavage of the Arg 446–Ile 447 peptide bond generating a heavy chain containing two CUB, one EGF-like and two sushi/CCP/SCR domains (see Chapter 4) and two modifications, an OH-Asn (aa 150) and a P-Ser (aa 189), and a light chain comprising the serine protease domain. The two chains are linked by a single interchain disulfide bridge (Cys 434 heavy chain–Cys 114 light chain). C1r exhibits an elongated shape and is very similar in structure and function to C1s. The activation of C1r is regulated by the main inhibitor of the C1 component, the plasma protease C1 inhibitor (see Chapter 11).

The 3D structure of a shortened (mutated) single-chain proenzyme form of C1r (399 aa: Ile 290–asp 688) comprising the two sushi domains (in green) and the catalytic chain (in yellow) was determined by X-ray diffraction (PDB: 1GPZ) and is shown in Figure 8.4. The homodimeric elongated structure is in head-to-tail interaction between one sushi domain of one

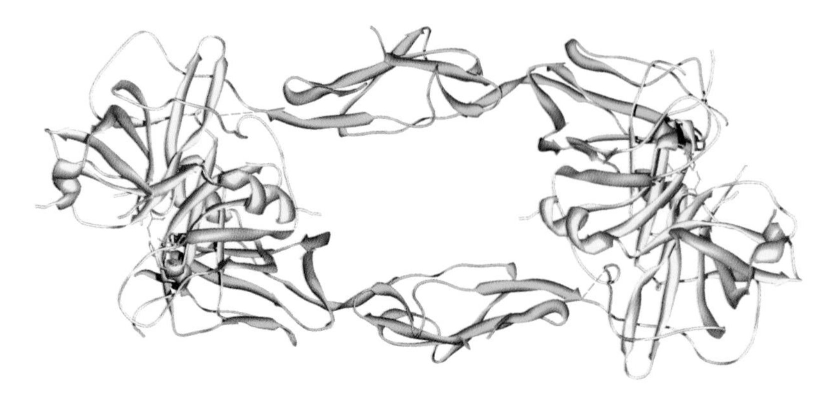

**Figure 8.4 3D structure of an N-terminally truncated form of C1r (PDB: 1GPZ)**
In the homodimer of C1r the two sushi domains (in green) and the catalytic chain (in yellow) are connected by a disulfide bridge (drawn schematically: dashed blue line) and the active site residues (in pink) are indicated.

monomer and the catalytic domain of the other monomer. The disulfide bridge between the catalytic chain and the sushi domain is drawn schematically (dashed blue line) and the active site residues of the catalytic triad (in pink) are indicated. The structural data support the hypothesis that the activation of C1r in C1 is triggered by mechanical stress caused by target recognition, resulting in the disruption of the interface between the sushi domain and the catalytic chain and a subsequent conformational change.

**Complement C1s subcomponent** (C1s, EC 3.4.21.42) (see Data Sheet) is a single-chain multidomain glycoprotein that is present in plasma as a zymogen and belongs to the peptidase S1 family. Activated C1r activates C1s, resulting in the active two-chain C1s, by cleaving the Arg 422–Ile 423 peptide bond generating a heavy chain containing two CUB, one EGF-like and two sushi/CCP/SCR domains (see Chapter 4) and an OH-Asn residue (aa 134), and a light chain comprising the serine protease domain. The two chains are linked by a single interchain disulfide bridge (Cys 410 heavy chain–Cys 112 light chain). C1s is very similar in structure and function to C1r. The activation of C1s is regulated by the main inhibitor of the C1 component, the plasma protease C1 inhibitor (see Chapter 11).

The 3D structure of the second sushi domain (in green) with the catalytic chain of C1s (333 aa: Asp 343–Asp 673 + N-terminal dipeptide, in yellow) was determined by X-ray diffraction (PDB: 1ELV) and is shown in Figure 8.5. The ellipsoidal sushi domain is oriented perpendicularly to the catalytic domain, which exhibits a chymotrypsin-like structure. The disulfide

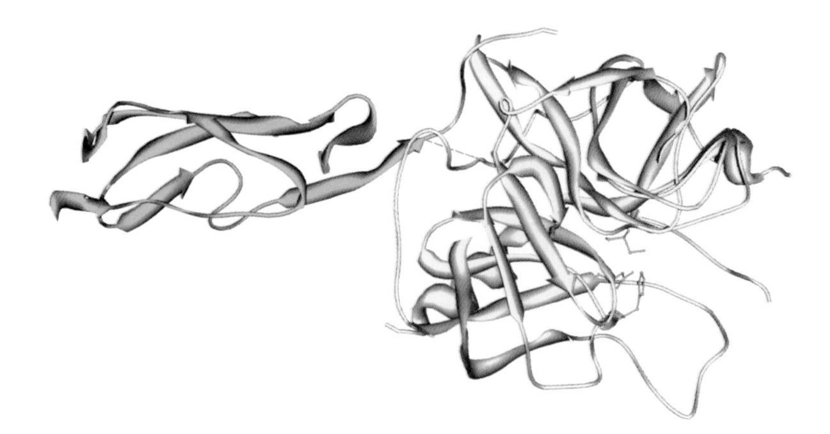

**Figure 8.5 3D structure of an N-terminally truncated form of C1s (PDB: 1ELV)**
In truncated C1s the second sushi domain (in green) and the catalytic chain (in yellow) are connected by a single disulfide bridge (drawn schematically: dashed blue line) and the active site residues (in pink) are indicated.

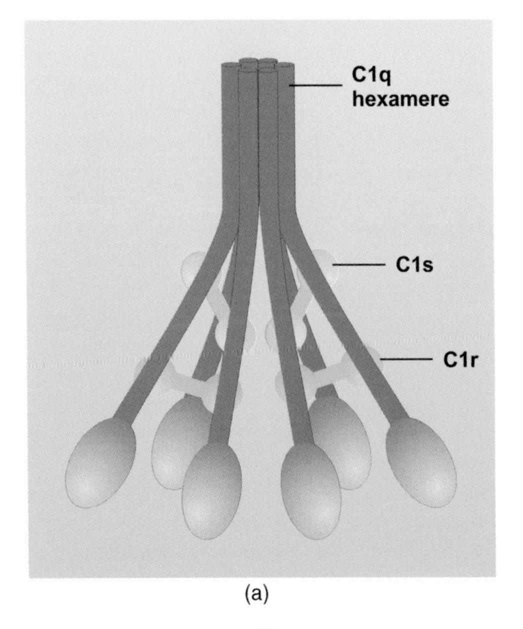

**Figure 8.6 Complement component C1**
(a) Schematic representation and (b) electron microscopic picture of the first component of the classical pathway of the complement system $C1qC1r_2C1s_2$ complex. The globular heads of C1q can be clearly seen. Two C1r and two C1s molecules are bound to the collagen-like stem of C1q. (Adapted from Janeway *et al.*, Immunobiology. Copyright (2005), Garland. Adapted from Colomb, *et al.*, Philos. Trans. R. Soc. London Biol., Issue 306, p. 283 Activation of C1. Copyright (1984), with permission from Royal Society Publishing.)

bridge between the catalytic chain and the sushi domain is drawn schematically (dashed blue line) and the active site residues of the catalytic triad (in pink) are indicated.

C1r and C1s are very similar in structure and function and contain the same number of the same domains. Two C1r and two C1s molecules are noncovalently bound to the fibril-like stem of the C1q molecule. A schematic representation of the complement component C1 and the structure determined by electron microscopy (EM) are shown in Figures 8.6(a) and (b). The six globular heads of C1q are clearly visible. Two C1r and two C1s molecules are bound to the collagen-like stem of C1q.

After the generation of the active two-chain C1s the complement components C4 and C2 are cleaved by activated C1s into two large fragments, C4b and C2b, and two smaller fragments, C4a, an anaphylatoxin, and C2a. The C4b fragment binds covalently via its active thioester bond to amino or hydroxyl groups on the surface of the pathogens. The formation and the binding of the reactive thioester bond in complement component C4 to the surface of a pathogen is depicted in Figure 8.7. The release of C4a by activated C1s exposes the reactive thioester bond in C4b, which is rapidly hydrolysed by a water molecule if it is not immediately bound to the surface of a pathogen, thus inactivating C4b, and in consequence preventing the diffusion from its site of activation on the surface of a pathogen to the surface of host cells. The structurally and functionally related complement component C3 contains an identical reactive thioester bond, which becomes exposed if C3a is released by the C3 convertase.

C4b covalently linked to the surface of a pathogen can bind a C2 molecule, which also becomes susceptible to cleavage by activated C1s. The two larger fragments, C4b and C2b, an active serine protease, form a complex, the C3 convertase (C4b2b) that remains on the surface of the pathogen. As the name suggests, the main function of the C3 convertase is the generation of the large fragment C3b and the small fragment C3a anaphylatoxin from native C3. Some of the C3b molecules remain on the surface of the pathogen and the released C3a causes local inflammation. The covalent fixation of C4b on the surface of the pathogen is essential for the activation of C3 on the surface of a pathogen and not on the surface of host cells.

**Complement C4** (C4) (see Data Sheet) is a large single-chain precursor glycoprotein (7 % CHO) that belongs to the protease inhibitor I39 family and is first processed by removal of two propeptides (4 aa and 7 aa, respectively), thus generating a three-chain molecule in the order of β-chain (95 kDa), α-chain (75 kDa) and γ-chain (33 kDa) linked by interchain disulfide bridges, one between the α′–β-chains (Cys 64 α′-chain–Cys 548 β-chain) and two between the α′–γ-chains. The C-terminal end of the α-chain contains three S-Tyr residues that increase the hemolytic activity of C4 by a factor

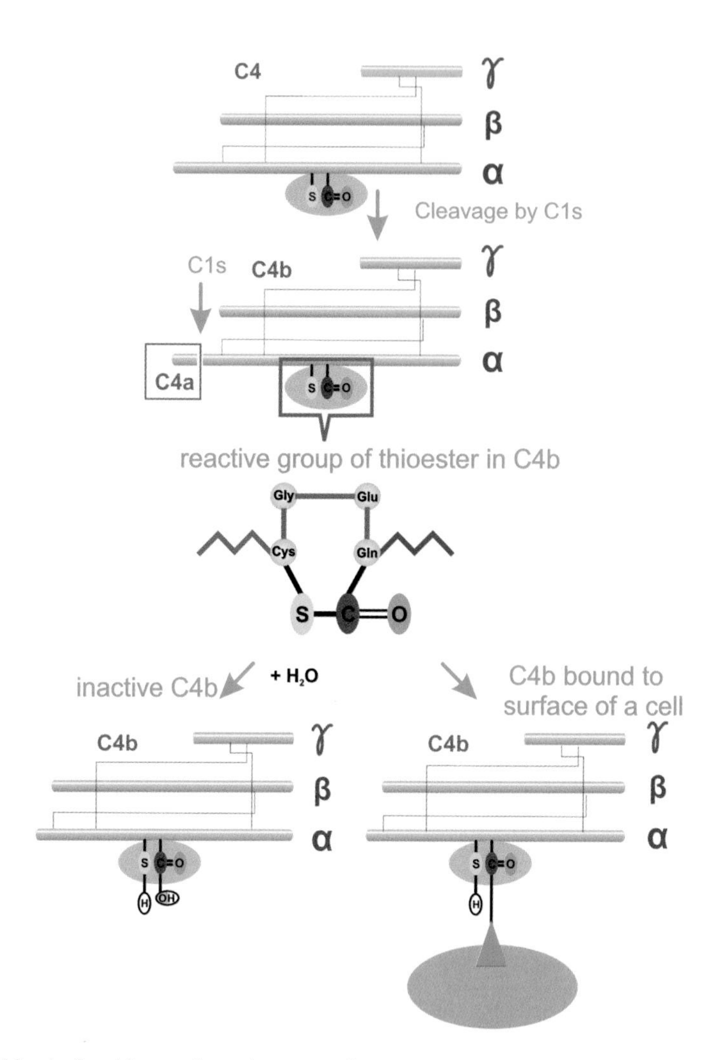


**Figure 8.7 Reactive thioester bond in complement components**
The formation and binding of the reactive thioester in complement component C4 is shown. The cleavage of C4 by activated C1s releases the C4a anaphylatoxin and the active three-chain C4b with an exposed and reactive thioester bond, which is rapidly inactivated by a water molecule if it is not bound to the surface of a pathogen. C3, similar in structure and function, contains an identical reactive thioester bond. (Adapted from Janeway *et al.*, Immunobiology. Copyright (2005), with permission from Garland.)

of about three. In a second step C4a anaphylatoxin is released from the N-terminus of the α-chain, resulting in the C4b three-chain fragment. C4b contains the reactive isoglutamyl cysteine thioester bond (**Cys 991–Gly–Glu–Gln 994**). C4 exhibits structural similarity with complement components C3 and C5 and $\alpha_2$-macroglobulin. C4 is a globular molecule that appears as a flat disc by X-ray diffraction.

The 3D structure of the factor I-generated C4d fragment (380 aa: Thr 938–Arg 1317) of complement C4 (N-terminally truncated form: 335 aa: Ser 983–Arg 1317) was determined by X-ray diffraction (PDB: 1HZF) and is shown in Figure 8.8. The structure exhibits the α–α six barrel fold characterised by six parallel α-helices forming the core of the barrel, surrounded by a second set of six parallel α-helices in antiparallel orientation. C4d shares this structure with the corresponding complement C3d fragment.

**Complement C2** (C2, EC 3.4.21.43) (see Data Sheet) or C3/C5 convertase is a single-chain glycoprotein that belongs to the peptidase S1 family and is cleaved by active C1s at the Arg 223–Lys 224 peptide bond, resulting in the larger C2b

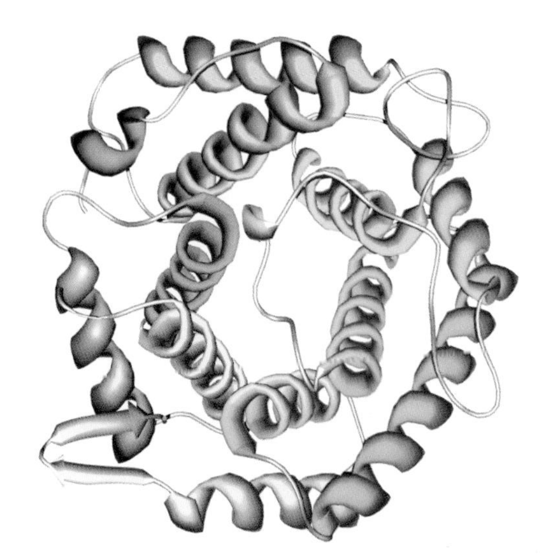

**Figure 8.8 3D structure of the C4d fragment of complement C4 (PDB: 1HZF)**
The structure of the C4d fragment is characterised by a α-α six barrel fold, characterised by six parallel α-helices forming the core of the barrel, surrounded by a second set of six parallel α-helices in antiparallel orientation.

fragment containing one VWFA domain and the serine protease domain and the smaller C2a fragment containing three sushi/CCP/SCR domains. C2 is very similar in structure and function to factor B of the alternative pathway.

C2 deficiency, type I and type II, are autosomal recessive diseases, which are the cause of systemic lupus erythematosus (SLE), disorders of the immune system and autoimmune reactions leading to inflammation and damage of body tissues, SLE-like syndromes, glomerulonephritis (inflammation of the kidneys usually caused by autoimmune reactions), vasculitis (inflammation of the blood vessel wall) and pyogenic infections by bacteria. Type I is characterised by the complete loss of C2 while in type II the secretion of C2 is selectively blocked (for details see MIM: 217000).

**Complement C3** (C3) (see Data Sheet) is a large single-chain precursor glycoprotein (2 % CHO) that belongs to the protease inhibitor I39 family and is first processed by removal of a tetrapeptide (four consecutive Arg residues, aa 646–649), resulting in an α-chain (115 kDa) and a β-chain (75 kDa) linked by a single interchain disulfide bridge (Cys 67 α′-chain–Cys 537 β-chain). The C3 convertase releases from the N-terminus of the α-chain the C3a anaphylatoxin (see Chapter 4) generating the C3b fragment (α′- and β-chains) that contains the reactive isoglutamyl cysteine thioester bond (**Cys 988–Gly–Glu–Gln 991**). C3b is rapidly inactivated by complement factor I (see Section 8.6) by cleaving the Arg 1281–Ser 1282 and Arg 1298–Ser 1299 peptide bonds, releasing C3f (17 aa) and generating inactivated iC3b, which is further processed to even smaller fragments. C3 exhibits structural similarity with complement components C4 and C5 and $\alpha_2$-macroglobulin. The tertiary structure of C3 exhibits a two-domain shape characterised by a flat ellipsoid and a second smaller flat domain.

The 3D structure of two-chain C3 was determined by X-ray diffraction (PDB: 2A73) and is shown in Figure 8.9. C3 consists of the β-chain (643 aa: Ser 1–Pro 643, in green) and the α-chain (991 aa: Val 651–Asn 1641, in yellow) linked by a single disulfide bridge (Cys 537 in the β-chain, Cys 67 in the α-chain, in blue and circled in pink). C3a anaphylatoxin (in magenta) is located in the N-terminal region of the α-chain.

The C3 convertase bound to the surface of a pathogen can cleave up to 1000 native C3 molecules into C3b and C3a. The main consequence of the complement activation is the generation and deposition of large numbers of C3b molecules on the surface of a pathogen. It is therefore not astonishing that C3 has the highest concentration of all complement proteins in plasma (1–2 mg/ml). This covering of the surface of a pathogen with C3b molecules is actually the signal for the destruction of the pathogen by phagocytes. The large number of C3b molecules deposited on the surface of a pathogen are recognised by receptors of the complement system on phagocytic cells, which are stimulated to devour the pathogen.

The binding of C3b to the C3 convertase (C4b2b) is the basis of the formation of the C4b2b3b complex, the C5 convertase, which leads further down the complement cascade. The C5 convertase cleaves complement component C5 into a large

**Figure 8.9 3D structure of two-chain complement C3 (PDB: 2A73)**
Two-chain C3 consists of the β-chain (in green) and the α-chain (in yellow) linked by a single disulfide bridge (in blue and circled in pink). C3a anaphylatoxin (in magenta) is in the α-chain.

fragment C5b and the small fragment C5a anaphylatoxin. This step initiates the assembly of the terminal components of the complement system that finally results in the formation of the membrane attack complex (MAC).

### 8.2.1 The anaphylatoxins C3a, C4a and C5a

Many molecules that are released during the immune response induce local inflammatory reactions. Under certain conditions mast cells release local inflammatory mediators and activated macrophages can cause similar reactions.

The small complement fragments C3a, C4a and C5a, termed 'anaphylatoxins', interact with specific receptors and cause similar local inflammations (anaphylaxis is an acute systemic inflammation). All three anaphylatoxins are able to induce the contraction of smooth muscles and can increase the vascular permeability. In addition, C3a and C5a are able to activate mast cells releasing mediators with the same implications. C5a is the most stable anaphylatoxin and exhibits the highest specific biological activity. As a consequence, antibodies, complement components and phagocytic cells are attracted by the local inflammation.

## 8.3 THE ALTERNATIVE PATHWAY

The large number of C3b molecules bound to the surface of a pathogen are enhancing the effects of the classical pathway by the activation of the alternative pathway. With the exception of the initiative step the cascade of the alternative pathway corresponds to that of the classical pathway. The covalent fixation of C3b on the surface of a pathogen leads to the activation of the alternative pathway corresponding to the covalent binding of C4b in the classical pathway. As seen previously, C3b and C4b are structurally and functionally very similar. In the second step of the alternative pathway C3b binds complement factor B, the structural and functional equivalent of C2 in the classical pathway. Factor B is the key enzyme of the alternative pathway. Upon binding to C3b factor B becomes accessible for cleavage by complement factor D, a serine protease, resulting in a smaller fragment Ba and a larger fragment Bb, an active serine protease, that remains bound to the C3b. Thus the C3bBb complex comes into existence as the C3 convertase of the alternative pathway, which is the structural and functional equivalent of C4b2b, the C3 convertase of the classical pathway. Similar to the classical pathway C4b2b, the C3 convertase of the alternative pathway generates many C3b molecules, which are covalently bound on the surface of the pathogen. A compilation of the components of the alternative pathway of the complement system is given in Table 8.2.

The activation of the classical pathway and its amplification by the alternative pathway results in the quick saturation of the surface of the pathogen with C3b molecules and the release of the inflammatory C3a anaphylatoxin. Some of the C3b molecules bind to the C3 convertase (C3bBb), resulting in the $C3b_2Bb$ complex, the C5 convertase of the alternative

**Table 8.2** Characteristics of the components of the alternative pathway of the complement system.

| Component | Mass (kDa)[a] | Concentration [mg/l][b] | Function |
|---|---|---|---|
| C3 | 185 | 1000–2000 | Key component complement activation |
| C3a anaphylatoxin | | | Inflammatory mediator |
| C3b | | | Binding to pathogen surface |
| Factor B (C3/C5 convertase) | 93 | 210 | Zymogen |
| Ba | | | Unknown |
| Bb | 60 | | Serine protease |
| Factor D (C3 convertase activator) | 24 | 2 | Zymogen |

[a]Values obtained by either SDS-PAGE or sedimentation equilibrium.
[b]Represent average values.

pathway. As in the case of the classical pathway the C5 convertase of the alternative pathway cleaves C5 into C5b and C5a. At this level the alternative and the classical pathway converge and initiate the assembly of the terminal components of the complement cascade, finally resulting in the generation of the membrane attack complex (MAC).

**Complement factor B** (B, EC 3.4.21.47) (see Data Sheet) or C3/C5 convertase is a single-chain precursor glycoprotein (7–10 % CHO) that belongs to the peptidase S1 family and becomes activated by factor D upon binding to C3b by cleaving the Arg 234–Lys 235 peptide bond, resulting in the Ba fragment containing three sushi/CCP/SCR domains and the Bb fragment containing one VWFA domain (see Chapter 4) and the serine protease domain. Factor B is very similar in structure and function to C2 of the classical pathway. In the EM factor B appears as a tightly packed three-domain molecule with the shape of a triangle. Activated Bb appears as a dumbbell-shaped molecule of 60 kDa.

The 3D structure of the Bb fragment of factor B (497 aa: Ser 243–Leu 739) comprising the VWFA domain and the inactive catalytic domain was determined by X-ray diffraction (PDB: 1RRK) and is shown in Figure 8.10. The Bb fragment consists of two compact domains with a linker region and has the shape of a distorted dumbbell, a very similar appearance as in the EM. The VWFA domain in Bb (in green) exhibits the same structure as the corresponding domain in vWF with a central β-sheet surrounded by α-helices and the catalytic domain (in yellow) exhibits the expected structure of a serine protease domain. The active site residues of the catalytic triad are shown in pink.

**Complement factor D** (D, EC 3.4.21.46) (see Data Sheet) or C3 convertase activator is a single-chain protein that belongs to the peptidase S1 family and is present in plasma already in its active form with a buried and inaccessible N-terminus inside the molecule. A potential propeptide (5 aa) seems already to have been removed. Factor D has the lowest concentration of all complement components in plasma with approximately 2 mg/l. It has a similar function to active C1s in the classical pathway that cleaves C2.

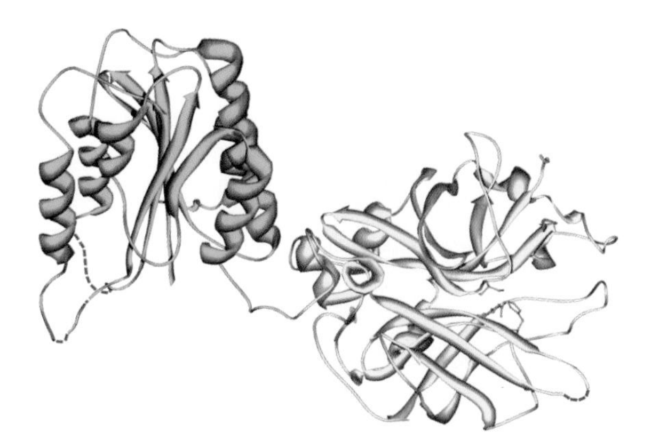

**Figure 8.10 3D structure of the Bb fragment of factor B (PDB: 1RRK)**
The Bb fragment consists of two compact domains with the shape of a distorted dumbbell. The VWFA domain (in green) contains a central β-sheet surrounded by α-helices and the catalytic domain (in yellow) exhibits the expected structure of a serine protease domain. The active site residues are shown in pink.

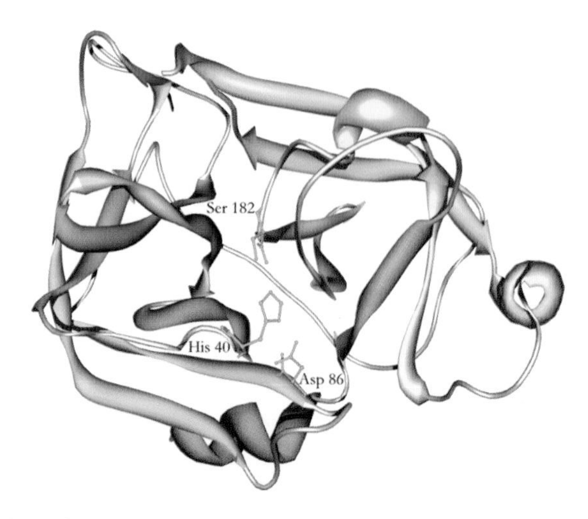


**Figure 8.11 3D structure of the proform of factor D (PDB: 1FDP)**
The structure of the proform of D is similar as in trypsinogen and chymotrypsinogen. The active site residues His 40, Asp 86 and Ser 182 are shown in pink.

The 3D structure of a proform of factor D (228 aa: Ile 1–Ala 228 + N-terminal propeptide (5aa) + 2 Ala from the C-terminus of the signal peptide) was determined by X-ray diffraction (PDB: 1FDP) and is shown in Figure 8.11. The structure of the profactor D is similar to that in trypsinogen and chymotrypsinogen. Comparison of the zymogen–active enzyme pairs of profactor D, trypsinogen and chymotrypsinogen revealed a similar distribution of the flexible regions, but profactor D is the most flexible. The active site residues His 40, Asp 86 and Ser 182 are shown in pink. Mature factor D displays an inactive, self-inhibited active site conformation.

## 8.4 THE TERMINAL COMPONENTS OF THE COMPLEMENT SYSTEM

The generation of the C5 convertase (C4b2b3b) in the classical pathway and the C5 convertase ($C3b_2Bb$) in the alternative pathway leads to the common terminal part of the complement cascade of both pathways, resulting in the cleavage of complement component C5. The characteristics of the terminal components of the complement system are summarised in Table 8.3.

C5 binds to C3b in the C4b2b3b and $C3b_2Bb$ complexes, where it is cleaved by the active serine proteases C2b of the classical pathway and Bb of the alternative pathway into the large fragment C5b and the small fragment C5a anaphylatoxin. C5b is the initiator of the formation of the MAC. A schematic depiction of the structural organisation of the pore-forming proteins of the complement system and the related protein perforin is shown in Figure 8.12.

C5b binds to the C-terminal region of complement component C6, resulting in the soluble C5b,6 complex that is the nucleus of the assembly of the lytic complex. The binding of complement component C7 to the C5b,6 complex and the

**Table 8.3** Characteristics of the terminal components of the complement system.

| Component | Mass (kDa)[a] | Concentration (mg/l)[b] | Function |
|---|---|---|---|
| C5 | 190 | 75 | Precursor |
| C5a anaphylatoxin | | | Inflammatory mediator |
| C5b | 179 | | Initiation membrane attack complex |
| C6 | 105 | 50 | Binding C5b |
| C7 | 97 | 55 | Binding C5b,6 |
| C8 | 151 | 80 | Binding C5b,6,7 |
| C9 | 71 | 60 | Binding C5b, 6, 7, 8 and polymerisation |

[a]Values obtained by either SDS-PAGE or sedimentation equilibrium.
[b]Represent average values.

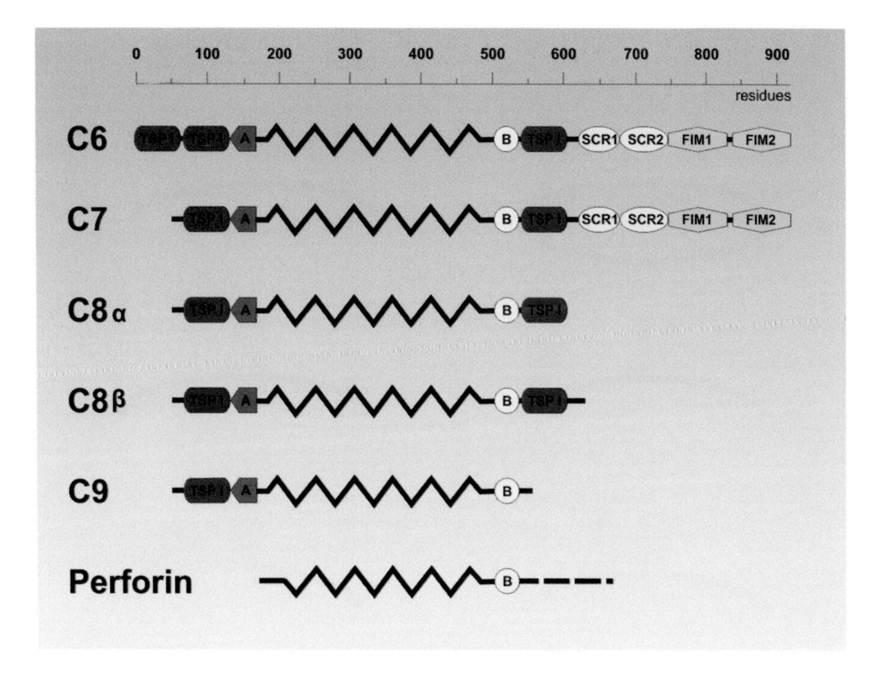


**Figure 8.12 Complement components C6, C7, C8, C9 and perforin**
Schematic depiction of the structural organisation at the pore forming proteins of the complement system and perforin. TSP 1: thrombospondin type I repeat; A and B, low-density lipoprotein receptor class A and class B domains, respectively; SCR: sushi/complement control protein (CCP)/short consensus repeat; FIM: factor I module. (Adapted from Haefliger *et al.*, J Biol Chem 264: 18'041–18'051. Copyright (1989) The American Society for Biochemistry and Molecular Biology.)

formation of the C5b,6,7 complex induce a conformational change in the complex, upon which a hydrophobic region in C7 becomes accessible for binding to membranes. The C5b,6,7 complex binds via C7 to the lipid bilayer of a pathogen and is the nucleus for the formation of the MAC C5b–C9. Complement component C8 binds via its β-chain to C5b of the membrane-associated C5b,6,7 complex, thus forming the C5b,6,7,8 complex. Similar to the case of C7, this complex formation induces a conformational change in the C5b,6,7,8 complex, exposing a hydrophobic binding region in the α-chain of C8 and enabling the penetration of the α–γ heterodimer into the lipid bilayer of the target membrane. The γ-chain seems not to be essential for the lytic activity of C8 and its function is unclear. The α–γ heterodimer of the C5b,6,7,8 complex acts as a catalyst for the polymerisation reaction of complement component C9. The GPI-anchored membrane glycoprotein CD59 is a potent regulator of complement activation by binding to the α-chain of C8 and protecting different types of host cells from complement-mediated lysis (see also Section 8.6). The C5b,6,7,8 complex itself already has some slight capability to destroy membrane structures. The binding of C9 to the α–γ heterodimer in the C5b,6,7,8 complex induces a conformational change in C9, exposing a hydrophobic region with the subsequent induction of the polymerisation of up to 18 C9 monomers to a ring-like structure termed 'membrane attack complex' (MAC), the C5b–C9 complex. Figures 8.13(a) and (b) show a schematic depiction of the formation of the membrane attack complex and an electron microscopic picture of the MAC.

The outer surface of the MAC is hydrophobic, which enables its interaction and penetration of the lipid bilayer. The inner surface of the channel is hydrophilic and the pore diameter is approximately 100 Å, thus allowing the passage of water and solutions. The penetration of the MAC into the lipid bilayer results in

- the loss of cellular homeostasis;
- the breakdown of the proton gradient;
- the penetration of enzymes like lysozyme into the cell;
- the destruction of the pathogens by phagocytes.

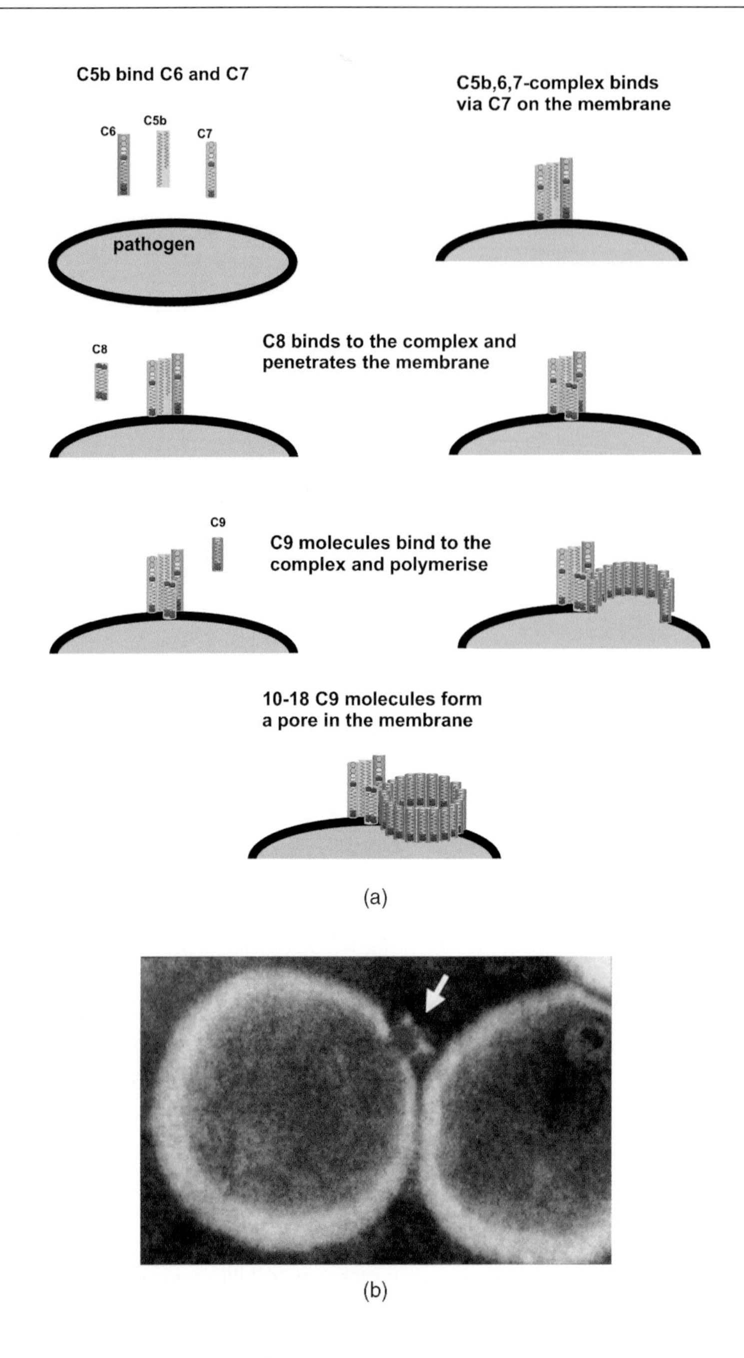


**Figure 8.13 The membrane attack complex (MAC)**
(a) Schematic representation of the events leading to the formation of the MAC and (b) an electron microscopic picture of the MAC. The membrane lesions can be clearly seen in the electron microscopic picture. (Adapted from Janeway *et al.*, Immunobiology. Copyright (2005), Garland. Reproduced from Bhakdi S, Tranum-Jensen J. Immunol Today 12: 318–320. Copyright (1991), with permission from Elsevier.)

The nascent C5b–C9 complex formed in solution and in the absence of membranes is complexed by vitronectin and clusterin (soluble terminal complex).

There is a remarkable similarity between the MAC and the pores formed by perforin, a protein with structural and functional similarity to complement components C6, C7, C8 and C9. The larger diameter of the pore formed by perforin molecules with approximately 160 Å seems to be required for the direct access of granzymes to the inside of target cells, thus enabling apoptosis.

**Complement C5** (C5) (see Data Sheet) is a single-chain precursor glycoprotein (3 % CHO) that belongs to the protease inhibitor I39 family and is first processed by removal of a tetrapeptide (Arg 656–Pro–Arg–Arg 659) generating an α-chain (115 kDa) containing a C-terminal NTR domain and a β-chain (75 kDa) linked together by a single interchain disulfide bridge. The C5-convertase releases from the N-terminus of the α-chain the C5a anaphylatoxin (see Chapter 4) resulting in C5b fragment (α′- and β-chains). C5 exhibits structural similarity with complement components C3 and C4 and $\alpha_2$-macroglobulin. EM shows an irregular multidomain and asymmetric structure of 104 Å × 140 Å × 168 Å with 17 % α-helix and 20 % β-sheet structure.

**Complement component C6** (C6) (see Data Sheet) is a single-chain multidomain glycoprotein (3 % CHO) that belongs to the complement C6/C7/C8/C9 family containing three TSP1, two sushi/CCP/SCR, one EGF-like and one LDLRA domain and two complement control factor I modules. C6 has two potential transmembrane segments, Asp 310–Tyr 328 and Tyr 333–Gly 352. C6 exhibits structural and functional similarity with complement components C7, C8 and C9 and perforin. C6 exhibits an elongated structure of approximately 180 Å length and 60 Å diameter in the centre with 12 % α-helix, 29 % β-sheet and 21 % β-turn structure.

**Complement component C7** (C7) (see Data Sheet) is a single-chain multidomain glycoprotein (5 % CHO) that belongs to the complement C6/C7/C8/C9 family containing two TSP1, two sushi/CCP/SCR, one EGF-like and one LDLRA domain and two complement control factor I modules. C7 has two potential transmembrane segments, Gln 249–Tyr 265 and Tyr 270–Gly 289. C7 exhibits structural and functional similarity with complement components C6, C8 and C9 and perforin. C7 exhibits an irregular ellipsoid structure of 151 Å length with bulbous ends of 59 Å and 42 Å diameters, respectively.

**Complement component C8** (C8) (see Data Sheet) is a single-chain multidomain glycoprotein that belongs to the complement C6/C7/C8/C9 family and is composed of three chains: an α-chain (64 kDa), a β-chain (64 kDa) and a γ-chain (22 kDa). The α- and γ-chains are linked by a single disulfide bridge (Cys 164 α-chain–Cys 40 γ-chain) and the β-chain is noncovalently attached to the α-chain. The α- and β-chains each contain two TSP1, one EGF-like and one LDLRA domain. Both chains have two potential transmembrane segments: in the α-chain, Met 279–Tyr 295 and Tyr 300–Gly 319 and in the β-chain, Met 258–Tyr 274 and Tyr 279–Gly 298. Both chains exhibit structural and functional similarity with complement components C6, C7 and C9 and perforin. The γ-chain carries an N-terminal pyro-Glu residue and belongs to the lipocalin family.

**Complement component C9** (C9) (see Data Sheet) is a single-chain multidomain glycoprotein (8–15 % CHO) that belongs to the complement C6/C7/C8/C9 family containing one TSP1, one EGF-like and one LDLRA domain. C9 contains two potential transmembrane segments, Val 293–Tyr 309 and Tyr 314–Gly 333. C9 exhibits a structural and functional similarity to complement components C6, C7 and C8 and perforin. C9 molecules form a cylindrical polymer of approximately 160 Å height and 160 Å outer and 100 Å inner diameters.

Deficiencies in C7, C8 or C9 are the cause of recurrent bacterial infections, predominantly from *Neisseria meningitidis* (for details see MIM: 217070, 120950, 120960, 120940).

**Perforin** (see Data Sheet) or cytolysin is a single-chain glycoprotein (70 kDa) that belongs to the complement C6/C7/C8/C9 family containing one EGF-like and one C2 domain. Perforin has two potential transmembrane segments, Pro 167–Phe 183 and Tyr 191–Gly 210. Perforin is the major cytolytic protein of killer lymphocytes and lyses cell membranes nonspecifically. Perforin exhibits a structural and functional similarity to complement components C6, C7, C8 and C9. Perforin polymerises in the presence of calcium ions to a cylindrical complex containing about 20 molecules with a hollow cylindrical structure of 160 Å height and approximately 160 Å inner diameter in EM.

## 8.5 COMPONENTS OF COMPLEMENT ACTIVATION

There are a series of proteins involved directly or indirectly in the activation of the complement cascade. Antibody-independent activation of the complement system is an important way to cope with pathogens, especially important in very young children who have not yet acquired a well-developed immune system or in immunodeficient individuals, and is thought to play an important role in innate immunity.

Mannose-binding protein C (MBP-C) is a calcium-dependent acute phase protein involved in complement activation by binding with its lectin-like carbohydrate recognition domain (CRD) to carbohydrate structures on the surface of a pathogen and subsequent activation of the complement system using its inflammatory, opsonisation and killing potential. It is thought to play an important role in innate immunity, especially in young children and individuals with an immunodeficiency. MBP-C is associated with two closely related glycoproteins that are also involved in complement activation: complement-activating component (MASP-1) and mannan-binding lectin serine protease 2 (MASP-2). MASP-1 and MASP-2 are very similar in structure and function and are also closely related with C1r and C1s involved in the activation phase of the classical pathway. Like C1s, MASP-1 and MASP-2 can activate C4 and C2, respectively. MASP-1 and MASP-2 form with MBP-C a complex, the MBP-C : MASP-1: MASP-2 complex, that is similar in the overall structure and function as the C1q : C1r$_2$ : C1s$_2$ complex.

Although the exact biological function of C-reactive protein (CRP) is unknown, it seems to be involved in complement activation and opsonisation. It is an acute phase protein resulting in an up to a thousand-fold increase in plasma and is therefore used as a marker for the extent of tissue damage and infection.

**Mannose-binding protein C** (MBP-C) (see Data Sheet) or mannose-binding lectin is a single-chain protein with a mass of 32 kDa consisting of an N-terminal region containing Cys residues involved in interchain disulfide bonds (probably Cys 5, 12 and 18), a collagen-like domain containing five OH-Pro residues and repeating triplets of **Gly–Xaa–Yaa** involved in the formation of the collagen-like triple helical structure, an α-helical neck region and a characteristic C-terminal C-type lectin CRD domain with two intrachain disulfide bridges (Cys 135–Cys 224, Cys 202–Cys 216). MBP-C forms multimers: the largest oligomeric form contains 18 monomers arranged as a hexamer of trimers with an approximate mass of 570 kDa exhibiting a bouquet-like structure with six globular heads connected by collagen-like strands to the central core.

The 3D structure of the neck and the carbohydrate recognition domain (CRD) of MBP-C (141 aa: Ala 88–Ile 228) was determined by X-ray diffraction (PDB: 1HUP) and is shown in Figure 8.14. The neck of this trimeric structure forms a triple α-helical coiled coil central region (in blue) and each α-helix interacts with a neighbouring CRD.

**Complement-activating component** (MASP-1, EC 3.4.21.-) (see Data Sheet) or mannan-binding lectin serine protease 1 and **mannan-binding lectin serine protease 2** (MASP-2, EC 3.4.21.104) (see Data Sheet) are single-chain glycoproteins very similar in structure and function both belonging to the peptidase S1 family and each containing one potential OH-Asn

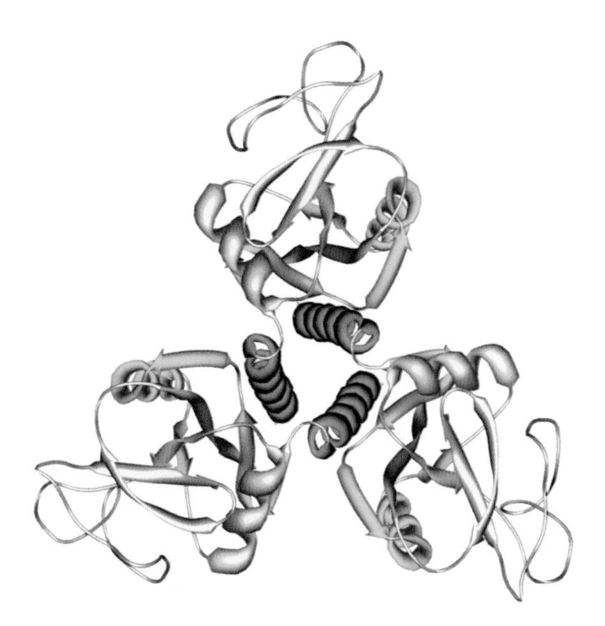

**Figure 8.14 3D structure the carbohydrate recognition domain (CRD) of mannose-binding protein C (PDB: 1HUP)**
The trimeric structure of the CRD domain of the mannose-binding protein C forms a triple α-helical coiled-coil central region (in blue) and each α-helix interacts with a neighbouring CRD.

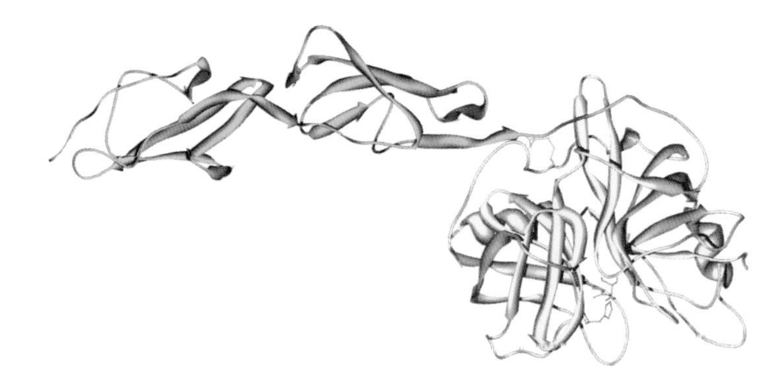

**Figure 8.15 3D structure of the proenzyme form of an MASP-2 fragment (PDB: 1ZJK)**
The MASP-2 fragment comprising two sushi domains (in green) and the catalytic chain (in yellow) are connected by a single disulfide bridge (in blue) and the residues of the catalytic triad (in pink) are indicated.

residue (aa 140 in MASP-1 and aa 143 in MASP-2). MASP-1 and MASP-2 are activated by cleaving their Arg 429–Ile 430 peptide bonds. Activated MASP-1 and MASP-2 are composed of a heavy chain containing two sushi/CCP/SCR, two CUB and one EGF-like domain and a light chain comprising the serine protease domain. Both chains are linked by an interchain disulfide bridge (Cys 417 heavy chain–Cys 104 light chain in MASP-1 and Cys 419 heavy chain–Cys 92 light chain in MASP-2). MASP-2 forms disulfide-linked homodimers. MASP-1 and MASP-2 exhibit structural and functional similarity with C1r and C1s.

The 3D structure of the proenzyme form of a MASP-2 fragment (403 aa: Thr 272–Phe 671 + N-terminal tripeptide) comprising the two sushi domains (in green) and the catalytic chain (in yellow) was determined by X-ray diffraction (PDB: 1ZJK) and is shown in Figure 8.15. As expected, the 3D structure is similar to the corresponding regions in C1r and C1s. The MASP-2 proenzyme form is capable of cleaving its natural substrate complement C4 with an efficiency of approximately 10 % compared with active MASP-2. The disulfide bridge between the catalytic chain and the two sushi domains (in blue) and the residues of the catalytic triad (in pink) are indicated.

**C-reactive protein** (CRP) (see Data Sheet) is a single-chain protein that belongs to the pentaxin family containing one pentaxin domain, an N-terminal pyro-Glu residue and one intrachain disulfide bridge (Cys 36–Cys 97). CRP is present in plasma as a noncovalently linked homopentamer with an approximate mass of 120 kDa, and each monomer binds two calcium ions. CRP is a classical acute phase protein and is an indicator for the extent of tissue damage and infection. The concentration in serum may increase within 24 hours by a factor of 1000 from normal levels of 100–500 μg/l to acute phase levels of 200–500 mg/l.

The 3D structure of the entire CRP (206 aa: Gln 1–Pro 206) was determined by X-ray diffraction (PDB: 1B09) and is shown in Figure 8.16. The monomeric structure exhibits a flattened β-sheet jelly-roll fold similar to lectins (concanavalin A) and contains two calcium ions (pink balls). The homopentamer is arranged as a spherical regular pentamer.

## 8.6 REGULATION OF THE COMPLEMENT SYSTEM

It is quite evident that such a pathway as the complement system with strong destructive and inflammatory power and an inbuilt amplification step is potentially dangerous for an organism and therefore has to be regulated precisely. There are several precautionary measures that guarantee an efficient regulation of the complement cascade. First of all, the activated key components are rapidly inactivated if they are not bound on the surface of an activating pathogen. In addition, the complement cascade is regulated at several levels in such a way that regulatory proteins interact with complement proteins to prevent an undesired activation on the surface of host cells and as a consequence to prevent their destruction.

Usually, the activated complement components are bound to the surface of a pathogen and therefore remain limited to the activating pathogen. However, sometimes activated complement components can also bind to the surface of host cells. In addition, as in the case of coagulation and fibrinolysis, complement components are activated spontaneously to a small extent and thus represent a potential danger for the destruction of host cells.

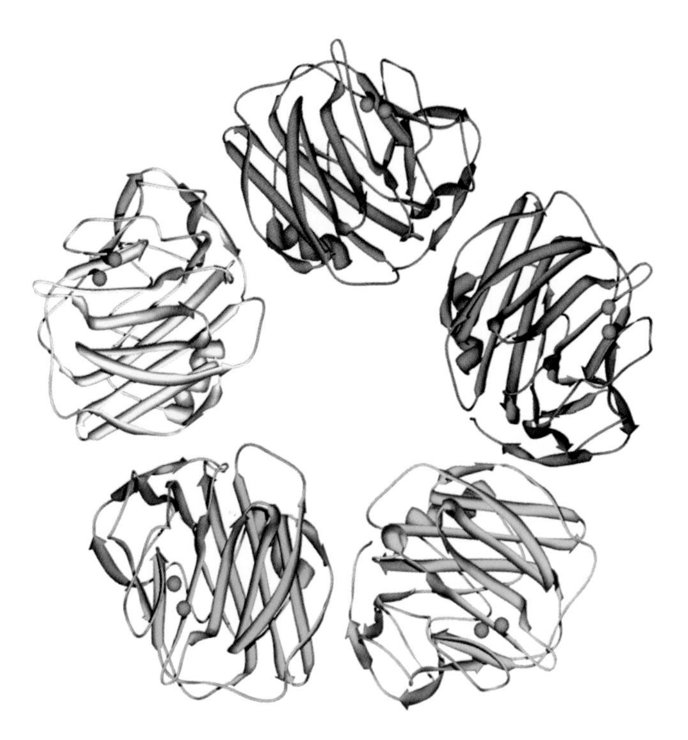

**Figure 8.16 3D structure of the homopentamer of C-reactive protein (PDB: 1B09)**
The monomer of C-reactive protein exhibits a flattened β-sheet jelly-roll fold with two calcium ions (pink balls) and the pentamer exhibits a spherical regular fold.

In Table 8.4 the major regulatory proteins of the complement system are tabulated. The complement system is primarily regulated at four levels:

**Table 8.4** Characteristics of the major control proteins of the complement system.

| Component | Mass (kDa)[a] | Concentration (mg/l)[b] | Function |
|---|---|---|---|
| Plasma protease C1 inhibitor (C1Inh) | 104 | 200 | Serpin |
| Factor H | 155 | 200–600 | Major regulator |
| Factor I (C3b/C4b inactivator) | 88 | 34 | Zymogen |
| Properdin (factor P) | 53 | 5–25 | Regulator alternative pathway |
| C4b-binding protein (Pro-rich protein) | 570 | 150 | C4b binding |
| Decay-accelerating factor (CD55 antigen) | 74 | 0.04–0.4 | Membrane protein |
| Membrane cofactor protein (CD46 antigen) | 51–68 | | Membrane protein |
| CD59 glycoprotein | 18–25 | | Cell surface glycoprotein |
| Complement receptor type 1 | 160–250 | 0.01–0.08 | Integral membrane glycoprotein |
| Complement receptor type 2 | 120 | | Integral membrane glycoprotein |

[a]Values obtained by either SDS-PAGE or sedimentation equilibrium.
[b]Represent average values.

1. The active C1 complex is regulated by the plasma protease C1 inhibitor, a serpin.
2. The active C3 convertase (C4b2b) is regulated by several complement proteins: the decay-accelerating factor (DAF) or CD55 antigen, the C4b-binding protein (C4BP), the membrane cofactor protein (MCP) or CD46 antigen, complement receptors type 1 and type 2 (CR1 and CR2) and factor I.
3. The C5 convertase of both pathways, C4b2b3b and $C3b_2Bb$, are regulated by CR1, factor H and factor I.

4. The formation of the MAC is regulated by the CD59 glycoprotein.
5. Properdin is a positive regulator of the alternative pathway.

The activation of the C1 complex is regulated by the plasma protease C1 inhibitor (C1Inh), a serpin (for details see Chapter 11). Two molecules of C1Inh bind covalently to activated C1r and C1s, respectively, leading to the disintegration of the pentameric C1 complex. C1Inh limits the time during which C4 and C2 can be activated by active C1s and additionally also limits the spontaneous activation of C1 in the fluid phase.

The type I or common form of hereditary angio(neurotic) edema is a disease with a C1Inh deficiency caused by insufficient synthesis or elevated turnover of C1Inh, resulting in a level of only 5–30 % of the normal C1Inh content. A chronic and spontaneous activation of the complement system leads to an elevated cleavage of C4 and C2 (for details see MIM: 106100).

The large fragments of C4 and C2, C4b and C2b, normally form the C3 convertase. The active thioester bond in C4b is rapidly inactivated by hydrolysis with a water molecule if the thioester is not bound to the surface of a pathogen, thus preventing the attachment to the surface of host cells. However, if by chance the C3 convertase is generated on the surface of a host cell the control mechanisms at the C3 convertase level limit the damage. Two proteins, the C4b-binding protein (C4BP) and the cell surface protein complement decay-accelerating factor (DAF) are able to compete with C2b for binding to C4b. Upon binding of C4BP to C4b the latter becomes accessible for cleavage by complement factor I, a serine protease, cleaving the C4b into C4c and C4d and thus inactivating C4b. C4BP acts as a cofactor for the factor I-mediated proteolysis of C4b.

The formation of the C3 convertase, C3bBb, cleaves more C3. To prevent the complete consumption of C3, several regulatory proteins control the formation and stability of C3bBb and the generation of C3b. Complement factor H is the major regulatory protein of this process. In addition, membrane-bound regulatory proteins such as the complement receptors type 1 and type 2 (CR1 and CR2), DAF and membrane cofactor protein (MCP) also regulate the consumption of C3 in a similar way to factor H. Once bound to C3b, C3b becomes accessible for the serine protease factor I and is cleaved at two sites (Arg 1281–Ser 1282, Arg 1298–Ser 1299), creating inactivated C3b, termed iC3b. All these proteins involved in the regulation of the C3b production act as cofactors for factor I. They all share structural and functional similarity and they all contain several sushi/CCP/SCR domains, 30 in the case of CR1. The sushi/CCP/SCR domains are responsible for the binding to C3b (see Chapter 4).

Finally, the activity of the terminal complement components is regulated by proteins of the cell surface membrane. A major inhibitor of the formation of the membrane attack complex is CD59 glycoprotein. CD59 binds to C8 and C9 upon their incorporation into the C5b–C9 complex and inhibits the assembly of the C9 homopolymer in the membrane.

Properdin is a positive regulator of the alternative pathway of the complement system. It binds to the labile C3 and C5 convertase complexes, C3bBb and $C3b_2Bb$, thus stabilising these complexes.

**Complement decay-accelerating factor** (DAF) (see Data Sheet) or CD55 antigen is a single-chain membrane glycoprotein (41 % CHO) that belongs to the receptors complement activation (RCA) family containing four sushi/CCP/SCR domains and a Ser/Thr-rich C-terminal region, which is heavily glycosylated. The C-terminal Ser 319 carries a GPI anchor that provides the membrane attachment. A C-terminal 28 residue propeptide is removed in the mature form of DAF. DAF exhibits structural and functional similarity to C4BP, MCP, CR1 and CR2, and factor H.

The 3D structure of the four sushi domains of DAF (254 aa: Asp 1–Lys252Cys + N-terminal dipeptide) was determined by X-ray diffraction (PDB: 1OK1) and is shown in Figure 8.17. The four sushi domains each containing two disulfide bridges (in pink) are arranged in a rod-like structure and a hydrophobic patch between the sushi 2 and 3 domain in DAF seems to be involved in the regulation of the C3- convertase.

**C4b-binding protein** (C4BP) (see Data Sheet) or proline-rich protein is a multimeric glycoprotein with an approximate mass of 570 kDa composed of α-chains (70 kDa) and β-chains (45 kDa). The major form ($\alpha_7\beta_1$) is composed of seven α-chains and one β-chain linked by two disulfide bridges (Cys 498, Cys 510 in α-chains–Cys 185, Cys 199 in β-chains). Other forms consist of six α-chains and one β-chain or of seven α-chains (homoheptamer) all linked by disulfide bridges. The α-chain contains eight and the β-chain three sushi/CCP/SCR domains. High-resolution electron microscopy reveals an octopus- or spider-like structure with seven thin tentacles, identified as the seven α-chains linked to a ring-like central core region. The binding site for C4b seems to be located at the peripheral end of each tentacle. C4BP exhibits structural and functional similarity with DAF, MCP, CR1 and CR2, and factor H.

**Membrane cofactor protein** (MCP) (see Data Sheet) or CD46 antigen is a single-chain type I membrane glycoprotein containing four sushi/CCP/SCR domains, a Ser/Thr/Pro-rich region (STP region) that is heavily glycosylated and a potential transmembrane segment, Val 310–Tyr 332. At least 15 splice variants exist: alternatively spliced STP regions resulting in the common higher molecular weight form of MCP containing the B + C sequence (29 aa) or the lower molecular weight form containing only the C sequence (14 aa): B: **VSTSSTTKSPASSAS** and C: **GPRPTYKPPVSNYP**.

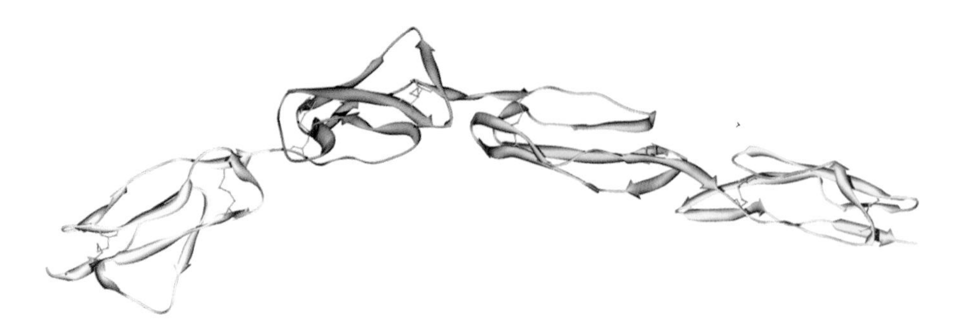

**Figure 8.17 3D structure of the four sushi domains of the complement decay-accelerating factor (PDB: 1OK1)**
The four sushi domains of the complement decay-accelerating factor each containing two disulphide bridges (in pink) exhibit a rod-like structure.

MCP isoforms have one of two alternatively spliced cytoplasmic tails: CYT-1 (16 aa) or CYT-2 (23 aa):

CYT−1: RYLQRRKKKG→ **TYLTDETHREVKFTSL**

CYT−2: RYLQRRKKKG→ **KADGGAEYATYQTKSTTPAEQRG**

MCP exhibits structural and functional similarity to DAF, C4BP, CR1 and CR2, and factor H.

**Complement receptor type 1** (CR1) (see Data Sheet) or C3b/C4b receptor is a large single-chain integral membrane glycoprotein (6–9 % CHO) that belongs to the receptors complement activation (RCA) family containing 30 sushi/CCP/SCR domains, a potential transmembrane segment (Ala 1931–Leu 1955) in the very C-terminal region of the molecule and probably an N-terminal pyro-Glu residue. Every eighth sushi/CCP/SCR domain is a highly similar repeat (65–100 % identity) and thus seven sushi/CCP/SCR repeats from a so-called long homologous repeat (LHR). A soluble form of CR1 has been identified in plasma and there exist at least four polymorphic variants in the mass range of 160–250 kDa that differ in the number of sushi/CCP/SCR domains but seem not to differ in their function. CR1 is a mosaic protein that shows an elongated shape in EM.

**Complement receptor type 2** (CR2) (see Data Sheet) is a single-chain membrane glycoprotein with an approximate mass of 120 kDa and also belongs to the receptors complement activation (RCA) family containing 15 sushi/CCP/SCR domains and a potential C-terminal transmembrane segment (Ser 952–Ser 979). Four splice variants have been described.

CR1 and CR2 exhibit a structural and functional similarity to DAF, C4BP, MCP and factor H.

**Complement factor I** (I, EC 3.4.21.45) (see Data Sheet) or C3b/C4b inactivator is a two-chain glycoprotein linked by disulfide bridge(s) and belongs to the peptidase S1 family consisting of a heavy chain (50 kDa) containing two LDLRA and one SRCR domains and a light chain (38 kDa) comprising the serine protease domain. The scavenger cysteine-rich (SRCR) domain is approximately 110 residues long and is characterised by the following consensus pattern:

SRCR_1 (PS00420): [GNRVM]−x(5)−[GLKA]−x(2)−[EQ]−x(6)−[WPS]−[GLKH]−x(2)−C−x(3)−
[FYW]−x(8)−[CM]−x(3)−G

The SRCR domain is thought to mediate protein–protein interactions and ligand binding. It is an important regulator of the complement cascade by cleaving C3b and C4b or their breakdown products at the C-terminal side of Arg residues.

**Complement factor H** (H) (see Data Sheet) is a single-chain glycoprotein (12 % CHO) with a mass of 155 kDa containing 20 sushi/CCP/SCR domains. The two C-terminal basic residues, Lys 1212 and Arg 1213, are released in circulation. Two well-defined truncated forms exist, a non-glycosylated 49 kDa form with seven sushi/CCP/SCR domains that are identical to the first seven in the 155 kDa form and an unique four aa C-terminal tail (**SFTL**) and a glycosylated 39–43 kDa form containing five sushi/CCP/SCR domains, of which three are identical to the sushi domains 18–20 in the 155 kDa form. The independently folding sushi domains form an elongated thread-like structure of approximately 80 nm, like beads on a string. Factor H exhibits structural and functional similarity with DAF, C4BP, MCP, and CR1 and CR2.

Defects in factor H are the cause of several diseases:

(a) The hemolytic-uremic syndrome is a disorder of the microvasculature leading to microangiopathic hemolytic anemia associated with distorted erythrocytes ('burr cells'), thrombocytopenia (low number of platelets) and acute renal failure (for details see MIM: 235400).

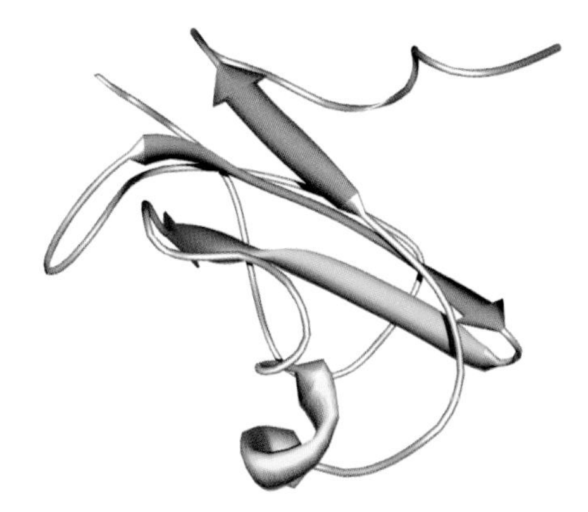

**Figure 8.18 3D structure of CD59 glycoprotein (PDB: 1CDQ)**
CD 59 glycoprotein is characterised by two antiparallel β-sheets and a short helix.

(b) The chronic hypocomplementemic nephropathy. Defects of many proteins of the complement system lead to an increased susceptibility to infections or are associated with autoimmune disorders (for details see MIM: 609496).
(c) The age-related macular degeneration 1 (ARMD1) is the most common cause of the irreversible loss of vision in the developed world, characterised by the progressive destruction and dysfunction of the central retina by accumulation of protein and lipid within an elastin-containing structure known as Bruch membrane (for details see MIM: 603075).
(d) Factor H deficiency is the cause of uncontrolled activation of the alternative pathway with concomitant consumption of C3 and terminal complement components (for details MIM: 609814).

**CD59 glycoprotein** (CD59) (see Data Sheet) is a single-chain cell surface membrane glycoprotein of 18–25 kDa mass containing one UPAR/LY6 domain. CD59 has five intrachain disulfide bridges and a GPI anchor linked to the C-terminal amidated Asn 77. A C-terminal 26 aa propeptide is removed in the mature form and the soluble form of CD59 lacks the GPI anchor.

The 3D structure of the entire CD59 (77 aa: Leu 1–Asn 77) was determined by NMR spectroscopy (PDB: 1CDQ) and is shown in Figure 8.18. The structure is characterised by two antiparallel β-sheets and a short helix.

**Properdin** (P) (see Data Sheet) or factor P is a single-chain glycoprotein (10 % CHO) containing six TSP1 domains present in plasma as a mixture of dimers, trimers and tetramers with the ratio of 20 : 54 : 26, formed from the asymmetric monomer by head-to-tail interactions. Structure prediction and infrared (IR) spectroscopy indicate a high degree of β-turn (56–67 %) and β-sheet (19–38 %) structures.

## REFERENCES

Colomb *et al.*, 1984, Activation of C1, *Philos. trans. R. Soc. London Biol.* Issue **306**, p. 283, Royal Society Publishing, London.
Gadjeva *et al.*, 2001, The mannan-binding-lectin pathway of the innate immune response, *Curr. Opin. Immunol.*, **13**, 74–78.
Gal *et al.*, 2001, Structure and function of complement activating enzyme complexes: C1 and MBL-MASPs, *Curr. Protein. Pept. Sci.*, **2**, 43–59.
Haefliger *et al.*, 1989, Complete primary structure and functional characterization action of the sixth component of the human complement system. Identification of the C5b-binding domain in complement C6, *J. Biol. Chem.*, **264**, 18 041–18 051.
Janeway *et al.*, 2005, *Immunobiology*, Garland Science Publishing.
Kirkitadze and Barlow, 2001, Structure and flexibility of the multiple domain proteins that regulate complement activation, *Immunol. Rev.*, **180**, 146–161.
Mastellos *et al.*, 2003, Complement: structure, function, evolution, and viral molecular mimicry, *Immunol. Res.*, **27**, 367–386.
Morgan, 2000, Complement. Methods and protocols, *Methods Mol. Biol.*, **150**.
Nanoka *et al.*, 2004, Evolution of the complement system, *Mol. Immunol.*, **40**, 897–902.
Reid, 1996, The complement system, in *Molecular Immunology* (eds. Hames and Glover), IRL Press, pp. 326–381.
Sim *et al.*, 2004, Proteases of the complement system, *Biochem. Soc. Trans.*, **32**, 21–27.
Sjoholm *et al.*, 2006, Complement deficiency and diseases: an update, *Mol. Immunol.*, **43**, 78–85.

DATA SHEETS

## C4b-binding protein (Proline-rich protein)

### Fact Sheet

| | | | |
|---|---|---|---|
| ***Classification*** | | ***Abbreviations*** | C4BP |
| ***Structures/motifs*** | α-chain: 8 sushi/CCP/SCR<br>β-chain: 3 sushi/CCP/SCR | ***DB entries*** | C4BP_HUMAN/P04003<br>C4BB_HUMAN/P20851 |
| ***MW/length*** | α-chain: 61 671 Da/549 aa<br>β-chain: 26 351 Da/235 aa | ***PDB entries*** | |
| ***Concentration*** | 100–300 mg/l | ***Half-life*** | |
| ***PTMs*** | α-chain: 3 N-CHO; β-chain: 5 N-CHO (potential) | | |
| ***References*** | Matsuguchi *et al.*, 1989, *Biochem. Biophys. Res. Commun.*, **165**, 138–144.<br>Hillarp *et al.*, 1993, *J. Biol. Chem.*, **268**, 15017–15023. | | |

### Description

**C4b-binding protein** is a multimeric glycoprotein synthesised in the liver and circulates in blood in complex with serum amyloid P-component and protein S. C4b-binding protein is involved in the regulation of the classical pathway of the complement system and exhibits structural and functional similarity with other C3/C4 binding proteins such as complement factor H, MCP, DAF, CR1 and CR2.

### Structure

C4b-binding protein exists in several multimeric forms: the major form is composed of seven α-chains and one β-chain linked by two disulfide bridges; other forms contain six α-chains and one β-chain or a homoheptamer of α-chains all linked by disulfide bridges. The α-chain contains eight and the β-chain three sushi/CCP/SCR domains.

### Biological Functions

C4b-binding protein regulates the classical pathway of complement activation by binding to component C4b inhibiting the formation of the C3 convertase complex (C4b2b) and by accelerating its decay. It also serves as a cofactor in the complement factor I-mediated proteolysis of C4b.

# *CD59 glycoprotein*

## Fact Sheet

| | | | |
|---|---|---|---|
| ***Classification*** | Cell surface glycoprotein | ***Abbreviations*** | CD59 |
| ***Structures/motifs*** | 1 UPAR/LY6 | ***DB entries*** | CD59_HUMAN/P13987 |
| ***MW/length*** | 8961 Da/77 aa | ***PDB entries*** | 1CDQ |
| ***Concentration*** | approx. 150 000–200 000/cell (umbilical vein) | ***Half-life*** | |
| ***PTMs*** | 3 N-CHO; 2 O-CHO (probable); 1 GPI-anchor (Asn 77) | | |
| ***References*** | Davies *et al.*, 1989, *J. Exp. Med.*, **170**, 637–654.<br>Fletcher *et al.*, 1994, *Structure*, **2**, 185–199. | | |

## Description

**CD59 glycoprotein** is a GPI-anchored plasma membrane glycoprotein involved in the regulation of the complement system.

## Structure

CD59 is a single-chain plasma membrane glycoprotein containing a UPAR/LY6 domain and a GPI anchor linked to amidated Asn 77. A C-terminal 26 residue propeptide is removed in the mature form. The soluble form lacks the GPI anchor. The 3D structure is characterised by two antiparallel β-sheets and a short α-helix (**1CDQ**).

## Biological Functions

CD59 is a potent inhibitor of the complement membrane attack complex by binding to complement C8 and/or C9, preventing the incorporation of multiple copies of C9 required for the formation of the osmolytic pore.

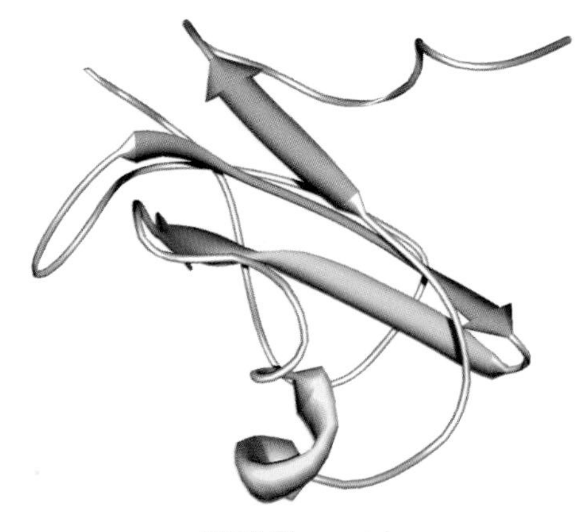

CD59 Glycoprotein

## *Complement-activating component (Mannan-binding lectin serine protease 1, EC 3.4.21.-)*

### Fact Sheet

| | | | |
|---|---|---|---|
| ***Classification*** | Peptidase S1 | ***Abbreviations*** | MASP-1 |
| ***Structures/motifs*** | 2 Sushi/CCP/SCR; 1 EGF-like; 2 CUB | ***DB entries*** | MASP1_HUMAN/P48740 |
| ***MW/length*** | 77 040 Da/680 aa | ***PDB entries*** | |
| ***Concentration*** | | ***Half-life*** | |
| ***PTMs*** | 4 N-CHO (potential); 1 OH-Asn (aa 140, potential) | | |
| ***References*** | Sato *et al.*, 1994, *Int. Immunol.*, **6**, 665–669. | | |

### Description

**Complement-activating component** (MASP-1) triggers the activation of the complement cascade. MASP-1 exhibits structural similarity with MASP-2 and complement components C1r and C1s.

### Structure

MASP-1 is a single-chain glycoprotein that is activated to a two-chain molecule composed of a heavy chain and a light chain linked by an interchain disulfide bridge (Cys 417 heavy chain–Cys 104 light chain). The heavy chain contains two sushi/CCP/SCR, two CUB and one EGF-like domain and one potential OH-Asn (140 aa) and the light chain comprises the serine protease part.

### Biological Functions

MASP-1 plays an important role in complement activation by cleaving complement components C2 and C4, thus triggering the activation of the complement cascade.

## *Complement C1q subcomponent*

### Fact Sheet

| | | | |
|---|---|---|---|
| ***Classification*** | | ***Abbreviations*** | C1q |
| ***Structures/motifs*** | Each chain: 1 collagen-like; 1 C1q | ***DB entries*** | C1QA_HUMAN/P02745<br>C1QB_HUMAN/P02746<br>C1QC_HUMAN/P02747 |
| ***MW/length*** | A: 23 688 Da/223 aa<br>B: 23 742 Da/226 aa<br>C: 22 813 Da/217 aa | ***PDB entries*** | 1PK6 |
| ***Concentration*** | 70–180 mg/l | ***Half-life*** | 3.5–4.5 days |
| ***PTMs*** | A-chain: 5 OH-Lys; 7 OH-Pro; 5 O-CHO, 1 N-CHO<br>B-chain: 6 OH-Lys; 12 OH-Pro; 4 O-CHO; 1 Pyro-Glu<br>C-chain: 2 OH-Lys; 11 OH-Pro; 1 O-CHO | | |
| ***References*** | Gaboriaud *et al.*, 2003, *J. Biol. Chem.*, **278**, 46974–46982.<br>Sellar *et al.*, 1991, *Biochem. J.*, **274**, 481–490. | | |

### Description

**Complement C1q** is a subcomponent of the complement component C1. C1 is a trimolecular complex of C1q, C1r and C1s with the molar ratio of 1 : 2 : 2 ($C1qC1r_2C1s_2$) and an approximate mass of 740 kDa. C1q initiates the classical pathway of complement activation and is one of the most cationic proteins in plasma.

### Structure

The C1q subcomponent is a plasma glycoprotein with an approximate mass of 400 kDa and is composed of six heterodimers containing six A-chains and six B-chains linked by disulfide bridges and three homodimers containing six C-chains linked by disulfide bridges. The A-, B- and C-chains are very similar in structure, containing an N-terminal collageneous domain of approximate 80 aa and a globular C1q domain of approximate 130 aa. The structure of C1q resembles a 'bunch of tulips' with six globular heads connected to six collagen-like stalks forming a central fibril-like stem. The 3D structure of the globular C1q domain of C1q exhibits a compact, almost spherical heterotrimeric assembly (A-chain: yellow; B-chain: green; C-chain: blue) (**1PK6**).

### Biological Functions

C1q initiates the classical pathway of complement activation by binding via their globular heads to IgG and IgM leading to the activation of the C1r and C1s subcomponents of C1 complement component.

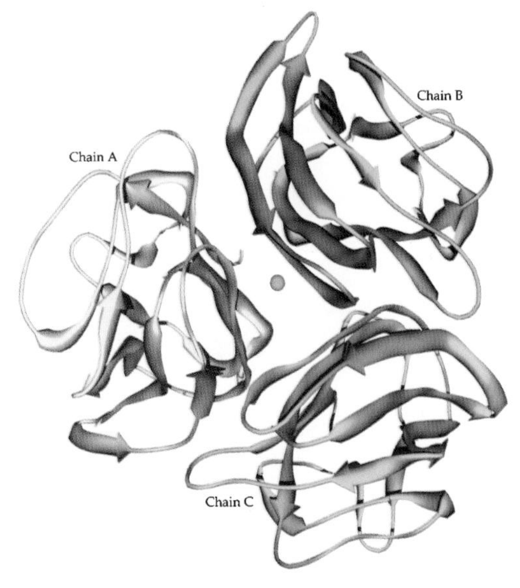


Complement Clq

## *Complement C1r subcomponent (EC 3.4.21.41)*

### Fact Sheet

| | | | |
|---|---|---|---|
| ***Classification*** | Peptidase S1 | ***Abbreviations*** | C1r |
| ***Structures/motifs*** | 2 CUB; 1 EGF-like; 2 sushi/CCP/SCR | ***DB entries*** | C1R_HUMAN/P00736 |
| ***MW/length*** | 78 268 Da/688 aa | ***PDB entries*** | 1GPZ |
| ***Concentration*** | approx. 100 mg/l | ***Half-life*** | |
| ***PTMs*** | 4 N-CHO; 1 OH-Asn (150 aa); 1 P-Ser (189 aa) | | |
| ***References*** | Budayova-Spano *et al.*, 2002, *EMBO J.*, **21**, 231–239.<br>Leytus *et al.*, 1986, *Biochemistry*, **25**, 4855–4863. | | |

### Description

**Complement C1r** is a subcomponent of the complement component C1. C1 is a trimolecular complex of C1q, C1r and C1s with the molar ratio of 1:2:2 ($C1qC1r_2C1s_2$) and an approximate mass of 740 kDa. C1r initiates the classical pathway of complement activation. C1r exhibits structural similarity with C1s and MASP-1 and MASP-2.

### Structure

The C1r subcomponent is a single-chain multidomain glycoprotein (9 % CHO) present in plasma as proenzyme. C1r is activated to the active two-chain C1r (cleavage of Arg 446–Ile 447) generating a heavy chain containing two CUB, one EGF-like and two sushi/CCP/SCR domains and two modifications, an OH-Asn (150 aa) and a P-Ser (189 aa), and a light chain comprising the serine protease part. The two chains are linked by a single interchain disulfide bridge (Cys 434 heavy chain–Cys 114 light chain). The 3D structure of a shortened proenzyme form of C1r comprising the two sushi domains (in green) and the catalytic domain (in yellow) adopts a homodimeric, elongated head-to-tail structure (**1GPZ**). The disulfide bridge connecting the catalytic chain and the sushi domain is drawn schematically (dashed blue line).

### Biological Functions

C1r interacts with C1s to form a $Ca^{2+}$-dependent tetramer (C1s–C1r–C1r–C1s) which interacts with C1q to form C1 ($C1qC1r_2C1s_2$). The binding of C1 to an activator leads to self-activation of C1r and to the activation of C1s by C1r. The main physiological inhibitor is the serpin plasma protease C1 inhibitor.

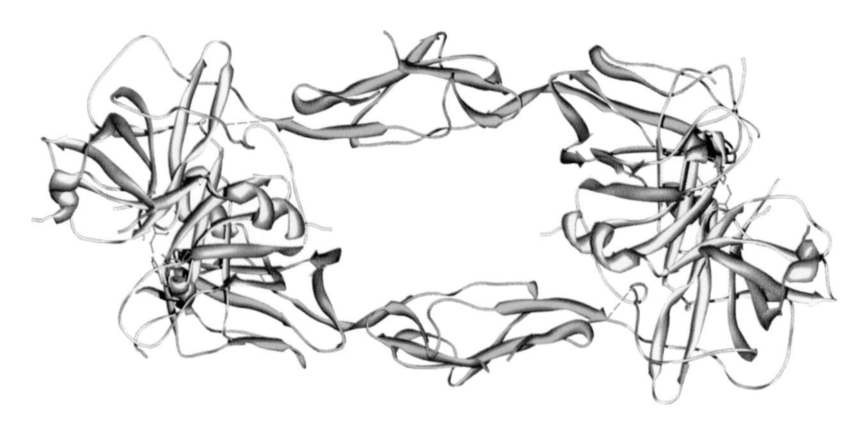

Complement Clr

# *Complement C1s subcomponent (C1 esterase; EC 3.4.21.42)*

## Fact Sheet

| | | | |
|---|---|---|---|
| ***Classification*** | Peptidase S1 | ***Abbreviations*** | C1s |
| ***Structures/motifs*** | 2 CUB; 1 EGF-like; 2 sushi/CCP/SCR | ***DB entries*** | C1S_HUMAN/P09871 |
| ***MW/length*** | 74 887 Da/673 aa | ***PDB entries*** | 1ELV |
| ***Concentration*** | approx. 30 mg/l | ***Half-life*** | |
| ***PTMs*** | 2 N-CHO; 1 OH-Asn (134 aa) | | |
| ***References*** | Gaboriaud *et al.*, 2000, *EMBO J.*, **19**, 1755–1765. | | |
| | Kusumoto *et al.*, 1988, *Proc. Natl. Acad. Sci. USA*, **85**, 7307–7311. | | |

## Description

**Complement C1s** is a subcomponent of the complement component C1. C1 is a trimolecular complex of C1q, C1r and C1s with the molar ratio of 1 : 2 : 2 ($C1qC1r_2C1s_2$) and an approximate mass of 740 kDa. C1s is activated by C1r and is involved in the early phase of the classical pathway of complement activation. C1s exhibits structural similarity with C1r and MASP-1 and MASP-2.

## Structure

The C1s subcomponent is a single-chain multidomain glycoprotein present in plasma as proenzyme. C1s is activated by C1r to the active two-chain C1s (cleavage of Arg 422–Ile 423) generating a heavy chain containing two CUB, one EGF-like and two sushi/CCP/SCR domains and an OH-Asn residue (134 aa) and a light chain comprising the serine protease part. The two chains are linked by a single interchain disulfide bridge (Cys 410 heavy chain–Cys 112 light chain). The 3D structure shows the second sushi domain (in green) linked to the chymotrypsin-like catalytic chain (in yellow) of C1s (**1ELV**). The interdomain disulfide bridge is drawn schematically (dashed blue line). The ellipsoidal sushi domain is oriented perpendicularly to the catalytic chain.

## Biological Functions

C1s forms with C1r a $Ca^{2+}$-dependent tetramer (C1s–C1r–C1r–C1s) which interacts with C1q to form C1 ($C1qC1r_2C1s_2$). Activated C1s activates complement C2 and C4. C1s activity is regulated by the plasma protease C1 inhibitor.

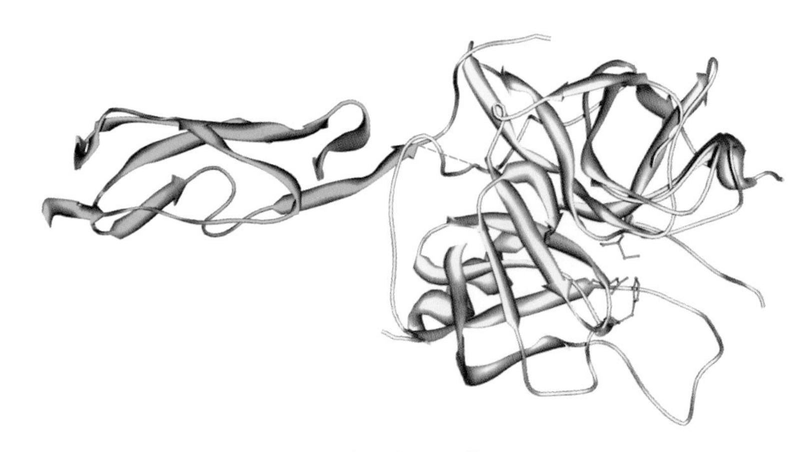

Complement Cls

## *Complement C2 (C3/C5 convertase; EC 3.4.21.43)*

### Fact Sheet

| | | | |
|---|---|---|---|
| ***Classification*** | Peptidase S1 | ***Abbreviations*** | C2 |
| ***Structures/motifs*** | 3 Sushi/CCP/SCR; 1 VWFA | ***DB entries*** | CO2_HUMAN/P06681 |
| ***MW/length*** | 81 085 Da/732 aa | ***PDB entries*** | |
| ***Concentration*** | approx. 15 mg/l | ***Half-life*** | |
| ***PTMs*** | 7 N-CHO (5 potential) | | |
| ***References*** | Bently, 1986, *Biochem. J.*, **239**, 339–345. | | |

### Description

**Complement component C2** is synthesised in the liver and in macrophages and is part of the initial phase of the classical pathway of complement activation. C2b forms together with C4b the C3/C5 convertase. C2 exhibits structural similarity with complement factor B of the alternative pathway of complement activation.

### Structure

C2 is a single-chain glycoprotein that is activated by C1s (cleavage of Arg 223–Lys 224), resulting in the C2a fragment containing three sushi/CCP/SCR domains and the C2b fragment that contains one VWFA domain and the serine protease part.

### Biological Functions

Activated C2 cleaves the component C3 α-chain into C3a and C3b and the component C5 α-chain into C5a and C5b.

## *Complement C3/C3a anaphylatoxin*

### Fact Sheet

| | | | |
|---|---|---|---|
| *Classification* | Protease inhibitor I39 | *Abbreviations* | C3 |
| *Structures/motifs* | 8 macroglobulin; 1 anaphylatoxin-like; 1 CUB; 1 NTR | *DB entries* | CO3_HUMAN/P01024 |
| *MW/length* | C3: 184967 Da/1641 aa<br>C3a: 9095 Da/77 aa | *PDB entries* | 2A73 |
| *Concentration* | 1000–2000 mg/l | *Half-life* | 2.5 days |
| *PTMs* | 3 N-CHO (1 potential) | | |
| *References* | Janssen *et al.*, 2005, *Nature*, **437**, 505–511<br>de Bruijn *et al.*, 1985, *Proc. Natl. Acad. Sci. USA.*, **82**, 708–712. | | |

### Description

**Complement component C3** is a large glycoprotein mainly synthesised in the liver and plays a key role in complement activation. C3 is the parent protein of the C3a anaphylatoxin. The processing of C3 by C3 convertase is the central reaction in both the classical and the alternative complement activation. C3 exhibits structural similarity with complement components C4 and C5 and $\alpha_2$-macroglobulin.

### Structure

The large single-chain precursor C3 glycoprotein (2 % CHO) is first processed by removal of four consecutive Arg residues (646–649 aa) resulting in an α-chain and a β-chain linked by a single disulfide bridge (Cys 67 α′-chain–Cys 537 β-chain). The C3 convertase releases from the N-terminus of the α-chain the 77-residue C3a anaphylatoxin, resulting in C3b (α′- and β-chain). C3b contains the reactive isoglutamyl cysteine thioester bond (**Cys 988–Gly–Glu–Gln 991**). C3b is rapidly processed by factor I (cleavage of Arg 1281–Ser 1282 and Arg 1298–Ser 1299) releasing the 17 residue C3f and generating inactivated iC3b. iC3b is further cleaved to smaller fragments. The 3D structure of two-chain C3 consisting of the β-chain (in green) linked by a single disulfide bridge (in blue) to the α-chain (in yellow) and C3a (in magenta) is shown (**2A73**).

### Biological Functions

Via its reactive thioester bond active C3b can bind covalently to carbohydrates on cell surfaces and immune aggregates. C3b can trigger the formation of the membrane attack complex C5b–C9 (MAC). C3a anaphylatoxin is a mediator of local inflammatory processes. It induces the contraction of smooth muscle, increases the vascular permeability and causes the release of histamine from mast cells.

Complement C3

# *Complement C4/C4a anaphylatoxin*

## Fact Sheet

| | | | |
|---|---|---|---|
| ***Classification*** | Protease inhibitor I39 | ***Abbreviations*** | C4 |
| ***Structures/motifs*** | 1 Anaphylatoxin-like; 1 NTR | ***DB entries*** | CO4A_HUMAN/P0COL4<br>CO4B_HUMAN/P0COL5 |
| ***MW/length*** | C4: 190 520 Da/1725 aa<br>C4a: 8764 Da/77 aa | ***PDB entries*** | 1HZF |
| ***Concentration*** | 200–600 mg/l | ***Half-life*** | |
| ***PTMs*** | 4 N-CHO (2 potential); 3 S-Tyr (1398, 1401, 1403 aa) | | |
| ***References*** | Belt *et al.*, 1984, *Cell*, **36**, 907–914.<br>van den Elsen, 2002, *J. Mol. Biol.*, **322**, 1103–1115. | | |

## Description

**Complement component C4** is a large glycoprotein mainly synthesised in the liver and is important in the activation of the classical pathway. C4 is the parent protein of C4a anaphylatoxin. C4 exhibits structural similarity with complement components C3 and C5 and $\alpha_2$-macroglobulin.

## Structure

The large single-chain precursor C4 glycoprotein (7 % CHO) is first processed by removal of two propeptides of 4 and 7 residues, resulting in a three-chain molecule consisting of an $\alpha$-chain, a $\beta$-chain and a $\gamma$-chain linked by disulfide bridges. Activated C1s releases from the N-terminus of the $\alpha$-chain the 77-residue C4a anaphylatoxin, resulting in C4b linked to the $\beta$- and $\gamma$- chains. C4b contains the reactive isoglutamyl cysteine thioester bond (**Cys 991–Gly–Glu–Gln 994**). The shown 3D structure of the C4d fragment (N-terminally truncated form; 335 aa: Ser 983–Arg 1317) exhibits the $\alpha$-$\alpha$ six-barrel fold (**1HZF**).

## Biological Functions

C4b is an essential subunit of the C3 convertase (C4b2b) and the C5 convertase (C3b4b2b) enzymes of the classical complement pathway. Via its reactive thioester bond active C4b can bind covalently to activated cell surfaces. C3a anaphylatoxin is a mediator of local inflammatory processes. It induces the contraction of smooth muscle, increases the vascular permeability and causes the release of histamine from mast cells.

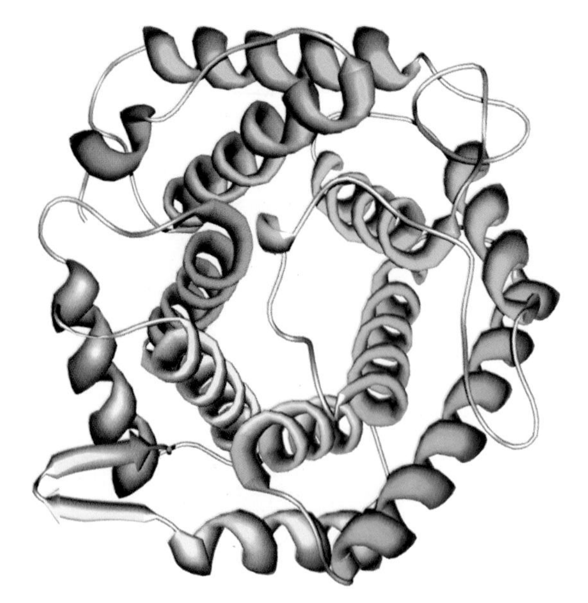

Complement C4d fragment

# *Complement C5/C5a anaphylatoxin*

## Fact Sheet

| | | | |
|---|---|---|---|
| ***Classification*** | Protease inhibitor I39 | ***Abbreviations*** | C5 |
| ***Structures/motifs*** | 1 Anaphylatoxin-like; 1 NTR | ***DB entries*** | CO5_HUMAN/P01031 |
| ***MW/length*** | C5: 186 369 Da/1658 aa;<br>C5a: 8274 Da/74 aa | ***PDB entries*** | |
| ***Concentration*** | 70–100 mg/l | ***Half-life*** | 1.9%/h |
| ***PTMs*** | 4 N-CHO (3 potential) | | |
| ***References*** | Haviland *et al.*, 1991, *J. Immunol.*, **146**, 362–368. | | |

## Description

**Complement C5** is a large glycoprotein mainly synthesised in hepatocytes and is part of the classical and the alternative pathway. C5 is the parent protein of C5a anaphylatoxin. C5b initiates the formation of the membrane attack complex. C5 exhibits structural similarity with complement components C3 and C4 and $\alpha_2$-macroglobulin.

## Structure

The large single-chain precursor C5 glycoprotein (3 % CHO) is first processed by removal of a propeptide (656–659 aa), resulting in an α-chain containing a C-terminal NTR domain and a β-chain linked by a single disulfide bridge. The C5 convertase releases from the N-terminus of the α-chain the 74-residue C5a anaphylatoxin, resulting in C5b (α′- and β-chains).

## Biological Functions

The formation of C5b initiates the spontaneous assembly of the late complement components C5b–C9 into the membrane attack complex (MAC). The C5b–C6 complex is the base for the assembly of the lytic complex. C5a anaphylatoxin is a mediator of local inflammatory processes. It induces the contraction of smooth muscle, increases the vascular permeability and causes the release of histamine from mast cells. C5a is involved in chemokinesis and chemotaxis.

## *Complement component C6*

### Fact Sheet

| | | | |
|---|---|---|---|
| ***Classification*** | Complement C6/C7/C8/C9 | ***Abbreviations*** | C6 |
| ***Structures/motifs*** | 1 LDLRA; 2 sushi/CCP/SCR; 1 EGF-like; 3 TSP1 | ***DB entries*** | CO6_HUMAN/P13671 |
| ***MW/length*** | 102 470 Da/913 aa | ***PDB entries*** | |
| ***Concentration*** | approx. 50 mg/l | ***Half-life*** | Cleared in 30–50 min |
| ***PTMs*** | 2 N-CHO (potential); 6 C-CHO (Trp) | | |
| ***References*** | Haefliger *et al.*, 1989, *J. Biol. Chem.*, **264**, 18041–18051. | | |

### Description

**Complement component C6** is a large glycoprotein mainly synthesised in the liver. C6 is part of the lytic complex C5b–C9. C6 exhibits structural similarity with complement components C7, C8, C9 and perforin.

### Structure

C6 is a single-chain multidomain glycoprotein (3 % CHO) containing three TSP1, two sushi/CCP/SCR, one EGF-like and one LDLRA domains and two complement control factor I modules. C6 has two potential transmembrane segments (Asp 310–Tyr 328, Tyr 333–Gly 352).

### Biological Functions

C5b binds to the C-terminal region of C6 containing two sushi/CCP/SCR and two complement control factor I modules, resulting in the soluble C5b,6 complex that is the nucleus for the assembly of the lytic complex.

## *Complement component C7*

### Fact Sheet

| | | | |
|---|---|---|---|
| ***Classification*** | Complement C6/C7/C8/C9 | ***Abbreviations*** | C7 |
| ***Structures/motifs*** | 1 LDLRA; 2 sushi/CCP/SCR; 1 EGF-like; 2 TSP1 | ***DB entries*** | CO7_HUMAN/P10643 |
| ***MW/length*** | 91 115 Da/821 aa | ***PDB entries*** | |
| ***Concentration*** | 40–70 mg/l | ***Half-life*** | 61 h |
| ***PTMs*** | 2 N-CHO (1 potential); 4 C-CHO (Trp) | | |
| ***References*** | DiScipio *et al.*, 1988, *J. Biol. Chem.*, **263**, 549–560. | | |

### Description

**Complement component C7** is a large glycoprotein mainly synthesised in phagocytes. C7 is part of the lytic complex C5b–C9. C7 exhibits structural similarity with complement components C6, C8, C9 and perforin.

### Structure

C7 is a single-chain multidomain glycoprotein (5 % CHO) containing two TSP1, two sushi/CPP/SCR, one EGF-like and one LDLRA domain and two complement control factor I modules. C7 has two potential transmembrane segments (Gln 249–Tyr 265, Tyr 270–Gly 289).

### Biological Functions

C7 combines with the C5b,6 complex to form the C5b,6,7 complex which has the capability to attach to cell membranes. This complex is the base for the formation of the membrane attack complex C5b–C9.

# *Complement component C8*

## Fact Sheet

| | | | |
|---|---|---|---|
| ***Classification*** | Complement C6/C7/C8/C9; Lipocalin | ***Abbreviations*** | C8 (C8A, C8B, C8G) |
| ***Structures/motifs*** | Each α-/β-chain:<br>1 LDLRA; 1 EGF-like; 2 TSP1 | ***DB entries*** | CO8A_HUMAN/P07357<br>CO8B_HUMAN/P07358<br>CO8G_HUMAN/P07360 |
| ***MW/length*** | α-chain: 61 711 Da/554 aa<br>β-chain: 61 043 Da/537 aa<br>γ-chain: 20 327/182 aa | ***PDB entries*** | |
| ***Concentration*** | 70–90 mg/l | ***Half-life*** | |
| ***PTMs*** | α-chain: 1 N-CHO (potential); 4 C-CHO (Trp)<br>β-chain: 2 N-CHO (potential); 4 C-CHO (Trp)<br>γ-chain: 1 Pyro-Glu | | |
| ***References*** | Howard *et al.*, 1987, *Biochemistry*, **26**, 3565–3570.<br>Ng *et al.*, 1987, *Biochemistry*, **26**, 5229–5233.<br>Rao *et al.*, 1987, *Biochemistry*, **26**, 3556–3564. | | |

## Description

**Complement component C8** is a large glycoprotein composed of three nonidentical chains, the α- β- and γ-chains. C8 is synthesised in the liver, monocytes and fibroblasts and binds to the C5b,6,7 complex, forming the C5b,6,7,8 complex that is part of the lytic complex C5b–C9. C8α and C8β exhibit structural similarity with complement components C6, C7, C9 and perforin.

## Structure

C8 is a multidomain glycoprotein composed of three chains: α-chain, β-chain and γ-chain. The α- and the γ-chains are linked by a single disulfide bridge and the β-chain is noncovalently attached. The α- and the β-chains each contain two TSP1, one EGF-like and one LDLRA domains. Both chains have two potential transmembrane segments. The γ-chain carries a pyro-Glu residue and belongs to the lipocalin family.

## Biological Functions

C8 combines with the C5b,6,7 complex to form the C5b,6,7,8 complex, which in turn binds C9 and thus acts as a catalyst in the polymerisation reaction of C9. The CD59 GPI-anchored membrane glycoprotein binds to C8 and/or C9, protecting the host cell from complement-mediated lysis.

# *Complement component C9*

## Fact Sheet

| | | | |
|---|---|---|---|
| ***Classification*** | Complement C6/C7/C8/C9 | ***Abbreviations*** | C9 |
| ***Structures/motifs*** | 1 LDLRA; 1 EGF-like; 1 TSP1 | ***DB entries*** | CO9_HUMAN/P02748 |
| ***MW/length*** | 60 979 Da/538 aa | ***PDB entries*** | |
| ***Concentration*** | 60 mg/l | ***Half-life*** | |
| ***PTMs*** | 2 N-CHO (1 probable); 2 C-CHO (Trp) | | |
| ***References*** | Stanley *et al*., 1985, *EMBO J*., **4**, 375–382. | | |

## Description

**Complement component C9** is a plasma glycoprotein synthesised in the liver and in monocytes. C9 is the final component to be incorporated into the C5b,6,7,8 complex, forming the lytic complex C5b–C9. C9 exhibits structural similarity with complement components C6, C7, C8 and perforin.

## Structure

C9 is a single-chain multidomain glycoprotein (8–15 % CHO) containing one TSP1, one EGF-like and one LDLRA domain. C9 has two potential transmembrane segments (Val 293–Tyr 309, Tyr 314–Gly 333).

## Biological Functions

Up to 18 C9 molecules combine with the C5b,6,7,8 complex to form the membrane attack complex (MAC). This cylindrical structure is able to enter the lipid bilayer resulting in a transmembrane channel and subsequent cell death. The nascent C5b–C9 complex formed in the fluid phase is inhibited by clusterin and vitronectin.

# *Complement decay-accelerating factor (CD55 antigen)*

## Fact Sheet

| | | | |
|---|---|---|---|
| ***Classification*** | Receptors complement activation (RCA) | ***Abbreviations*** | DAF |
| ***Structures/motifs*** | 4 Sushi/CCP/SCR | ***DB entries*** | DAF_HUMAN/P08174 |
| ***MW/length*** | 34 964 Da/319 aa | ***PDB entries*** | 1OK1 |
| ***Concentration*** | 40–400 μg/l | ***Half-life*** | approx. 7 h (tissue) |
| ***PTMs*** | 1 N-CHO (potential); many O-CHO (Ser/Thr-rich); 1 GPI-anchor (Ser 319) | | |
| ***References*** | Caras *et al.*, 1987, *Nature*, **325**, 545–549. Lukacik *et al.*, 2004, *Proc. Natl. Acad. Sci. USA*, **101**, 1279–1284. | | |

## Description

**Complement decay-accelerating factor** (DAF) is a membrane glycoprotein involved in the inhibition of the complement cascade and is synthesised in many cell types. DAF exhibits structural and functional similarity with other C3/C4 binding proteins such as C4b binding protein, complement factor H, MCP, CR1 and CR2.

## Structure

DAF is a single-chain membrane glycoprotein (41 % CHO) containing four sushi/CCP/SCR domains, a Ser/Thr-rich region which is heavily glycosylated and a GPI anchor attached to C-terminal Ser 319. A C-terminal 28 residue propeptide is removed in the mature form. The four sushi/CCP/SCR domains are arranged in a rod-like structure (**1OK1**).

## Biological Functions

DAF regulates the central enzyme of the complement system, the C3 convertase, by dissociation (decay) of C2b from C4b2b and Bb from C3bBb thus protecting the cells from damage.

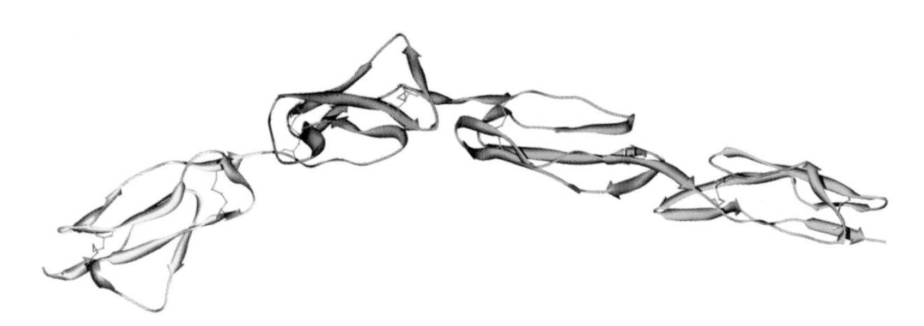

Complement Decay-accelerating Factor

## *Complement factor B (C3/C5 convertase; EC 3.4.21.47)*

### Fact Sheet

| | | | |
|---|---|---|---|
| ***Classification*** | Peptidase S1 | ***Abbreviations*** | B |
| ***Structures/motifs*** | 3 Sushi/CCP/SCR; 1 VWFA | ***DB entries*** | CFAB_HUMAN/P00751 |
| ***MW/length*** | 88 001 Da/739 aa | ***PDB entries*** | 1RRK |
| ***Concentration*** | 170–260 mg/l | ***Half-life*** | approx. 3 days |
| ***PTMs*** | 5 N-CHO | | |
| ***References*** | Mole *et al.*, 1984, *J. Biol. Chem.*, **259**, 3407–3412.<br>Ponnuraj *et al.*, 2004, *Mol. Cell.*, **14**, 17–28. | | |

### Description

**Complement factor B** is part of the alternative pathway of the complement system and is present in plasma as a zymogen. Together with complement factor C3b, Bb forms the C3/C5 convertase of the alternative pathway. Factor B exhibits structural similarity with complement factor C2.

### Structure

Factor B is a single-chain glycoprotein (7–10 % CHO). Upon binding to complement C3b it is activated by complement factor D (cleavage of Arg 234–Lys 235), resulting in the Ba fragment containing three sushi/CCP/SCR domains and the Bb fragment that contains one VWFA domain and the serine protease part. The 3D structure of fragment Bb consisting of the VWFA domain (in green) and the catalytic chain (in yellow) exhibits the structure as the individual domains (**1RRK**).

### Biological Functions

Factor B is the central enzyme of the alternative pathway of complement activation. The C3/C5 convertase (C3bBb) cleaves the complement components C3 and C5, releasing the anaphylatoxins C3a and C5a, respectively. Complement factor H inhibits the biological activity of C3bBb by accelerating the release of Bb from the complex.

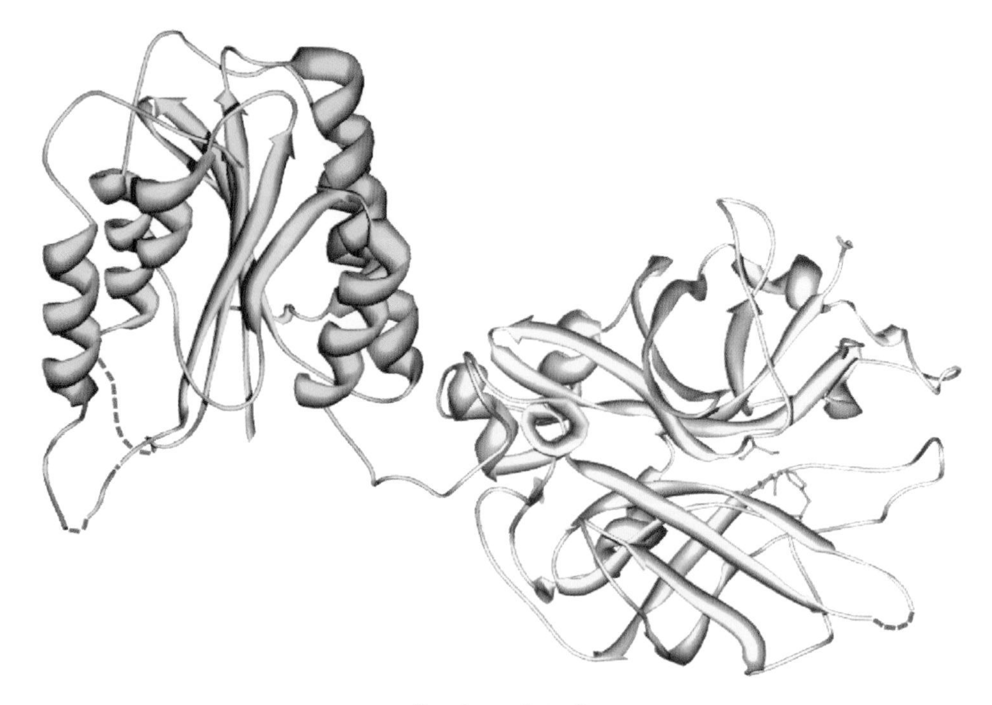

Complement factor B

# *Complement factor D (C3 convertase activator; EC 3.4.21.46)*

## Fact Sheet

| | | | |
|---|---|---|---|
| ***Classification*** | Peptidase S1 | ***Abbreviations*** | D |
| ***Structures/motifs*** | | ***DB entries*** | CFAD_HUMAN/P00746 |
| ***MW/length*** | 24 405 Da/229 aa | ***PDB entries*** | 1FDP |
| ***Concentration*** | 1.4–2.2 mg/l | ***Half-life*** | 1 h |
| ***PTMs*** | | | |
| ***References*** | Jing *et al.*, 1999, *EMBO J.*, **18**, 804–814.<br>White *et al.*, 1992, *J. Biol. Chem.*, **267**, 9210–9213. | | |

## Description

**Complement factor D** is part of the alternative pathway of complement activation and is present in plasma in its active form (no zymogen in plasma). Factor D has the lowest plasma concentration of all complement components.

## Structure

Factor D is a single-chain protein belonging to the peptidase S1 family. A potential propeptide (5 aa) seems to be removed. The 3D structure of the proform of factor D was determined (**1FDP**). Active site residues are in pink (numbering according to active factor D).

## Biological Functions

Factor D cleaves factor B when the latter is bound to factor C3b, which then becomes the C3 convertase (C3bBb) of the alternative pathway. The function of factor B is analogous to that of the complement subcomponent C1s of the classical pathway.

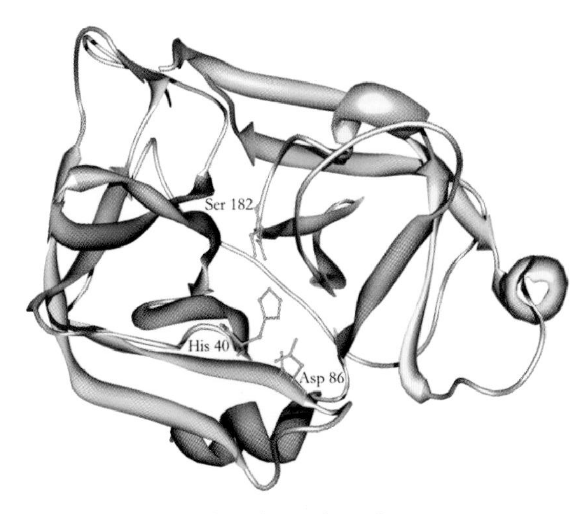


Complement factor D

## *Complement factor H*

### Fact Sheet

| | | | |
|---|---|---|---|
| ***Classification*** | | ***Abbreviations*** | H |
| ***Structures/motifs*** | 20 Sushi/CCP/SCR | ***DB entries*** | CFAH_HUMAN/P08603 |
| ***MW/length*** | 137 082 Da/1213 aa | ***PDB entries*** | |
| ***Concentration*** | 200–600 mg/l | ***Half-life*** | 4–5 days |
| ***PTMs*** | 8 N-CHO (3 potential) | | |
| ***References*** | Ripoche *et al.*, 1988, *Biochem. J.*, **249**, 593–602. | | |

### Description

**Complement factor H** is a large glycoprotein of the alternative pathway mainly synthesised in the liver. Factor H is a major regulatory protein of the complement system and exhibits structural and functional similarity with other C3/C4 binding proteins such as C4b binding protein, MCP, DAF, CR1 and CR2.

### Structure

Factor H is a single-chain glycoprotein (12 % CHO) containing 20 sushi/CCP/SCR domains. In circulation the two C-terminal residues, Lys 1212 and Arg 1213, are released.

### Biological Functions

Factor H is the major soluble protein regulating the formation and the stability of C3bBb protease complex (C3 convertase) and is involved as a cofactor in the inactivation of C3b by complement factor I.

## *Complement factor I (C3b/C4b inactivator; EC 3.4.21.45)*

### Fact Sheet

| | | | |
|---|---|---|---|
| ***Classification*** | Peptidase S1 | ***Abbreviations*** | I |
| ***Structures/motifs*** | 2 LDLRA; 1 SRCR | ***DB entries*** | CFAI_HUMAN/P05156 |
| ***MW/length*** | 63 457 Da/565 aa | ***PDB entries*** | |
| ***Concentration*** | approx. 35 mg/l | ***Half-life*** | approx. 3 days |
| ***PTMs*** | 6 N-CHO (5 potential) | | |
| ***References*** | Catterall *et al.*, 1987, *Biochem. J.*, **242**, 849–856. | | |

### Description

**Complement factor I** is a regulatory protein of the complement system that is present in plasma as an active serine protease.

### Structure

Factor I is a two-chain glycoprotein linked by disulfide bridge(s). The heavy chain (317 aa) consists of two LDLRA and one SRCR domain, the light chain (244 aa) comprises the serine protease part.

### Biological Functions

Factor I cleaves and inactivates the complement subcomponents C3b and C4b and their breakdown products, thus stopping complement activation. The presence of the cofactors C4-binding protein and complement factor H is required for the inactivation process.

## *Complement receptor type 1 (C3b/C4b receptor)*

### Fact Sheet

| | | | |
|---|---|---|---|
| *Classification* | Receptors complement activation (RCA) | *Abbreviations* | CR1 |
| *Structures/motifs* | 30 Sushi/CCP/SCR | *DB entries* | CR1_HUMAN/P17927 |
| *MW/length* | 219 637 Da/1998 aa | *PDB entries* | |
| *Concentration* | 10–80 μg/l | *Half-life* | 11–32 days |
| *PTMs* | 20 N-CHO (3 potential); 1 Pyro-Glu (potential) | | |
| *References* | Vik and Wong, 1993, *J. Immunol.*, **151**, 6214–6224. | | |

### Description

**Complement receptor type 1** (CR1) is a large integral membrane glycoprotein involved in the regulation of the complement system and is synthesised by many different cell types. CR1 exhibits structural and functional similarity with other C3/C4 binding proteins such as C4b binding protein, complement factor H, MCP, DAF and CR2.

### Structure

CR1 is a single-chain membrane glycoprotein (6–9 % CHO) containing 30 sushi/CCP/SCR domains, a potential C-terminal transmembrane segment (Ala 1931–Leu 1955) and most likely a pyro-Glu residue.

### Biological Functions

CR1 plays a key role in the regulation of the complement cascade. CR1 binds C3b and C4b reversibly and acts as a cofactor for their breakdown and is predominantly involved in the clearance of immune complexes.

## *Complement receptor type 2*

### Fact Sheet

| | | | |
|---|---|---|---|
| *Classification* | Receptors complement activation (RCA) | *Abbreviations* | CR2 |
| *Structures/motifs* | 15 Sushi/CCP/SCR | *DB entries* | CR2_HUMAN/P20023 |
| *MW/length* | 111 166 Da/1013 aa | *PDB entries* | |
| *Concentration* | | *Half-life* | |
| *PTMs* | 11 N-CHO (potential) | | |
| *References* | Weis *et al.*, 1988, *J. Exp. Med.*, **167**, 1047–1066. | | |

### Description

**Complement receptor type 2** (CR2) is a large integral membrane glycoprotein involved in the regulation of the complement system and is synthesised by many different cell types. CR2 exhibits structural and functional similarity with other C3/C4 binding proteins such as C4b binding protein, complement factor H, MCP, DAF and CR1.

### Structure

CR2 is a single-chain membrane glycoprotein containing 15 sushi/CCP/SCR domains and a potential C-terminal transmembrane segment (Ser 952–Ser 979).

### Biological Functions

CR2 is a receptor for the C3d region of complement component C3 and plays a certain role in the regulation of the complement cascade.

# *C-reactive protein*

## Fact Sheet

| | | | |
|---|---|---|---|
| ***Classification*** | Pentaxin | ***Abbreviations*** | CRP |
| ***Structures/motifs*** | 1 Pentaxin | ***DB entries*** | CRP_HUMAN/P02741 |
| ***MW/length*** | 23 047 Da/206 aa | ***PDB entries*** | 1B09 |
| ***Concentration*** | 100–500 μg/l | ***Half-life*** | 13–16 h |
| ***PTMs*** | 1 Pyro-Glu | | |
| ***References*** | Lei *et al.*, 1985, *J. Biol. Chem.*, **260**, 13377–13383. | | |
| | Thompson *et al.*, 1999, *Structure Fold Des.*, **7**, 169–177. | | |

## Description

**C-reactive protein** (CRP) is an acute-phase plasma protein synthesised in the liver and is involved in complement activation among other biological functions.

## Structure

CRP is a single-chain protein containing one pentaxin domain, an N-terminal pyro-Glu residue and one intrachain disulfide bridge (Cys 36–Cys 97). CRP is present in plasma as a noncovalently linked homopentamer and each subunit binds two calcium ions. The 3D structure of monomeric CRP exhibits a β-sheet jelly-roll fold containing two calcium ions (pink balls) and the homopentamer exhibits a spherical regular pentameric structure (**1B09**).

## Biological Functions

CRP exhibits several functions associated with host defence, among them complement activation. CRP is an acute phase reactant upon tissue damage and infection resulting in up to a thousand-fold increase in plasma.

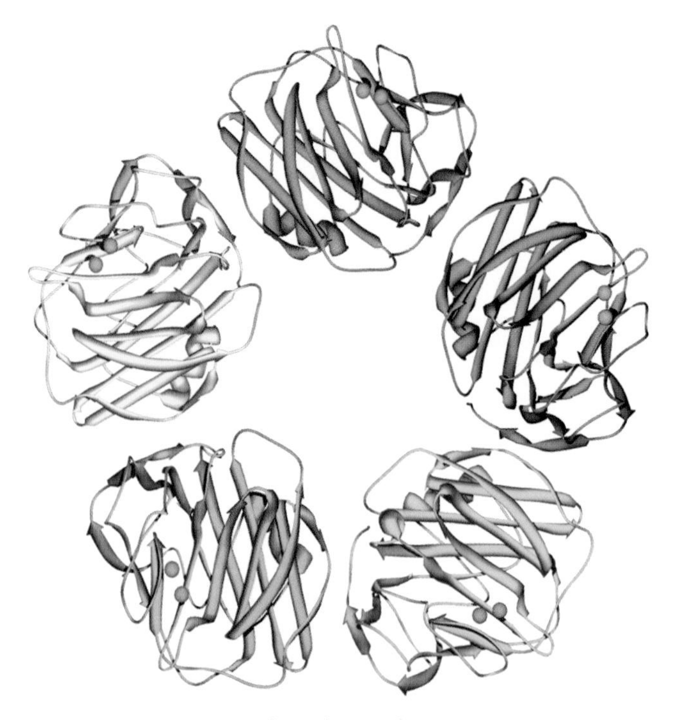

C-reactive protein

# *Mannan-binding lectin serine protease 2 (EC 3.4.21.-)*

## Fact Sheet

| | | | |
|---|---|---|---|
| ***Classification*** | Peptidase S1 | ***Abbreviations*** | MASP-2 |
| ***Structures/motifs*** | 2 Sushi/CCP/SCR; 1 EGF-like; 2 CUB | ***DB entries*** | MASP2_HUMAN/O00187 |
| ***MW/length*** | 74 143 Da/671 aa | ***PDB entries*** | 1ZJK |
| ***Concentration*** | | ***Half-life*** | |
| ***PTMs*** | 1 OH-Asn (143 aa, potential) | | |
| ***References*** | Gal *et al.*, 2005, *J. Biol. Chem.*, **280**, 33435–33444.<br>Thiel *et al.*, 1997, *Nature*, **386**, 506–510. | | |

## Description

**Mannan-binding lectin serine protease 2** (MASP-2) plays an important role in the activation process of the complement system. MASP-2 exhibits structural similarity with MASP-1 and complement components C1r and C1s.

## Structure

MASP-2 is a single-chain protein that is activated autocatalytically (cleavage of Arg 429–Ile 430) resulting in a heavy and a light chain linked by an interchain disulfide bridge (Cys 419 heavy chain–Cys 92 light chain). The heavy chain contains two sushi/CCP/SCR, two CUB and one EGF-like domain and one potential OH-Asn (143 aa) and the light chain comprises the serine protease part. MASP-2 forms disulfide-linked homodimers. The 3D structure of the proenzyme form of a MASP-2 fragment comprising the two sushi domains (in green) and the catalytic chain (in yellow) is comparable as in C1r and C1s (**1ZJK**).

## Biological Functions

MASP-2 plays an important role in complement activation via mannose-binding lectin. Activated MASP-2 cleaves complement components C2 and C4, resulting in their activation and the subsequent formation of the C3 convertase.

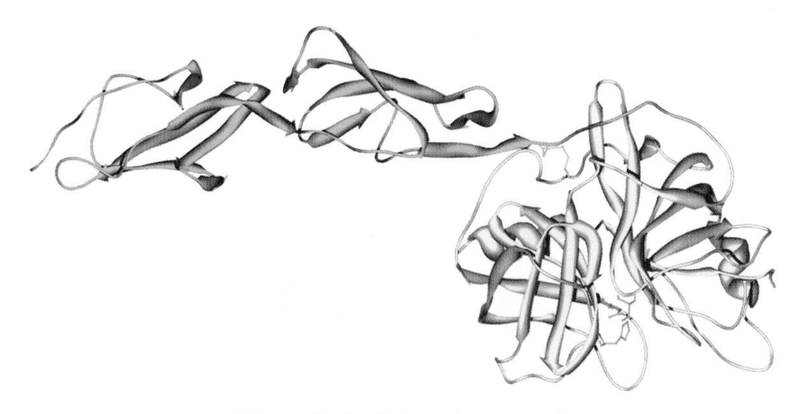

Mannan-binding lectin serine protease 2

## *Mannose-binding protein C (Mannose-binding lectin)*

### Fact Sheet

| | | | |
|---|---|---|---|
| ***Classification*** | | ***Abbreviations*** | MBP-C |
| ***Structures/motifs*** | 1 C-type lectin; 1 collagen-like | ***DB entries*** | MABC_HUMAN/P11226 |
| ***MW/length*** | 24 021 Da/228 aa | ***PDB entries*** | 1HUP |
| ***Concentration*** | approx. 1 mg/l | ***Half-life*** | approx. 5 days |
| ***PTMs*** | 5 OH-Pro | | |
| ***References*** | Ezekowitz *et al.*, 1988, *J. Exp. Med.*, **167**, 1034–1046.<br>Sheriff *et al.*, 1994, *Nat. Struct. Biol.*, **1**, 789–794. | | |

### Description

**Mannose-binding protein C** (MBP-C) is primarily synthesised in the liver and is capable of activating the classical complement pathway independently of antibody.

### Structure

MBP-C is a single-chain protein containing one C-type lectin with two intrachain disulfide bridges (Cys 135–Cys 224, Cys 202–Cys 216) and one collagen-like domain with five OH-Pro residues. MBP-C forms an oligomeric complex of six homotrimers linked by interchain disulfide bridges (probably Cys 5, 12 and 18 in the N-terminal region). The carbohydrate recognition domain (CRD) forms trimers with a triple α-helical coiled coil central region (**1HUP**).

### Biological Functions

MBP-C binds mannose and *N*-acetylglucosamine in the presence of calcium. It is capable of host defence against pathogens by activation of the classical complement system independently of antibody.

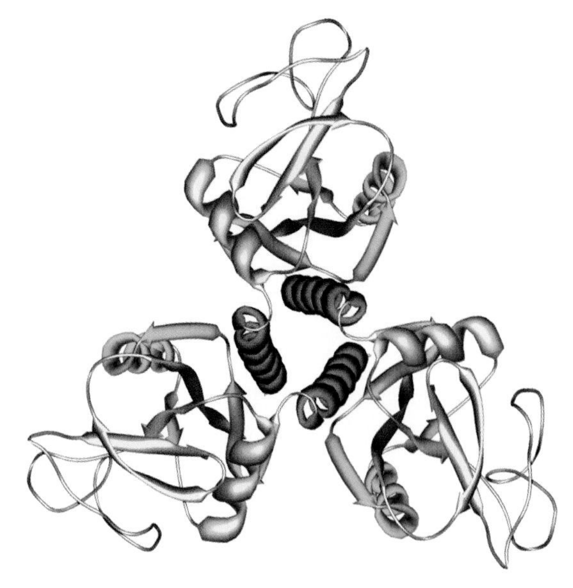

Mannose-binding protein C

# *Membrane cofactor protein (CD46 antigen)*

## Fact Sheet

| | | | |
|---|---|---|---|
| ***Classification*** | Type I membrane | ***Abbreviations*** | MCP |
| ***Structures/motifs*** | 4 Sushi/CCP/SCR | ***DB entries*** | MCP_HUMAN/P15529 |
| ***MW/length*** | 39 897 Da/358 aa | ***PDB entries*** | |
| ***Concentration*** | approx. 20 000–50 000/cell | ***Half-life*** | approx. 12 h (transfected cells) |
| ***PTMs*** | 3 N-CHO (potential); many O-CHO (Ser/Thr-rich) | | |
| ***References*** | Lublin *et al.*, 1988, *J. Exp. Med.*, **168**, 181–194. | | |

## Description

**Membrane cofactor protein** (MCP) is a membrane glycoprotein expressed by many cell types. It is involved in the protection of host cells from damage by the complement system. MCP exhibits structural and functional similarity with other C3/C4 binding proteins such as C4b binding protein, complement factor H, DAF, CR1 and CR2.

## Structure

MCP is a single-chain membrane glycoprotein containing four sushi/CCP/SCR domains, a Ser/Thr/Pro-rich STP region which is heavily glycosylated and a potential transmembrane segment (Val 310–Tyr 332).

## Biological Functions

MCP possesses cofactor activity in the inactivation process of C3b and C4b by factor I and thus is involved in the regulation of the complement system at the C3 convertase level.

# *Perforin 1 (Cytolysin)*

## Fact Sheet

| | | | |
|---|---|---|---|
| ***Classification*** | Complement C6/C7/C8/C9 | ***Abbreviations*** | P1 |
| ***Structures/motifs*** | 1 EGF-like; 1 C2 | ***DB entries*** | PERF_HUMAN/P14222 |
| ***MW/length*** | 59 184 Da/534 aa | ***PDB entries*** | |
| ***Concentration*** | | ***Half-life*** | approx. 10 h (intracellular) |
| ***PTMs*** | 2 N-CHO (potential) | | |
| ***References*** | Lichtenheld *et al.*, 1988, *Nature*, **335**, 448–451. | | |

## Description

**Perforin** is exclusively synthesised in cytotoxic T lymphocytes and natural killer cells and is capable of lysing cell membranes nonspecifically. It exhibits structural similarity with complement components C6, C7, C8 and C9.

## Structure

Perforin is a single-chain glycoprotein containing one EGF-like and one C2 domain. Perforin has two potential transmembrane segments (Pro 167–Phe 183, Tyr 191–Gly 210).

## Biological Functions

In the presence of calcium ions perforin is able to polymerise into transmembrane tubules and as a consequence is able to lyse membrane of target cells nonspecifically.

# *Properdin (Factor P)*

## Fact Sheet

| | | | |
|---|---|---|---|
| ***Classification*** | | ***Abbreviations*** | P |
| ***Structures/motifs*** | 6 TSP1 | ***DB entries*** | PROP_HUMAN/P27918 |
| ***MW/length*** | 48 494 Da/442 aa | ***PDB entries*** | |
| ***Concentration*** | 5–25 mg/l | ***Half-life*** | |
| ***PTMs*** | 1 N-CHO (potential); 4 O-CHO; 14 C-CHO (Trp) | | |
| ***References*** | Nolan *et al.*, 1986, *Biochem. J.*, **287**, 291–297. | | |

## Description

**Properdin** is a glycoprotein of the alternative pathway of the complement system mainly synthesised in the spleen, lung and in macrophages and circulates in plasma as a mixture of cyclic polymers, mainly as dimers, trimers and tetramers.

## Structure

Properdin is a single-chain glycoprotein (10 % CHO) containing six TSP1 domains. Properdin is observed under dissociating conditions as monomer but under physiological conditions as a mixture of dimers, trimers and tetramers (20 : 54 : 26).

## Biological Functions

Properdin is a positive regulator of the alternative pathway of the complement system. It binds to and stabilises the C3- and the C5-convertase enzyme complexes.

# 9

# The Immune System

## 9.1 INTRODUCTION

Maintaining the integrity of a living organism is a vital process for its prosperity and survival. The immune system composed of specialised cells and organs protects an organism from attack by hostile pathogens such as bacteria and viruses and fends off foreign substances. In addition, it plays a crucial role in the recognition and elimination process of transformed and/or malignant cells such as cancer cells.

The immune system is composed of a variety of cellular and soluble components:

1. lymphatic tissues such as the lymph nodes, spleen, thymus and mucosa;
2. epithelial cells of the skin and mucous tissues representing a natural barrier against pathogens;
3. mobile and resident nucleated cells of the hematopoietic system originating from the bone marrow;
4. soluble components such as antibodies, cytokines and components of the complement system.

Functionally, the immune system can be divided into two systems:

1. ***Innate immunity.*** This is a natural, innate, nonadaptive immune defence, which is an immediate defence to fend off any type of pathogen.
2. ***Adaptive or acquired immunity.*** The body can develop a specific immunity to particular pathogens by producing specific antibodies to a certain pathogen and by expressing T cells as specific targets to a particular pathogen.

Both systems are composed of a whole set of cellular and soluble components, which act synergistically in a coordinated effort resulting in a specific immune response.

In the nonadaptive immune defence monocytes/macrophages and neutrophilic granulocytes ingest foreign particulate material by phagocytosis, but also older own body cells, dead cells and tissue fragments are removed in the same way. The material ingested by phagocytosis is then degraded intracellularly by lysosomal enzymes. Mast cells and basophilic granulocytes bind IgE via high-affinity surface receptors and are the effector cells for the IgE-mediated allergic immune response. Natural killer (NK) cells are cytotoxic cells that are able to lyse virus-infected cells and tumor cells.

In the adaptive or humoral immune response B cells produced in the bone marrow carry specific surface proteins, which recognise and react with antigens, resulting in the differentiation of the lymphocytes capable of expression and secretion of antibodies specifically designed to a certain invader. The antigen–antibody complex acts as a marker for macrophages, resulting in the destruction of the unwanted particle. In addition, T cells from the thymus gland also carry cell surface proteins, which recognise specific antigens and support their destruction.

Activated B lymphocytes either differentiate into plasma cells, which are short-lived and return to the bone marrow where they secrete antibodies, or into memory B cells, which remain in the body for longer periods, primarily in the lymphoid tissue.

*Human Blood Plasma Proteins: Structure and Function* Johann Schaller, Simon Gerber, Urs Kämpfer, Sofia Lejon and Christian Trachsel

The mobile cells of the immune system are developed from common progenitor cells in the bone marrow, the pluripotent stem cells, under the control of growth factors and differentiation factors. The blood circulatory system and the lymphatic system transport the white blood cells such as B cells, T cells, natural killer cells and macrophages through the body. Although all these cells exhibit a different and specialised function and task they work together in a coordinated effort to destroy pathogens and to fend off all components recognised as foreign. Without such a coordinated effort a living organism would not be able to survive for a long period.

## 9.2 IMMUNOGLOBULINS

The recognition of antigens by B lymphocytes is based on ***immunoglobulins***, which are present either as membrane-bound receptors or in soluble form. Five classes or isotypes of immunoglobulins exist, which are all composed of two heavy chains (mass range: ~53–75 kDa) and two light chains (~23 kDa):

1. Immunoglobulin G (IgG) or $\gamma$-globulin exists in four subclasses: IgG1 (60%), IgG2 (30%), IgG3 (4%) and IgG4 (6%).
2. Immunoglobulin D (IgD).
3. Immunoglobulin E (IgE).
4. Immunoglobulin A (IgA) exists in two forms: IgA1 (~90%) and IgA2 (~10%).
5. Immunoglobulin M (IgM) or $\gamma$-macroglobulin.

The five classes of immunoglobulins with their subunit structure and some physical-chemical parameters are summarised in Table 9.1.

Structurally, the five classes of immunoglobulins are organised in a very similar way: two identical light chains (L) and two identical heavy chains (H) are linked covalently by disulfide bridges and in addition associate noncovalently to form a symmetrical, Y-shaped dimer. Each light chain consists of two homologous domains, a variable region ($V_L$) and a constant region ($C_L$). Each domain consists of approximately 110 aa characterised by two $\beta$-sheet structures containing three to five antiparallel $\beta$-strands stabilised by a disulfide bridge. The heavy chains consist of four homologous domains, a variable region ($V_H$) and three constant regions ($C_H1$, $C_H2$, $C_H3$). IgM and IgE contain four constant regions. The schematic structural organisation of the five classes and the subclasses of the immunoglobulins is shown in Figures 9.1(a)–(c).

The schematic structural organisation of immunoglobulin IgG is depicted in Figure 9.2. The variable domains of IgG ($V_L$ and $C_L$) contain at the exposed end of the $\beta$-barrel three small regions with a much higher variability than the rest of the domain, termed 'hypervariable regions' or 'complementarity determining regions' (CDR1–CDR3 in IgG). The CDR regions of the light and heavy chains in native immunoglobulins are in close proximity to each other and form the specific binding sites for antigens. The light and heavy chains as well as the two heavy chains are connected covalently by disulfide bridges and the heavy chains usually carry N-linked CHO and occasionally also O-linked CHO. The hinge region between the constant domains $C_H1$ and $C_H2$ in the heavy chain of IgG is very flexible and accessible to enzymes. Limited

**Table 9.1** Classes of immunoglobulins.

| Class | Subunit structure | Mass (kDa) | Concentration range (g/l) | Half-life (days) |
|---|---|---|---|---|
| IgA | $\alpha_2\kappa_2/\lambda_2$ | 150 | 0.5–3.5 | 4.5–5.9 |
| | $(\alpha_2\kappa_2/\lambda_2)_nJ$ $_{(n=2-4)}$ | 390–690 | | |
| IgD | $\delta_2\kappa_2/\lambda_2$ | 176 | 0.03 | 2.8 |
| IgE | $\varepsilon_2\kappa_2/\lambda_2$ | 190 | 0.0001–0.0004 | 2.3 |
| IgG | $\gamma_2\kappa_2/\lambda_2$ | 150 | | |
| IgG1 (60%) | | 150 | 4.2–13.0 | 21 |
| IgG2 (30%) | | 150 | 1.2–7.5 | 21 |
| IgG3 (4%) | | 170 | 0.4–1.3 | 7 |
| IgG4 (6%) | | 150 | 0.01–2.9 | 21 |
| IgM | $(\mu_2\kappa_2/\lambda_2)_5J$ | 960 | 0.5–2.0 | 5 |

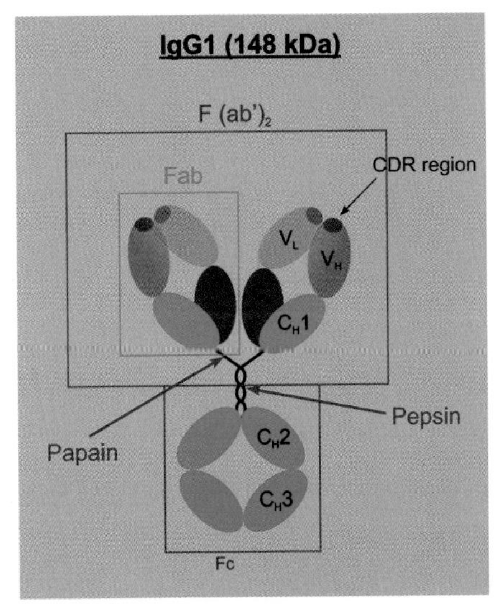


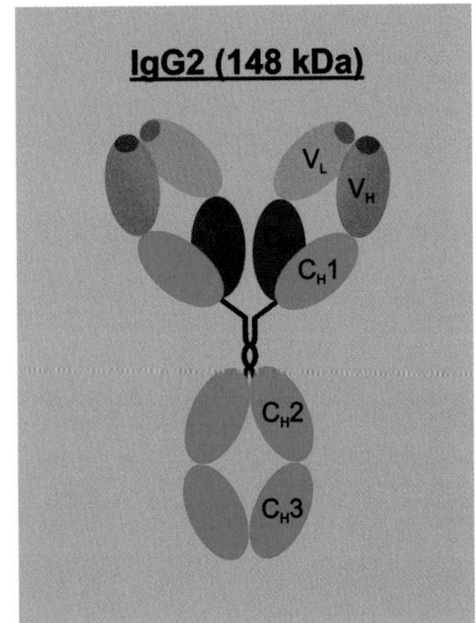


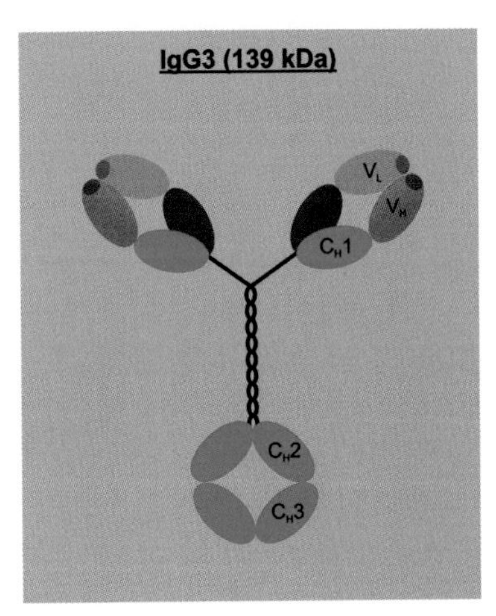


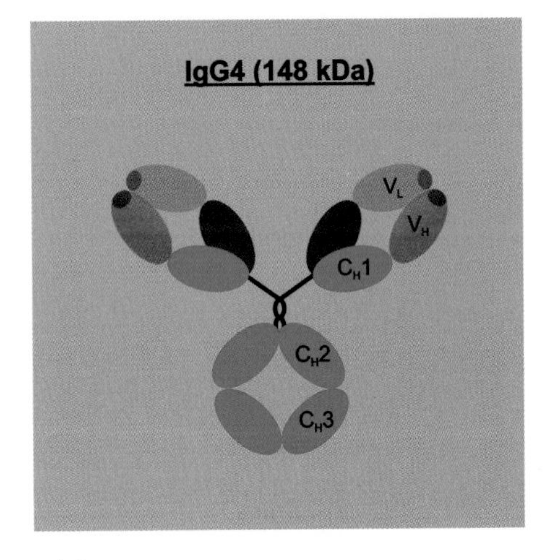


(a)

**Figure 9.1 The five classes of Immunoglobulins (a) immunoglobulin G (IgG) and its four subclasses, IgG1, IgG2, IgG3 and IgG4** Schematic structural organisation of the five classes and its subclasses of immunoglobulins. Two identical light chains (L) and two identical heavy chains (H) are linked covalently by disulfide bridges and associate also non-covalently to form a symmetrical, Y-shaped dimer. Each light chain consists of two homologous domains, a variable region ($V_L$) and a constant region ($C_L$). The heavy chains consist of four homologous domains, a variable region ($V_H$) and three constant regions ($C_H1$, $C_H2$, $C_H3$). IgM and IgE contain four constant regions.

proteolysis with papain occurs in the hinge region N-terminal to the interchain disulfide bridges generating two identical Fab (fragment, antigen binding) fragments and one Fc (fragment, crystallisable) fragment. Limited proteolysis with pepsin cleaves the heavy chain C-terminal to the interchain disulfide bridges, resulting in a $(Fab')_2$ fragment, which is still able to bind antigens but has otherwise lost the biological functions of native IgG.

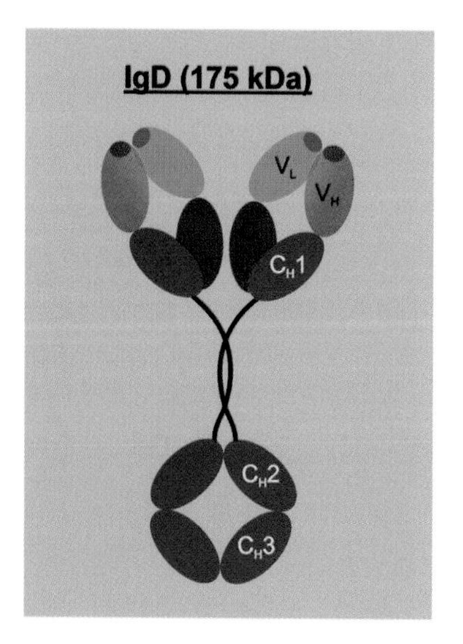


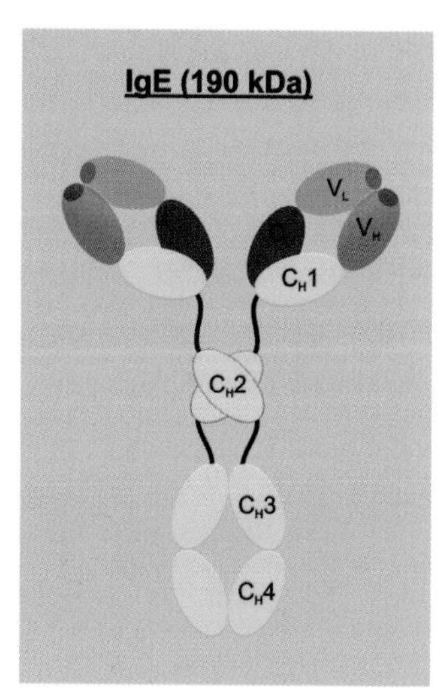


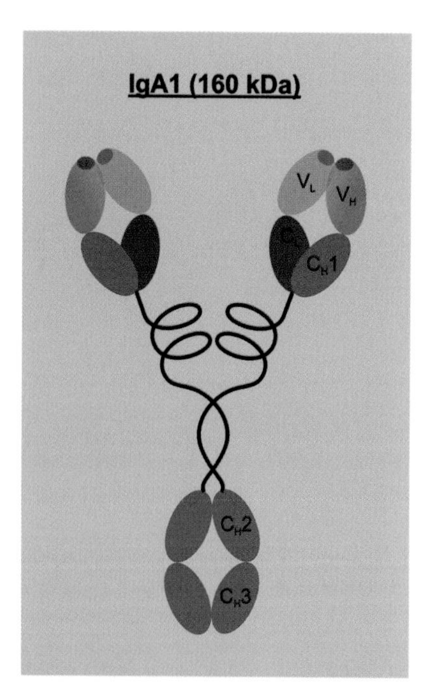


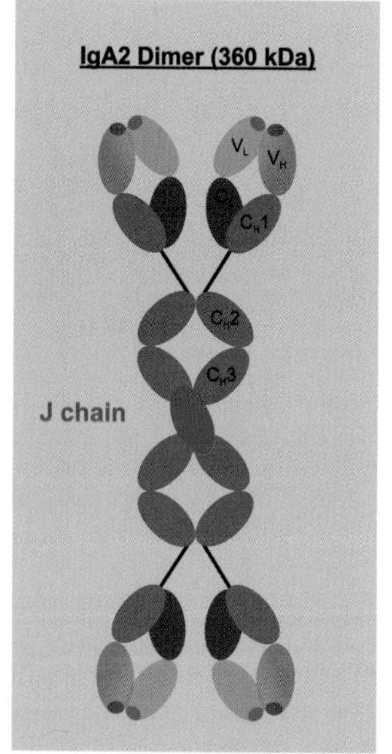


(b)

**Figure 9.1** (b) Immunoglobulins IgD, IgE, IgA1 and IgA2

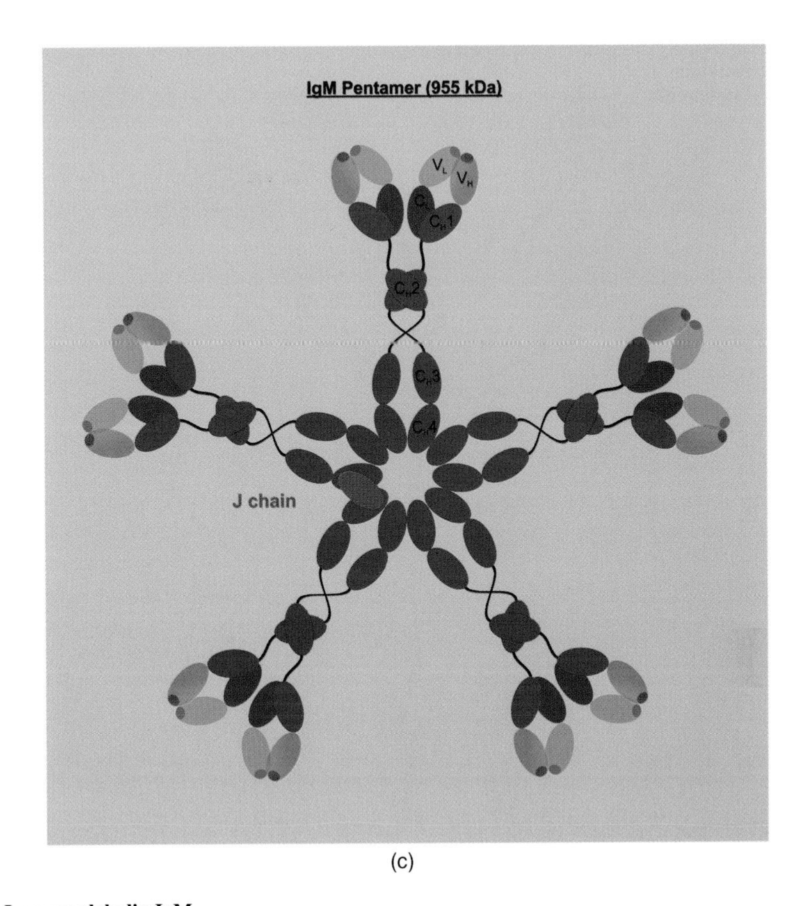


(c)

**Figure 9.1** **(c) Immunoglobulin IgM**

Although the constant regions in the light and the heavy chains are similar in length and structure, a comparison of both constant regions of IgG1λ is quite useful. The 3D structures of the constant region of the Fab fragment of the light chain (95 aa: Ala 108–Val 202) containing a single intrachain disulfide bridge (Cys 130–Cys 189) and the corresponding domain in the Fab fragment of the heavy chain (95 aa: Pro 123–Pro 217) also containing a single intrachain disulfide bridge (Cys 144–Cys 200) was determined by X-ray diffraction and is shown in Figures 9.3(a) and (b) (PDB: 7FAB). The structures are characterised by a compressed antiparallel β-barrel composed of a three-stranded β-sheet (in green) packed against a four-stranded β-sheet (in blue) connected by loops and stabilised by a single disulfide bridge (in pink), shown in the side view and the top view.

The variable regions of the light and heavy chains exhibit similar structures as the corresponding constant regions. The 3D structure of the variable region of the Fab fragment of the light chain (102 aa: Gln1Ala–Leu 102) containing a single intrachain disulfide bridge (Cys 22–Cys 83) and the corresponding domain in the Fab fragment of the heavy chain (116 aa: Ala 1–Ser 116) also containing a single intrachain disulfide bridge (Cys 22–Cys 95) was determined by X-ray diffraction and is shown in Figures 9.4(a) and (b) (PDB: 7FAB). As in the domains of the constant regions the structures represent an antiparallel β-barrel composed of a three-stranded (in green) and a four-stranded β-sheet (in blue) connected by loops and stabilised by a single disulfide bridge (in pink). Furthermore, the variable regions contain an additional region usually composed of two β-strands (in yellow). The CDR1–3 regions (in red), which determine the specificity of the antigen–antibody interaction, located in the loops connecting the β-strands are clearly exposed on the surface, as can be seen in the side view as well as in the top view of the domains.

The 3D structure of the entire Fab fragment of IgG1 consisting of the light chain ($V_L + C_L$ (208 aa): Pyro-Glu 1–Ser 208) and the heavy chain ($V_H + C_H1$ (217 aa): Pyro-Glu 1–Pro 217) determined by X-ray diffraction (PDB: 7FAB) is shown in

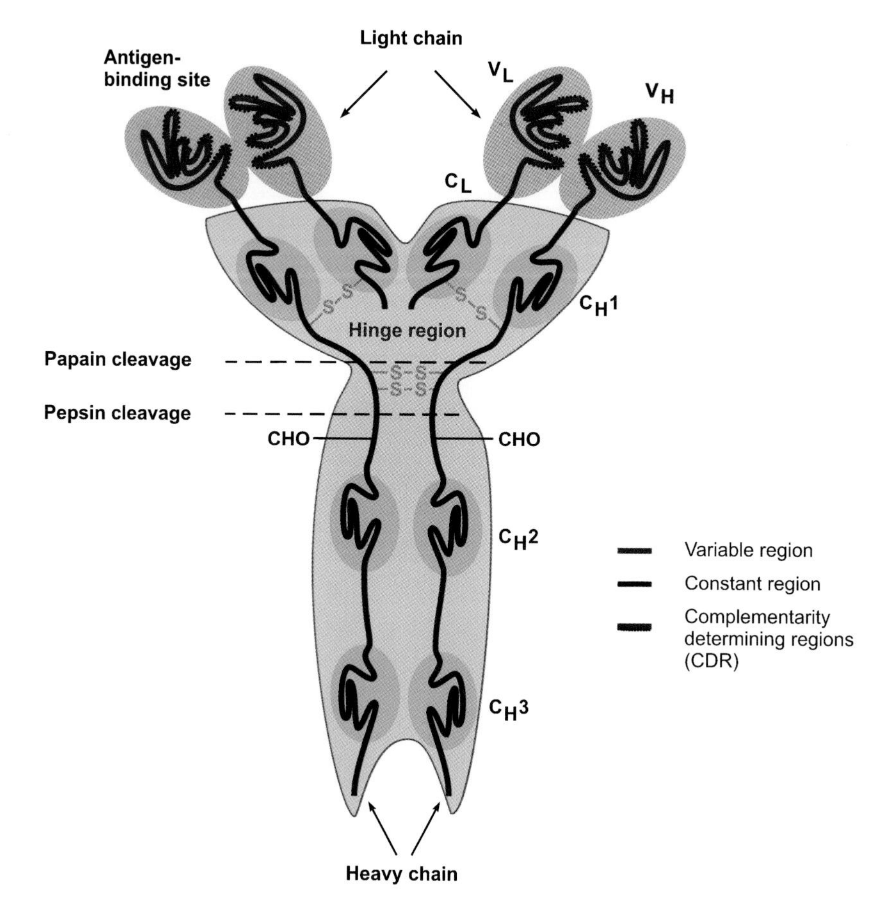


**Figure 9.2 Immunoglobulin IgG**
Schematic structural organisation of immunoglobulin IgG. Two light chains (L) and two heavy chains (H) are linked covalently by disulfide bridges and the heavy chains usually carry N-linked CHO. The variable domains $V_L$ and $C_L$ contain the complementarity determining regions (CDR). Limited proteolysis with papain generates two identical Fab fragments and one Fc fragment and limited proteolysis with pepsin generates the $(Fab')_2$ fragment.

Figure 9.5. The structure is characterised predominantly by β-sheet folding within the $V_L$, $V_H$, $C_L$ and $C_H1$ domains of the Fab fragment, typical for the immunoglobulin fold.

The 3D structure of the Fc fragment of IgG1 consisting of the two C-terminal constant regions of the heavy chain ($C_H2 + C_H3$ (223 aa): Thr 225–Lys 447) determined by X-ray diffraction (PDB: 2DTQ) is shown in Figure 9.6. The Fc fragment is a homodimer linked by two interchain disulfide bridges in the N-terminal $C_H2$ regions (not shown) and noncovalent interactions in the C-terminal $C_H3$ domains. Each chain carries a CHO moiety (in pink) attached to Asn 297 and each domain ($C_H2 + C_H3$) contains two intrachain disulfide bridges (in blue). The overall structure resembles that of a horseshoe and the space between the two chains is primarily filled with the CHO chains attached to Asn 297.

**Immunoglobulin G** (IgG) (see Data Sheet) is a large glycoprotein synthesised by plasma cells in the bone marrow, lymph nodes and spleen and is the most important plasma antibody. Each IgG molecule ($\gamma_2\kappa_2/\lambda_2$) consists of two heavy chains (γ-chains) of approximately 420 aa and two light chains (κ- or λ-chains) of about 210 aa linked together by interchain disulfide bridges between the heavy chains on one side and the light and heavy chains on the other side. Four subclasses exist with an approximate distribution of 60% IgG1, 30% IgG2, 4% IgG3 and 6% IgG4. The constant regions of the four

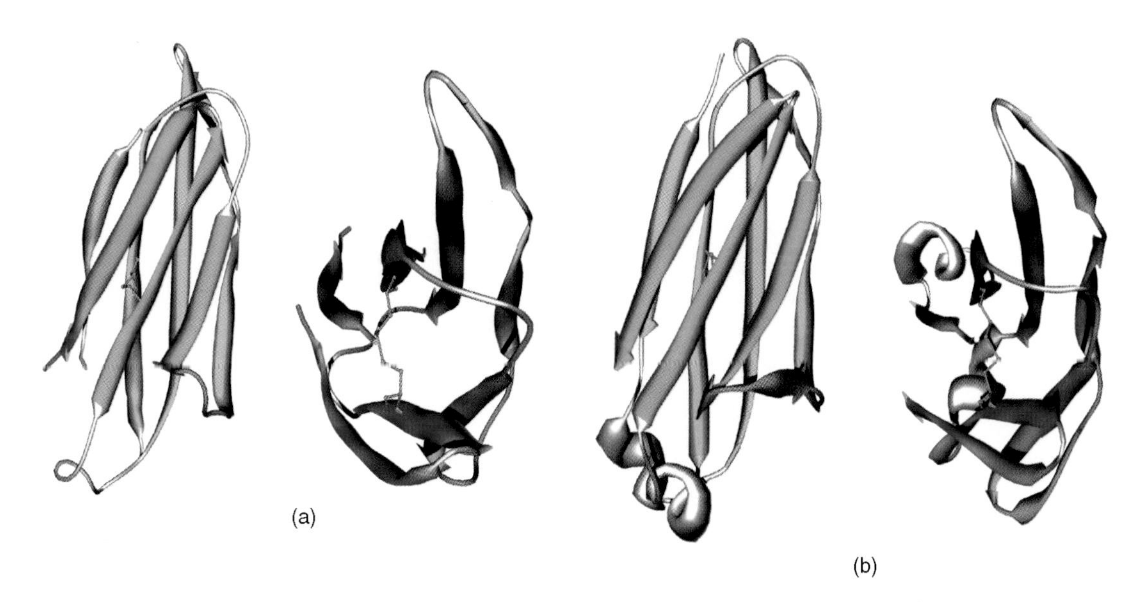


**Figure 9.3 The 3D structures of the constant region of the IgG (a) heavy chain and (b) light chain (side and top views) (Fab fragment, PDB: 7FAB)**
The structures represent an antiparallel β-barrel composed of a three-stranded β-sheet (in green) and a four-stranded β-sheet (in blue). The single disulfide bridge is shown in pink.

subclasses are very similar in sequence, with the greatest differences in the hinge region: IgG1, 15aa, IgG2, 12 aa, IgG3, 62 aa and IgG4, 12 aa. They have different antigenic, structural and biological properties located in the heavy chains. IgG neutralises viruses, bacterial antigens and bacteria, fungi and yeasts. IgG binds to mononuclear cells, neutrophils, mast cells and basophils. In addition IgG1, IgG2 and IgG3 can activate the classical pathway of the complement system, but not IgG4.

**Immunoglobulin A** (IgA) (see Data Sheet) is a large glycoprotein synthesised by plasma cells in mucosal tissues, secretory glands, bone marrow, spleen and lymph nodes. Approximately 90 % of the plasma IgA is monomeric ($\alpha_2\kappa_2/\lambda_2$) consisting of two heavy chains (α-chains) and two light chains (κ- or λ-chains) connected by interchain disulfide bridges. In addition, oligomeric forms of IgA exist in the mass range of 390–690 kDa with two to four IgA molecules linked by disulfide bridges and a disulfide-linked joining chain J (J-chain) ($(\alpha_2\kappa_2/\lambda_2)_{n=2-4}J$). The J-chain will be discussed below. A secretory IgA (S-IgA) exists composed of two to four disulfide-linked IgA molecules, a J-chain and a secretory component (SC). The secretory component (585 aa: Lys 1–Arg 585) is derived from the polymeric-immunoglobulin receptor (746 aa: P01833) by proteolytic processing. IgA antibodies neutralise antigens such as viruses, enzymes and toxins.

**Immunoglobulin M** (IgM) (see Data Sheet) or γ-macroglobulin is a very large pentameric plasma glycoprotein with an approximate mass of 970 kDa, representing about 7–10 % of the immunoglobulins in plasma. The monomer is composed of two heavy chains (μ-chains) and two light chains (κ- or λ-chains) connected by interchain disulfide bridges. Each pentamer $(\mu_2\kappa_2/\lambda_2)_5J$ contains one J-chain and adjacent IgM monomers are linked by two intermonomer disulfide bridges (Cys 414–Cys 414 and Cys 575–Cys 575). IgM is produced early in the immune response before IgG and is an effective initial defence against bacteria. IgM binds to antigens and as a result activates the classical pathway of the complement system (see Chapter 8) and macrophages. In antigen-bound IgM the arms of the $(Fab')_2$ domain are able to bend out of the plane, resulting in a 'staple-like' structure in close contact with the antigen.

**Immunoglobulin D** (IgD) (see Data Sheet) is a large glycoprotein consisting of two heavy chains (δ-chains) and two light chains (κ- or λ-chains) connected by interchain disulfide bridges ($\delta_2\kappa_2/\lambda_2$). The hinge region (58 aa) is susceptible to proteolytic cleavage. IgD acts as membrane antigen receptor resulting in endocytosis and cell activation; otherwise it functions as a typical antibody.

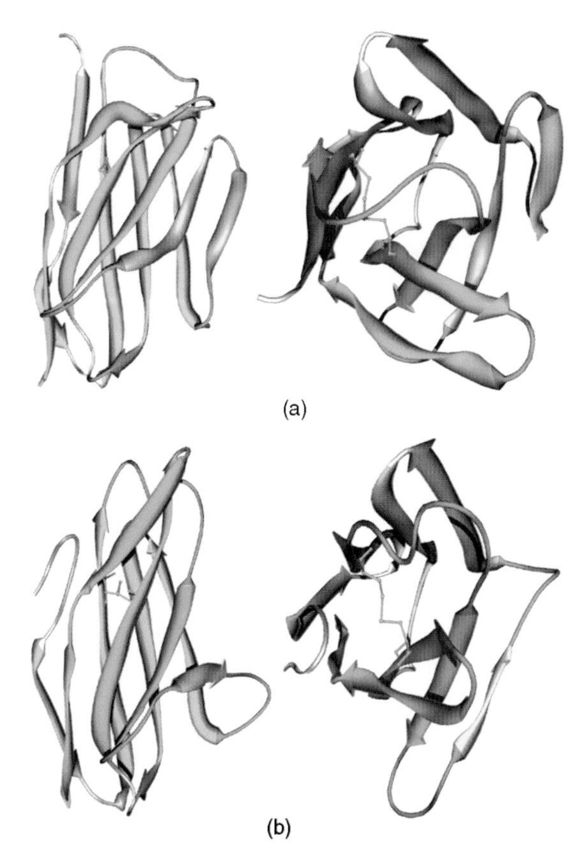


**Figure 9.4 The 3D structures of the variable region of the IgG (a) heavy chain and (b) light chain (side and top views) (Fab fragment, PDB: 7FAB)**

The structures represent an antiparallel β-barrel composed of a three-stranded β-sheet (in green) and a four-stranded β-sheet (in blue). The single disulfide bridge is shown in pink. The additional part of the variable region contains one β-strand (in yellow). The CDR1 – 3 regions (in red) located in the loops connecting the strands are exposed.

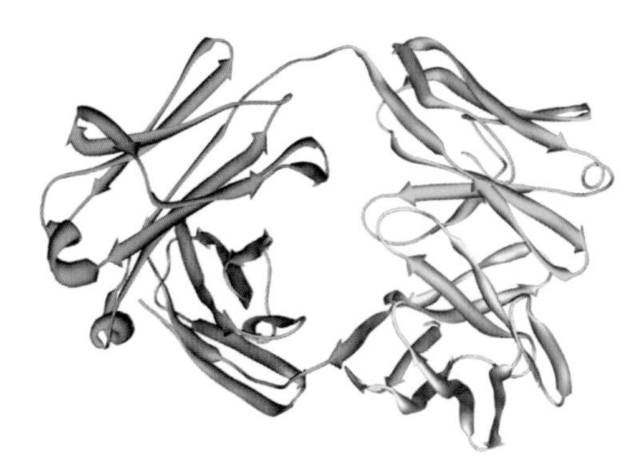

**Figure 9.5 3D structure of the Fab fragment of IgG1 (PDB: 7FAB)**

The 3D structure of the Fab fragment of IgG1 consisting of two domains of the light chain ($V_L + C_L$) and two domains of the heavy chain ($V_H + C_H1$) is typical for the immunoglobulin fold characterised predominantly by β-sheet folding.

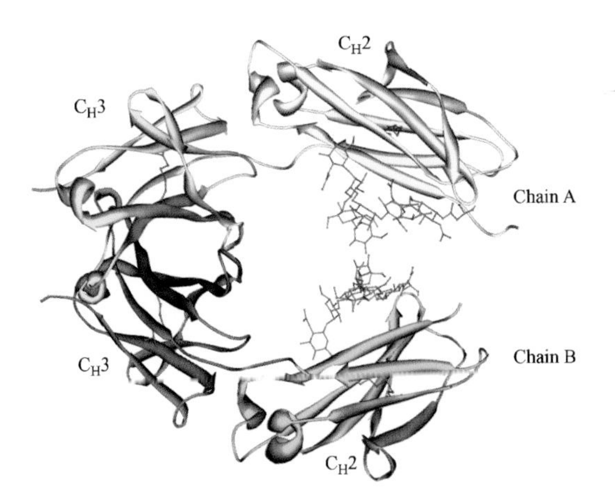


**Figure 9.6 3D structure of the Fc fragment of IgG1 (PDB: 2DTQ)**
The 3D structure of the Fc fragment consisting of the two C-terminal domains of the constant heavy chain region ($C_H2 + C_H3$) resembles a horseshoe and the space between the two chains is primarily filled with the CHO moieties (in pink).

**Immunoglobulin E** (IgE) (see Data Sheet) is a large glycoprotein (approximately 12 % CHO) consisting of two heavy chains (ε-chains) and two light chains (κ- or λ-chains) connected by interchain disulfide bridges ($\varepsilon_2\kappa_2/\lambda_2$). IgE carries unique antigenic determinants associated with the ε-chains. IgE antibodies are the cause of immediate-type allergic reactions called reaginic hypersensitivity, which are only found with IgE.

**Immunoglobulin J-chain** (J-chain) (see Data Sheet) produced by plasma cells is a small component associated with polymeric IgA and IgM. The J-chain is a single-chain glycoprotein (approximately 8 % CHO) containing three intrachain disulfide bridges (Cys 13–Cys 101, Cys 72–Cys 92, Cys 109–Cys 134) and two Cys residues (Cys 15, Cys 69) forming interchain disulfide bridges with the penultimate Cys residue in the α- and μ-chains of oligomeric IgA and IgM, respectively. Additionally, it carries an N-terminal pyro-Glu residue.

## 9.3 ANTIBODY DIVERSITY

The immunoglobulin class of proteins is characterised by an incredible diversity consisting of 10 Mio or even more different sequences as estimated by Anderson and Anderson in their article 'The human plasma proteome: history, character and diagnostic prospects' in 2002.

The human body produces approximately 50 Mio antibody-producing B cells every day, of which most produce an antibody unique to these cells. On the genetic level more than 1000 small segments of DNA exist where this immense diversity is located. The sequence information is clustered in three gene pools, one for the heavy chain and one for each of the two isotypes of light chains. The variable domain in the heavy chain gene pool is characterised by three regions, the V, D and J segments. The V part is approximately 90 residues long, the D part covers the hypervariable region CDR3 and the J part consists of the rest of the variable domain (approximately 15 residues). The variable domain of the light chain gene pool is organised in a similar way, except that there is no D segment.

The number of functional gene segments in a human immunoglobulin locus is large. As an example the germ line organisation of the immunoglobulin heavy chain in the human genome is shown in Figure 9.7. The heavy chain locus

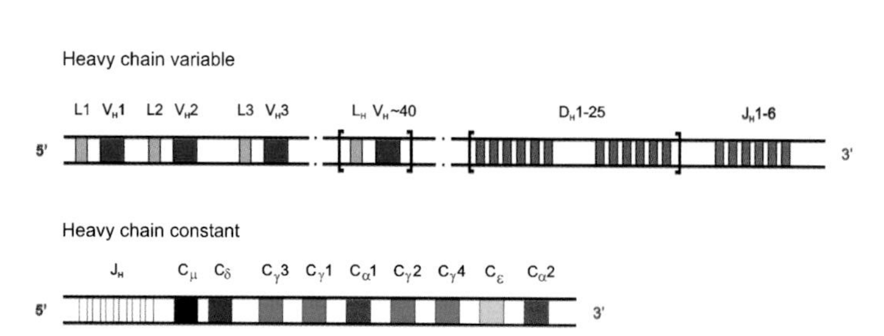


**Figure 9.7 Germ line organisation of the immunoglobulin heavy chain locus in the human genome**
The heavy chain locus of the variable region consists of approx. 40 functional $V_H$ gene segments, followed by about 25 $D_H$ gene segments and six $J_H$ segments. In addition, the constant region contains the $C_H$ gene segments of all classes and subclasses of immunoglobulins: $C_\mu$, $C_\delta$, $C_{\gamma 1\text{-}4}$, $C_{\alpha 1,2}$, $C_\varepsilon$. For reasons of simplicity all pseudogens are omitted. L: leader sequence.

(located on chromosome 14) of the variable region is composed of approximately 40 functional $V_H$ gene segments, followed by a cluster of about 25 $D_H$ gene segments and six $J_H$ gene segments. In addition, the constant region contains the $C_H$ gene segments of all five classes and their subclasses of immunoglobulins: $C_\mu$, $C_\delta$, $C_{\gamma 1-4}$, $C_{\alpha 1,2}$, $C_\varepsilon$ (L represents the leader sequences). For reasons of simplicity all pseudogens are omitted and the details of the exon organisation in the C genes are not shown. The variable region of the immunoglobulin light chain is organised in a similar way with the exception that the light chains contain no D gene segments.

The enormous antibody diversity is generated by several mechanisms. As an example, the generation of an immunoglobulin heavy chain from different gene segments by somatic recombination is depicted in Figure 9.8. In a first step the D and J segments in the germ line DNA are joined, followed by joining of the V segments to the joined DJ segment resulting in the $V_H$ exon. After transcription of the DNA to the primary RNA transcript and splicing the generated mRNA is translated to the corresponding polypeptide chain.

The various mechanisms that generate the huge number of antibodies are briefly summarised:

1. ***Combinatorial joining of gene segments.*** In the gene pool of the variable domain of the heavy chain there are approximately 1000 different V segments, approximately 10 different D segments of variable length and about four different J segments. These segments are assembled randomly into a single exon in a process termed 'combinatorial joining', allowing approximately 40 000 different combinations.
2. ***Junction diversity.*** The joining of the V, D and J segments does not take place exactly at the junction sites, thus enabling additional diversity termed 'junction diversity'. The variable domain of the light chain contains no D segments, thus the diversity is mainly generated in the joining process of the V and J segments.
3. ***Class-switching.*** The vast number of the V-D-J exons is combined to one of the eight C segments encoding the constant regions of heavy chain, and usually the variable regions are first expressed in IgM molecules. In a genetic rearrangement process the μ exon, which encodes for the constant region of IgM, is deleted and the variable region will be recombined to exons encoding the constant regions of IgG, IgA, IgD and IgE. This process is termed 'class-switching'.
4. ***Combination of heavy and light chains.*** Of course, the various heavy and light chains are combined, thus generating the vast number of antibodies.
5. ***Somatic point mutations.*** In antigen-activated B lymphocytes a process termed 'somatic point mutation' leads to point mutations in the variable regions of the antibodies, creating a 'somatic hypermutation', which is orders of magnitude more frequent than in other genes. This results in an even higher affinity of an antibody for a certain antigen during an immune response, so-called 'affinity maturation'.

From these genetic mechanisms it is possible to estimate roughly the theoretically possible number of combinations. There are approximately 90 000 different heavy chains and 3 000 different light chains and in principle all combinations are possible. By combinatorial association the vast number of 270 Mio different antibody molecules are possible. Somatic hypermutation will lead to an even higher number of antibody molecules with specific antigen-binding sites.

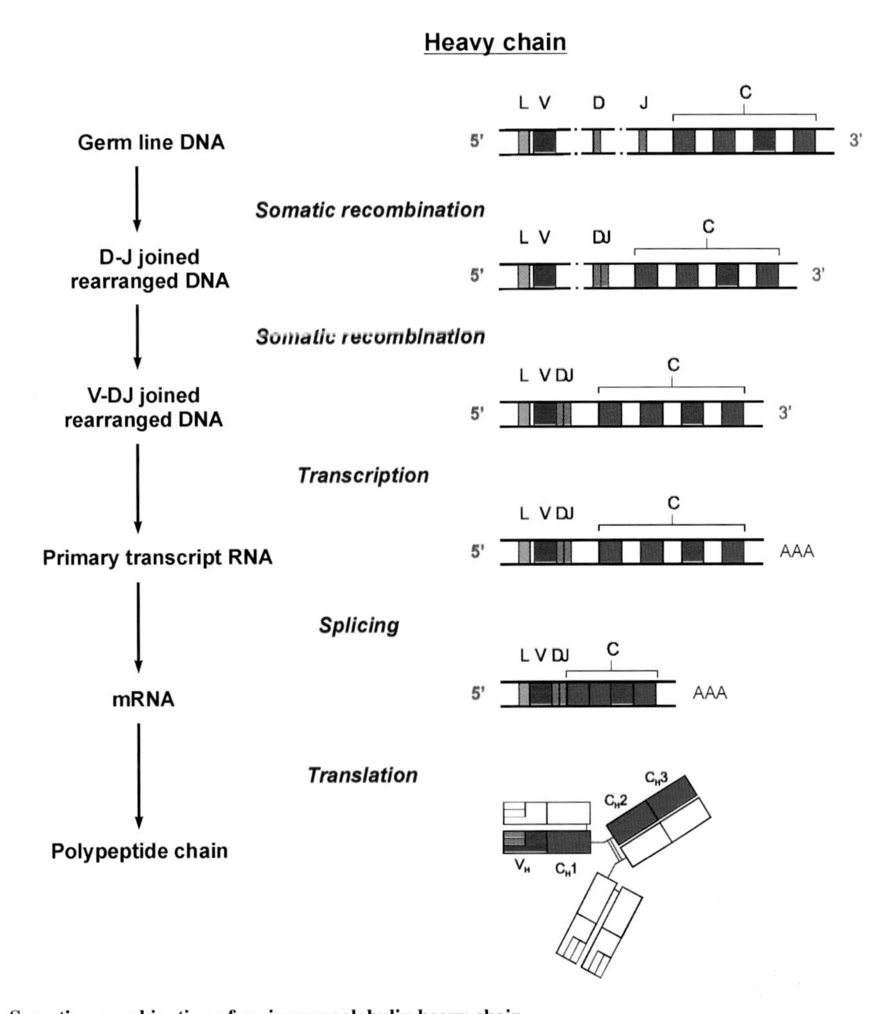


**Figure 9.8 Somatic recombination of an immunoglobulin heavy chain**
In somatic recombination in a first step the D and J segments are joined followed by the joining of the V segments to the joined DJ segment resulting in the $V_H$ exon. After transcription of the DNA to the primary RNA transcript and splicing the generated mRNA is translated to the corresponding polypeptide chain.

## 9.4 MAJOR HISTOCOMPATIBILITY COMPLEX (MHC) PROTEINS

The main function of the major histocompatibility complex (MHC) proteins is the binding of pathogen-derived peptide fragments and to present them on cell surfaces for recognition by the corresponding T cells, resulting in most cases in the death of the pathogen-infected cells, in the activation of macrophages able to kill bacteria and the activation of B cells for production of antibodies to inactivate and eliminate extracellular pathogens.

Two classes of MHC proteins exist:

1. The MHC class I molecules are composed of an $\alpha$-chain consisting of three extracellular domains ($\alpha_1$, $\alpha_2$, $\alpha_3$), a transmembrane segment, a cytoplasmic tail and a noncovalently attached $\beta$-chain consisting of $\beta_2$-microglobulin containing a single extracellular domain. MHC class I proteins are expressed by all nucleated cells. MHC class I molecules present peptides originating from cytosolic proteins, and the MHC class I pathway is often termed 'cytosolic' or 'endogenous pathway'.
2. The MHC class II molecules are composed of an $\alpha$-chain and a $\beta$-chain, both composed of two extracellular domains, a transmembrane segment and a cytoplasmic tail. MHC class II proteins are expressed by specialised antigen-presenting cells

such as dendritic cells, macrophages and B cells. MHC class II molecules present peptides originating from extracellular proteins and the MHC class II pathway is often called 'endocytic' or 'exogenous pathway'.

In addition, proteins exist that exhibit a significant similarity with MHC class molecules, among them Zn-$\alpha_2$-glycoprotein (Zn-$\alpha_2$-GP).

**HLA class I histocompatibility antigen A-1 α-chain** (HLA class I) (see Data Sheet) or MHC class I antigen A1 expressed by most nucleated cells is a single-chain type I membrane glycoprotein composed of a large N-terminal extracellular domain, a potential transmembrane segment (24 aa: Val 285–Trp 308) and a short C-terminal cytoplasmic domain. HLA class I α-chain consists of an Ig-like C1-type domain and three regions of approximately 90 residues: $\alpha_1$, $\alpha_2$ and $\alpha_3$. In addition, the molecule contains two intrachain disulfide bridges (Cys 101–Cys 164, Cys 203–Cys 259) and is sulfated at Tyr 59.

The 3D structure of the extracellular domain of the MHC class I molecule HLA-B53 α-chain (276 aa: Gly 1–Pro 276) and the β-chain consisting of $\beta_2$-microglobulin (99 aa: Ile 1–Met 99, in red) in complex with an antigen nonapeptide from the gag protein of HIV2 was determined by X-ray diffraction (PDB: 1A1M) and is shown in Figure 9.9(a). The α-chain is composed of three regions: $\alpha_1$ (yellow), $\alpha_2$ (green) and $\alpha_3$ (blue). The $\alpha_1$ and $\alpha_2$ domains contain the antigen-binding site containing the nonapeptide TPYDINQML (in pink) and faces away from the membrane surface. Both domains are very similar in structure composed of four up and down antiparallel β-strands and a helical region located across the β-strands. The $\alpha_3$ region, which

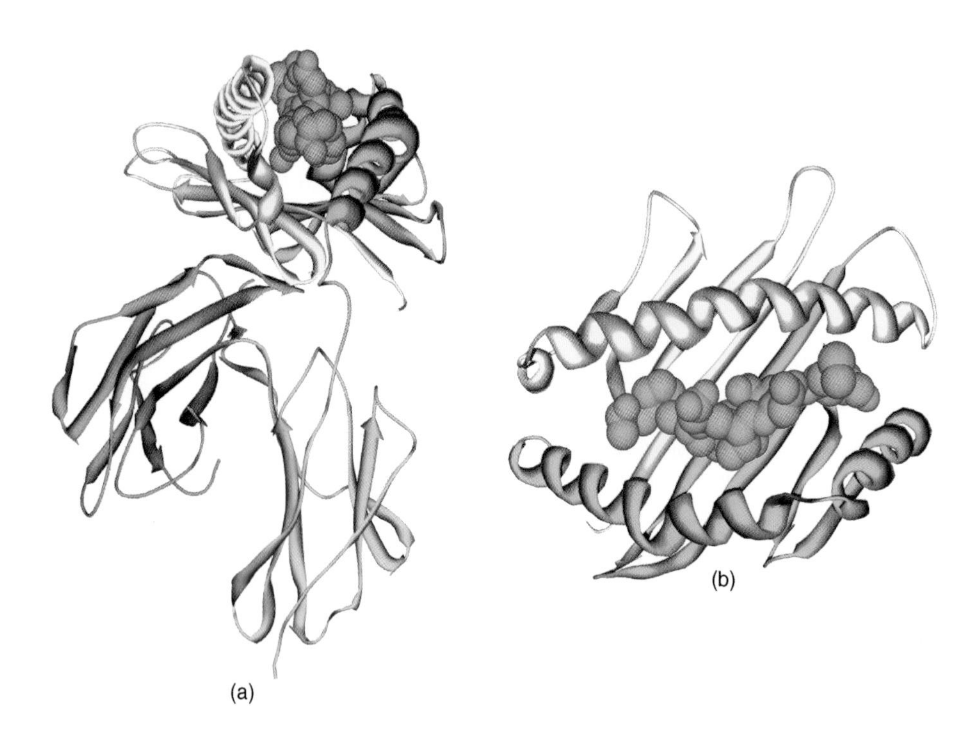


**Figure 9.9 (a) 3D structure of the extracellular domain of the MHC class I molecule HLA-B53 α-chain and the β-chain consisting of $\beta_2$-microglobulin (PDB: 1A1M). (b) MHC class I binding region containing a nonapeptide (PDB: 1A1M)**

(a) The 3D structure of the extracellular domain of the MHC class I molecule HLA-B53 α-chain is composed of three regions: $\alpha_1$ (yellow), $\alpha_2$ (green) and $\alpha_3$ (blue), where the $\alpha_1$ and $\alpha_2$ domains contain the antigen binding site containing the nonapeptide. The β-chain consists of $\beta_2$-microglobulin (in red).

(b) The MHC class I binding region consists of two domains, $\alpha_1$ and the $\alpha_2$ characterised by eight antiparallel β-strands forming a base structure and two extended helical segments across the β-sheet structure, engulfing the antigen, a nonapeptide (in pink).

is connected to the transmembrane segment, exhibits the Ig-like domain fold containing four-stranded β-sheets. The β-chain consisting of $\beta_2$-M (in red) exhibits the expected structure of a sandwich of two β-sheets.

The peptide-binding region of the MHC class I antigen consisting of the $\alpha_1$ (in yellow) and the $\alpha_2$ (green) domains containing a nonapeptide (TPYDINQML, in pink) is shown in Figure 9.9(b). The $\alpha_1$ and the $\alpha_2$ domains are characterised by eight antiparallel β-strands forming a base structure and two extended helical segments across the β-sheet structure, thus engulfing the antigen.

**$\beta_2$-Microglobulin** ($\beta_2$-M) (see Data Sheet) synthesised by all nucleated cells is a single-chain protein containing one Ig-like C1-type domain stabilised by a single intrachain disulfide bridge (Cys 25–Cys 80) and an N-terminal pyro-Glu residue. $\beta_2$-M represents the β-chain of the MHC class I molecules and other related proteins, but is also present as monomer and in oligomeric form.

The 3D structure of entire $\beta_2$-microglobulin (99 aa: Ile 1–Met 99) in complex with the α-chain of the MHC class I antigen was determined by X-ray diffraction (PDB: 1A1M) and the structure of $\beta_2$-M is shown in Figure 9.10. The structure is characterised by a sandwich of two β-sheets consisting of four (in blue) and three (in green) antiparallel β-strands stabilised by a single intrachain disulfide bridge (in pink) representing the classical Ig-like C1-type domain with the arrangement of the β-strands ABED in sheet 1 (in blue) and CFG in sheet 2 (in green).

A pattern exists that recognises a short sequence located in the related domains of immunoglobulins and MHC proteins. The consensus sequence is located around a Cys residue involved in a conserved intrachain disulfide bridge in these domains:

IG_MHC (PS00290): [FY]–{L}–C–{PGAD}–[VA]–{LC}–H

**HLA class II histocompatibility antigen DR α-chain and DR-1 β-chain** (HLA class II) (see Data Sheet) or MHC class II antigen, expressed in specialised cells such as macrophages, dendritic cells, activated T cell and B cells, is composed of an α-chain and a β-chain, which are very similar in structure. They are both single-chain type I membrane glycoproteins composed of a large N-terminal extracellular domain, a potential transmembrane segment of 23 residues and a short C-terminal cytoplasmic region. The α-chain contains an Ig-like C1-type domain and two regions, $\alpha_1$ and $\alpha_2$, and a single intrachain disulfide bridge (Cys 107–Cys 163). The β-chain contains an Ig-like

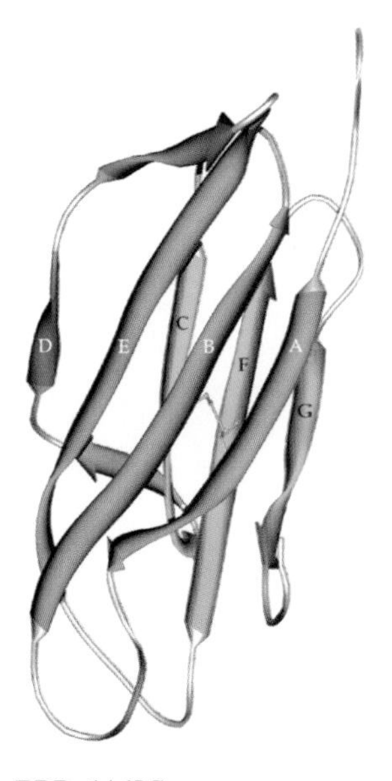


**Figure 9.10 3D structure of $\beta_2$-microglobulin (PDB: 1A1M)**
The 3D structure of $\beta_2$-microglobulin is characterised by a sandwich of two β-sheets consisting of four (in blue) and three (in green) antiparallel β-strands stabilised by a single intrachain disulfide bridge (in pink).

C1-type domain and two regions, $\beta_1$ and $\beta_2$, and two intrachain disulfide bridges (Cys 15–Cys 79, Cys 117–Cys 173).

The 3D structure of the extracellular domain of the HLA class II antigen consisting of the $\alpha$-chain (176 aa: His 5–Phe 180) and the $\beta$-chain (187 aa: Pro 5–Arg 191) in complex with a fragment of the HLA class II antigen invariant $\gamma$-chain (24 aa: Leu 81–Met 104) was determined by X-ray diffraction (PDB: 1A6A) and is shown in Figure 9.11. The $\alpha$- and $\beta$-chains are both composed of two structurally similar regions: in the $\alpha$-chain, $\alpha_1$ (in green) and $\alpha_2$ (in blue), and in the $\beta$-chain, $\beta_1$ (in yellow) and $\beta_2$ (in red). The $\alpha_1$ and $\beta_1$ domains form the antigen-binding site and the peptide of the invariant $\gamma$-chain (in pink) binds in an almost identical way as an antigenic peptide and stabilises the peptide-free HLA class II antigen. Both domains are very similar in structure, composed of four up and down antiparallel $\beta$-strands and a helical region located across the $\beta$-strands. The $\alpha_2$ and $\beta_2$ regions, which are connected to the transmembrane segment, exhibit the Ig-like domain fold containing four-stranded $\beta$-sheets.

**Zn-$\alpha_2$-glycoprotein** (Zn-$\alpha_2$-GP) (see Data Sheet) present in many body fluids and various types of cells is a single-chain glycoprotein (18 % CHO) containing one Ig-like domain and two intrachain disulfide bridges (Cys 103–Cys 166, Cys 205–Cys 260) and an N-terminal pyro-Glu residue. Structural similarity with HLA class I molecules is suggestive that Zn-$\alpha_2$-GP is involved in immune response. It stimulates lipid degradation in adipocytes and causes extensive fat loss in some types of cancer.

**Figure 9.11 3D structure of the HLA class II antigen (PDB: 1A6A)**
The 3D structure of the extracellular domain of the HLA class II antigen consisting of the $\alpha$-chain (in green and blue) and the $\beta$-chain (in yellow and red) in complex with an invariant peptide of HLA class II antigen $\gamma$-chain (in pink).

**Figure 9.12 (a) 3D structure of Zn-$\alpha_2$-glycoprotein (PDB: 1T80). (b) Overlay of Zn-$\alpha_2$-glycoprotein (PDB: 1T80) with an MHC class I molecule (PDB: 1A1M)**

(a) Zn-$\alpha_2$-glycoprotein [GP] consists of three regions: The $\alpha_1$ and $\alpha_2$ domains (in blue) are characterised by four antiparallel β-strands and a helical region located across the β-strands molecules and the $\alpha_3$ domain (in yellow) contains two four-stranded β-sheets.

(b) Overlay of Zn-$\alpha_2$-glycoprotein (in pink) with a MHC class I molecule (in green). The β-chain ($\beta_2$-microglobulin, in yellow) is only present in the MHC class I molecules.

The 3D structure of Zn-$\alpha_2$-glycoprotein (278 aa: Gln 1–Ser 278) was determined by X-ray diffraction (PDB: 1T80) and is shown in Figure 9.12(a). Zn-$\alpha_2$-GP consists of three regions: the $\alpha_1$ and $\alpha_2$ domains (in blue) are characterised by four antiparallel β-strands and a helical region located across the β-strands, and represent the counterpart of the $\alpha_1$ and $\alpha_2$ peptide-binding regions in the MHC class I molecules. The $\alpha_3$ domain (in yellow) exhibits the Ig-like domain fold containing two four-stranded β-sheets.

Zn-$\alpha_2$-GP exhibits structural similarity with the α-chain of MHC class I molecules and the overlay of Zn-$\alpha_2$-GP (in pink) with the MHC class I molecule (in green, see Figure 9.9(a)) is evidenced in Figure 9.12(b). The β-chain consisting of $\beta_2$-microglobulin (in yellow) is only present in MHC class I molecules.

## 9.5 INNATE IMMUNE SYSTEM

The innate immune system is thought to be an older immune defence strategy than the adaptive immune system and represents the immediate defence against hostile pathogens. It is found throughout the animal kingdom as well as in plants and is the major immune system present in insects, plants, fungi and most primitive organisms. The innate immune system consists of cells and mechanisms to defend an organism in a nonspecific way from hostile attacks by pathogens. The cells recognise, respond and destroy the pathogens, but this process does not result in a long-lasting or even protective immunity for an organism. The innate immune system in mammals is characterised by several major functions:

- recruiting immune cells to the sites of infection and inflammation by producing chemical factors such as cytokines;
- activation of the complement system, resulting in the identification of pathogens and the subsequent clearance of dead cells and antibody complexes;
- identification and clearance of foreign components in organs, tissues and in the blood and lymph system by white blood cells;
- activation of the adaptive immune system.

Inflammation is one of the primary reactions of the immune system to infection or irritation and helps to establish an initial physical barrier against the spreading of an infection. The leukocytes of the innate immune response include a series of specialised cells such as natural killer cells, mast cells, eosinophils and basophils. The phagocytic cells include macrophages, neutrophils and dendritic cells (for more details see Chapter 2). The innate immune system contains a series of proteins, which are either described in this part of the chapter or have been described previously (see Chapter 8).

The mannose-binding protein C (mannose-binding lectin), which can activate the classical pathway of the complement system independent from antibodies, has been described previously (see Chapter 8). Lysozyme C is a glycolytic enzyme with bacteriolytic activity and is referred to as the body's antibiotic (for details see Chapter 10).

In addition, there are antimicrobial peptides, which are thought to be the oldest form of defence against infections. A well-studied group of such antimicrobial peptides are the defensins, which exhibit biological activity against bacteria, viruses and fungi. They are thought to attach to the cell membrane of microbes by electrostatic interaction, followed by an embedding process into the cell membrane resulting in pore formation and the subsequent efflux of the cell content. Defensins are small cysteine-rich cationic peptides usually containing three intrachain disulfide bridges, and can be divided into two main groups, the α-defensins and the β-defensins:

1. α-Defensins in humans are a family of six peptides, neutrophil defensins 1–4 and intestinal defensins 5 and 6. α-Defensins are characterised by a consensus sequence containing all six invariant Cys residues:

   DEFENSIN (PS00269): **C**–x–**C**–x(3,5)–**C**–x(7)–G–x–**C**–x(9)–**C**–**C**

   All six Cys residues are involved in three intrachain disulfide bridges arranged in the following pattern:

   **Cys 1–Cys 6, Cys 2–Cys 4, Cys 3–Cys 5**

   Although arthropod defensins exhibit a similar biological function as mammalian defensins, the 3D structures and the arrangement of disulfide bridges are distinct in the two families.
2. β-Defensins in humans are a family of over 30 peptides, which are structurally and functionally related. Most of them share a common disulfide bridge pattern arranged in the following way:

   **Cys 1–Cys 5, Cys 2–Cys 4, Cys 3–Cys 6**

**Serum amyloid P-component** (SAP) (see Data Sheet), synthesised exclusively by hepatocytes, is a single-chain glycoprotein containing one pentaxin domain with a single intrachain disulfide bridge (Cys 36–Cys 95). In the presence of calcium ions, SAP exhibits a highly symmetrical pentameric structure with each monomer carrying two calcium ions, in the absence of calcium two pentameric discs form a face-to-face decamer. Calcium 1 is coordinated to Asp 58, Asn 59, Glu 136, Gln 137 and Asp 138, calcium 2 to Glu 136, Asp 138 and Gln 148. SAP exhibits structural similarity with C-reactive protein and is universally present in amyloid deposits. SAP seems to interact with DNA and histones and may act as a scavenger for nuclear material released from damaged cells.

The 3D structure of the pentameric serum amyloid P-component (204 aa: His 1–Val 204) was determined by X-ray diffraction (PDB: 2A3W) and is shown in Figure 9.13 (a). Each monomer consists of a sandwich of two large β-sheets with a single α-helix on one face and a ligand binding site for two $Ca^{2+}$ (pink balls) on the other face. The pentamer has a five-fold symmetry axis. The binding of two calcium ions (pink balls) on the surface of each monomer is depicted in the surface representation shown in Figure 9.13(b).

SAP is a precursor of amyloid component P, which is found in basement membranes and is associated with amyloid deposits, e.g. the most ill-famous cerebral amyloid deposits found in the brain of Alzheimer disease patients.

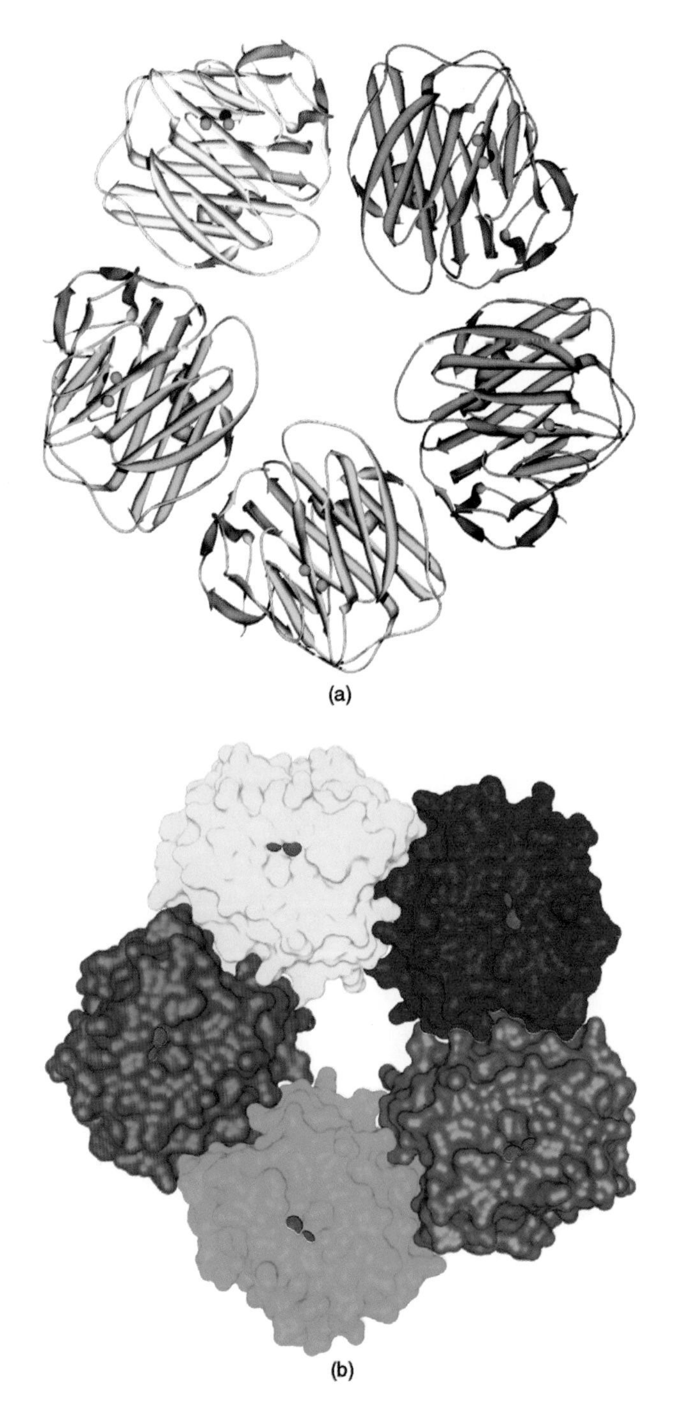


**Figure 9.13 (a) 3D structure of serum amyloid P-component (PDB: 2A3W)**
(a) The 3D structure of monomeric serum amyloid P-component consists of a sandwich of two large β-sheets with a single α-helix on one face and a ligand binding site for two $Ca^{2+}$ (pink balls) on the other face. The pentamer has a fivefold symmetry axis.
**(b) The binding of two calcium ions to serum amyloid P-component (PDB: 2A3W)**
(b) The binding of two calcium ions (pink balls) on the surface of each monomer of the amyloid P-component.

**Monocyte differentiation antigen CD14** (see Data Sheet), expressed mainly on the surface of monocytes, exists in two forms, a urinary form (348 aa) and a membrane-bound form (326 aa) with a potential GPI anchor linked to the C-terminal amidated Asn residue. The urinary form is generated from the proform by removing a C-terminal 30 residues propeptide and is a single-chain glycoprotein containing 11 leucine-rich repeats (LRR) and four intrachain disulfide bridges (Cys 6–Cys 17, Cys 15–Cys 32, Cys 168–Cys 198, Cys 222–Cys 253). CD14 is part of the innate immune response and cooperates with lymphocyte antigen 96 (MD-2) and toll-like receptor 4 (TLR4) to trigger a response to bacterial lipopolysaccharides (LPSs).

There are no 3D structural data available of human CD14; therefore the 3D structure of mouse CD 14 (312 aa: Ala 3–Glu 314, P10810) determined by X-ray diffraction (PDB: 1WWL) is shown in Figure 9.14. The sequence identity between the human and the mouse species amounts to 62 %. CD14 dimer exhibits a horseshoe-like structure with a concave surface and each monomer (in yellow and blue) consists of a large β-sheet composed of 11 parallel and two antiparallel β-strands. The convex surface contains both helices and loops in no regular pattern. The N-terminal region, thought to be the main LPS binding site, contains a large hydrophobic pocket (indicated by red arrows).

**Neutrophil defensin 1** (DEF1) (see Data Sheet) present in cells of the immune system is a single-chain cationic peptide (30 aa) containing three intrachain disulfide bridges (Cys 2–Cys 30, Cys 4–Cys 19, Cys 9–Cys 29). **Neutrophil defensin 2** (29 aa) lacks the N-terminal Ala residue of DEF1, which is generated from the precursor by releasing an N-terminal propeptide (19 aa) and by proteolytic processing from HP 1–56 (56 aa: Asp 1–Cys 56), releasing the 26 N-terminal residues.

The 3D structure of neutrophil defensin 3 (30 aa: Asp 1–Cys 30) was determined by X-ray diffraction (PDB: 1DFN) and is shown in Figure 9.15. Dimeric defensin exhibits an elongated, ellipsoidal structure and the monomer is characterised by a three-stranded antiparallel β-sheet stabilised by three intrachain disulfide bridges (in pink) arranged as mentioned above.

**β-Defensin 1** (BD1) (see Data Sheet) present in cells of the immune system is a single-chain cationic peptide (36 aa) containing three intrachain disulfide bridges (Cys 52–Cys 34, Cys 12–Cys 27, Cys 17–Cys 35) and is generated from the precursor form by releasing an N-terminal 11 residues propeptide.

The 3D structure of β-defensin 1 (36 aa: Asp 1–Lys 36) was determined by X-ray diffraction (PDB: 1IJV) and is shown in Figure 9.16. The 3D structure is characterised by a three-stranded antiparallel β-sheet flanked by an α-helix and stabilised by three intrachain disulfide bridges (in pink) arranged as mentioned above.

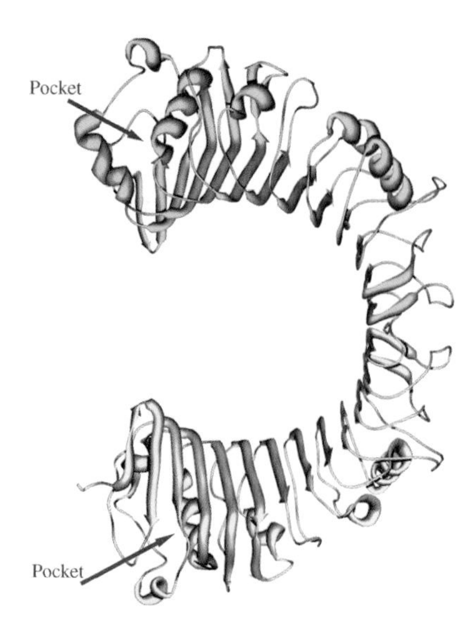


**Figure 9.14 3D structure of dimeric mouse monocyte differentiation antigen CD 14 (PDB: 1WWL)**
Dimeric CD 14 exhibits a horseshoe-like structure and each monomer (in yellow and blue) consists of a large β-sheet composed of 11 parallel and two antiparallel β-strands and a large hydrophobic pocket in the N-terminal region (indicated by red arrow).

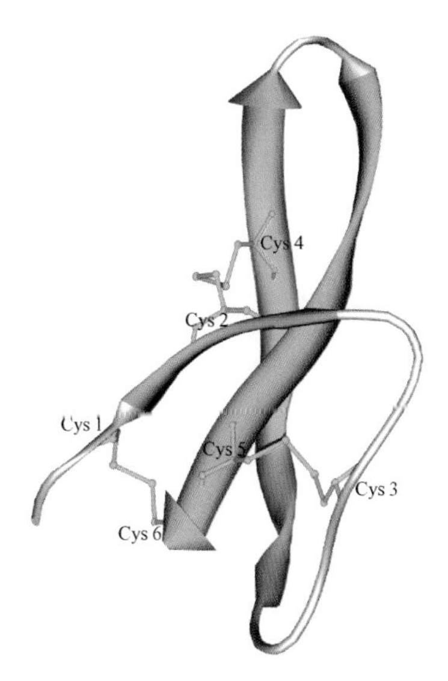


**Figure 9.15 3D structure of neutrophil defensin 3 (PDB: 1DFN)**
Neutrophil defensin 3 is characterised by a three-stranded antiparallel β-sheet stabilised by three intrachain disulfide bridges (in pink).

## 9.6 OTHER PROTEINS

Several proteins are involved in the modulation and regulation of the activity of the immune system but are not assigned to one of the afore-mentioned protein families.

**$\alpha_1$-Acid glycoprotein 1 and 2** (AGP1, AGP2) (see Data Sheet) or orosomucoid 1 and 2, synthesised in the liver, are single-chain glycoproteins containing two intrachain disulfide bridges (Cys 5–Cys 147, Cys 72–Cys 165) and an N-terminal

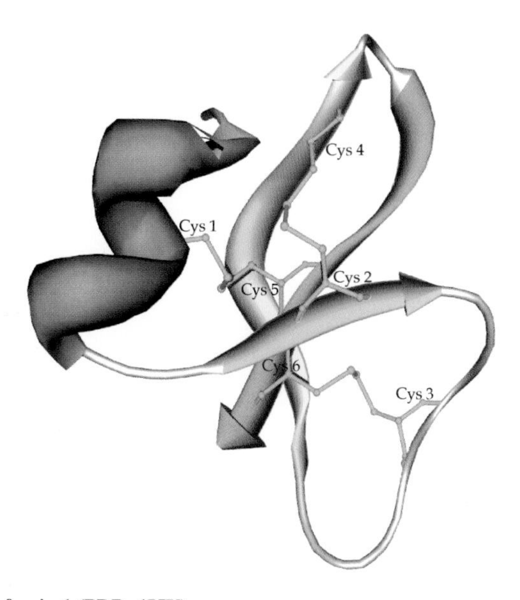


**Figure 9.16 3D structure of β-defensin 1 (PDB: 1IJV)**
β-Defensin 1 is characterised by a three-stranded antiparallel β-sheet flanked by an α-helix and stabilised by three intrachain disulfide bridges (in pink).

pyro-Glu residue. AGP1 and AGP2 appear to be involved in the modulation of the immune system during the acute-phase reaction. Their synthesis is controlled by glucocorticoids and interleukin-1 and interleukin-6.

**$\alpha_1$-Microglobulin** ($\alpha_1$-M) (see Data Sheet) is synthesised in the liver from the same mRNA as diprotein AMBP, which is processed before secretion into $\alpha_1$-microglobulin and the inter-$\alpha$-trypsin inhibitor light chain or bikunin (for details see Chapter 11). $\alpha_1$-M is present in plasma in free form or covalently linked to monomeric IgA or serum albumin. $\alpha_1$-M is a single-chain glycoprotein containing a single intrachain disulfide bridge (Cys 72–Cys 169) and a free Cys residue (aa 34) of unknown function. $\alpha_1$-M structurally belongs to the lipocalin family characterised by two $\beta$-sheets with eight to nine $\beta$-strands, forming a hollow cone with a hydrophobic interior. $\alpha_1$-M is heterogeneous in structure and has a yellow–brown colour caused by an array of small chromophore groups attached to amino acids at the entrance of the lipocalin pocket. $\alpha_1$-M seems to be involved in immunoregulation and exhibits immunosuppressive properties.

## REFERENCES


Anderson and Anderson, 2002, The human plasma proteome: history, character and diagnostic prospects, *Mol. Cell. Proteomics*, **1**, 845–867.

Cushley and Owen, 1983, Structural and genetic similarities between immunoglobulins and class-1 histocompatibility antigens, *Immunol. Today*, **4**, 88–92.

Halaby *et al.*, 1999, The immunoglobulin fold family: sequence analysis and 3D structure comparisons, *Protein Eng.*, **12**, 563–571.

Janeway, 2005, *Immunobiology*, Garland Science, New York.

Li *et al.*, 2004, The generation of antibody diversity through somatic hypermuation and class switch recombination, *Genes Dev.*, **18**, 1–11.

Saul and Poljak, 1992, Crystal structure of human immunoglobulin fragment Fab New refined at 2.0 Å resolution, *Proteins*, **14**, 363–371.

White *et al.*, 1995, Structure, function, and membrane integration of defensins, *Curr. Opin. Struct. Biol.*, **5**, 521–527.

DATA SHEETS

# $\alpha_1$-*Acid glycoprotein* 1 + 2 (*Orosomucoid* 1 + 2)

## Fact Sheet

| | | | |
|---|---|---|---|
| ***Classification*** | Lipocalin | ***Abbreviations*** | AGP1; AGP2 |
| ***Structures/motifs*** | | ***DB entries*** | A1AG1_HUMAN/P02763<br>A1AG2_HUMAN/P19652 |
| ***MW/length*** | AGP1: 21 560 Da/183 aa<br>AGP2: 21 651 Da/183 aa | ***PDB entries*** | |
| ***Concentration*** | 870 mg/l | ***Half-life*** | |
| ***PTMs*** | AGP1/AGP2: 5 N-CHO; 1 Pyro-Glu | | |
| ***References*** | Dente *et al.*, 1987, *EMBO J.*, **6**, 2289–2296.<br>Merritt *et al.*, 1988, *Gene*, **66**, 97–106. | | |

## Description

**$\alpha_1$-Acid glycoproteins** (AGP 1 + 2) are synthesised in the liver and seem to be involved in the modulation of the activity of the immune system.

## Structure

AGP 1 + 2 are single-chain glycoproteins containing two intrachain disulfide bridges (Cys 5–Cys 147, Cys 72–Cys 165) and are N-terminally modified (pyro-Glu).

## Biological Functions

AGP 1 + 2 appear to function in the modulation of the immune system during the acute-phase reaction. Their synthesis is controlled by glucocorticoids and interleukin-1 and interleukin-6.

# β-*Defensin* 1

## Fact Sheet

| | | | |
|---|---|---|---|
| ***Classification*** | β-Defensin | ***Abbreviations*** | BD1 |
| ***Structures/motifs*** | | ***DB entries*** | BD01_HUMAN/P60022 |
| ***MW/length*** | 3935 Da/36 aa | ***PDB entries*** | 1IJV |
| ***Concentration*** | | ***Half-life*** | |
| ***PTMs*** | | | |
| ***References*** | Hoover *et al.*, 2001, *J. Biol. Chem.*, **276**, 39021–39026.<br>Liu *et al.*, 1997, *Genomics*, **43**, 316–320. | | |

## Description

**β-Defensin** (BD1) is a secreted antimicrobial peptide of the innate immune system.

## Structure

BD1 is a peptide containing three intrachain disulfide bridges (Cys 5–Cys 34, Cys 12–Cys 27, Cys 17–Cys 35) belonging to the β-defensin family. The 3D structure is characterised by a three-stranded antiparallel β-sheet flanked by an α-helix and stabilised by three intrachain disulfide bridges (in pink) (**1IJV**).

## Biological Functions

BD1 exhibits antimicrobial activity against bacteria, viruses and fungi. They are thought to form pores in the cell membranes of microbes.

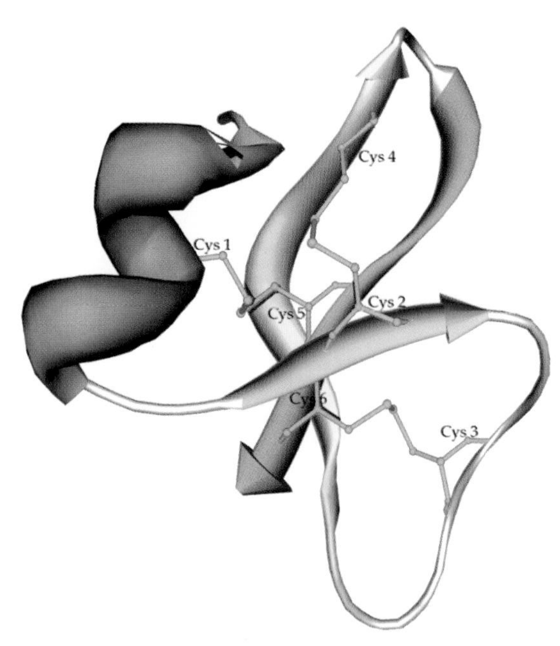


β-Defensin1

# *HLA class I histocompatibility antigen (MHC class I antigen): A-1 α-chain*

## Fact Sheet

| | | | |
|---|---|---|---|
| ***Classification*** | MHC class I | ***Abbreviations*** | HLA class I |
| ***Structures/motifs*** | 1 Ig-like C1 type | ***DB entries*** | 1A01_HUMAN/P30443 |
| ***MW/length*** | 38 322 Da/341 aa | ***PDB entries*** | 1A1M |
| ***Concentration*** | | ***Half-life*** | |
| ***PTMs*** | 1 S-Tyr (59 aa); 1 N-CHO | | |
| ***References*** | Parham *et al.*, 1988, *Proc. Natl. Acad. Sci. USA*, **85**, 4005–4009.<br>Smith *et al.*, 1996, *Immunity*, **4**, 215–228. | | |

## Description

**HLA class I histocompatibility antigen** (HLA I) is expressed by almost all types of nucleated cells and is involved in the presentation of pathogen-derived peptide fragments on cell surfaces.

## Structure

The HLA class I antigen, A-1 α-chain is a single-chain glycoprotein composed of a large N-terminal extracellular domain, a potential transmembrane segment (24 aa: Val 285–Trp 308) and a short C-terminal cytoplasmic domain. The molecule contains an Ig-like C1 type domain and three regions, $\alpha_1$, $\alpha_2$ and $\alpha_3$, two intrachain disulfide bridges (Cys 101–Cys 164, Cys 203–Cys 259) and an S-Tyr residue (59 aa). The MHC class I molecules are noncovalently linked heterodimers composed of an α-chain and a β-chain ($\beta_2$-microglobulin). The 3D structure of the extracellular domain of HLA class I α- and β-chains in complex with a nonapeptide (in pink) clearly reveals the three regions $\alpha_1$(in yellow), $\alpha_2$ (in green) and $\alpha_3$ (in blue) and $\beta_2$-microglobulin (in red) (**1A1M**).

## Biological Functions

HLA class I antigens play a key role in the immune system by presenting peptides derived from cytosolic proteins on cell surfaces for recognition by the corresponding T cells.

MHC class I antigen

## *HLA class II histocompatibility antigen (MHC class II antigen): DR α-chain/DR-1 β-chain*

### Fact Sheet

| | | | |
|---|---|---|---|
| ***Classification*** | MHC class II | ***Abbreviations*** | HLA class II |
| ***Structures/motifs*** | 1 Ig-like C1 type | ***DB entries*** | 2DRA_HUMAN/P01903<br>HB2B_HUMAN/P01912 |
| ***MW/length*** | α-chain: 25 987 Da/229 aa<br>β-chain: 27 190 Da/237 aa | ***PDB entries*** | 1A6A |
| ***Concentration*** | | ***Half-life*** | |
| ***PTMs*** | α-chain: 2 N-CHO; β-chain: 1 N-CHO (potential) | | |
| ***References*** | Das *et al.*, 1983, *Proc. Natl. Acad. Sci. USA*, **80**, 3543–3547.<br>Ghosh *et al.*, 1995, *Nature*, **378**, 457–462.<br>Peterson *et al.*, 1984, *EMBO J.*, **3**, 1655–1660. | | |

### Description

**HLA class II histocompatibility antigen** (HLA II) is expressed in specialised cells such as macrophages, dendritic cells, activated T cells and B cells, and is involved in the presentation of pathogen-derived peptide fragments on cell surfaces.

### Structure

The HLA class II antigen, consisting of the DR α-chain and DR-1 β-chain, are both very similar in structure and are single-chain glycoproteins composed of a large N-terminal extracellular domain, a potential transmembrane segment of 23 residues and a short C-terminal cytoplasmic region. The α-chain contains an Ig-like C1 type domain and two regions, $\alpha_1$ and $\alpha_2$, and a single intrachain disulfide bridge (Cys 107–Cys 163); the β-chain contains an Ig-like C1 type domain and two regions, $\beta_1$ and $\beta_2$, and two intrachain disulfide bridges (Cys 15–Cys 79, Cys 117–Cys 173). The $\alpha_1$ and $\beta_1$ domains form the antigen-binding site and a fragment of the HLA class II antigen invariant γ-chain binds in an almost identical way as an antigenic peptide (**1A6A**).

### Biological Functions

HLA class II antigens play a key role in the immune system by presenting peptides derived from extracellular proteins on cell surfaces for recognition by the corresponding T cells.

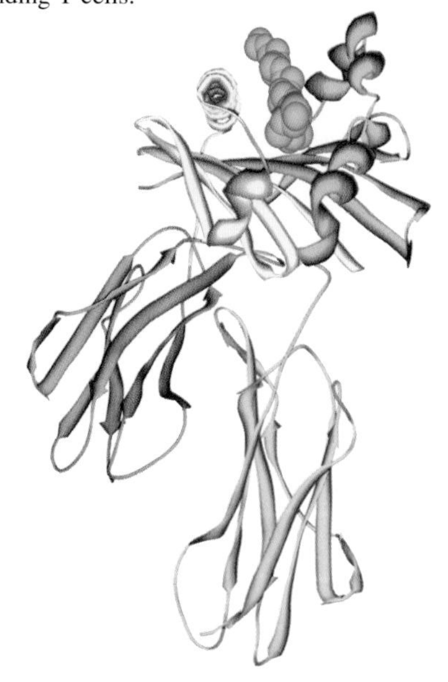

MHC class II antigen

# *Immunoglobulin A/α-Chain C regions (IgA1, IgA2)*

## Fact Sheet

| | | | |
|---|---|---|---|
| ***Classification*** | Immunoglobulin A | ***Abbreviations*** | IgA |
| ***Structures/motifs*** | IGHA1, IGHA2: Each 3 Ig-like | ***DB entries*** | IGHA1_HUMAN/P01876<br>IGHA2_HUMAN/P01877 |
| ***MW/length*** | IGHA1: 37 655 Da/353 aa<br>IGHA2: 36 508 Da/340 aa | ***PDB entries*** | |
| ***Concentration*** | 0.5–3.5 g/l | ***Half-life*** | 4.5–5.9 days |
| ***PTMs*** | IGHA1: 5 N-CHO, 2 O-CHO; IGHA2: 5 N-CHO | | |
| ***References*** | Flanagan *et al.*, 1984, *Cell*, **36**, 681–688. | | |

## Description

**Immunoglobulin A** (IgA) is synthesised by plasma cells in various tissues and glands and exists in two subclasses: IgA1 and IgA2.

## Structure

Monomeric IgA molecules consist of two heavy chains (α-chains) and two light chains (κ- or λ-chains) connected by interchain disulfide bridges exhibiting the classical structure of immunoglobulins. Oligomeric forms contain 2–4 IgA molecules disulfide-linked to a joining J-chain. Secretory IgA (S-IgA) is composed as the oligomeric forms and additionally contains a secretory component (SC).

## Biological Functions

IgA neutralises active antigens such as viruses, enzymes and toxins. IgA does not activate the classical pathway of the complement system.

## *Immunoglobulin D/δ-Chain C regions (IgD)*

### Fact Sheet

| | | | |
|---|---|---|---|
| ***Classification*** | Immunoglobulin D | ***Abbreviations*** | IgD |
| ***Structures/motifs*** | IGHD: 3 Ig-like | ***DB entries*** | IGHD_HUMAN/P01880 |
| ***MW/length*** | IGHD: 42 126 Da/383 aa | ***PDB entries*** | |
| ***Concentration*** | approx. 30 mg/l | ***Half-life*** | 2.8 days |
| ***PTMs*** | IGHD: 3 N-CHO, 7 O-CHO | | |
| ***References*** | Putnam *et al.*, 1981, *Proc. Natl. Acad. Sci. USA*, **78**, 6168–6172.<br>Lin and Putnam, 1981, *Proc. Natl. Acad. Sci. USA*, **78**, 504–508. | | |

### Description

**Immunoglobulin D** (IgD) exists as secreted and membrane-bound immunoglobulin.

### Structure

IgD consists of two heavy chains (δ-chains) and two light chains (κ- or λ-chains) connected by interchain disulfide bridges. IgD exhibits the typical immunoglobulin structure.

### Biological Functions

At low concentration IgD has the typical function of an antibody. Its membrane form acts as an antigen receptor resulting in endocytosis and cell activation.

## *Immunoglobulin E/ε-chain C regions (IgE)*

### Fact Sheet

| | | | |
|---|---|---|---|
| ***Classification*** | Immunoglobulin E | ***Abbreviations*** | IgE |
| ***Structures/motifs*** | IGHE: 4 Ig-like | ***DB entries*** | IGHE_HUMAN/P01854 |
| ***MW/length*** | IGHE: 47 019 Da/428 aa | ***PDB entries*** | |
| ***Concentration*** | 100–400 μg/l | ***Half-life*** | 2.3 days |
| ***PTMs*** | IGHE: 6 N-CHO | | |
| ***References*** | Seno *et al.*, 1983, *Nucleic Acids Res.*, **11**, 719–726. | | |

### Description

Immunoglobulin E (IgE) contains unique antigenic determinants and causes allergic reactions.

### Structure

IgE consists of two heavy chains (ε-chains) and two light chains (κ- or λ-chains) connected by interchain disulfide bridges. As monomeric IgM the heavy chains contain four constant regions.

### Biological Functions

IgE has specific antigenic determinants associated with IgE ε-chains and causes immediate allergic reactions called reaginic hypersensitivity.

## *Immunoglobulin G/γ-Chain C regions (IgG1, IgG2, IgG3, IgG4)*

### Fact Sheet

| | | | |
|---|---|---|---|
| ***Classification*** | Immunoglobulin G | ***Abbreviations*** | IgG |
| ***Structures/motifs*** | IGHG1, IGHG2, IGHG3, IGHG4:<br>Each 1 CH1, CH2, CH3 | ***DB entries*** | IGHG1_HUMAN/P01857<br>IGHG2_HUMAN/P01859<br>IGHG3_HUMAN/P01860<br>IGHG4_HUMAN/P01861 |
| ***MW/length*** | IGHG1: 36 110 Da/330 aa<br>IGHG2: 35 885 Da/326 aa<br>IGHG3: 32 331 Da/290 aa<br>IGHG4: 35 941 Da/327 aa | ***PDB entries*** | Fab: 7FAB<br>Fc: 2DTQ |
| ***Concentration*** | IgG1: 4.2–13 g/l; IgG2: 1.2–7.5 g/l;<br>IgG3: 0.4–1.3 g/l; IgG4: 0.01 g–2. 9 g/l | ***Half-life*** | IgG1: 21 days; IgG2: 21 days;<br>IgG3: 7 days; IgG4: 21 days |
| ***PTMs*** | IGHG1: 1 N-CHO; IGHG3: 1 Pyro-Glu, 2 N-CHO | | |
| ***References*** | Alexander *et al.*, 1982, *Proc. Natl. Acad. Sci. USA*, **79**, 3260–3264.<br>Ellison *et al.*, 1981, *DNA*, **1**, 11–18.<br>Ellison *et al.*, 1982, *Nucleic Acids Res.*, **10**, 4071–4079.<br>Matsumiya *et al.*, 2007, *J. Mol. Biol.*, **368**, 767–779.<br>Saul and Poljak, 1992, *Proteins*, **14**, 363–371.<br>Takahashi *et al.*, 1982, *Cell*, **29**, 671–679. | | |

### Description

**Immunoglobulin G** (IgG) is produced by plasma cells of the bone marrow, spleen and lymph nodes and is the most important immunoglobulin. There are four subclasses of IgG: IgG1 (60 %), IgG2 (30 %), IgG3 (4 %) and IgG4 (6 %).

### Structure

IgG molecules consist of two heavy chains (γ-chains) and two light chains (κ- or λ-chains) connected by interchain disulfide bridges. The heavy chains (HC) are approx. 420 aa long consisting of one variable (VH) and three constant (CH1, CH2 and CH3) domains and the light chains are approximately 210 aa long, consisting of a variable (VL) and a constant (CL) domain. The hinge regions in IgG1, IgG2 and IgG4 contain 12 aa, in IgG3 62 aa. The 3D structures of the Fab fragment (**7FAB**) and the Fc fragment (**2DTQ**) exhibit the typical immunoglobulin folds.

### Biological Functions

IgG is the most important antibody in plasma. It neutralises viruses, bacteria, fungi and yeasts. The four subclasses exhibit different antigenic, biological and structural properties.

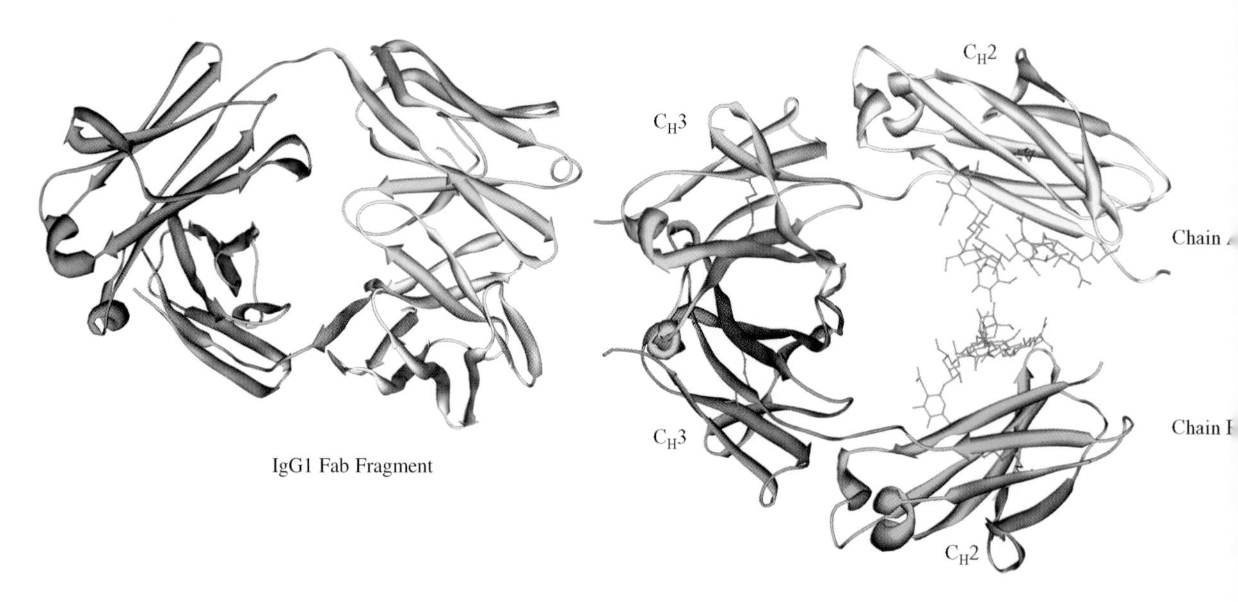


IgG1 Fab Fragment

IgG1 FC Fragment

# Immunoglobulin J-chain

## Fact Sheet

| | | | |
|---|---|---|---|
| *Classification* | | *Abbreviations* | J-chain |
| *Structures/motifs* | 1 Ig-like | *DB entries* | IGJ_HUMAN/P01591 |
| *MW/length* | 15 594 Da/137 aa | *PDB entries* | |
| *Concentration* | Not found in free form | *Half-life* | |
| *PTMs* | 1 Pyro-Glu; 1 N-CHO | | |
| *References* | Max and Korsmeyer, 1985, *J. Exp. Med.*, **161,** 832–849. | | |

## Description

**Immunoglobulin J-chain** (J-chain) is produced by plasma cells and is a component of polymeric immunoglobulins.

## Structure

J-chain is a single-chain glycoprotein (approx. 8 % CHO) containing three intrachain disulfide bridges (Cys 13–Cys 101, Cys 72–Cys 92, Cys 109–Cys 134) usually not found in free form. It is linked to oligomeric IgA and IgM via two Cys residues (Cys 15, Cys 69) forming interchain disulfide bridges with the penultimate Cys residue in the α- and μ-chains of IgA and IgM, respectively. Additionally, it is N-terminally modified (pyro-Glu).

## Biological Functions

It seems to participate in the intracellular assembly of polymeric Ig.

# Immunoglobulin M (γ-Macroglobulin)/μ-Chain C regions (IgM)

## Fact Sheet

| | | | |
|---|---|---|---|
| *Classification* | Immunoglobulin M | *Abbreviations* | IgM |
| *Structures/motifs* | MUC: 1 CH1, CH2, CH3, CH4 | *DB entries* | MUC_HUMAN/P01871 |
| *MW/length* | MUC: 49 557 Da/454 aa | *PDB entries* | |
| *Concentration* | 0.5–2.0 g/l | *Half-life* | 5 days |
| *PTMs* | MUC: 5 N-CHO | | |
| *References* | Friedlander *et al.*, 1990, *Nucleic Acids Res.*, **18**, 4278. | | |

## Description

**Immunoglobulin M** (IgM) is produced very early in the immune response, even before IgG, and is a primary defence against pathogens. It exists as secreted and membrane-bound immunoglobulin.

## Structure

Monomeric IgM consists of two heavy chains (μ-chains) and two light chains (κ- or λ-chains) connected by interchain disulfide bridges and the heavy chain contains four constant regions instead of the usual three. IgM exists predominantly as a pentamer connected to a J-chain (compare IgA) and adjacent monomers are linked by intermonomer disulfide bridges.

## Biological Functions

IgM binds to antigens resulting in the activation of the classical pathway of the complement system and macrophages.

## $\alpha_1$-*Microglobulin*

### Fact Sheet

| | | | |
|---|---|---|---|
| ***Classification*** | Lipocalin | ***Abbreviations*** | $\alpha_1$-M |
| ***Structures/motifs*** | 2 BPTI/Kunitz inhibitor | ***DB entries*** | AMBP_HUMAN/P02760 |
| ***MW/length*** | 20 847 Da/184 aa | ***PDB entries*** | |
| ***Concentration*** | 20–100 mg/l | ***Half-life*** | |
| ***PTMs*** | 2 N-CHO; 1 O-CHO | | |
| ***References*** | Åkerström *et al.*, 2000, *Biochim. Biophys. Acta*, **1482**, 172–184. | | |
| | Kaumeyer *et al.*, 1986, *Nucleic Acids Res.*, **14**, 7839–7850. | | |

### Description

**$\alpha_1$-Microglobulin** ($\alpha_1$-M) is synthesised in the liver as a diprotein (AMBP) that is processed before secretion into $\alpha_1$-microglobulin ($\alpha_1$-M) and the inter-$\alpha$-trypsin inhibitor light chain (bikunin). It is present in plasma in free form or covalently linked to monomeric IgA or serum albumin.

### Structure

$\alpha_1$-M is a single-chain glycoprotein containing an array of chromophore groups, a single intrachain disulfide bridge (Cys 72–Cys 169) and a free Cys residue (34 aa) of unknown function.

### Biological Functions

$\alpha_1$-M seems to be involved in immunoregulation and exhibits immunosuppressive properties.

# $\beta_2$-*Microglobulin*

## Fact Sheet

| | | | |
|---|---|---|---|
| ***Classification*** | | ***Abbreviations*** | $\beta_2$-M |
| ***Structures/motifs*** | 1 Ig-like C1-type | ***DB entries*** | B2MG_HUMAN/P61769 |
| ***MW/length*** | 11 731 Da/99 aa | ***PDB entries*** | 1A1M |
| ***Concentration*** | 16 mg/l | ***Half-life*** | 1.2 – 2.8 hrs |
| ***PTMs*** | 1 Pyro-Glu | | |
| ***References*** | Guessow *et al.*, 1987, *J. Immunol.*, **139**, 3132–3138.<br>Smith *et al.*, 1996, *Immunity*, **4**, 215–228. | | |

## Description

**$\beta_2$-Microglobulin** ($\beta_2$-M) is expressed in all nucleated cells and plays an important role in the immune system.

## Structure

$\beta_2$-M is a single-chain protein containing an Ig-like domain with a single intrachain disulfide bridge (Cys 25–Cys 80) and is N-terminally modified (pyro-Glu). The 3D structure of $\beta_2$-M is characterised by a sandwich of two $\beta$-sheets consisting of three and four antiparallel $\beta$-strands stabilised by the single disulfide bridge and resulting in a compact globular fold (**1A1M**).

## Biological Functions

$\beta_2$-M plays a key role in the immune system and is required for a proper cell surface expression of the MHC class I antigen.

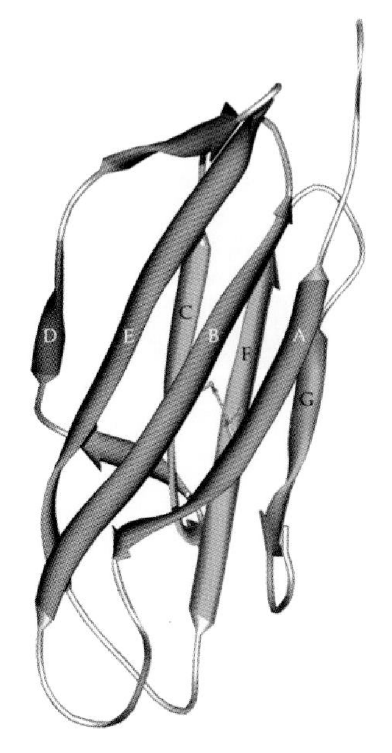


$\beta_2$-Microglobulin

# *Monocyte differentiation antigen CD14*

## Fact Sheet

| | | | |
|---|---|---|---|
| ***Classification*** | | ***Abbreviations*** | |
| ***Structures/motifs*** | 11 LRR (leucine-rich) | ***DB entries*** | CD14_HUMAN/P08571 |
| ***MW/length*** | Urinary form: 37 215 Da/348 aa<br>Membrane form: 35 159 Da/326 aa | ***PDB entries*** | 1WWL |
| ***Concentration*** | | ***Half-life*** | |
| ***PTMs*** | 4 N-CHO (2 potential); membrane form: 1 GPI-anchor (potential, Asn 326) | | |
| ***References*** | Kim *et al.*, 2005, *J. Biol. Chem.*, **280**, 11347–11351.<br>Simmons *et al.*, 1989, *Blood*, **73**, 284–289. | | |

## Description

**Monocyte differentiation antigen** CD14 is heavily expressed on the surface of monocytes and is part of the innate immune response.

## Structure

CD14 exists in two forms, a membrane-bound form with a potential GPI anchor attached to C-terminal amidated Asn 326 and a urinary form. The urinary form is a single-chain glycoprotein containing 11 LRR repeats (leucine-rich) and four intrachain disulfide bridges. A 30 residue propeptide is removed in the mature form. Mouse dimeric CD14 exhibits a horseshoe-like structure with a large hydrophobic pocket in the N-terminal region, thought to be the main LPS binding site (**1WWL**).

## Biological Functions

CD14 cooperates with lymphocyte antigen 96 (MD-2) and toll-like receptor 4 (TLR4) to mediate the innate immune response to bacterial lipopolysaccharides.

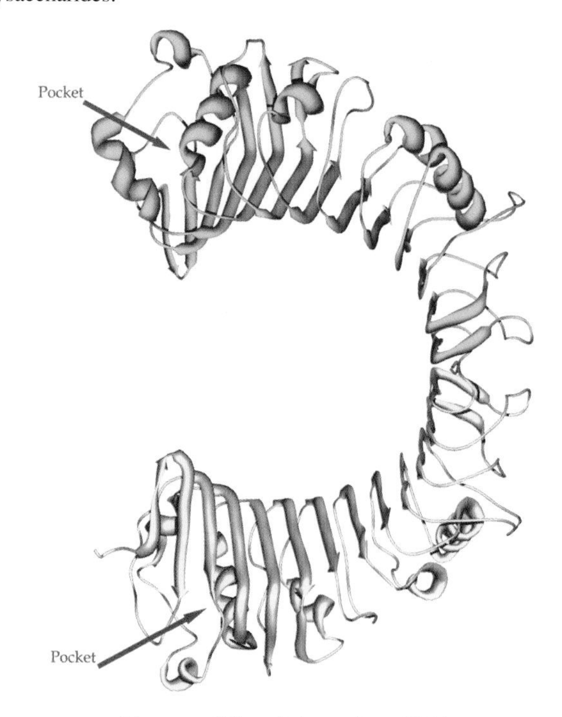


Monocyte differentiation antigen CD14

# *Neutrophil defensin 1 + 2 (α-Defensins)*

## Fact Sheet

| | | | |
|---|---|---|---|
| ***Classification*** | α-Defensin | ***Abbreviations*** | DEF1; DEF2 |
| ***Structures/motifs*** | | ***DB entries*** | DEF1_HUMAN/P59665 |
| ***MW/length*** | DEF1: 3448 Da/30 aa<br>DEF2: 3377 Da/29 aa<br>HP 1–56: 6306 Da/56 aa | ***PDB entries*** | 1DFN |
| ***Concentration*** | | ***Half-life*** | |
| ***PTMs*** | | | |
| ***References*** | Daher *et al.*, 1988, *Proc. Natl. Acad. Sci. USA*, **85**, 7327–7331.<br>Hill *et al.*, 1991, *Science*, **251**, 1481–1485. | | |

## Description

**Neutrophil defensins 1 + 2** (DEF 1 + 2) are antimicrobial peptides of the innate immune system.

## Structure

DEF 1 + 2 are peptides containing three intrachain disulfide bridges (DEF 1: Cys 2–Cys 30, Cys 4–Cys 19, Cys 9–Cys 29) and DEF 2 lacks the N-terminal Ala residue of DEF 1. The 3D structure of neutrophil defensin 3 is characterised by a three-stranded β-sheet stabilised by three intrachain disulfide bridges (in pink) (**1DFN**).

## Biological Functions

DEF 1 + 2 exhibit antimicrobial activity against bacteria, viruses and fungi. They are thought to form pores into the cell membranes of microbes.

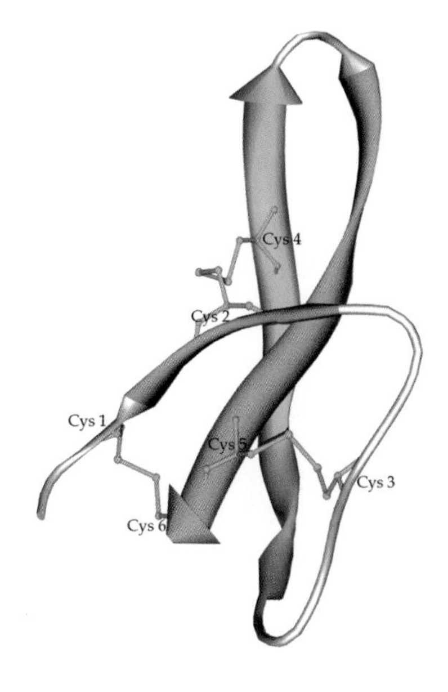


Neutrophil defensin3

# *Serum amyloid P-component*

## Fact Sheet

| | | | |
|---|---|---|---|
| ***Classification*** | Pentaxin | ***Abbreviations*** | SAP |
| ***Structures/motifs*** | 1 pentaxin | ***DB entries*** | SAMP_HUMAN/P02743 |
| ***MW/length*** | 23 259 Da/204 aa | ***PDB entries*** | 2A3W |
| ***Concentration*** | 10–50 mg/l | ***Half-life*** | 24 hrs |
| ***PTMs*** | 1 N-CHO | | |
| ***References*** | Ho *et al.*, 2005, *J. Biol. Chem.*, **280**, 31999–32008. | | |
| | Mantzouranis *et al.*, 1985, *J. Biol. Chem.*, **260**, 7752–7756. | | |

## Description

**Serum amyloid P-component** (SAP) is exclusively synthesised by hepatocytes and exhibits a high sequence similarity with C-reactive protein.

## Structure

SAP is a single-chain glycoprotein containing a single intrachain disulfide bridge (Cys 36–Cys 95) and in the presence of calcium exhibits a pentameric structure. Each monomer consists of a sandwich of two large β-sheets with a single α-helix on one face and a ligand binding site for two $Ca^{2+}$ (pink balls) on the other face (**2AW3**).

## Biological Functions

SAP is universally present in amyloid deposits such as the cerebral amyloid deposits found in the brain of Alzheimer disease patients. SAP seems to interact with DNA and histones and may act as a scavenger for nuclear material released from damaged cells.

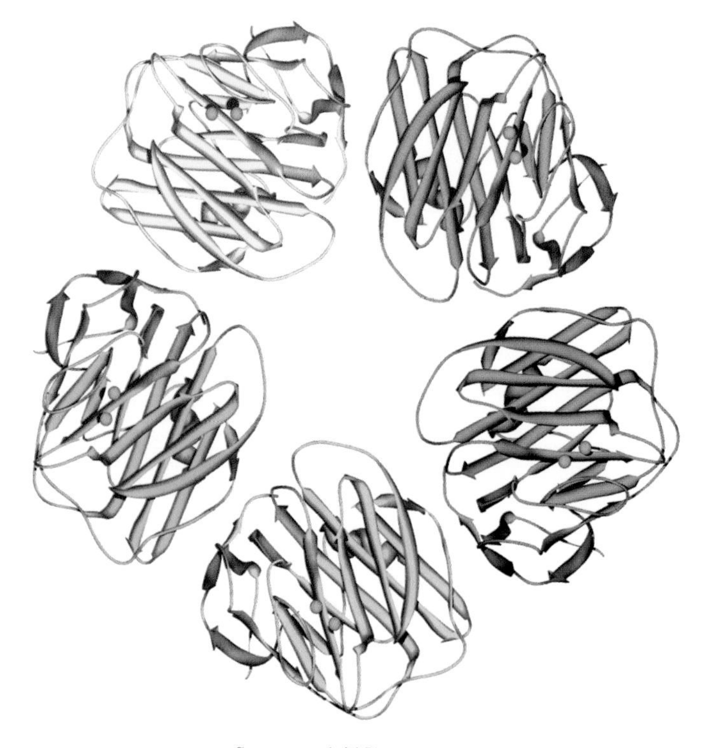

Serum amyloid P-component

# *Zn-$\alpha_2$-Glycoprotein*

## Fact Sheet

| | | | |
|---|---|---|---|
| ***Classification*** | MHC class I | ***Abbreviations*** | Zn-$\alpha_2$-GP |
| ***Structures/motifs*** | 1 Ig-like | ***DB entries*** | ZA2G_HUMAN/P25311 |
| ***MW/length*** | 32 145 Da/278 aa | ***PDB entries*** | 1T80 |
| ***Concentration*** | 40–75 mg/l | ***Half-life*** | |
| ***PTMs*** | 3 N-CHO; 1 Pyro-Glu | | |
| ***References*** | Araki *et al.*, 1988, *Proc. Natl. Acad. Sci. USA*, **85**, 679–683.<br>Delker *et al.*, 2004, *J. Struct. Biol.*, **148**, 205–213. | | |

## Description

**Zn-$\alpha_2$-glycoprotein** (Zn-$\alpha_2$-GP) is present in many body fluids and various types of cells and binds zinc. Zn-$\alpha_2$-GP exhibits similarity to the MHC class I molecules.

## Structure

Zn-$\alpha_2$-GP is a single-chain glycoprotein (18 % CHO) containing an Ig-like domain and two intrachain disulfide bridges (Cys 103–Cys 166, Cys 205–Cys 260) and carries an N-terminal pyro-Glu residue. The 3D structure clearly demonstrates the structural similarity with MHC class I molecules (**1T80**).

## Biological Functions

Although the exact function of Zn-$\alpha_2$-GP is unknown it seems to be involved in the immune response, in lipid degradation in adipocytes and cell adhesion processes.

Zn-$\alpha_2$-glycoprotein

# 10

# Enzymes

## 10.1 INTRODUCTION

Enzymes are proteins that accelerate or catalyse chemical reactions. Enzymes occupy key positions in the synthesis of a wide variety of compounds in biological systems. Almost all processes in the cells of a living organism require enzymes. Enzymes are known to catalyse approximately 4000 chemical reactions but not all biological catalysts are enzymes since ribozymes (RNA molecules) can also catalyse reactions. Enzymes are usually named according to the reaction they catalyse. The International Union of Biochemistry and Molecular Biology (IUBMB) and the International Union of Pure and Applied Chemistry (IUPAC) have developed a nomenclature for enzymes, the *EC number system*. Each enzyme is described by a sequence of four numbers preceded by EC. The first number broadly classifies the enzymes into six groups based on their catalysing mechanism followed by three numbers separated by periods representing a progressively finer classification of each enzyme. The six enzyme groups ***oxidoreductases***, ***transferases***, ***hydrolases***, ***lyases***, ***isomerases***, and ***ligases*** are summarised in Table 10.1 together with the reactions they catalyse.

Many biochemical pathways and cascades such as *coagulation and fibrinolysis* (Chapter 7) and *the complement system* (Chapter 8) are governed, controlled and regulated by enzymes where peptidases, especially serine proteases, play a crucial role. Many of these enzymes are described in detail in the corresponding chapters. The enzymes involved in these cascades together with their main function are compiled in Table 10.2.

Normally, enzymes are present in plasma in their proenzyme form and require a specific activation step with a concomitant large conformational change to obtain their full activity. They are often multidomain proteins composed of the catalytic domain linked to other types of domains, such as EGF-like, kringle, fibronectin and sushi/CCP/SCR, which are required for correct recognition and binding (for details see Chapter 9).

Of course, plasma contains many more enzymes than those compiled in Table 10.2. A selection of these 'remaining' enzymes are discussed in this chapter according to the EC classification system. The groups EC 4: Lyases and EC 6: Ligases are not discussed with examples in this chapter because they are not considered as 'classical' blood plasma proteins.

## 10.2 EC 1: OXIDOREDUCTASES

An oxidoreductase is an enzyme that catalyses the transfer of electrons from one molecule to another:

$$A^- + B \Rightarrow A + B^-$$

Here, A is the electron donor (the reductant, also called the hydrogen acceptor) and B is the electron acceptor (the oxidant, also called the hydrogen donor). In biochemical reactions, the redox reaction system is not always evident, as can be seen in the following example:

$$P_i + \text{glyceraldehyde-3-phosphate} + NAD^+ \Rightarrow NADH{+}H^+ + 1,3\text{-bisphosphoglycerate}$$

Here, glyceraldehyde-3-phosphate is the oxidant (hydrogen donor) and $NAD^+$ is the reductant (hydrogen acceptor).

Oxidoreductases are further classified into 22 subclasses, from EC 1.1 to EC 1.21 and EC 1.97.

*Human Blood Plasma Proteins: Structure and Function* Johann Schaller, Simon Gerber, Urs Kämpfer, Sofia Lejon and Christian Trachsel

**Table 10.1** Enzyme nomenclature and chemical reactions catalysed.

| Group | Reaction catalysed | Typical reaction | Examples |
|---|---|---|---|
| EC 1: Oxidoreductases | Oxidation/reduction: transfer of H and O atoms or electrons | Reduced: AH + B ⇒ A + BH<br>Oxidised: A + O ⇒ AO | Dehydrogenase, oxidase |
| EC 2: Transferases | Transfer of functional groups: methyl, acyl, amino, phosphate groups | AB + C ⇒ A + BC | Transaminase, kinase |
| EC 3: Hydrolases | Substrate hydrolysis: formation of two products | AB + $H_2O$ ⇒ AOH + BH | Lipase, amylase, peptidase |
| EC 4: Lyases | Nonhydrolytic addition/removal of groups: C–C, C–N, C–O, C–S bonds may be cleaved | RCOCOOH ⇒ RCOH + $CO_2$ | |
| EC 5: Isomerases | Intramolecular rearrangement, e.g. isomerisation | AB ⇒ BA | Isomerase, mutase |
| EC 6: Ligases | Linking two molecules: synthesis of new C–O, C–S, C–N, C–C bonds (requires ATP) | X + Y + ATP ⇒ XY + ADP + $P_i$ | Synthetase |

**Table 10.2** Enzymes involved in coagulation and fibrinolysis, the complement system and lipoprotein metabolisms.

| Enzyme | EC Number | Code | Function |
|---|---|---|---|
| ***Coagulation and fibrinolysis*** | | | |
| Factor XIII | 2.3.2.13 | P00488, P05160 | Crosslinks fibrin |
| Angiotensin-converting enzyme | 3.4.15.1 | P12812 | Converts angiotensin I ⇒ angiotensin II |
| Carboxypeptidase B | 3.4.17.20 | Q96IY4 | Removes C-terminal basic residues |
| Prothrombin | 3.4.21.5 | P00734 | Cleaves fibrinogen, releasing fibrinopeptides |
| Factor X | 3.4.21.6 | P00742 | Complex FVa/FXa: prothrombin ⇒ thrombin |
| Plasminogen | 3.4.21.7 | P00747 | Cleaves fibrin polymers |
| Factor VII | 3.4.21.21 | P08709 | Complex TF/FVIIa: FX ⇒ FXa; FIX ⇒ FIXa |
| Factor IX | 3.4.21.22 | P00740 | Complex FVIIIa/FIXa: FX ⇒ FXa |
| Factor XI | 3.4.21.27 | P03951 | Contact-phase activation |
| Plasma kallikrein | 3.4.21.34 | P03952 | Initiation intrinsic pathway |
| Factor XII | 3.4.21.38 | P00748 | FXI ⇒ FXIa; prekallikrein ⇒ kallikrein |
| Tissue-type plasminogen activator | 3.4.21.68 | P00750 | Activation plasminogen |
| Protein C | 3.4.21.69 | P04070 | Regulation blood coagulation |
| Urokinase-type plasminogen activator | 3.4.21.73 | P00749 | Activation plasminogen |
| Renin | 3.4.23.15 | P00797 | Generates angiotensin I from angiotensinogen |
| ***Complement system*** | | | |
| Complement activating component | 3.4.21.- | P48740 | Complement activation: activation C2, C4 |
| Mannan-binding lectin serine protease 2 | 3.4.21.- | O00187 | Complement activation: activation C2, C4 |
| C1r subcomponent | 3.4.21.41 | P00736 | Activation C1s |
| C1s subcomponent | 3.4.21.42 | P09871 | Activation C2, C4 |
| C2 | 3.4.21.43 | P06681 | Cleavage C3 |
| Factor I | 3.4.21.45 | P05156 | Inactivation C3b, C4b |
| Factor D | 3.4.21.46 | P00746 | Activation factor B |
| Factor B | 3.4.21.47 | P00751 | Activation alternative pathway |
| ***Lipoprotein metabolism*** | | | |
| Phosphatidylcholine-sterol acyltransferase | 2.3.1.43 | P04180 | Synthesis acylglycerophosphocholine |
| Hepatic triacylglycerol lipase | 3.1.1.3 | P11150 | Hydrolysis phospholipids, acylglycerols |
| Lipoprotein lipase | 3.1.1.34 | P06858 | Hydrolysis triacylglycerol |
| Apolipoprotein(a) | 3.4.21.- | P08519 | Regulation fibrinolysis |

### 10.2.1 EC 1.11: peroxidases

Peroxidases (EC 1.11) are oxidoreductases that act on peroxide as an acceptor. Free peroxides such as hydrogen peroxide, which is a reaction product of various oxidases (e.g. in peroxisomes, in the endoplasmic reticulum) are very dangerous to cells and peroxidases protect the cells, enzymes and heme-containing proteins from oxidative damage. There are several types of peroxidases, of which three examples present in plasma, at least in traces, are discussed in this chapter:

1. **Catalase** (CAT) is a peroxidase that is widely distributed in many types of tissues with the highest concentrations in the liver and kidney, where it is localised in peroxisomes. Catalase protects the cells from the toxic effects of $H_2O_2$ by decomposing it to oxygen and water:

   $$2\,H_2O_2 \Rightarrow O_2 + 2\,H_2O$$

   Catalase has a turnover rate of approximately 5 million molecules of $H_2O_2$ per minute per enzyme molecule, which is one of the highest of all enzymes.

   Catalases share two consensus sequences, one around a conserved Tyr residue that is the proximal heme-binding ligand and one around a conserved His residue that is part of the active site:

   CATALASE_1 (PS00437): R–[LIVMFSTAN] –F–[GASTNP]–**Y**–x–D–[AST]–[QEH]

   CATALASE_2 (PS00438): [IF]–x–[RH]–x(4) –[EQ]–R–x(2)–**H**–x(2)–[GAS]–[GASTFY]–[GAST]

   **Catalase** (CAT, EC 1.11.1.6) (see Data Sheet) is found in many types of tissues localised in peroxisomes, with the highest concentrations in the liver and kidney and the lowest in connective tissue. The active site of single-chain catalase contains His 74 and Asn 147 and the single heme group is coordinated to Tyr 357. Catalase is present as noncovalently linked homotetramer.

   The 3D structure of erythrocyte catalase (498 aa: Ser 3–Asn 500) was determined by X-ray diffraction (PDB: 1DGB) and the noncovalently linked homotetramer is shown in Figure 10.1(a). The monomer is characterised by a central eight-stranded antiparallel β-barrel surrounded by a α-helical domain, an N-terminal threading arm and a wrapping loop and is given in Figure 10.1(b). The heme group (in pink) is located in the central β-barrel together with the active site residues His and Asn (in blue) and the NADPH group (in green) is located on the far side of the molecule.

   Defects in the CAT gene are the cause of acatalasia (acatalasemia) characterised by the absence of catalase activity in erythrocytes, resulting in ulcerating oral lesions (for details see MIM: 115500).

2. **Myeloperoxidase** (MPO) is a peroxidase most prominently present in neutrophil granulocytes, where it is part of the host defence system of polymorphonuclear leukocytes and is responsible for the microbicidal activity against many organisms. MPO catalyses the following reactions and the production of hypochloric acid:

   $$\text{Donor} + H_2O_2 \Rightarrow \text{oxidised donor} + 2\,H_2O$$
   $$Cl^- + H_2O_2 \Rightarrow HOCl + 2\,H_2O$$

   Heme is required as a cofactor.

   **Myeloperoxidase** (MPO, EC 1.11.1.7) (see Data Sheet) is primarily contained in the granulocytes of neutrophils. Mature MPO (581 aa: Val 1–Ser 581) is generated from the precursor by releasing an N-terminal propeptide (116 aa) and a subsequent proteolytic cleavage of the Gly 114–Val 115 peptide bond, thus generating a light chain (114 aa: Val 1–Gly 114) and a heavy chain (467 aa: Val 115–Ser 581). MPO exists as a tetramer of two light and two heavy chains, where the two heavy chains are covalently linked by a single interchain disulfide bridge (Cys 155 in MPO). The light chain contains one intrachain disulfide bridge (Cys 3–Cys 16) and the heavy chain five intrachain disulfide bridges (Cys 117–Cys 127, Cys 121–Cys 145, Cys 223–Cys 234, Cys 442–Cys 499, Cys 540–Cys 566). MPO contains a rare Cys-sulfenic acid residue (aa 152). MPO binds two calcium ions and covalently two heme B groups (iron-protoporphyrin IX) containing two iron ions, one per heterodimer. The calcium-binding site in the heterodimer is (MPO numbering):

   $Ca^{2+}$: **His 97, Arg 98** (both MPO light chain), **Thr 170, Phe 172** (carbonyl oxygen), **Asp 174, Ser 176** (all MPO heavy chain). The Heme (iron)-binding site in the heterodimer is (MPO numbering):

   Heme (covalently): **Asp 96, Glu 244, Met 245** and **$Fe^{3+}$: His 338**

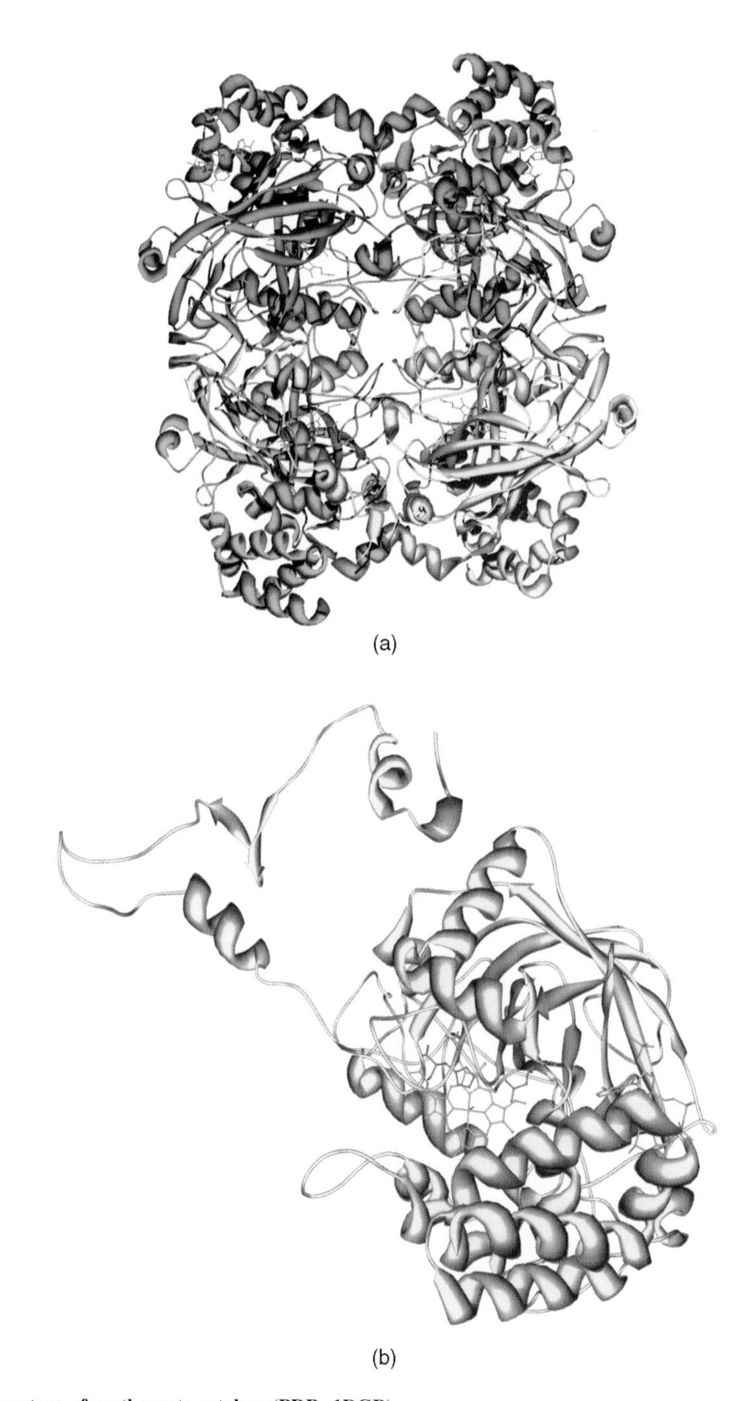


**Figure 10.1 3D structure of erythrocyte catalase (PDB: 1DGB)**
(a) Catalase forms a non-covalently linked homotetramer
(b) The monomer of catalase consists of a central eight-stranded antiparallel β-barrel with the heme group (in pink), which is surrounded by a α-helical domain, a N-terminal threading arm and a wrapping loop and the NADPH group (in green) is located on the far side of the molecule. The active site residues His and Asn are shown in blue.

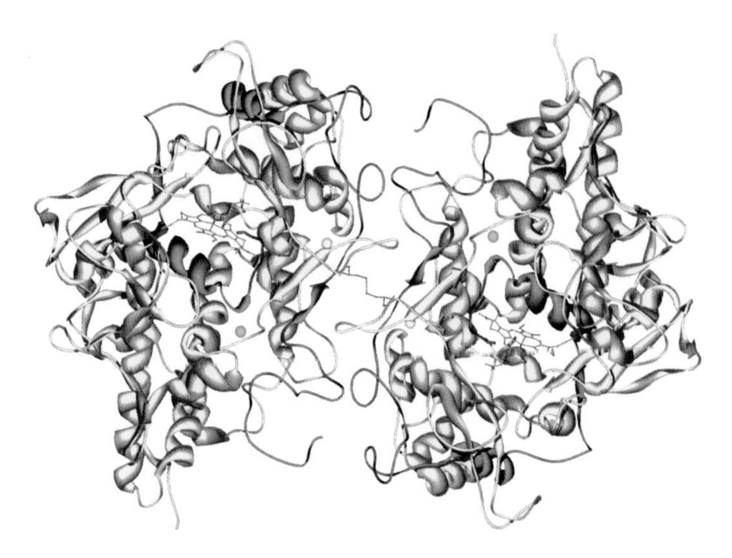

**Figure 10.2 3D structure of the dimer of myeloperoxidase (PDB: 1MHL)**
The dimer of myeloperoxidase is linked by a single interchain disulphide bridge (in red) and each monomer consists of a non-covalently linked light chain (in blue) and heavy chain (in yellow). The heme group (in pink), the unusual Cys-sulfenic acid residue (in green) and the calcium and chloride ions (green and orange balls, respectively) are indicated.

The 3D structure of the dimer of MPO, where each monomer consists of a noncovalently linked light chain (108 aa: Val 1–Arg 108, in blue) and heavy chain (466 aa: Val 115–Ala 580, in yellow), was determined by X-ray diffraction (PDB: 1MHL) and is shown in Figure 10.2. The structure is largely $\alpha$-helical with very little $\beta$-sheet and the heme group (in pink) is covalently linked via two ester bonds to the carboxyl groups of Glu 244 and Asp 96 and via a sulfonium ion linkage to the sulfur of Met 245. The unusual Cys-sulfenic acid residue (aa 152, in green) and the interchain disulfide bridge (in red) as well as the calcium and chloride ions (green and orange balls, respectively) are indicated.

Defects in the MPO gene are the cause of myeloperoxidase deficiency, which is an autosomal recessive disease resulting in disseminated candidiasis (for details see MIM: 254600).

3. There are seven **glutathione peroxidases** (GPX1–GPX7) in different cells and body fluids that protect the cells and enzymes from oxidative damage by catalysing the reduction of free hydrogen peroxide to water, lipid peroxides and organic hydroperoxides to their corresponding alcohols using glutathione (GSH), which is transformed into oxidised glutathione (GSSG):

$$H_2O_2 + 2\,GSH \Rightarrow GSSG + 2\,H_2O$$

$$ROOH + 2\,GSH \Rightarrow GSSG + ROH + H_2O$$

Glutathione peroxidases share two consensus sequences, one around the active site residue, which is either a Cys residue or in many cases a Se-Cys residue, and a conserved octapeptide sequence in the central region:

GLUTATHIONE_PEROXIDASE_1 (PS00460): [GNDR]–[RKHNQFYCS]–x–[LIVMFCS]–[LIVMF]–x–N–[VT]–x–[STCA]–x–**C**–[GA]–x–[TA]

GLUTATHIONE_PEROXIDASE_2 (PS00763): [LIV]–[AGD]–F–P–[CS]–[NG]–Q–F

Oxidised glutathione is reduced to monomeric glutathione by the enzyme glutathione reductase:

$$GSSG + NADPH + H^+ \Rightarrow 2\,GSH + NADP^+$$

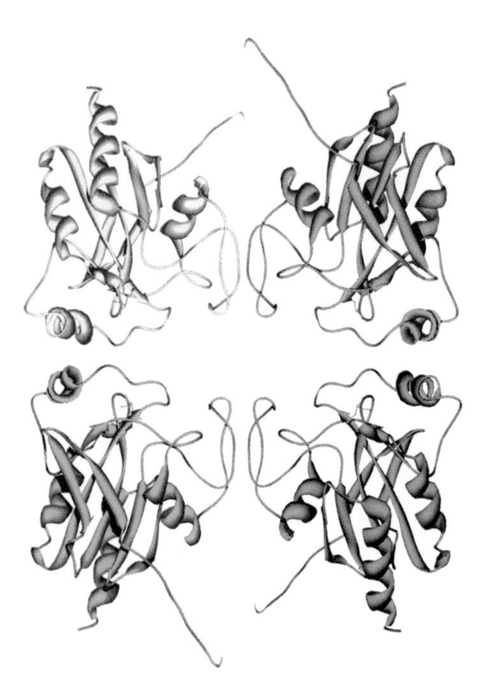

**Figure 10.3 3D structure of epididymal glutathione peroxide 5 (PDB: 2I3Y)**
Glutathione peroxidase 5 forms a non-covalently linked homotetramer and each monomer consists of a central β-sheet core structure surrounded by α-helices and the active site Se-Cys (in pink) is located in the N-terminal region.

**Glutathione peroxidase 3** (GPX3, EC 1.11.1.9) (see Data Sheet) is mainly synthesised in the kidney as a single-chain protein present in plasma as a noncovalently linked homotetramer. The active site consists of a Se-Cys residue and the N-terminus is modified (pyro-Glu). It protects extracellular fluid components and the cell surface against peroxide-mediated damage. GPX3 plasma levels are used as a clinical marker for selenium deficiency.

The 3D structure of epididymal glutathione peroxide 5 (GPX5: 193 aa: Met 7–Lys 199 + His-tag), exhibiting a sequence identity of 71 % with GPX3, was determined by X-ray diffraction (PDB: 2I3Y) and forms a noncovalently linked homotetramer as shown in Figure 10.3. The monomer is characterised by a central β-sheet core structure surrounded by α-helices and the active site Se-Cys 53 (in pink) is located in the N-terminal region.

## 10.3 EC 2: TRANSFERASES

A transferase is an enzyme that catalyses the transfer of a functional group from one molecule to another according to the following reaction scheme:

$$A{-}X + B \Rightarrow A + B{-}X$$

Here, A is the donor (e.g. a methyl or a phosphate group), often a coenzyme, and B is the acceptor. Transferases are further classified into nine subclasses, from EC 2.1 to EC 2.9.

### 10.3.1 EC 2.3: acyltransferases

γ-Glutamyltranspeptidase 1 (GGT) is a member of aminoacyltransferases (EC 2.3.2) and is part of the cell antioxidant defence mechanism that initiates the extracellular breakdown of glutathione and provides the cell with Cys supply. It catalyses the transfer of a glutamyl group of glutathione to amino acids and dipeptides and plays a key role in the $\gamma$-glutamyl cycle, a pathway for the synthesis and degradation of glutathione:

$$\text{5-Glutamyl-peptide} + \text{amino acid} \Rightarrow \text{peptide} + \text{5-glutamyl amino acid}$$

GGT is a good example of an enzyme following the mechanism of so-called ping-pong bi-bi reactions, also known as double-displacement reactions.

$\gamma$-Glutamyltranspeptidases share a conserved sequence pattern in the N-terminal region of the light chain:

G_GLU_TRANSPEPTIDASE (PS00462): T–[STA]–H–x–[ST]–[LIVMA]–x(4)–G–[SN]–x–V–[STA]
–x–T–x – T–[LIVM]–[NE]–x(1, 2)–[FY]–G

**$\gamma$-Glutamyltranspeptidase 1** (EC 2.3.2.2) (see Data Sheet) or $\gamma$-Glutamyltransferase 1 (GGT) is present in many tissues and organs and the small amounts present in plasma originate mainly from the liver. GGT is a type II membrane glycoprotein composed of two noncovalently associated chains, a heavy chain (380 aa: Met 1–Gly 380) and a light chain (189 aa: Thr 1–Tyr 189). The heavy chain contains an N-terminal potential membrane-anchor sequence (22 aa: Leu 5–Leu 26) and the active site is located in the light chain. GGT catalyses the first reaction in the glutathione breakdown and is involved in the transport of amino acids across the cell membrane.

### 10.3.2 EC 2.6: transfer of nitrogenous group

Transaminases or aminotransferases (EC 2.6.1) such as aspartate aminotransferase (AspAT) or alanine transaminase transfer nitrogenous groups. AspAT catalyses the reaction:

$$\text{Aspartate} + \text{2-oxoglutarate} \Longleftrightarrow \text{oxaloacetate} + \text{glutamate}$$

In the case of tissue, heart or liver damage the released amount of AST into the bloodstream is directly related to the extent of tissue damage.

**Aspartate aminotransferase** (AspAT, EC 2.6.1.1) (see Data Sheet) is an ubiquitous pyridoxal-5′-phosphate (vitamin $B_6$)-dependent enzyme present as noncovalently linked homodimer that carries the covalently-linked pyridoxal-5′-phosphate in the form of a Schiffs base at the active site Lys 258. AspAT plays an important role in the amino acid metabolism, linking it to the urea and the tricarboxylic acid cycles.

The 3D structure of human AspAT is not yet determined. Therefore, the 3D structure of a mutant porcine AspAT (412 aa: Ala 1–Gln 412, P00503) in complex with the substrate analogue 2-methyl-Asp was determined by X-ray diffraction (PDB: 1AJS) and is shown in Figure 10.4. Porcine AspAT exhibiting a sequence identity of 93 % with the human species is a homodimer. Each monomer consists of a large (in blue) and a small domain (in yellow) with the active site located at the interface of the two domains. The coenzyme pyridoxal 5′-phosphate (in pink) is covalently linked to Lys 258 at the active site. The presence of the substrate analogue 2-methyl-Asp results in a ligand-induced conformational change in subunit 1 with a large movement of the small domain to form a 'closed' conformation.

### 10.3.3 EC 2.7: transfer of phosphorous-containing groups

The transfer of phosphorous-containing groups, especially the transfer of phosphate from ATP to various compounds, is an important reaction. Creatine kinase (CK) catalyses the transfer of phosphate from ATP to creatine:

$$\text{Creatine} + \text{ATP} \Longleftrightarrow \text{Phosphocreatine} + \text{ADP}$$

Two nonidentical types of creatine kinase exist, the M type and the B type, resulting in two homodimeric (MM, BB) and a heterodimeric (MB) form. Creatine kinase plays a central role in the energy transduction in tissues and is involved in the restoration of the cellular energy reservoir. The measurement of CK activity in serum is a useful diagnostic tool of heart, muscle and brain diseases.

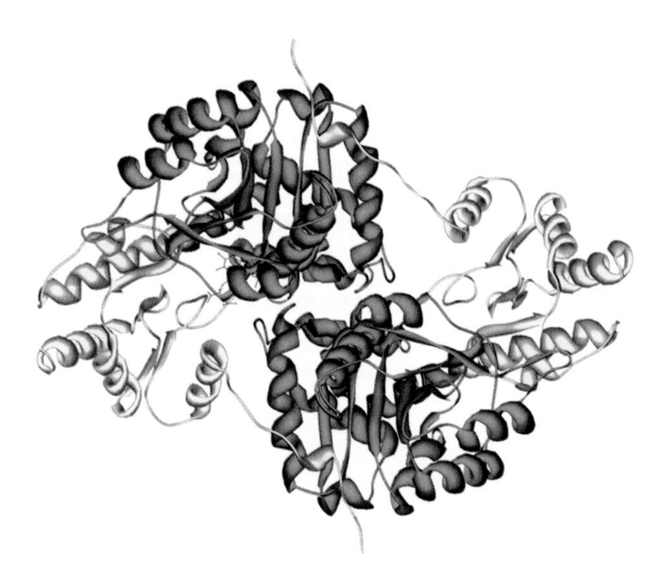

**Figure 10.4 3D structure of porcine aspartate aminotransferase (PDB: 1AJS)**
Porcine aspartate aminotransferase exhibits a sequence identity of 93% with the human species and consists of a large (in blue) and a small domain (in yellow) with the active site located at the interface of the two domains. The coenzyme pyridoxal 5′-phosphate (in pink) is covalently linked to Lys 258 at the active site.

**Creatine kinase** (CK, EC 2.7.3.2) (see Data Sheet) has its highest concentration in high-energy tissues, primarily in the brain, heart and muscle. There are two forms, the M type and B type, which gives rise to homo- and heterodimers. The MM form is mainly found in skeletal muscle, the BB form primarily in the brain and the MB form in the heart. The M and B type CK share an 81 % sequence identity and in both chains the active site consists of Cys 283.

The 3D structure of muscle creatine kinase (381 aa: Met 1–Lys 381) was determined by X-ray diffraction (PDB: 1I0E) and is shown in Figure 10.5. The dimers form an infinite 'double-helix'-like structure along an unusual long crystallographic axis.

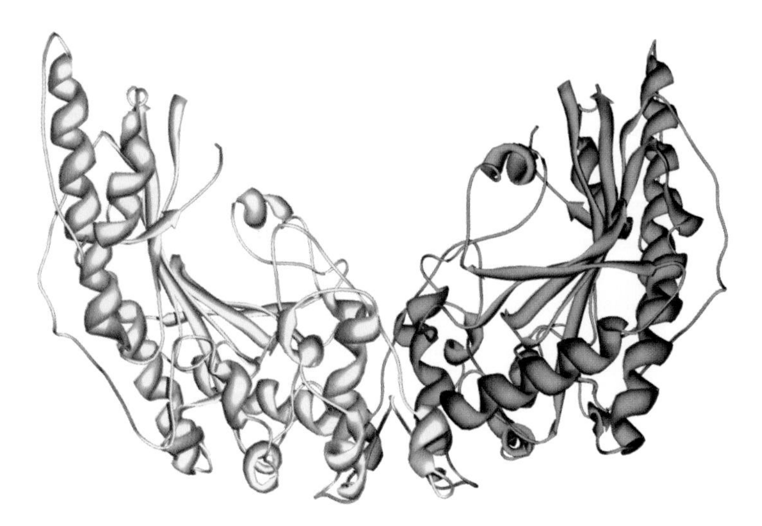

**Figure 10.5 3D structure of muscle creatine kinase (PDB: 1I0E)**
Creatine kinase forms dimers with an infinite 'double-helix'-like structure.

## 10.4 EC 3: HYDROLASES

Hydrolases are enzymes that catalyse the hydrolysis of a chemical bond:

$$A{-}B + H_2O \Rightarrow A{-}OH + B{-}H$$

Hydrolases are classified into 13 subclasses, based on the type of bond they catalyse for hydrolysis.

### 10.4.1 EC 3.1: esterases (ester bond)

Carboxylic ester hydrolases (EC 3.1.1) cleave ester bonds. The hepatic triacylglycerol lipase catalyses the hydrolysis of acylglycerols and phospholipids and lipoprotein lipase triacylglycerols (for details see Chapter 12). Cholinesterase is a serine esterase that catalysis the hydrolysis of acylcholines, mainly the neurotransmitter acetylcholine:

$$\text{Acylcholine} + H_2O \Rightarrow \text{choline} + \text{carboxylate}$$

Cholinesterase is clinically important for the diagnosis of poisoning by carbamates and organophosphates and seems to play a role in cell proliferation and cell differentiation. It is thought to have a protective role against ingested poisons.

Serum paraoxonase/arylesterase 1 is able to hydrolyse a series of toxic metabolites such as organophosphates and aromatic carboxylic acid esters:

$$\text{Aryl dialkyl phosphate} + H_2O \Rightarrow \text{dialkyl phosphate} + \text{aryl alcohol}$$

$$\text{Phenyl acetate} + H_2O \Rightarrow \text{phenol} + \text{acetate}$$

It is associated with high-density lipoprotein and seems to protect low-density lipoprotein from oxidative damage.

Angiogenin exhibits a similar enzymatic activity to pancreatic ribonuclease A but with a $10^5$–$10^6$ times lower efficiency. Angiogenin is involved in angiogenesis by promoting the endothelial invasiveness required for blood vessel formation. Angiogenin induces vascularisation of normal as well as malignant tissues.

**Cholinesterase** (EC 3.1.1.8) (see Data Sheet) or Butyrylcholinesterase (BChE) is primarily synthesised in the liver and is present in plasma as a tetramer of dimers, where the dimer is covalently linked by a single disulfide bridge (Cys 571) and two

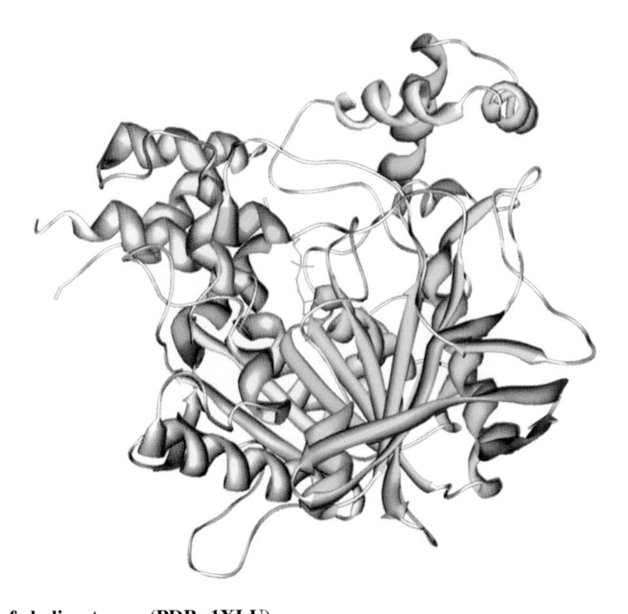

**Figure 10.6 3D structure of cholinesterase (PDB: 1XLU)**
Monomeric cholinesterase consists of a central propeller-like β-sheet structure surrounded by a series of α-helices. An organophosphate molecule (in green) binds to the reactive site Ser (in pink).

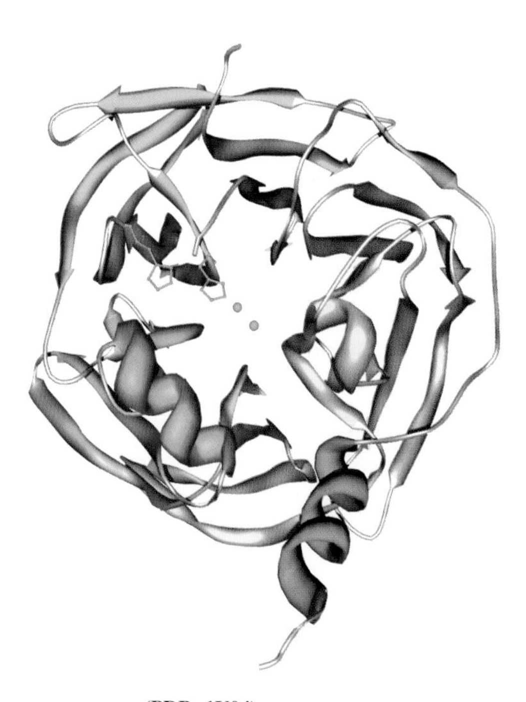

**Figure 10.7 3D structure of serum paraoxonase (PDB: 1V04)**
Serum paraoxonase consists of a six-bladed β-propeller, which forms a central tunnel containing two calcium ions (green balls) and the two His residues of the putative catalytic dyad are indicated in pink.

non covalently dimers form the tetramer. The monomer is a single-chain glycoprotein (24 % CHO) containing three intrachain disulfide bridges (Cys 65–Cys 95, Cys 252–Cys 263, Cys 400–Cys 519) and one inaccessible, free Cys 66 and the active site consists of Ser 198, Glu 325 and His 438.

The 3D structure of a C-terminally truncated form of monomeric cholinesterase (529 aa: Glu 1–Val 529) was determined by X-ray diffraction (PDB: 1XLU) and is shown in Figure 10.6. Cholinesterase is characterised by a central propeller-like β-sheet structure surrounded by a series of α-helices. Poisonous organophosphates (monoisopropyl ester phosphonic acid in green) bind to the reactive site Ser 198 (in pink), which can usually be reactivated by strong nucleophiles such as oximes. The covalently bound organophosphates can undergo a suicide reaction termed 'ageing' involving the reactive site His 438 and Glu 197, which are adjacent to Ser 198.

Mutant alleles of cholinesterase are responsible for hypocholinesterasemia.

**Serum paraoxonase/arylesterase 1** (PON1, EC 3.1.1.2, EC 3.1.8.1) (see Data Sheet) is primarily synthesised in the liver as a single-chain glycoprotein (15 % CHO) containing a single disulfide bridge (Cys 41–Cys 352) and a free, reactive Cys 283. Mature PON1 only loses the N-terminal Met residue but retains the signal peptide, which is important for the association with HDL particles. PON1 is capable of hydrolysing a broad spectrum of toxic organophosphate substrates and confers resistance to poisoning from a wide variety of insecticides.

The 3D structure of the entire PON1 including the N-terminal Met (355 aa: Met 0–Leu 354) was determined by X-ray diffraction (PDB: 1V04) and is shown in Figure 10.7. The structure is characterised by a six-bladed β-propeller, which forms a central tunnel containing two calcium ions (green balls); the putative catalytic His 114–His 133 dyad in the mature form is indicated in pink.

**Angiogenin** (ANG, EC 3.1.27.-) (see Data Sheet) or ribonuclease 5 is predominantly synthesised in the liver and belongs to the pancreatic ribonuclease family. ANG is a single-chain monomer containing three intrachain disulfide bridges (Cys 26–Cys 81, Cys 39–Cys 92, Cys 57–Cys 107) and an N-terminal pyro-Glu and the active site consists of His 13 and His 114 and the substrate-binding region is located at the peptide stretch Lys 40–Thr 44.

The 3D structure of angiogenin (123 aa: Gln 1–Pro 123) in complex with the ribonuclease inhibitor (460 aa: Ser 1–Ser 460, P13489) was determined by X-ray diffraction and is shown in Figure 10.8(a) (PDB: 1A4Y). The 15 Leu-rich repeats (28–29 aa) of the ribonuclease inhibitor form β–α hairpin units arranged in the nonglobular, symmetric shape of a horseshoe (in

**Figure 10.8 3D structure of angiogenin in complex with the ribonuclease inhibitor (PDB: 1A4Y)**
(a) The 15 Leu-rich repeats of the ribonuclease inhibitor form β-α hairpin units arranged in the non-globular, symmetric shape of a horseshoe (in violet) around angiogenin (in yellow) shown in the top and side view.
(b) Monomeric angiogenin PDB: 1B1I exhibits a V-shaped structure with three α-helices and seven β-strands and an active site cleft in the middle with two His residues (in pink).

violet) around angiogenin (in yellow). The molecular fold is generated by an extended right-handed superhelix with alternating β-strands and α-helices shown in the top and side view. The V-shaped structure of angiogenin is characterised by three α-helices and seven β-strands and an active site cleft in the middle with two His residues (in pink), as shown in Figure 10.8(b) (PDB: 1B1I).

### 10.4.2 EC 3.2: glycosylases (sugars)

Glycosidases (EC 3.2.1) are enzymes that hydrolyse *O*-and *S*-glycosyl compounds. Lysozymes are evolutionary and structurally related but functionally unrelated to α-lactalbumin and belong to family 22 in the classification of glycosyl hydrolases. They share a common disulfide bridge pattern:

**Cys 1**–**Cys 8**, **Cys 2**–**Cys 7**, **Cys 3**–**Cys 5**, **Cys 4**–**Cys 6**

and are characterised by a common consensus pattern, where the three Cys residues are involved in intrachain disulfide bridges:

LACTALBUMIN_LYSOZYME (PS00128): **C**–x (3)–**C**–x (2)–[LMF]–x (3)–[DEN]–[LI]–x (5)–**C**

Different classes of lysozymes exist in different species and act as bacteriolytic enzymes hydrolysing β-1,4-linkages between *N*-acetylglucosamine and *N*-acetylmuramic acid residues in peptidoglycans and between *N*-acetylglucosamines in chitodextrins. The reaction mechanism resembles that of a serine protease. Since lysozymes kill bacteria they are also referred to as the body's antibiotic. In addition, they are a potent inhibitor of chemotaxis and of the production of toxic oxygen radicals.

**Lysozyme C** (EC 3.2.1.17) (see Data Sheet) or muramidase is widely distributed in several tissues and in secretions such as milk, tears and saliva. Lysozyme C is a highly cationic single-chain protein containing four intrachain disulfide bridges (Cys 6–Cys 128, Cys 30–Cys 116, Cys 65–Cys 81, Cys 77–Cys 95) and the active site consists of Glu 35 and Asp 53. Glu acts as a proton donor to the glycosidic bond, cleaving the C–O bond, and Asp acts as a nucleophile to generate the glycosyl-enzyme intermediate, which in turn will react with a water molecule, resulting in the hydrolysis of the glycosidic bond and the free enzyme.

The 3D structure of entire lysozyme (130 aa: Lys 1–Val 130) was determined by X-ray diffraction (PDB: 1B5U) and is shown in Figure 10.9. The structure is characterised by four α-helices and three antiparallel β-strands, with the active site residues Glu and Asp indicated in pink and the four intrachain disulfide bridges shown in green.

Defects in the lysozyme gene are the cause of amyloidosis VIII, also called familial visceral or Ostertag-type amyloidosis (for details see MIM: 105200).

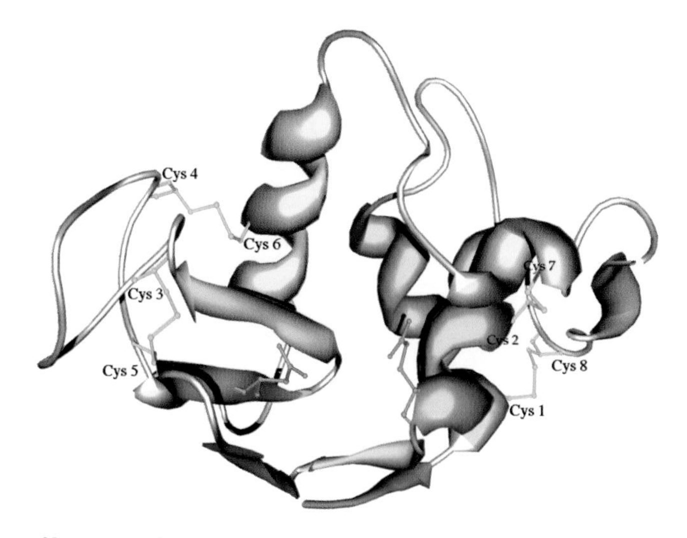


**Figure 10.9 3D structure of lysozyme (PDB: 1B5U)**
Lysozyme contains four α-helices and three antiparallel β-strands and the active site consists of Glu 35 and Asp 53 (in pink) and the four disulfide bridges are shown in green.

### 10.4.3 EC 3.4: peptidases (peptide bonds)

Peptidases are widely spread in living organisms and 15 subcategories are classified (EC 3.4.11–3.4.19, EC 3.4.21–3.4.25, EC 3.4.99). Many peptidases, especially serine proteases, are discussed in detail in Chapters 7 and 8. There are many more peptidases that are not directly assigned to a certain cascade, but nevertheless they are of the utmost importance for balance in the vascular system. A selection of these peptidases is discussed in this section.

*EC 3.4.17: metallocarboxypeptidases*

There are different types of zinc-dependent carboxypeptidases (EC 3.4.17.-) that are structurally and functionally related. The zinc ion is indispensable for the catalytic activity of these enzymes and is coordinated to three conserved amino acids, two His and one Glu residue. Two consensus patterns are located around the zinc-binding region:

CARBOXYPEPT_ZN_1 (PS00132): [PK]–x–[LIVMFY]–x(2)–{E}–x–**H**–[STAG]–x–**E**–x–[LIVM]–[STAG]–{L}–x(5)–[LIVMFYTA]

CARBOXYPEPT_ZN_2 (PS00133): **H**–[STAG]–{ADNV}–{VGFI}–{YAR}–[LIVME]–{SDEP}–x–[LIVMFYW]–P–[FYW]

There are two main types of pancreatic carboxy-peptidases:

1. Carboxypeptidase A1 (309 aa: P15085) cleaves all C-terminal amino acids except Lys, Arg and Pro and the closely related carboxypeptidase A2 (305 aa: P48052), which has a similar specificity but with a preference for bulkier side chains.
2. Carboxypeptidase B and carboxypeptidase B2 or thrombin-activatable fibrinolysis inhibitor (TAFI, for details see Chapter 7) preferentially release C-terminal Arg/Lys residues.

Carboxypeptidase N (CPN) belongs to the peptidase M14 family and consists of two catalytic chains and two heavy chains, forming a noncovalently linked tetramer. CPN protects the body from vasoactive and inflammatory peptides containing C-terminal Arg/Lys residues such as kinins or anaphylatoxins.

**Carboxypeptidase B** (CPB, EC 3.4.17.2) (see Data Sheet) is a secreted pancreatic enzyme generated from the proform by releasing an N-terminal 95 residue activation peptide. Single-chain CPB contains three intrachain disulfide bridges (Cys 63–Cys 76, Cys 135–Cys 158, Cys 149–Cys 163) and the active site consists of Glu 268 and a $Zn^{2+}$ ion coordinated to His 66 and His 194 and to Glu 69.

The 3D structure of entire procarboxypeptidase B consisting of the activation peptide (95 aa: His 1–Arg 95) and mature carboxypeptidase B (307 aa: Ala 1–Tyr 307) was determined by X-ray diffraction (PDB: 1KWM) and is shown in Figure 10.10. The N-terminal globular prodomain (in red), which covers the active site of the enzyme, thus shielding it from substrates, is linked to the catalytic domain by a connecting region. The catalytic domain (in blue) exhibits the typical central mixed eight-stranded twisted β-sheet surrounded by eight α-helices. The active site consists of a Glu residue (in pink) and the catalytically active zinc ion (green ball), which is coordinated to a Glu and two His residues and a water molecule.

**Carboxypeptidase N, catalytic chain and subunit 2** (CPN, EC 3.4.17.3) (see Data Sheet) is synthesised in the liver and circulates in blood as a noncovalently linked tetramer consisting of two catalytic (light) chains and two heavy chains. The light chain bears the catalytic activity with the active site consisting of Glu 288 and a $Zn^{2+}$ ion coordinated to His 66 and His 196 and to Glu 69. The heavy chain is highly glycosylated and contains 13 LRR (leucine-rich) repeats.

*EC 3.4.21: serine endopeptidases (serine proteases)*

Serine proteases are usually present in plasma as inactive precursors or proenzymes, also termed 'zymogens', which require a specific activation step to obtain full enzymatic activity from the usually inactive proform with a distorted active site. The activation step is usually a limited proteolysis with a concomitant large conformational change of the enzyme, which finally leads to the correct 3D structure of the active site. This process is often accompanied by the release of an activation peptide or

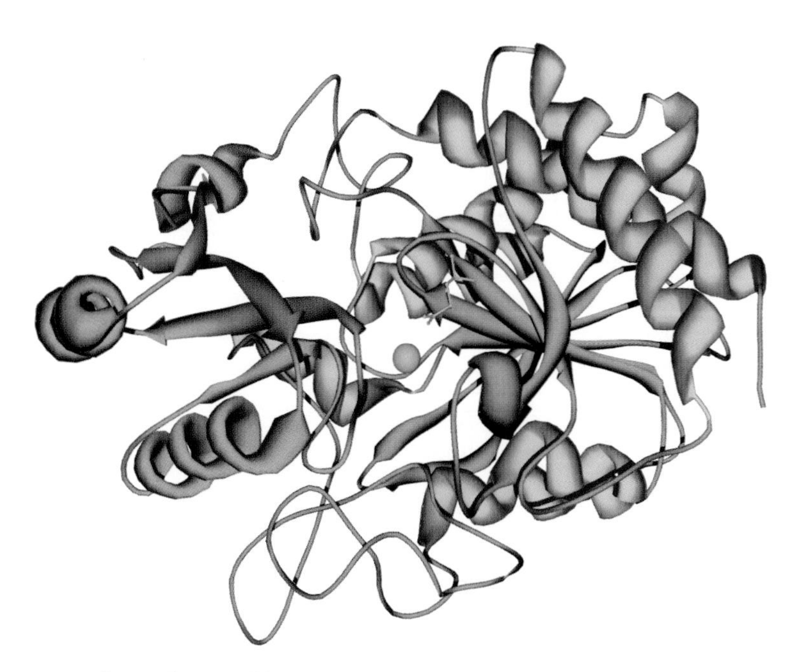

**Figure 10.10 3D structure of procarboxypeptidase B (PDB: 1KWM)**
Procarboxypeptidase B contains the N-terminal globular pro-domain (in red) and the catalytic domain (in blue) with the typical central mixed eight-stranded twisted β-sheet surrounded by eight α-helices. The active site consists of a Glu residue (in pink) and the catalytically active zinc ion (green ball).

several activation peptides, the cleavage of one or several peptide bonds and the generation of primarily two-chain, but also three-chain (e.g. chymotrypsin), active enzymes.

Many important serine proteases that are part of the coagulation cascade and fibrinolysis and the complement system are discussed in the corresponding chapters and are not included in this chapter (for a summary see Table 10.2).

A preserved particular structure called catalytic triad consisting of His 57, Asp 102 and Ser 195 (chymotrypsin numbering) is essential for the catalytic activity of a serine protease. The catalytic mechanism can be divided into two parts, an acylation and a deacylation step, and is shown in Figure 10.11.

Acylation involves:

1. Formation of a covalent bond between the hydroxyl group of the Ser residue in the active site of the protease and the carbonyl carbon of the peptide bond with the concomitant transfer of the hydrogen of the Ser-OH group to the free electron pair of the His nitrogen.
2. Formation of a negatively charged transition state with a tetrahedral coordination.
3. Cleavage of the peptide bond and formation of the acyl-enzyme intermediate and transfer of the proton from the active site His to the released first peptide product.

Deacylation involves:

1. Formation of a covalent bond between the hydroxyl group of a water molecule and the concomitant transfer of the proton from the water molecule to the His residue in the active site.
2. Formation of a negatively charged transition state with a tetrahedral coordination.
3. Release of the second peptide product and transfer of the His proton to the Ser residue in the active site, thus regenerating the original state of the active site.

A serine protease possesses four important structural elements required for its catalytic activity. Figure 10.12 shows the case of chymotrypsin in complex with the bound inhibitor acetyl–Pro–Ala–Pro–Tyr–$NH_2$, where hydrogen bonds are indicated by red dashed lines:

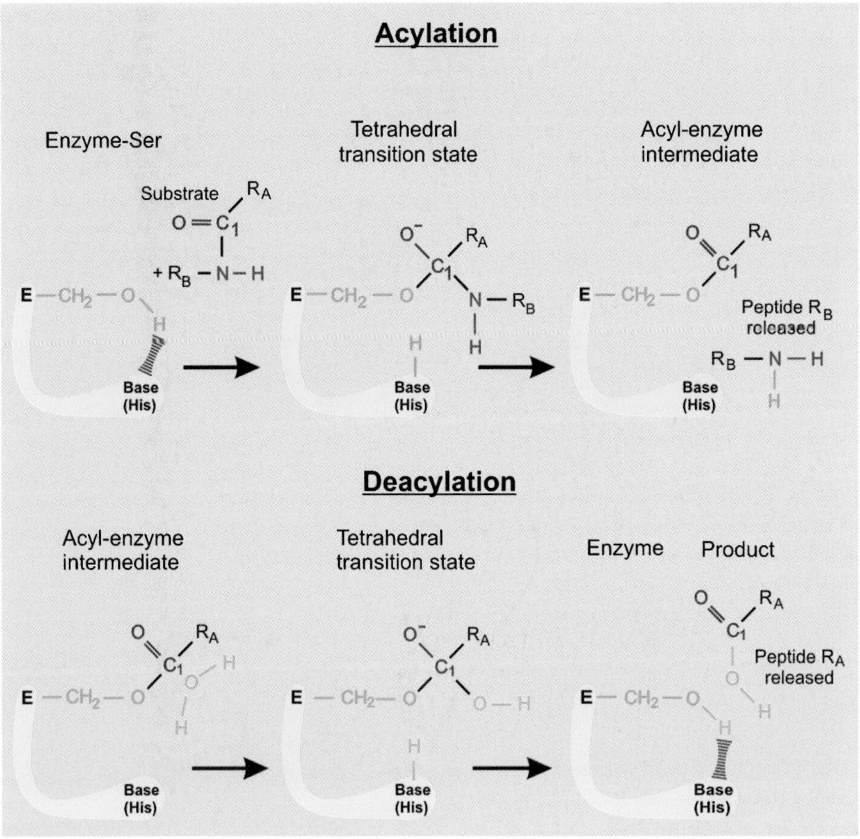

**Figure 10.11 Catalytic mechanism of a serine protease**
The reaction can be divided into two parts, an **acylation** and a **deacylation** step.
**Acylation:** Formation of a covalent bond between the hydroxyl group of the Ser residue in the active site of the protease and the carbonyl carbon of the peptide bond and formation of a negatively charged tetrahedral transition state. Cleavage of the peptide bond and formation of the acyl-enzyme intermediate and release of the first peptide product.
**Deacylation:** Formation of a covalent bond between the hydroxyl group of a water molecule and formation of a negatively charged tetrahedral transition state. Release of the second peptide product and regeneration of the original state of the active site.

1. **Catalytic triad: His 57, Asp 102, Ser 195.** His is the acceptor of the proton and the tetrahedral intermediate transition state is stabilised by hydrogen bonds.
2. **Oxyanion hole: Gly 193, Ser 195.** The carboxy oxygen of the substrate in the oxyanion hole forms hydrogen bonds with the NH groups (backbone) of Gly 193 and Ser 195.
3. **Nonspecific binding site: Gly 216.** The NH and CO groups of Gly 216 form hydrogen bonds with NH and CO groups of the substrate (backbone).
4. **Specificity pocket: Ser 189, Gly 216 and 226.** Specific binding site for the side chains of the substrate, resulting in the specificity of a serine protease.

Serine proteases are characterised by two signatures of the trypsin family, covering the sequences in the vicinity of the active site residues His and Ser:

TRYPSIN_HIS (PS00134): [LIVM]−[ST]−A−[STAG]−**H**−C

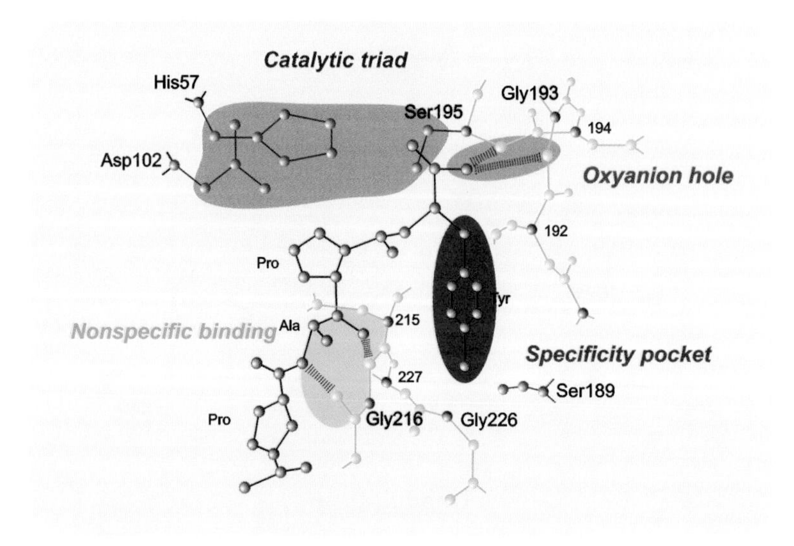


**Figure 10.12 Essential structural elements for the catalytic reaction of chymotrypsin**
1. **Catalytic triad**: His 57, Asp 102, Ser 195. 2. **Oxyanion hole**: Gly 193, Ser 195. 3. **Non-specific binding site**: Gly 216. 4. **Specificity pocket**: Ser 189, Gly 216 and 226. Hydrogen bonds are indicated in red dashed lines.

TRYPSIN_SER (PS00135): [DNSTAGC]–[GSTAPIMVQH]–x(2)–G–[DE]–**S**–G–[GS]–[SAPHV]–[LIVMFYWH]
– [LIVMFYSTANQH]

The group of the chymotrypsin-like serine proteases with chymotrypsin, trypsin and leukocyte elastase are structurally closely related and are among the best studied enzymes.

Chymotrypsin cleaves peptide bonds C-terminal to amino acids with bulky side chains, preferably aromatic residues Phe, Trp and Tyr, but also Leu and Met. The specificity pocket of chymotrypsin consists of Ser 189, Gly 216 and Gly 226, allowing the penetration of the substrate into the binding site and enabling hydrophobic interactions of the bulky, aromatic side chains. Chymotrypsin is a proteolytic enzyme of the digestive tract inhibited by $\alpha_2$-macroglobulin and by $\alpha_1$-antichymotrypsin.

Trypsin cleaves peptide bonds C-terminal to amino acids with positively charged side chains, namely Lys and Arg residues, but also other basic amino acids such as ornithine. The specificity pocket of trypsin consists of Asp 189, Gly 216 and Gly 226, allowing the penetration of the substrate into the binding site and enabling an electrostatic interaction of the charged Arg and Lys residues with Asp 189. Trypsin is a proteolytic enzyme of the small intestine and is essential for the activation of all pancreatic zymogens such as chymotrypsinogen, proelastase, prekallikrein, phospholipase $A_2$ and procarboxypeptidases. Trypsin is inhibited by Kunitz-type trypsin inhibitors, $\alpha_1$-antitrypsin and $\alpha_2$-macroglobulin.

Elastase cleaves peptide bonds with C-terminal smaller neutral side chains such as Ala and Val. The specificity pocket of elastase consists of Val 216 and Thr 226; thus the 'binding pocket' is just a shallow depression accommodating smaller residues. Elastase is involved in the degradation of elastin, collagen, proteoglycans and certain plasma proteins. Elastase is inhibited by $\alpha_2$-macroglobulin and serpins like leukocyte elastase inhibitor and $\alpha_1$-antitrypsin.

**Chymotrypsin B** (EC 3.4.21.1) (see Data Sheet) is synthesised in the pancreas as proenzyme chymotrypsinogen, which is cleaved to two-chain chymotrypsinogen by trypsin cleaving the peptide bond Arg 15–Ile 16. In an autocatalytic activation process two-chain chymotrypsinogen is activated to three-chain chymotrypsin by releasing two dipeptides, resulting in chain A (13 aa: Cys 1–Leu 13), chain B (131 aa: Ile 16–Tyr 146) and chain C (97 aa: Asn 149–Asn 245) linked by two interchain disulfide bridges: A–B (Cys 1–Cys 122) and B–C (Cys 136–Cys 201).

The 3D structure of human chymotrypsin has not yet been determined. Therefore, the 3D structure of bovine chymotrypsinogen A (245 aa: Cys 1–Asn 245, P00766) exhibiting a sequence identity of 82 % with the human species determined by X-ray diffraction (PDB: 1CHG) is given in Figure 10.13. Chymotrypsinogen exhibits a very similar fold to active $\alpha$-chymotrypsin and contains two domains, which are characterised by large sections of distorted antiparallel $\beta$-sheets

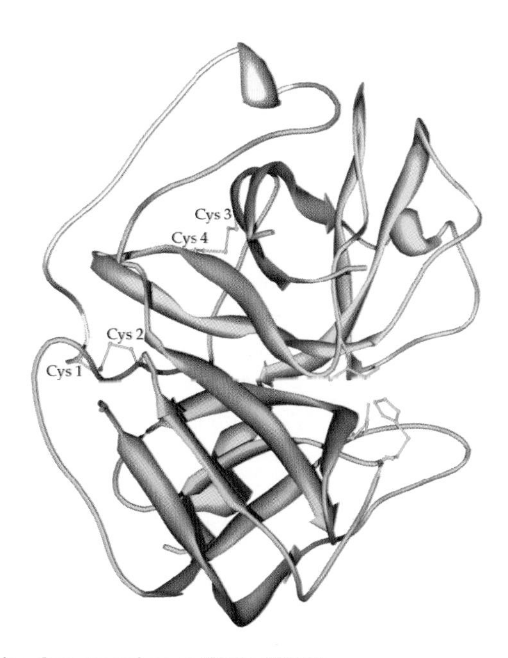


**Figure 10.13 3D structure of bovine chymotrypsinogen (PDB: 1CHG)**
Active chymotrypsin consists of three chains: Chain A in yellow, chain B in green and chain C in blue and the chains are linked by two interchain disulphide bridges (in red). The residues of the active site are marked in pink.

but few α-helices. The three chains in active chymotrypsin are indicated in different colours, chain A in yellow, chain B in green and chain C in blue, and the chains are linked by two interchain disulfide bridges (in red). The residues of the active site are marked in pink.

**Trypsin I and II** (EC 3.4.21.4) (see Data Sheet) are synthesised in the pancreas as proenzymes, cationic and anionic trypsinogens, respectively. Initially, enterokinase activates trypsinogen to trypsin by releasing an N-terminal octapeptide, and trypsin itself can activate trypsinogen autocatalytically. Trypsin I belongs to the peptidase S1 family, contains five intrachain disulfide bridges (Cys 7–Cys 137, Cys 25–Cys 41, Cys 116–Cys 183, Cys 148–Cys 162, Cys 173–Cys 197), carries a P-Tyr residue (aa 131) and binds one calcium ion. Trypsin II is very similar to trypsin I (sequence identity: 90 %), but contains only four disulfide bridges (Cys 116–Cys 183 is missing) and is not phosphorylated.

The 3D structure of entire trypsin I (224 aa: Ile 1–Ser 224) was determined by X-ray diffraction (PDB: 1TRN) and is shown in Figure 10.14. Trypsin exhibits the typical fold of a serine protease domain (for details see Chapter 5) and the structure is very similar to trypsins of other mammalian species. Ser, Asp and His of the catalytic triad are indicated in pink.

Defects in the trypsinogen gene are the cause of hereditary pancreatitis, also known as chronic pancreatitis. It is an autosomal dominant disease characterised by calculi in the pancreatic duct causing severe abdominal pain attacks (for details see MIM: 167800).

**Leukocyte elastase** (EC 3.4.21.37) (see Data Sheet) or neutrophil elastase (Medullasin) is synthesised in the bone marrow as proelastase, which is converted to active elastase by releasing an N-terminal dipeptide and a C-terminal 20 residues peptide. Elastase is a single-chain glycoprotein (approximately 22 % CHO) containing four intrachain disulfide bridges (Cys 26–Cys 42, Cys 122–Cys 179, Cys 152–Cys 158, Cys 169–Cys 194).

The 3D structure of entire active elastase (218 aa: Ile 1–Gln 218) in complex with a peptide inhibitor was determined by X-ray diffraction (PDB: 1HNE) and is shown in Figure 10.15. The inhibitory mechanism of the peptide inhibitor methoxysuccinyl–Ala–Ala–Pro–Ala chloromethyl ketone (in red) involves cross-links with the active site residues His 57 and Ser 195 of elastase. Ser, Asp and His of the catalytic triad are marked in pink.

Defects in the elastase gene are the cause of cyclic haematopoiesis, also known as cyclic neutropenia, which is an autosomal dominant disease. The production of blood cells oscillates with a 21-day periodicity, resulting in a concentration of circulating neutrophils between almost normal and zero and leading to risks of opportunistic infections (for details see MIM: 162800).

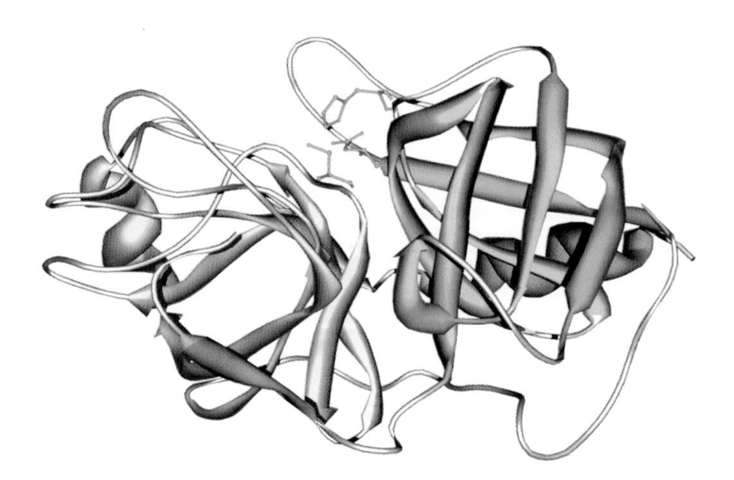

**Figure 10.14 3D structure of trypsin I (PDB: 1TRN)**
Trypsin exhibits the typical fold of a serine protease domain and Ser, Asp and His of the catalytic triad are indicated in pink.

Of the many enzymes belonging to the serine protease category EC 3.4.21 a selection is discussed in some detail in this section of the chapter. Cathepsin G and chymase are two closely related serine proteases with a chymotrypsin-like specificity. Cathepsin G degrades connective tissues, glycoproteins and collagen and therefore causes damage to tissues during an inflammatory response. Together with leukocyte elastase it seems to participate in the killing and digestion of pathogens during phagocytosis. Chymase together with tryptase is the major protein of mast cells granules and is thought to be involved in the generation of vasoactive peptides, the degradation of the extracellular matrix and the regulation of gland secretion. Tryptase exists in several forms:

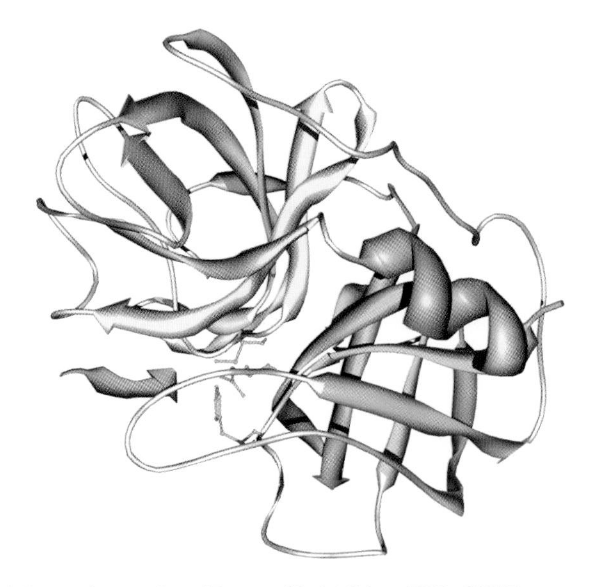

**Figure 10.15 3D structure of elastase in complex with a peptide inhibitor (PDB: 1HNE)**
The peptide inhibitor methoxysuccinyl – Ala – Ala – Pro – Ala chloromethyl ketone (in red) and Ser, Asp and His of the catalytic triad (in pink) are indicated.

1. Tryptase $\alpha_1$ (245 aa: Ile 1–Pro 245, TRYA1_HUMAN/P15157)
2. Tryptase $\beta_1$ (245 aa: Ile 1–Pro 245, TRYB1_HUMAN/Q15661)

3. Tryptase $\beta_2$ (245 aa: Ile 1–Pro 245, TRYB2_HUMAN/P20231)
4. Tryptase $\delta$ (205 aa: Ile 1–Thr 205, TRYD_HUMAN/Q9BZJ3)

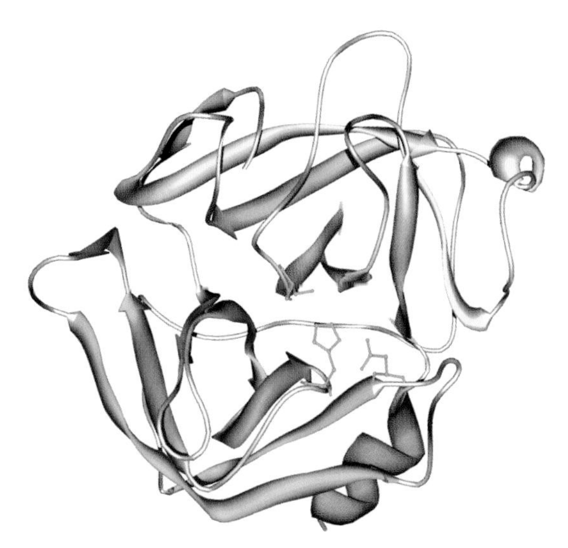

**Figure 10.16 3D structure of cathepsin G (PDB: 1KYN)**
Cathepsin G is characterised by two six-stranded $\beta$-barrels held together by three trans-domain segments. Ser, Asp and His of the catalytic triad are indicated in pink.

Together with chymase, tryptases are the major neutral proteases in mast cells granules with a trypsin-like specificity. Although their biological function is not clear tryptases seem to be involved in tissue inflammation and remodelling.

In addition, prostate-specific antigen (PSA) is a kallikrein-like serine protease with a chymotrypsin-like specificity and a preference for the cleavage of Tyr–Xaa peptide bonds. PSA cleaves the main gel-forming proteins in semen, semenogelin-1 and -2 and fibronectin, leading to the liquefaction of the seminal coagulum. Elevated levels of PSA in blood is currently the most reliable marker for the detection of prostate cancer and other prostate disorders.

**Cathepsin G** (EC 3.4.21.20) (see Data Sheet) is mainly contained in granules of neutrophils and mature cathepsin G is generated from the proform by releasing an N-terminal activation dipeptide and a C-terminal dodecapeptide. Cathepsin G is a single-chain glycoprotein containing three intrachain disulfide bridges (Cys 29–Cys 45, Cys 122–Cys 187, Cys 152–Cys 166) cleaving C-terminally bulky aromatic and aliphatic residues such as Phe or Leu, thus exhibiting a similar specificity as chymotrypsin, and is inhibited by the serpins $\alpha_1$-antichymotrypsin and $\alpha_1$-antitrypsin and by $\alpha_2$-macroglobulin.

The 3D structure of cathepsin G (235 aa: Ile 1–Leu 235) was determined by X-ray diffraction (PDB: 1KYN) and is shown in Figure 10.16. The structure is characterised by two six-stranded $\beta$-barrels held together by three transdomain segments. The residues of the catalytic triad Ser, Asp and His are indicated in pink.

**Chymase** (EC 3.4.21.39) (see Data Sheet) is secreted by mast cells in complex with a carboxypeptidase and sulfated proteoglycans and is generated from the proform by dipeptidyl peptidase releasing an acidic N-terminal activation dipeptide. Chymase is a single-chain glycoprotein containing three intrachain disulfide bridges (Cys 30–Cys 46, Cys 123–Cys 188, Cys 154–Cys 167) and a single Cys 7, which seems not to participate in an interchain disulfide bridge. Chymase is closely related to cathepsin G and exhibits a chymotrypsin-like specificity and is inhibited by the serpins $\alpha_1$-antichymotrypsin and $\alpha_1$-antitrypsin and to a lesser extent by $\alpha_2$-macroglobulin.

The 3D structure of entire chymase (226 aa: Ile 1–Asn 226) in complex with the low molecular weight inhibitor phenylmethylsulfonyl fluoride (PMSF) was determined by X-ray diffraction (PDB: 1KLT) and is shown in Figure 10.17. The structure is characterised by two six-stranded $\beta$-barrels, which form a cleft containing the active site region consisting of His 45, Asp 89 and Ser 182 (in pink), to which the inhibitor PMSF (in green) is covalently linked.

**Tryptase $\beta_2$** (EC 3.4.21.59) (see Data Sheet) is secreted from mast cells as a noncovalently linked tetramer in complex with heparin. Tryptase is generated from the proform by releasing an N-terminal dodecapeptide, but the monomer is inactive.

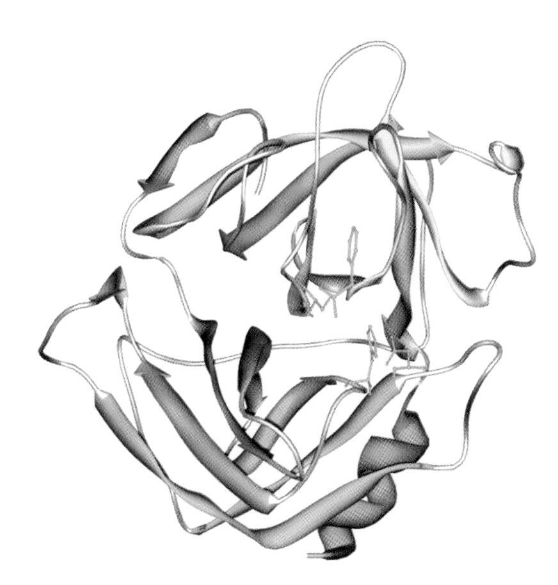

**Figure 10.17 3D structure of chymase in complex with the low molecular weight inhibitor phenylmethylsulfonyl fluoride (PDB: 1KLT)**

Chymase is characterised by two six-stranded $\beta$-barrels, which form a cleft with the active site region consisting of His 45, Asp 89 and Ser 182 (in pink) and the inhibitor PMSF (in green).

Tryptase is a single-chain glycoprotein containing four intrachain disulfide bridges (Cys 29–Cys 45, Cys 125–Cys 200, Cys 158–Cys 181, Cys 190–Cys 218). Tryptase is not inhibited by high molecular weight inhibitors such as serpins or $\alpha_2$-macroglobulin but by low molecular weight inhibitors.

The 3D structure of tetrameric tryptase $\beta_2$ (244 aa: Ile 1–Lys 244) was determined by X-ray diffraction (PDB: 1A0L) and is shown in Figure 10.18(a). Tryptase forms a ring-like tetrameric structure with the active sites of each monomer (in pink) facing an oval central pore, thus restricting the access for macromolecular substrates and high molecular weight inhibitors. The nature of this unique tetrameric structure helps to explain the rather unusual biochemical properties such as the resistance to inhibition by endogenous high molecular weight inhibitors or the fact that monomeric tryptase is inactive. The 3D structure of monomeric tryptase $\beta_2$ with the residues of the catalytic triad (in pink) is shown in Figure 10.18(b).

**Prostate-specific antigen** (PSA, EC 3.4.21.77) (see Data Sheet) is synthesised in the prostate gland and is generated from the proform releasing an N-terminal activation heptapeptide. PSA is a single-chain glycoprotein containing five intrachain disulfide bridges (Cys 7–Cys 149, Cys 26–Cys 42, Cys 128–Cys 195, Cys 160–Cys 174, Cys 185–Cys 210). PSA is inhibited by serpins such as plasma serine protease inhibitor and $\alpha_1$-antichymotrypsin, and by $\alpha_2$-macroglobulin and pregnancy-zone protein.

*EC 3.4.22: cysteine endopeptidases (cysteine proteases)*

There are several groups of cysteine proteases (thiol proteases); important ones are cathepsins and caspases. Probably one of the best-studied cysteine protease is papain present in papaya (212 aa: P00784). The reactive site of a cysteine protease consists of a Cys, a His and an Asn residue (in the case of papain: Cys 25, His 159, Asn 175) and the reaction mechanism resembles that of serine proteases. The deprotonation of the reactive site Cys by the adjacent His enables the nucleophilic attack of Cys-S$^-$ at the carbonyl carbon of the peptide bond, resulting in the cleavage of the peptide bond and the formation of the acyl-enzyme intermediate. The deacylation reaction is carried out by a water molecule, resulting in the release of the second peptide product.

Cysteine proteases are characterised by a consensus sequence pattern around the reactive site Cys residue:

THIOL_PROTEASE_CYS (PS00139): Q–{V}–x–{DE}–[GE]–{F}–**C**–[YW]–{DN}–x–[STAGC]–[STAGCV]

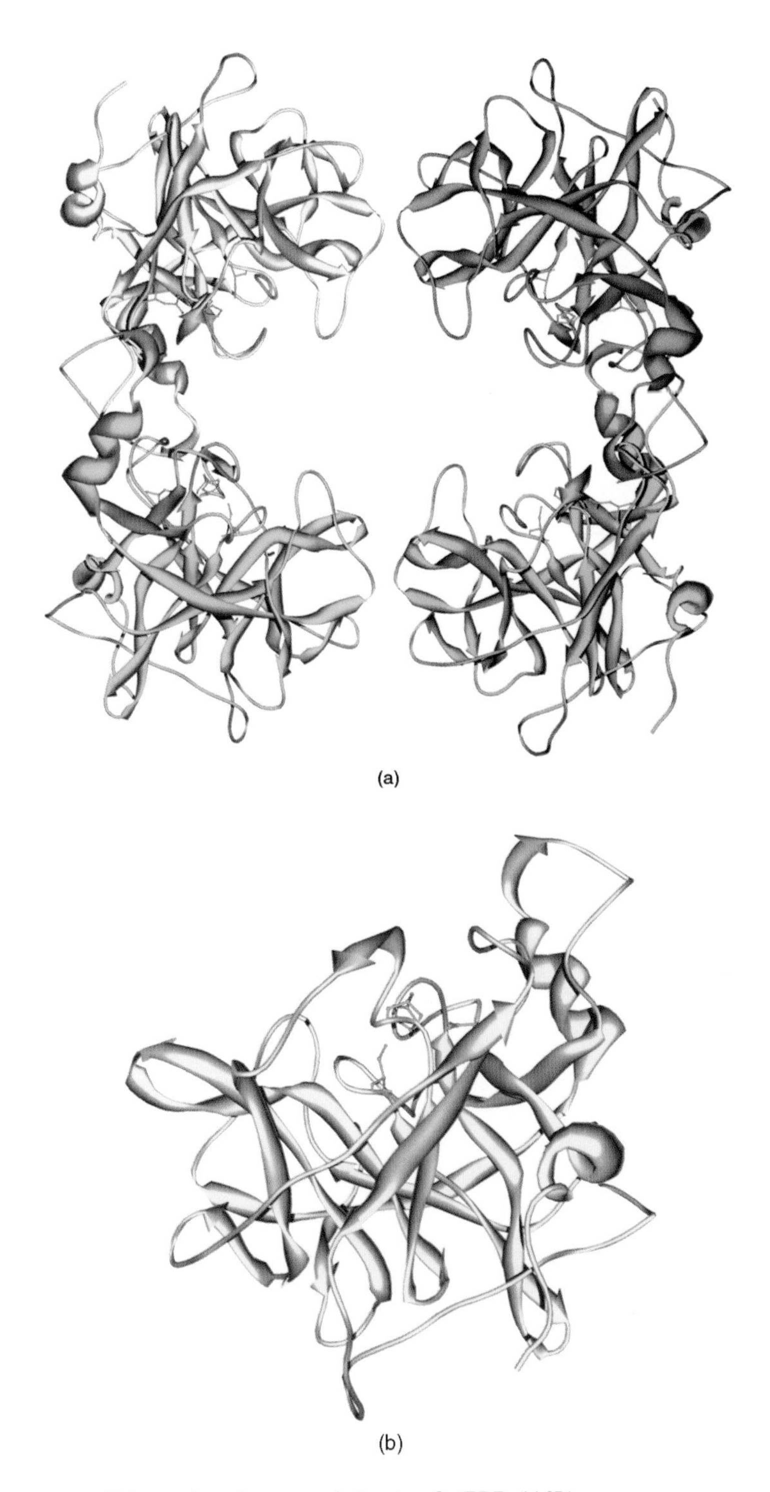


**Figure 10.18 3D structures of tetrameric and monomeric tryptase $\beta_2$ (PDB: 1A0L)**
(a) Tryptase $\beta_2$ forms a ring-like tetrameric structure with the active sites of each monomer (in pink) facing an oval central pore.
(b) Monomeric tryptase $\beta_2$ exhibits the expected serine protease conformation with the residues of the catalytic triad shown in pink.

Cathepsins are a large family of lysosomal cysteine proteases that belong to the peptidase C1 family, which are involved in cellular turnover such as bone resorption and cartilage breakdown. Cathepsins can cleave many proteins such as elastin and collagen and are often involved in a more general degradation of proteins.

Caspases are a large group of cysteine proteases with a rather unusual specificity cleaving after Asp residues. The name caspase is derived from the molecular function: cysteine-*asp*artic-acid-prote*ase*. Caspases are essential in apoptosis, the programmed cell death, and some of them are required in the immune system for cytokine maturation. The failure of apoptosis contributes to tumor development and to autoimmune diseases. There are two consensus sequence patterns around the two active site residues Cys and Asp in caspases:

CASPASE_CYS (PS01122): K–P–K–[LIVMF] –[LIVMFY] –[LIVMF](2)–[QP]–[AF]– **C**–[RQG]–[GE]

CASPASE_HIS (PS01121): H–x(2,4)–[SC]–x(2)–{A}–x–[LIVMF](2)–[ST]–**H**–G

**Cathepsin L** (EC 3.4.22.15) (see Data Sheet) is present in lysosomes and is important for the general degradation of proteins in lysosomes. It plays an important role in the degradation of the invariant chain of MHC class II molecules. Cathepsin L is generated from the proform by releasing an N-terminal propeptide (96 aa) and an internal tripeptide, resulting in active two-chain cathepsin L. The heavy chain (175 aa: Ala 1–Thr 175) containing two intrachain disulfide bridges (Cys 22–Cys 65, Cys 56–Cys 98) and the light chain (42 aa: Asn 1–Val 42) are connected by a single interchain disulfide bridge (Cys 156 in the heavy chain and Cys 31 in the light chain).

The 3D structure of cathepsin L (heavy chain (175 aa): Ala 1–Thr 175; light chain (42 aa): Asn 1–Val 42) in complex with a fragment of the HLA class II histocompatibility antigen (65 aa): Leu 210–Ser 274) was determined by X-ray diffraction (PDB: 1ICF) and is shown in Figure 10.19. The heavy chain (in green) and the light chain (in yellow-green) are linked by a single disulfide bond (in blue). The fragment of the invariant chain of the MHC class II molecules (in red) has a homologous structure as the thyroglobulin type 1 domain and specifically inhibits cathepsin L but not cathepsin S and binds to the active site cleft of cathepsin L (residues of the active site are in pink).

**Caspase-1** (EC 3.4.22.36) (see Data Sheet) is primarily synthesised in the spleen and the lung and plays a critical role in the inflammatory response by activating the proforms of cytokines such as interleukin-1β and interleukin-18. It activates the precursor of interleukin-1β to the mature active cytokine by cleaving the peptide bonds Asp 27–Gly 28 and Asp 116–Ala 117. Active caspase-1 is generated from its proform (404 aa) either autocatalytically or by other caspases, releasing an

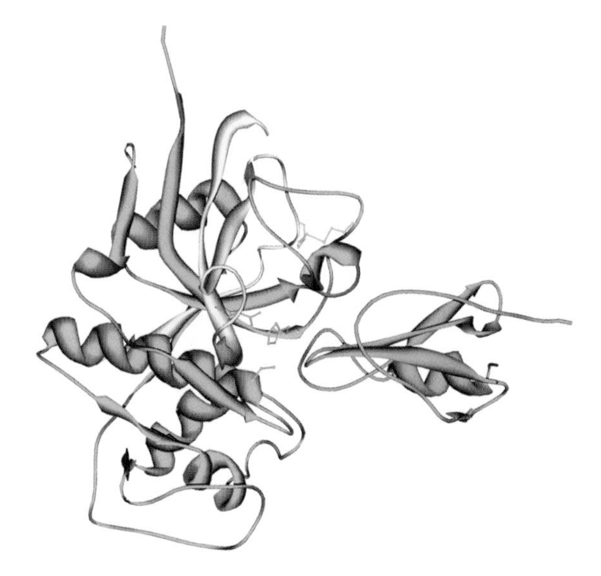

**Figure 10.19 3D structure of cathepsin L in complex with a fragment of the HLA class II histocompatibility antigen (PDB: 1ICF)** Cathepsin L consists of a heavy chain (in green) and a light chain (in yellow) linked by a single interchain disulfide bond (in blue). The fragment of the HLA class II histocompatibility antigen (in red) binds to the active site cleft of cathepsin L (residues of the active site in pink).

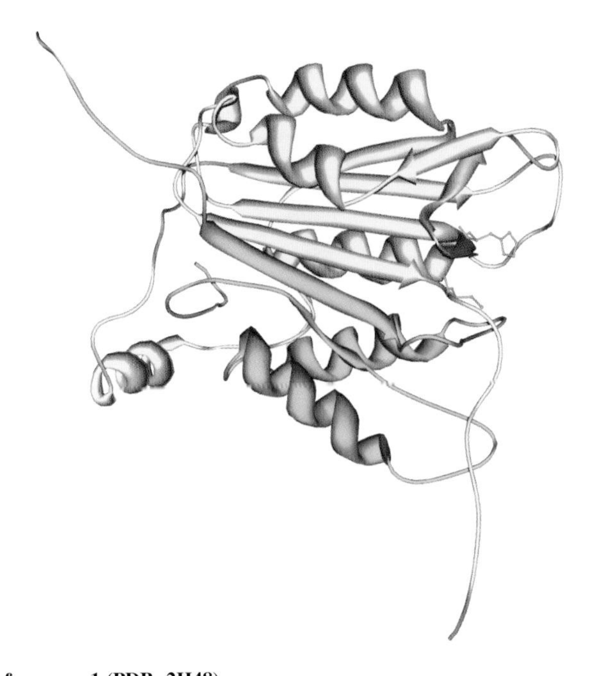

**Figure 10.20 3D structure of caspase-1 (PDB: 2H48)**
Caspase-1 is a heterodimer of the p20 subunit (in yellow) containing the active site residues His 118 and Cys 166 (in pink) and the p10 subunit (in blue).

N-terminal propeptide (119 aa) and an internal propeptide (19 aa) and resulting in p20 and p10 subunits. Caspase-1 is a tetramer of two non-covalently-linked heterodimers consisting of the p20 subunit (178 aa) and the p10 subunit (88 aa). The active site in the p20 subunit consists of His 118 and Cys 166.

The 3D structure of caspase-1 consisting of the p20 subunit (178 aa: Asn 1–Asp 178) and the p10 subunit (88 aa: Ala 1–His 88) was determined by X-ray diffraction (PDB: 2H48) and is shown in Figure 10.20. The p20 (in yellow)/p10 (in blue) heterodimer exhibits the structural motif of an $\alpha/\beta$ protein with a single central six-stranded $\beta$-sheet core flanked by $\alpha$-helices on both sides. The residues His 118 and Cys 166 of the active site in the p20 subunit are shown in pink.

**Caspase-3** (EC 3.4.22.56) (see Data Sheet) is primarily synthesised in the spleen, lung, liver, kidney and heart and is involved in the activation cascade of caspases. At the onset of apoptosis it cleaves the poly [ADP-ribose] polymerase (1724 aa: Q9UKK3) at the Asp 216–Gly 217 peptide bond. It cleaves and activates various proteins, among them several caspases. Active caspase-3 is generated from its proform (277 aa) by other caspases and autocatalytically by releasing two N-terminal propeptides (9 + 19 aa) and by an internal cleavage of an Asp–Ser peptide bond, resulting in p17 and p12 subunits. Caspase-3 is a non-covalently-linked heterodimer consisting of the p17 subunit (147 aa) and the p12 subunit (102 aa). The active site in the p17 subunit consists of His 93 and Cys 135.

The 3D structure of caspase-3 consisting of the p17 subunit (147 aa: Ser 1–Asp 147) and the p12 subunit (97 aa: Asp 6–His 102) was determined by X-ray diffraction (PDB: 1GFW) and is shown in Figure 10.21. The p17 (in yellow)/p12 (in blue) heterodimer exhibits the structural motif of an $\alpha/\beta$ protein with a single central six-stranded $\beta$-sheet core flanked by $\alpha$-helices on both sides. The residues His 93 and Cys 135 of the active site in the p17 subunit are shown in pink.

### *EC 3.4.23: aspartic endopeptidases (aspartic proteases)*

Aspartyl proteases or acid proteases (EC 3.4.23) carry in their reactive site two Asp residues. The reaction mechanism involves the $\gamma$-carboxyl groups of the Asp residues and a water molecule. One deprotonated Asp residue activates a water molecule, which enables a nucleophilic attack of the water molecule at the carbonyl carbon of the peptide bond. The released proton protonates the Asp residue and reacts with the amide nitrogen of the peptide bond resulting in its cleavage. The reaction mechanism of aspartyl proteases is depicted in Figure 10.22.

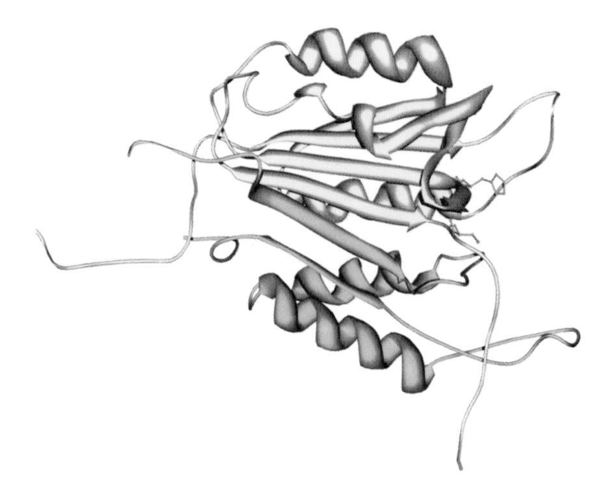

**Figure 10.21 3D structure of caspase-3 (PDB: 1GFW)**
Caspase-3 is a heterodimer of the p17 subunit (in yellow) containing the active site residues His 93 and Cys 135 (in pink) and the p12 subunit (in blue).

Aspartyl proteases are characterised by a consensus sequence pattern around one Asp residue of the active site, which is identical with the single active site residue in viral proteases:

ASP_PROTEASE (PS00141): [LIVMFGAC] –[LIVMTADN] –[LIVFSA]–**D**–[ST]–G–[STAV]–[STAPDENQ]–{GQ}–[LIVMFSTNC]–{EGK}–[LIVMFGTA]

The best known aspartyl proteases are the pepsins and gastricins, which are the main digestive enzymes in the stomach degrading the proteins in food into peptides. In addition, renin, whose function is the generation of angiotensin I from angiotensinogen in plasma (for details see Chapter 7), and cathepsin D from the cathepsin family, which is involved in the general intracellular protein degradation, are other important aspartyl proteases.

**Pepsin A** (EC 3.4.23.1) (see Data Sheet) is released by the gastric mucosa and is the main digestive enzyme in the stomach. The acidic pH in the stomach, created by the release of HCl from the stomach lining, triggers an autocatalytic generation of pepsin from the proenzyme pepsinogen by releasing an N-terminal activation peptide (47 aa). Pepsin is a single-chain enzyme containing three intrachain disulfide bridges (Cys 45–Cys 50, Cys 208–Cys 210, Cys 249–Cys 282) and a potential

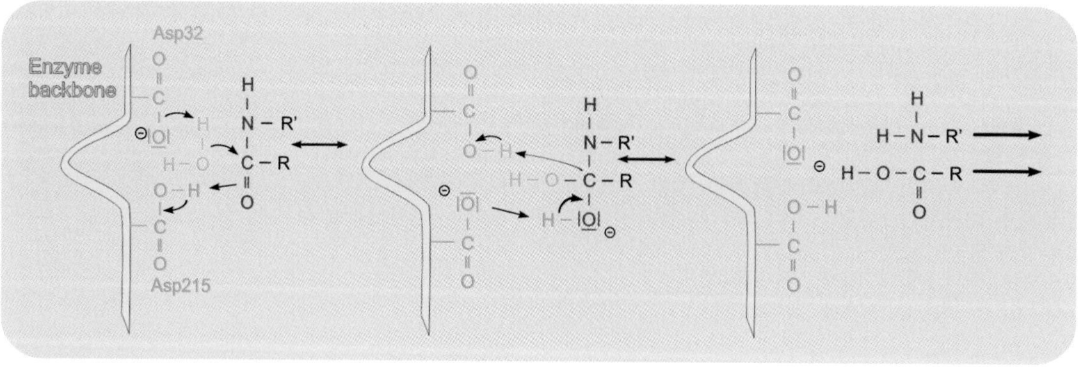

**Figure 10.22 Catalytic mechanism of an aspartyl protease**
A deprotonated Asp residue activates a water molecule, which enables a nucleophilic attack of the water molecule at the carbonyl carbon of the peptide bond. The protonated Asp residue reacts with the amide nitrogen of the peptide bond and is cleaved.

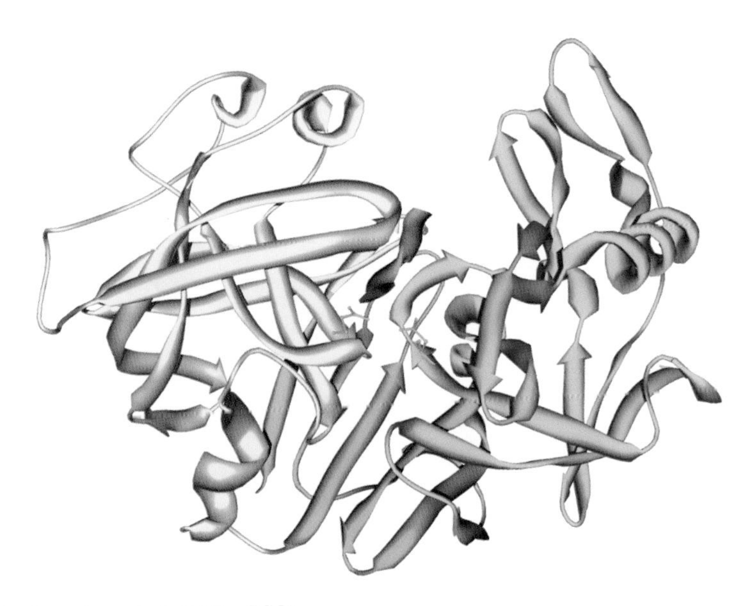

**Figure 10.23 3D structure of pepsin A (PDB: 1PSO)**
Pepsin consists of a central domain composed of a six-stranded antiparallel β-sheet (in green) and N-terminal (in yellow) and C-terminal (blue) lobes containing orthogonally packed β-sheets. The two Asp residues of the active site (in pink) and the inhibitor pepstatin (in red) are indicated.

P-Ser residue (aa 68) and the active site consists of Asp 32 and 215. Pepsin has a rather broad specificity with a preference for aromatic or bulky aliphatic side chains. A main inhibitor is pepstatin (isovaleryl–Val–Val–AHMHA–Ala–AHMHA. AHMHA: 4-amino-3-hydroxy-6-methyl heptanoic acid).

The 3D structure of entire pepsin A (326 aa: Val 1–Ala 326) in complex with the inhibitor pepstatin was determined by X-ray diffraction (PDB: 1PSO) and is shown in Figure 10.23. Pepsin consists of a central domain composed of a six-stranded antiparallel β-sheet (in green) that serves as a backbone to the active site region with two Asp residues (in pink) and two terminal lobes, an N-terminal (Glu 7–Gln 148, in yellow) and C-terminal (Thr 185–Arg 307, in blue) consisting of orthogonally packed β-sheets. Upon binding, the inhibitor pepstatin (in red) induces a confo-rmational change in pepsin, resulting in a more enclosed and tighter binding and pepstatin adopts an extended conformation.

**Cathepsin D** (EC 3.4.23.5) (Data Sheet) is present in all nucleated cells and is primarily involved in the general protein breakdown and turnover in cells. Cathepsin D, a two-chain glycoprotein, consisting of a light chain (97 aa: Gly 1–Gln 97) containing two intrachain disulfide bridges (Cys 27–Cys 96, Cys 46–Cys 53) and a heavy chain (244 aa: Leu 1–Leu 244) also containing two intrachain disulfide bridges (Cys 117–Cys 121, Cys 160–Cys 197) are noncovalently associated. Active single-chain cathepsin D is generated from procathepsin D by releasing an N-terminal activation peptide (46 aa) and two-chain cathepsin D is generated from the single-chain form by releasing an internal heptapeptide.

The 3D structure of two-chain cathepsin D (light chain (97 aa): Gly 1–Gln 97; heavy chain (241 aa): Gly 2–Ala 242) was determined by X-ray diffraction (PDB: 1LYA) and is shown in Figure 10.24. The light chain (in yellow-green) and the heavy chain (in green) are noncovalently linked. The structure consists of two related domains rich in β-sheets connected by an interdomain linker peptide and is very similar to other aspartyl proteases. Residues Asp 33 and Asp 231 of the active site are indicated in pink.

## 10.5 EC 5: ISOMERASES

Isomerases catalyse the interconversion of isomers:

$$A \Rightarrow B$$

Isomerases are classified into six subclasses, but only oxidoreductases are discussed.

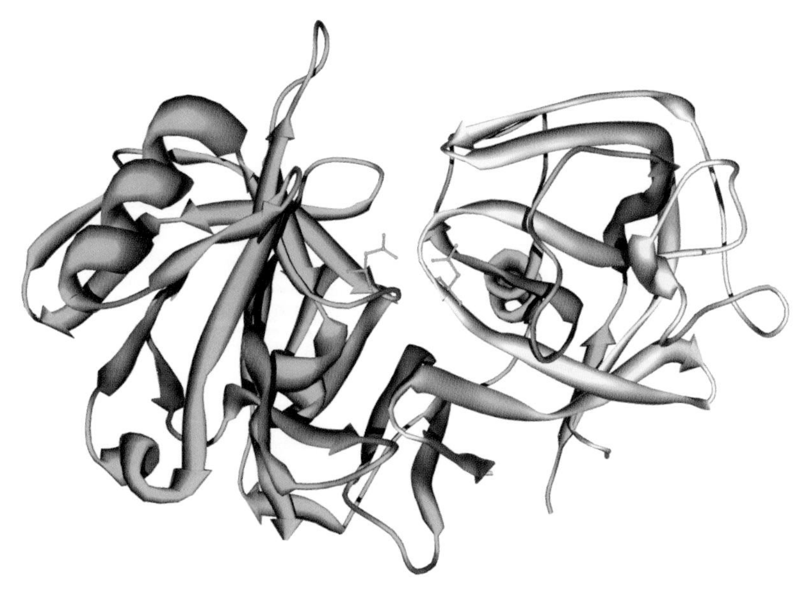

**Figure 10.24 3D structure of two-chain cathepsin D (PDB: 1LYA)**
Cathepsin D consists of two related domains rich in β-sheets connected by an interdomain linker peptide. The light chain (in yellow) and the heavy chain (in green) are non-covalently associated and the active site residues Asp 33 and Asp 231 are shown in pink.

### 10.5.1 EC 5.3: oxidoreductases (intramolecular)

Protein disulfide isomerases (PDIs) are important enzymes in protein folding, catalysing the formation and breakage of disulfide bonds, thus allowing the proteins to find the correct arrangement of the disulfide bonds rapidly during the (re-) folding process. Reduced PDIs are able to catalyse the reduction of mispaired disulfide bonds, thus catalysing disulfide exchange, leading to an intramolecular rearrangement of disulfide bonds in the same protein. In addition, PDIs act as chaperones, enabling incorrectly folded proteins to obtain a correctly folded structure.

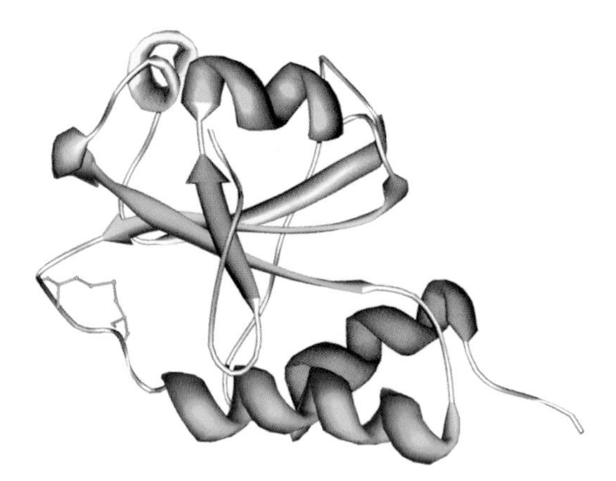

**Figure 10.25 3D structure of the first thioredoxin domain in protein disulfide isomerase (PDB: 1MEK)**
The thioredoxin domain consists of a central β-sheet surrounded by four α-helices typical for a thioredoxin fold and the redox-active disulfide bond of the active site is indicated (in pink).

All protein disulfide isomerases contain at least two thioredoxin domains. The sequence around the redox-reactive disulfide bond is well conserved and is characterised by the following consensus pattern, where the two Cys residues form the redox-reactive bond:

THIOREDOXIN (PS00194): [LIVMF]–[LIVMSTA]–x–[LIVMFYC]–[FYWSTHE]–x(2)–[FYWGTN] –**C**–[GATPLVE]–[PHYWSTA]–**C**–{I}–x–{A}–x(3)–[LIVMFYWT]

**Protein disulfide isomerase** (PDI, EC 5.3.4.1) (see Data Sheet) or prolyl 4-hydroxylase beta subunit is contained in the endoplasmic reticulum and catalyses the formation, breakage and rearrangement of disulfide bonds. PDI is a single-chain protein containing two thioredoxin domains and two redox-active disulfide bonds: Cys 36–Cys 39 and Cys 380–Cys 383. These four Cys residues act as nucleophiles in the isomerisation reaction and adjacent Gly 37 and His 38 in the first active site and Gly 381 and His 382 in the second active site contribute to the required redox potential and Arg 103 lowers the p$K_a$ value of the second Cys in the first active site and Arg 444 lowers the p$K_a$ of the second Cys in the second active site.

PDI, which is identical to the beta subunit of prolyl 4-hydroxylase, also catalyses the hydroxylation of Pro residue:

$$\text{Procollagen-Pro} + \text{2-oxoglutarate} + O_2 \Longleftrightarrow \text{procollagen-4-OH-Pro} + \text{succinate} + CO_2$$

The 3D structure of the first thioredoxin domain in protein disulfide isomerase (120 aa: Asp 1–Ala 120) was determined by X-ray diffraction (PDB: 1MEK) and is shown in Figure 10.25. The structure is characterised by a central β-sheet surrounded by four α-helices and exhibits the expected typical thioredoxin fold. The redox-active disulfide bond Cys 36–Cys 39 of the active site is shown in pink.

## REFERENCES


Bairoch, 2000, The enzyme database in 2000, *Nucleic Acids Res.*, **28**, 304–305.

Freedman *et al.*, 1994, Protein disulfide isomerase: building bridges in protein folding, *Trends Biochem. Sci.*, **19**, 331–336.

Garcia-Viloca *et al.*, 2004, How enzymes work: analysis by modern rate theory and computer simulation, *Science*, **303**,186–195.

Henrissat, 1991, A classification of glycosyl hydrolases based on amino acid sequence similarities, *Biochem. J.*, **280**,309–316.

Nicholson and Thornberry, 1997, Caspases: killer proteases, *Trends Biochem. Sci.*, **22**, 299–306.

Rawlings and Barrett, 1994, Families of cysteine peptidases, *Meth. Enzymol.*, **244**, 461–486.

Rawlings and Barrett, 1995, Evolutionary families of metallopeptidases, *Meth. Enzymol.*, **248**,183–228.

Stadtman, 1990, Selenium biochemistry, *Annu. Rev. Biochem.*, **59**, 111–127.

DATA SHEETS

## *Angiogenin (Ribonuclease 5; EC 3.1.27.-)*

### Fact Sheet

| | | | |
|---|---|---|---|
| ***Classification*** | Pancreatic ribonuclease | ***Abbreviations*** | ANG |
| ***Structures/motifs*** | | ***DB entries*** | ANGI_HUMAN/P03950 |
| ***MW/length*** | 14 143 Da/123 aa | ***PDB entries*** | 1B1I |
| ***Concentration*** | 270–500 μg/l | ***Half-life*** | |
| ***PTMs*** | 1 Pyro-Glu | | |
| ***References*** | Kurachi *et al.*, 1985, *Biochemistry*, **24**, 5494–5499.<br>Leonidas *et al.*, 1999, *J. Mol. Biol.*, **285**, 1209–1233.<br>Papageorgiou *et al.*, 1997, *EMBO J.*, **16**, 5162–5177. | | |

### Description

**Angiogenin** (ANG) is predominantly expressed in the liver and is a potent inducer of blood vessel formation.

### Structure

ANG is a single-chain plasma protein carrying a pyro-Glu residue with His 13 and 114 in the active site. The 3D structure of angiogenin exhibits a V-shaped structure containing three α-helices and seven β-sheets with the active site cleft (in pink) in the middle (**1B1I**).

### Biological Functions

ANG induces the vascularisation of normal and malignant tissues. ANG promotes the endothelial invasiveness necessary for blood vessel formation.

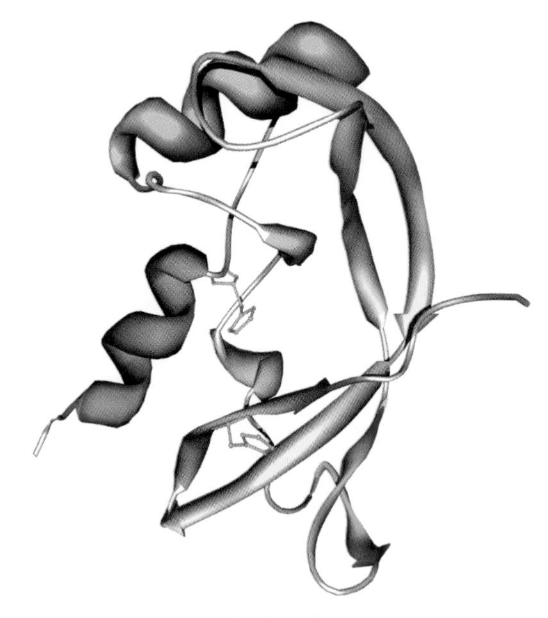

Angiogenin

## *Aspartate aminotransferase (Transaminase A; EC 2.6.1.1)*

### Fact Sheet

| | | | |
|---|---|---|---|
| ***Classification*** | Class I aminotransferase | ***Abbreviations*** | AspAT |
| ***Structures/motifs*** | | ***DB entries*** | AATC_HUMAN/P17174 |
| ***MW/length*** | 46 116 Da/412 aa | ***PDB entries*** | 1AJS |
| ***Concentration*** | approx. 100 μg/l | ***Half-life*** | approx. 3 days |
| ***PTMs*** | 1 Pyridoxal 5'-phosphate (Lys 258) | | |
| ***References*** | Doyle *et al.*, 1990, *Biochem. J.*, **270**, 651–657. | | |

### Description

**Aspartate aminotransferase** (AspAT) is an ubiquitous enzyme that is important in the amino acid metabolism.

### Structure

AspAT is a single-chain protein present as homodimer. It carries a pyridoxal 5'-phosphate covalently linked as a Schiffs base to the active site Lys 258. The complex formation of porcine AspAT with the substrate analogue 2-methyl-Asp results in a large ligand-induced conformational change (**1AJS**).

### Biological Functions

AspAT catalyses the reaction aspartate + 2-oxoglutarate ⇔ oxaloacetate + glutamate. AspAT plays an important role in the amino acid metabolism, linking it to the urea and the tricarboxylic acid cycles.

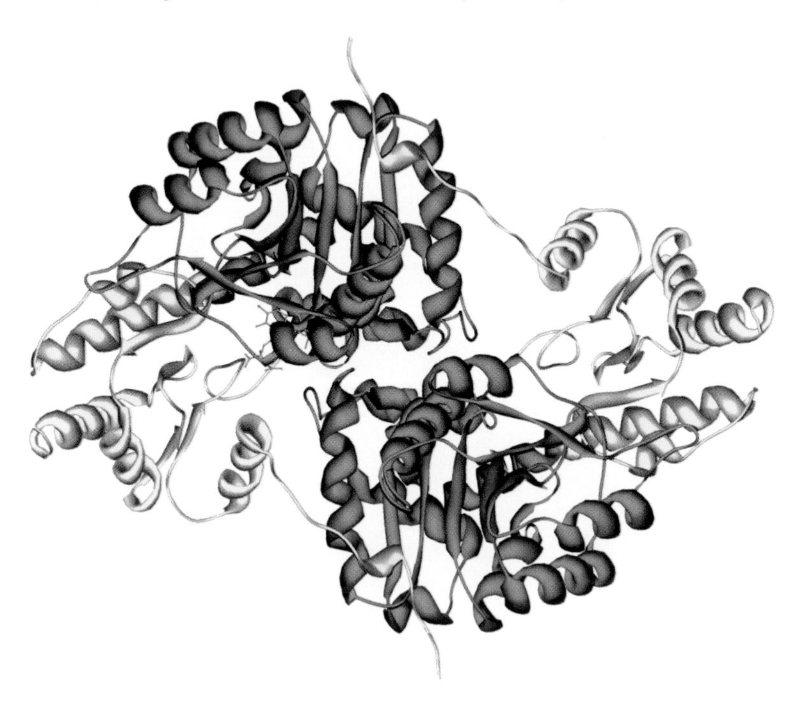

Aspartate aminotransferase

## *Carboxypeptidase B (EC 3.4.17.2)*

### Fact Sheet

| | | | |
|---|---|---|---|
| ***Classification*** | Peptidase M14 | ***Abbreviations*** | CPB |
| ***Structures/motifs*** | | ***DB entries*** | CBPB1_HUMAN/P15086 |
| ***MW/length*** | 34 848 Da/307 aa | ***PDB entries*** | 1KWM |
| ***Concentration*** | | ***Half-life*** | |
| ***PTMs*** | | | |
| ***References*** | Barbosa Pereira *et al.*, 2002, *J. Mol. Biol.*, **321**, 537–547.<br>Yamamoto *et al.*, 1992, *J. Biol. Chem.*, **267**, 2575–2581. | | |

### Description

**Carboxypeptidase B** (CPB) is a pancreatic peptidase releasing C-terminal basic residues.

### Structure

CPB is generated from its proform by releasing an N-terminal 95 residue activation peptide. Single-chain CPB contains three intrachain disulfide bridges (Cys 63–Cys 76, Cys 135–Cys 158, Cys 149–Cys 163) and the active site is composed of Glu 268 (in pink) and a $Zn^{2+}$ (green ball) ion coordinated to His 63 and His 194 and to Glu 69. The globular prodomain is linked via a connecting region to the catalytic domain, which consists of a central mixed eight-stranded twisted β-sheet surrounded by eight α-helices (**1KWM**).

### Biological Functions

CPB preferentially releases C-terminal Lys and Arg residues from its target proteins.

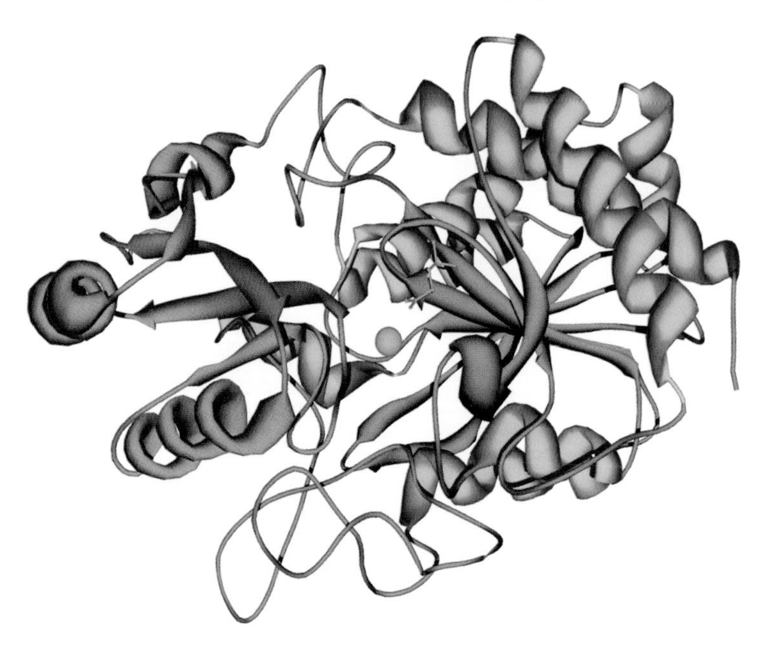

Procarboxypeptidase B

## *Carboxypeptidase N: Catalytic chain + subunit 2 (EC 3.4.17.3)*

### Fact Sheet

| | | | |
|---|---|---|---|
| ***Classification*** | Peptidase M14 | ***Abbreviations*** | CPN |
| ***Structures/motifs*** | Subunit 2: 13 LRR (leucine-rich) | ***DB entries*** | CBPN_HUMAN/P15169<br>CPN2_HUMAN/P22792 |
| ***MW/length*** | Catalytic chain: 50 034 Da/438 aa<br>Subunit 2: 58 285 Da/524 aa | ***PDB entries*** | |
| ***Concentration*** | 30 mg/l | ***Half-life*** | approx. 24 h |
| ***PTMs*** | Subunit 2: 8 N-CHO (6 potential) | | |
| ***References*** | Gebhard *et al.*, 1989, *Eur. J. Biochem.*,**178**, 603–607.<br>Tan *et al.*, 1990, *J. Biol. Chem.*, **265**, 13–19. | | |

### Description

**Carboxypeptidase N** (CPN) is synthesised in the liver and is present in plasma as a tetramer of two light chains (catalytic chain) and two heavy chains.

### Structure

CPN is a glycoprotein composed of two catalytic chains (light chain), each containing Glu 288 and one $Zn^{2+}$ coordinated to His 66 and His 196 and to Glu 69 in the active site, and two heavy chains, each containing 13 LRR (leucine-rich repeats).

### Biological Functions

CPN protects the body from vasoactive and inflammatory peptides containing C-terminal Arg/Lys residues such as kinins or anaphylatoxins.

## Caspase-1 (EC 3.4.22.36)

### Fact Sheet

| | | | |
|---|---|---|---|
| ***Classification*** | Peptidase C14 | ***Abbreviations*** | CASP1 |
| ***Structures/motifs*** | | ***DB entries*** | CASP1_HUMAN/P29466 |
| ***MW/length*** | p20 subunit: 19 844 Da/178 aa<br>p10 subunit: 10 244 Da/88 aa | ***PDB entries*** | 2H48 |
| ***Concentration*** | | ***Half-life*** | 3 h |
| ***PTMs*** | | | |
| ***References*** | Thomberry *et al.*, 1992, *Nature,* **356**, 768–774.<br>Scheer *et al.*, 2006, *Proc. Natl. Acad. Sci. USA*, **103**, 7595–7600. | | |

### Description

**Caspase-1** is primarily expressed in the spleen and lung and plays a critical role in the inflammatory response by processing the proforms of cytokines.

### Structure

Caspase-1 is a tetramer of two heterodimers consisting of non-covalently-linked p20 and p10 subunits. The active site in the p20 subunit consists of His 118 and Cys 166 (in pink). Caspase-1 is generated from its proform autocatalytically by releasing an N-terminal propeptide (119 aa) and an internal propeptide (19 aa), resulting in the two subunits. The 3D structure of the heterodimer consisting of the p20 (in yellow) and the p10 (in blue) subunits is characterised by a central six-stranded β-sheet core flanked by α-helices on both sides (**2H48**).

### Biological Functions

Caspase-1 activates the precursor of interleukin-1β by specifically cleaving the Asp 116–Ala 117 and Asp 27–Gly 28 peptide bonds in the proform.

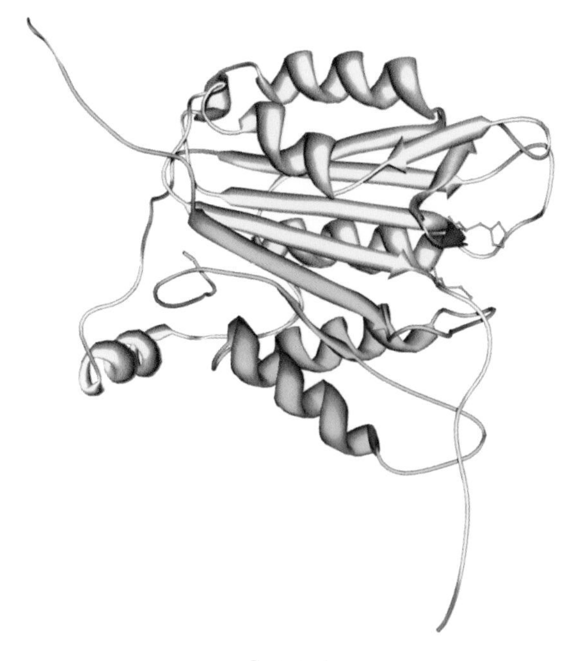

Caspase-1

# *Caspase-3 (EC 3.4.22.-)*

## Fact Sheet

| | | | |
|---|---|---|---|
| ***Classification*** | Peptidase C14 | ***Abbreviations*** | CASP3 |
| ***Structures/motifs*** | | ***DB entries*** | CASP3 HUMAN/P42574 |
| ***MW/length*** | P17 subunit: 16 615 Da/147 aa<br>p12 subunit: 11 896 Da/102 aa | ***PDB entries*** | 1GFW |
| ***Concentration*** | | ***Half-life*** | |
| ***PTMs*** | | | |
| ***References*** | Fernandes-Alnemri *et al.*, 1994, *J. Biol. Chem.*, **269**, 30761–30764.<br>Feeney *et al.*, 2006, *Biochemistry*, **45**, 13249–13263.<br>Lee *et al.*, 2000, *J. Biol. Chem.*, **275**, 16007–16014. | | |

## Description

**Caspase-3** is primarily expressed in the lung, spleen, heart, liver and kidney and is involved in the activation cascade of caspases responsible for apoptosis.

## Structure

Caspase-3 is a heterodimer consisting of non-covalently-linked p17 and p12 subunits. The active site in the p17 subunit consists of His 93 and Cys 135 (in pink). Caspase-3 is generated from its proform by other caspases and also autocatalytically by releasing two N-terminal propeptides (9 + 19 aa) and by cleaving an internal peptide bond (Asp 147–Ser 148), resulting in the two subunits. The 3D structure of the heterodimer consisting of the p17 (in yellow) and the p12 (in blue) subunits is characterised by a central six-stranded β-sheet core flanked by α-helices on both sides (**1GFW**).

## Biological Functions

Capase-3 is involved in the activation cascade of caspases. At the onset of apoptosis it cleaves the Asp 216–Gly 217 peptide bond in poly(ADP-ribose) polymerase and activates several caspases.

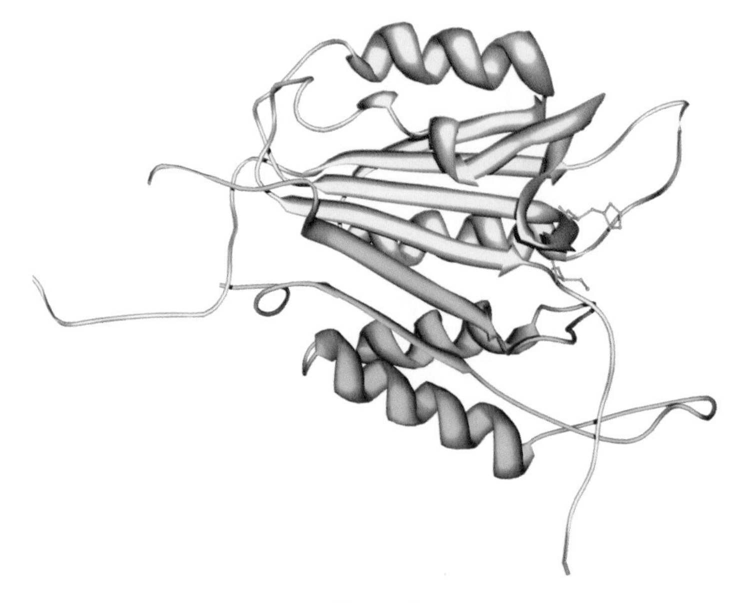

Caspase-3

## *Catalase (EC 1.11.1.6)*

### Fact Sheet

| | | | |
|---|---|---|---|
| ***Classification*** | Catalase | ***Abbreviations*** | CAT |
| ***Structures/motifs*** | | ***DB entries*** | CATA_HUMAN/P04040 |
| ***MW/length*** | 59 625 Da/526 aa | ***PDB entries*** | 1DGB |
| ***Concentration*** | Traces; abundant in erythrocytes | ***Half-life*** | |
| ***PTMs*** | | | |
| ***References*** | Quan *et al.*, 1986, *Nucleic Acids Res.*, **14**, 5321–5335.<br>Putnam *et al.*, 2000, *J. Mol. Biol.*, **296**, 295–309. | | |

### Description

**Catalase** (CAT) is widely distributed in different tissues, the highest being liver and kidney, where it is localised in peroxisomes.

### Structure

CAT is a single-chain protein that is present as non-covalently-linked homotetramer. Each monomer contains one heme group coordinated to Tyr 357 and the active site is composed of His 74 and Asn 147 (in blue). The monomer is characterised by a central eight-stranded antiparallel β-barrel surrounded by an α-helical domain, an N-terminal threading arm and a wrapping loop. The heme group (in pink) is located in the central β-barrel (**1DGB**).

### Biological Functions

CAT catalyses the decomposition of hydrogen peroxide: $2\ H_2O_2 \Rightarrow O_2 + 2\ H_2O$ and protects the cells from oxidative damage.

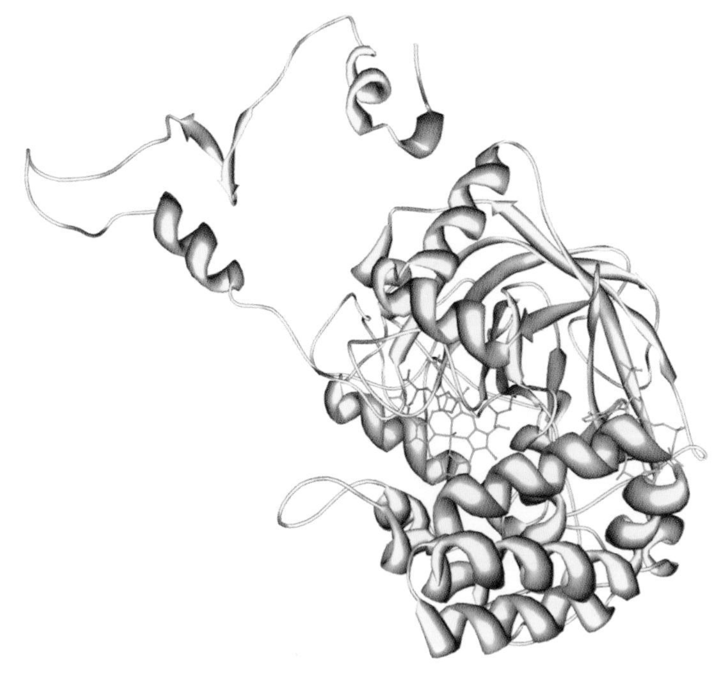

Catalase monomer

## *Cathepsin D (EC 3.4.23.5)*

### Fact Sheet

| | | | |
|---|---|---|---|
| ***Classification*** | Peptidase A1 | ***Abbreviations*** | CatD |
| ***Structures/motifs*** | | ***DB entries*** | CATD HUMAN/P07339 |
| ***MW/length*** | 37 852 Da/348 aa<br>Light chain: 10 680 Da/97 aa<br>Heavy chain: 26 629 Da/244 aa | ***PDB entries*** | 1LYA |
| ***Concentration*** | | ***Half-life*** | Several days |
| ***PTMs*** | 2 N-CHO | | |
| ***References*** | Faust *et al.*, 1985, *Proc. Natl. Acad. Sci. USA*, **82**, 4910–4914.<br>Baldwin *et al.*, 1993, *Proc. Natl Acad. Sci. USA*, **90**, 6796–6800. | | |

### Description

**Cathepsin D** (CatD) is an aspartyl protease present in almost all tissues and organs and is involved in the intracellular breakdown of proteins.

### Structure

Procathepsin D (394 aa) is activated by releasing an N-terminal 46 residue activation peptide generating single-chain active cathepsin D (348 aa), which is further processed to active two-chain cathepsin D consisting of a non-covalently-linked light (97 aa) and heavy chain (244 aa). Both chains contain two intrachain disulfide bridges, Cys 27–Cys 96 and Cys 46–Cys 53 in the light chain and Cys 117–Cys 121 and Cys 160–Cys 197 in the heavy chain. The 3D structure is characterised by two related domains rich in β-sheets connected by a linker peptide and the structure is very similar to other aspartyl proteases (**1LYA**).

### Biological Functions

CatD is an aspartyl protease with a specificity similar as pepsin. CatD is involved in the general degradation of proteins inside lysosomes.

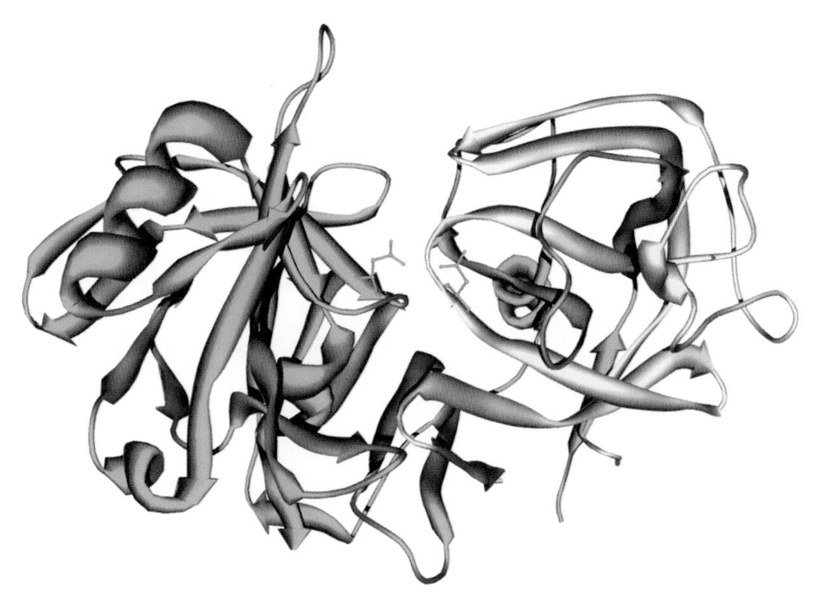

Cathepsin D

## *Cathepsin G (EC 3.4.21.20)*

### Fact Sheet

| | | | |
|---|---|---|---|
| ***Classification*** | Peptidase S1 | ***Abbreviations*** | CatG |
| ***Structures/motifs*** | | ***DB entries*** | CATG_HUMAN/P08311 |
| ***MW/length*** | 26 758 Da/235 aa | ***PDB entries*** | 1KYN |
| ***Concentration*** | Traces | ***Half-life*** | approx. 12 min |
| ***PTMs*** | 1 N-CHO | | |
| ***References*** | Greco *et al.*, 2002, *J. Am. Chem. Soc.*, **124**, 3810–3811.<br>Hohn *et al.*, 1989, *J. Biol. Chem.*, **264**, 13412–13419. | | |

### Description

**Cathepsin G** (CatG) is a neutral protease mainly contained in granules of leukocytes and is present in plasma only in traces.

### Structure

CatG is a single-chain glycoprotein that is activated by releasing an N-terminal dipeptide and a C-terminal dodecapeptide not present in mature CatG. The 3D structure is characterised by two six-stranded β-barrels connected by three transdomain segments and the active site residues are indicated in pink (**1KYN**).

### Biological Functions

CatG is a chymotrypsin-like serine protease that preferentially cleaves at the C-terminal side of bulky aliphatic and aromatic amino acids like Phe or Leu. CatG degrades connective tissues, glycoproteins and collagen and participates in the killing and digestion of pathogens during phagocytosis. CatG is inhibited by $\alpha_1$-antichymotrypsin, $\alpha_1$-antitrypsin and $\alpha_2$-macroglobulin.

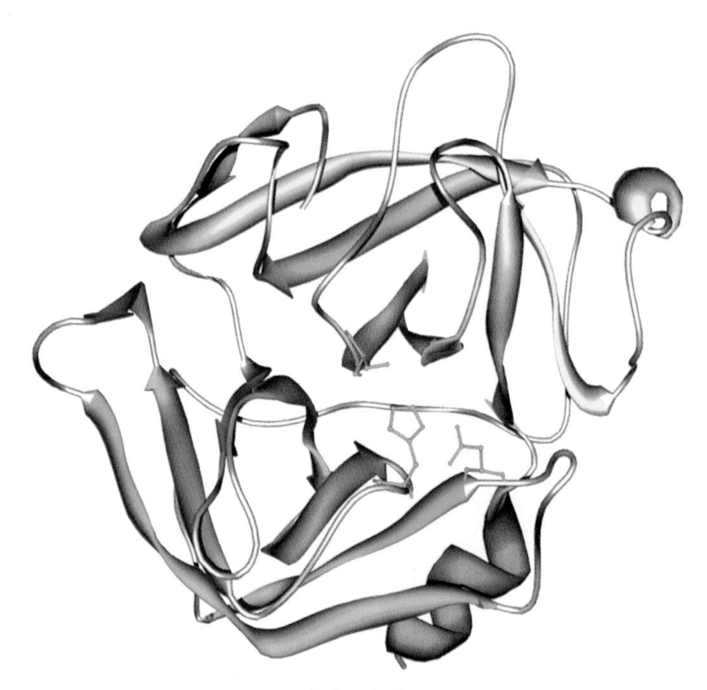

Cathepsin G

# Cathepsin L (EC 3.4.22.15)

## Fact Sheet

| | | | |
|---|---|---|---|
| ***Classification*** | Peptidase C1 | ***Abbreviations*** | |
| ***Structures/motifs*** | | ***DB entries*** | CATL_HUMAN/P07711 |
| ***MW/length*** | Heavy chain: 19 080 Da/175 aa<br>Light chain: 4776 Da/42 aa | ***PDB entries*** | 1ICF |
| ***Concentration*** | | ***Half-life*** | |
| ***PTMs*** | 1 N-CHO | | |
| ***References*** | Gal *et al.*, 1988, *Biochem. J.*, **253**, 303–306.<br>Guncar *et al.*, 1999, *EMBO J.*, **18**, 793–803. | | |

## Description

**Cathepsin L** is a lysosomal cysteine protease important for the degradation process of proteins in the lysosomes.

## Structure

Procathepsin L is activated to two-chain cathepsin L by releasing an N-terminal 96 residue activation peptide and an internal tripeptide, thus generating a heavy and a light chain linked by a single interchain disulfide bridge (Cys 156 in the heavy chain and Cys 31 in the light chain). The heavy chain contains two intrachain disulfide bridges (Cys 22–Cys 65, Cys 56–Cys 98). A fragment of the invariant chain of the HLA class II histocompatibility antigen (in red) binds specifically to the residues of active site cleft (in pink) of cathepsin L consisting of the heavy chain (in green) and the light chain (in yellow) linked by a single disulfide bridge (in blue) (**1ICF**).

## Biological Functions

Cathepsin L exhibits a rather broad specificity similar to that of papain. Cathepsin L is important for the overall degradation of proteins in lysosomes.

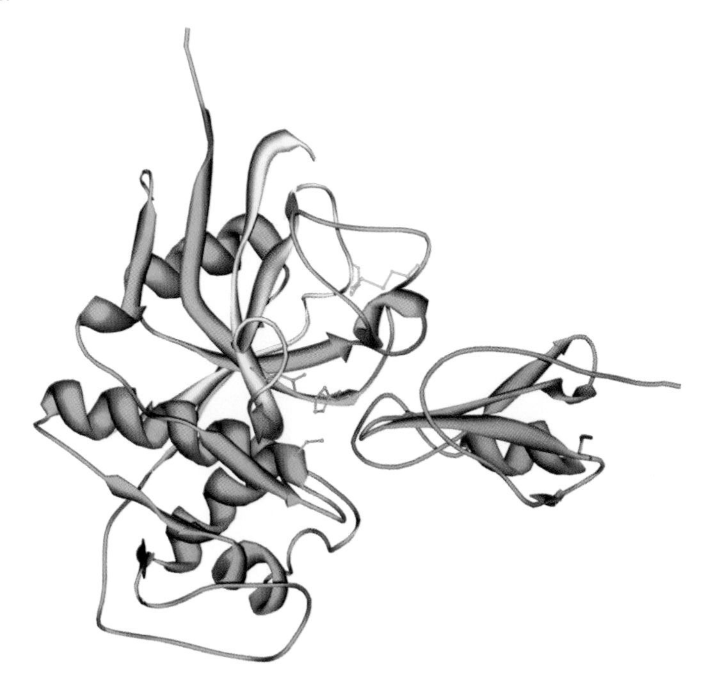

Cathepsin L

## *Cholinesterase (Butyrylcholinesterase; EC 3.1.1.8)*

### Fact Sheet

| | | | |
|---|---|---|---|
| ***Classification*** | Type-B carboxylesterase/lipase | ***Abbreviations*** | BChE |
| ***Structures/motifs*** | | ***DB entries*** | CHLE_HUMAN/P06276 |
| ***MW/length*** | 65 084 Da/574 aa | ***PDB entries*** | 1XLU |
| ***Concentration*** | 5 mg/l | ***Half-life*** | 11 days |
| ***PTMs*** | 9 N-CHO | | |
| ***References*** | Lockridge *et al.*, 1987, *J. Biol. Chem.*, **262**, 549–557.<br>Nachon *et al.*, 2005, *Biochemistry,* **44**, 1154–1162. | | |

### Description

**Cholinesterase** (BChE) is synthesised in the liver and is present in most cells except in erythrocytes. It is clinically important for the diagnosis of poisoning by carbamate and organophosphate insecticides.

### Structure

BChE is a single-chain plasma glycoprotein (24 % CHO) that circulates in plasma as homotetramer (dimers of dimer). Two monomers form a dimer linked by a single interchain disulfide bridge (Cys 571) and two dimers form the non-covalently-linked tetramer. Covalently bound organophosphates (in green) are linked to the active site Ser 198 (in pink) and a suicide reaction termed 'aging' involves the adjacent active site His 438 and Glu 197 (**1XLU**).

### Biological Functions

BChE catalyses the reaction acylcholine + $H_2O \Rightarrow$ choline + carboxylate. BChE seems to play a role in cell proliferation and cell differentiation and is thought to have a protective role against ingested poisons.

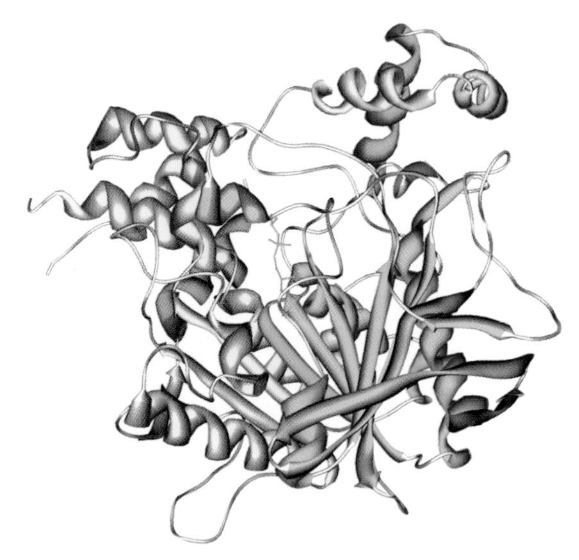

Cholinesterase

## *Chymase (EC 3.4.21.39)*

### Fact Sheet

| | | | |
|---|---|---|---|
| ***Classification*** | Peptidase S1 | ***Abbreviations*** | |
| ***Structures/motifs*** | | ***DB entries*** | MCT1_HUMAN/P23946 |
| ***MW/length*** | 25 030 Da/226 aa | ***PDB entries*** | 1KLT |
| ***Concentration*** | | ***Half-life*** | |
| ***PTMs*** | 2 N-CHO (1 potential) | | |
| ***References*** | Caughey *et al.*, 1991, *J. Biol. Chem.*, **266**, 12956–12963.<br>McGrath *et al.*, 1997, *Biochemistry*, **36**, 14318–14324. | | |

### Description

**Chymase** is synthesised in mast cells and is secreted in complex with mast cell carboxypeptidase and sulfated proteoglycans.

### Structure

Chymase is activated by a dipeptidyl peptidase releasing an acidic N-terminal activation dipeptide. Chymase is a single-chain glycoprotein containing three intrachain disulfide bridges (Cys 30–Cys 46, Cys 123–Cys 188, Cys 154–Cys 167) and a single Cys residue (aa 7). The 3D structure is characterised by two six-stranded β-barrels forming a cleft containing the active site residues (in pink) and the covalently linked inhibitor PMSF (in green) (**1KLT**).

### Biological Functions

Chymase is a chymotrypsin-like serine protease that preferentially cleaves at the C-terminal side of aromatic amino acids. Chymase is inactivated by serpins like $\alpha_1$-antichymotrypsin and $\alpha_1$-antitrypsin. Chymase is thought to be involved in the generation of vasoactive peptides, the degradation of the extracellular matrix and the regulation of gland secretion.

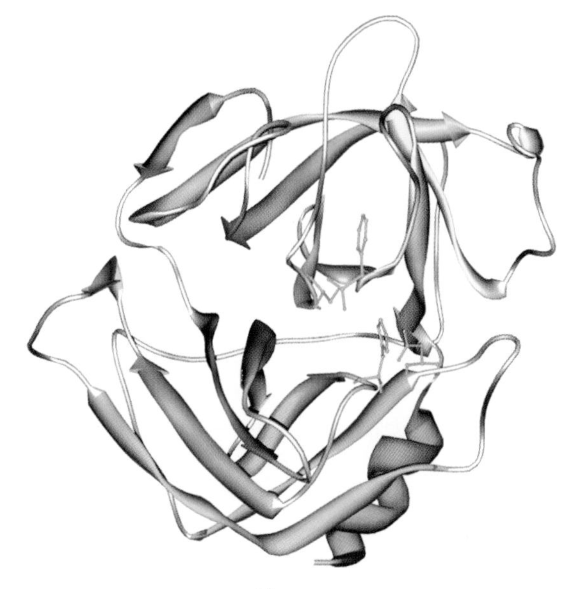

Chymase

## *Chymotrypsinogen B (EC 3.4.21.1)*

### Fact Sheet

| | | | |
|---|---|---|---|
| ***Classification*** | Peptidase S1 | ***Abbreviations*** | HLE |
| ***Structures/motifs*** | | ***DB entries*** | CTRB1_HUMAN/P17538 |
| ***MW/length*** | 25 858 Da/245 aa | ***PDB entries*** | 1CHG |
| ***Concentration*** | | ***Half-life*** | |
| ***PTMs*** | | | |
| ***References*** | Tomita *et al.*, 1989, *Biochem. Biophys. Res. Commun.*, **158**, 569–575.<br>Freer *et al.*, 1970, *Biochemistry*, 9, 1997–2009. | | |

### Description

**Chymotrypsinogen B** is synthesised in the pancreas and acts in the digestive tract.

### Structure

Chymotrypsinogen B is initially cleaved by trypsin (Arg 15–Ile 16) resulting in a two-chain chymotrypsinogen linked by a disulfide bridge. In an autocatalytic process two-chain chymotrypsinogen is activated to three-chain chymotrypsin connected by two disulfide bridges (in red): A–B (Cys 1–Cys 122), B–C (Cys 136–Cys 201). Chain B contains one intrachain disulfide bridge (Cys 42–Cys 58) and chain C two intrachain disulfide bridges (Cys 168–Cys 182, Cys 191–Cys 220). Bovine chymotrypsinogen contains two domains characterised by large sections of distorted antiparallel β-sheets (**1CHG**).

### Biological Functions

Chymotrypsin cleaves peptide bonds with C-terminal bulky side chains, preferably aromatic residues Phe, Trp and Tyr, but also Leu and Met. Chymotrypsin is inhibited by $\alpha_2$-macroglobulin and the serpin $\alpha_1$-antichymotrypsin.

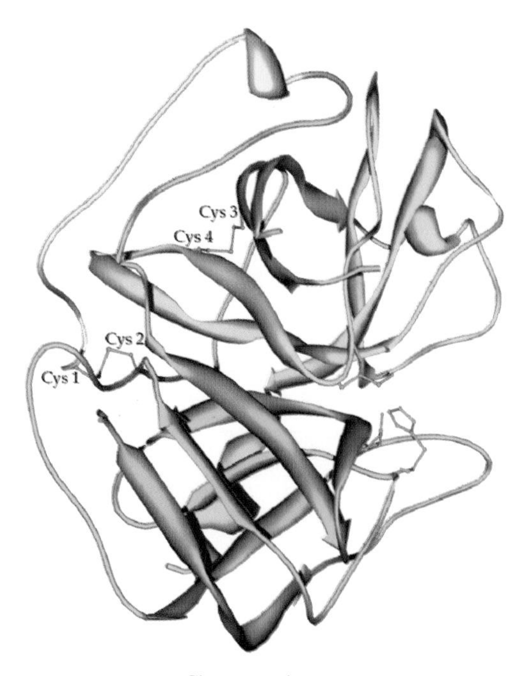


Chymotrypsinogen

# *Creatine kinase: M-type and B-type (EC 2.7.3.2)*

## Fact Sheet

| | | | |
|---|---|---|---|
| ***Classification*** | ATP guanidino phosphotransferase | ***Abbreviations*** | M-CK; B-CK |
| ***Structures/motifs*** | | ***DB entries*** | KCRM_HUMAN/P06732<br>KCRB_HUMAN/P12277 |
| ***MW/length*** | M-type: 43 101 Da/381 aa<br>B-type: 42 644 Da/381 aa | ***PDB entries*** | 1I0E |
| ***Concentration*** | | ***Half-life*** | 3–15 h |
| ***PTMs*** | | | |
| ***References*** | Mariman *et al.*, 1987, *Genomics*, **1**, 126–137.<br>Trask *et al.*, 1988, *J. Biol. Chem.*, **263**, 17142–17149.<br>Shen *et al.*, 2001, *Acta Crystallogr. Sect. D*, **57**, 1196–1200. | | |

## Description

**Creatine kinase** (CK) is found primarily in high-energy tissues such as brain, heart and skeletal muscle.

## Structure

M-CK and B-CK are single-chain proteins that are present as non-covalently-linked homo- or heterodimers: MM, BB and MB. They require divalent ions as activators and the active site consists of Cys 283. The dimers form an infinite 'double-helix'-like structure along a long crystallographic axis (**1I0E**).

## Biological Functions

CK catalyses the reaction ATP + creatine ⇔ ADP + phosphocreatine. CK plays a central role in the energy transduction in tissues and is involved in the restoration of the cellular energy reservoir. The measurement of CK activity in serum is useful for the diagnosis of diseases.

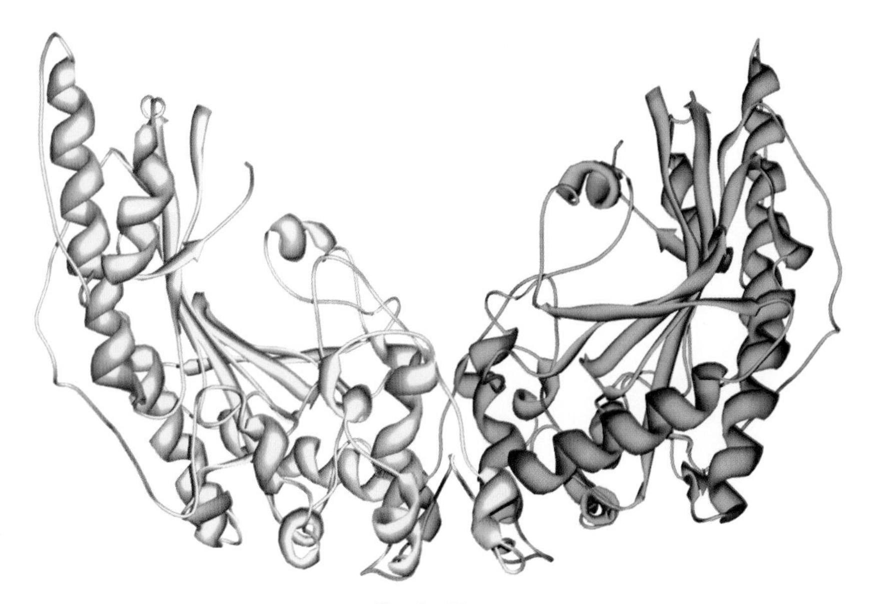

Creatine kinase

# *γ-Glutamyltranspeptidase 1(γ-Glutamyltransferase 1) (EC 2.3.2.2)*

## Fact Sheet

| | | | |
|---|---|---|---|
| ***Classification*** | γ-Glutamyltransferase | ***Abbreviations*** | GGT |
| ***Structures/motifs*** | | ***DB entries*** | GGT1_HUMAN/P19440 |
| ***MW/length*** | Heavy chain: 41 403 Da/380 aa<br>Light chain: 19 998 Da/189 aa | ***PDB entries*** | |
| ***Concentration*** | Traces | ***Half-life*** | 9 h |
| ***PTMs*** | Heavy chain: 6 N-CHO (potential); light chain: 1 N-CHO (potential) | | |
| **References** | Sakamuro *et al.*, 1988, *Gene*, **73**, 1–9. | | |

## Description

**γ-Glutamyltransferase** (GGT) in blood mainly originates from the liver. GGT plays a key role in the gamma-glutamyl cycle.

## Structure

GGT is a type II membrane glycoprotein present as a heterodimer of two non-covalently-linked chains: a heavy chain (380 aa) with a potential 22 residue transmembrane segment (Leu 5–Leu 26) and a light chain (189 aa) carrying the active site.

## Biological Functions

GGT catalyses the reaction 5-glutamyl-peptide + amino acid ↔ peptide + 5-glutamyl amino acid. GGT follows the mechanism of 'ping-pong bi-bi' reactions. GGT catalyses the first reaction in the glutathione breakdown. GGT is involved in the transport of amino acids across the cell membrane.

# *Glutathione peroxidase 3 (EC 1.11.1.9)*

## Fact Sheet

| | | | |
|---|---|---|---|
| ***Classification*** | Glutathione peroxidase | ***Abbreviations*** | GPX3 |
| ***Structures/motifs*** | | ***DB entries*** | GPX3_HUMAN/P22352 |
| ***MW/length*** | 23 417 Da/206 aa | ***PDB entries*** | 2I3Y |
| **Concentration** | 15–20 mg/l | ***Half-life*** | |
| **PTMs** | 1 Pyro-Glu; 1 Se-Cys (53 aa) | | |
| **References** | Epp *et al.*, 1983, *Eur. J. Biochem.*, **133**, 51–69.<br>Takahashi *et al.*, 1990, *J. Biochem.*, **108**, 145–148. | | |

## Description

**Glutathione peroxidase 3** (GPX3) is primarily synthesised in the kidney and protects cells, enzymes and heme-containing proteins from oxidative damage.

## Structure

GPX3 is a single-chain protein with an N-terminal pyro-Glu residue and a Se-Cys residue at the active site. The 3D structure of the epididymal GPX5 is characterised by a central β-sheet core surrounded by α-helices and the active site Se-Cys (in pink) is located in the N-terminal region (**2I3Y**).

## Biological Functions

GPX3 catalyses the reaction 2 glutathione + $H_2O_2$ ↔ glutathione disulfide + 2 $H_2O$. It protects extracellular fluid components and the cell surface against peroxide-mediated damage. GPX3 plasma levels are used as a clinical marker for selenium deficiency.

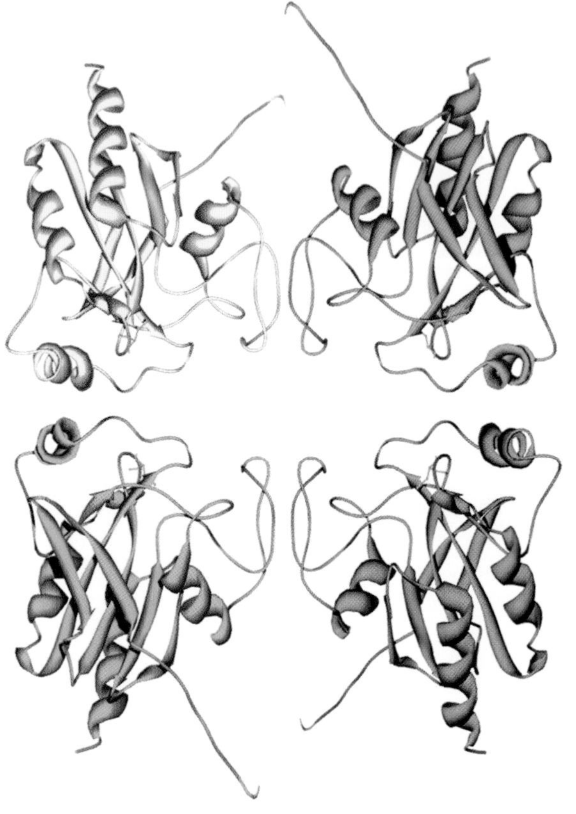

Glutathione peroxidase 3

## *Leukocyte elastase (Medullasin; EC 3.4.21.37)*

### Fact Sheet

| | | | |
|---|---|---|---|
| ***Classification*** | Peptidase S1 | **Abbreviations** | HLE |
| ***Structures/motifs*** | | ***DB entries*** | ELNE_HUMAN/P08246 |
| ***MW/length*** | 23 295 Da/218 aa | ***PDB entries*** | 1HNE |
| ***Concentration*** | | ***Half-life*** | approx. 5 min |
| ***PTMs*** | 3 N-CHO (1 potential) | | |
| ***References*** | Navia *et al.*, 1989, *Proc. Natl. Acad. Sci. USA*, **86**, 7–11.<br>Sinha *et al.*, 1987, *Proc. Natl. Acad. Sci. USA*, **84**, 2228–2232. | | |

### Description

**Leukocyte elastase** (HLE) is synthesised in the bone marrow and is involved in the degradation of various proteins.

### Structure

HLE is a single-chain glycoprotein (22 % CHO) that is activated by the release of an N-terminal dipeptide. Active HLE (218 aa) is devoid of a C-terminal 20 residue peptide and contains four intrachain disulfide bridges (Cys 26–Cys 42, Cys 122–Cys 179, Cys 152–Cys 158, Cys 169–Cys 194). The catalytic mechanism of the peptide inhibitor methoxysuccinyl-Ala-Ala-Pro-Ala chloromethyl ketone (in red) involves cross-links with the active site residues His 57 and Ser 195 (in pink) of elastase (**1HNE**).

### Biological Functions

HLE is a serine protease that preferentially cleaves at the C-terminal side of small, nonpolar amino acids like Ala and Val. HLE is involved in the degradation of elastin, collagen, proteoglycans and certain plasma proteins. HLE is inhibited by $\alpha_2$-macroglobulin and serpins like leukocyte elastase inhibitor and $\alpha_1$-antitrypsin.

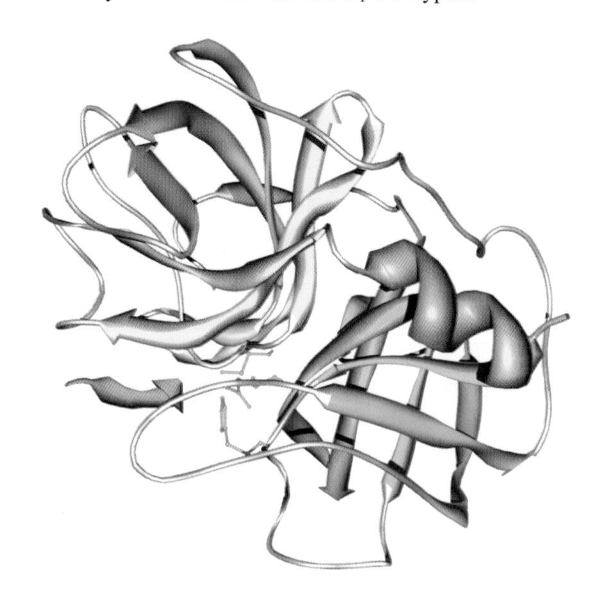

Elastase

## *Lysozyme C (Muramidase; EC 3.2.1.17)*

### Fact Sheet

| | | | |
|---|---|---|---|
| ***Classification*** | Glycosyl hydrolyse 22 | ***Abbreviations*** | LYZ |
| ***Structures/motifs*** | | ***DB entries*** | LYSC_HUMAN/P61626 |
| ***MW/length*** | 14 701 Da/130 aa | ***PDB entries*** | 1B5U |
| ***Concentration*** | 7–13 mg/l | ***Half-life*** | |
| ***PTMs*** | | | |
| ***References*** | Castanon *et al.*, 1988, *Gene*, **66**, 223–234.<br>Takano *et al.*, 1999, *Biochemistry*, **38**, 6623–6629. | | |

### Description

**Lysozyme** (LYZ) is a highly cationic protein widely distributed in various types of tissues and has primarily a bacteriolytic function.

### Structure

LYZ is a single-chain protein with Glu 35 and Asp 53 in its active site. LYZ shows structural similarity with α-lactalbumin. The 3D structure is characterised by four α-helices and three β-sheets (**1B5U**).

### Biological Functions

LYZ hydrolyses the β-1,4-linkages between *N*-acetylmuramic acid and *N*-acetylglucosamine residues in peptidoglycans and between *N*-acetylglucosamines in chitodextrins. LYZ dissolves living cells of *Micrococcus lysodeikticus* and is a potent inhibitor of chemotaxis and the production of toxic oxygen radicals.

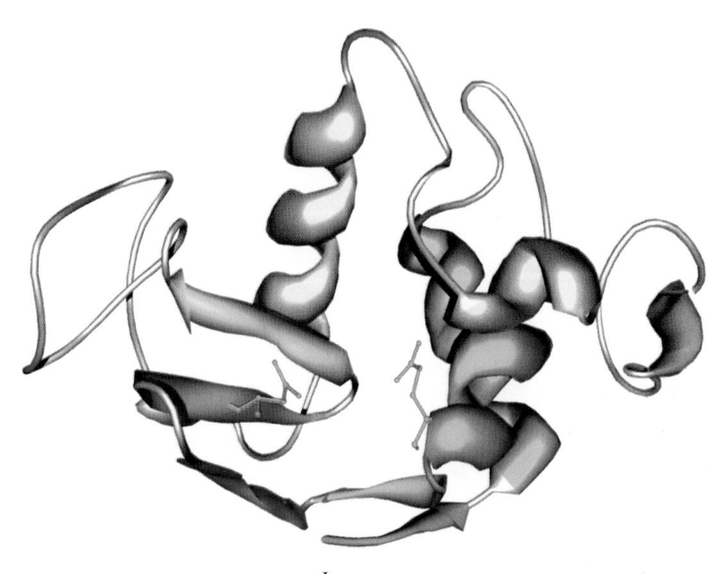

Lysozyme

## *Myeloperoxidase (EC 1.11.1.7)*

### Fact Sheet

| | | | |
|---|---|---|---|
| ***Classification*** | Peroxidase | ***Abbreviations*** | MPO |
| ***Structures/motifs*** | | ***DB entries*** | PERM_HUMAN/P05164 |
| ***MW/length*** | 66 107 Da/581 aa | ***PDB entries*** | 1MHL |
| ***Concentration*** | Traces | ***Half-life*** | |
| ***PTMs*** | 3 N-CHO; 1 Cys-sulfenic acid (aa 152) | | |
| ***References*** | Fiedler *et al.*, 2000, *J. Biol. Chem.*, **275**, 11964–11971.<br>Morishita *et al.*, 1987, *J. Biol. Chem.*, **262**, 3844–3851. | | |

### Description

**Myoeloperoxidase** (MPO) is predominantly present in the granulocytes of neutrophils as part of the host defence system of polymorphonuclear leukocytes (PMNs).

### Structure

MPO is a glycoprotein present as a tetramer of two heterodimers. Each heterodimer consists of a light chain (114 aa) and a heavy chain (467 aa) generated from the proform by the release of a 116 residue propeptide and the cleavage of a single peptide bond (Gly 114–Val 115). The two heterodimers are linked by a single disulfide bridge (in red) between the two heavy chains (Cys 155). The heavy chain carries an unusual Cys-sulfenic acid (aa 152, in green). Each heterodimer binds one calcium ion and one heme B group. MPO exhibits primarily an $\alpha$-helical structure and the heme group (in pink) is covalently bound to Asp 96, Glu 244 and Met 245; the calcium and chloride ions are shown as orange and green balls, respectively (**1MHL**).

### Biological Functions

MPO catalyses the reaction donor + $H_2O_2 \Rightarrow$ oxidised donor + 2 $H_2O$, especially $Cl^-$ + $H_2O_2 \Rightarrow$ HOCl + 2 $H_2O$. As part of the host defence system it is responsible for microbicidal activity against a wide range of organisms.

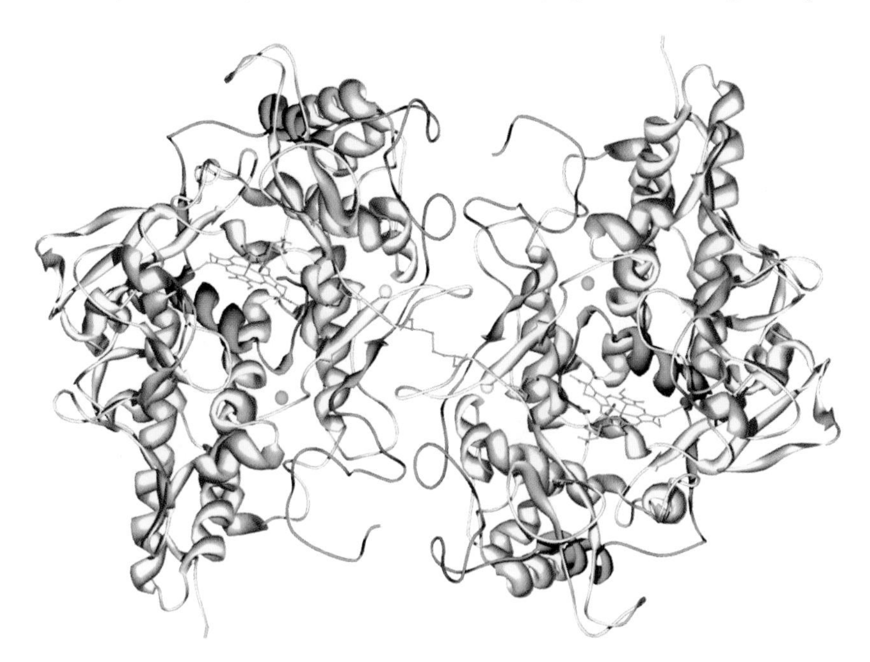

Myeloperoxidase

# *Pepsin A (EC 3.4.23.1)*

## Fact Sheet

| | | | |
|---|---|---|---|
| *Classification* | Peptidase A1 | *Abbreviations* | |
| *Structures/motifs* | | *DB entries* | PEPA_HUMAN/P00790 |
| *MW/length* | 34 628 Da/326 aa | *PDB entries* | 1PSO |
| *Concentration* | | *Half-life* | |
| *PTMs* | 1 P-Ser (potential) | | |
| *References* | Fujinaga *et al.*, 1995, *Prot. Sci.*, **4**, 960–972.<br>Sogawa *et al.*, 1983, *J. Biol. Chem.*, **258**, 5306–5311. | | |

## Description

**Pepsin** is synthesised in the gastric mucosa and is the main enzyme of the stomach responsible for food digestion.

## Structure

Pepsin is generated autocatalytically from its proform pepsinogen A by releasing an N-terminal 47 residue activation peptide. Pepsin is a single-chain protein containing three intrachain disulfide bridges (Cys 45–Cys 50, Cys 208–Cys 210, Cys 249–Cys 288) and the active site consists of two Asp residues (aa 32, 215.) and Ser 68 seems to be phosphorylated. The 3D structure of pepsin consists of a central domain composed of a six-stranded β-sheet and N-terminal and C-terminal lobes consisting of orthogonally packed β-sheets (**1PSO**).

## Biological Functions

Pepsin is an aspartic protease that cleaves C-terminal to hydrophobic, preferentially aromatic residues.

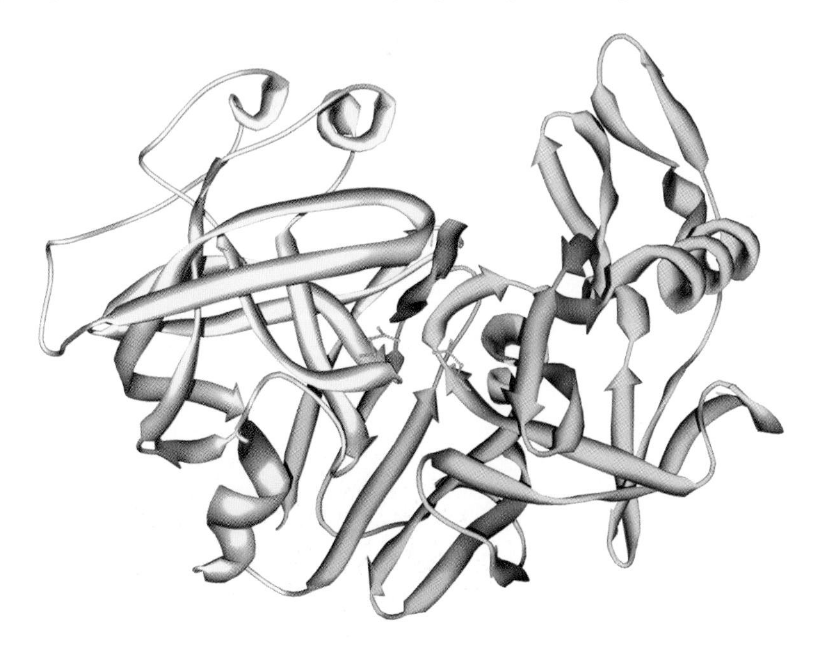

Pepsin A

# *Prostate-specific antigen (EC 3.4.21.77)*

## Fact Sheet

| | | | |
|---|---|---|---|
| ***Classification*** | Peptidase S1 | ***Abbreviations*** | PSA |
| ***Structures/motifs*** | | ***DB entries*** | KLK3_HUMAN/P07288 |
| ***MW/length*** | 26 089 Da/237 aa | ***PDB entries*** | |
| ***Concentration*** | < 3 μg/l | ***Half-life*** | 12–14 hrs |
| ***PTMs*** | 1 N-CHO | | |
| ***References*** | Lundwall and Lilja, 1987, *FEBS Lett.*, **214**, 317–322.<br>Schaller *et al.*, 1987, *Eur. J. Biochem.*, **170**, 111–120. | | |

## Description

**Prostate-specific antigen** (PSA) is synthesised in the epithelial cells of the prostate and is involved in the liquefaction of the seminal coagulum. PSA in serum is used for the detection and monitoring of prostate cancer.

## Structure

PSA is a single-chain glycoprotein containing five intrachain disulfide bridges (Cys 7–Cys 149, Cys 26–Cys 42, Cys 128–Cys 195, Cys 160–Cys 174, Cys 185–Cys 210) that is activated by the release of an N-terminal activation heptapeptide.

## Biological Functions

PSA is a kallikrein-like serine protease that preferentially cleaves at the C-terminal side of Tyr. PSA cleaves the major gel-forming proteins in semen: semenogelin I and II and fibronectin. PSA is inhibited by the serpins plasma serine protease inhibitor and $\alpha_1$-antichymotrypsin and by $\alpha_2$-macroglobulin and pregnancy zone protein.

# *Protein disulfide isomerase (EC 5.3.4.1)*

## Fact Sheet

| | | | |
|---|---|---|---|
| ***Classification*** | Protein disulfide isomerase | ***Abbreviations*** | PDI |
| ***Structures/motifs*** | 2 Thioredoxin | ***DB entries*** | PDIA1_HUMAN/P07237 |
| ***MW/length*** | 55 294 Da/491 aa | ***PDB entries*** | 1MEK |
| ***Concentration*** | | ***Half-life*** | |
| ***PTMs*** | | | |
| ***References*** | Kemmink *et al.*, 1996, *Biochemistry*, **35**, 7684–7691.<br>Pihlajaniemi *et al.*, 1987, *EMBO J.*, **6**, 643–649. | | |

## Description

**Protein disulfide isomerase** (PDI) is identical with the beta subunit of prolyl 4-hydroxylase and both originate from the same gene. PDI catalyses the formation, breakage and rearrangement of disulfide bridges.

## Structure

PDI is a single-chain protein containing two thioredoxin domains each containing two redox-active disulfide bonds: Cys 36–Cys 39 and Cys 380–Cys 383. The Cys residues act as nuclophiles in the isomerisation reaction. The 3D of the N-terminal thioredoxin domain is characterised by a central β-sheet surrounded by four α-helices typical for thioredoxin domains (**1MEK**).

## Biological Functions

PDI catalyses the rearrangement of disulfide bonds in proteins. In addition, it catalyses the reaction procollagen-Pro + 2-oxoglutarate + $O_2$ ⇔ procollagen-4-OH-Pro + succinate + $CO_2$.

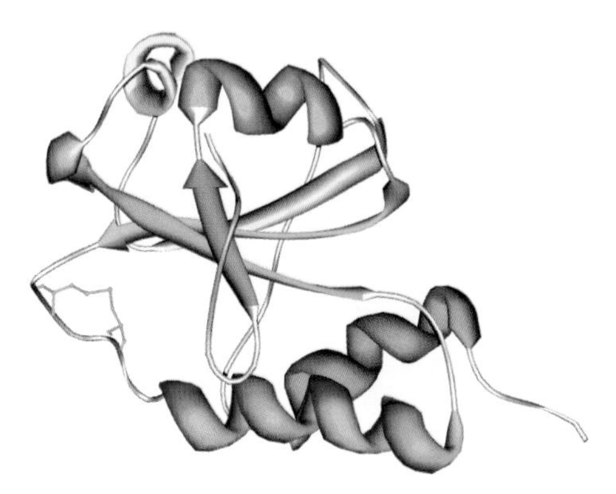

Protein disulfide isomerase thio redoxin domain 1

## *Serum paraoxonase/Arylesterase 1 (EC 3.1.1.2, EC 3.1.8.1)*

### Fact Sheet

| | | | |
|---|---|---|---|
| ***Classification*** | Paraoxonase | ***Abbreviations*** | PON1 |
| ***Structures/motifs*** | | ***DB entries*** | PON1_HUMAN/P27169 |
| ***MW/length*** | 39 618 Da/354 aa | ***PDB entries*** | 1V04 |
| ***Concentration*** | approx. 50 mg/l | ***Half-life*** | |
| ***PTMs*** | 3 N-CHO (2 potential) | | |
| ***References*** | Harel *et al.*, 2004, *Nat. Struct. Mol. Biol.*, **11**, 412–419.<br>Hassett *et al.*, 1991, *Biochemistry*, **30**, 10141–10149. | | |

### Description

**Serum paraoxonase** (PON1) is synthesised in the liver and circulates in plasma associated with high-density lipoprotein (HDL).

### Structure

PON1 is a single-chain glycoprotein (15 % CHO) that only loses its N-terminal Met upon maturation but retains its signal peptide which seems to be important for complex formation with HDL. Mature PON1 contains a single disulfide bridge (Cys 41–Cys 352) and one free, reactive Cys 283. The 3D structure is characterised by a six-bladed β-propeller forming a central tunnel, which contains the active site (in pink) and two calcium ions (green balls) (**1V04**).

### Biological Functions

PON1 catalyses the reaction aryl dialkyl phosphate + $H_2O \Rightarrow$ dialkyl phosphate + aryl alcohol and phenyl acetate + $H_2O \Rightarrow$ phenol + acetate. PON1 is capable of hydrolysing a broad spectrum of toxic organophosphate substrates. It also confers resistance to poisoning from a wide variety of insecticides.

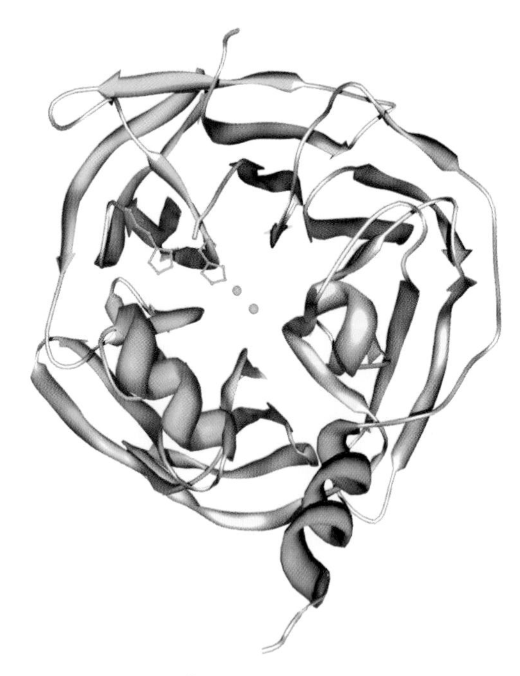

Serum paraoxonase

# *Trypsin I + II (Cationic + anionic trypsinogen; EC 3.4.21.4)*

## Fact Sheet

| | | | |
|---|---|---|---|
| ***Classification*** | Peptidase S1 | ***Abbreviations*** | Tg1; Tg2 |
| ***Structures/motifs*** | | ***DB entries*** | TRY1_HUMAN/P07477<br>TRY2_HUMAN/P07478 |
| ***MW/length*** | Tg1: 24 114 Da/224 aa<br>Tg2: 24 042 Da/224 aa | ***PDB entries*** | 1TRN |
| ***Concentration*** | 15–30 μg/l | ***Half-life*** | |
| ***PTMs*** | Tg1: 1 P-Tyr (131 aa) | | |
| ***References*** | Emi *et al.*, 1986, *Gene*, **41**, 305–310.<br>Gaboriaud *et al.*, 1996, *J. Mol. Biol.*, **259**, 995–1010. | | |

## Description

**Cationic** and **anionic trypsinogen** (Tg1 + Tg2) are synthesised in the pancreas as inactive precursors. Tg1 and Tg2 are essential for the activation of pancreatic zymogens.

## Structure

Tg1 and Tg2 are single-chain proteins that are activated initially by enterokinase and afterwards autocatalytically by trypsin, releasing an N-terminal octapeptide. Tg1 is phosphorylated at Tyr 131 and both bind one calcium ion. Tg1 contains five intrachain disulfide bridges (Cys 7–Cys 137, Cys 25–Cys 41, Cys 116–Cys 183, Cys 148–Cys 162, Cys 173–Cys 197), Tg2 only four (Cys 116–Cys 183 is missing). The 3D structure exhibits the classical fold of a serine protease domain and the residues of the active site are marked in pink (**1TRN**).

## Biological Functions

Trypsin is essential for the activation of pancreatic zymogens like chymotrypsinogen, proelastase, prekallikrein, prophospholipase $A_2$ and procarboxypeptidases. Trypsin preferentially cleaves at the C-terminal side of basic residues. Trypsin is inhibited by Kunitz-type trypsin inhibitors, $\alpha_1$-antitrypsin and $\alpha_2$-macroglobulin.

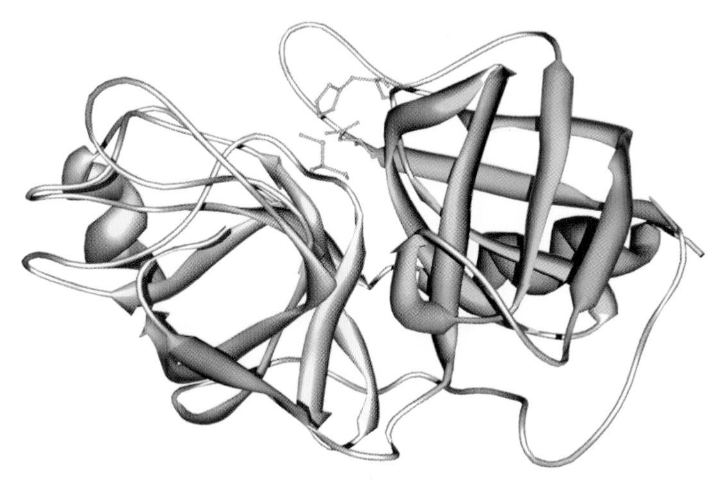

Trypsin

## *Tryptase* $\beta_2$*(EC 3.4.21.59)*

### Fact Sheet

| | | | |
|---|---|---|---|
| ***Classification*** | Peptidase S1 | ***Abbreviations*** | |
| ***Structures/motifs*** | | ***DB entries*** | TRYB2_HUMAN/P20231 |
| ***MW/length*** | 27 458 Da/245 aa | ***PDB entries*** | 1A0L |
| ***Concentration*** | < 10 µg/l | ***Half-life*** | approx. 2 h |
| ***PTMs*** | 1 N-CHO (potential) | | |
| ***References*** | Miller *et al.*, 1990, *J. Clin. Invest.*, **86**, 864–870.<br>Pereira *et al.*, 1998, *Nature*, **392**, 306–311. | | |

### Description

**Tryptase** is the major protein of the secretory granules of mast cells and is present in plasma in small amounts, usually as tetrameric complex with heparin.

### Structure

Tryptase is generated from the proenzyme form by releasing an N-terminal 12 residue activation peptide, but monomeric tryptase is inactive. Monomeric tryptase is a single-chain glycoprotein containing four intrachain disulfide bridges (Cys 29–Cys 45, Cys 125–Cys 200, Cys 158–Cys 181, Cys 190–Cys 218). Tryptase forms a ring-like tetrameric structure, where the active sites of each monomer (in pink) are facing towards a central oval pore (**1A0L**).

### Biological Functions

Tryptase is a trypsin-like serine protease that preferentially cleaves at the C-terminal side of Lys and Arg residues. Tryptase is secreted by mast cells upon the activation–degranulation response of mast cells.

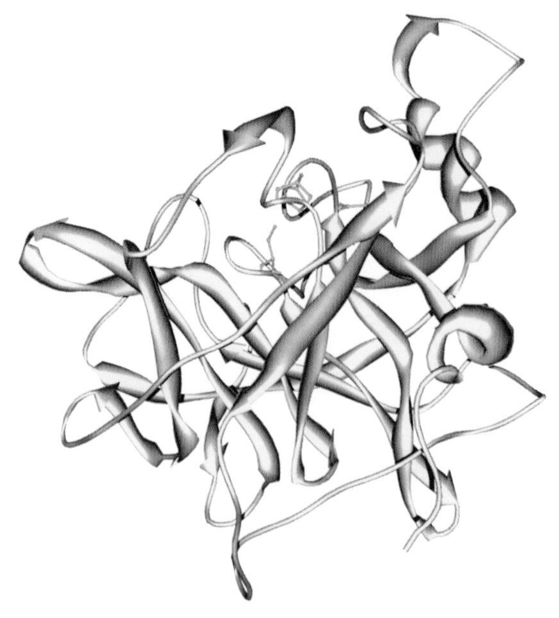

Tryptase

# 11

# Inhibitors

## 11.1 INTRODUCTION

Biochemical pathways and cascades such as blood coagulation and fibrinolysis and the complement system have to be controlled and regulated very precisely in order to obtain and maintain a well-balanced system in an organism. Especially, active enzymes such as proteases in these pathways and other enzymes have to be tightly controlled and their action has to be limited to their physiological substrates and to the location (often surfaces) where they are really required. For the control and regulation, especially of proteases, various types of inhibitors exist, which can be divided into several classes of protease inhibitors:

1. The serine protease inhibitors (serpins)
2. The Kunitz- and Kazal-type inhibitors
3. General protease inhibitors like $\alpha_2$-macroglobulin
4. The cystatins
5. Other inhibitors such as inter-$\alpha$-trypsin inhibitors

The importance of the various types of inhibitors in an organism is reason enough to discuss this group of proteins in a separate chapter.

## 11.2 SERINE PROTEASE INHIBITORS (SERPINS)

### 11.2.1 Introduction

The regulation of proteolytic activity is a fundamental process for maintaining the integrity of a living organism. During evolution various mechanisms have evolved to obtain and maintain an equilibrated balance within an organism itself and between the host and a pathogen. The family of serine protease inhibitors (serpins) represents such a regulative system. Serpins are high molecular weight inhibitors in the mass range of 40–70 kDa, which regulate many physiological processes like coagulation, fibrinolysis, complement activation, inflammation, angiogenesis, apoptosis, neoplasia and viral pathogenesis.

Serpins are suicide inhibitors of serine proteases, which have been identified in species as diverse as vertebrates, insects, plants and viruses. Serpins act like a 'sink' for serine proteases, removing them before they cause serious damage to the surrounding cells and tissues. The loss of inhibitory activity of a serpin leads to an imbalance between serine proteases and their inhibitors. Several pathological disorders have been correlated with serpin dysfunction, such as thrombosis, emphysema, liver cirrhosis, immune hypersensitivity and mental disorder.

In the bloodstream of vertebrates haemostasis is a well-balanced equilibrium between limited proteolytic activity, often consigned to a surface, and its regulatory control. Serpins play a key role in these regulatory processes. The coagulation cascade and fibrinolysis and complement activation are main events in the bloodstream and are therefore heavily regulated by inhibitors, mainly by serpins.

*Human Blood Plasma Proteins: Structure and Function* Johann Schaller, Simon Gerber, Urs Kämpfer, Sofia Lejon and Christian Trachsel

### 11.2.2 Structural properties

Basically, there are two different types of serpins:

1. those with inhibitory activity
2. those without any known inhibitory activity

Based on structural similarity, a series of proteins without inhibitory activity have been assigned to the family of serpins such as angiotensinogen from the blood coagulation system and ovalbumin, probably the best studied and best known serpin.

Inhibitory serpins are fundamentally different from other types of inhibitors in the sense that they are consumed in a one-to-one reaction termed 'suicide inhibition'. Serpins are globular proteins usually containing nine α-helices, three β-sheets, A, B and C, and a reactive-centre loop peptide of approximately 20 residues located on the surface of the inhibitor. Serpins exist in three major conformational states: native, latent and cleaved. The conformational change upon cleavage of a serpin is known as the stressed → relaxed (S → R) transition. Upon proteolytic cleavage by a serine protease the reactive-centre loop peptide is inserted as strand 4 into β-sheet A with a concomitant profound conformational change. In the case of $\alpha_1$-antitrypsin, the residues adjacent to the cleaved peptide bond in the reactive loop centre are approximately 70 Å apart after cleavage and the 3D structure of the $\alpha_1$-antitrypsin–trypsin complex shows inhibition by deformation, as can be seen in Figure 11.1.

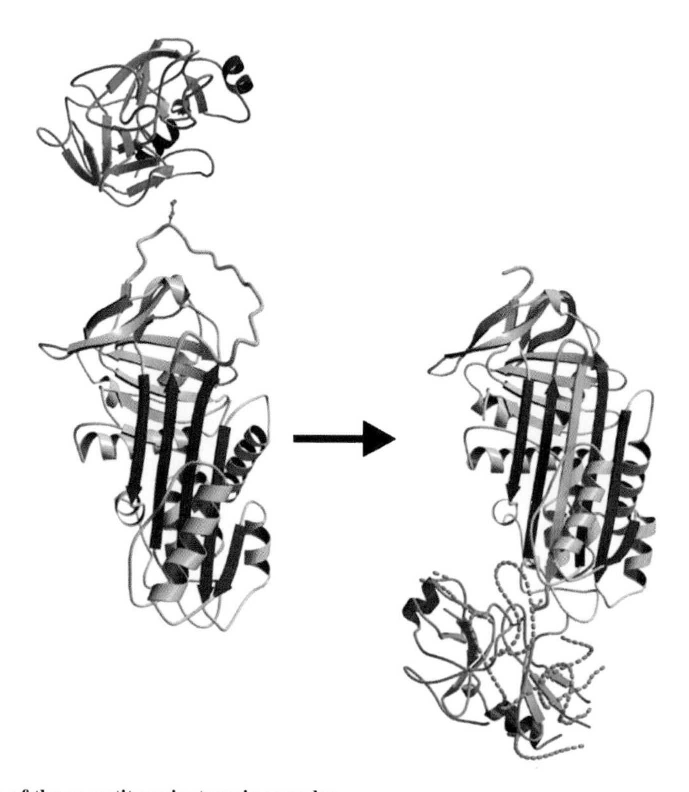

**Figure 11.1 Formation of the $\alpha_1$-antitrypsin–trypsin complex**
Insertion of the reactive loop peptide (in yellow) located on the surface of native $\alpha_1$-antitrypsin into the β-sheet A (in red) of cleaved $\alpha_1$-antitrypsin. In the course of the inhibitory reaction trypsin is drawn from one side of $\alpha_1$-antitrypsin to the other side and the C- and N-terminal ends of the cleaved reactive bond are now on either side of cleaved and inactivated $\alpha_1$-antitrypsin, approx. 70 Å apart. Trypsin is shown in magenta. (Reproduced from Huntington *et al.*, Nature 407: 923–926. Copyright (2000), with permission from Nature Publishing Group.)

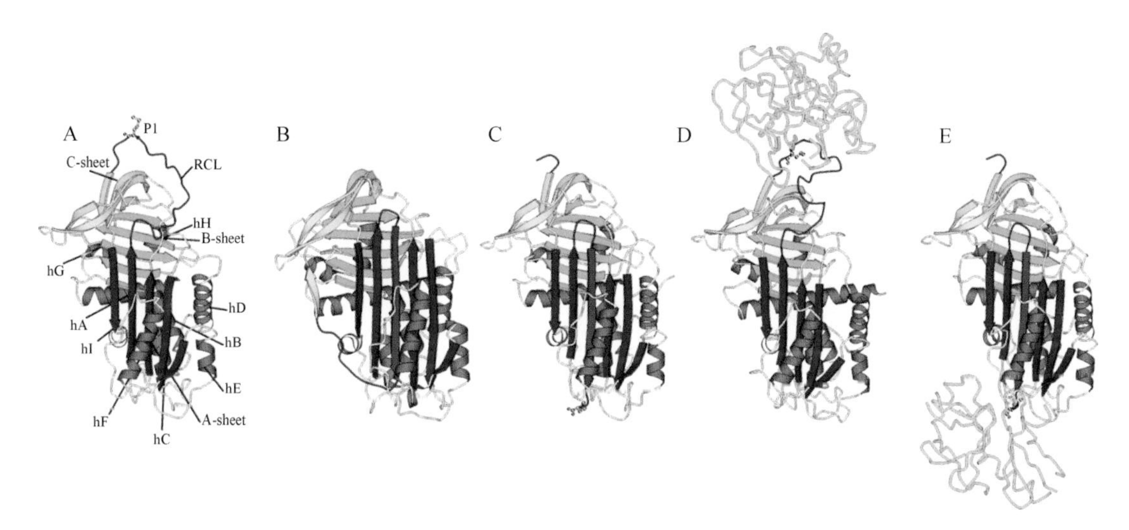


**Figure 11.2 Conformational states of a serpin**
A: **Native state** of human $\alpha_1$-antitrypsin (**PDB: 1QLP**) with the reactive-centre loop peptide containing the reactive bond on the surface of the serpin. B: **Latent, inserted state** of human antithrombin III (**PDB: 2ANT**) with the reactive loop peptide inserted into β-sheet A representing the uncleaved stressed state of a serpin. C: **Cleaved state** of human α-antitrypsin (**PDB: 7API**) with the ends of the cleaved peptide bond on either side of the serpin. D: **Non-covalent, Michaelis complex** of alaserpin (*Manduca Sexta*) with rat trypsin (**PDB: 1I99**). E: **Cleaved state**. Covalent complex of human α-antitrypsin with bovine trypsin (**PDB: 1EZX**). The serpin is covalently linked to the side chain of the Ser residue in the catalytic triad of the serine protease. (Reproduced from Silverman *et al.*, J Biol Chem 276: 33293–33296. Copyright (2001) The American Society for Biochemistry and Molecular Biology.)

There are many conformations and forms of serpins. Several conformational states of them have been described in detail and the major ones are depicted in Figure 11.2.

1. *The native state.* The native state is usually characterised by three β-sheets, nine α-helices and an approximately 20 residues long reactive-centre loop (RCL) peptide located on the surface of the molecule.
2. *The latent, inserted state.* In the inserted state the reactive loop peptide is inserted into β-sheet A. This uncleaved stressed state is metastable in its nature and is often compared with a spring under tension before the stress is released.
3. *The cleaved state.* In the cleaved state the tension of the spring was released (to remain with the spring comparison) and the C- and N-terminal ends of the cleaved peptide bond in the reactive centre are now on either side of the molecule, in many cases as far apart as 70 Å.
4. *The serpin–serine protease complex.* The first step of the inhibitory action of a serpin is the formation of a noncovalent Michaelis-like complex between the serine protease and the serpin, characterised by an interaction of the reactive loop peptide with the catalytic triad region of the serine protease.
5. *Cleaved serpin with inactivated serine protease.* Upon insertion of the reactive loop peptide into β-sheet A the serine protease is drawn from one side of the molecule to the other side and the cleavage of the reactive bond releases the tension and the C- and N-terminal ends of the cleaved peptide bond end up on either side of the inhibitor. The serpin is now covalently linked to the side chain of the Ser residue in the catalytic triad of the serine protease.

Multiple forms of serpins have been identified in biological fluids such as native inhibitory, inactivated and latent forms and noninhibitory forms like those cleaved, polymerised and oxidised. A selection of the conformational polymorphisms of serpins is shown in Figure 11.3.

The biological function of the various forms of serpins and their relation to pathological states is not yet very well understood and remains to be investigated.

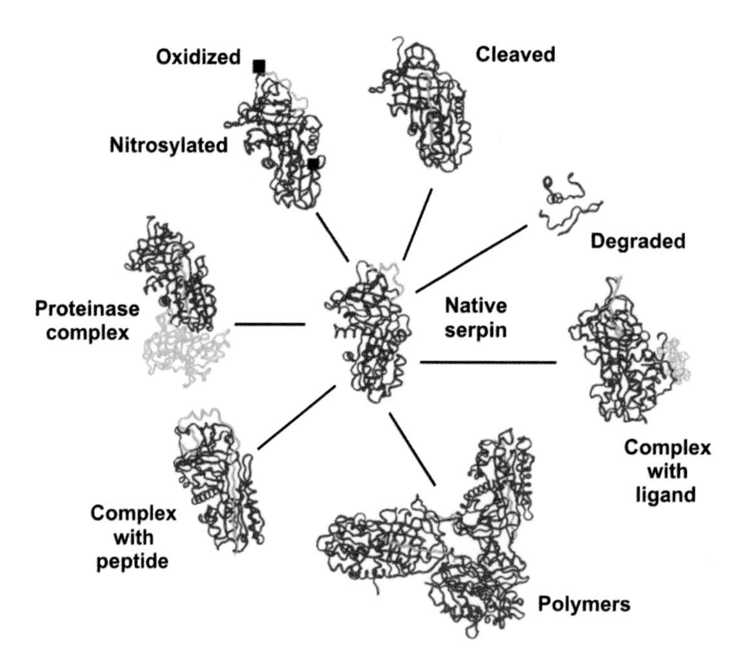


**Figure 11.3 Conformational polymorphisms of serpins**
There exist many conformational polymorphisms of serpins in biological fluids. Beside the native form there exist complexes with serine proteases, peptides and ligands, oxidised and nitrosylated forms, polymers, cleaved and degraded forms. The serpins are shown in lilac and the reactive loop peptide in yellow. (Reproduced from Janciauskiene S. Biochim Biophys Acta 1535: 221–235. Copyright (2001), with permission from Elsevier.)

### 11.2.3 Inhibitory reaction

The inhibition of a serine protease by a serpin is a two-stage process:

1. The first step is a fast, reversible second-order reaction leading to a noncovalent 1 : 1 complex of the serpin with the serine protease.
2. The second step is a slower, irreversible first-order reaction with the insertion of the reactive loop peptide into β-sheet A and the subsequent cleavage of the reactive bond in the reactive loop peptide, resulting in the covalent attachment of the serpin to the side chain of the Ser residue in the catalytic triad of the serine protease.

The sequence of events with the corresponding constants is depicted in Figure 11.4, illustrated with the serpin $\alpha_1$-antitrypsin and the serine protease trypsin. The serpin (inhibitor, I) forms with the serine protease (E) a noncovalent Michaelis-like complex (EI). The substrate-like cleavage occurs via the formation of the acyl enzyme intermediate ($EI^{\#}$) and its subsequent hydrolysis, releasing the cleaved serpin ($I^*$) and the inactivated serine protease (E). The inhibition involves the insertion of the cleaved reactive loop centre into β-sheet A of the serpin and the covalently bound serine protease is dragged from the top to the bottom of the serpin, giving rise to a kinetically trapped covalent serpin–serine protease complex ($EI^+$).

## 11.3 BLOOD COAGULATION AND FIBRINOLYSIS

Many proteins in blood coagulation and fibrinolysis are present in plasma as zymogens that are activated by specific activators, usually by limited proteolysis. The active proteases have to be controlled and regulated very precisely, because they represent a potential danger for proteins in the fluid phase as well as for those proteins present on cell surfaces and tissues. A list of the main inhibitors involved in blood coagulation and fibrinolysis has already been compiled in Table 7.6 in Chapter 7. Many factors in blood coagulation and fibrinolysis are serine proteases and for almost all of them there are specific

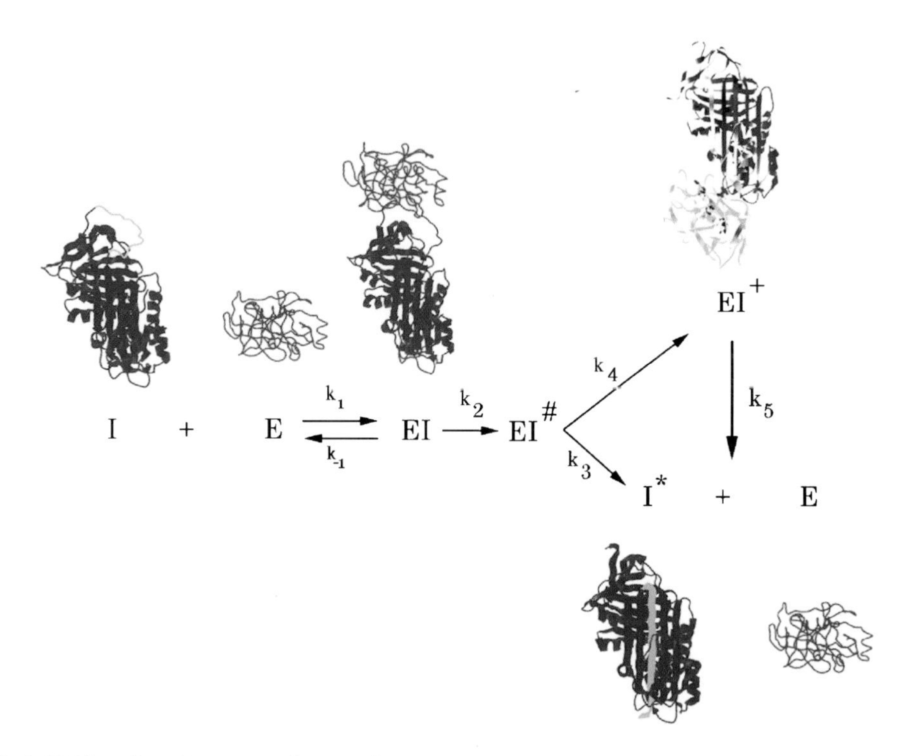


**Figure 11.4 Inhibition of a serine protease by a serpin**
The inhibition reaction is a two-stage process with a fast and reversible second-order reaction leading to a non-covalent Michaelis-like complex (**EI**) between the serpin (**I**) and the serine protease (**E**) followed by a slower, irreversible first-order reaction resulting in the cleaved serpin (**I***) and the inactivated serine protease (**E**). **EI**$^{\#}$: acyl enzyme intermediate. **EI**$^{+}$: kinetically trapped covalent serpin – serine protease complex. (Reproduced from Gettins PG. Genome Res 10: 1833–1835. Copyright (2000), with permission from Cold Spring Harbor Laboratory Press.)

physiological inhibitors, very often serpins. The concentration of a serpin is usually of the same order as the corresponding serine protease.

### 11.3.1 Blood coagulation

Probably the most important inhibitor in blood coagulation is the serpin antithrombin-III (ATIII), the main inhibitor of the key component of blood coagulation, thrombin, but ATIII also inhibits coagulation factors FIXa, FXa and FXIa. A unique heparin-binding site in ATIII modulates the heparin-enhanced ATIII–thrombin interaction, resulting in an approximately thousand-fold enhancement of the inhibitory activity (heparin cofactor activity).

Heparin cofactor II (HCII) is another serpin that specifically regulates the activity of thrombin. The rate of the HCII–thrombin complex formation is dramatically increased (approximately thousand-fold) by polyanions such as heparin or dermatan sulfate. In the presence of dermatan sulfate HCII becomes the predominant inhibitor of thrombin. HCII also inhibits chymotrypsin and cathepsin G.

$\alpha_1$-Antitrypsin ($\alpha_1$AT) is the main inhibitor of neutrophil elastase and is essential to protect the lung tissue from damage by elastase, but $\alpha_1$AT also inhibits components of blood coagulation such as thrombin, FXa, FXIa and protein C. The inhibition of trypsin, as the name might suggest, is only of minor importance. The 3D structure of $\alpha_1$AT is well studied and comprises eight $\alpha$-helices (in yellow–green) and three $\beta$-sheets (in blue), and the characteristic reactive loop peptide (in pink) containing the reactive peptide bond P1–P1′ (Met 358–Ser 359) is located on the surface and is shown in Figure 11.5.

Plasma protease C1 inhibitor (C1Inh), a serpin, is an important inhibitor in the early phases of the activation of the complement system as well as the activation of the blood coagulation cascade, where C1Inh inhibits coagulation factor FXIIa and kallikrein (for details see Section 11.4). Kallistatin, another serpin, inhibits tissue kallikrein and heparin prevents the complex formation of kallistatin with kallikrein.

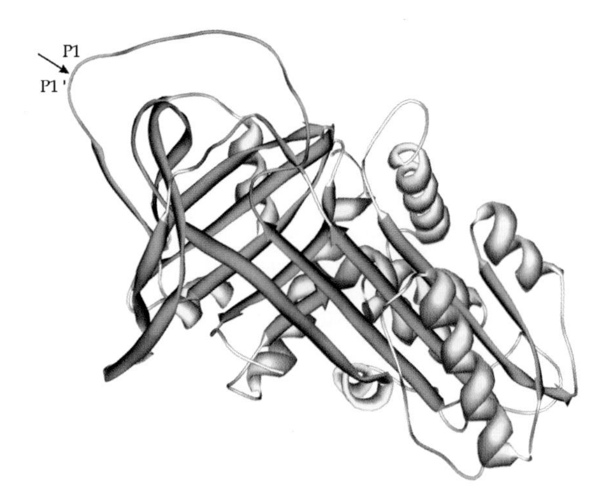


**Figure 11.5 3D structure of $\alpha_1$-antitrypsin (PDB: 1QLP)**
The 3D structure is characterised by eight α-helices (in yellow-green) and the three β-sheets (in blue) and the characteristic reactive loop peptide (in pink) containing the reactive peptide bond P1 – P1′ (Met 358 – Ser 359) is located on the surface.

Plasma serine protease inhibitor (PAI-3) is a major inhibitor of activated protein C, which is involved in the regulation of blood coagulation by selective inactivation of coagulation factors FVa and FVIIa. In addition, PAI-3 also inhibits plasma kallikrein, thrombin, coagulation factors FXa and FXIa and uPA.

Tissue factor pathway inhibitor (TFPI), a Kunitz-type inhibitor, is an important inhibitor of coagulation factor FXa and in an FXa-dependent way produces feedback inhibition of the TF-FVIIa complex. The second Kunitz-type domain is involved in FXa inhibition and the first Kunitz-type domain is required for the inhibition of the TF–FVIIa complex.

Protein Z-dependent protease inhibitor (PZI) is a serpin that is involved in the inhibition of coagulation factor FXa in the presence of protein Z, calcium ions and phospholipids.

**Antithrombin-III** (ATIII) (see Data Sheet) is a single-chain glycoprotein that belongs to the serpin family with the RCL peptide containing the reactive bond Arg 393–Ser 394, three disulfide bridges (Cys 8–Cys 128, Cys 21–Cys 95, Cys 247–Cys 430) that are essential for its activity and a heparin-binding site composed of a cluster of basic amino acids including Arg 47, Lys 125, Arg 129, Lys 136 and Arg 415.

The 3D structure of entire ATIII (432 aa: His 1–Lys 432) present as a dimer of one active and one inactive ATIII molecule was determined by X-ray diffraction (PDB: 1ANT) and is shown in Figure 11.6. The RCL peptide of active ATIII exhibits the predicted conformation with the expected initial entry of two residues into the main β-sheet. In inactive ATIII the reactive loop peptide (in pink) is completely incorporated into the main β-sheet, common in latent serpin conformations. The cleaved C-terminal region of ATIII is shown in red and the cleaved peptide bond in the RCL peptide is marked with pink circles. The mechanism of dimerisation is relevant in the polymerisation process in diseases associated with variant serpins.

Antithrombin-III deficiency is an important risk factor for hereditary thrombophilia, classified into four types:

(a) Type I: 50 % decrease in antigenic and functional levels;
(b) Type II: defects influencing the thrombin-binding domain;
(c) Type III: alterations of the heparin-binding domain;
(d) Type IV: mixed group of nonclassifiable mutations. (For details see MIM: 107300, 188050.)

**Heparin cofactor II** (HCII) (see Data Sheet) or protease inhibitor leuserpin 2 is a single-chain glycoprotein (10 % CHO) that belongs to the serpin family with no disulfide bridges but three sulfhydryl groups of unknown function and the RCL peptide containing the reactive bond Leu 444–Ser 445. The N-terminal region of HCII contains two hirudin-like repeats of approximately 13 aa (Asp/Glu-rich) each with an S-Tyr residue (aa 60 and 73) and the N-terminus is modified (most likely pyro-Glu). Several basic residues including Lys 173 and 185 and Arg 189, 192 and 193 are involved in the binding of polyanions such as heparin and dermatan sulfate.

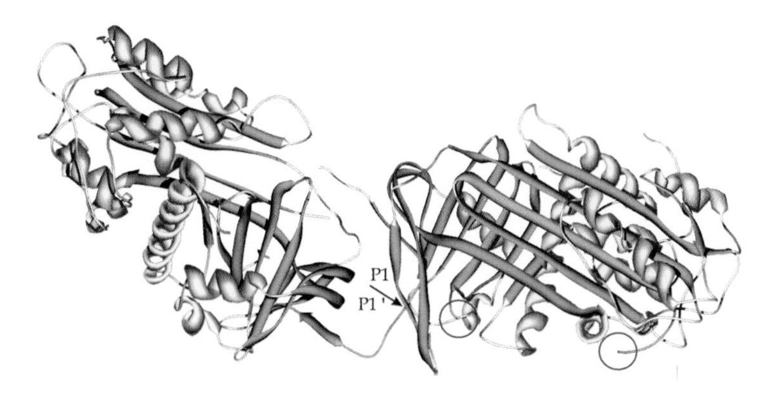


**Figure 11.6 3D structure of a dimer of an active and inactive antithrombin-III molecule (PDB: 1ANT)**
Active ATIII exhibits the predicted conformation with the exposed RCL peptide. In inactive ATIII the reactive loop peptide (in pink) is completely incorporated into the main β-sheet. The cleaved C-terminal region of ATIII is shown in red and the cleaved peptide bond in the RCL peptide is marked with pink circles.

The 3D structure of native HCII (480 aa: Gly 1–Ser 480) and of the HCII–thrombin complex was determined by X-ray diffraction (PDB: 1JMJ, 1JMO) and the complex is shown in Figure 11.7. The native structure of HCII resembles that of antithrombin III. The structure of the HCII–thrombin complex (thrombin in green, the RCL peptide in pink, with the reactive bond P1–P1′ (Leu 444–Ser 445) of HCII) together with native HCII reveals a multistep allosteric mechanism of inhibition that relies on sequential contraction and expansion of the central β-sheet in HCII.

Heparin cofactor II deficiency is an important risk factor for hereditary thrombophilia characterised by recurrent thrombosis and abnormal platelet aggregation (for details see MIM: 142360 and 188050).

**$\alpha_1$-Antitrypsin** ($\alpha_1$AT) (see Data Sheet) or $\alpha_1$-protease inhibitor is a single-chain glycoprotein (13 % CHO) that belongs to the serpin family with no disulfide bridges and a single sulfhydryl group of unknown function with the RCL peptide

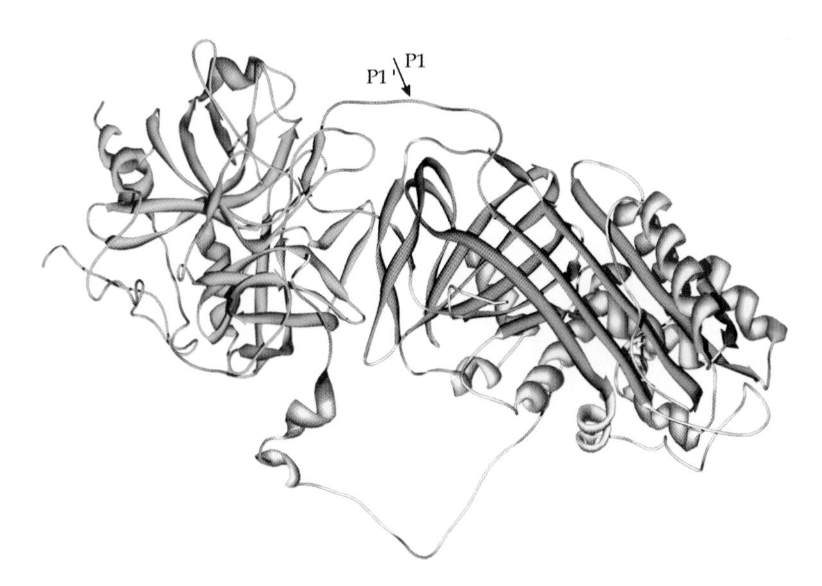


**Figure 11.7 3D structure of heparin cofactor II in complex with thrombin (PDB: 1JMJ, 1JMO)**
HCII exhibits the expected serpin structure with the exposed RCL peptide (in pink) containing the reactive bond P1 – P1′ (Leu 444 – Ser 445) and thrombin (in green).

containing the reactive bond (Met 358–Ser 359). $\alpha_1$AT is susceptible to inactivation by proteolytic cleavage near the reactive centre.

The 3D structure of entire $\alpha_1$-antitrypsin (394 aa: Glu1/Met–Lys 394) was determined by X-ray diffraction (PDB: 1QLP) and is shown in Figure 11.5. The structure reveals the classical features of serpins but is characterised by novel conformations in the flexible RCL peptide region (in pink) with the reactive bond P1–P1′: Met 358–Ser 359.

The major physiological function of $\alpha_1$AT is the protection of the lower respiratory tract against the proteolytic destruction by human leukocyte elastase. A hereditary deficiency of $\alpha_1$AT increases the risk of the development of chronic obstructive pulmonary disease by a factor of 20–30.

**Kallistatin** (see Data Sheet) or kallikrein inhibitor is a single-chain glycoprotein that belongs to the serpin family with no disulfide bridges and three sulfhydryl groups of unknown function with the RCL peptide containing an unique reactive bond P2–P1–P1′, Phe 367–Phe–Ser 369. The P1 position (Phe 368) is crucial for the specificity of kallistatin; the P2 position (Phe 367) is essential for the required hydrophobic surrounding in the formation of the kallistatin–tissue kallikrein complex. Kallistatin is susceptible to inactivation by proteolytic cleavage in the reactive centre and exhibits the typical serpin structure with eight helices and three long β-sheets.

**Plasma serine protease inhibitor** (PAI-3) (see Data Sheet) or plasminogen activator inhibitor-3 or protein C inhibitor is a single-chain glycoprotein that belongs to the serpin family with the RCL peptide containing the reactive bond Arg 354–Ser 355.

The 3D structure of PAI-3 cleaved in the RCL peptide (chain A (346 aa), Arg 11–Ala 356; chain B (31 aa), Arg 357–Pro 387) was determined by X-ray diffraction (PDB: 1LQ8) and is shown in Figure 11.8. The structure is typical for cleaved serpins with the cleaved regions of the RCL peptide on either side of the molecule (pink circles) and part of the RCL peptide inserted into β-sheet A (in pink); the cleaved chain B is shown in red.

**Tissue factor pathway inhibitor** (TFPI) (see Data Sheet) or lipoprotein-associated coagulation inhibitor is a single-chain Kunitz-type serine protease inhibitor with an acidic N-terminal region, three central BPTI/Kunitz inhibitor domains and a basic C-terminal region and a P-Ser residue in position 2. Each BPTI/Kunitz inhibitor domain contains 51 aa with three intrachain disulfide bridges arranged in the same way as in kringle domains, creating a triple-loop structure similar in structure to the kringle domains (for details see Chapter 5):

**Cys 1–Cys 6, Cys 2–Cys 4, Cys 3–Cys 5**

The 3D structure of the TFPI Kunitz II domain (coagulation factor FXa binding domain) was determined by X-ray diffraction (PDB: 1ADZ) (for details see Chapter 5).

**Protein Z-dependent protease inhibitor** (PZI) (see Data Sheet) is a single-chain glycoprotein that belongs to the serpin family with the RCL peptide containing the reactive bond Tyr 387–Ser 388. PZI has the typical structural features of a serpin.

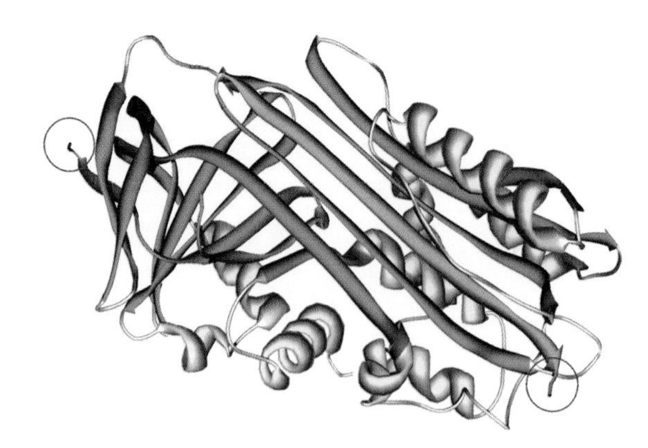

**Figure 11.8 3D structure of cleaved plasma serine protease inhibitor (PDB: 1LQ8)**
The cleaved regions of the RCL peptide are located on either side of the molecule (pink circles) and part of the RCL peptide is inserted into β-sheet A (in pink). The cleaved chain B is shown in red.

### 11.3.2 Fibrinolysis

The most important inhibitor of the key enzyme of the fibrinolytic system, plasmin, is the serpin $\alpha_2$-antiplasmin ($\alpha_2$AP). $\alpha_2$AP is a major serpin in plasma with a high affinity for plasmin(ogen). $\alpha_2$AP is the only serpin that contains an unique C-terminal extension of approximately 50 aa involved in the binding to the kringle domains of plasmin(ogen). In addition, $\alpha_2$AP also inhibits chymotrypsin where the reactive bond is moved by one position from Arg 364–Met 365 to Met 365–Ser 366.

Activators of plasminogen, tissue-type plasminogen activator (tPA) and urokinase-type plasminogen activator (uPA), also play an important role in the fibrinolytic system. It is therefore quite evident that specific physiological inhibitors also exist for these activators, plasminogen activator inhibitor-1 (PAI-1) and plasminogen activator inhibitor-2 (PAI-2), both serpins. PAI-1 primarily inhibits tPA and uPA, but it also inhibits thrombin, plasmin, protein C and trypsin. PAI-1 is biologically unstable and is prone to proteolytic cleavage in the reactive loop peptide. In plasma PAI-1 is bound to vitronectin (see Chapter 7), which stabilises the active conformation of PAI-1. PAI-2 is an efficient inhibitor of free as well as cell-surface-bound uPA and to a lesser extent also of tPA. Since uPA plays a key role in extracellular proteolytic processes such as tissue remodelling, cell migration, inflammation, cancer invasion and metastasis it is quite likely that PAI-2 as an efficient inhibitor of uPA acts as a regulator of these processes. During pregnancy the level of PAI-2 increases from normal concentration of $<10\ \mu g/l$ to levels as high as $300\ \mu g/l$.

Neuroserpin seems to protect neurons from cell damage by tPA via its inhibition. In addition, it also inhibits plasmin.

**$\alpha_2$-Antiplasmin** ($\alpha_2$AP) (see Data Sheet) or $\alpha_2$-plasmin inhibitor is a single-chain glycoprotein that belongs to the serpin family. The main form in plasma (Asn-form, 452 aa, 14 % CHO) contains one disulfide bridge (Cys 31–Cys 104) and two Cys of undefined state, the RCL peptide containing the reactive bond Arg 364–Met 365, the cross-linking site (Gln 2) that forms an isopeptide bond with the $\alpha$-chain of fibrinogen (Lys 303) and the plasminogen-binding region (Asn 398–Lys 452) with an S-Tyr residue (aa 445) of unknown function. The minor form in plasma still carries an N-terminal 12 aa propeptide (Met-form, 464 aa).

The deficiency of $\alpha_2$AP is the cause of severe hemorrhagic diathesis, an unusual susceptibility to bleeding (for details see MIM: 262850).

**Plasminogen activator inhibitor-1** (PAI-1) (see Data Sheet) is a single-chain glycoprotein (13 % CHO) that belongs to the serpin family with the RCL peptide containing the reactive bond Arg 346–Met 347.

The 3D structure of a mutated form (Ala335Glu) of PAI-1 (379 aa: Val 1–Pro 379) in complex with two inhibitory RCL pentapeptides (**Ac-TVASS-NH$_2$**) was determined by X-ray diffraction (PDB: 1A7C) and is shown in Figure 11.9. The two RCL pentapeptides (in pink) bind between the space of $\beta$-strands 3 and 5 in $\beta$-sheet A, thus preventing the insertion of the reactive-centre loop into $\beta$-sheet A and thereby abolishing the ability of the serpin to irreversibly inactivate the target enzyme. Otherwise, PAI-1 exhibits the typical structural features of cleaved serpins with the cleaved regions of the RCL peptide on either side of the molecule (pink circles). The cleaved C-terminal region is shown in red. Due to the lack of interpretable data the region of the RCL peptide was removed from the final structure.

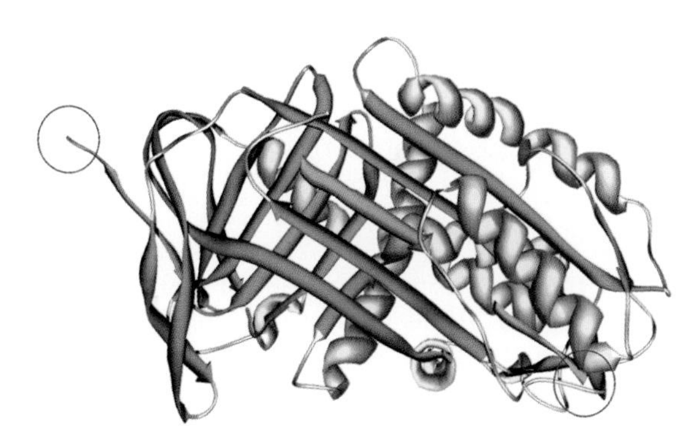

**Figure 11.9 3D structure of cleaved plasminogen activator inhibitor-1 in complex with two inhibitory RCL pentapeptides (PDB: 1A7C)**

PAI-1 exhibits the typical structure of cleaved serpins with the cleaved regions of the RCL peptide on either side of the molecule (pink circles). The cleaved C-terminal region is shown in red. The two RCL pentapeptides (in pink) are located between $\beta$-strands 3 and 5 in $\beta$-sheet A preventing the insertion of the reactive-centre loop into $\beta$-sheet A.

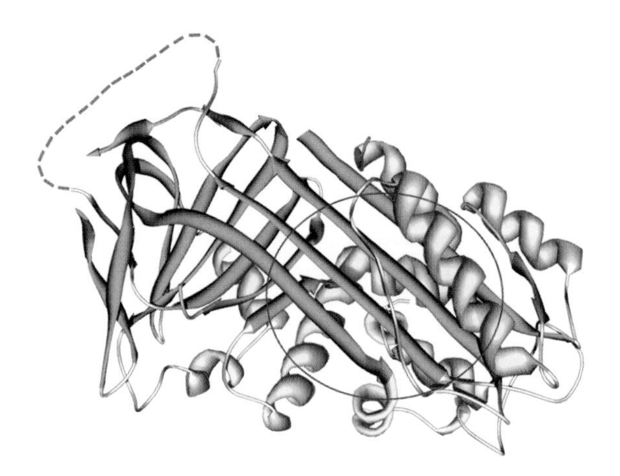

**Figure 11.10 3D structure of a deletion mutant of plasminogen activator inhibitor-2 (PDB: 1BY7)**
PAI-2 represents the stressed state of a serpin with a disordered structure of the RCL (reconstructed RCL peptide shown in grey dashed line). The shutter region is indicated by a pink circle.

The deficiency of PAI-1 is characterised by an abnormal bleeding tendency. Increased levels of PAI-1 are associated with thrombophilia (for details see MIM: 173360, 188050).

**Plasminogen activator inhibitor-2** (PAI-2) (see Data Sheet) is a single-chain glycoprotein that belongs to the serpin family with an uncleaved signal peptide, two potential intrachain disulfide bridges (Cys 5–Cys 405, second bridge unassigned) and the RCL peptide containing the reactive bond Arg 380–Thr 381.

The 3D structure of a deletion mutant of PAI-2 (382 aa: Met 1–Pro 415, deletion of loop Asn 66–Gln 98) was determined by X-ray diffraction (PDB: 1BY7) and is shown in Figure 11.10. The deletion mutant represents the stressed state of a serpin with a disordered structure of the RCL (reconstructed RCL peptide is shown in grey dashed line). A buried cluster of polar amino acids beneath β-sheet A, the 'shutter' region (pink circle), stabilises the stressed as well as the relaxed forms. The insertion of the RCL into β-sheet A is important for protease inhibition.

**Neuroserpin** (see Data Sheet) or protease inhibitor 12 is a single-chain glycoprotein that belongs to the serpin family with the RCL peptide containing the reactive bond Arg 346–Met 347. Neuroserpin has the typical structure of serpins.

Defects in neuroserpin are the cause of familial encephalopathy characterised by unique neuronal inclusion bodies with neuroserpin polymers (for details see MIM: 604218).

## 11.4 THE COMPLEMENT SYSTEM

Like blood coagulation and fibrinolysis the complement cascade has to be controlled and regulated. However, in contrast to blood coagulation and fibrinolysis the complement system is not mainly regulated by serpins or other protease inhibitors but by a series of other regulatory proteins (for details see Chapter 8). Nevertheless, the plasma protease C1 inhibitor (C1Inh), a serpin, is an important inhibitor in the early phases of the activation of the complement system, as well as the activation of the blood coagulation cascade. C1Inh is the only inhibitor with a high efficacy for the regulation of the activation of the classical pathway of the complement system. C1Inh regulates the activity of C1 by forming a covalent complex, C1Inh:C1r:C1s:C1Inh, with activated C1r and C1s. In addition, C1Inh inhibits FXIIa and kallikrein in the blood coagulation cascade.

**Plasma protease C1 inhibitor** (C1Inh) (see Data Sheet) or C1-inhibitor is a single-chain heavily glycosylated glycoprotein (at least 26 % CHO) that belongs to the serpin family with an approximately mass of 105 kDa with two disulfide bridges (Cys 108–Cys 183, Cys 101–Cys 406) and the RCL peptide containing the reactive bond Arg 444–Thr 445. In addition to the typical serpin part (365 aa), C1Inh contains an N-terminal 113 aa long, hydrophilic and rod-like domain that is heavily glycosylated and contains 14 tandem repeats of the tetrapeptide **[QE]-P-T-[TQ]**.

Defects in C1Inh are the cause of hereditary angioedema characterised by rapid swelling of subcutaneous and submucosal tissues. In the type I disorder (85 % of patients) the level of C1Inh is less than 35 %, while in type II the amounts are normal but the protein is nonfunctional (for details see MIM: 106100).

## 11.5 GENERAL INHIBITORS

In addition to the specific inhibitors for proteases there are general inhibitors in plasma such as $\alpha_2$-macroglobulin ($\alpha_2$M) and pregnancy zone protein (PZP) that can regulate the activity of all four types of proteases: serine proteases, cysteine proteases, aspartic proteases and metallo proteases. Together with the other protease inhibitors they control the activity of proteases in circulation and tissues. $\alpha_2$M and PZP act as scavengers for proteases; limited proteolysis in the 'bait' region of $\alpha_2$M at specific sites (cleavage sites for each type of protease) by the active protease induces a large conformational change, resulting in the trapping of the protease in a large central cavity (trapping mechanism). The trapping mechanism of $\alpha_2$M is depicted in Figure 11.11. As a consequence, the internal thioester bond is cleaved, which mediates covalent binding of $\alpha_2$M to the protease. The protease is still active towards low molecular weight substrates but is protected from reaction with high molecular weight substrates and inhibitors.

**$\alpha_2$-Macroglobulin** ($\alpha_2$M) (see Data Sheet) is a large single-chain glycoprotein (10 % CHO) comprising 8–10 % of the total serum protein content. $\alpha_2$M belongs to the protease inhibitor I39 family and contains eleven intrachain and two interchain disulfide bridges. $\alpha_2$M is present in plasma as a tetramer with an approximate mass of 720 kDa: two monomers are linked by two interchain disulfide bridges (Cys 447–Cys 540, Cys 540–Cys 447) with an antiparallel orientation and two monomers are noncovalently associated. $\alpha_2$M contains a 'bait' region of 39 aa (Pro 667–Thr 705) with three inhibitory regions (Arg 681–Glu 686, Arg 696–Val 700, Thr 707–Phe 712) containing specific cleavage sites for the four types of proteases. Like complement components C3 and C4 and PZP, $\alpha_2$M contains a reactive isoglutamyl cysteine thioester bond (**Cys 949–Gly–Glu–Gln 952**) and, in addition, Gln 670 and Gln 671 are potential cross-linking sites with the side chains of Lys residues in other proteins. Cleavage in the 'bait' region induces a large conformational change followed by hydrolysis of the thioester bond and covalent binding to the protease, resulting in the elimination from the circulation within minutes. $\alpha_2$M exhibits a high similarity to PZP (approximately 70 % sequence identity) and to a lower extent to complement components C3, C4 and C5.

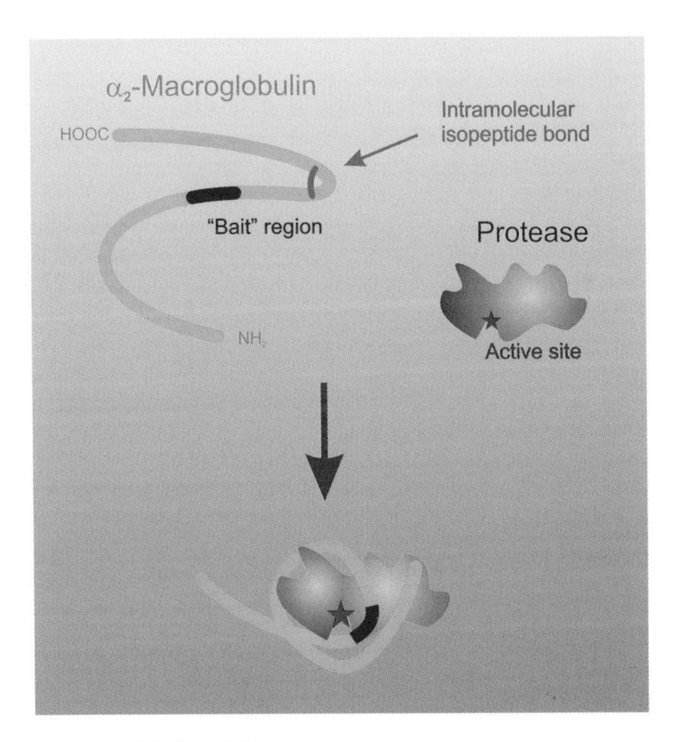


**Figure 11.11 Inhibition by $\alpha_2$-macroglobulin ($\alpha_2$M)**

The inhibitory reaction of $\alpha_2$M is characterised by a trapping mechanism. Cleavage in the 'bait' region of $\alpha_2$M by an active protease leads to a large conformational change in $\alpha_2$M resulting in the trapping of the active protease followed by the hydrolysis of the thioester bond and the covalent binding to the protease.

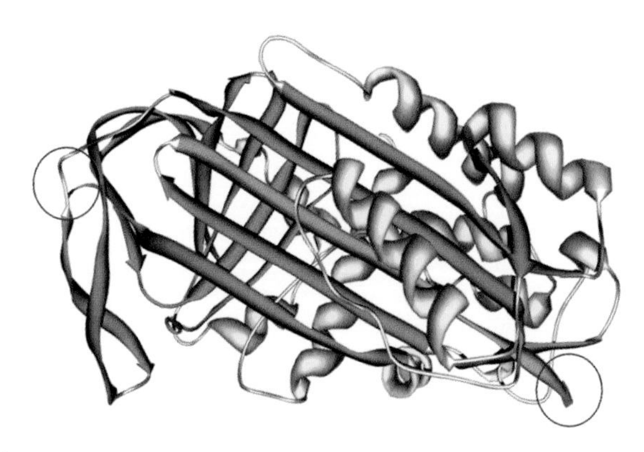

**Figure 11.12 3D structure of cleaved α-antichymotrypsin (PDB: 2ACH)**
The RCL peptide is inserted as the middle strand in β-sheet A (in pink). The cleaved regions of the RCL peptide are on either side of the molecule (pink circles) and the cleaved C-terminal region (40 aa) is shown in red.

**Pregnancy zone protein** (PZP) (see Data Sheet) is a large single-chain glycoprotein that belongs to the protease inhibitor I39 family. PZP contains a 'bait' region of 51 aa (Cys 660–Glu 710) with specific cleavage sites for the four types of proteases. Like complement components C3 and C4 and $\alpha_2$M, PZP contains a reactive isoglutamyl cysteine thioester bond (**Cys 953–Gly–Glu–Gln 956**). PZP exhibits a high similarity to $\alpha_2$M (approximately 70 % sequence identity) and to a lower extent to complement components C3, C4 and C5.

## 11.6 OTHER SERPINS

There are several important serpins in plasma that are not directly assigned to a certain cascade and are briefly discussed in this part of the chapter. $\alpha_1$-Antichymotrypsin (ACT) is a major serpin in plasma that inhibits various proteases, primarily chymotrypsin, cathepsin G, chymase and prostate specific antigen. ACT is susceptible to inactivation by proteolytic cleavage in the exposed RCL peptide. Leukocyte elastase inhibitor (LEI), a serpin, regulates primarily the activity of leukocyte elastase and cathepsin G.

**$\alpha_1$-Antichymotrypsin** (ACT) (see Data Sheet) is a single-chain glycoprotein (25 % CHO) that belongs to the serpin family with an approximate mass of 68 kDa and a single Cys residue (aa 236) of unknown function and the RCL peptide containing the reactive bond Leu 360–Ser 361. A form exists that is devoid of the N-terminal dipeptide His–Pro.

The 3D structure of cleaved ACT (360 aa: His 1–Leu 360; 40 aa: Ser 361–Ala 400) was determined by X-ray diffraction (PDB: 2ACH) and is shown in Figure 11.12. The structure is very similar to cleaved $\alpha_1$-antitrypsin and after cleavage of the RCL peptide the loop is inserted as the middle strand in β-sheet A (in pink). The cleaved regions of the RCL peptide are on either side of the molecule (pink circles) and the cleaved C-terminal region (40 aa) is shown in red.

The concentration of ACT in serum is about 10 % of that of $\alpha_1$-antitrypsin and defects in ACT are the cause of chronic obstructive pulmonary disease (for details see MIM: 107280).

**Leukocyte elastase inhibitor** (LEI) (see Data Sheet) is a single-chain protein that belongs to the serpin family with an N-terminal acetyl-Met residue and the RCL peptide containing the reactive bond Cys 344–Met 345. LEI lacks a typical cleavable N-terminal signal sequence.

## 11.7 OTHER INHIBITORS

Other types of inhibitors exist that have not been discussed so far. Some of them are only present in plasma in trace amounts. However, their function justifies a brief discussion in this part of the chapter.

Pancreatic secretory trypsin inhibitor (PSTI) is a specific inhibitor of trypsin in the pancreas and the pancreatic juices and is present in plasma only in traces. PSTI prevents the autoactivation of trypsinogen and as a consequence prevents the pancreas from autodigestion. PSTI inhibits the enzymatic activity of trypsin by complex formation. In a second step the Arg 42–Lys 43 peptide bond in PSTI is cleaved by trypsin, but PSTI remains bound to trypsin and retains its inhibitory activity. In a third step the Arg 44–Gln 45 peptide bond is cleaved, releasing the Lys–Arg dipeptide and resulting in a conformational change of PSTI and the disruption of the PSTI–trypsin complex.

Inter-$\alpha$-trypsin inhibitor light chain or bikunin circulates in plasma in complex with inter-$\alpha$-trypsin inhibitor heavy chains H1 and H2. Bikunin is synthesised from the same mRNA in the form of the precursor protein $\alpha_1$-microglobulin–bikunin. Proteolytic cleavage in the central region of the precursor molecule releases a basic tetrapeptide RVRR with the concomitant secretion of $\alpha_1$-microglobulin (for details see Chapter 9) and bikunin. Bikunin binds with a lower affinity to proteases than other protease inhibitors and exhibits inhibitory activity for trypsin, chymotrypsin, plasmin, neutrophil elastase and cathepsin G.

Together with bikunin, inter-$\alpha$-trypsin inhibitor heavy chains H1, H2, H3 and H4 form several types of inhibitors:

1. Inter-$\alpha$-plasma protease inhibitors are composed of one or two heavy chains (H1, H2, H3) and the light chain (bikunin).
2. Inter-$\alpha$-inhibitor contains H1, H2 and bikunin.
3. Inter-$\alpha$-like inhibitor is composed of H2 and bikunin.
4. Pre-$\alpha$-inhibitor contains H3 and bikunin.

Cystatins are thiol protease inhibitors that are present in many different types of tissues and in body fluids and to some extent also in plasma. The most important cystatins are compiled in Table 11.1 and their typical characteristics are given.

Cystatin A and cystatin B belong to the family 1 of the cystatin superfamily containing no disulfide bridges. The other cystatins contain two invariant disulfide bridges in the C-terminal region of the molecule with the disulfide bridge pattern **Cys 1–Cys 2, Cys 3–Cys 4** and belong to the cystatin family 2. Cystatins inhibit cysteine proteases of the papain type such as cathepsin B, H, L and S. Inhibition of cysteine proteases by cystatins is thought to be a tight but reversible binding, forming an equimolar enzyme–inhibitor complex.

**Pancreatic secretory trypsin inhibitor** (PSTI) (see Data Sheet) is a small single-chain Kazal-type serine protease inhibitor with a Kazal-like domain containing three intrachain disulfide bridges (Cys 9–Cys 38, Cys 16–Cys 35, Cys 24–Cys 56) and the reactive site Lys 18–Ile 19 (for details see Chapter 5). The Kazal-like domain exhibits structural similarity with the EGF-like domain, but their exon–intron junctions are not related and both domains seem not to have evolved from a common ancestral gene.

The 3D structure of a mutant recombinant PSTI (56 aa: Asp 1–Cys 56) was determined by X-ray diffraction (PDB: 1HPT) and is shown in Figure 11.13. PSTI is characterised by a central $\alpha$-helix and a short antiparallel $\beta$-sheet and the structure is stabilised by three disulfide bridges (in pink). The reactive site region (Lys18-lee19) is shown in green. Differences with other known PSTI structures mainly occur within the flexible N-terminal region.

Defects in PSTI are the cause of hereditary or chronic pancreatitis characterised by the presence of calculi (stones) in the pancreatic ducts that cause severe abdominal pain attacks (for details see MIM: 167800).

**Inter-$\alpha$-trypsin inhibitor light chain** (see Data Sheet) or bikunin is a single-chain Kunitz-type protease inhibitor glycoprotein containing two BPTI/Kunitz inhibitor domains each with three intrachain disulfide bridges (for details see Chapter 5). Together with $\alpha_1$-microglobulin, bikunin is processed and released from the parent diprotein AMBP (see also Chapter 9). Bikunin is linked to the inter-$\alpha$-trypsin inhibitor heavy chains 1 and 2 by chondroitin-4-sulfate.

**Table 11.1** Characteristics of various cystatins.

| Type | Code | Length (aa) | Reactive areas | Comments |
|---|---|---|---|---|
| A | CYTA_HUMAN/P01040 | 98 | Gly 4, Gln 46–Gly 50 | No Cys; cytoplasmic; traces plasma |
| B | CYTB_HUMAN/P04080 | 98 | Gly 4, Gln 46–Gly 50 | Free Cys 3; *N*-Ac-Met (aa 1); dimerisation; cytoplasmic; $\sim$0.6 mg/l plasma |
| C | CYTC_HUMAN/P01034 | 120 | Gly 11, Gln 55–Gly 59 | Cys 73–Cys 83, Cys 97–Cys 117; OH-Pro (aa 3, partial) $\sim$1.1 mg/l plasma |
| D | CYTD_HUMAN/P28325 | 122 | Gly 12, Gln 50–Ala 54 | Cys 75–Cys 85, Cys 99–Cys 119; $<0.1$ mg/l plasma |
| S | CYTS_HUMAN/P01036 | 121 | Gly 12, Gln 56–Gly 60 | Cys 74–Cys 84, Cys 98–Cys 118; salivary |
| SA | CYTT_HUMAN/P09228 | | | |
| SN | CYTN_HUMAN/P01037 | | | |
| F | CYTF_HUMAN/O76096 | 126 | Gly 18, Gln 62–Gly 66 | Cys 80–Cys 91, Cys 105–Cys 125; secreted, cytoplasmic |
| M | CYTM_HUMAN/Q15828 | 121 | Gly 8, Gln 52–Gly 56 | Cys 70–Cys 85, Cys 98–Cys 118; secreted |
| 11 | CST11_HUMAN/Q9H112 | 112 | Xaa ?, Gln 50–His 54 | Cys 68–Cys 76, Cys 89–Cys 109; cytoplasmic |

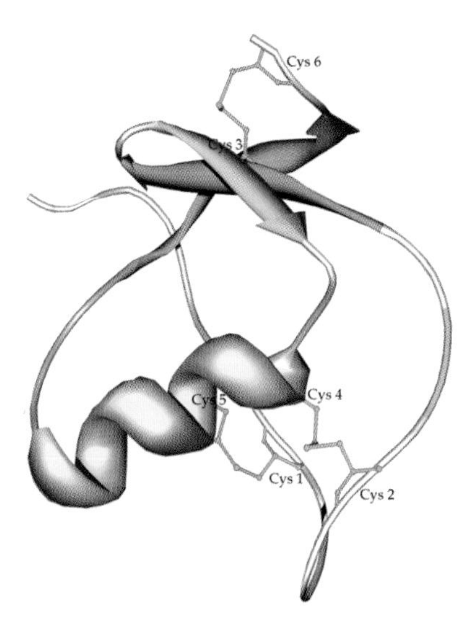


**Figure 11.13 3D structure of a mutant recombinant pancreatic secretory trypsin inhibitor (PDB: 1HPT)**
PSTI is characterised by a central α-helix and a short antiparallel β-sheet and the structure is stabilised by three disulfide bridges (in pink).

The 3D structure of **bikunin** (147 aa: Ala 1–Asn 147) was determined by X-ray diffraction (PDB: 1BIK) and is shown in Figure 11.14. The two domains (in green and yellow) are very similar to each other and to other proteins containing the BPTI/Kunitz inhibitor domain. Each domain is stabilised by three disulfide bridges (in pink and blue) arranged in the pattern **Cys 1–Cys 6, Cys 2–Cys 4, Cys 3–Cys 5** (for details see Chapter 5).

**Inter-α-trypsin inhibitor heavy chains H1, H2, H3 and H4** (see Data Sheet) are single-chain glycoproteins that belong to the ITIH family each containing one VWFA domain. They represent the heavy chain part of several inhibitors and exhibit structural and functional similarity to each other. H1 contains an unusual S-linked glycan (Cys 26). After proteolytic processing of the C-terminal propeptides in the precursor proteins the H1, H2 and H3 heavy chains are

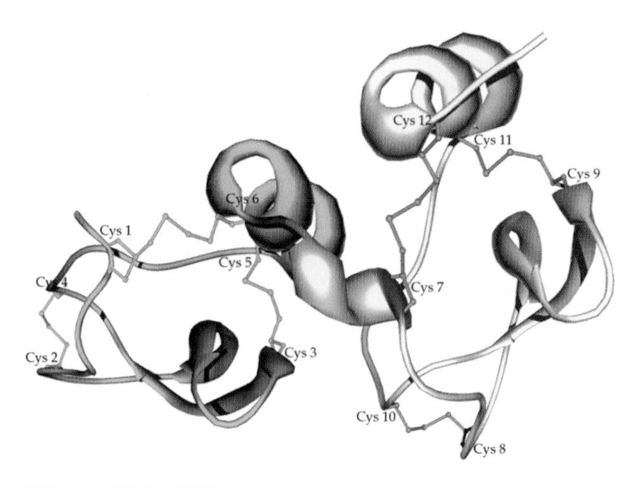


**Figure 11.14 3D structure of bikunin (PDB: 1BIK)**
The two BPTI/Kunitz inhibitor domains are very similar in structure and each domain is stabilised by three disulfide bridges (in pink and blue).

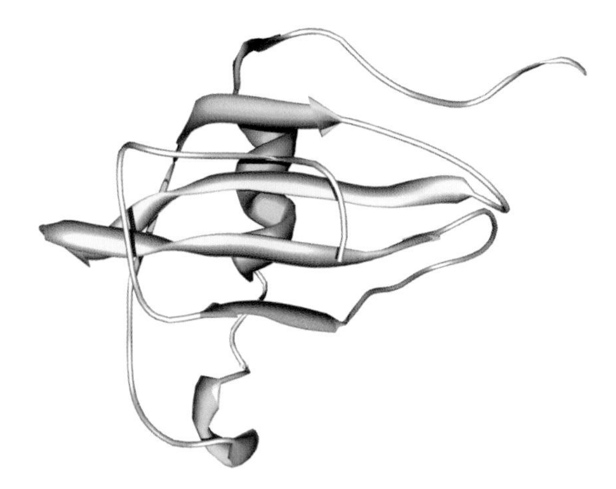

**Figure 11.15 3D structure of recombinant cystatin A (PDB: 1DVC)**
The structure consists of five antiparallel β-strands wrapped around a central five-turn α-helix.

covalently linked via the α-carboxyl group of their C-terminal Asp residue to chondroitin 4-sulfate at Ser 10 in bikunin, the inter-α-trypsin inhibitor light chain.

**Cystatin A** or stefin A and **Cystatin B** or stefin B (see Data Sheet) are single-chain cysteine protease inhibitors with Gly 4 as essential residue for the inhibitory activity and the sequence Gln 46–Val–Val–Ala–Gly 50 is important as a contact area for the activity. Cystatin A and B contain no disulfide bridges but cystatin B contains one free Cys and is capable of forming dimers. Cystatin A and B keep their biological activity over a wide pH range (pH 2–11) and cystatin A is stable at elevated temperatures (100 °C, 10 min). They specifically form tight but reversible equimolar enzyme–inhibitor complexes with cysteine proteases of the papain type, such as cathepsin B, H, L and S, but they do not inhibit any other types of proteases.

The 3D solution structure of recombinant cystatin A (98aa: Met 1–Phe 98) was determined by NMR spectroscopy (PDB: 1DVC) and is shown in Figure 11.15. The structure has a well-defined fold consisting of five antiparallel β-strands wrapped around a central five-turn α-helix. There is a considerable similarity to cystatin B.

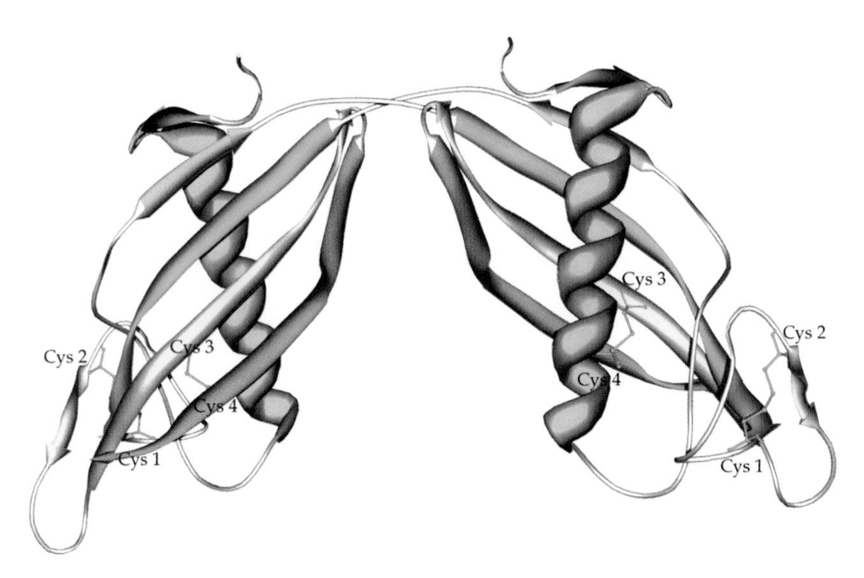


**Figure 11.16 3D structure of cystatin C as a dimer (PDB: 1G96)**
Each monomer is characterised by five antiparallel β-strands stabilised by two disulfide bridges (in pink) and a single detached α-helix.

Defects in cystatin B are the cause of progressive myoclonus epilepsy characterised by brief shock-like jerks of muscle(s) and convulsions with an early onset at the age of 6–13 years (for details see MIM: 254800).

**Cystatin C** (see Data Sheet) is a single-chain cysteine protease inhibitor containing two disulfide bridges (Cys 73–Cys 83, Cys 97–Cys 117) and a partially hydroxylated Pro residue (aa 3). Gly 11 is essential for its inhibitory activity and the sequence Gln 55–Ile–Val–Ala–Gly 59 is important as a contact area.

The 3D structure of dimeric cystatin C (120aa: Ser 1–Ala 120) was determined by X-ray diffraction (PDB: 1G96) and is shown in Figure 11.16. The structure of each monomer is characterised by five antiparallel β-strands stabilised by two disulfide bridges (in pink) and a single detached α-helix. Cystatin C has the tendency to refold in very tight two-fold symmetric dimers.

Hereditary cystatin C amyloid angiopathy, also called hereditary cerebral hemorrhage with amyloidosis, is characterised by the deposition of cystatin C amyloid in tissues, particularly in the cerebral vasculature. The disease is caused by a mutation of Leu 68 → Gln in the cystatin C polypeptide chain (for details see MIM: 105150).

The multifunctional kininogens, parent protein of the vasoactive nonapeptide bradykinin, is a cysteine protease inhibitor containing three consecutive cystatin-like type 2 domains (see Chapter 7). The cystatin structure is characterised by the following consensus sequence:

CYSTATIN (PS00287): [GSTEQKRV]–Q–[LIVT]–[VAF]–[SAGQ]–G–{DG}–[LIVMNK]–{TK}–x–[LIVMFY]
–{s}–[LIVMFYA]–[DENQKRHSIV]

## REFERENCES


Abrahamson *et al.*, 2003, Cystatins, *Biochem. Soc. Symp.*, **70**, 179–199.

Akerstroem *et al.*, 2000, Alpha(1)-microglobulin: a yellow–brown lipocalin, *Biochim. Biophys. Acta*, **1482**, 172–184.

Borth, 1992, Alpha 2-macroglobulin a multifunctional binding protien with targeting characteristics, *FASEB J*, **6**, 3345–3353.

Coughlin, 2005, Antiplasmin. The forgotten serpin?, *FEBS J.*, **272**, 4852–4857.

Gettins, 2000, Keeping the serpin machine running smoothly, *Genome*, **10**, 1833–1835.

Huntington *et al.*, 2000, Structure of a serpin-protease complex shows inhibition by deformation, *Nature*, **407**, 923–926.

Janciauskine, 2001, Conformational properties of serine protease inhibitors (serpins) confer multiple pathophysiological roles, *Biochim. Biophys. Acta.*, **1535**, 221–235.

Pike *et al.*, 2005, Control of the coagulation system by serpins. Getting by with a little help from glycosaminoglycans, *FEBS J.*, **272**, 4842–4851.

Rawlings *et al.*, 1990, Evolution of proteins of the cystatin superfamily, *J. Mol. Evol.*, **30**, 60–71.

Silverman *et al.*, 2001, The serpins are an expanding superfamily of structurally similar but functionally diverse proteins, *J. Biol. Chem.*, **276**, 33293–33296.

Stein and Carrell, 1995, What do dysfunctional serpins tell us about molecular mobility and disease?, *Nat. Struct. Biol.*, **2**, 96–113.

van Gent *et al.*, 2003, Serpins: structure, function and molecular evolution, *Int. J. Biochem. Cell. Biol.*, **35**, 1536–1547.

Whisstock *et al.*, 1998, An atlas of serpin conformations, *Trends Biochem. Sci.*, **23**, 63–67.

Whisstock *et al.*, 2005, Serpins 2005–fun between the β-sheets, *FEBS J.*, **272**, 4868–4873.

DATA SHEETS

# $\alpha_1$-*Antichymotrypsin*

## Fact Sheet

| | | | |
|---|---|---|---|
| *Classification* | Serpin | *Abbreviations* | ACT |
| *Structures/motifs* | | *DB entries* | AACT_HUMAN/P01011 |
| *MW/length* | 45 266 Da/400 aa | *PDB entries* | 2ACH |
| *Concentration* | 300–600 mg/l | *Half-life* | |
| *PTMs* | 3 N-CHO | | |
| *References* | Baumann *et al.*, 1991, *J. Mol. Biol.*, **218**, 595–606.<br>Chandra *et al.*, 1983, *Biochemistry*, **22**, 5055–5061. | | |

## Description

**$\alpha_1$-Antichymotrypsin** (ACT) is a serpin predominantly synthesised in the liver that primarily inhibits chymotrypsin and cathepsin G and exhibits structural and functional similarity with other serpins.

## Structure

ACT (400 aa) is a single-chain plasma glycoprotein containing a 26 residue RCL peptide (Gly 346–Arg 371) with the reactive bond Leu 360–Ser 361. A 398 aa form exists that is devoid of the N-terminal dipeptide His–Pro. The 3D structure of cleaved ACT is similar to cleaved $\alpha_1$-antitrypsin, with the cleaved strand inserted into β-sheet A (in pink). The cleaved RCL peptide regions are on either side of the molecule (pink circles) and the cleaved C-terminal region is shown in red (**2ACH**).

## Biological Functions

ACT inhibits primarily the serine proteases chymotrypsin, cathepsin G, chymase and prostate specific antigen. ACT is susceptible to inactivation by proteolytic cleavage in the reactive loop peptide.

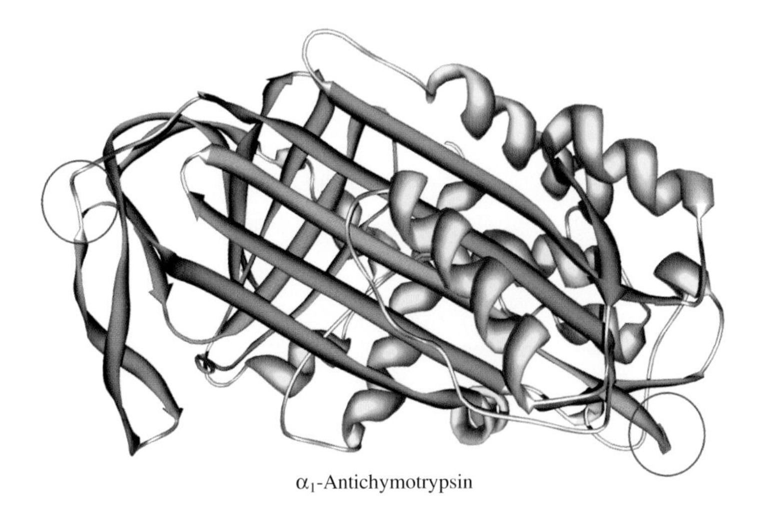

$\alpha_1$-Antichymotrypsin

## *$\alpha_2$-Antiplasmin ($\alpha_2$-Plasmin inhibitor)*

### Fact Sheet

| | | | |
|---|---|---|---|
| ***Classification*** | Serpin | ***Abbreviations*** | $\alpha_2$AP; $\alpha_2$PI |
| ***Structures/motifs*** | | ***DB entries*** | A2AP_HUMAN/P08697 |
| ***MW/length*** | 50 451 Da/452 aa | ***PDB entries*** | |
| ***Concentration*** | approx. 70 mg/l | ***Half-life*** | 2.6–3.3 days |
| ***PTMs*** | 4 N-CHO; 1 S-Tyr (aa 445) | | |
| ***References*** | Tone *et al.*, 1987, *J. Biochem.*, **102**, 1033–1041. | | |

### Description

**$\alpha_2$-Antiplasmin** ($\alpha_2$AP) is a serpin synthesised in the liver that primarily inhibits plasmin and exhibits structural and functional similarity with other serpins.

### Structure

The main form of $\alpha_2$AP (Asn-form, 452 aa) is a single-chain plasma glycoprotein (14 % CHO) containing the RCL peptide with the reactive bond Arg 364–Met 365, the cross-linking site (Gln 2) forming an isopeptide bond with the $\alpha$-chain of fibrinogen (Lys 303) and the plasminogen-binding region (Asn 398–Lys 452) with an S-Tyr residue (445 aa). About one-third of $\alpha_2$AP in plasma still carries the 12 residue propeptide (Met-form, 464 aa).

### Biological Functions

$\alpha_2$AP is the main inhibitor of the fibrinolytic system inhibiting plasmin in a two-stage reaction: a fast, reversible second-order reaction followed by a slower, irreversible first-order reaction. $\alpha_2$AP also inhibits chymotrypsin (reactive bond: Met 365–Ser 366).

# *Antithrombin-III*

## Fact Sheet

| | | | |
|---|---|---|---|
| ***Classification*** | Serpin | ***Abbreviations*** | ATIII |
| ***Structures/motifs*** | | ***DB entries*** | ANT3_HUMAN/P01008 |
| ***MW/length*** | 49 039 Da/432 aa | ***PDB entries*** | 1ANT |
| ***Concentration*** | 115–160 mg/l | ***Half-life*** | about 3 days |
| ***PTMs*** | 4 N-CHO | | |
| ***References*** | Bock *et al.*, 1982, *Nucleic Acids Res.*, **10**, 8113–8125.<br>Carrell *et al.*, 1994, *Structure*, **2**, 257–270. | | |

## Description

**Antithrombin-III** (ATIII) is the most important serpin in plasma synthesised in the liver and primarily inhibits thrombin and exhibits structural and functional similarity with other serpins.

## Structure

ATIII is a single-chain plasma glycoprotein with the reactive bond Arg 393–Ser 394, three disulfide bridges (Cys 8–Cys 128, Cys 21–Cys 95, Cys 247–Cys 430) that are essential for its activity and a heparin-binding site. The 3D of a dimer containing an active and an inactive ATIII molecule exhibits the expected conformation (**1ANT**).

## Biological Functions

ATIII is the most important serpin in plasma and regulates the coagulation cascade by inhibition of thrombin and also coagulation factors FIXa, FXa and FXIa. In the presence of heparin its inhibitory activity is greatly enhanced (up to a thousand-fold).

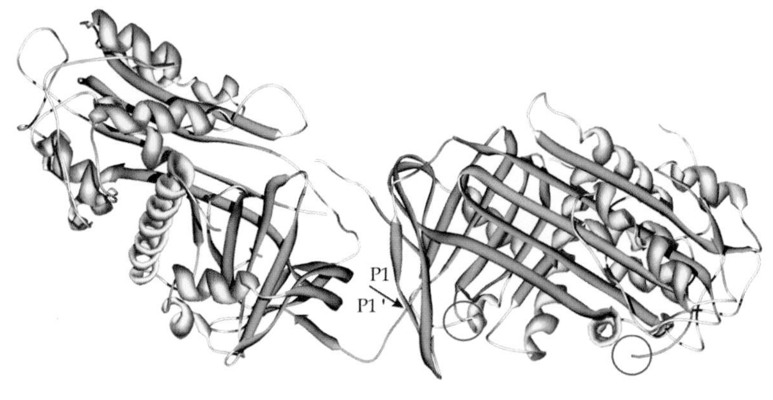


Antithrombin III dimer

## *$\alpha_1$-Antitrypsin ($\alpha_1$-Protease inhibitor)*

### Fact Sheet

| | | | |
|---|---|---|---|
| ***Classification*** | Serpin | ***Abbreviations*** | $\alpha_1$AT; $\alpha_1$PI |
| ***Structures/motifs*** | | ***DB entries*** | A1AT_HUMAN/P01009 |
| ***MW/length*** | 44 325 Da/394 aa | ***PDB entries*** | 1QLP |
| ***Concentration*** | 1300 mg/l | ***Half-life*** | 4–5 days |
| ***PTMs*** | 3 N-CHO | | |
| ***References*** | Elliot *et al.*, 1998, *J. Mol. Biol.*, **275**, 419–425.<br>Jacobs *et al.*, 1983, *DNA*, **2**, 255–264. | | |

### Description

**$\alpha_1$-Antitrypsin** ($\alpha_1$AT) is a serpin synthesised in the liver that primarily inhibits elastase and exhibits structural and functional similarity with other serpins.

### Structure

$\alpha_1$AT is a single-chain plasma glycoprotein (13 % CHO) containing a 25 residue RCL peptide (Gly 344–Lys 368) with the reactive bond Met 358–Ser 359. The 3D structure reveals the characteristic features of serpins with eight $\alpha$-helices and three $\beta$-sheets and the reactive loop peptide is shown in pink (**1QLP**).

### Biological Functions

$\alpha_1$AT inhibits the serine protease elastase and thus protects the lung from proteolytic damage. To a lesser extent it also inhibits other serine protease like plasmin and thrombin. $\alpha_1$AT is susceptible to inactivation by proteolytic cleavage near the reactive centre.

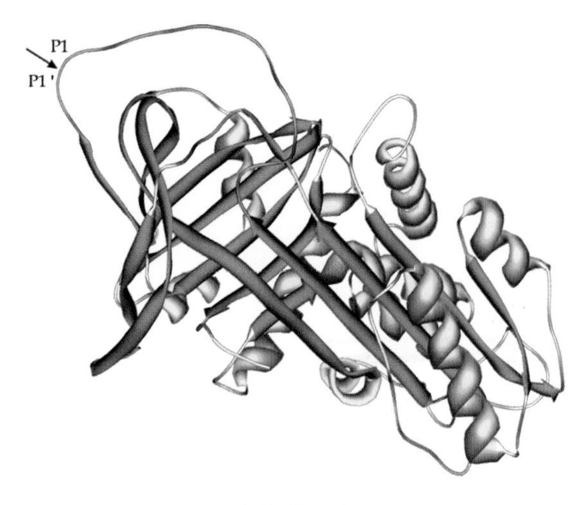


$\alpha_1$-Antitrypsin

# *Cystatin A (Stefin A)/Cystatin B (Stefin B)*

## Fact Sheet

| | | | |
|---|---|---|---|
| ***Classification*** | Cystatin | ***Abbreviations*** | |
| ***Structures/motifs*** | | ***DB entries*** | CYTA_HUMAN/P01040<br>CYTB_HUMAN/P04080 |
| ***MW/length*** | Cystatin A: 11 006 Da/98 aa<br>Cystatin B: 11 140 Da/98 aa | ***PDB entries*** | 1DVC |
| ***Concentration*** | Cystatin A: traces in plasma;<br>Cystatin B: approx. 0.6 mg/l | ***Half-life*** | |
| ***PTMs*** | Cystatin B: N-Ac-Met (aa 1, probable) | | |
| ***References*** | Machleidt *et al.*, 1983, *Hopp-Seyler's Z. Physiol. Chem.*, **364**, 1481–1486.<br>Martin *et al.*, 1995, *J. Mol. Biol.*, **246**, 331–343.<br>Ritonja *et al.*, 1985, *Biochem. Biophys. Res. Commun.*, **131**, 1187–1192. | | |

## Description

**Cystatin A** and **B** are intracellular thiol protease inhibitors and belong to the family I of the cystatin superfamily.

## Structure

Cystatin A and B are single-chain cysteine protease inhibitors without disulfide bridges, where Gly 4 is essential for their inhibitory activity and a 5 residue sequence (Gln 46–Gly 50) is important as the contact area. The 3D structure of cystatin A is characterised by five antiparallel β-strands wrapped around a central α-helix (**1DVC**).

## Biological Functions

Cystatin A and B inhibit cysteine proteases of the papain type and are thought to form a tight but reversible 1:1 (molar) enzyme–inhibitor complex. They inhibit cysteine proteases such as cathepsin B, H, L and S.

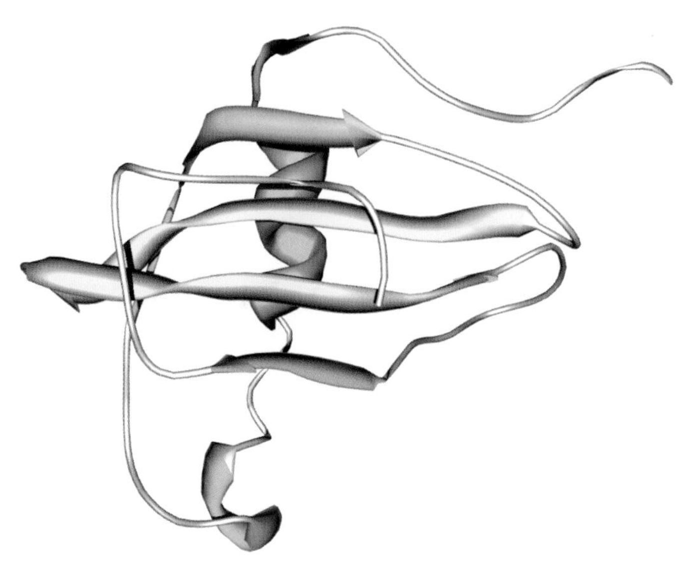

Cystatin A

# Cystatin C

## Fact Sheet

| | | | |
|---|---|---|---|
| ***Classification*** | Cystatin | ***Abbreviations*** | |
| ***Structures/motifs*** | | ***DB entries*** | CYTC_HUMAN/P01034 |
| ***MW/length*** | 13 347 Da/120 aa | ***PDB entries*** | 1G96 |
| ***Concentration*** | 1.1 mg/l | ***Half-life*** | 2 min |
| ***PTMs*** | OH-Pro (aa 3, partial) | | |
| ***References*** | Grub *et al.*, 1982, *Proc. Natl. Acad. Sci. USA*, **79**, 3024–3027. | | |
| | Janowski *et al.*, 2001, *Nat. Struct. Biol.*, **8**, 316–320. | | |

## Description

**Cystatin C** is a thiol protease inhibitor synthesised in almost all cell types and belongs to the family 2 of the cystatin superfamily.

## Structure

Cystatin C is a single-chain cysteine protease inhibitor containing two disulfide bridges (Cys 73–Cys 83, Cys 97–Cys 117) and a partially hydroxylated Pro residue (aa 3). Gly 11 is essential for its inhibitory activity and a 5 residue sequence (Gln 55–Gly 59) is important as the contact area. The 3D structure is characterised by five antiparallel β-strands and a detached α-helix (**1G96**).

## Biological Functions

Cystatin C inhibits cysteine proteases of the papain type such as cathepsin B, L and S.

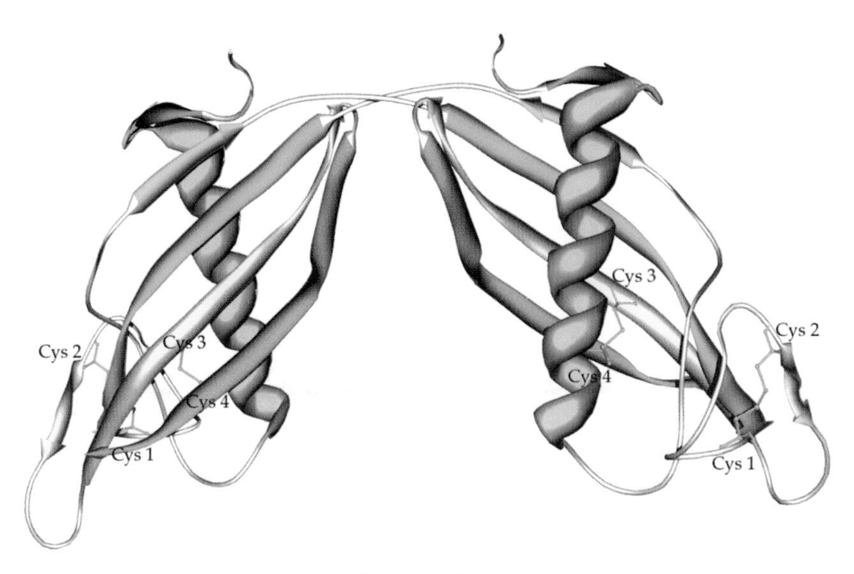


Cystatin C dimer

## *Heparin cofactor II (Protease inhibitor leuserpin 2)*

### Fact Sheet

| | | | |
|---|---|---|---|
| ***Classification*** | Serpin | ***Abbreviations*** | HCII |
| ***Structures/motifs*** | | ***DB entries*** | HEP2_HUMAN/P05546 |
| ***MW/length*** | 54 960 Da/480 aa | ***PDB entries*** | 1JMJ; 1JMO |
| ***Concentration*** | 60–90 mg/l | ***Half-life*** | 2.5 days |
| ***PTMs*** | 3 N-CHO (potential); 2 S-Tyr (aa 60 and 73); 1 Pyro-Glu (potential) | | |
| ***References*** | Baglin *et al.*, 2002, *Proc. Natl. Acad. Sci. USA*, **99**, 11079–11084. Blinder *et al.*, 1988, *Biochemistry*, **27**, 752–759. | | |

### Description

**Heparin cofactor II** (HCII) is predominantly synthesised in the liver and is an important inhibitor of thrombin and exhibits structural and functional similarity with other serpins.

### Structure

HCII is a single-chain plasma glycoprotein (10 % CHO) containing the RCL peptide with the reactive bond Leu 444–Ser 445. HCII contains in its N-terminal region two hirudin-like repeats (approx. 13 aa, Asp/Glu-rich) each with an S-Tyr residue (aa 60 and 73) and the N-terminus is modified (most likely pyro-Glu). The 3D structure of the HCII–thrombin complex reveals a multistep allosteric mechanism of inhibition (**1JMJ; 1JMO**) with thrombin in green and the RCL peptide with the reactive bond in pink.

### Biological Functions

In the presence of glycosaminoglycans, heparin and dermatan sulfate the inhibition of thrombin is increased approximately a thousand-fold. In the presence of dermatan sulfate HCII becomes the predominant thrombin inhibitor.

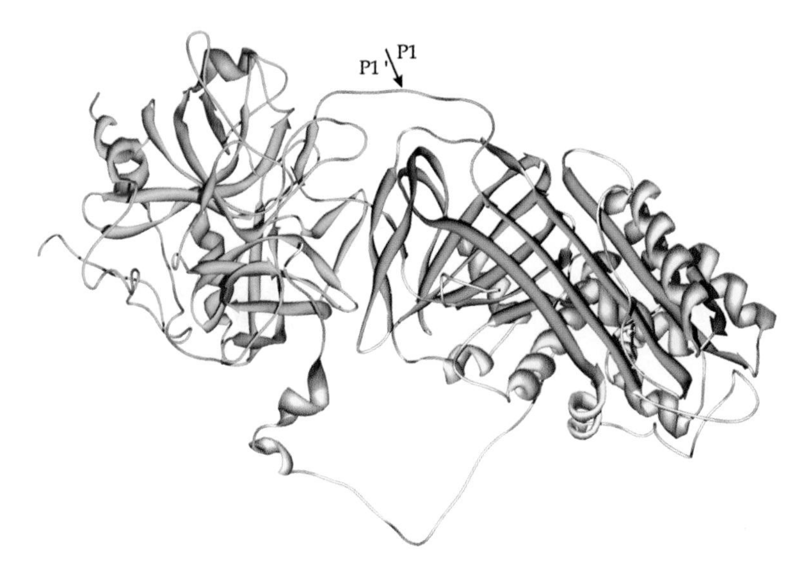


Heparin cofactor II in complex with thrombin

## *AMBP protein ($\alpha_1$-microglobulin/Inter-$\alpha$-trypsin inhibitor light chain (Bikunin))*

### Fact Sheet

| | | | |
|---|---|---|---|
| ***Classification*** | Lipocalin | ***Abbreviations*** | AMBP |
| ***Structures/motifs*** | 2 BPTI/Kunitz inhibitor | ***DB entries*** | AMBP_HUMAN/P02760 |
| ***MW/length*** | $\alpha_1$-Microglobulin: 20 847 Da/184 aa<br>Inter-$\alpha$-trypsin inhibitor light chain: 15 974 Da/147 aa | ***PDB entries***<br>***Half-life*** | 1BIK |
| ***Concentration*** | $\alpha_1$-Microglobulin: 20–100 mg/l<br>Inter-$\alpha$-trypsin inhibitor light chain: 500 mg/l | | |
| ***PTMs*** | $\alpha_1$-Microglobulin: 2 N-CHO; 1 O-CHO;<br>inter-$\alpha$-trypsin inhibitor light chain: 1 N-CHO; 1 O-CHO | | |
| ***References*** | Kaumeyer *et al.*, 1986, *Nucleic Acids Res.*, **14**, 7839–7850. | | |

### Description

The **AMBP protein** is synthesised in the liver as a diprotein that is processed before secretion into $\alpha_1$-microglobulin ($\alpha_1$M) and the inter-$\alpha$-trypsin inhibitor light chain (bikunin).

### Structure

$\alpha_1$M is a single-chain glycoprotein containing a chromophore. The inter-$\alpha$-trypsin inhibitor light chain (bikunin) is a single-chain glycoprotein containing two BPTI/Kunitz inhibitor domains and bikunin is linked to the inter-$\alpha$-trypsin inhibitor heavy chains via chondroitin 4-sulfate. The 3D structure of bikunin is very similar to other proteins containing BPTI/Kunitz inhibitor domains (**1BIK**).

### Biological Functions

$\alpha_1$M seems to be involved in immunoregulation. In addition to the free monomer form $\alpha_1$-M is also present in plasma in complex with IgA and albumin. Inter-$\alpha$-trypsin inhibitor exhibits inhibitory activity for trypsin, chymotrypsin, plasmin, neutrophil elastase and cathepsin G.

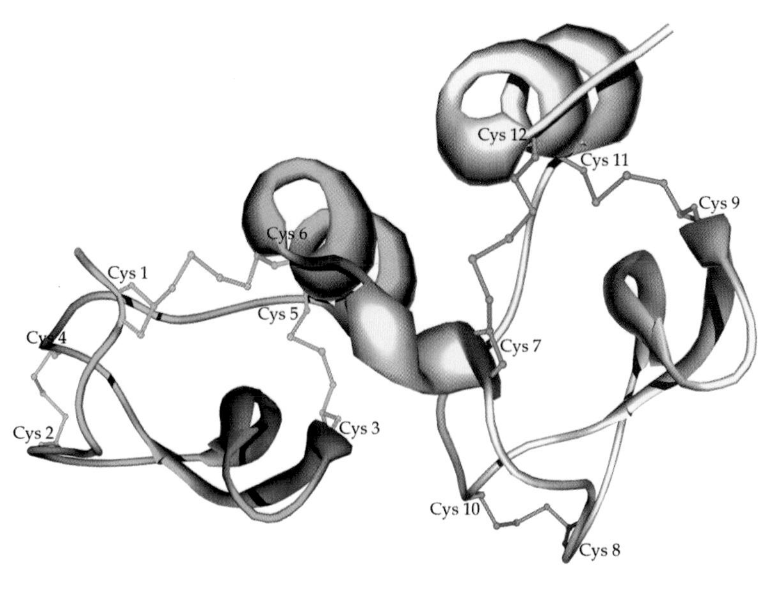


Bikunin

## *Inter-α-trypsin inhibitor: heavy chains H1, H2, H3, H4*

### Fact Sheet

| | | | |
|---|---|---|---|
| ***Classification*** | ITIH | ***Abbreviations*** | ITIH |
| ***Structures/motifs*** | H1, H2, H3, H4: each 1 VWFA | | |
| | | ***DB entries*** | ITIH1_HUMAN/P19 827<br>ITIH2_HUMAN/P19 823<br>ITIH3_HUMAN/Q06 033<br>ITIH4_HUMAN/Q14 624 |
| ***MW/length*** | H1: 71 415 Da/638 aa<br>H2: 72 425 Da/648 aa<br>H3: 69 360 Da/617 aa<br>H4: 70 587 Da/633 aa | ***PDB entries*** | 1BIK |
| ***Concentration*** | | ***Half-life*** | |
| ***PTMs*** | H1: 2 N-CHO; 1 O-CHO; 1 S-CHO; 1 chondroitin 4-sulfate<br>H2: 2 N-CHO; 4 O-CHO; 2 γ-Gla; 1 chondroitin 4-sulfate<br>H3: 2 N-CHO (potential); 1 chondroitin 4-sulfate<br>H4: 4 N-CHO (2 potential); 3 O-CHO (probable) | | |
| ***References*** | H1: Bost *et al.*, 1993, *Eur. J. Biochem*, **218**, 283–291.<br>H2: Gebhard *et al.*, 1988, *FEBS Lett*, **229**, 63–67.<br>H3: Bourguignon *et al.*, 1993, *Eur. J. Biochem*, **212**, 771–776.<br>H4: Saguchi *et al.*, 1995, *J. Biochem*, **117**, 14–18. | | |

### Description

The **inter-α-trypsin inhibitor heavy chains H1, H2, H3 and H4** are synthesised in the liver and belong to the ITIH family that exhibit structural and functional similarity with each other and are part of various plasma protease inhibitors.

### Structure

H1, H2, H3 and H4 heavy chains are glycoproteins each containing one VWFA domain. H1 contains an unusual S-linked glycan (Cys 26). The inter-α-plasma protease inhibitors are composed of one or two heavy chains (H1, H2 or H3) and the light chain (bikunin), the inter-α-inhibitor of H1, H2 and bikunin, the inter-α-like inhibitor of H2 and bikunin and the pre-α-inhibitor of H3 and bikunin. After proteolytic processing of the C-terminal propeptides in the precursor proteins the H1, H2 and H3 heavy chains are covalently linked via the α-carboxyl group of their C-terminal Asp residues to chondroitin 4-sulpfate at Ser 10 in bikunin. The 3D structure of bikunin is very similar to other proteins containing the BPTI/Kunitz inhibitor domain (**1BIK**).

### Biological Functions

The inter-α-trypsin inhibitor seems to act as a carrier of hyaluronan and as the binding protein between hyaluronan and other matrix proteins.

## *Kallistatin (Kallikrein inhibitor)*

### Fact Sheet

| | | | |
|---|---|---|---|
| ***Classification*** | Serpin | ***Abbreviations*** | |
| ***Structures/motifs*** | | ***DB entries*** | KAIN_HUMAN/P29622 |
| ***MW/length*** | 46 369 Da/407 aa | ***PDB entries*** | |
| ***Concentration*** | 19–23 mg/l | ***Half-life*** | |
| ***PTMs*** | 4 N-CHO (3 potential) | | |
| ***References*** | Zhou *et al.*, 1992, *J. Biol. Chem.*, **267**, 25873–25880. | | |

### Description

**Kallistatin** is mainly synthesised in the liver and apart from plasma is found in various tissues. It is an important inhibitor of kallikrein and exhibits structural and functional similarity with other serpins.

### Structure

Kallistatin is a single-chain plasma glycoprotein containing the RCL peptide with the reactive bond Phe 368–Ser 369. Kallistatin exhibits the typical serpin structure with eight helices and three long β-sheets.

### Biological Functions

Kallistatin inhibits tissue kallikrein and heparin prevents the complex formation of kallistatin with kallikrein. Kallistatin is susceptible to inactivation by proteolytic cleavage in the reactive centre.

## *Leukocyte elastase inhibitor*

### Fact Sheet

| | | | |
|---|---|---|---|
| ***Classification*** | Serpin | ***Abbreviations*** | LEI |
| ***Structures/motifs*** | | ***DB entries*** | ILEU_HUMAN/P30740 |
| ***MW/length*** | 42 742 Da/379 aa | ***PDB entries*** | |
| ***Concentration*** | | ***Half-life*** | |
| ***PTMs*** | 1 *N*-acetyl-Met | | |
| ***References*** | Remold-O'Donnell *et al.*, 1992, *Proc. Natl. Acad. Sci. USA*, **89**, 5635–5639. | | |

### Description

**Leukocyte elastase inhibitor** (LEI) primarily regulates the activity of neutrophil elastase and exhibits structural and functional similarity with other serpins.

### Structure

LEI is a single-chain serpin containing the RCL peptide with the reactive bond Cys 344–Met 345 and an N-terminal acetyl-Met. LEI lacks a typical cleavable N-terminal signal sequence.

### Biological Functions

LEI primarily regulates the activity of neutrophil elastase and cathepsin G.

## *$\alpha_2$-Macroglobulin*

### Fact Sheet

| | | | |
|---|---|---|---|
| ***Classification*** | Protease inhibitor I39 | ***Abbreviations*** | $\alpha_2$M |
| ***Structures/motifs*** | | ***DB entries*** | A2MG_HUMAN/P01023 |
| ***MW/length*** | 160 797 Da/1451 aa | ***PDB entries*** | |
| ***Concentration*** | approx. 1200 mg/l | ***Half-life*** | approx. 10 days |
| ***PTMs*** | 8 N-CHO | | |
| ***References*** | Sottrup-Jensen *et al.*, 1984, *J. Biol. Chem.*, **259**, 8318–8327. | | |

### Description

**$\alpha_2$-Macroglobulin** ($\alpha_2$M) is primarily synthesised in the liver and circulates in blood as a tetramer. $\alpha_2$M is able to inhibit all four types of proteases.

### Structure

$\alpha_2$M is a large single-chain glycoprotein (10 % CHO) with a 39 residue bait region (Pro 667–Thr 705) with three inhibitory regions (Arg 681–Glu 686, Arg 696–Val 700, Thr 707–Phe 712) containing specific cleavage sites for the different proteases. Gln 670 and Gln 671 are potential crosslinking sites with Lys residues in other proteins and Cys 949 and Gln 952 form an internal isoglutamyl cysteine thioester bond. In the tetramer the two monomers are linked by two interchain disulfide bridges (Cys 447–Cys 540, Cys 540–Cys 447) and two monomers are noncovalently linked.

### Biological Functions

$\alpha_2$M inhibits all four classes of proteases by a unique trapping mechanism. A peptide stretch called the 'bait' region contains specific cleavage sites for the different proteases. The cleavage-induced conformational change enables $\alpha_2$M to trap the proteases. In consequence, the internal thioester bond is cleaved and this mediates the covalent binding of $\alpha_2$M to the protease.

## *Neuroserpin (Protease inhibitor 12)*

### Fact Sheet

| | | | |
|---|---|---|---|
| ***Classification*** | Serpin | ***Abbreviations*** | |
| ***Structures/motifs*** | | ***DB entries*** | NEUS_HUMAN/Q99574 |
| ***MW/length*** | 44 704 Da/394 aa | ***PDB entries*** | |
| ***Concentration*** | | ***Half-life*** | |
| ***PTMs*** | 3 N-CHO (potential) | | |
| ***References*** | Schrimpf *et al.*, 1997, *Genomics*, **40**, 55–62. | | |

### Description

**Neuroserpin** is mainly expressed in the brain and inhibits plasminogen activators and plasmin and exhibits structural and functional similarity with other serpins.

### Structure

Neuroserpin is a single-chain plasma glycoprotein containing the RCL peptide with the reactive bond Arg 346–Met 347.

### Biological Functions

Neuroserpin seems to protect neurons from cell damage by tissue-type plasminogen activator.

## *Pancreatic secretory trypsin inhibitor*

### Fact Sheet

| | | | |
|---|---|---|---|
| ***Classification*** | Kazal-type inhibitor | ***Abbreviations*** | PSTI |
| ***Structures/motifs*** | 1 Kazal-like | ***DB entries*** | IPK1_HUMAN/P00995 |
| ***MW/length*** | 6247 Da/56 aa | ***PDB entries*** | 1HPT |
| ***Concentration*** | approx. 12 μg/l | ***Half-life*** | 6 min |
| ***PTMs*** | | | |
| ***References*** | Hecht *et al.*, 1992, *J. Mol. Biol.*, **225**, 1095–1103.<br>Horii *et al.*, 1987, *Biochem. Biophys. Res. Commun.*, **149**, 635–641. | | |

### Description

**Pancreatic secretory trypsin inhibitor** (PSTI) is mainly found in pancreas but traces are present in blood. PSTI is a specific inhibitor of trypsin.

### Structure

PSTI is a small single-chain Kazal-type serine protease inhibitor with a Kazal-like domain containing three intrachain disulfide bridges (Cys 9–Cys 38, Cys 16–Cys 35, Cys 24–Cys 50) and the reactive bond Lys 18–Ile 19. The domain exhibits structural similarity with the EGF-like domain but their exon–intron junctions are not related and they seem not to have evolved from a common ancestral gene. The 3D structure of a mutant human PSTI is similar to known PSTI structures of other species (**1HPT**).

### Biological Functions

PSTI is a specific trypsin inhibitor that prevents the autoactivation of trypsinogen in pancreas and pancreatic juices and thus protects the pancreas from autodigestion.

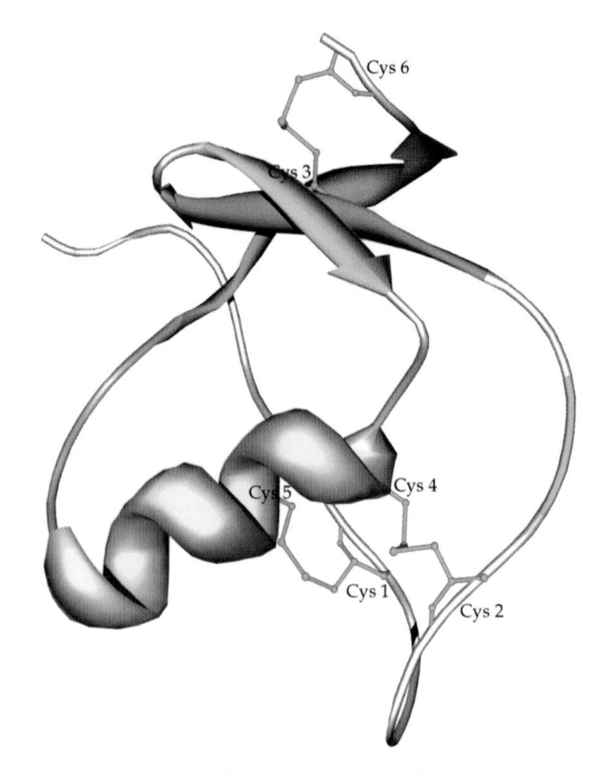


Pancreatic secretory trypsin inhibitor

## *Plasma protease C1 inhibitor (C1-inhibitor)*

### Fact Sheet

| | | | |
|---|---|---|---|
| ***Classification*** | Serpin | ***Abbreviations*** | C1Inh |
| ***Structures/motifs*** | | ***DB entries*** | IC1_HUMAN/P05155 |
| ***MW/length*** | 52 843 Da/478 aa | ***PDB entries*** | |
| ***Concentration*** | 100–260 mg/l | ***Half-life*** | 67–72 h |
| ***PTMs*** | 7 N-CHO; 7 O-CHO | | |
| ***References*** | Bock *et al.*, 1986, *Biochemistry,* **25**, 4292–4301. | | |

### Description

**Plasma protease C1 inhibitor** (C1Inh) plays a key role in the regulation of the classical pathway of complement activation and exhibits structural and functional similarity with other serpins.

### Structure

C1Inh is a single-chain plasma glycoprotein (at least 26 % CHO) containing the RCL peptide with the reactive bond Arg 444–Thr 445. In addition to the typical serpin part C1Inh contains an N-terminal approx. 110 residue long rod-like domain, which is highly glycosylated (*N*- and *O*-glycosylation) and contains 14 tandem repeats of the tetrapeptide **[QE]-P-T-[TQ]**.

### Biological Functions

C1Inh regulates the activation of the C1 complex by forming an inactive complex with C1r or C1s. Beside its key function in the regulation of complement activation it also plays an important role in the regulation of blood coagulation and fibrinolysis and the kinin system.

# *Plasma serine protease inhibitor (Protein C inhibitor, Plasminogen activator inhibitor-3)*

## Fact Sheet

| | | | |
|---|---|---|---|
| ***Classification*** | Serpin | ***Abbreviations*** | PAI-3 |
| ***Structures/motifs*** | | ***DB entries*** | IPSP_HUMAN/P05154 |
| ***MW/length*** | 43 660 Da/387 aa | ***PDB entries*** | 1LQ8 |
| ***Concentration*** | 3–7 mg/l | ***Half-life*** | 24 h |
| ***PTMs*** | 3 N-CHO (2 potential) | | |
| ***References*** | Huntington *et al.*, 2003, *Structure*, **11**, 205–215.<br>Suzuki *et al.*, 1987, *J. Biol. Chem.*, **262**, 611–616. | | |

## Description

**Plasma serine protease inhibitor** (PAI-3) is mainly synthesised in the liver and is a major inhibitor of protein C and exhibits structural and functional similarity with other serpins.

## Structure

PAI-3 is a single-chain plasma glycoprotein containing the reactive loop peptide with the reactive bond Arg 354–Ser 355 and exhibits the typical serpin structure. The 3D structure of PAI-3 cleaved in the reactive loop peptide exhibits the typical structure of cleaved serpins with the cleaved region of the RCL peptide on either side of the molecule (pink circle) and part of the RCL peptide is inserted into β-sheet A (in pink) and cleaved chain B is shown in red (**1LQ8**).

## Biological Functions

Besides protein C, PAI-3 also inhibits kallikrein, thrombin, FXa, FXIa and uPA.

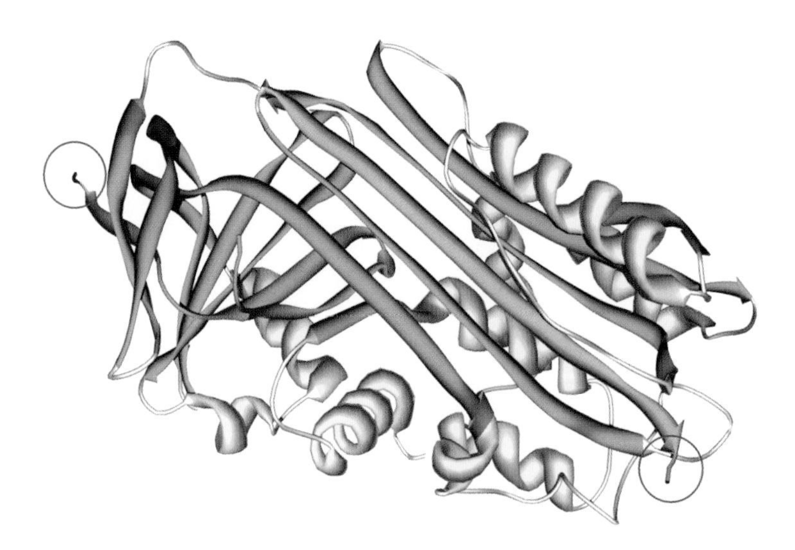

Plasma serine protease inhibitor

# *Plasminogen activator inhibitor-1*

## Fact Sheet

| | | | |
|---|---|---|---|
| ***Classification*** | Serpin | ***Abbreviations*** | PAI-1 |
| ***Structures/motifs*** | | ***DB entries*** | PAI1_HUMAN/P05121 |
| ***MW/length*** | 42 769 Da/379 aa | ***PDB entries*** | 1A7C |
| ***Concentration*** | Variable: 24 μg/l | ***Half-life*** | 6–7 min |
| ***PTMs*** | 3 N-CHO (1 potential) | | |
| ***References*** | Pannekoek *et al.*, 1986, *EMBO J.*, **5**, 2539–2544.<br>Xue *et al.*, 1998, *Structure*, **6**, 627–636. | | |

## Description

Active **plasminogen activator inhibitor-1** (PAI-1) circulates in plasma bound to vitronectin and mainly inhibits tissue-type (tPA) and urokinase-type (uPA) plasminogen activators and exhibits structural and functional similarity with other serpins.

## Structure

PAI-1 is a single-chain plasma glycoprotein (13 % CHO) containing the RCL peptide with the reactive bond Arg 346–Met 347. The 3D structure of PAI-1 in complex with two inhibitory RCL pentapeptides (in pink) is shown **(1A7C)**. The cleaved regions of the RCL peptide are on either side of the molecule (pink circles) and the cleaved C-terminal region is shown in red.

## Biological Functions

Besides tPA and uPA, PAI-1 inhibits other serine proteases like protein C, thrombin, plasmin and trypsin. PAI-1 is sensitive to inactivation by proteolytic cleavage in the reactive loop peptide.

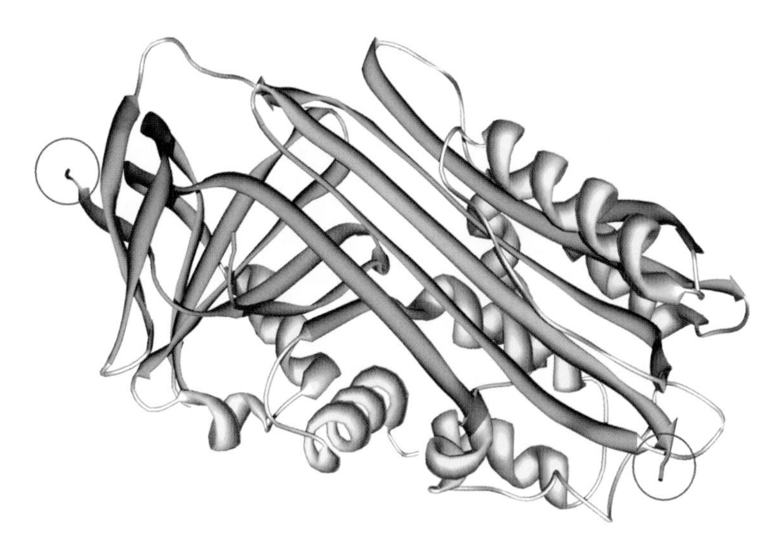

Plasminogen activator inhibitor-1

## *Plasminogen activator inhibitor-2*

### Fact Sheet

| | | | |
|---|---|---|---|
| ***Classification*** | Serpin | ***Abbreviations*** | PAI-2 |
| ***Structures/motifs*** | | ***DB entries*** | PAI2_HUMAN/P05120 |
| ***MW/length*** | 46 596 Da/415 aa | ***PDB entries*** | 1BY7 |
| ***Concentration*** | < 10 μg/l | ***Half-life*** | |
| ***PTMs*** | 3 N-CHO (potential) | | |
| ***References*** | Harrop *et al.*, 1999, *Structure*, **7**, 43–54.<br>Ye *et al.*, 1987, *J. Biol. Chem.*, **262**, 3718–3725. | | |

### Description

**Plasminogen activator inhibitor-2** (PAI-2) is the main inhibitor of urokinase-type (uPA) plasminogen activator and exhibits structural and functional similarity with other serpins.

### Structure

PAI-2 is a single-chain plasma glycoprotein containing the RCL peptide with the reactive bond Arg 380–Thr 381 and an uncleaved signal peptide. The 3D structure of a deletion mutant (Asn 66–Gln 98) of PAI-2 represents the stressed state (**1BY7**).

### Biological Functions

Beside uPA PAI-2 also inhibits tPA to a lesser extent. During pregnancy the level of PAI-2 may be as high as 300 μg/l.

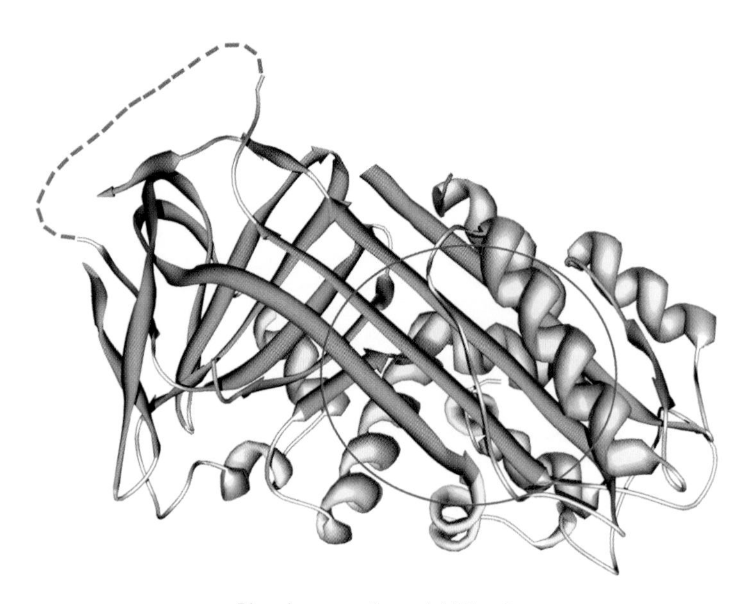

Plasminogen activator inhibitor-2

# *Pregnancy zone protein*

## Fact Sheet

| | | | |
|---|---|---|---|
| ***Classification*** | Protease inhibitor I39 | ***Abbreviations*** | PZP |
| ***Structures/motifs*** | | ***DB entries*** | PZP_HUMAN/P20742 |
| ***MW/length*** | 161 030 Da/1457 aa | ***PDB entries*** | |
| ***Concentration*** | | ***Half-life*** | |
| ***PTMs*** | 11 N-CHO (potential) | | |
| ***References*** | Devriendt *et al.*, 1991, *Biochim. Biophys. Acta*, **1088**, 95–103. | | |

## Description

**Pregnancy zone protein** (PZP) circulates in blood as a tetramer. PZP is able to inhibit all four types of proteases.

## Structure

PZP is a large single-chain glycoprotein with a 51 residue bait region (Cys 660–Glu 710) that contains specific cleavage sites for different proteases. Cys 953 and Gln 956 form an internal isoglutamyl cysteine thioester bond.

## Biological Functions

PZP inhibits all four classes of proteases by a unique trapping mechanism. A peptide stretch called the bait region contains specific cleavage sites for the different proteases. The cleavage-induced conformational change enables PZP to trap the proteases. In consequence, the internal thioester bond is cleaved and this mediates the covalent binding of PZP to the protease.

# *Protein Z-dependent protease inhibitor*

## Fact Sheet

| | | | |
|---|---|---|---|
| ***Classification*** | Serpin | ***Abbreviations*** | PZI |
| ***Structures/motifs*** | | ***DB entries*** | ZPI_HUMAN/Q9UK55 |
| ***MW/length*** | 48 426 Da/423 aa | ***PDB entries*** | |
| ***Concentration*** | | ***Half-life*** | |
| ***PTMs*** | 4 N-CHO (potential) | | |
| ***References*** | Han *et al.*, 1999, *Biochemistry*, **38**, 11073–11078. | | |

## Description

**Protein Z-dependent protease inhibitor** (PZI) is synthesised in the liver and inhibits coagulation factor FXa and exhibits structural and functional similarity with other serpins.

## Structure

PZI is a single-chain plasma glycoprotein containing the reactive loop peptide with the reactive bond Tyr 387–Ser 388.

## Biological Functions

PZI inhibits coagulation factor FXa in the presence of protein Z, calcium and phospholipids.

## *Tissue factor pathway inhibitor (Lipoprotein-associated coagulation inhibitor)*

### Fact Sheet

| | | | |
|---|---|---|---|
| ***Classification*** | Kunitz-type inhibitor | ***Abbreviations*** | TFPI/LCAI |
| ***Structures/motifs*** | 3 BPTI/Kunitz inhibitor | ***DB entries*** | TFPI1_HUMAN/P10646 |
| ***MW/length*** | 31 950 Da/276 aa | ***PDB entries*** | 1ADZ |
| ***Concentration*** | 16 μg/l | ***Half-life*** | 30–60 min |
| ***PTMs*** | 2 N-CHO; 2 O-CHO; P-Ser (aa 2) | | |
| ***References*** | Maurits *et al.*, 1997, *J. Mol. Biol.*, **269**, 395–407.<br>Wun *et al.*, 1988, *J. Biol. Chem.*, **263**, 6001–6004. | | |

### Description

**Tissue factor pathway inhibitor** (TFPI) circulates in plasma in a free and lipoprotein-associated form. TFPI is involved in the inhibition of the coagulation cascade.

### Structure

TFPI is a Kunitz-type serine protease inhibitor with an acidic N-terminal region, three BPTI/Kunitz inhibitor domains and a basic C-terminal region and is phosphorylated at Ser 2. The 3D structure of the Kunitz II domain was determined (see Chapter 5) (**1ADZ**).

### Biological Functions

TFPI inhibits coagulation factor FXa directly and in an FXa-dependent way the activity of the FVIIa/TF complex.

# 12

# Lipoproteins

## 12.1 INTRODUCTION

Lipids are linked to proteins in two ways:

1. Covalently, known as lipid-linked proteins
2. Noncovalently, known as lipoproteins

The covalently-linked lipids in lipid-linked proteins belong to three classes:

- Isoprenoid ($C_5$ hydrocarbon) group such as farnesyl ($C_{15}$) and geranylgeranyl ($C_{20}$)
- Fatty acid groups such as myristoyl ($C_{14}$) and palmitoyl ($C_{16}$)
- Glycosylphosphatidylinositol (GPI) anchor

All these covalently-linked lipids are thought to function as membrane anchors for the corresponding protein.

Lipoproteins are highmolecular weight aggregates or complexes of lipids and apolipoproteins. Lipoproteins exhibit structural functions or act as enzymes, pump protons and ions or transport lipids. Plasma lipoproteins contain lipids such as cholesterol in free or bound form, triacylglycerols and phospholipids and, of course, proteins termed 'apolipoproteins'. Plasma lipoproteins execute a variety of functions such as the transport of water-insoluble lipids via the vascular system to the target cells either for consumption or for storage, and they play a key role in homeostasis of cholesterol.

Lipoproteins are classified into five categories listed according to their size and density, from larger and less dense to smaller and denser particles with an increase of the protein and a decrease of the lipid content. The characteristics of the various lipoprotein classes are summarised in Table 12.1.

1. *Chylomicrons* (<0.95 g/cm$^3$) are large lipoprotein particles that are assembled in the intestinal mucosa and transport exogenous triacylglycerols and cholesterol from the intestine to the peripheral tissues and the liver.
2. *Very-low-density lipoproteins* (*VLDLs*) (<1.006 g/cm$^3$) are assembled in the liver and transport endogenous triacylglycerols and cholesterol to the peripheral tissues. VLDLs are the main internal transport system of endogenous lipids.
3. *Intermediate-density lipoproteins* (IDLs) (1.006–1.019 g/cm$^3$) are generated from the degradation of VLDLs and are usually not detectable in blood. IDLs have a similar function to VLDLs.
4. *Low-density lipoproteins* (*LDLs*) (1.019–1.063 g/cm$^3$) are generated in circulation from VLDLs by lipoprotein lipase. LDLs transport triacylglycerols and cholesterol away from cells and tissues where they exist in abundance to cells and tissues that are taking them up. It is referred to as '***bad cholesterol***', because there is a link between high levels of LDLs and cardiovascular diseases.
5. *High-density lipoproteins* (*HDLs*) (1.063–1.21 g/cm$^3$) are synthesised in the liver and the small intestine and transport endogenous cholesterol from tissues to the liver. Because HDLs can remove cholesterol from arteries to the liver for excretion or reutilisation HDLs are considered as '***good lipoproteins***'.

*Human Blood Plasma Proteins: Structure and Function* Johann Schaller, Simon Gerber, Urs Kämpfer, Sofia Lejon and Christian Trachsel

**Table 12.1** Characterisation of the major classes of lipoproteins in human plasma.

| Parameter | Chylomicrons | VLDL | IDL | LDL | HDL |
|---|---|---|---|---|---|
| Density (g/cm$^3$) | <0.95 | <1.006 | 1.006–1.019 | 1.019–1.063 | 1.063–1.210 |
| Particle diameter (Å) | 750–12 000 | 300–800 | 250–350 | 180–250 | 50–120 |
| Particle mass (kDa) | 400 000 | 10 000–80 000 | 5000–10 000 | 2300 | 175–360 |
| Protein (%) | 1.5–2.5 | 5–10 | 15–20 | 20–25 | 40–55 |
| Phospholipids (%) | 7–9 | 15–20 | 22 | 15–20 | 20–35 |
| Free cholesterol (%) | 1–3 | 5–10 | 8 | 7–10 | 3–4 |
| Cholesteryl esters (%) | 3–5 | 10–15 | 30 | 35–40 | 12 |
| Triacylglycerols (%) | 84–89 | 50–65 | 22 | 7–10 | 3–5 |
| Apolipoproteins | A-I, A-II, B-48 C-I, C-II, C-III, E | B-100 C-I, C-II, C-III, E | B-100, C-III, E | B-100 | A-I, A-II, D, E C-I, C-II, C-III |

Lipoprotein particles are permanently processed and therefore their composition and properties are variable to some extent. The general structure of a lipoprotein particle is depicted in Figure 12.1(a). The characteristic components of a lipoprotein particle are apolipoproteins, triacylglycerols, free cholesterol, cholesteryl esters and phospholipids. Figure 12.1 (b) shows an LDL particle, the major carrier of cholesterol, consisting of approximately 1500 cholesteryl ester molecules surrounded by some 800 phospholipid molecules, 500 cholesterol molecules and a single copy of the large apolipoprotein B-100. In cholesteryl esters cholesterol primarily forms an ester linkage with linoleic acid (18 : 2). The distribution of the apolipoproteins in the various lipoprotein particles is given in Table 12.2.

## 12.2 ENZYMES IN LIPOPROTEIN METABOLISM

The processing of the lipoprotein particles by hydrolysis of the triacylglycerols in circulating chylomicrons and VLDLs by lipoprotein lipase or the esterification of free cholesterol transported in plasma lipoproteins by phosphatidylcholine-sterol acyltransferase or the reesterification of excess intracellular cholesterol for storage are important processes catalysed by specific enzymes.

Lipases and phosphatidylcholine-sterol acyltransferase share a conserved region centred around a Ser residue that participates in the catalytic triad:

LIPASE _SER (PS 00120): [LIV]–{KG}–[LIVFY]– [LIVMST]–G–[HYWV]–**S**–{YAG}–G–[GSTAC]

**Lipoprotein lipase** (LPL, EC 3.1.1.34) (see Data Sheet) is synthesised in parenchymal cells of various tissues and exists in its functional form as homodimer in plasma. LPL is a single-chain glycolipoprotein (8 % CHO) containing five intrachain disulfide bridges (Cys 27–Cys 40, Cys 216–Cys 239, Cys 264–Cys 283, Cys 275–Cys 278, Cys 418–Cys 438), a PLAT domain and a potential heparin-binding region (Lys 292–Lys 304) and is attached to membranes by a GPI anchor. The active site consists of Ser 132, Asp 156 and His 241. The primary function of LPL is the hydrolysis of triacylglycerols of circulating chylomicrons and VLDLs on the luminal surface of the vascular endothelium:

$$\text{Triacylglycerol} + H_2O \Longleftrightarrow \text{diacylglycerol} + \text{fatty acid}$$

LPL obtains maximal activity with Apo-CII as cofactor and Apo-CIII seems to be an important biological inhibitor.

Defects in the LPL gene are the cause of:

1. Familial chylomicronemia, also known as hyperlipoproteinemia type I, is a recessive disorder already manifested in childhood and characterised by various symptoms such as abdominal pain and massive hypertriglyceridemia.
2. Lipoprotein lipase deficiency leading to hypertrigly ceridemia.

(For both diseases see details in MIM: 238600.)

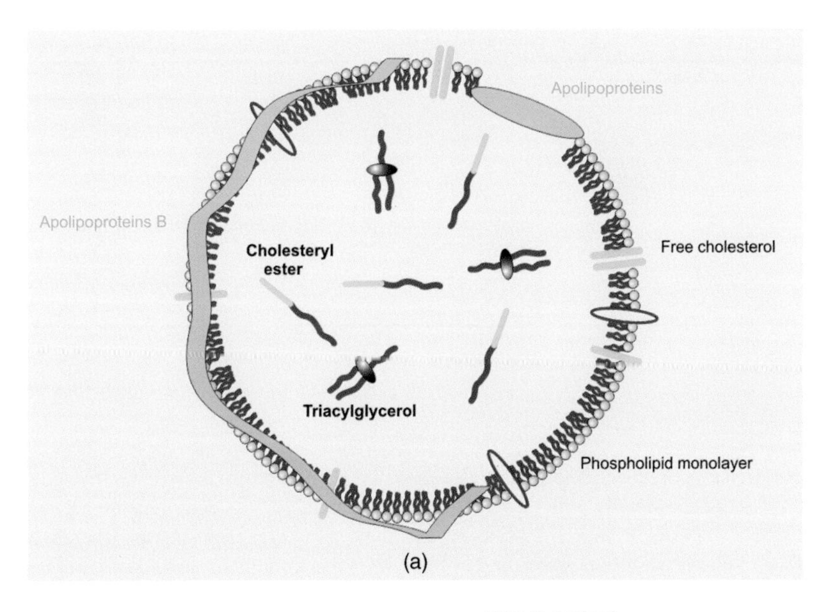


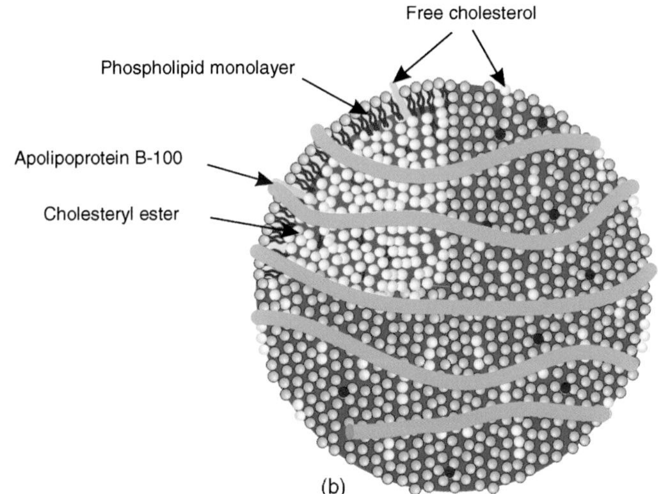


**Figure 12.1 Schematic representations of lipoprotein particles and a low-density lipoprotein (LDL) particle**
(a) Lipoprotein particles consist of apolipoproteins, triacylglycerols, free cholesterol, cholesteryl esters and phospholipids.
(b) LDL, the major carrier of cholesterol, is composed of approx. 1500 cholesteryl ester molecules, some 800 phospholipid and 500 cholesterol molecules and a single copy of the large apolipoprotein B-100.

**Hepatic triacylglycerol lipase** (HL, EC 3.1.1.3) (see Data Sheet) or **hepatic lipase** is synthesised exclusively by hepatocytes and is usually not present in plasma except in post-heparin plasma. HL is a single-chain glycolipoprotein (18 % CHO) containing five intrachain disulfide bridges (Cys 40–Cys 53, Cys 232–Cys 255, Cys 280–Cys 299, Cys 291–Cys 294, Cys 445–Cys 465, assigned by sequence comparison with lipoprotein lipase), a PLAT domain and a potential heparin-binding region (Ile 159–Leu 171). The active site consists of Ser 146, Asp 172 and His 257. HL seems to be involved in the metabolism of chylomicron remnants, IDLs and HDLs. HL has the capacity to catalyse the hydrolysis of phospholipids, mono-, di-, and triacylglycerols and acyl-CoA thioesters.

**Table 12.2** Distribution of apolipoproteins in lipoprotein classes (%).

| Apolipoprotein | Chylomicrons | VLDL | LDL | $HDL_2$[a] | $HDL_3$ |
|---|---|---|---|---|---|
| A-I | 33 | Traces | Traces | 65 | 62 |
| A-II | Traces | Traces | Traces | 10 | 23 |
| A-IV | 14 | — | — | ? | Traces |
| B | 5 | 25 | >95 | Traces | Traces |
| C-I, C-II, C-III | 32 | 55 | 0–2 | 13 | — |
| D | ? | ? | ? | 2 | 4 |
| E | 10 | 15 | 0–3 | 3 | 1 |
| Others | 6 | 5 | 0–5 | 4 | 5 |

[a]HDL, discoid shape; $HDL_3$, spherical HDL; $HDL_2$, derived form of $HDL_3$ by additional incorporation of cholesterol and triacylglycerols. Apolipoprotein(a) is not included (variable mass).

The 3D structure of HL has not yet been determined. However, the 3D structure of the structurally and functionally related pancreatic lipase (34 % sequence identity with HL) in complex with colipase was determined by X-ray diffraction (PDB: 1N8S). Pancreatic lipase (449 aa: P16233) consists of the same structural elements as HL: an N-terminal domain (approximately 340 aa) with the active site consisting of Ser 153, Asp 117 and His 254 and a C-terminal PLAT domain.

Defects in the HL gene are the cause of hepatic lipase deficiency (for details see MIM: 151670).

**Phosphatidylcholine-sterol acyltransferase** (EC 2.3.1.43) (see Data Sheet) or lecithin-cholesterol acyltransferase (LCAT) is synthesised primarily in the liver and is the central enzyme in the extracellular metabolism of plasma lipoproteins. LCAT is a single-chain glycolipoprotein (20 % CHO) containing two intrachain disulfide bridges (Cys 50–Cys 74, Cys 313–Cys 356) and two free Cys (aa 31 and 184) located near the active site, which have to be in the reduced state for optimal activity of LCAT. The active site consists of Ser 181, His 377 and an unidentified Asp/Glu residue and Apo-AI is a potent activator. Among other substrates it esterifies free cholesterol in transported plasma lipoproteins:

$$\text{Phosphatidylcholine} + \text{sterol} \Longleftrightarrow \text{1-acylglycerophosphocholine} + \text{sterol ester}$$

LCAT plays a key role in the reverse transport of cholesterol by HDLs from peripheral tissues to the liver.

Defects in the LCAT gene are the cause of

1. Fish-eye disease, an autosomal recessive disorder characterised by the accumulation of unesterified cholesterol in certain tissues, resulting in corneal opacities (for details see MIM: 136120).
2. LCAT deficiency also called Norum disease, a disorder in the lipoprotein metabolism resulting in corneal opacities, target cell haemolytic anemia and proteinuria with renal failure (for details see MIM: 245900).

## 12.3 APOLIPOPROTEINS

**Apolipoprotein A-I** (Apo-AI) (see Data Sheet) is primarily synthesised in the liver and the small intestine and is the major protein of HDL particles, but is also found in chylomicrons. Apo-AI is a single-chain protein generated from the proform by releasing an N-terminal propeptide (6 aa) and contains eight tandem repeats (approximately 22 aa) and two repeats of half length (11 aa) and is palmitoylated. Apo-AI is involved in the reverse transport of cholesterol from tissues to the liver for excretion by promoting cholesterol efflux from tissues and acts as a cofactor of the lecithin cholesterol acyltransferase (LCAT), a central enzyme in the extracellular metabolism of plasma lipoproteins (see LCAT). Apo-AI seems to be protective against the development of atherosclerosis and its plasma concentration is negatively correlated with atherosclerosis.

The 3D structure of an N-terminally truncated Apo-AI (201 aa: Asn43Met–Gln 243) was determined by X-ray diffraction (PDB: 1AV1) and is shown in Figure 12.2. The structure consists almost entirely of an amphipathic α-helix and adopts a shape similar to a horseshoe. The tetramer adopts an antiparallel four-helix bundle with an elliptical ring shape.

Defects in the Apo-AI gene are the cause of:

1. HDL deficiency type 1, also known as analphalipoproteinemia or Tangier disease, is a recessive disorder characterised by the absence of HDL in plasma and the accumulation of cholesteryl esters (for details see MIM: 205400).

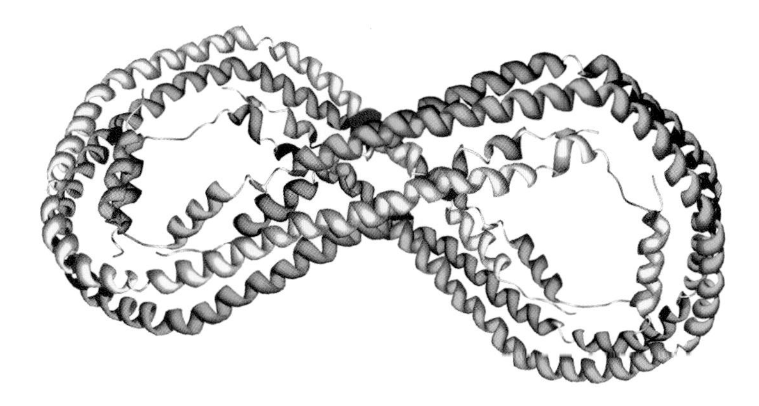

**Figure 12.2 3D structure of tetrameric apolipoprotein A-I (PDB: 1AV1)**
The tetramer of apolipoprotein A-I adopts an antiparallel four-helix bundle with an elliptical ring shape.

2. HDL efficiency type 2, also known as familial hypoalphalipoproteinemia, is an inherited autosomal dominant disease (for details see MIM: 604091).
3. Systemic nonneuropathic amyloidosis, also known as amyloidosis VIII or Ostertag-type amyloidosis, is an autosomal dominant disorder characterised by a generalised deposition of amyloid protein (for details see MIM: 105200).

**Apolipoprotein A-II** (Apo-AII) (see Data Sheet) is primarily synthesised in the liver and also in the intestine and is a major component of HDLs. Apo-AII exists in plasma as disulfide-linked homodimer (Cys 6–Cys 6) but also forms disulfide-linked heterodimers with apolipoprotein D (Apo-AII: Cys 6; Apo-D: Cys 116). Apo-AII is generated from its proform by releasing an N-terminal propeptide (5 aa) and contains 11 aa and 22 aa repeats and carries an N-terminal pyro-Glu residue. Beside lipid transport Apo-AII enhances hepatic lipase activity and seems to be important for the HDL metabolism by stabilising the HDL particles.

The 3D structure of entire Apo-AII (77 aa: Gln 1–Gln 77) was determined by X-ray diffraction (PDB: 2OU1) and is shown in Figure 12.3. The disulfide-linked homodimers (the disulfide bridges are in pink) form amphipathic $\alpha$-helices, which aggregate into non-covalently-linked dimers.

**Apolipoprotein A-IV** (Apo-AIV) (see Data Sheet) is primarily synthesised in the intestine and is a major component of chylomicrons. Apo-AIV is a single-chain glycoprotein (6 % CHO) containing 13 approximately 22 aa long tandem repeats

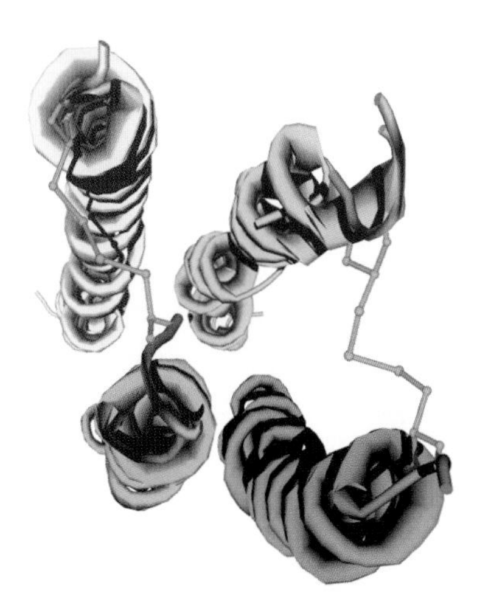

**Figure 12.3 3D structure of dimeric apolipoprotein A-II (PDB: 2OU1)**
The tetramer of apolipoprotein A-II consists of disulphide-linked homodimers (disulphide bridges in pink) containing amphipathic $\alpha$-helices.

and an approximately 20 aa long C-terminal Glu/Gln-rich segment. Apo-AIV is required for the efficient activation of lipoprotein lipase and is a potent activator of lecithin cholesterol acyltransferase (LCAT). In the presence of lipids Apo-AIV adopts a structure with an α-helical content of over 80 %. The structure is similar to Apo-AI.

Familial combined hyperlipidemia is an inherited disorder with a high level of serum cholesterol or high blood triacylglycerols that leads to an increased risk of cardiovascular diseases.

**Apolipoprotein B-100** (Apo B-100) (see Data Sheet) is synthesised in the liver and is the main protein of the LDL particles, but is also found in VLDLs and chylomicrons. Apo B-100 is one of the largest known monomeric single-chain glycoproteins containing an N-terminal vitellogenin domain and is palmitoylated at Cys 1085. It contains several heparin-binding regions and two basic regions thought to be possible receptor binding regions: Lys 3147–His 3157 **(KAQYKKNKHRH)** and Arg 3359–Lys 3367 **(RLTRKRGLK)**. Apo B-100 is linked to Apo(a) via an interchain disulfide bridge (Apo B-100: Cys 4326; Apo(a): Cys 4059). Apo B-100 is a recognition signal for cellular binding and internalisation of LDL particles by the Apo B/E receptor. Elevated levels of Apo B-100 are associated with premature atherosclerosis and it is thought to be a better indicator as a risk factor for heart diseases than total cholesterol or LDL. Native Apo B-100 in LDLs is characterised by approximately 43 % α-helix, 21 % β-sheet, 20 % random coil and 16 % β-turn structures.

**Apolipoprotein B-48** is synthesised in the intestine and is exclusively found in chylomicrons. Apo B-48 is the shortened form of Apo B-100 comprising the 2152 N-terminal aa of Apo B-100 created by the introduction of a stop codon (UAA) at the corresponding position (Gln 2153) by RNA editing. Apo B-48 lacks the receptor-binding region of Apo B-100.

Defects in the Apo B gene are the cause of:

1. Familial hypobetalipoproteinemia is a genetically heterogeneous autosomal co-dominant disorder associated with reduced concentrations of LDLs, VLDLs and Apo B in plasma (for details see MIM: 107730).
2. Familial ligand-defective apolipoprotein B-100 is a dominantly inherited disorder of lipoprotein metabolism resulting in hypercholesterolemia due to an impaired clearance of LDL particles by defective Apo B/E receptors and an increased proneness to coronary artery disease (for details see MIM: 144010).
3. Hypocholesterolemia is associated with other genetic defects.

**Apolipoprotein C-I** (Apo-CI) (see Data Sheet) is primarily synthesised in the liver and also in the intestine and is a major component of VLDLs, but is also contained in HDLs. Apo-CI is a small single-chain lipoprotein carrying many charged residues: 11 Asp/Glu and 12 Arg/Lys. Apo-CI inhibits the binding of ApoE-containing β-VLDL to the LDL receptor and to the LDL receptor-related protein. In addition, it moderately activates the lecithin cholesterol acyltransferase (LCAT).

The 3D solution structure of entire Apo-CI (57 aa: Thr 1–Ser 57) was determined by NMR spectroscopy (PDB: 1IOJ) and is shown in Figure 12.4. The structure is characterised by two α-helices linked by a short unstructured region.

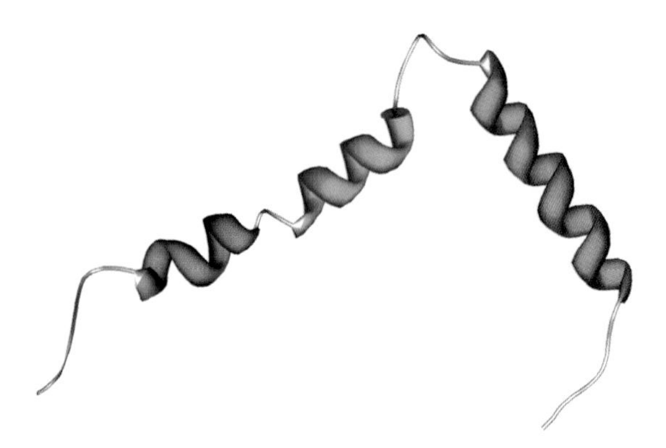

**Figure 12.4 3D structure of apolipoprotein C-I (PDB: 1IOJ)**
The structure contains two α-helices linked by a short unstructured region.

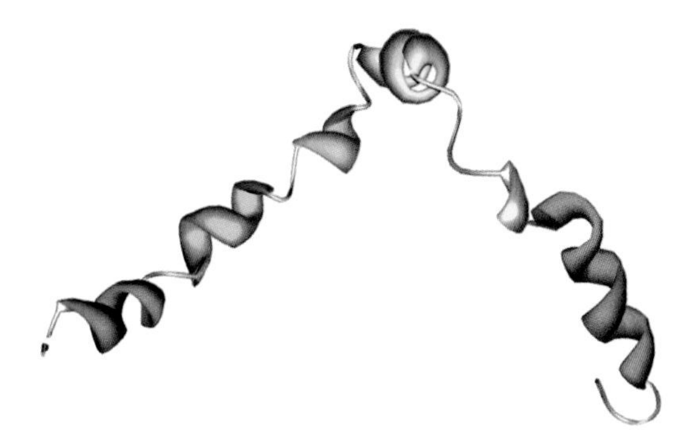

**Figure 12.5 3D structure of apolipoprotein C-II (PDB: 1SOH)**
The structure of apolipoprotein C-II in dodecyl phosphocholine micelles contains amphipathic α-helices connected by β-turns.

**Apolipoprotein C-II** (Apo-CII) (see Data Sheet) is primarily synthesised in the liver and to a smaller extent in the intestine and is a major component of VLDLs and chylomicrons but is also contained in HDLs. Apo-CII is a small single-chain lipoprotein containing a lipoprotein lipase cofactor region. Apo-CII is a potent activator of the enzyme lipoprotein lipase responsible for the hydrolysis of triacylglycerols.

The 3D solution structure of entire Apo-CII (79 aa: Thr 1–Glu 79) in the presence of dodecyl phosphocholine micelles was determined by NMR spectroscopy (PDB: 1SOH) and is shown in Figure 12.5. The structure is characterised by amphipathic α-helices connected by β-turns. The structural data are an experimental proof of the 'snorkel hypothesis' for the interaction of apolipoproteins with phospholipids first introduced by Segrest *et al.* (1974).

Defects in the Apo-CII gene are the cause of hyperlipoproteinemia type IB, which is an autosomal recessive disease characterised by hypertriglyceridemia (for details see MIM: 207750).

**Apolipoprotein C-III** (Apo-CIII) (see Data Sheet) is primarily synthesised in the liver and to a smaller extent in the intestine and is a major component of VLDLs (approximately 50 % of total protein) and chylomicrons, but is also contained in HDLs. Apo-CIII is a small single-chain glycolipoprotein containing a C-terminal lipid-binding region. Apo-CIII inhibits lipoprotein lipase and hepatic lipase and seems to delay the catabolism of triacylglycerol-rich particles.

**Apolipoprotein C-IV** (Apo-CIV) (see Data Sheet) is primarily expressed in the liver and seems to participate in the lipoprotein metabolism. Apo-CIV is a small single-chain glycolipoprotein.

**Apolipoprotein D** (Apo-D) (see Data Sheet) is expressed in many different types of tissues and exists in plasma as a homodimer and is primarily localised in HDLs. Apo-D is a glycolipoprotein (18 % CHO) containing two intrachain disulfide bridges (Cys 8–Cys 114, Cys 41–Cys 165) and an N-terminal pyro-Glu residue and belongs to the lipocalin family. Apo-D forms disulfide-linked heterodimers with Apo-AII (Apo-D: Cys 116; Apo-AII: Cys 6). Although its role in plasma is not clear, as a member of the lipocalin family it seems to transport small hydrophobic molecules such as cholesteryl esters, progesterone, arachidonic acid and heme-related components.

**Apolipoprotein E** (Apo-E) (see Data Sheet) is produced in almost all organs and is an important component of chylomicrons and VLDLs, but is also present to some extent in LDLs and HDLs. Apo-E is a single-chain glycolipoprotein with three major isoforms, E2 (15 %), E3 (77 %) and E4 (8 %), differing at two positions (112/158): Apo-E2 (Cys/Cys), Apo-E3 (Cys/Arg) and Apo-E4 (Arg/Arg). Like Apo-AII and Apo-AIV, Apo-E also contains tandem repeats (eight 22 aa long repeats) and, in addition, two heparin-binding regions and an LDL receptor-binding region (Thr 130–Ala 160) with basic residues that are crucial for receptor binding. Apo-E mediates the binding, internalisation and catabolism of lipoprotein particles and serves as a ligand for the LDL receptor and the LDL receptor-related protein. Apo-E is a major risk factor in heart diseases and in neurodegenerative diseases such as Alzheimer.

The 3D structure of the N-terminal domain of Apo-E (191 aa: Lys 1–Arg 191) containing the major heparin binding site in complex with a heparin-derived oligosaccharide was determined by X-ray diffraction (PDB: 1B68) and is shown in Figure 12.6. The structure is characterised by four α-helical segments and the heparin-binding site was assigned to a basic residue-rich region (indicated in pink) in helix 4 (Arg 142–Arg 150).

Defects in the Apo-E gene are the cause of hyperlipoproteinemia type III, also known as familial dysbetalipoproteinemia, characterised by xanthomas, yellowish lipid deposits in the palmar crease and on tendons and elbows (for details see MIM: 107741). The Apo-E4 allele is associated with the late onset of Alzheimer's disease (for details see MIM: 104310).

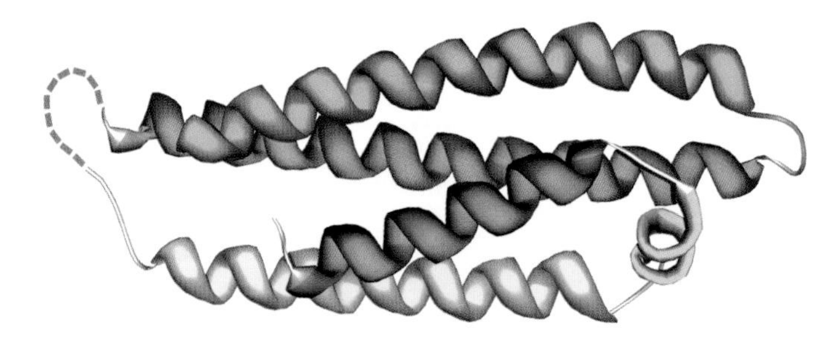

**Figure 12.6 3D structure of the N-terminal domain of apolipoprotein E (PDB: 1B68)**
Apolipoprotein E contains four α-helical segments and the heparin-binding site is located in a basic residues-rich region (in pink) in helix 4 (Unstructured region (grey dashed line) was modelled).

**Apolipoprotein F** (Apo-F) (see Data Sheet) is a minor apolipoprotein that associates with LDLs and to a lesser extent with VLDLs. Apo-F is a single-chain lipoprotein that is generated from a proform by releasing an N-terminal propeptide (128 aa). Apo-F inhibits the activity of the cholesteryl ester transfer protein and seems to be an important regulator of cholesterol transport.

**Apolipoprotein L1** (Apo-L1) (see Data Sheet) is expressed in many different tissues and is primarily associated with the large HDLs and interacts with Apo-AI. Apo-L1 is a single-chain glycolipoprotein that seems to play a role in the lipid exchange and transport in the body. It also seems to be involved in the reverse transport of cholesterol from peripheral tissues to the liver.

**Apolipoprotein(a)** (Apo(a), EC 3.4.21.-) (see Data Sheet) is synthesised in the liver and mainly circulates in blood linked to Apo B-100. It represents the protein component of the cholesteryl ester-rich lipoprotein Lp(a). Together with Apo B-100, Apo(a) is one of the largest known single-chain plasma glycoproteins (23 % CHO) containing 38 kringle domains and a serine protease domain. Apo(a) is linked to Apo B-100 via an interchain disulfide bridge (Apo(a): Cys 4059, Apo B-100: Cys 4326). Dozens of variants have been described with masses ranging from 240 to 880 kDa and kringle numbers from 12 to 51. Lp(a) seems to participate in the regulation of fibrinolysis by interaction with fibrin, thus competing with plasminogen. Apo (a) is known to be proteolytically degraded, resulting in the so-called mini-Lp(a). Apo(a) fragments accumulate in atherosclerotic lesions and thus may promote thrombogenesis.

The 3D structure of several kringle domains of Apo(a) are known and resemble those of other kringle-containing proteins (for details see Chapter 4). Increased levels of Apo(a) and its naturally occurring proteolytic fragments are the cause of atherosclerosis.

## 12.4 LDL AND VLDL RECEPTORS

Exogenous cholesterol is mainly internalised into cells via receptor-mediated endocytosis of lipoprotein particles. Receptor-mediated endocytosis is a general way in which cells can take up large molecules by an invagination process into the plasma membrane through corresponding and specific receptors. LDLs and to a lesser extent also VLDLs are the main carriers of cholesterol in plasma. Before the internalisation process can take place the formed receptor–lipoprotein particle complex must cluster into clathrin-coated pits that invaginate the plasma membrane to form coated vesicles, which fuse with lysosomes in the cells.

**Low-density lipoprotein receptor** (LDLR) (see Data Sheet) is a large single-chain type I membrane glycoprotein containing seven LDLRA, six LDLRB and three EGF-like domains (for details see Chapter 4), a potential C-terminal transmembrane segment (Ala 768–Trp 789) and a clustered O-CHO-linked segment (Thr 700–Ser 747). LDLR specifically binds to Apo B-100 and Apo-E in the LDLs, the major cholesterol-carrying lipoprotein in plasma, and transports it into the cells by endocytosis and it discharges its ligand into the endosome at pH < 6.

The 3D structure of the extracellular domain of LDLR (699 aa: Ala 1–Ala 699) was determined by X-ray diffraction (PDB: 1N7D) and is shown in Figure 12.7. The ligand-binding region comprising the LDLRR2–LDLRR7 domains folds back as an arc over the EGF-like A + B domains, the β-propeller region comprising the LDLRB1–LDLRB6 domains and the EGF-like C domain. The calcium ions are shown as orange balls (poor backbone connectivities are shown with dashed lines). The β-propeller region comprising the LDLRB1–LDLRB6 domains is given in pink. The LDLRR4 + R5 domains, which are critical for lipoprotein binding, associate with the β-propeller via their calcium-binding loop.

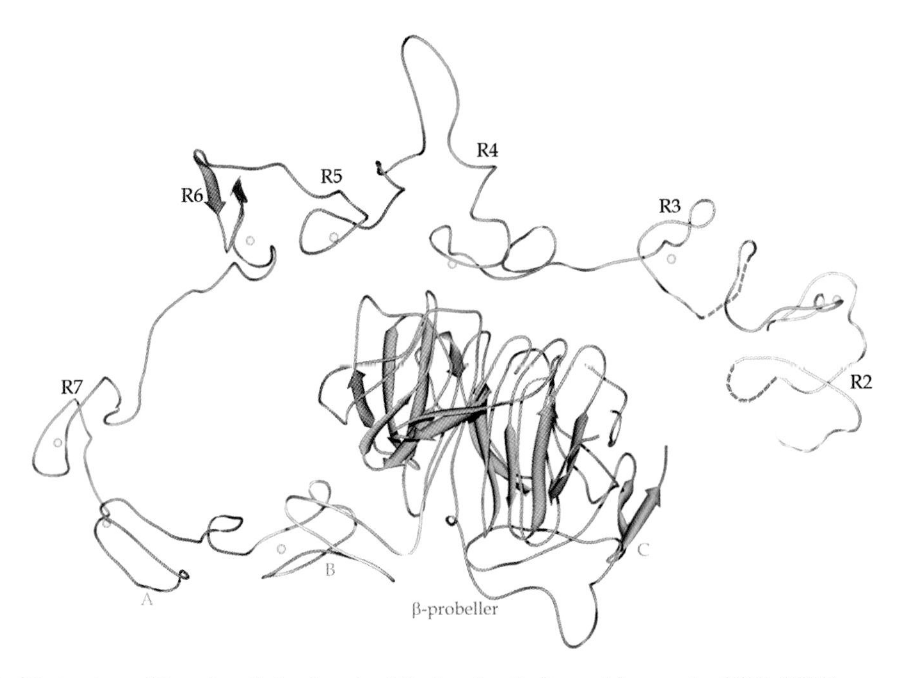


**Figure 12.7 3D structure of the extracellular domain of the low-density lipoprotein receptor (PDB: 1N7D)**
The extracellular domain of the low-density lipoprotein receptor contains six LDLRA domains R2 – R7 and three EGF-like domains A – C shown in different colours and the calcium ions are included as orange balls. The six LDLRBB1 – B6 domains (in pink) are contained in a β-propeller region.

Defects in the LDLR gene are the cause of familial hypercholesterolemia, a common (1 in 500 individuals) autosomal semi-dominant disease resulting in deposition of cholesterol in the skin (xanthelasma), tendons (xanthomas) and coronary arteries (atherosclerosis) (for details see MIM: 143890).

**Very-low-density lipoprotein receptor** (VLDLR) (see Data Sheet) is a large single-chain type I membrane glycoprotein that is abundant in heart and skeletal muscle, less in the ovaries and kidneys but not in the liver. VLDLR contains eight LDLRA, six LDLRB and three EGF-like domains (for details see Chapter 4), a potential C-terminal transmembrane segment (Ala 771–Trp 792) and a clustered O-CHO-linked segment (Thr 724–Ser 763). VLDLR binds VLDLs and transports it into the cells by endocytosis.

## 12.5 SERUM AMYLOID A PROTEINS

Serum amyloid A (SAA) proteins are a family of apolipoproteins predominantly associated with HDLs. SAAs are primarily synthesised in the liver but have been documented in various tissues. SAAs are a major acute phase marker during inflammation and their concentration can increase up to a thousand-fold within 24 h. SAAs are linked to inflammation, pathogen defence, HDL metabolism and cholesterol transport and are related to atherosclerosis, rheumatoid arthritis, Alzheimer's disease and cancer.

**Serum amyloid A protein** (SAA) (see Data Sheet) is primarily expressed in the liver and the precursor protein of the **amyloid protein A** (AA) by removing a C-terminal propeptide (28 aa) and is primarily contained in HDL particles. SAA is a single-chain lipoprotein with a probable and rare *N,N*-dimethyl-Asn (aa 83). Several well-documented isoforms exist where one, two or three residues at the N- or C-terminal end are removed, resulting in SAA(N-1), SAA(N-2), SAA(N-1/C-1), SAA(N-1/C-2) and SAA(N-3/C-3). There is a consensus pattern for the SAA protein family located in the central part of the sequence:

SAA (PS00992): A–R–G–N–Y–[ED]–A–x–[QKR]–R–G–x–G–G–x–W–A

In amyloidosis the AA protein frequently forms amyloid deposits in the liver, spleen and kidney. SAA is a major acute phase protein in inflammatory processes. EM pictures suggest a doughnut-shaped structure with a central cavity or a putative pore.

Reactive, secondary amyloidosis is caused by a chronic infection or inflammatory disease such as rheumatoid arthritis, fever, osteomyelitis (infection of the bone) or granulomatous ileitis (infection of the ileum, a portion of the small intestine) and is characterised by the extracellular accumulation of the SAA protein in various types of tissues. As these deposits (often fibrils) are highly insoluble and also resistant to proteolytic degradation they disrupt the tissue structure and impair its function.

**Serum amyloid A-4 protein** (SAA-4: 112 aa: P35542) belongs to the SAA family and is very similar in structure and function to SAA and exhibits a 55 % sequence identity with SAA.

## REFERENCES

Alaupovic, 1996, Significance of apolipoproteins for structure, function and classification of plasma lipoproteins, *Meth. Enzymol.*, **263**, 32–60.

Gursky, 2005 Apolipoprotein structure and dynamics, *Curr. Opin. Lipidol.*, **16**, 287–294.

Persson *et al.*, 1989, Structural features of lipoprotein lipase. Lipase family relationships, binding interactions, non-equivalence of lipase cofactors, vitellogenin similarities and functional subdivision of lipoprotein lipase, *Eur. J. Biochem.*, **179**, 39–45.

Segrest *et al.*, 1974, A molecular theory of lipid–protein interactions in the plasma lipoproteins, *FEBS lett.*, **38**, 247–258.

DATA SHEETS

## *Apolipoprotein(a) (EC 3.4.21.-)*

### Fact Sheet

| | | | |
|---|---|---|---|
| ***Classification*** | Peptidase S1 | ***Abbreviations*** | Apo(a) |
| ***Structures/motifs*** | 38 kringle | ***DB entries*** | APOA_HUMAN/P08519 |
| ***MW/length*** | 499 185 Da/4529 aa | ***PDB entries*** | |
| ***Concentration*** | 100–200 mg/l | ***Half-life*** | approx. 6 h |
| ***PTMs*** | Heavily glycosylated | | |
| ***References*** | McLean *et al.*, 1987, *Nature*, **330**, 132–137. | | |

### Description

**Apolipoprotein(a)** (Apo(a)) is the main component of lipoprotein(a) (Lp(a)) and is synthesised in the liver. It circulates in blood linked to Apo B-100.

### Structure

Apo(a) is, together with Apo B-100, one of the largest known single-chain plasma glycoproteins. Apo(a) is a single-chain glycoprotein (23 % CHO) containing 38 kringle domains and a serine protease domain. Variants containing from 12 to 51 kringles have been described. Apo(a) is disulfide-linked to Apo B-100 (Apo(a): Cys 4059, Apo B-100: Cys 4326).

### Biological Functions

The biological function of Apo(a) is not very well understood. However, Apo(a) seems to participate in the regulation of fibrinolysis by inhibiting tissue-type plasminogen activator and thus regulating plasmin activity.

# *Apolipoprotein A-I*

## Fact Sheet

| | | | |
|---|---|---|---|
| ***Classification*** | Apolipoprotein A1/A4/E | ***Abbreviations*** | Apo-AI |
| ***Structures/motifs*** | | ***DB entries*** | APOA1_HUMAN/P02647 |
| ***MW/length*** | 28 079 Da/243 aa | ***PDB entries*** | 1AV1 |
| ***Concentration*** | 1300–1500 mg/l | ***Half-life*** | 3–5 days |
| ***PTMs*** | 1 Palmitoyl (site unknown) | | |
| ***References*** | Borhani *et al.*, 1997, *Proc. Natl. Acad. Sci. USA*, **94**, 12291–12296.<br>Shoulders *et al.*, 1983, *Nucleic Acids Res.*, **11**, 2827–2837. | | |

## Description

**Apolipoprotein A-I** (Apo-AI) is primarily synthesised in the liver and in the small intestine and is the major apolipoprotein in the high-density lipoproteins (HDLs).

## Structure

Apo-AI is a single-chain plasma protein containing eight tandem repeats (22 aa) and two half-length repeats (11 aa) and is palmitoylated. A six residue N-terminal propeptide is released from the proform to generate mature Apo-AI. The 3D structure of tetrameric N-terminal truncated Apo-AI (201 aa: Asn43Met–Gln 243) consists almost entirely of an amphipathic α-helix and adopts a shape similar to a horseshoe (**1AV1**).

## Biological Functions

Apo-AI participates in the transport of cholesterol from tissues to the liver for excretion by promoting the cholesterol efflux from tissues and by acting as a cofactor for the lecithin cholesterol acyltransferase (LCAT). Apo-AI seems to be protective against the development of atherosclerosis.

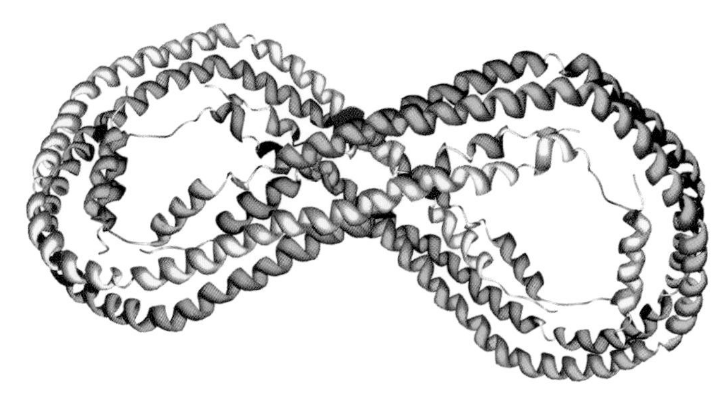

Apolipoprotein A-I

## *Apolipoprotein A-II*

### Fact Sheet

| | | | |
|---|---|---|---|
| ***Classification*** | Apolipoprotein A2 | ***Abbreviations*** | Apo-AII |
| ***Structures/motifs*** | | ***DB entries*** | APOA2_HUMAN/P02652 |
| ***MW/length*** | 8708 Da/77 aa | ***PDB entries*** | 2OU1 |
| ***Concentration*** | 450–750 mg/l | ***Half-life*** | 3–6 days |
| ***PTMs*** | 1 Pyro-Glu | | |
| ***References*** | Knott *et al.*, 1985, *Nucleic Acids Res.*, **13**, 6387–6398.<br>Kumar *et al.*, 2002, *Biochemistry*, **41**, 11681–11691. | | |

### Description

**Apolipoprotein A-II** (Apo-AII) is primarily synthesised in the liver and exists in plasma as a homodimer and is a major apolipoprotein in the high-density lipoproteins (HDLs).

### Structure

Apo-AII is a homodimer of two disulfide-linked chains (Cys 6–Cys 6) containing 11 aa and 22 aa repeats and an N-terminal pyro-Glu residue. In addition, Apo-AII forms heterodimers with Apo-D (Apo-AII: Cys 6; Apo-D: Cys 116). A five residue N-terminal propeptide is released from the proform to generate mature Apo-AII. The disulfide-linked homodimers (disulfide bridges in pink) aggregate into tetramers characterised by amphipathic α-helices (**2OU1**).

### Biological Functions

Apo-AII is a major component of HDLs and seems to be important for its stabilisation.

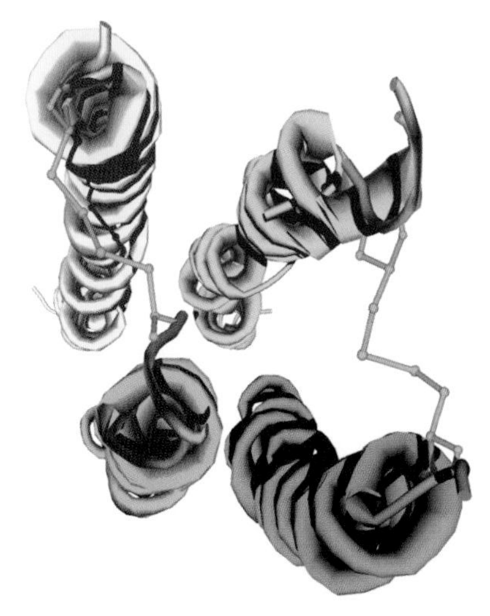

Apolipoprotein A-II

# *Apolipoprotein A-IV*

## Fact Sheet

| | | | |
|---|---|---|---|
| ***Classification*** | Apolipoprotein A1/A4/E | ***Abbreviations*** | Apo-AIV |
| ***Structures/motifs*** | 13 Tandem repeats (approx. 22 aa) | ***DB entries*** | APOA4_HUMAN/P06727 |
| ***MW/length*** | 43 375 Da/376 aa | ***PDB entries*** | |
| ***Concentration*** | 120–150 mg/l | ***Half-life*** | 16–20 h |
| ***PTMs*** | | | |
| ***References*** | Karathanasis *et al.*, 1986, *Proc. Natl. Acad. Sci. USA*, **83**, 8457–8461. | | |

## Description

**Apolipoprotein A-IV** (Apo-AIV) is synthesised in the enterocytes of the small intestine and is a major component of the high-density lipoproteins (HDLs) and chylomicrons.

## Structure

Apo-AII is a single-chain glycoprotein (6 % CHO) containing 13 approximately 22 aa long tandem repeats and an approx. 20 aa long C-terminal region rich in Gln/Glu.

## Biological Functions

Apo-AIV is a major component of chylomicrons and seems to play a role in the secretion and catabolism of chylomicrons and very-low-density lipoproteins (VLDLs).

# *Apolipoprotein B-100/Apolipoprotein B-48*

## Fact Sheet

| | | | |
|---|---|---|---|
| ***Classification*** | | ***Abbreviations*** | Apo B-100/Apo B-48 |
| ***Structures/motifs*** | 1 Vitellogenin | ***DB entries*** | APOB_HUMAN/P04114 |
| ***MW/length*** | Apo B-100: 512 816 Da/4536 aa<br>Apo B-48: 240 748 Da/2152 aa | ***PDB entries*** | |
| ***Concentration*** | approx. 1000 mg/l | ***Half-life*** | 36 h |
| ***PTMs*** | 19 N-CHO (potential);<br>1 S-palmitoyl (Cys 1085) | | |
| ***References*** | Knott *et al.*, 1986, *Nucleic Acids Res.*, **14**, 7501–7503. | | |

## Description

**Apolipoprotein B-100** (Apo B-100) is the major apolipoprotein of low-density lipoproteins (LDLs) but is also present in very-low-density lipoproteins (VLDLs) and chylomicrons and is synthesised in the liver. Apolipoprotein B-48 (Apo B-48) is derived from Apo B-100 by RNA editing and is found in chylomicrons.

## Structure

Apo B-100 is one of the largest known monomeric proteins. Apo B-100 is a single-chain glycoprotein containing a vitellogenin domain and is palmitoylated at Cys 1085. Apo B-100 contains several heparin-binding regions and two basic regions (11 aa: **KAQYKKNKHRH**, 9 aa: **RLTRKRGLK**), thought to be possible receptor binding regions. Apo B-48 (2152 aa) comprises the N-terminal part of Apo B-100 and is created by RNA editing (introduction of a stop codon UAA at Gln 2153).

## Biological Functions

Apo B-100 plays a role in the internalisation of LDL particles by the ApoB/E receptor and functions as a recognition signal for cellular binding.

# *Apolipoprotein C-I*

## Fact Sheet

| | | | |
|---|---|---|---|
| ***Classification*** | Apolipoprotein C1 | ***Abbreviations*** | Apo-CI |
| ***Structures/motifs*** | | ***DB entries*** | APOC1_HUMAN/P02654 |
| ***MW/length*** | 6631 Da/57 aa | ***PDB entries*** | 1IOJ |
| ***Concentration*** | 40–60 mg/l | ***Half-life*** | 3–3.5 days |
| ***PTMs*** | | | |
| ***References*** | Knott *et al.*, 1984, *Nucleic Acids Res.*, **12**, 3909–3915.<br>Rozek *et al.*, 1998, *Biochem. Cell. Biol.*, **76**, 267–275. | | |

## Description

**Apolipoprotein C-I** (Apo-CI) is primarily synthesised in the liver and is a major component of the very-low-density lipoproteins (VLDLs).

## Structure

Apo-ACI is a small single-chain plasma lipoprotein carrying many charged residues: 11 Asp/Glu, 12 Arg/Lys. The 3D structure is characterised by two α-helices linked by a short unstructured region (**1IOJ**).

## Biological Functions

Apo-CI is a major component of VLDLs and a minor of HDLs and inhibits the binding of Apo-E-containing β-VLDL to the LDL receptor-related protein and the LDL receptor. In addition, Apo-CI moderately activates the lecithin-cholesterol acyltransferase (LCAT).

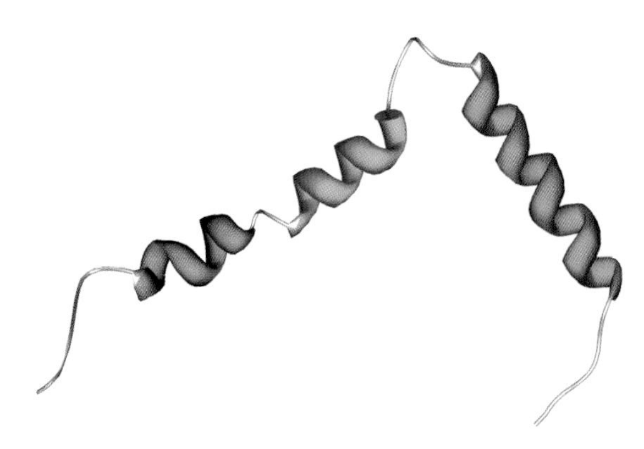

Apolipoprotein C-I

# *Apolipoprotein C-II*

## Fact Sheet

| | | | |
|---|---|---|---|
| ***Classification*** | Apolipoprotein C2 | ***Abbreviations*** | Apo-CII |
| ***Structures/motifs*** | | ***DB entries*** | APOC2_HUMAN/P02655 |
| ***MW/length*** | 8915 Da/79 aa | ***PDB entries*** | 1SOH |
| ***Concentration*** | 30–50 mg/l | ***Half-life*** | 2.6–3.2 days |
| ***PTMs*** | | | |
| ***References*** | Fojo *et al.*, 1984, *Proc. Natl. Acad. Sci. USA*, **81**, 6354–6357.<br>MacRalid *et al.*, 2004, *Biochemistry*, **43**, 8084–8093. | | |

## Description

**Apolipoprotein C-II** (Apo-CII) is primarily synthesised in the liver and is a major component of the very-low-density lipoproteins (VLDL).

## Structure

Apo-ACII is a small single-chain plasma lipoprotein containing a lipoprotein lipase cofactor region. The 3D structure is characterised by amphipathic α-helices connected by β-turns (**1SOH**).

## Biological Functions

Apo-CII is a major component of VLDL and a minor of HDL and is a potent activator of lipoprotein lipase.

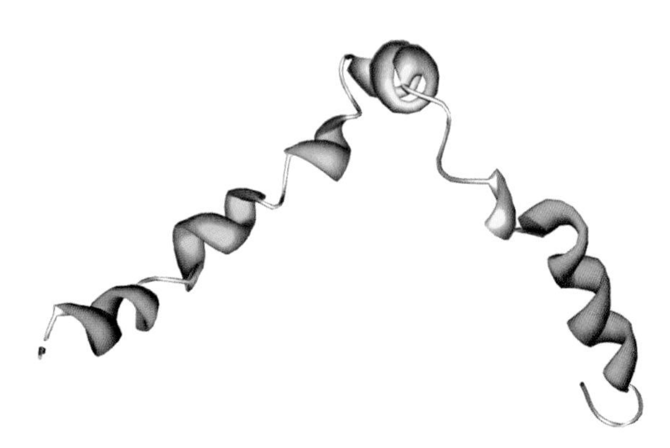

Apolipoprotein C-II

# *Apolipoprotein C-III*

## Fact Sheet

| | | | |
|---|---|---|---|
| ***Classification*** | Apolipoprotein C3 | ***Abbreviations*** | Apo-CIII |
| ***Structures/motifs*** | | ***DB entries*** | APOC3_HUMAN/P02656 |
| ***MW/length*** | 8765 Da/79 aa | ***PDB entries*** | |
| ***Concentration*** | 120–140 mg/l | ***Half-life*** | 2.1–2.8 days |
| ***PTMs*** | 1 O-CHO | | |
| ***References*** | Protter *et al.*, 1984, *DNA*, **3**, 449–456. | | |

## Description

**Apolipoprotein C-III** (Apo-CIII) is mainly synthesised in the liver and is the main protein component of the very-low-density lipoproteins (VLDLs).

## Structure

Apo-ACIII is a small single-chain plasma glycolipoprotein containing a C-terminal lipid-binding region.

## Biological Functions

At 50 % Apo-CIII is the main component of the protein fraction of VLDLs and a minor of HDLs. Apo-CIII inhibits lipoprotein lipase and hepatic triacylglycerol lipase.

# *Apolipoprotein C-IV*

## Fact Sheet

| | | | |
|---|---|---|---|
| ***Classification*** | Apolipoprotein C4 | ***Abbreviations*** | Apo-CIV |
| ***Structures/motifs*** | | ***DB entries*** | APOC4_HUMAN/P55056 |
| ***MW/length*** | 11 816 Da/101 aa | ***PDB entries*** | |
| ***Concentration*** | | ***Half-life*** | |
| ***PTMs*** | 1 N-CHO (potential) | | |
| ***References*** | Allan *et al.*, 1995, *Genomics*, **28**, 291–300. | | |

## Description

**Apolipoprotein C-IV** (Apo-CIV) is primarily expressed in the liver and secreted into plasma.

## Structure

Apo-CIV is a single-chain plasma glycolipoprotein.

## Biological Functions

Apo-CIV seems to participate in the lipoprotein metabolism.

# *Apolipoprotein D*

## Fact Sheet

| | | | |
|---|---|---|---|
| ***Classification*** | Lipocalin | ***Abbreviations*** | Apo-D |
| ***Structures/motifs*** | | ***DB entries*** | APOD_HUMAN/P05090 |
| ***MW/length*** | 19 303 Da/169 aa | ***PDB entries*** | |
| ***Concentration*** | 60–120 mg/l | ***Half-life*** | |
| ***PTMs*** | 2 N-CHO; 1 Pyro-Glu | | |
| ***References*** | Drayna *et al.*, 1986, *J. Biol. Chem.*, **261**, 16535–16539. | | |

## Description

**Apolipoprotein D** (Apo-D) is synthesised in many different types of tissues and exists in plasma as a homodimer. Apo-D is primarily localised in high-density lipoproteins (HDLs).

## Structure

Apo-D is a glycolipoprotein (18 % CHO) with an N-terminal pyro-Glu and forms disulfide-linked heterodimers with Apo-AII (Apo-D: Cys 116; Apo-AII: Cys 6).

## Biological Functions

The exact role of Apo-D in the lipid metabolism remains unclear. However, as a member of the lipocalin family Apo-D seems to be involved in the binding and transport of small hydrophobic molecules in plasma.

## *Hepatic triacylglycerol lipase (Hepatic lipase; EC 3.1.1.3)*

### Fact Sheet

| | | | |
|---|---|---|---|
| ***Classification*** | Lipase | ***Abbreviations*** | HL |
| ***Structures/motifs*** | 1 PLAT | ***DB entries*** | LIPC_HUMAN/P11150 |
| ***MW/length*** | 53 497 Da/477 aa | ***PDB entries*** | |
| ***Concentration*** | 120–150 μg/l (post-heparin plasma) | ***Half-life*** | |
| ***PTMs*** | 4 N-CHO (potential) | | |
| ***References*** | Datta *et al.*, 1988, *J. Biol. Chem.*, **263**, 1107–1110. | | |

### Description

**Hepatic triacylglycerol lipase** (HL) is exclusively synthesised in hepatocytes and is an important enzyme in the high-density lipoprotein (HDL) metabolism.

### Structure

HL is a single-chain plasma glycolipoprotein (18 % CHO) containing five intrachain disulfide bridges, a PLAT domain, a potential heparin-binding region (Ile 159–Leu 171) and the active site consists of Ser 146, Asp 172 and His 257.

### Biological Functions

HL has the capacity to catalyse the hydrolysis of phospholipids, mono-, di-, and triacylglycerols and acyl-CoA thioesters.

## *Lipoprotein lipase (EC 3.1.1.34)*

### Fact Sheet

| | | | |
|---|---|---|---|
| ***Classification*** | Lipase | ***Abbreviations*** | LPL |
| ***Structures/motifs*** | 1 PLAT | ***DB entries*** | LIPL_HUMAN/P06858 |
| ***MW/length*** | 50 394 Da/448 aa | ***PDB entries*** | |
| ***Concentration*** | 150–300 μg/l | ***Half-life*** | 30–40 min |
| ***PTMs*** | 2 N-CHO (potential) | | |
| ***References*** | Wion *et al.*, 1987, *Science*, **235**, 1638–1641. | | |

### Description

**Lipoprotein lipase** (LPL) is synthesised in parenchymal cells and exists as a homodimer in its functional form. LPL hydrolysis triacylglycerols of chylomicrons and very-low-density lipoproteins (VLDL).

### Structure

LPL is a single-chain plasma glycolipoprotein (8 % CHO) containing five intrachain disulfide bridges, a PLAT domain and a heparin-binding region (Lys 292–Lys 304) and the active site consists of Ser 132, Asp 156 and His 241.

### Biological Functions

LPL catalyses the reaction triacylglycerol + $H_2O \Leftrightarrow$ diacylglycerol + fatty acid. LPL requires apolipoprotein C-II as a cofactor for maximal activity and apolipoprotein C-III seems to be an important biological inhibitor.

## *Low-density lipoprotein receptor*

### Fact Sheet

| | | | |
|---|---|---|---|
| ***Classification*** | LDLR | ***Abbreviations*** | LDLR |
| ***Structures/motifs*** | 7 LDLRA; 6 LDLRB; 3 EGF-like | ***DB entries*** | LDLR_HUMAN/P01130 |
| ***MW/length*** | 93 095 Da/839 aa | ***PDB entries*** | 1N7D |
| ***Concentration*** | | ***Half-life*** | |
| ***PTMs*** | 5 N-CHO (potential); clustered O-CHO (Thr 700–Ser 747) | | |
| ***References*** | Rudenko *et al.*, 2002, *Science*, **298**, 2353–2358. | | |
| | Yamamoto *et al.*, 1984, *Cell*, **39**, 27–38. | | |

### Description

**Low-density lipoprotein receptor** (LDLR) binds to low-density lipoproteins (LDLs), the major cholesterol-carrying lipoprotein in plasma.

### Structure

LDLR is a single-chain plasma glycolipoprotein containing seven LDLRA, six LDLRB and three EGF-like domains and in the C-terminal region a potential 22 residue transmembrane segment (Ala 768–Trp 789). In the 3D structure of the extracellular domain of the LDLR the ligand-binding region comprising the LDLRR2–R7 domains folds back as an arc to the EGF-like A + B domains, the β-propeller region comprising the LDLRB1–B6 domains and the EGF-like C domain (**1N7D**).

### Biological Functions

LDLR forms a complex with LDLs, the major-cholesterol-carrying lipoprotein, and thus transports cholesterol via endocytosis into the cells.

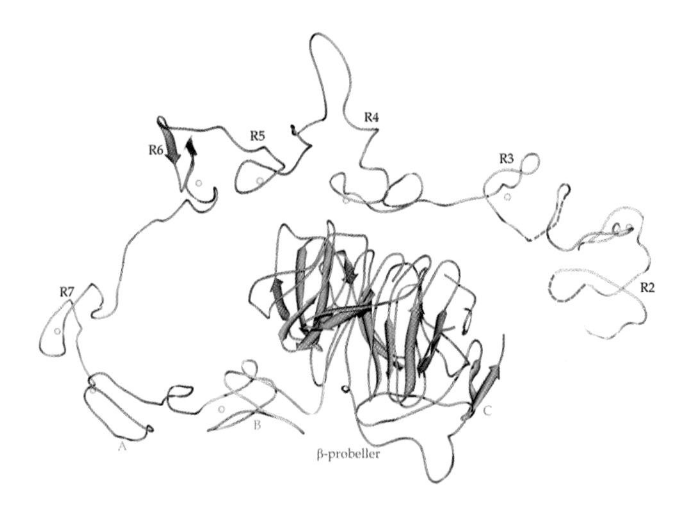


Low-density lipoprotein receptor

## Phosphatidylcholine-sterol acyltransferase (Lecithin-cholesterol acyltransferase; *EC 2.3.1.43*)

### Fact Sheet

| | | | |
|---|---|---|---|
| ***Classification*** | Lipase | ***Abbreviations*** | LCAT |
| ***Structures/motifs*** | | ***DB entries*** | LCAT_HUMAN/P04180 |
| ***MW/length*** | 47 084 Da/416 aa | ***PDB entries*** | |
| ***Concentration*** | 5–7 mg/l | ***Half-life*** | |
| ***PTMs*** | 4 N-CHO; 2 O-CHO | | |
| ***References*** | McLean *et al.*, 1986, *Proc. Natl. Acad. Sci. USA*, **83**, 2335–2339. | | |

### Description

**Phosphatidylcholine-sterol acyltransferase** (LCAT) is primarily synthesised in the liver and is the central enzyme in the extracellular metabolism of plasma lipoproteins.

### Structure

LCAT is a single-chain glycolipoprotein (20 % CHO) containing two intrachain disulfide bridges (Cys 50–Cys 74, Cys 313–Cys 356) and two free Cys (aa 31 and 184) adjacent to the active site, which have to be in the reduced state for optimal activity of LCAT. The active site consists of Ser 181, His 377 and an unidentified Asp/Glu residue.

### Biological Functions

LCAT plays a key role in the reverse transport of cholesterol from peripheral tissues to the liver. It catalyses the reaction phosphatidylcholine + sterol ⇔ 1-acylglycerophosphocholine + sterol ester. Apolipoprotein A-I is an efficient activator of LCAT.

# *Serum amyloid A protein (SAA)/Amyloid protein A (AA)*

## Fact Sheet

| | | | |
|---|---|---|---|
| ***Classification*** | SAA | ***Abbreviations*** | SAA |
| ***Structures/motifs*** | | ***DB entries*** | SAA_HUMAN/P02735 |
| ***MW/length*** | AA: 8575 Da/76 aa<br>SAA: 11 683 Da/104 aa | ***PDB entries*** | |
| ***Concentration*** | approx. 10 mg/l | ***Half-life*** | Mouse: approx. 1 h |
| ***PTMs*** | *N,N*-dimethyl-Asn (probable, aa 83) | | |
| ***References*** | Sipe *et al.*, 1985, *Biochemistry*, **24**, 2931–2936. | | |

## Description

**Serum amyloid A protein** (SAA) is synthesized primarily in the liver and is associated with high-density lipoproteins (HDLs). SAA is the precursor of amyloid protein A (AA).

## Structure

SAA is a single-chain plasma lipoprotein containing a C-terminal 28 residue propeptide and a rare *N,N*-dimethyl-Asn (aa 83). During amyloidogenesis the propeptide is often cleaved from the precursor SAA, resulting in amyloid protein A (76 aa). There are several forms of serum amyloid protein A.

## Biological Functions

SAA is a major acute-phase protein in inflammatory processes. The biological function of SAA is not clear. However, the SAA protein can accumulate in extracellular tissues, leading to highly insoluble deposits that are resistant to proteolysis and resulting in the disrupture of tissue structures and loss of their function.

# *Very-low-density lipoprotein receptor*

## Fact Sheet

| | | | |
|---|---|---|---|
| ***Classification*** | VLDLR | ***Abbreviations*** | VLDLR |
| ***Structures/motifs*** | 8 LDLRA; 6 LDLRB; 3 EGF-like | ***DB entries*** | VLDLR_HUMAN/P98155 |
| ***MW/length*** | 93 240 Da/846 aa | ***PDB entries*** | |
| ***Concentration*** | | ***Half-life*** | |
| ***PTMs*** | 3 N-CHO (potential); clustered O-CHO (Ser 724–Ser 763) | | |
| ***References*** | Sakai *et al.*, 1994, *J. Biol. Chem.*, **269**, 2173–2182. | | |

## Description

**Very-low-density lipoprotein receptor** (VLDLR) binds to very-low-density lipoproteins (VLDLs).

## Structure

VLDLR is a single-chain plasma glycolipoprotein containing eight LDLRA, six LDLRB and three EGF-like domains and in the C-terminal region a potential 22 residue transmembrane segment (Ala 771–Trp 792).

## Biological Functions

VLDLR forms a complex with VLDLs and transports it into the cell via endocytosis.

# 13

# Hormones

## 13.1 INTRODUCTION

The coordination of the signalling pathways and the signal transduction is of utmost importance in every living organism at all organisational levels. Intracellular signalling occurs via synthesis and modifications of various types of molecules within the cell such as the feedback control mechanisms in blood coagulation (see Chapter 7) or by covalent modifications (see Chapter 6). Intercellular signalling occurs either via chemical messengers (*hormones*) or via electrochemical impulses through the neuronal system. Hormones are divided into three classes depending on the distance over which they act:

1. ***Autocrine hormones*** act on the same cell that has released them. Interleukin-2, for example, is required for stimulation of T cell proliferation.
2. ***Paracrine hormones*** are local mediators that act on cells close to the site of the releasing cell. Many growth factors are paracrine hormones.
3. ***Endocrine hormones*** act on cells in a distance from the site of the releasing cell. Endocrine hormones are often released by endocrine glands such as the islets of Langerhans in the case of the release of insulin and are transported via the bloodstream to the target cells.

Most hormones are either peptides/proteins, modified amino acids or steroids. Hormones are usually specifically addressed to a certain site in the body and only cells that carry a specific receptor for that type of hormone will respond to the signal. In this chapter only the class of (poly)peptide hormones will be discussed. Most peptide hormones are present in plasma at very low concentrations (pg/ml level); nevertheless, their presence in plasma is of great importance. Many glands of the endocrine system such as the hypothalamus, the pituitary, the parathyroid, the thyroid and pancreas as well as many tissues such as the stomach, intestine, liver and placenta release peptide hormones. In Table 13.1 a selection of (poly)peptide hormones with their characteristics is summarised.

Many peptide hormones are synthesised as preprohormones in the ribosomes, then processed in the endoplasmic reticulum and the Golgi apparatus to their mature form, stored in secretory granules and then released by exocytosis upon the corresponding signal.

## 13.2 PANCREATIC HORMONES

The pancreas is a large gland most of which is an exocrine gland that synthesises many digestive enzymes such as trypsin and α-amylase. A small part of the pancreas is an endocrine gland with cells called the islets of Langerhans. They contain three different types of cells, which produce three different hormones:

1. The α cells produce *glucagon*, which is secreted in response to a low level of glucose in blood. Glucagon stimulates the liver to release glucose and regulates the blood glucose level by decreasing glycolysis and increasing gluconeogenesis. In addition, it stimulates adipose tissue to release fatty acids via lipolysis. Glucagon is the counterregulatory hormone of insulin.

---

*Human Blood Plasma Proteins: Structure and Function* Johann Schaller, Simon Gerber, Urs Kämpfer, Sofia Lejon and Christian Trachsel

**Table 13.1** Characteristics of a selection of hormones in human plasma.

| Hormone | Origin | Size (# aa) | Major effects |
|---|---|---|---|
| Glucagon | Pancreas | 29 | Stimulation glucose release |
| Insulin | Pancreas | 21 + 30 | Stimulation glucose uptake |
| Somatostatins | Hypothalamus/pancreas | 28/14 | Inhibition growth hormone release |
| Pancreatic hormone | Pancreas | 30 | Antagonist cholecystokinin |
| Gastrins | Stomach | 34/17 | Stimulation secretion HCl and pepsinogen |
| Gastrin-releasing peptide | Vagus nerve | 27 | Gastrin release |
| Secretin | Intestine | 27 | Stimulation secretion $HCO_3^-$ |
| Cholecystokinins | Intestine | Variable | Stimulation secretion digestive enzymes and $HCO_3^-$ from pancreas; stimulation gallbladder emptying |
| Gastric inhibitory polypeptide | Intestine | 42 | Stimulation insulin release; inhibition gastric acid secretion |
| Motilin | Small intestine | 22 | Gastrointestinal motility |
| Parathyroid hormone | Parathyroid | 84 | Stimulation $Ca^{2+}$ uptake |
| Calcitonin | Thyroid | 32 | Inhibition $Ca^{2+}$ uptake |
| Corticoliberin | Hypothalamus | 41 | Stimulation corticotropin release |
| Corticotropin | Adenohypophysis | 39 | Stimulation release corticosteroids |
| Thyroliberin | Hypothalamus | 3 | Stimulation thyrotropin release |
| Thyrotropin | Adenohypophysis | 92 + 112 | Stimulation release $T_3$ and $T_4$[a] |
| Thyroglobulin | Thyroid gland | 2749 | Precursor thyroid hormones |
| Gonadoliberin I + II | Hypothalamus/kidney | 10 | Stimulation gonadotropin release |
| Follitropin | Adenohypophysis | 92 + 111 | Ovaries: stimulation development, follicles, ovulation, estrogen synthesis<br>Testes: stimulation spermatogenesis |
| Lutropin | Adenohypophysis | 92 + 121 | Ovaries: stimulation maturation oocytes; synthesis estrogen and progesterone<br>Testes: stimulation androgen synthesis |
| Somatoliberin | Hypothalamus | 44 | Stimulation growth hormone release |
| Somatotropin | Adenohypophysis | 191 | Stimulation somatomedin synthesis |
| Somatomedins | Liver | 70/67 | Stimulation cartilage growth |
| Choriogonadotropin | Placenta | 92 + 145 | Stimulation progesterone release |
| Prolactin | Adenohypophysis | 199 | Promotion lactation |
| Prolactin-releasing peptide | Hypothalamus | 31/20 | Release of prolactin |
| Arg-vasopressin | Neurohypophysis | 9 | Stimulation water resorption by kidneys; increase blood pressure |
| Oxytocin | Neurohypophysis | 9 | Stimulation uterus contraction |
| β-Endorphin | Adenohypophysis | 31 | Opoid effects, central nervous system |
| Met-enkephalin/Leu-enkephalin | Adenohypophysis | 5 | Opoid effects, central nervous system |
| Dynorphin | Brain | 17 | Endogenous opioid peptide |
| Atrial natriuretic factor | Atria of heart | 28 | Homeostatic control of water and sodium |
| Vasoactive intestinal peptide | Pancreas | 28 | Vasodilatation; decreases blood pressure |
| Calcitonin gene-related peptides | Central nervous system | 37 | Vasodilatation |
| Substance P | Central nervous system | 11 | Neurotransmission/vasodilatation |
| Erythropoietin | Kidney/liver | 166 | Differentiation erythrocytes |
| Thrombopoietin | Hepatocytes | 332 | Regulation production of megakaryocytes and platelets |
| Islet amyloid polypeptide | Pancreas | 37 | Inhibition glucose utilisation and glycogen deposition |

[a] $T_3$: triiodothyronine; $T_4$: thyroxine.

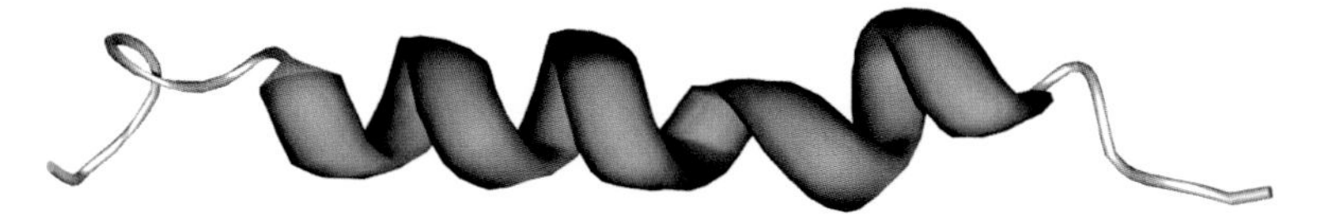

**Figure 13.1 3D structure of the glucagon-like peptide 1 (PDB: 1D0R)**
The glucagon-like peptide 1 assumes a stable, single-stranded helical structure in a water/trifluoroethanol solution.

2. The β cells produce *insulin*, which is secreted in response to a high level of glucose in the bloodstream. Insulin stimulates the liver to store glucose for later use via the synthesis of glycogen. Insulin accelerates glycolysis and the pentose phosphate cycle and increases the cell permeability for monosaccharides, amino acids and fatty acids. In addition, it promotes the protein and fat metabolism. The effects of insulin revert those of glucagon.
3. The δ cells (as well as the hypothalamus) produce *somatostatin*, which inhibits the release of somatotropin, a hormone that is important for growth control and the control of the release of glucagon and insulin.

**(Pro)-glucagon** (see Data Sheet) is the precursor and the parent protein of several physiologically important peptide hormones. Proglucagon is proteolytically processed by the prohormone convertases PCSK1 and PCSK2 to the corresponding, biologically active peptides: **glucagon** (29 aa), **glicentin** (69 aa), **oxyntomodulin** (37 aa), three **glucagon-like peptides 1** (37aa, 31 aa, 30 aa, respectively) and a **glucagon-like peptide 2** (33 aa). Glucagon regulates the blood glucose level and plays a key role in glucose metabolism and homeostasis. Glicentin modulates the gastric acid secretion and the gastro-pyloro-duodenal activity and seems to play an important role in the growth of the intestinal mucosa in early life. Oxyntomodulin significantly reduces food intake and inhibits gastric emptying. Glucagon-like peptide 1 (37 aa) is a potent stimulator of the glucose-dependent release of insulin and plays an important role in the suppression of the glucagon level in plasma. Glucagon-like peptide 2 plays a key role in nutrient homoestasis, stimulates the intestinal transport of glucose and decreases the permeability of the mucosa.

Glucagon exhibits an extended structure with an α-helical segment (PDB: 1BH0; see Chapter 5). Charged residues (Arg 17, Arg 18, Asp 21) and the ability to form salt bridges are important for the biological activity of glucagon. The glucagon-like peptide 1 (30 aa: His 7–Arg-$NH_2$ 36) gradually assumes a stable, single-stranded helical structure in an aqueous solution when the concentration of trifluoroethanol is increased from 0 to 35 % (v/v) (PDB: 1D0R), as can be seen in Figure 13.1.

Deficiency of glucagon results in hypoglycemia, a decreased level of the blood sugar content. The lower limit is thought to be 0.7 mg/ml (3.9 mM) for normal glucose levels.

**(Pro)-insulin** (see Data Sheet) is the precursor of **insulin** that is generated by limited proteolysis of proinsulin by the prohormone-converting enzyme, releasing a propeptide that is further processed to the insulin C-peptide by cleavage of two terminal dibasic dipeptides. Insulin is a two-chain (A- and B-chain) peptide hormone linked by two interchain disulfide bridges (Cys A7–Cys B7, Cys A20–Cys B19). The A-chain contains one intrachain disulfide bridge (Cys 6–Cys 11).

Insulin is primarily involved in the carbohydrate and secondarily the protein and fat metabolism. It is responsible for the decrease of the glucose concentration in blood. It increases the permeability of cells for monosaccharides, amino acids and fatty acids. It accelerates the synthesis of glycogen in the liver, glycolysis and the pentose phosphate cycle. Insulin is the counterregulatory hormone of glucagon that rises the blood glucose level in response to hypoglycemia.

The 3D structure of insulin is highly flexible and the hexameric structure is symmetrical (PDB: 3AIY; see Chapter 5) and is shown in Figure 13.2. The A-chain (21 aa: Gly 1–Asn 21) has two short α-helical segments and the B-chain (30 aa: Phe 1–Thr 30) a longer α-helix and a short β-strand. Each monomer is shown in a different colour.

Diabetes mellitus is characterised by persistent hyperglycemia resulting from insufficient secretion of insulin or an inadequate response of target cells to insulin. The three most common forms of diabetes are type 1, type 2 and gestational diabetes.

Type 1 diabetes mellitus is characterised by the loss of insulin-producing β cells of the islets of Langerhans of the pancreas and is most commonly diagnosed in children and adolescents but can also occur in adults. Type 2 diabetes mellitus results from a defective insulin secretion and a defective responsiveness to insulin and may remain undiagnosed for many years because of typically milder symptoms. Gestational diabetes is a combination of an inadequate insulin secretion and responsiveness, thus resembling type 2 diabetes. It occurs in 2–5 % of all pregnancies, is of temporary nature and completely treatable.

Defects in the insulin gene are the cause of familial hyperproinsulinemia (for details see: MIM: 176730).

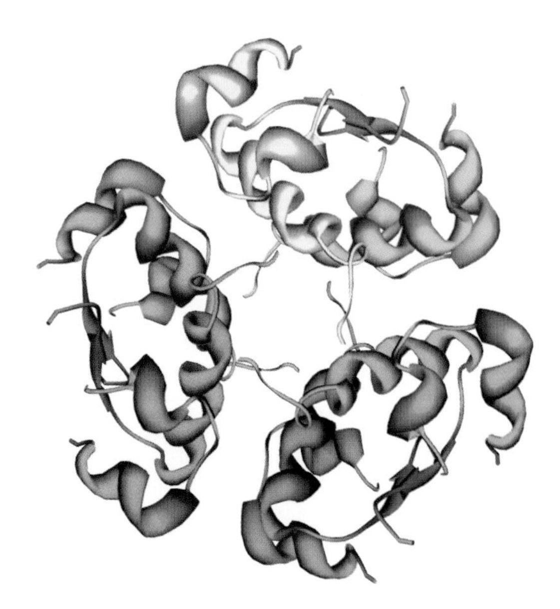

**Figure 13.2 3D structure of hexameric insulin (PDB: 3AIY)**
Hexameric insulin exhibits a symmetrical structure and each monomer is shown in a different colour.

**Somatostatin** (see Data Sheet) or growth hormone release-inhibiting factor (GRIF) is generated from the proform by the release of an N-terminal propeptide (64 aa), resulting in **somatostatin-28** containing a single disulfide bridge (Cys 17–Cys 28). **Somatostatin-14** is the C-terminal half of **somatostatin-28** with the same single disulfide bridge (Cys 3–Cys 14). Somatostatin inhibits the release of somatotropin, which plays an important role in growth control.

In addition, the islets of Langerhans synthesise the pancreatic prohormone (66 aa: P01298) or pancreatic polypeptide (PP) and from this proform the **pancreatic hormone** (36 aa: Ala 1–Tyr-$NH_2$ 36) and the **pancreatic icosapeptide** (20 aa: His 1–Arg 20) are generated by the release of a C-terminal propeptide (7 aa). The Gly residue from the basic tripeptide (**GKR**) C-terminal to the Tyr residue provides the amide group for the amidation of Tyr. Pancreatic hormone is an antagonist of cholecystokinin by suppression of pancreatic secretion and stimulation of gastric secretion.

## 13.3 GASTROINTESTINAL HORMONES

The uptake, digestion and absorption of nutrients into the body are complicated processes with the gastric release of HCl and pepsinogen, the zymogen of the digestive enzyme pepsin, into the stomach. On one hand this process is regulated by the autonomous nervous system and on the other several peptide hormones are involved in the regulation. The gastrointestinal tract contains specialised cells that secrete the gastrointestinal enzymes into the bloodstream. The gastric mucosa and the small intestine produce four main gastrointestinal hormones:

1. The *gastrins* are a family of biologically active peptide hormones that are produced by the gastric mucosa with three major forms: gastrin 34 or big gastrin, gastrin 17 or little gastrin and gastrin 14 or minigastrin. They stimulate the stomach mucosa to produce and secrete HCl and the pancreas to secrete digestive enzymes such as pepsinogen. Gastrins stimulate the smooth muscle contraction and increase blood circulation and water secretion in the stomach and the intestine.
2. *Secretin* is produced by the upper small intestinal mucosa and stimulates the pancreas to secrete bicarbonate in order to neutralise the released gastric acid.
3. *Cholecystokinins* (CCKs) are produced by the upper small intestine and stimulate the pancreas to secrete digestive enzymes and bicarbonate, stimulate the emptying of the gallbladder and inhibit gastric emptying.
4. *Gastric inhibitory polypeptide* (GIP, glucose-dependent insulinotropic polypeptide) is secreted by K cells in the duodenum of the gastrointestinal tract. Its major function is the stimulation of the insulin release from the pancreas. In addition, it is an inhibitor of gastric acid secretion, gastric mobility and gastric emptying.
5. *Motilin* is secreted by the small intestine, increasing the gastrointestinal motility and stimulating the production of pepsin.

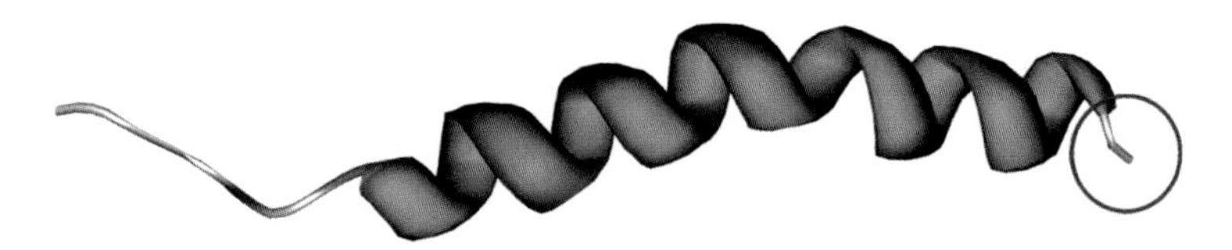

**Figure 13.3 3D structure of the gastric inhibitory polypeptide (PDB: 1T5Q)**
The major biologically active fragment of the gastric inhibitory polypeptide exhibits an elongated α-helical conformation with a C-terminally amidated Lys (pink circle).

**Gastrins** (see Data Sheet) are a family of peptide hormones that are generated by proteolytic cleavage from the proform (80 aa), the removal of the C-terminal propeptide (6 aa: Ser 75–Asn 80) and the basic tripeptide (**GRR**) required for amidation of the C-terminal Phe residue (see Chapter 6). The numbers refer to the amino acid count: **gastrin 71** (Ser 1–Phe-$NH_2$ 71), **gastrin 52** (Asp 20–Phe-$NH_2$ 71), **gastrin 34/big gastrin** (Gln 38–Phe-$NH_2$ 71), **gastrin 17/little gastrin** (Gln 55–Phe-$NH_2$ 71), **gastrin 14/minigastrin** (Trp 58–Phe-$NH_2$ 71) and **gastrin 6** (Tyr 66–Phe-$NH_2$ 71). All gastrins carry a partially sulfated Tyr residue and are C-terminally amidated (Phe-$NH_2$). The two major biologically active gastrins, gastrin 34 and gastrin 17, which stimulate the stomach mucosa to produce and secrete gastric HCl and the pancreas to secrete its digestive enzymes, carry an N-terminal pyro-Glu residue. Sulfation of the Tyr residue increases the biological activity of gastrins and reduces their degradation. **Gastrin-releasing peptide** (GRP, 27 aa: Val 1–Met-$NH_2$ 27) is generated from the proform (125 aa: P07492) by releasing a C-terminal propeptide (95 aa). The C-terminal region of GRP is known as **neuromedin C** (10 aa: Gly 18–Met-$NH_2$ 27). GRP stimulates the release of gastrin and other gastrointestinal hormones.

**Secretin** (see Data Sheet) is a 27 residue peptide hormone generated by proteolytic cleavage from the proform (103 aa) via removal of N-terminal (8 aa) and C-terminal (64 aa) propeptides. The C-terminal Val residue is amidated (Val-$NH_2$). The Gly residue from the basic tripeptide (**GKR**) C-terminal to the Val residue provides the amide group for the amidation reaction of Val. Secretin regulates the pH of the duodenum via controlling gastric acid secretion and buffering with bicarbonate. Secretin reduces acid secretion from the stomach by inhibiting the release of gastrin, which helps to neutralise the digestive products from the stomach entering the duodenum; thus the digestive pancreatic enzymes amylase and lipase have optimal neutral pH conditions.

**Cholecystokinins** (CCKs) (see Data Sheet) are a family of peptide hormones that are generated by proteolytic cleavage from the proform (95 aa) by removing N-terminal (24 aa) and C-terminal (9 aa) propeptides and the basic tripeptide (**GRR**) required for amidation of the C-terminal Phe residue (see Chapter 6). The numbers refer to the amino acid count: **cholecystokinin 58** (Val 1–Phe-$NH_2$ 58), **cholecystokinin 39** (Tyr 20–Phe-$NH_2$ 58), **cholecystokinin 33** (Lys 26–Phe-$NH_2$ 58), **cholecystokinin 12** (Ile 47–Phe-$NH_2$ 58), **cholecystokinin 8** (Asp 51–Phe-$NH_2$ 58). All CCKs carry a sulfated Tyr residue and are C-terminally amidated (Phe-$NH_2$). CCK 33 is the major cholecystokinin and enhances the effects of secretin. CCKs are released in response to digestive products such as fatty acids, monoacylglycerols, amino acids and peptides.

**Gastric inhibitory polypeptide** (GIP) (see Data Sheet) or glucose-dependent insulinotropic polypeptide is a 42 residue peptide hormone generated by proteolytic cleavage from the proform (132 aa) via removal of N-terminal (29 aa) and C-terminal (59 aa) propeptides. After the detection of glucose in the small intestine GIP is thought to induce the secretion of insulin.

The 3D structure of the major biologically active fragment of GIP (30 aa: Tyr 1–Lys-$NH_2$ 30) was determined by NMR spectroscopy (PDB: 1T5Q) and is shown in Figure 13.3. GIP exhibits an elongated structure with a full-length α-helical conformation from Phe 6 to Ala 28 and a C-terminally amidated Lys residue (pink circle).

**Motilin** (22 aa: Phe 1–Gln 22) and **motilin-associated peptide** (66 aa: Ser 1–Lys 66) are generated from the same precursor (90 aa: P12872). They are secreted by the small intestine and play an important role in the regulation of gastrointestinal motility and indirectly cause the rhythmic contraction of duodenal and colonic smooth muscle.

## 13.4 CALCIUM-RELATED HORMONES

Calcium is important in many biological processes and is essential for the function of Ca-containing proteins. $Ca^{2+}$ is part of hydroxyapatite ($Ca_5(PO_4)_3OH$), the main component of bones. Calcium ions are involved as a second messenger in hormone signalling and in the transmission of nerve impulses, they trigger muscle contraction and are essential for several proteins of the blood coagulation cascade. Therefore, the extracellular $Ca^{2+}$ concentration of approximately 1.2 m*M* is strictly regulated, mainly by two peptide hormones and vitamin D, which acts synergistically with parathyroid hormone to increase the calcium level in blood:

1. *Parathyroid hormone* (PTH) is secreted by the parathyroid gland and is responsible for the increase in the $Ca^{2+}$ level in blood by stimulating its resorption from the bone and kidney and by increasing its absorption in the intestine. PTH primarily controls the calcium and phosphate homeostasis in humans. PTH exhibits structural similarity to the PTH-related protein (141 aa: P12272) and is characterised by a conserved sequence in the N-terminal region:

   PARATHYROID (PS00335): V–S–E–x–Q–x(2)–H–x(2)–G

2. *Calcitonin* is produced primarily by the C cells in the thyroid gland and is responsible for the decrease in the $Ca^{2+}$ level in blood by decreasing its absorption by the intestines and its reabsorption by the kidney and reducing the activity of osteoclasts in bones.

**Parathyroid hormone** (PTH) (see Data Sheet) or parathormone is an 84 residue peptide hormone generated from a proform by the release of an N-terminal hexapeptide. It is involved in the regulation of a constant blood calcium concentration by increasing its level. PTH increases the excretion rate of phosphate in the kidneys, resulting in the leach of $Ca_5(PO_4)_3OH$ from the bones and, as a consequence, the increase in the $Ca^{2+}$ level in blood. In addition, PTH increases the production of vitamin D and upregulates the hydroxylase that catalyses the hydroxylation reaction to form active vitamin D (1$\alpha$,25-dihydroxycholecalciferol).

The 3D structure of biologically active N-terminal fragments of parathyroid hormone (39 aa: Ser 1–Ala 39) was determined by NMR spectroscopy (PDB: 1BWX) and is shown in Figure 13.4. The structure is characterised by a short N-terminal and a longer C-terminal $\alpha$-helix connected by a defined loop region.

Defects in the PTH gene are the cause of familial isolated hypoparathyroidism existing as autosomal dominant and recessive forms (for details see MIM: 146200).

**(Pro)-calcitonin** (see Data Sheet) is a 32 residue peptide hormone (Cys 1–Pro-$NH_2$ 32) generated from a proform (116 aa) by releasing an N-terminal propeptide (57 aa) and two basic peptides, an N-terminal dipeptide (**KR**) and a C-terminal tetrapeptide (**GKKR**), the latter being required for amidation of the C-terminal Pro residue (see Chapter 6). In addition, from the C-terminal region of the proform the 21 residue long **katacalcin** (Asp 1–Asn 21) is generated. Calcitonin contains a single intrachain disulfide bridge (Cys 1–Cys 7) and is C-terminally amidated (Pro-$NH_2$). Calcitonin is involved in the calcium and phosphate metabolism, but the main physiological regulator of calcium and phosphate homeostasis is parathyroid hormone. Katacalcin is involved in the decrease of the $Ca^{2+}$ level in blood.

The 3D structure of eel calcitonin (32 aa: Cys 1–Pro-$NH_2$ 32, PO1232) in micelles was determined by NMR spectroscopy (PDB: 1BKU). Eel calcitonin exhibits 50 % sequence identity with human calcitonin and due to the lack of an appropriate human 3D structure the structure of eel calcitonin is shown in Figure 13.5. The structure is characterised by an amphipathic $\alpha$-helix extending from Ser 5–Leu 19 followed by an unstructured C-terminal region and C-terminal Pro 32 is amidated (pink circle).

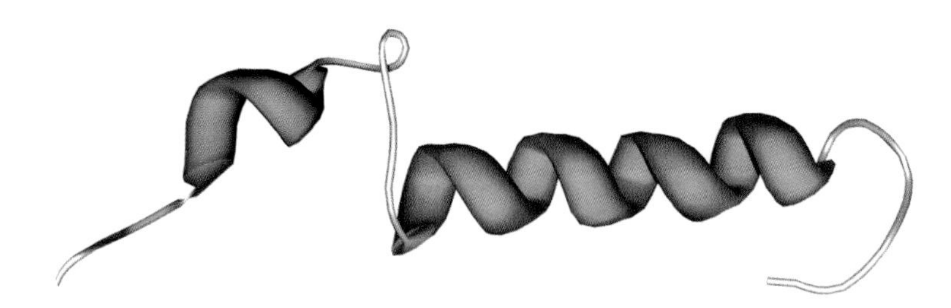

**Figure 13.4 3D structure of biologically active N-terminal fragments of parathyroid hormone (PDB: 1BWX)**
The structures of these biologically active fragments are characterised by short N-terminal and longer C-terminal $\alpha$-helix segments connected by a defined loop region.

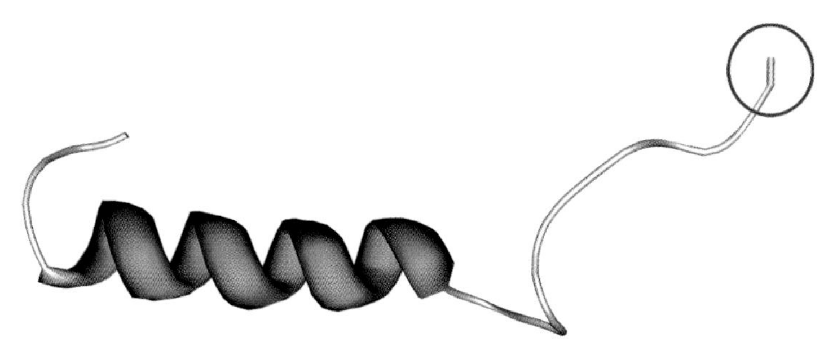

**Figure 13.5 3D structure of eel calcitonin in micelles (PDB: 1BKU)**
Eel calcitonin exhibits 50% sequence identity with the human species. The structure is characterised by an amphipathic α-helical segment and C-terminal Pro 32 is amidated (pink circle).

## 13.5 HORMONE-RELEASING FACTORS, THEIR TROPHIC HORMONES AND RELATED HORMONES

The hypothalamus synthesises a series of polypeptide hormones known as releasing factors and releasing-inhibiting factors. They stimulate or inhibit the release of trophic hormones into the bloodstream, which in turn stimulate their target cells to secrete the corresponding hormones. The involved releasing factors, the trophic hormones and the released hormones are summarised in Table 13.2 and detailed as follows:

1. *Corticoliberin* or corticotropin-releasing factor (CRF) is a releasing factor that is synthesised in the hypothalamus and causes the adenohypophysis to release corticotropin or adrenocorticotropic hormone (ACTH), which in turn stimulates the release of corticosteroids.
2. *Thyroliberin* or thyrotropin-releasing factor (TRF) is a releasing factor that stimulates the adenohypophysis to release thyrotropin or thyroid-stimulating hormone (TSH), which in turn stimulates the thyroid gland to synthesise and release $T_3$ (triiodothyronine) and $T_4$ (thyroxine). Thyroglobulin is the precursor of $T_3$ and $T_4$.
3. *Gonadoliberin I* and *II* or gonadotropin-releasing factor (GnRF) is a releasing factor from the hypothalamus that stimulates the adenohypophysis to release the gonadotropins lutropin or luteinizing hormone (LH) and follitropin or follicle-stimulating hormone (FSH). In males LH stimulates the testes to secrete androgens and triggers ovulation in females. FSH promotes spermatogenesis in males and stimulates the development of ovarian follicles in females.
4. *Somatoliberin* or growth hormone-releasing factor (GRF) is a releasing factor from the hypothalamus that stimulates the release of growth hormone (GH) or somatotropin. GH accelerates the growth of various tissues directly and induces the synthesis of growth factors known as somatomedins in the liver.
5. *Somatostatin* or growth hormone release-inhibiting factor (GRIF) is a release-inhibiting factor from the hypothalamus that inhibits the release of growth hormone (GH) or somatotropin.

**Corticoliberin** (see Data Sheet) or corticotropin-releasing factor (CRF) is a 41 residue peptide hormone generated from a proform (172 aa) by releasing an N-terminal propeptide (129 aa) and a C-terminal dipeptide (**GK**) required for the amidation of the C-terminus. Corticoliberin carries a C-terminal Ile-$NH_2$ residue (see Chapter 6).

**Table 13.2** Releasing factors, the corresponding trophic hormones and the released hormones.

| Releasing factor (code)/aa | Trophic hormone (code)/aa | Released hormone |
|---|---|---|
| Corticoliberin (P06850)/41 | Corticotropin (P01189)/39 | Corticosteroids |
| Thyroliberin (P20396)/3 | Thyrotropin (P01215/P01222)/92 + 112 | $T_3$: triiodothyronine $T_4$: thyroxine |
| Gonadoliberin (P01148)/10 | Lutropin (P01215/P01229)/92 + 121 | Androgens |
| | Follitropin (P01215/P01225)/92 + 111 | |
| Somatoliberin (P01286)/44 | Somatotropin (P01241)/191 | Somatomedins |

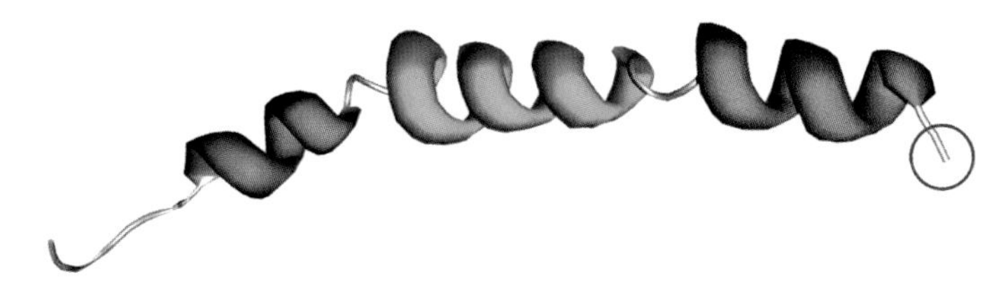

**Figure 13.6 3D structure of corticoliberin (PDB: 1GO9)**
Corticoliberin is characterised by a linear helical structure and C-terminal Ile is amidated (pink circle).

The 3D solution structure of corticoliberin (41 aa: Ser 1–Ile-$NH_2$ 41) was determined by NMR spectroscopy (PDB: 1GO9) and is shown in Figure 13.6. Corticoliberin adopts an almost liner helical structure with a helical content of over 80 % and is C-terminally amidated (Ile-$NH_2$) (pink circle).

Corticoliberin regulates the release of corticotropin (see Data Sheet) from the adenohypophysis. Together with many other biologically active peptides corticotropin is generated from a proform (241 aa) known as **proopiomelanocortin** (POMC) by proteolytic cleavage at paired basic residues (listed according to their position in the proform from the N- to the C-terminus).

**Melanotropin gamma** (11 aa: Tyr 1–Phe-$NH_2$ 11) and **melanotropin alpha** (13 aa: Ser 1–Val-$NH_2$ 13) or melanocyte-stimulating hormones (MSHs) are C-terminally amidated peptide hormones where the Gly residues from the adjacent tripeptides (**GRR/GKK**) provide the amide groups. Melanotropin alpha comprises the 13 N-terminal residues of corticotropin. **Melanotropin beta** (18 aa: Asp 1–Asp 18) is derived from lipotropin gamma (see below) by cleavage at a basic dipeptide and corresponds to its 18 C-terminal residues. MSHs increase skin pigmentation by an increase in the melanin production in melanocytes.

**Corticotropin** (39 aa: Ser 1–Phe 39) stimulates the adrenal gland to release corticosteroids.

**Lipotropin gamma** (56 aa: Glu 1–Asp 56) and **lipotropin beta** (89 aa: Glu 1–Glu 89) are involved in the mobilisation of fat from adipose tissue; lipotropin gamma comprises the 56 N-terminal residues of lipotropin beta.

In addition, the opioid peptides **β-endorphin** (31 aa: Tyr 1–Glu 31) comprising the 31 C-terminal residues of proopiomelanocortin and **Met-enkephalin** (5 aa: Tyr 1–Met 5) comprising the five N-terminal residues of β-endorphin are also released from the proform proopiomelanocortin (more details are given in Section 13.7).

**(Pro)-thyroliberin** (see Data Sheet) or thyrotropin- releasing factor (TRF) is a tripeptide (Pyro-Glu–His–Pro-$NH_2$) generated from the proform prothyroliberin (218 aa) in six copies. The Gly residue from the basic tripeptides (**GKR/GRR**) C-terminal to the Pro residue provides the amide group for the amidation reaction at Pro. TRF is a releasing factor that stimulates the adenohypophysis to release thyrotropin.

**Thyrotropin** (see Data Sheet) or thyroid-stimulating hormone belongs to the glycoprotein hormone (gonadotropin) family together with follitropin, lutropin and choriogonadotropin. Gonadotropins consist of a common α-chain and a different but structurally related β-chain (for details see Chapter 5) that are noncovalently associated. The α-chain (approximately 21 % CHO) comprises 92 residues with five intrachain disulfide bridges and the β-chain (112 aa: Phe 1–Tyr 112; approximately 12 % CHO) containing six intrachain disulfide bridges is processed from the proform by releasing a C-terminal hexapeptide (for structural data see follitropin later). Thyrotropin is essential for the proper control of thyroid function and causes the release of $T_3$ and $T_4$.

**Thyroglobulin** (TG) (see Data Sheet) is a large glycoprotein (approximately 10 % CHO) usually present as homodimer containing eleven thyroglobulin type I, three type II and five type III domains. The thyroglobulin type I domain contains three intrachain disulfide bridges arranged in the following pattern (in some variants the second bridge is missing):

**Cys 1–Cys 2, Cys 3–Cys 4, Cys 5–Cys 6**

In addition, the thyroglobulin type I domain is characterised by a consensus sequence in the central part of the domain:

THYROGLOBULIN_1_1 (PS00484): [FYWHPVAS]–x(3)–C–x(3,4)–[SG]–x–[FYW]–x(3)–Q–x(5,12)–[FYW]–C–[VA]–x(3,4)–[SG]

Thyroglobulin is essential for thyroid hormone synthesis and is the precursor of the thyroid hormones $T_3$ and $T_4$, which are rather unusual iodinated amino acids. In a series of quite unique biochemical reactions in the thyroid gland two neighbouring Tyr residues on the surface of thyroglobulin are iodinated in the presence of $H_2O_2$ by the enzyme thyroid peroxidase (EC 1.11.1.8, P07202) at the 3 and 5 positions in the benzene ring, leading to 3,5-diiodo-Tyr. Furthermore, two adjacent 3, 5-diiodo-Tyr residues are linked via an ether bond forming thyroxine, which at this stage is still part of thyroglobulin, and finally proteolysis of thyroglobulin results in the release of free thyroxine. Thyroglobulin contains four major thyroxine sites (modified Tyr 5, 1291, 2554 and 2568) and one triiodothyronine site (modified Tyr 2747) and several minor sites. In addition, Tyr 5 is sulfated. In plasma TG is a useful marker for differentiated thyroid cancer.

Defects in the TG gene are the cause of certain forms of goiter (enlargement of the thyroid gland), which might be linked to hypothyroidism (for details see MIM: 188450).

**(Pro)-gonadoliberin-1 and -2** (see Data Sheet) or gonadotropin-releasing factor (GnRF) are decapeptides (Pyro-Glu 1–Gly-$NH_2$ 10) that are generated from progonadoliberin-1 (69 aa) and –2 (97 aa) together with GnRH-associated peptide 1 (56 aa) and GnRH-associated peptide 2 (84 aa). The Gly residue from the basic tripeptide (**GKR**) C-terminal to the amidated Gly residue provides the amide group for the amidation reaction of Gly. GnRFs are releasing factors that stimulate the adenohypophysis to release the gonadotropins follitropin and lutropin.

**Follitropin** (see Data Sheet) or follicle-stimulating hormone (FSH) belongs to the glycoprotein hormone (gonadotropin) family consisting of the common α-chain and the β-chain (111 aa: Asn 1–Glu 111). The main members of this hormone family are follitropin, lutropin, thyrotropin and choriogonadotropin. FSH promotes spermatogenesis and stimulates the development of the ovarian follicles.

The 3D structure of follitropin in complex with the extracellular hormone-binding domain of its receptor (252 aa: Gly 1–Ser 252, P23945) was determined by X-ray diffraction (PDB: 1XWD) and is shown in Figure 13.7. Follitropin consists of the α-chain (92 aa: Ala1–Ser 92) (in yellow) and the β-chain (111 aa: Asn 1–Gly 111) (in blue) associated in a head-to-tail arrangement and adopts the expected structure (see Chapter 5), being bound in a handclasp fashion to the elongated, curved receptor domain (in red).

**Lutropin** (see Data Sheet) or luteinizing hormone (LH) also belongs to the glycoprotein hormone (gonadotropin) family consisting of the common α-chain and the β-chain (121 aa: Ser 1–Leu 121). LH stimulates the testes to secrete androgens and triggers ovulation.

**Somatoliberin** (see Data Sheet) or growth hormone-releasing factor (GRF) is a 44 residue releasing factor (Tyr 1–Leu-$NH_2$ 44) generated from a proform (88 aa) by cleaving two propeptides, an N-terminal one (10 aa) and a C-terminal one (31 aa). The Gly residue from the basic dipeptide (**GR**) C-terminal to the amidated Leu residue provides the amide group for the amidation reaction of Leu. GRF stimulates the adenohypophysis to release somatotropin.

**Somatotropin** (see Data Sheet) or growth hormone (GH) is a peptide hormone (191 aa: Phe 1–Phe 191) containing two disulfide bridges (Cys 53–Cys 165, Cys 182–Cys 189), two phosphorylation sites (Ser 106 and 150) and two deamidation

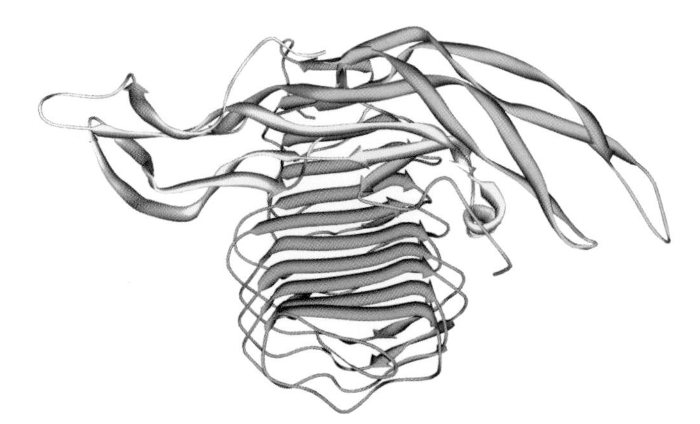

**Figure 13.7 3D structure of follitropin in complex with the extracellular hormone-binding receptor domain (PDB: 1XWD)**
The two chains of follitropin (α-chain: yellow, β-chain: blue) are associated in a head-to-tail arrangement resulting in an elongated, slightly curved structure and are bound in a handclasp fashion to the elongated, curved receptor domain (in red).

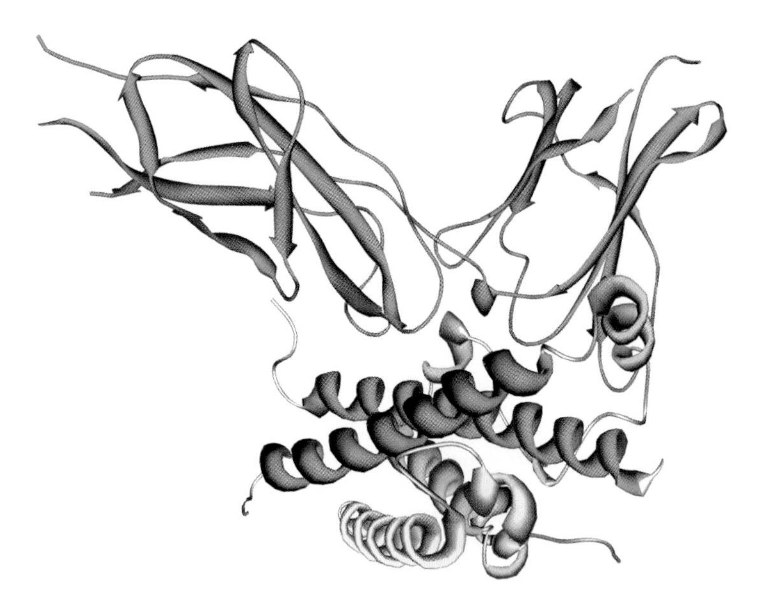

**Figure 13.8 3D structure of somatotropin in complex with the extracellular receptor domain (PDB: 1A22)**
Somatotropin is highly helical (rainbow) and the receptor domain contains two β-sheets (in red).

sites (Gln 137 and Asn 152) and at least four splice variants exist. Somatotropin, which is structurally related to prolactin, is characterised by two consensus sequences located around conserved Cys residues involved in disulfide bridges:

SOMATOTROPIN_1 (PS00266): **C**–x–[STN]–x(2)–[LIVMFYS]–x–[LIVMSTA]–P–x(5)–[TALIV]–x(7)–[LIVMFY]–x(6)–[LIVMFY]–x(2)–[STACV]–W

SOMATOTROPIN_2 (PS00338): **C**–[LIVMFY]–{PT}–x–D–[LIVMFYSTA]–x–{S}–{RK}–{A}–x–[LIVMFY]–x(2)–[LIFMFYT]–x(2)–**C**

GH exists as monomer and also as oligomers up to pentamers, either noncovalently associated or disulfide-linked. GH is essential for postnatal growth and development and plays an important role in growth control and directly accelerates the growth of various tissues. It stimulates the liver to synthesise growth factors known as somatomedins.

The 3D structure of somatotropin (191 aa: Phe 1–Phe 191) in complex with the extracellular domain of its receptor (238 aa: Phe 1–Gln 238, P10912) was determined by X-ray diffraction (PDB: 1A22) and is shown in Figure 13.8. The structure of somatotropin is highly helical (rainbow) and the receptor domain contains two β-sheets (in red). The complex formation does not induce major conformational changes in somatotropin.

Defects in the growth hormone gene are the cause of various diseases:

1. Isolated growth hormone deficiency type IB or pituitary dwarfism I is an autosomal recessive disease causing short stature (for details see MIM: 262400).
2. Kowarski syndrome or pituitary dwarfism VI (for details see MIM: 262650).
3. Isolated growth hormone deficiency type II is an autosomal dominant disease causing short stature (for details see MIM: 173100).

**Somatomedins** (see Data Sheet) are generated from insulin-like growth factor (IGF) proforms IA and IB by releasing N-terminal (27 aa) and C-terminal (35 aa, 77 aa) propeptides resulting in **somatomedin C** (70 aa: Gly 1–Ala 70) and from insulin-like growth factor proform II by releasing a C-terminal (89 aa) propeptide resulting in **somatomedin A** (67 aa: Ala 1–Glu 67). Somatomedins are structurally and functionally related to insulin and contain three disulfide bridges arranged in the following pattern:

**Cys 1–Cys 4, Cys 2–Cys 6, Cys 3–Cys 5**

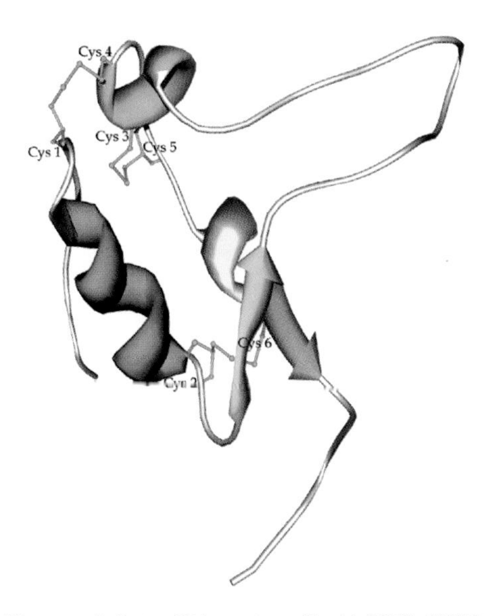


**Figure 13.9 3D structure of insulin-like growth factor II (somatomedin A) (PDB: 1IGL)**
Insulin-like growth factor II is mainly characterised by α-helices and a small antiparallel β-sheet. The three intrachain disulfide bridges are shown in pink.

Somatomedins are growth factors that have insulin-like activities but exhibit a much higher activity than insulin and stimulate cartilage growth.

The 3D solution structure of insulin-like growth factor II (somatomedin A) (67 aa: Ala 1–Glu 67) was determined by NMR spectroscopy (PDB: 1IGL) and is shown in Figure 13.9. The structure contains mainly α-helices and in addition a small antiparallel β-sheet. The three intrachain disulfide bridges are arranged in the indicated pattern (in pink).

**Somatostatin** (see Data Sheet) or growth hormone release-inhibiting factor inhibits the release of somatotropin (for details see Section 13.2).

In addition to the hormone-releasing factors and their trophic hormones, two important hormones exist that are structurally related and also functionally involved in the hormonal circuits of trophic hormones:

1. *Choriogonadotropin* (CG) belongs to the glycoprotein hormone family together with follitropin, lutropin and thyrotropin and is essential for maintaining pregnancy by stimulating the production of progesterone.
2. *Prolactin* belongs to the somatotropin/prolactin family and is structurally related to somatotropin and induces the mammary gland to promote lactation.

**Choriogonadotropin** (CG) (see Data Sheet) belongs to the glycoprotein (33 % CHO) hormone (gonadotropin) family together with follitropin, lutropin and thyrotropin and consists of the common α-chain and the individual β-chain (145 aa: Ser 1–Gln 145). CG is synthesised in the syncytiotrophoblasts of the placenta and stimulates the ovaries to synthesise the steroid progesterone and to maintain the progesterone level, thus ensuring the continuation of pregnancy.

The 3D structure of recombinant Se-Met-choriogonadotropin was determined by X-ray diffraction (PDB: 1HCN) and is shown in Figure 13.10. Despite a low sequence similarity (10 % identity) the α-chain (92 aa: Ala 1–Ser 92) and the β-chain (145 aa: Ser 1–Glu 145) have similar tertiary folds characterised by β-sheets. The α-chain (in yellow) contains five intrachain disulfide bridges (in green) and the β-chain (in blue) six intrachain disulfide bridges (in pink). There is a remote but distinct structural similarity between the glycoprotein hormone chains and other Cys-knot proteins such as the platelet-derived growth factor (see also Chapter 14).

**Prolactin** (see Data Sheet) is synthesised in the adenohypophysis and is structurally related to somatotropin. Prolactin (199 aa: Leu 1–Cys 199) contains three disulfide bridges (Cys 4–Cys 11, Cys 58–Cys 174, Cys 191–Cys 199) arranged in a

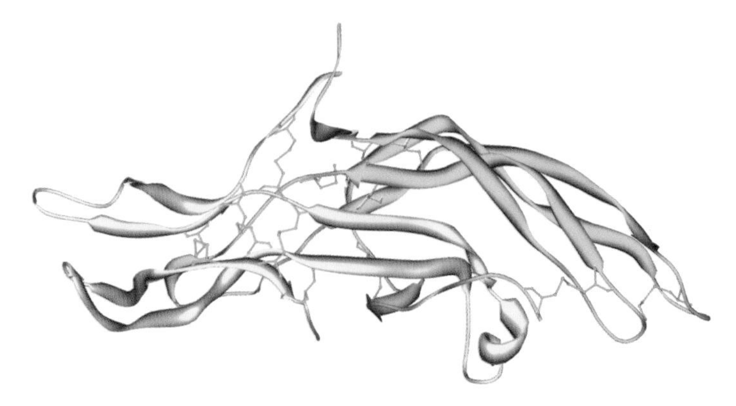

**Figure 13.10 3D structure of choriogonadotropin (PDB: 1HCN)**
The α-chain (in yellow) contains five intrachain disulfide bridges (in green) and the β-chain (in blue) six intrachain disulfide bridges (in pink). The two chains exhibit a low sequence similarity but have similar tertiary folds characterised by β-sheets and a Cys-knot structure.

**Cys 1–Cys 2, Cys 3–Cys 4, Cys 5–Cys 6** pattern. Prolactin causes many effects, primarily the stimulation of the mammary glands to produce milk during lactation. Prolactin also plays a role in osmoregulation, growth, morphogenesis and immunomodulation.

The 3D solution structure of entire prolactin (199 aa: Leu 1–Cys 199) was determined by NMR spectroscopy (PDB: 1N9D) and is shown in Figure 13.11. The structure is characterised by an 'up–up down–down' four helical bundle topology that resembles other members of the family of hematopoietic cytokines. The structure is stabilised with disulfide bridges (in pink).

**Prolactin-releasing peptide** (PrRP) or prolactin-releasing hormone is generated from the proform (65 aa: P81277) by releasing a C-terminal propeptide (30 aa) resulting in **PrRP31** (31 aa: Ser 1–Phe-$NH_2$ 31) and the N-terminally truncated **PrRP20** (20 aa: Thr 12–Phe-$NH_2$ 31). The Gly residue from the basic tetrapeptide (**GRRR**) C-terminal to the amidated Phe residue provides the amide group for the amidation of Phe. PrRP stimulates the release of prolactin and regulates the expression of prolactin via its receptor.

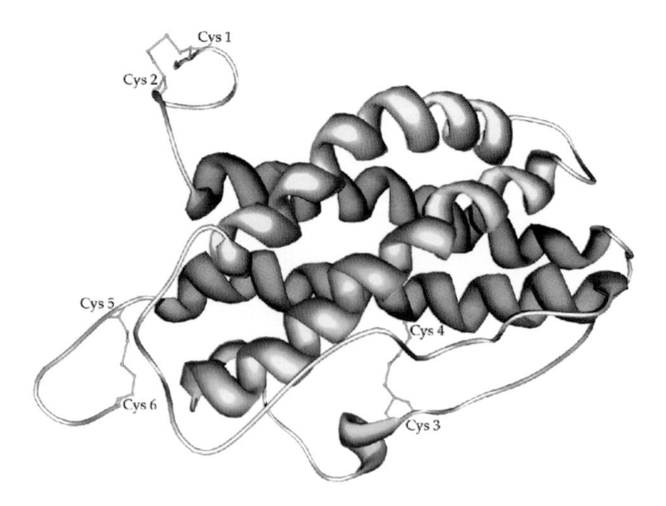


**Figure 13.11 3D structure of prolactin (PDB: 1N9D)**
Prolactin exhibits an 'up-up down-down' four helical bundle topology. The three intrachain disulfide bridges are shown in pink.

## 13.6 VASOPRESSIN AND OXYTOCIN

The neurohypophysis (posterior pituitary gland) produces and secretes two structurally related nonapeptides, vasopressin and oxytocin. Vasopressin is the main regulator of the osmolarity of body fluids and regulates the pressure of the vascular system. Oxytocin causes smooth muscles to contract, for example, uterus and mammary gland.

**(Arg)-vasopressin** (see Data Sheet) or antidiuretic hormone (ADH) is generated from the vasopressin-neurophysin 2-copeptin proform (145 aa) resulting in vasopressin, neurophysin 2 and copeptin (39aa: Ala 1–Tyr 39). Vasopressin contains a single disulfide bridge (Cys 1–Cys 6) and is C-terminally amidated (Gly-$NH_2$). The Gly residue from the basic tripeptide (**GKR**) C-terminal to the amidated Gly residue provides the amide group for the amidation of Gly. Vasopressin is secreted in the case of increased plasma osmolarity, causing the reabsorption of water in the kidneys and resulting in a more concentrated urine. In addition, vasopressin causes vasoconstriction in peripheral vessels.

**Neurophysin 2** (93 aa: Ala 1–Ala 93) contains seven disulfide bridges arranged in the following pattern:

**Cys 1–Cys 8, Cys 2–Cys 4, Cys 3–Cys 7, Cys 5–Cys 6, Cys 9–Cys 11, Cys 10–Cys 14, Cys 12–Cys 13**

Neurophysin 2 is a carrier of hormones along axons and specifically binds vasopressin.

The 3D structure of bovine oxytocin (9 aa: Cys 1–Gly-$NH_2$ 9, P01175) (in pink) in complex with its carrier protein neurophysin 2 (95 aa: Ala 1–Val 95, P01180) (in blue) was determined by X-ray diffraction (PDB: 1NPO) and is shown in Figure 13.12(a). The sequence of human and bovine oxytocin are identical and neurophysin 2 of the two species share a sequence identity of 94 % and contains seven intrachain disulfide bridges (in light green) which are arranged as expected. The structure confirms the fact that Cys 1 and Tyr 2 (in green) in oxytocin are the key residues involved in the neurophysin–oxytocin recognition. The surface representation in Figure 13.12(b) clearly shows the interaction of Tyr 2(in green) in oxytocin with the binding pocket in neurophysin 2.

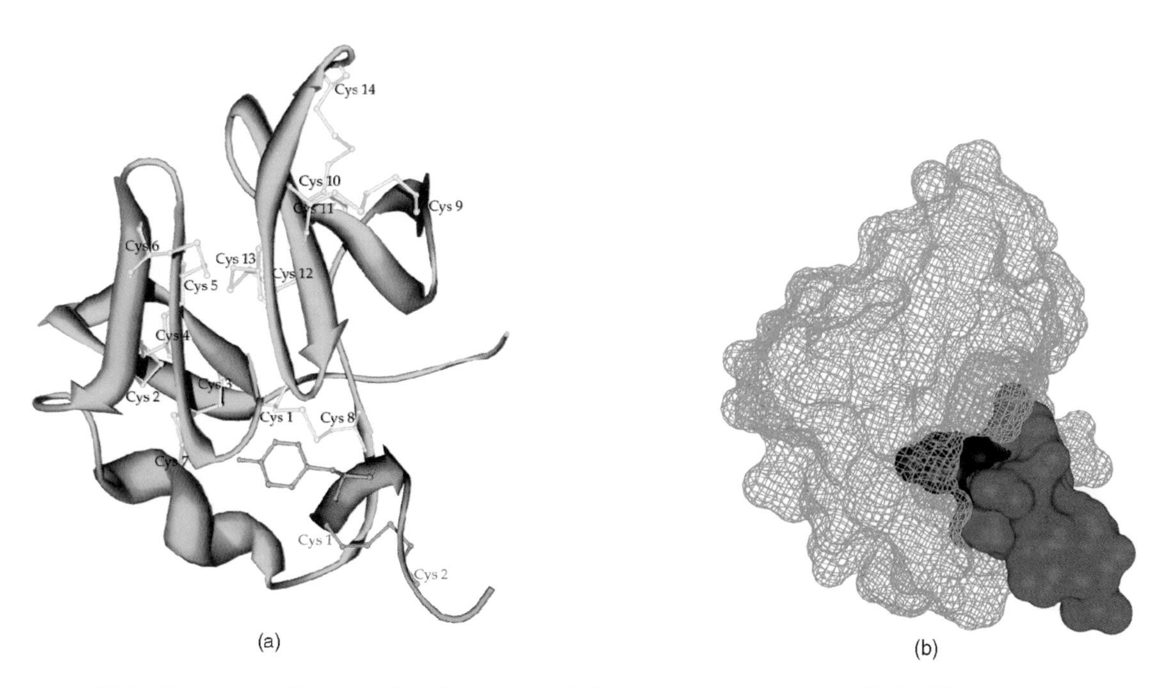


**Figure 13.12 3D structure of bovine oxytocin in complex with its carrier protein neurophysin (PDB: 1NPO).** (a) The sequence of human and bovine oxytocin (in pink) are identical and neurophysin 2 (in blue) of the two species share a sequence identity of 94% and contain seven intrachain disulphide bridges (in yellow). Cys 1 and Tyr 2 (in pink) in oxytocin are the key residues involved in the neurophysin – oxytocin recognition. (b) The surface representation shows the interaction of Tyr 2 (in green) in oxytocin with the binding pocket in neurophysin 2.

Defects in the vasopressin-neurophysin 2-copeptin gene are the cause of either autosomal dominant or autosomal recessive neurohypophyseal diabetes insipidus characterised by thirst, polydipsia (unusual intake of fluid by mouth) and polyuria (passage of large volumes of urine in a given period) (for details see MIM: 125700).

**Oxytocin** (see Data Sheet) is generated from the oxytocin–neurophysin 1 proform (106 aa) resulting in oxytocin and neurophysin 1. Oxytocin contains a single disulfide bridge (Cys 1–Cys 6) and is C-terminally amidated (Gly-$NH_2$). The Gly residue from the basic tripeptide (**GKR**) C-terminal to the amidated Gly residue provides the amide group for the amidation of Gly. Oxytocin causes the contraction of the smooth muscles of the uterus and the mammary gland.

**Melanoliberin (CYIQNC)** or melanotropin-releasing hormone (MRH) comprises the first six residues of oxytocin and is the releasing hormone of melanotropin and the antagonist of melanostatin.

**Melanostatin (PLG-$NH_2$)** or melanotropin-releasing inhibitory hormone (MIH) comprises the three last residues in oxytocin and inhibits the secretion of melanotropin and is the antagonist of melanoliberin.

**Neurophysin 1** (94 aa: Ala 1–Arg 94) contains seven disulfide bridges arranged in the same pattern as in neurophysin 2. Neurophysin 1 is a carrier of hormones along axons and specifically binds oxytocin.

## 13.7 NATURAL OPIOID PEPTIDES

Some of the most unusual peptide hormones are those with opiate-like effects on the central nervous system. They are secreted by the adenohypophysis and are the cause of analgesia (without pain) and a sense of well-being and behave like a kind of natural painkiller. There are three different families of opioid peptides:

1. *Endorphins* or endomorphines are endogenous opiates that are generated from proopiomelanocortin (POMC) such as α-, β- and γ-endorphins, of which β-endorphin appears to be the most important in pain relief.
2. *Enkephalins* (Met-enkephalin and Leu-enkephalin) exhibit opiate-like effects and are mainly generated from proenkephalin but also from proopiomelanocortin and β-neoendorphin-dynorphin (preprodynorphin).
3. *Dynorphins*, like endorphins and enkephalins, also exhibit opiate-like effects and are generated from β-neoendorphin-dynorphin (preprodynorphin).

**Endorphins** (see corticotropin/lipotropin Data Sheet) are generated by proteolytic cleavage at basic dipeptides in proopiomelanocortin (POMC) (see trophic hormones in Section 13.5), resulting in the main product **β-endorphin** (31 aa: Tyr 1–Glu 31) comprising the 31 C-terminal residues of proopiomelanocortin. **α-Endorphin** (16 aa: Tyr 1–Thr 16) and **γ-endorphin** (17 aa: Tyr 1–Leu 17) correspond to the N-terminal part of β-endorphin, which is an opioid peptide neurotransmitter and works as an analgesic (pain killer/relief). Although the exact function of α-endorphin and γ-endorphin are not very well understood, α-endorphin seems to stimulate the brain in a manner similar to amphetamines and γ-endorphin probably has antipsychotic effects and may be involved in the regulation of blood pressure.

**Enkephalins** (mainly **Met-enkephalin** and **Leu-enkephalin**, see Data Sheet, proenkephalin A) are generated and form various proforms:

1. Proteolytic cleavage at basic dipeptides in proenkephalin releases five propeptides of various lengths and generates the following enkephalin products:
   (a) **synenkephalin** (73 aa: Glu 1–Ala 73) containing three disulfide bridges (Cys 2–Cys 24, Cys 6–Cys 28, Cys 9–Cys 41);
   (b) four copies of **Met-enkephalin** (5aa: Tyr 1–Met 5);
   (c) **Met-enkephalin-Arg-Gly-Leu** (8 aa: Tyr 1–Leu 8);
   (d) **Leu-enkephalin** (5 aa: Tyr 1–Leu 5);
   (e) **Met-enkephalin-Arg-Phe** (7 aa: Tyr 1–Phe 7).
2. Met-enkephalin is also generated from proopiomelanocortin, corresponding to the five N-terminal residues in β-endorphin.
3. Leu-enkephalin is also generated from preprodynorphin, corresponding to the five N-terminal residues in leumorphin.

Enkephalins compete and mimic the effects of opiate drugs and they are involved in several physiological processes such as pain perception and response to stress. The conformation of Met-enkephalin in model membranes resembling natural membranes was determined by multidimensional NMR spectroscopy (PDB: 1PLW). Opioid peptides are able to adopt several conformations in a membrane environment consistent with the flexibility and poor selectivity of enkephalins.

**Dynorphins** (see proenkephalin B Data Sheet) are generated from preprodynorphin by releasing two propeptides (152 aa, 19 aa) and resulting in the following products:

a. **β-neoendorphin** (9 aa: Tyr 1–Pro 9);
b. **dynorphin** (17 aa: Tyr 1–Gln 17);
c. **leumorphin** (29 aa: Tyr 1–Ala 29);
d. **rimorphin** (13 aa: Tyr 1–Thr 13) identical with the N-terminal region of leumorphin.

Dynorphins are endogenous opioid peptides synthesised in different parts of the brain, such as the hypothalamus, the hippocampus and the spinal cord. Depending on the site of synthesis they have many different physiological actions. Dynorphin may act as an antidote to the effects of cocaine and may help some individuals against addiction.

## 13.8 VASOACTIVE PEPTIDES

There are several highly vasoactive peptides in body fluids, some of which have already been discussed in some detail:

1. *Bradykinin* is the major peptide hormone responsible for vasodilatation, thus lowering and regulating the blood pressure (see Chapter 7).
2. *Angiotensin II* is the major peptide hormone responsible for vasoconstriction and thus the increase of blood pressure. Angiotensin II and bradykinin are the main components responsible for the regulation of the vascular tonus (see Chapter 7).
3. *Vasopressin* is involved in the control of plasma osmolarity and causes vasoconstriction in peripheral vessels (see Section 13.6).

In addition, there are many other potent vasoactive peptides, which are involved in homeostatic control of the vascular system:

4. *Atrial natriuretic factor* (ANF) or atrial natriuretic peptide is primarily involved in the homeostatic control of body water and sodium.
5. *Vasoactive intestinal peptide* (VIP) is involved in many physiological processes such as water and electrolyte secretion in the intestine and vasodilatation in peripheral blood vessels.
6. *Calcitonin gene-related peptides I + II* (CGRP) are involved in vasodilatation and are very potent blood vessel-relaxing peptides.
7. *Substance P* is a neurotransmitter and neuromodulator but is also a potent vasodilatator.

**Atrial natriuretic factor** (ANF) (see Data Sheet) or atrial natriuretic peptide and **cardiodilatin-related peptide** are generated from prepronatriodilatin (128 aa), releasing an intermittent propeptide (67 aa) and two C-terminal Arg residues, resulting in cardiodilatin-related peptide (30 aa: Asn 1–Asp 30) and atrial natriuretic factor (28 aa: Ser 1–Tyr 28) containing a single disulfide bridge (Cys 7–Cys 23). ANF, which is structurally related to other natriuretic peptides, is characterised by a consensus sequence located around the single disulfide bridge and two conserved Gly residues:

NATRIURETIC_PEPTIDE (PS00263): **C**–F–**G**–x(3)–[DEA]–[RH]–I–x(3)–S–x(2)–**G**–**C**

ANF is involved in the homeostatic control of water and sodium in body fluids. ANF is released by monocytes in the atria of the heart in response to a raised blood pressure and thus counters the blood pressure-raising effects of the rennin–angiotensin system. Cardiodilatin-related peptide is thought to possess smooth muscle relaxant activity but no natriuretic or diuretic activity.

The 3D structure of atrial natriuretic factor (26 aa: Ser1-Phe 26) in complex with the extracellular domain of its receptor including the potential signal peptide (480 aa: Met 1–Glu 480, P17342) was determined by X-ray diffraction (PDB: 1YK0) and is shown in Figure 13.13. Although wild-type ANF (in yellow) is too flexible to adopt a definable structure, ANF,

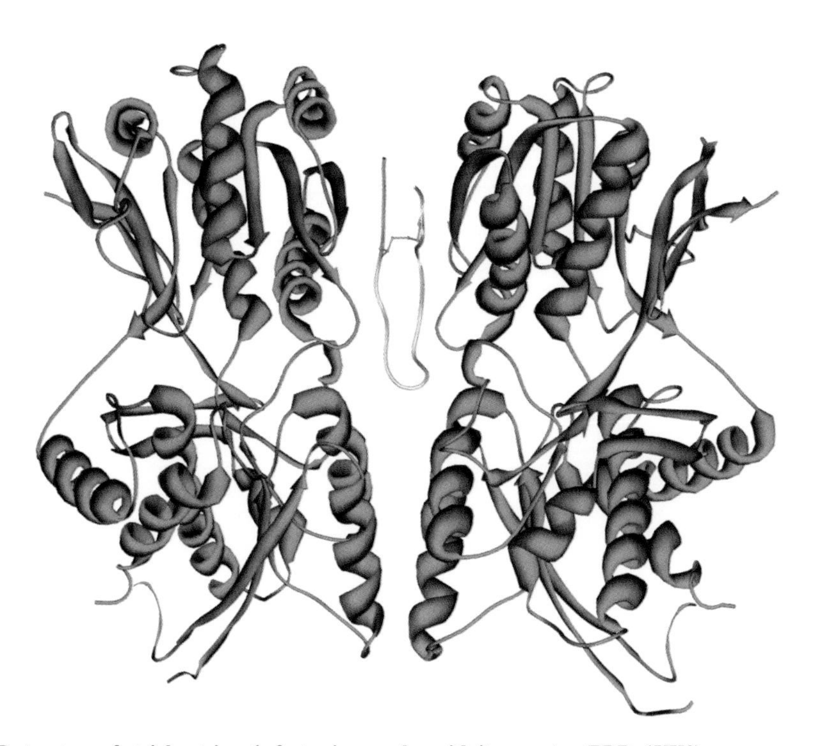

**Figure 13.13 3D structure of atrial natriuretic factor in complex with its receptor (PDB: 1YK0)**
In complex with two receptor molecules (in red) the otherwise very flexible atrial natriuretic factor (in yellow) adopts a stable structure stabilised by a single disulphide bridge (in pink).

stabilised by a single disulfide bridge (in pink), adopts a stable structure in complex with two receptor molecules (in red).

**Vasoactive intestinal peptide** (VIP) (see Data Sheet) and **intestinal peptides PHV-42/PHM-27** are generated from the proform (150 aa) by releasing two propeptides, an N-terminal one (59 aa) and a C-terminal one (15 aa). PHM-27 (27 aa: His 1–Met-$NH_2$ 27) corresponds to the N-terminal region of PHV-42 (42 aa: His 1–Val 42). The Gly residue from the basic tripeptide (**GKR**) C-terminal to the amidated Met residue provides the amide group for the amidation of Met in PHM-27. As in PHM-27 the amide group of the C-terminal Asn residue in the vasoactive intestinal peptide (28 aa: His 1–Asn-$NH_2$ 28) is provided from the adjacent Gly residue in a basic tripeptide (**GKR**). VIP has many biological functions such as the stimulation of water and electrolyte secretion, dilatation of intestinal smooth muscles, vasodilatation of peripheral blood vessels, stimulation of the pancreatic secretion of bicarbonate and inhibition of the gastrin-stimulated release of gastric acid.

**Calcitonin gene-related peptides I + II** (CGRP) (see Data Sheet) are generated from the proforms (I: 102 aa; II: 103 aa) by releasing two propeptides, an N-terminal one (I: 55 aa; II: 54 aa) and a C-terminal one (I + II: 4 aa). Calcitonin gene-related peptides I + II (37 aa: Ala 1–Phe-$NH_2$ 37) contain a single disulfide bridge (Cys 2–Cys 7) and share a sequence identity of 92 %. The amide group of the C-terminal Phe residue is provided from the adjacent Gly residue. CGRP induces vasodilatation in various types of vessels such as coronary, cerebral and systemic vasculature and is one of the most potent endogenous vasodilatators. In the central nervous system CGRP works as a neurotransmitter and neuromodulator.

**Substance P** (see Data Sheet) is generated from protachykinin 1 (110 aa) by releasing an N-terminal propeptide (37 aa), resulting in substance P (11 aa: Arg 1–Met-$NH_2$ 11) and **neuropeptide K** (36 aa: Asp 1–Met-$NH_2$ 36) and a C-terminal flanking peptide (16 aa). **Neuropeptide γ, first part** (2 aa: Asp 1–Ala 2) and **second part** (19 aa: Gly 18–Met-$NH_2$ 36) as well as **neurokinin A** (10 aa: His 27–Met-$NH_2$ 36) are contained in neuropeptide K. The Gly residue from the basic tripeptide (**GKR**) C-terminal to the Met residue provides the amide group for the amidation of Met. These peptides belong to the tachykinin family and are characterised by a consensus pattern covering the last five C-terminal residues, which are essential for their biological activity:

TACHYKININ (PS00267): F–[IVFY]–G–[LM]–M–[**G**>]

Substance P is a neurotransmitter and a neuromodulator. In addition, it is a potent vasodilatator by releasing nitric oxide from the endothelium, which may cause hypotension.

## 13.9 ERYTHROPOIETIN AND THROMBOPOIETIN

Erythropoietin (EPO) is primarily synthesised in the kidney and to a lesser extent in the liver and is the principal hormone involved in the regulation of the erythrocyte differentiation and in maintaining the physiological level of the erythrocyte mass. Thrombopoietin (TPO) is primarily synthesised in the liver and the kidney and regulates the production of platelets in the bone marrow and stimulates the production and differentiation of megakaryocytes. EPO and TPO are structurally related and belong to the EPO/TPO family characterised by a common signature (for details see Chapter 5).

**Erythropoietin** (EPO) (see Data Sheet) or epoietin is a single-chain glycoprotein containing two intrachain disulfide bridges (Cys 7–Cys 161, Cys 29–Cys 33) and is structurally related with the N-terminal region of TPO.

The 3D solution structure of a recombinant mutant EPO (166 aa: Ala 1–Arg 166) was determined by NMR spectroscopy (PDB: 1BUY) and is shown in Figure 13.14. The structure is characterised by an antiparallel four-helix bundle connected by two long and one short loop and by three short β-sheets (see also Chapter 5). The structure is stabilised by two intrachain disulfide bridges (in pink) arranged in a **Cys 1–Cys 4, Cys 2–Cys 3** pattern.

EPO is widely used for the treatment of anemic patients and several products are available on the market. Illegally, EPO is used by healthy athletes to increase the erythrocyte level in their blood in order to gain a competitive advantage (blood doping).

**Thrombopoietin** (TPO) (see Data Sheet) or megakaryocyte colony-stimulating factor is a single-chain glycoprotein containing two disulfide bridges (Cys 7–Cys 151, Cys 29–Cys 85) in its N-terminal region, which is structurally related with EPO. The C-terminal region is structurally unique and is rich in Pro/Ser/Thr, especially the segment Pro 264–Pro 316 (Pro/Ser/Thr content: 58 %) but is not related to EPO and seems not to be directly required for the biological activity of TPO.

The 3D structure of the receptor-binding domain corresponding to the N-terminal region in TPO (163 aa: Ser 1–Ser 163) in complex with a neutralising antibody fragment (not shown) was determined by X-ray diffraction (PDB: 1V7M) and is shown

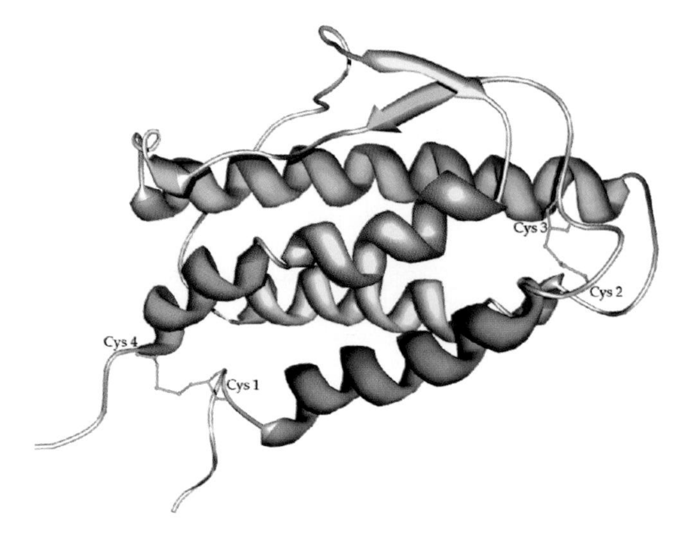


**Figure 13.14 3D structure of erythropoietin (PDB: 1BUY)**
The structure of erythropoietin contains an antiparallel four-helix bundle connected by two long and one short loop and three short β-sheets and is stabilised by two intrachain disulfide bridges (in pink).

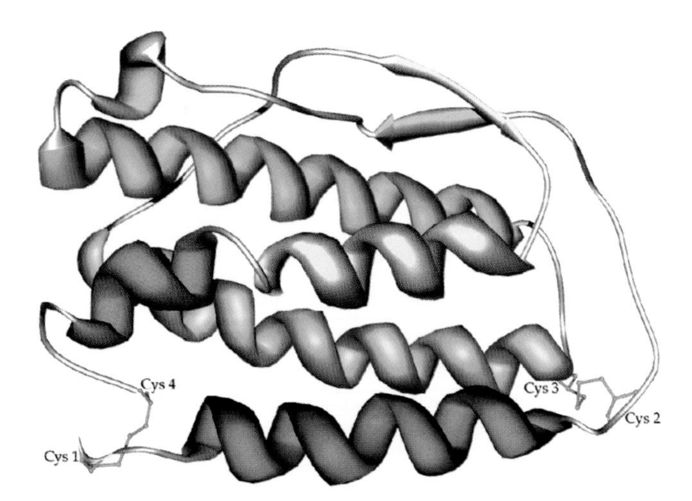


**Figure 13.15 3D structure of the N-terminal receptor-binding domain of thrombopoietin (PDB: 1V7M)**
The receptor-binding domain of thrombopoietin exhibits the expected antiparallel four-helix bundle fold stabilised by two intrachain disulfide bridges (in pink).

in Figure 13.15. The receptor-binding domain in TPO adopts the expected antiparallel four-helix bundle fold stabilised by two intrachain disulfide bridges (in pink) and the structure is similar to the related erythropoietin.

Defects in the thrombopoietin gene are the cause of primary (essential) thrombocythemia (overproduction of platelets), which is inherited as an autosomal dominant trait and is characterised by elevated platelet levels in blood, frequently leading to thrombotic and haemorrhagic complications (for details see MIM: 187950).

## 13.10 OTHERS

### 13.10.1 Adiponectin

Adiponectin is a protein hormone involved as modulator in several metabolic processes such as glucose regulation and fatty acid catabolism. Adiponectin is exclusively secreted from adipose tissue and is quite abundant in plasma (5–10 mg/l) compared with other hormones. Levels of adiponectin are inversely correlated with the body mass index (BMI) and are decreased in obesity and metabolic syndrome in contrast to most other adipokines.

**Adiponectin** (see Data Sheet) or gelatin-binding protein is a single-chain glycoprotein consisting of an N-terminal collagen-like domain that contains nine OH-Pro residues and Cys 18, which forms interchain disulfide bridges resulting in homooligomers and a C-terminal C1q domain with a globular fold. Adiponectin exhibits a distinct structural similarity with complement component C1q A-, B- and C- chains (sequence identity with C1qA: 33 %). Adiponectin promotes insulin sensitivity and has antiinflammatory effects.

Defects in the adiponectin gene are the cause of adiponectin deficiency, resulting in a low adiponectin concentration in plasma associated with obesity insulin resistance and diabetes type 2 (for details see MIM: 605441).

### 13.10.2 Islet amyloid polypeptide

Islet amyloid polypeptide (IAPP) or diabetes-associated peptide is generated from a proform (67 aa: P10997) releasing two propeptides, an N-terminal one (9 aa) and a C-terminal one (16 aa), resulting in the mature **islet amyloid polypeptide** (37 aa: Lys 1–Tyr-$NH_2$ 37) containing a single disulfide bridge (Cys 2–Cys 7) and belongs to the calcitonin family. The Gly residue from the basic tripeptide (**GKR**) C-terminal to the amidated Tyr residue provides the amide group for the amidation of Tyr. IAPP selectively inhibits insulin-stimulated glucose utilisation and glycogen deposition in muscle but does not affect the adipocytes glucose metabolism. IAPP is found in pancreatic islets of type 2 diabetes.

## REFERENCES

Baumann, 1999, Growth hormone heterogeneity in human pituitary and plasma, *Horm. Res.*, **51** (Suppl. 1), 2–6.

Hinuma *et al.*, 1998, A prolactin-releasing peptide in the brain, *Nature*, **393**, 272–276.

Jiang and Zhang, 2003, Glucagon and regulation of glucose metabolism, *Am. J. Physiol.*, **284**, E671–E678.

Maggio, 1988, Tachykinins, *Annu. Rev. Neurosci.* **11**, 13–28.

Molina *et al.*, 1996, Characterization of the type-1 repeat from thyroglobulin, a cysteine-rich module found in proteins from different families, *Eur. J. Biochem.*, **240**, 125–133.

Rosenzweig and Seidman, 1991, Atrial natriuretic factor and related peptide hormones, *Annu. Rev. Biochem.*, **60**, 229–255.

Sanke *et al.*,1988, An islet amyloid peptide is derived from an 89-amino acid precursor by proteolytic processing, *J. Biol. Chem.*, **263**, 17243–17246.

Seino *et al.*, 1987, Sequence of an intestinal cDNA encoding human motilin precursor, *FEBS Lett.*, **223**, 74–76.

Wallis, 2001, Episodic evolution of protein hormones in mammals, *J. Mol. Evol.*, **53**, 10–18.

DATA SHEETS

## *Adiponectin (Gelatin-binding protein)*

### Fact Sheet

| | | | |
|---|---|---|---|
| ***Classification*** | | ***Abbreviations*** | |
| ***Structures/motifs*** | 1 C1q; 1 collagen-like | ***DB entries*** | ADIPO_HUMAN/Q15848 |
| ***MW/length*** | 24 544 Da/226 aa | ***PDB entries*** | |
| ***Concentration*** | 5–10 mg/l | ***Half-life*** | |
| ***PTMs*** | 9 OH-Pro, 1 N-CHO (potential) | | |
| ***References*** | Saito *et al.*, 1999, *Gene*, **229**, 67–73. | | |

### Description

**Adiponectin** is exclusively synthesised in adipocytes and is a hormone regulator in hematopoiesis and the immune system.

### Structure

Adiponectin is a single-chain protein containing a collagen-like domain containing nine OH-Pro residues and a globular C1q domain. Cys 18 forms interchain disulfide bridges leading to homooligomers.

### Biological Functions

Adiponectin is an important negative regulator in hematopoiesis and the immune system.

## *Vasopressin-neurophysin 2-copeptin (Arg)-vasopressin*

### Fact Sheet

| | | | |
|---|---|---|---|
| ***Classification*** | Vasopressin/oxytocin | ***Abbreviations*** | AVP-NPII |
| ***Structures/motifs*** | | ***DB entries*** | NEU2_HUMAN/P01185 |
| ***MW/length*** | Arg-vasopressin: 1087 Da/9 aa<br>Neurophysin 2: 9787 Da/93 aa<br>Copeptin: 4021/39 aa | ***PDB entries*** | 1NPO |
| ***Concentration*** | 0.5–4.0 ng/l | ***Half-life*** | |
| ***PTMs*** | Vasopressin: Gly-$NH_2$ (aa 9);<br>Copeptin: 1 N-CHO | | |
| ***References*** | Rose *et al.*, 1996, *Nat. Struct. Biol.*, **3**, 163–169.<br>Sausville *et al.*, 1985, *J. Biol. Chem.*, **260**, 10236–10241. | | |

### Description

**Vasopressin** is secreted from the neurohypophysis, stimulates water resorption by the kidney and is involved in the increase of the blood pressure.

### Structure

Vasopressin is the N-terminal nonapeptide from the vasopressin–neurophysin 2–copeptin precursor with an intrachain disulfide bridge (Cys 1–Cys 6) and a C-terminal Gly-$NH_2$ residue, neurophysin 2 represents the central and copeptin the C-terminal part of the precursor.

### Biological Functions

Vasopressin has a direct influence on the antidiuretic action of the kidney and causes vasoconstriction of the peripheral vessels. Neurophysin 2 specifically binds vasopressin.

## *Atrial natriuretic factor (Atrial natriuretic peptide)*

### Fact Sheet

| | | | |
|---|---|---|---|
| ***Classification*** | Natriuretic peptide | ***Abbreviations*** | ANF |
| ***Structures/motifs*** | | ***DB entries*** | ANF_HUMAN/P01160 |
| ***MW/length*** | Cardiodilatin-related peptide: 3508 Da/30 aa<br>Atrial natriuretic peptide: 3082 Da/28 aa | ***PDB entries*** | 1YK0 |
| ***Concentration*** | | ***Half-life*** | |
| ***PTMs*** | | | |
| ***References*** | He *et al.*, 2006, *J. Mol. Biol.*, **361**, 698–714.<br>Oikawa *et al.*, 1984, *Nature*, **309**, 724–726. | | |

### Description

**Atrial natriuretic factor** (ANF) is present in the atria of the heart and is primarily involved in the homeostatic control of body water and sodium. In addition, it is a potent vasoactive peptide hormone.

### Structure

ANF (28 aa) containing a single disulfide bridge (Cys 7–Cys 23) and cardiodilatin-related peptide (30 aa) are generated from prepronatriodilatin by releasing an intermittent 67 residue propeptide and two C-terminal Arg residues. The 3D structure of ANF (in yellow) in complex with its receptor (in red) was determined by X-ray diffraction (**1YK0**).

### Biological Functions

ANF is a key player in the control of water and sodium concentrations in the body. It is also thought to play a key role in cardiovascular haemostasis. Cardiodilatin-related peptide exhibits vasorelaxant activity but has no natriuretic and diuretic activity.

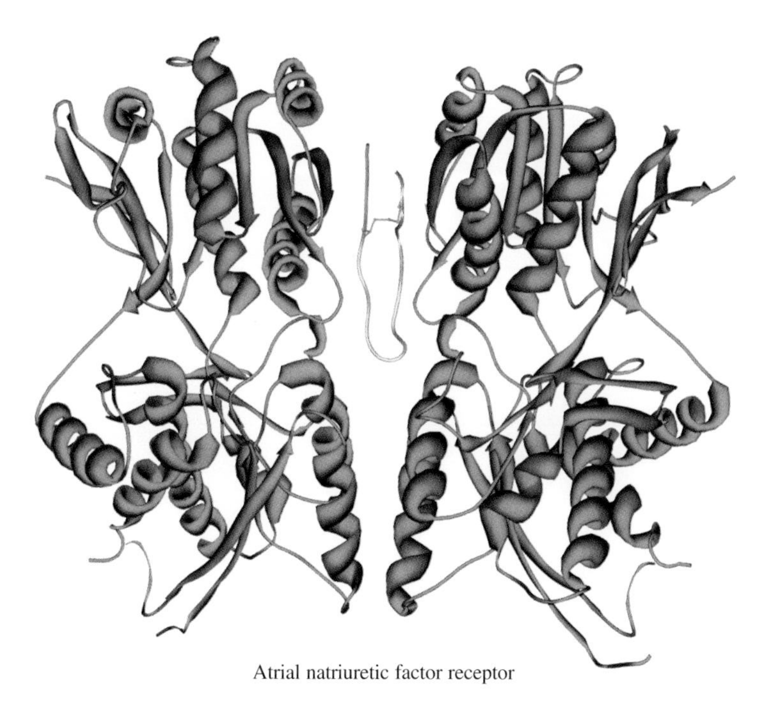

Atrial natriuretic factor receptor

## *Calcitonin gene-related peptides I + II*

### Fact Sheet

| | | | |
|---|---|---|---|
| *Classification* | Calcitonin | *Abbreviations* | CGRP |
| *Structures/motifs* | | *DB entries* | CALCA_HUMAN/P06881<br>CALCB_HUMAN/P10092 |
| *MW/length* | CGRP-I: 3792 Da/37 aa; | *PDB entries* | |
| *Concentration* | CGRP-II: 3796 Da/37 aa | *Half-life* | |
| *PTMs* | CGRP I + II: Phe-$NH_2$ | | |
| *References* | Nelkin *et al.*, 1984, *Biochem. Biophys. Res. Commun.*, **123**, 648–655.<br>Steenbergh *et al.*, 1986, *FEBS Lett.*, **209**, 97–103. | | |

### Description

**Calcitonin gene-related peptides I + II** (CGRPs) are abundant in the central nervous system and induce vasodilatation.

### Structure

CGRP I + II are generated from proforms by releasing N- and C-terminal propeptides. CGRP I + II (37 aa) contain a single disulfide bridge (Cys 2–Cys 7), are C-terminally amidated (Phe-$NH_2$) and share a sequence identity of 92 %.

### Biological Functions

CGRPs induce vasodilatation in various types of vessels and act as very potent endogenous peptides in the relaxation of blood vessels. Their abundance in the central nervous system suggests a role as neurotransmitter and neuromodulator.

# *Cholecystokinins*

## Fact Sheet

| | | | |
|---|---|---|---|
| ***Classification*** | Gastrin/cholecystokinin | ***Abbreviations*** | CCK |
| ***Structures/motifs*** | | ***DB entries*** | CCKN_HUMAN/P06307 |
| ***MW/length*** | Cholecystokinin: 10 750 Da/95 aa<br>Cholecystokinin 58: 6646 Da/58 aa<br>Cholecystokinin 39: 4626 Da/39 aa<br>Cholecystokinin 33: 3866 Da/33 aa<br>Cholecystokinin 12: 1536 Da/12 aa<br>Cholecystokinin 8: 1064 Da/8 aa | ***PDB entries*** | |
| ***Concentration*** | approx. 60 ng/l | ***Half-life*** | |
| ***PTMs*** | All cholecystokinin: 1 S-Tyr, Phe-$NH_2$ | | |
| ***References*** | Takahashi *et al.*, 1985, *Proc. Natl. Acad. Sci. USA*, **82**, 1931–1935. | | |

## Description

**Cholecystokinins** (CCKs) are produced by the upper small intestine and stimulate the emptying of the gallbladder and pancreatic secretion of digestive enzymes and bicarbonate.

## Structure

The various CCKs are generated by proteolytic cleavage from the proform (95 aa) releasing two propeptides, an N-terminal (24 aa) and a C-terminal (9 aa). CCKs carry a sulfated Tyr residue in their C-terminal segment and are C-terminally amidated (Phe-$NH_2$).

## Biological Functions

CCKs induce gallbladder contraction and the release of pancreatic enzymes and bicarbonate and inhibition of gastric emptying.

# *Choriogonadotropin*

## Fact Sheet

| | | | |
|---|---|---|---|
| ***Classification*** | Glycoprotein hormone beta-chain | ***Abbreviations*** | CG |
| ***Structures/motifs*** | | ***DB entries*** | GLHA_HUMAN/P01215<br>CGHB_HUMAN/P01233 |
| ***MW/length*** | α-chain: 10 205 Da/92 aa<br>β-chain: 15 532 Da/145 aa | ***PDB entries*** | 1HCN |
| ***Concentration*** | 16–24 mg/l | ***Half-life*** | 6–36 h |
| ***PTMs*** | α-chain: 2 N-CHO; 1 O-CHO;<br>β-chain: 2 N-CHO; 4 O-CHO | | |
| ***References*** | Fiddes *et al.*, 1979, *Nature*, **281**, 351–356.<br>Fiddes *et al.*, 1980, *Nature*, **286**, 684–687.<br>Wu *et al.*, 1994, *Structure*, **2**, 545–558. | | |

## Description

**Choriogonadotropin** (CG) is synthesised in the syncytiotrophoblasts of the placenta and is essential to maintain the maternal progesterone level.

## Structure

CG is a glycoprotein (33 % CHO) present in plasma as a heterodimer of two non-covalently-linked chains: a common α-chain (also present in follitropin, lutropin and thyrotropin) and a unique β-chain. The 3D structure of CG characterised by β-sheets resembles that of other Cys-knot proteins (**1HCN**). The α-chain (in yellow) contains five and the β-chain (in blue) six disulfide bridges.

## Biological Functions

CG maintains the functionality of the corpus luteum in early pregnancy and stimulates the ovaries to produce the steroids that are essential for the maintenance of pregnancy.

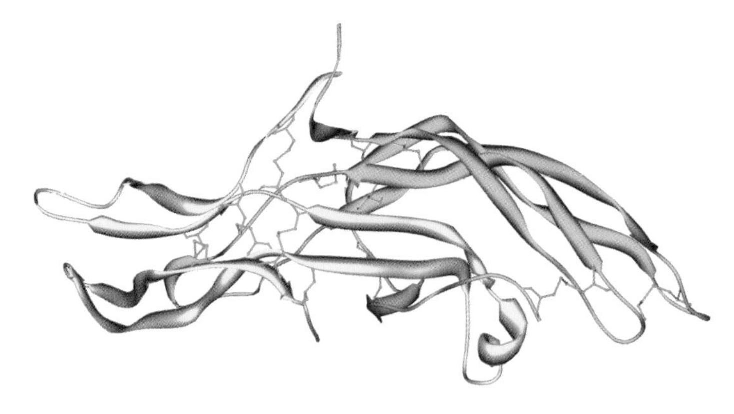

Choriogonadotropin

## *Corticoliberin (Corticotropin-releasing factor)*

### Fact Sheet

| | | | |
|---|---|---|---|
| ***Classification*** | Corticotropin-releasing factor/urotensin I | ***Abbreviations*** | CRF |
| ***Structures/motifs*** | | ***DB entries*** | CRF_HUMAN/P06850 |
| ***MW/length*** | Corticoliberin: 4758 Da/41 aa | ***PDB entries*** | 1GO9 |
| ***Concentration*** | | ***Half-life*** | |
| ***PTMs*** | Ile-$NH_2$ | | |
| ***References*** | Shibahara *et al.*, 1983, *EMBO J.*, **2**, 775–779.<br>Spyroullas *et al.*, 2002, *Eur. J. Biochem.*, **269**, 6009–6019. | | |

### Description

**Corticoliberin** is synthesised in the hypothalamus and is responsible for the release of corticotropin or adrenocorticotropic hormone (ACTH).

### Structure

CRF carrying a C-terminal Ile-$NH_2$ residue is generated from the proform by releasing an N-terminal 129 residue propeptide. CRF adopts an almost linear helical structure with over 80 % helix content (**1GO9**).

### Biological Functions

CRF causes the adenohypophysis to release corticotropin that stimulates the release of adrenocortical steroids.

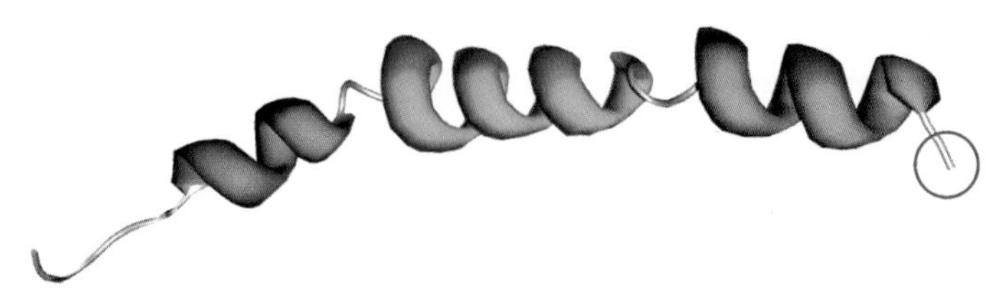

Corticoliberin

# *Corticotropin-Lipotropin (Proopiomelanocortin)*

## Fact Sheet

| | | | |
|---|---|---|---|
| ***Classification*** | POMC | ***Abbreviations*** | POMC |
| ***Structures/motifs*** | | ***DB entries*** | COLI_HUMAN/P01189 |
| ***MW/length*** | Melanotropin γ: 1514 Da/11 aa<br>Corticotropin: 4541 Da/39 aa<br>Melanotropin α: 1624/13 aa<br>Lipotropin β: 9806 Da/89 aa<br>Lipotropin γ: 6075 Da/56 aa<br>Melanotropin β: 2204 Da/18 aa<br>β-Endorphin: 3465 Da/31 aa<br>Met-enkephalin: 574 Da/5 aa | ***PDB entries*** | |
| ***Concentration*** | | ***Half-life*** | |
| ***PTMs*** | Melanotropin γ: Phe-$NH_2$; melanotropin α: Val-$NH_2$ | | |
| ***References*** | Takahashi *et al.*, 1981, *FEBS Lett.*, **135**, 97–102. | | |

## Description

All these peptide hormones with various functions are synthesised as a single proform (**proopiomelanocortin**) in the adenohypophysis.

## Structure

The various peptide hormones are generated by proteolytic cleavages at paired basic residues in the proform (248 aa). Melanotropin γ and α are C-terminally amidated (Phe-$NH_2$ and Val-$NH_2$, respectively).

## Biological Functions

Corticotropin (ACTH) stimulates the adrenal gland to release cortisol. β-Endorphin and Met-enkephalin are endogenous opiates effecting the central nervous system. Melanotropin γ increases the pigmentation of the skin by stimulating the melanin production.

## *Erythropoietin (Epoietin)*

### Fact Sheet

| | | | |
|---|---|---|---|
| ***Classification*** | EPO/TPO | ***Abbreviations*** | EPO |
| ***Structures/motifs*** | | ***DB entries*** | EPO_HUMAN/P01588 |
| ***MW/length*** | 18 396 Da/166 aa | ***PDB entries*** | 1BUY |
| ***Concentration*** | 0.5–2.5 μg/l | ***Half-life*** | 4–12 h |
| ***PTMs*** | 3 N-CHO; 1 O-CHO | | |
| ***References*** | Cheetham *et al.*, 1998, *Nat. Struct. Biol.*, **5**, 861–866.<br>Jacobs *et al.*, 1985, *Nature*, **313**, 806–810. | | |

### Description

**Erythropoietin** (EPO) is primarily synthesised in the kidney and to a lower extent in the liver. EPO is the main hormone for the erythrocyte differentiation.

### Structure

EPO is a single-chain monomeric glycoprotein with two intrachain disulfide bridges (Cys 7–Cys 161, Cys 29–Cys 33). The C-terminal tetrapeptide is probably removed in the mature form. The 3D structure is characterised by an antiparallel four-helix bundle connected by loops and three short β-sheets (**1BUY**).

### Biological Functions

EPO is the principal hormone involved in the regulation of the erythrocyte differentiation and the maintenance of the physiological level of the circulating erythrocytes. EPO is used for the treatment of anemia and illegally to increase the erythrocyte level in the blood of sportsmen (blood doping).

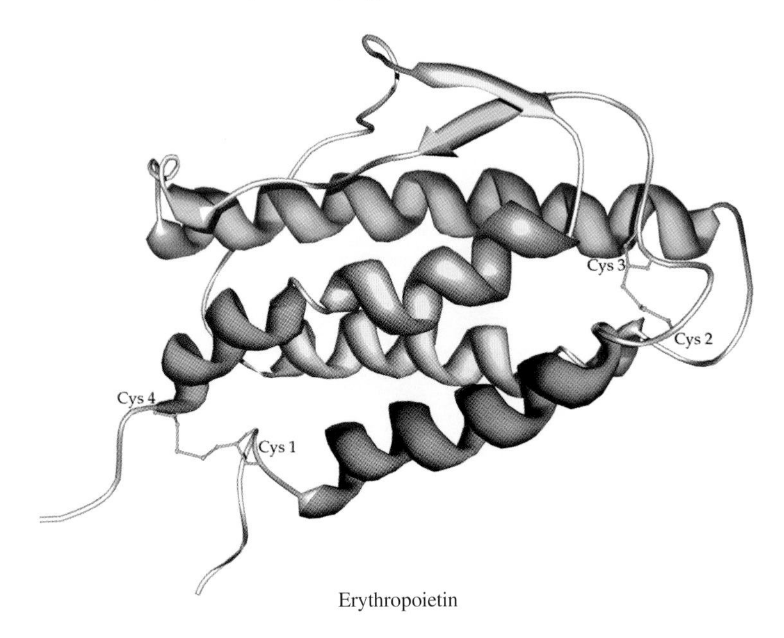


Erythropoietin

# *Follitropin (Follicle-stimulating hormone)*

## Fact Sheet

| | | | |
|---|---|---|---|
| ***Classification*** | Glycoprotein hormone beta-chain | ***Abbreviations*** | FSH |
| ***Structures/motifs*** | | ***DB entries*** | GLHA_HUMAN/P01215<br>FSHB_HUMAN/P01225 |
| ***MW/length*** | α-chain: 10 205 Da/92 aa<br>β-chain: 12 485 Da/111 aa | ***PDB entries*** | 1XWD |
| ***Concentration*** | | ***Half-life*** | |
| ***PTMs*** | α-chain: 2 N-CHO; 1 O-CHO;<br>β-chain: 2 N-CHO | | |
| ***References*** | Fan *et al.*, 2005, *Nature*, **433**, 269–277.<br>Fiddes *et al.*, 1979, *Nature*, **281**, 351–356.<br>Watkins *et al.*, 1987, *DNA*, **6**, 205–212. | | |

## Description

**Follitropin** (FSH) is synthesised in the adenohypophysis and stimulates the development of follicles and promotes spermatogenesis.

## Structure

FSH is a glycoprotein hormone present in plasma as a heterodimer of two non-covalently-linked chains: a common α-chain (also present in choriogonadotropin, lutropin and thyrotropin) and a unique β-chain. Follitropin (α-chain: yellow; β-chain: blue) adopts the expected structure and is bound in a hand-clasp fashion to its extracellular hormone-binding receptor domain (in red) (**1XWD**).

## Biological Functions

FSH stimulates in females the development of the ovarian follicles and in males it promotes spermatogenesis. Gonadoliberin is responsible for the release of FSH (and also lutropin) from the adenohyopophysis.

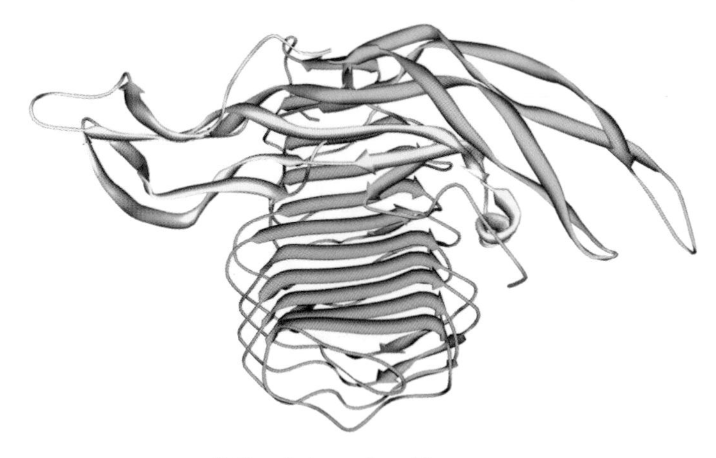

Follitropin incomplex with receptor

# *Gastric inhibitory polypeptide (glucose-dependent insulinotropic polypeptide)*

## Fact Sheet

| | | | |
|---|---|---|---|
| ***Classification*** | Glucagon | ***Abbreviations*** | GIP |
| ***Structures/motifs*** | | ***DB entries*** | GIP_HUMAN/P09681 |
| ***MW/length*** | 4984 Da/42 aa | ***PDB entries*** | 1T5Q |
| ***Concentration*** | approx. 100–400 ng/l | ***Half-life*** | |
| ***PTMs*** | | | |
| ***References*** | Alana *et al.*, 2004, *Biochem. Biophys. Res. Commun.*, **325**, 281–286.<br>Takeda *et al.*, 1987, *Proc. Natl. Acad. Sci. USA*, **84**, 7005–7008. | | |

## Description

**Gastric inhibitory polypeptide** (GIP) is produced in the small intestine and its major function is the stimulation of the insulin release.

## Structure

GIP is a single-chain peptide hormone that is generated from the proform by releasing two propeptides, an N-terminal (29 aa) and C-terminal (59 aa). The 3D structure of GIP exhibits an elongated structure with a pronounced α-helical segment (**1T5Q**).

## Biological Functions

The major physiological function of GIP is the stimulation of the insulin release from the pancreas and GIP is an important hormone involved in the inhibition of gastric acid secretion, gastric mobility and gastric emptying.

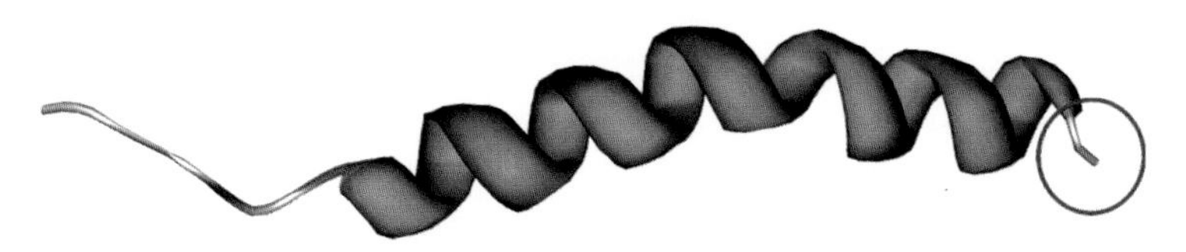

Gastric inhibitory polypeptide

# *Gastrins*

## Fact Sheet

| | | | |
|---|---|---|---|
| ***Classification*** | Gastrin/cholecystokinin | ***Abbreviations*** | |
| ***Structures/motifs*** | | ***DB entries*** | GAST_HUMAN/P01350 |
| ***MW/length*** | Gastrin 71: 8067 Da/71 aa<br>Gastrin 52: 6019 Da/52 aa<br>Gastrin 34/big gastrin: 3867 Da/34 aa<br>Gastrin 17/little gastrin: 2116 Da/17 aa<br>Gastrin 14/minigastrin: 1834 Da/14 aa<br>Gastrin 6: 818 Da/6 aa | ***PDB entries*** | |
| ***Concentration*** | approx. 70 ng/l | ***Half-life*** | |
| ***PTMs*** | All gastrins: 1 S-Tyr (partial), Phe-$NH_2$;<br>Pyro-Glu: Gastrin 34, gastrin 17 | | |
| ***References*** | Kato *et al.*, 1983, *Nucleic Acids Res.*, **11**, 8197–8203.<br>Spindel *et al.*, 1986, *Proc. Natl. Acad. Sci. USA*, **83**, 19–23. | | |

## Description

**Gastrins** are produced by the gastric mucosa and stimulates the gastric secretion of HCl and of pepsinogen.

## Structure

The various gastrins are generated by proteolytic cleavage from the proform (80 aa). All gastrins carry a partially sulfated Tyr residue and are C-terminally amidated (Phe-$NH_2$). Gastrin 34 and gastrin 17 carry an N-terminal pyro-Glu residue.

## Biological Functions

The two major bioactive gastrins, gastrin 34 and gastrin 17, stimulate the stomach mucosa to produce and secrete HCl and the pancreas to secrete digestive enzymes like pepsinogen. Gastrins also stimulate the smooth muscle contraction and increase blood circulation and water secretion in the stomach and intestine. The sulfation of Tyr increases the activity and reduces degradation.

# *Lutropin (Luteinizing hormone)*

## Fact Sheet

| | | | |
|---|---|---|---|
| ***Classification*** | Glycoprotein hormone beta-chain | ***Abbreviations*** | LH |
| ***Structures/motifs*** | | ***DB entries*** | GLHA_HUMAN/P01215<br>LHB_HUMAN/P01229 |
| ***MW/length*** | α-chain: 10 205 Da/92 aa<br>β-chain: 13 203 Da/121 aa | ***PDB entries*** | |
| ***Concentration*** | | ***Half-life*** | |
| ***PTMs*** | α-chain: 2 N-CHO; 1 O-CHO; β-chain: 1 N-CHO | | |
| ***References*** | Fiddes *et al.*, 1979, *Nature*, **281**, 351–356.<br>Talmadge *et al.*, 1984, *Nature*, **307**, 37–40. | | |

## Description

**Lutropin** (LH) is primarily present in the pituitary gland and promotes spermatogenesis and ovulation.

## Structure

LH is a glycoprotein hormone present in plasma as a heterodimer of two non-covalently-linked chains: a common α-chain (also present in choriogonadotropin, follitropin and thyrotropin) and a unique β-chain.

## Biological Functions

In females LH stimulates oocyte maturation and follicular synthesis of estrogens and progesterone and in males it stimulates androgen synthesis. Gonadoliberin is responsible for the release of LH (and also follitropin) from the adenohyopophysis.

## Fact Sheet

| | | | |
|---|---|---|---|
| ***Classification*** | Vasopressin/oxytocin | ***Abbreviations*** | OT-NP1 |
| ***Structures/motifs*** | | ***DB entries*** | NEU1_HUMAN/P01178 |
| ***MW/length*** | Oxytocin: 1010 Da/9 aa<br>Neurophysin 1: 9601 Da/94 aa<br>Melanoliberin: 743 Da/6 aa<br>Melanostatin: 285 Da/3 aa | ***PDB entries*** | 1NPO |
| ***Concentration*** | 2 ng/l; pregnancy: approx. 40 ng/l | ***Half-life*** | |
| ***PTMs*** | Oxytocin: Gly-$NH_2$ (aa 9) | | |
| ***References*** | Sausville *et al.*, 1985, *J. Biol. Chem.*, **260**, 10236–10241. | | |

## Description

**Oxytocin** is secreted from the neurohypophysis and is responsible for the contraction of the smooth muscle of the uterus and of the mammary gland.

## Structure

Oxytocin is the N-terminal nonapeptide from the oxytocin–neurophysin precursor with an intrachain disulfide bridge (Cys 1–Cys 6) and a C-terminal Gly-$NH_2$ residue and neurophysin 1 is the C-terminal part of the precursor. Melanoliberin comprises the N-terminal six residues of oxytocin and melanostatin the C-terminal three residues of oxytocin. The 3D structure of bovine oxytocin (in pink) in complex with its carrier protein neurophysin 2 (in blue) reveals Cys 1 and Tyr 2 (in green) of oxytocin as key residues of the recognition process (**1NPO**).

## Biological Functions

Oxytocin is released from the posterior lobe from the pituitary gland, the neurohypophysis, and causes the contraction of the smooth muscle of the uterus and of the mammary gland. Neurophysin 1 specifically binds to oxytocin. Melanoliberin releases melanotropin and is the antagonist of melanostatin. Melanostatin inhibits the release of melanotropin and is the antagonist of melanoliberin.

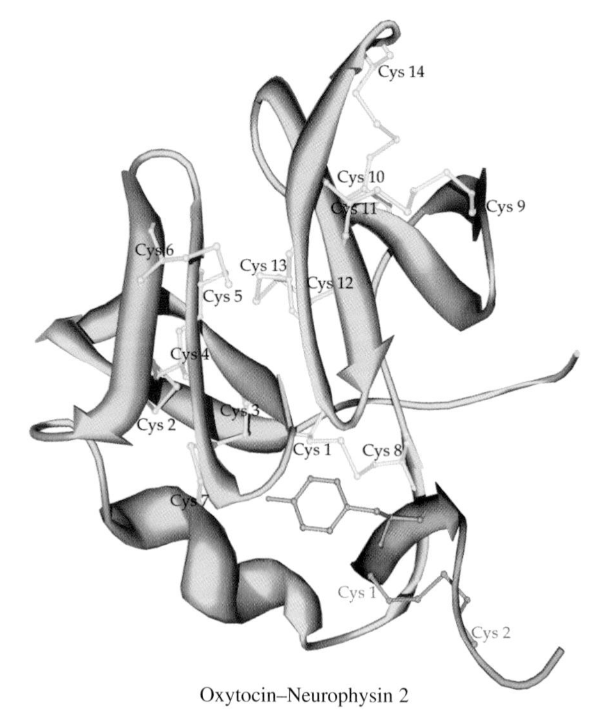


Oxytocin–Neurophysin 2

## *Parathyroid hormone (Parathormone)*

### Fact Sheet

| | | | |
|---|---|---|---|
| ***Classification*** | Parathyroid hormone | ***Abbreviations*** | PTH |
| ***Structures/motifs*** | | ***DB entries*** | PTHY_HUMAN/P01270 |
| ***MW/length*** | 9425 Da/84 aa | ***PDB entries*** | 1BWX |
| ***Concentration*** | approx. 10 ng/l | ***Half-life*** | |
| ***PTMs*** | | | |
| ***References*** | Hendy *et al.*, 1981, *Proc. Natl. Acad. Sci. USA*, **78**, 7365–7369.<br>Marx *et al.*, 2000, *Biochem. Biophys. Res. Commun.*, **267**, 213–220. | | |

### Description

**Parathyroid hormone** (PTH) is synthesised in the parathyroid gland and is the most important hypercalcaemic hormone.

### Structure

PTH is a single-chain protein hormone that is derived from the proform by removing an N-terminal hexapropeptide. The 3D structure of N-terminal fragments exhibits a short and a longer α-helix (**1BWX**).

### Biological Functions

PTH raises the serum calcium level by stimulating the renal reabsorption and the bone resorption and by increasing the dietary absorption of calcium from the intestine.

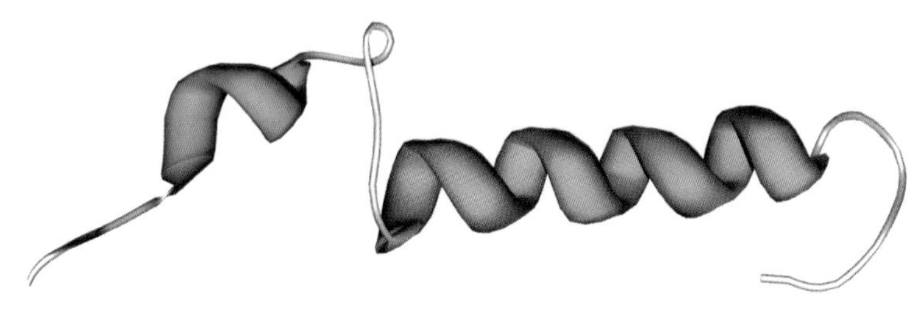

Parathyroid hormone

# *(Pro)-calcitonin*

## Fact Sheet

| | | | |
|---|---|---|---|
| ***Classification*** | Calcitonin | ***Abbreviations*** | CT |
| ***Structures/motifs*** | | ***DB entries*** | CALC_HUMAN/P01258 |
| ***MW/length*** | Calcitonin: 3421 Da/32 aa<br>Katacalcin: 2437 Da/21 aa | ***PDB entries*** | 1BKU |
| ***Concentration*** | approx. 20 ng/l | ***Half-life*** | |
| ***PTMs*** | Calcitonin: Pro-$NH_2$ | | |
| ***References*** | Hashimoto *et al.*, 1999, *Biochemistry*, **38**, 8377–8384.<br>le Moullec *et al.*, 1984, *FEBS Lett.*, **167**, 93–97. | | |

## Description

**Procalcitonin** is synthesised in the thyroid gland and is the most important hypocalcaemic hormone.

## Structure

Single-chain calcitonin is generated from procalcitonin by releasing an N-terminal propeptide (57 aa), a dibasic dipeptide located N-terminally and a tribasic tetrapeptide located C-terminally of calcitonin in procalcitonin. Calcitonin contains a single disulfide bridge (Cys 1–Cys 7) and the C-terminal Pro is amidated. Katacalcin (21 aa) comprises the C-terminal portion of procalcitonin. The corresponding 3D structure of eel calcitonin in micelles is characterised by an extended α-helix and an unstructured C-terminal region (**1BKU**).

## Biological Functions

Calcitonin lowers the calcium level in blood by inhibiting the resorption of calcium from bone and kidney. Katacalcin is a potent calcium-lowering peptide in plasma.

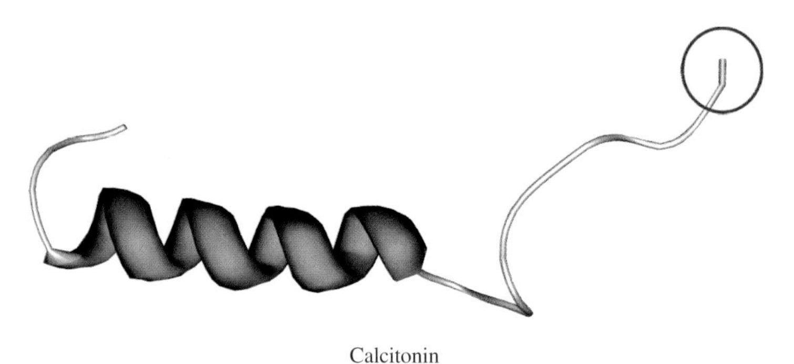

Calcitonin

# Proenkephalin A

## Fact Sheet

| | | | |
|---|---|---|---|
| *Classification* | Opoid neuropeptide precursor | *Abbreviations* | |
| *Structures/motifs* | | *DB entries* | PENK_HUMAN/P01210 |
| *MW/length* | Synenkephalin: 8198 Da/73 aa<br>Met-enkephalin: 574 Da/5 aa<br>Leu-enkephalin: 556 Da/5 aa | *PDB entries* | 1PLW |
| *Concentration* | | *Half-life* | |
| *PTMs* | | | |
| *References* | Marcotte *et al.*, 2004, *Biophys. J.*, **86**, 1587–1600.<br>Noda *et al.*, 1982, *Nature*, **297**, 431–434. | | |

## Description

**Met**- and **Leu-enkephalins** are present in the adenohypophysis and are endogenous opiates.

## Structure

Synenkephalin, four copies of Met-enkephalin, Met-enkephalin-Arg-Gly-Leu, Leu-enkephalin and Met-enkephalin-Arg-Phe, are generated by proteolytic processing from the precursor form (243 aa). The conformation of Met-enkephalin in model membranes resembling natural membranes was determined by multidimensional NMR spectroscopy (**1PLW**).

## Biological Functions

Met- and Leu-enkephalins are endogenous opiates that compete with and mimic the effects of opiate drugs. They play a role in pain perception and in responses to stress.

# Proenkephalin B/(β-Neoendorphin/Dynorphin)

## Fact Sheet

| | | | |
|---|---|---|---|
| *Classification* | Opoid neuropeptide precursor | *Abbreviations* | |
| *Structures/motifs* | | *DB entries* | PDYN_HUMAN/P01213 |
| *MW/length* | β-Neoendorphin: 1110 Da/9 aa<br>Dynorphin: 2148 Da/17 aa<br>Leumorphin: 3352/29 aa<br>Rimorphin: 1571 Da/13 aa<br>Leu-enkephalin: 556 Da/5 aa | *PDB entries* | |
| *Concentration* | | *Half-life* | |
| *PTMs* | | | |
| *References* | Hirokawa *et al.*, 1983, *Nature*, **306**, 611–614. | | |

## Description

All these peptide hormones are contained in the precursor **proenkephalin B** synthesised as a single polypeptide in the adenohypophysis.

## Structure

The various peptide hormones are generated by proteolytic cleavages at paired basic residues in proenkephalin B (234 aa).

## Biological Functions

All these peptide hormones have opiate-like effects on the central nervous system.

# (Pro)-glucagon

## Fact Sheet

| | | | |
|---|---|---|---|
| ***Classification*** | Glucagon | ***Abbreviations*** | |
| ***Structures/motifs*** | | ***DB entries*** | GLUC_HUMAN/P01275 |
| ***MW/length*** | Glucagon: 3483 Da/29 aa<br>Glicentin: 8101Da/69 aa<br>Oxyntomodulin: 4450 Da/37 aa<br>3 Glucagon-like peptides 1: 4170 Da/37 aa; 3356 Da/31aa; 3299 Da/30 aa<br>Glucagon-like peptide 2: 3766 Da/33 aa | ***PDB entries*** | 1BH0; 1D0R |
| ***Concentration*** | 50–150 ng/l | ***Half-life*** | |
| ***PTMs*** | Glucagon-like peptide 1 (30 aa): Arg-$NH_2$ | | |
| ***References*** | Chang *et al.*, 2001, *Mag. Reson. Chem.*, **39**, 477–483.<br>Kieffer *et al.*, 1999, *Endocr. Rev.*, **20**, 876–913.<br>Sturm *et al.*, 1998, *J. Med. Chem.*, **41**, 2693–2700.<br>White *et al.*, 1986, *Nucleic Acid Res.*, **14**, 4719–4730. | | |

## Description

**Glucagon** is secreted by the α cells in the pancreatic islets of Langerhans and plays a key role in glucose metabolism and haemostasis.

## Structure

Proglucagon (160 aa) is the precursor of several physiologically important peptide hormones.The prohormone convertases PCSK1 and PCSK2 process proglucagon by proteolytic cleavages to the corresponding biologically active peptides: glucagon (29 aa), glicentin (69 aa), oxyntomodulin (37 aa), three glucagon-like peptides 1 (37 aa, 31 aa and 30 aa (Arg-$NH_2$)) and glucagon-like peptide 2 (33 aa). The 3D structures of glucagon (**1BH0**) and glucagon-like peptide 1 (**1D0R**) exhibit an extended, partially helical structure.

## Biological Functions

Glucagon regulates the blood glucose level by increasing gluconeogenesis and by decreasing glycolysis and thus is a counterregulatory hormone of insulin. Glucagon-like peptide 1 (37 aa) is a potent stimulator of the glucose-dependent release of insulin. Glucagon-like peptide 2 plays a key role in nutrient haemostasis. Oxyntomodulin significantly reduces the food intake. Glicentin seems to modulate gastric acid secretion.

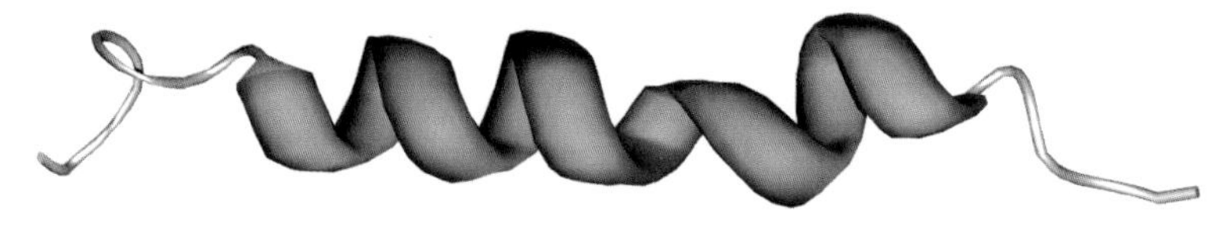

Glucagon - like peptide 1

## *(Pro)-gonadoliberin-1 and (pro)-gonadoliberin-2 (Gonadotropin-releasing factor I + II)*

### Fact Sheet

| | | | |
|---|---|---|---|
| ***Classification*** | GnRH | ***Abbreviations*** | GnRH |
| ***Structures/motifs*** | | ***DB entries*** | GON1_HUMAN/P01148<br>GON2_HUMAN/O43555 |
| ***MW/length*** | Progonadoliberin-1: 7894 Da/69 aa<br>Gonadoliberin-1: 1200 Da/10 aa<br>GnRH-associated peptide 1: 6370/56 aa<br>Progonadoliberin-2: 10 516 Da/97 aa<br>Gonadoliberin-2: 1254 Da/10 aa<br>GnRH-associated peptide 2: 8938 Da/84 aa | ***PDB entries*** | |
| ***Concentration*** | 1–80 ng/l | ***Half-life*** | |
| ***PTMs*** | Gonadoliberin 1 + 2: 1 Pyro-Glu, Gly-$NH_2$ | | |
| ***References*** | Adelman *et al.*, 1986, *Proc. Natl. Acad. Sci. USA*, **83**, 179–183.<br>White *et al.*, 1998, *Proc. Natl. Acad. Sci. USA*, **95**, 305–309. | | |

### Description

**Gonadoliberin-1** (GnRH) is produced in the hypothalamus and gonadoliberin-2 particularly in the kidney, bone marrow and the prostate; both stimulate the secretion of gonadotropins.

### Structure

Progonadoliberin-1 and -2 are the precursors of gonadoliberin-1 and -2 and the GnRH-associated peptides 1 and 2. Gonadoliberin-1 and -2 both carry an N-terminal pyro-Glu and a C-terminal Gly-$NH_2$.

### Biological Functions

GnRHs stimulate the adenohypophysis to release gonadotropins, luteinizing hormone (LH) and follicle-stimulating hormone (FSH). Trp 3 in gonadoliberin-1 and -2 seems to be essential for the biological activity.

# *(Pro)-insulin/Insulin*

## Fact Sheet

| | | | |
|---|---|---|---|
| ***Classification*** | Insulin | ***Abbreviations*** | HPI |
| ***Structures/motifs*** | | ***DB entries*** | INS_HUMAN/P01308 |
| ***MW/length*** | A chain: 2384 Da/21 aa; B chain: 3420 Da/30 aa | ***PDB entries*** | 3AIY |
| ***Concentration*** | 0.2–5.5 μg/l | ***Half-life*** | 5–8 min |
| ***PTMs*** | | | |
| ***References*** | Bell *et al.*, 1979, *Nature*, **282**, 525–527.<br>O'Donoghue *et al.*, 2000, *J. Biomol. NMR*, **16**, 93–108. | | |

## Description

**Pre-proinsulin** (HPI) is synthesised in the islets of Langerhans in the pancreas and insulin promotes the carbohydrate metabolism.

## Structure

Two-chain insulin (A- and B-chains) is generated from proinsulin by the prohormone-converting enzyme, releasing a propeptide that is processed to the C-peptide (31 aa) by cleavage of two terminal dibasic dipeptides. The A- and B-chains are linked by two interchain disulfide bridges (Cys A7–Cys B7, Cys A20–Cys B19) and the A-chain contains one intrachain disulfide bridge (Cys 6–Cys 11). The 3D structure of insulin is highly flexible and the hexameric form is symmetrical (**3AIY**).

## Biological Functions

Insulin promotes normal carbohydrate and secondary protein and fat metabolism. It increases the cell permeability for monosaccharides, amino acids and fatty acids. It accelerates glycolysis, the pentose phosphate cycle and glycogen synthesis in the liver. Absence or insufficiency of insulin leads to diabetes mellitus.

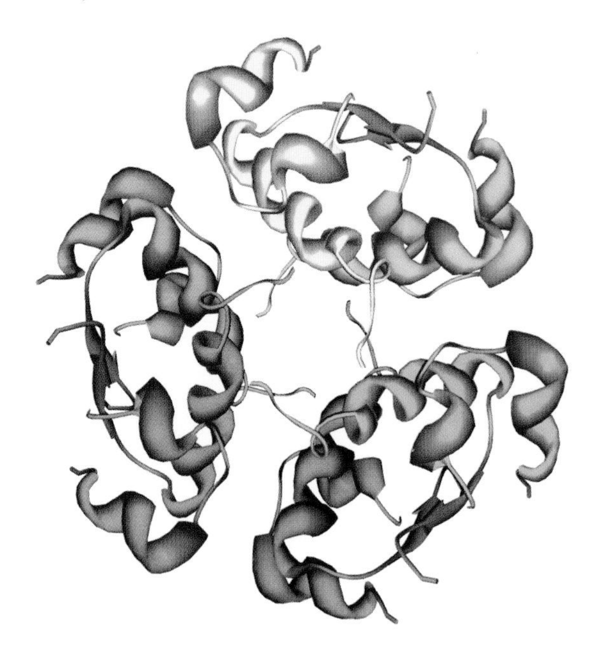

Insulin hexames

## *Prolactin (Luteotropic hormone)*

### Fact Sheet

| | | | |
|---|---|---|---|
| ***Classification*** | Somatotropin/prolactin | ***Abbreviations*** | PRL |
| ***Structures/motifs*** | | ***DB entries*** | PRL_HUMAN/P01236 |
| ***MW/length*** | 22 898 Da/199 aa | ***PDB entries*** | 1N9D |
| ***Concentration*** | <20–200 μg/l | ***Half-life*** | <1 h |
| ***PTMs*** | 1 N-CHO | | |
| ***References*** | Cooke *et al.*, 1981, *J. Biol. Chem.*, **256**, 4007–4016.<br>Keeler *et al.*, 2003, *J. Mol. Biol.* **328**, 1105–1121. | | |

### Description

**Prolactin** (PRL) is mainly synthesised in the adenohypophysis and is involved in lactation.

### Structure

PRL is a single-chain glycoprotein hormone that also exists as dimers and oligomers. The 3D structure of prolactin is characterised by an 'up–up down–down' four-helix bundle topology (**1N9D**).

### Biological Functions

PRL primarily acts on the mammary gland, promoting lactation. PRL also plays a role in osmoregulation, growth, morphogenesis and immunomodulation.

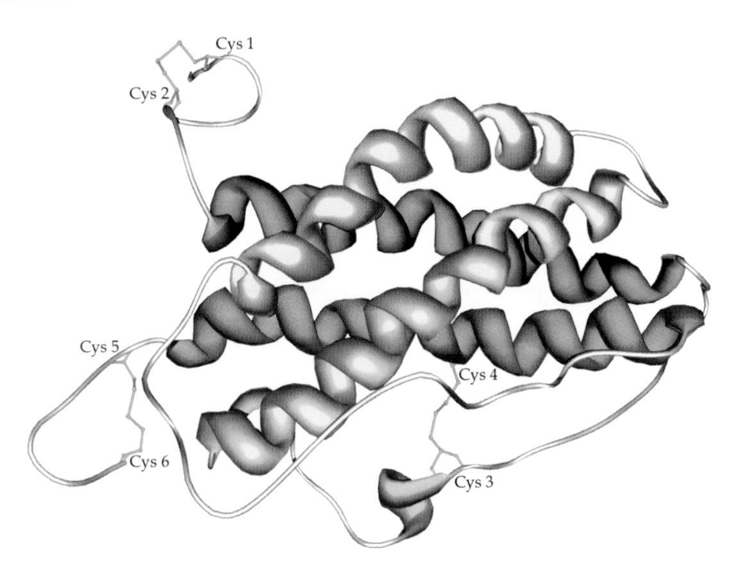


Prolactin

## *(Pro)-thyroliberin (thyrotropin-releasing factor)*

### Fact Sheet

| | | | |
|---|---|---|---|
| ***Classification*** | TRH | ***Abbreviations*** | TRF |
| ***Structures/motifs*** | | ***DB entries*** | TRH_HUMAN/P20396 |
| ***MW/length*** | Prothyroliberin: 24 990 Da/218 aa<br>Thyroliberin: 380 Da/3 aa | ***PDB entries*** | |
| ***Concentration*** | 5–50 ng/l | ***Half-life*** | |
| ***PTMs*** | Thyroliberin: 1 Pyro-Glu; Pro-$NH_2$ | | |
| ***References*** | Yamada *et al.*, 1990, *Mol. Endocrinol.*, **4**, 551–556. | | |

### Description

**Thyroliberin** (TRF) is primarily present in the hypothalamus and is responsible for the release of thyrotropin.

### Structure

TRF, a tripeptide with the sequence pyro-Glu-His-Pro-$NH_2$, is generated from prothyroliberin containing six copies.

### Biological Functions

TRF is a regulator of the synthesis of thyrotropin in the adenohypophysis and functions as a neurotransmitter/neuromodulator in the central and the peripheral nervous system.

## *Secretin*

### Fact Sheet

| | | | |
|---|---|---|---|
| ***Classification*** | Glucagon | ***Abbreviations*** | SCT |
| ***Structures/motifs*** | | ***DB entries*** | SECR_HUMAN/P09683 |
| ***MW/length*** | 3040 Da/27 aa | ***PDB entries*** | |
| ***Concentration*** | 30–50 ng/l | ***Half-life*** | |
| ***PTMs*** | Val-$NH_2$ | | |
| ***References*** | Whitmore *et al.*, 2000, *Cytogenet Cell. Genet.*, **90**, 47–52. | | |

### Description

**Secretin** (SCT) is produced by the upper small intestinal mucosa and stimulates the pancreatic secretion of bicarbonate.

### Structure

SCT is a single-chain peptide hormone that carries a C-terminal Val-$NH_2$ residue and is generated from the proform by releasing two propeptides, an N-terminal (8 aa) and C-terminal (64 aa).

### Biological Functions

SCT stimulates the production of bicarbonate-rich pancreatic juice and the secretion of bicarbonate-rich bile in order to inhibit the HCl production by the stomach.

# *Somatoliberin (Growth hormone-releasing factor)*

## Fact Sheet

| | | | |
|---|---|---|---|
| ***Classification*** | Glucagon | ***Abbreviations*** | GRF |
| ***Structures/motifs*** | | ***DB entries*** | SLIB_HUMAN/P01286 |
| ***MW/length*** | 5041 Da/44 aa | ***PDB entries*** | |
| ***Concentration*** | | ***Half-life*** | |
| ***PTMs*** | Leu-$NH_2$ | | |
| ***References*** | Mayo *et al.*, 1985, *Proc. Natl. Acad. Sci. USA*, **82**, 63–67. | | |

## Description

**Somatoliberin** or growth hormone-releasing factor is synthesised in the hypothalamus and stimulates the release of growth hormone.

## Structure

Somatoliberin is a single-chain peptide hormone that is C-terminally amidated (Leu-$NH_2$) and is generated from its precursor by releasing two propeptides, an N-terminal (10 aa) and a C-terminal (31 aa).

## Biological Functions

Somatoliberin is released by the hypothalamus and acts on the adenohypophysis to stimulate the secretion of growth hormone.

## (Insulin-like growth factors) Somatomedins

### Fact Sheet

| | | | |
|---|---|---|---|
| ***Classification*** | Insulin | ***Abbreviations*** | IGF |
| ***Structures/motifs*** | | ***DB entries*** | IGF1A_HUMAN/P01343<br>IGF1B_HUMAN/P05019<br>IGF2_HUMAN/P01344 |
| ***MW/length*** | Insulin-like growth factor IA/1B: 7655 Da/70 aa<br>Insulin-like growth factor II: 7475 Da/67 Da | ***PDB entries*** | 1IGL |
| ***Concentration*** | 100–500 μg/l | ***Half-life*** | |
| ***PTMs*** | | | |
| ***References*** | Dull *et al.*, 1984, *Nature*, **310**, 777–781.<br>Rotwein *et al.*, 1986, *J. Biol. Chem.*, **261**, 4828–4832. | | |

### Description

**Insulin-like growth factors** (IGFs) are synthesised in the liver and stimulate cartilage growth and have insulin-like activity.

### Structure

IGF IA/B are generated from the proform by releasing N-terminal and C-terminal propeptides and IGF II by releasing a C-terminal propeptide. The 3D structure of IGF2 is characterised mainly by α-helices and a short antiparallel β-sheet (**1IGL**).

### Biological Functions

IGFs possess growth-promoting activity and are structurally and functionally related to insulin, but have a much higher activity than insulin.

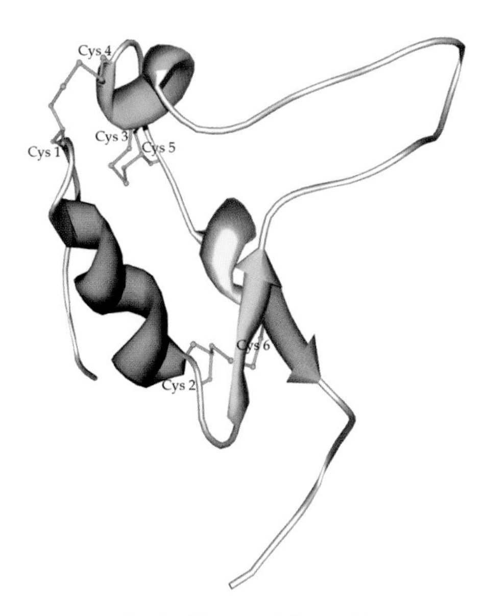


Insulin-like growth factors II

## *Somatostatin (Growth hormone release-inhibiting factor)*

### Fact Sheet

| | | | |
|---|---|---|---|
| ***Classification*** | Somatostatin | ***Abbreviations*** | |
| ***Structures/motifs*** | | ***DB entries*** | SMS_HUMAN/P61278 |
| ***MW/length*** | Somatostatin-28: 3151 Da/28 aa<br>Somatostatin-14: 1640 Da/14 aa | ***PDB entries*** | |
| ***Concentration*** | | ***Half-life*** | |
| ***PTMs*** | | | |
| ***References*** | Shen *et al.*, 1984, *Science*, **224**, 168–171. | | |

### Description

**Somatostatin** is released from the δ-cells in the pancreatic islets of Langerhans and inhibits the release of somatotropin, a hormone that is important for growth control.

### Structure

Somatostain-28 is generated from the proform by releasing an N-terminal 64 residue propeptide and contains a single disulfide bridge (Cys 17–Cys 28). Somatostatin-14 is the C-terminal half of somatostain-28 (disulfide bridge: Cys 3–Cys 14).

### Biological Functions

The main function of somatostatin is the inhibition of somatotropin, a growth hormone from the adenohypophysis.

# *Somatotropin (Growth hormone)*

## Fact Sheet

| | | | |
|---|---|---|---|
| ***Classification*** | Somatotropin/prolactin | ***Abbreviations*** | GH |
| ***Structures/motifs*** | | ***DB entries*** | SOMA_HUMAN/P01241 |
| ***MW/length*** | 22 129 Da/191 aa | ***PDB entries*** | 1A22 |
| ***Concentration*** | 20–70 ng/l | ***Half-life*** | 15–25 min |
| ***PTMs*** | 2 P-Ser (aa 106, 150); Gln 137 and Asn 152: deamidated | | |
| ***References*** | Clackson *et al.*, 1998, *J. Mol. Biol.*, **277**, 1111–1128.<br>Roskam and Rougeon, 1979, *Nucleic Acids Res.*, **7**, 305–320. | | |

## Description

**Somatotropin** (GH) is synthesised in the pituitary gland and plays an important role in growth control.

## Structure

GH is a single-chain protein hormone that exists as monomer and as oligomers up to pentamers, linked by either disulfide bridges or non-covalently associated. GH exhibits a great heterogeneity due to alternative splicing. It contains two disulfide bridges (Cys 53–Cys 165, Cys 182–Cys 189), two phosphorylation sites (Ser 106 and 150) and Gln 137 and Asn 152 are deamidated. The 3D structure of GH with the extracellular domain of its receptor is characterised by a highly helical structure of GH (rainbow) and two β-sheets of the receptor (in red) (**1A22**).

## Biological Functions

GH is an anabolic hormone that is important for postnatal growth and development. It stimulates the growth of a variety of tissues and induces the synthesis of somatomedins (growth factors) in the liver. Somatoliberin stimulates the release of somatotropin.

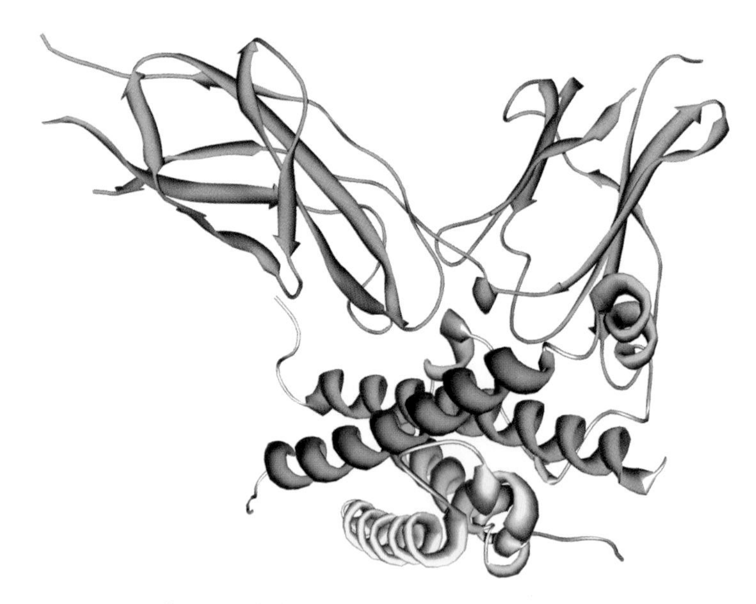

Somatotropin in complex with Somatotropin receptor

# *Substance P + Neurokinin A*

## Fact Sheet

| | | | |
|---|---|---|---|
| ***Classification*** | Tachykinin | ***Abbreviations*** | |
| ***Structures/motifs*** | | ***DB entries*** | TKN1_HUMAN/P20366 |
| ***MW/length*** | Substance P: 1349 Da/11 aa;<br>Neuropeptide K: 3982 Da/36 aa;<br>Neuropeptide γ, first part: 204 Da/2 aa;<br>Neuropeptide γ, second part: 2135 Da/19 aa;<br>Neurokinin A: 1134 Da/10 aa;<br>C-terminal flanking peptide: 1846 Da/16 aa | ***PDB entries*** | |
| ***Concentration*** | | ***Half-life*** | |
| ***PTMs*** | Substance P, neuropeptide K, neuropeptide γ, second part, neurokinin A: Met-$NH_2$ | | |
| ***References*** | Harmar *et al.*, 1986, *FEBS Lett.*, **208**, 67–72. | | |

## Description

**Substance P** is a neurotransmitter and a neuromodulator but is also a potent vasodilatator.

## Structure

Substance P (11 aa), neuropeptide K (36 aa) and neurokinin A (10 aa) are C-terminally amidated (Met-$NH_2$) and are derived from protachykinin.

## Biological Functions

Substance P is a neuropeptide that acts as a neurotransmitter and neuromodulator. In addition, substance P is a potent vasodilatator by releasing nitric oxide, which may cause hypotension.

## *Thrombopoietin (Megakaryocyte colony-stimulating factor)*

### Fact Sheet

| | | | |
|---|---|---|---|
| ***Classification*** | EPO/TPO | ***Abbreviations*** | TPO |
| ***Structures/motifs*** | | ***DB entries*** | TPO_HUMAN/P40225 |
| ***MW/length*** | 35 468 Da/332 aa | ***PDB entries*** | 1V7M |
| ***Concentration*** | 12–60 ng/l | ***Half-life*** | |
| ***PTMs*** | 6 N-CHO (2 potential); 8 O-CHO | | |
| ***References*** | de Sauvage *et al.*, 1994, *Nature*, **369**, 533–538. | | |
| | Feese *et al.*, 2004, *Proc. Natl. Acad. Sci. USA*, **101**, 1816–1821. | | |

### Description

**Thrombopoietin** (TPO) is mainly synthesised in hepatocytes and is the primary regulator of megakaryocyte and platelet production.

### Structure

TPO is a single-chain heavily glycosylated protein that is composed of two domains: an N-terminal domain with two intrachain disulfide bridges (Cys 7–Cys 151, Cys 29–Cys 85) that exhibits structural similarity with EPO and a unique C-terminal domain rich in Pro/Ser/Thr. The 3D structure of the receptor-binding domain of TPO in complex with a neutralising antibody fragment adopts the expect antiparallel four-helix bundle fold as in the related erythropoietin (**1V7M**).

### Biological Functions

TPO is a cytokine involved in the proliferation and regulation of megakaryocytes from the progenitor cells. It seems to be a major physiological regulator of circulating platelets.

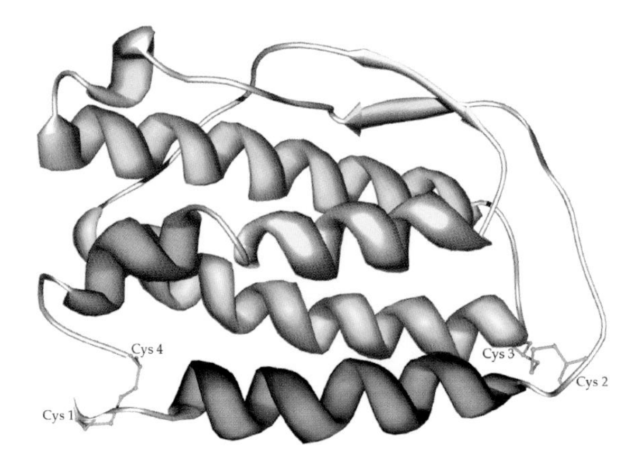


Thrombopoietin

# *Thyroglobulin*

## Fact Sheet

| | | | |
|---|---|---|---|
| ***Classification*** | Type-B carboxylesterase/lipase | ***Abbreviations*** | TG |
| ***Structures/motifs*** | 11 thyroglobulin type I, 3 type II, 5 type III | ***DB entries*** | THYG_HUMAN/P01266 |
| ***MW/length*** | 302 700 Da/2749 aa | ***PDB entries*** | |
| ***Concentration*** | 5 μg/l | ***Half-life*** | approx. 30 h |
| ***PTMs*** | 21 N-CHO (potential); 1 S-Tyr (aa 5); 4 thyroxine (Tyr 5, 1291, 2554, 2568); 1 triiodothyronine (Tyr 2747) | | |
| ***References*** | Malthiery *et al.*, 1987, *Eur. J. Biochem.*, **165**, 491–498. | | |

## Description

**Thyroglobulin** (TG) is synthesised in the thyroid cells and is the matrix for the formation of the thyroid hormones $T_3$ and $T_4$.

## Structure

TG is a large single-chain glycoprotein (approx. 10 % CHO) containing 11 thyroglobulin type I, 3 type II and 5 type III domains, a sulfated Tyr residue (aa 5) and is usually present as homodimer. TG carries four binding sites for thyroxine (Tyr 5, 1291, 2554 and 2568) and one for triiodothyronine (Tyr 2747).

## Biological Functions

TG is the precursor for the generation of thyroid hormones $T_3$ and $T_4$ and for their subsequent storage. TG is a useful marker in serum for thyroid cancer.

## *Thyrotropin (Thyroid-stimulating hormone)*

### Fact Sheet

| | | | |
|---|---|---|---|
| ***Classification*** | Glycoprotein hormone beta chain | ***Abbreviations*** | TSH |
| ***Structures/motifs*** | | ***DB entries*** | GLHA_HUMAN/P01215<br>TSHB_HUMAN/P01222 |
| ***MW/length*** | α-chain: 10 205 Da/92 aa<br>β-chain: 12 901 Da/112 aa | ***PDB entries*** | |
| ***Concentration*** | | ***Half-life*** | |
| ***PTMs*** | α-chain: 2 N-CHO; 1 O-CHO;<br>β-chain: 1 N-CHO | | |
| ***References*** | Fiddes *et al.*, 1979, *Nature*, **281**, 351–356.<br>Hayashizaki *et al.*, 1985, *FEBS Lett.*, **188**, 394–400. | | |

### Description

**Thyrotropin** (TSH) is primarily present in the adenohypophysis and stimulates the release of triiodothyronine ($T_3$) and thyroxine ($T_4$).

### Structure

TSH is a glycoprotein hormone present in plasma as a heterodimer of two non-covalently-linked chains: a common α-chain (approx. 21 % CHO) containing five intrachain disulfide bridges which is also present in choriogonadotropin, follitropin and lutropin, and a unique β-chain (approx. 12 % CHO) containing six intrachain disulfide bridges. The thyroid β-chain is generated from the proform by removing a C-terminal hexapeptide.

### Biological Functions

TSH stimulates the release of $T_3$ and $T_4$ and is indispensable for the thyroid structure and metabolism. Thyroliberin stimulates the adenohypophysis to release TSH.

# *Vasoactive intestinal peptides*

## Fact Sheet

| | | | |
|---|---|---|---|
| ***Classification*** | Glucagon | ***Abbreviations*** | VIP |
| ***Structures/motifs*** | | ***DB entries*** | VIP_HUMAN/P01282 |
| ***MW/length*** | Vasoactive intestinal peptide: 3327 Da/28 aa<br>Intestinal peptide PHV-42: 4552 Da/42 aa<br>Intestinal peptide PHM-27: 2986 Da/27 aa | ***PDB entries*** | |
| ***Concentration*** | 6–16 ng/l | ***Half-life*** | approx. 2 min |
| ***PTMs*** | Vasoactive intestinal peptide: Asn-$NH_2$; Intestinal peptide PHM-27: Met-$NH_2$ | | |
| ***References*** | Itoh *et al.*, 1983, *Nature*, **304**, 547–549. | | |

## Description

**Vasoactive intestinal peptides** (VIPs) are produced in the pancreas and are involved in several physiological processes.

## Structure

VIPs are generated from the proform (150 aa) by releasing two propeptides, an N-terminal (59 aa) and a C-terminal (15 aa). Vasoactive intestinal peptide (29 aa) contains a C-terminal Asn-$NH_2$, and intestinal peptide PHM-27 (27 aa) with a C-terminal Met-$NH_2$ corresponding to the N-terminal region of intestinal peptide PHV-42 (42 aa).

## Biological Functions

VIP causes vasodilation, lowers arterial blood pressure, stimulates myocardial contractility, increases glycogenolysis and relaxes smooth muscles.

# 14

# Cytokines and Growth Factors

## 14.1 INTRODUCTION

*Growth factors* are capable of stimulating cellular proliferation and differentiation and are important for various cellular processes. Growth factors act as signal transmission proteins between cells. *Cytokines*, like hormones, are short- or long-distance signalling molecules important in intercellular communication. A clear distinction between growth factors and cytokines and in some cases even hormones is often difficult or even impossible. Although historically cytokines were rather associated with hematopoietic cells and with cells of the immune system it is now clear that these signalling molecules are also used by other cells and tissues for various tasks in the whole body, during development and later in the adult state. A search in protein data bases such as UniProt for the keywords 'growth factors' and 'cytokines' often results in double entries, which is often the case for interleukins.

Cytokines are a group of smaller, soluble proteins, many of them glycoproteins, which are produced by a wide variety of cell types, hematopoietic and nonhematopoietic cells. Cytokines are responsible for intercellular communication via receptors on the surface of cells. Cytokines are involved in the development, differentiation and activation of cells. Cytokines, which are secreted from lymphocytes are termed 'lymphokines' and those secreted by monocytes and macrophages are termed 'monokines'. Traditionally, cytokines are divided into several groups and are compiled in Table 14.1:

1. ***Hematopoietins*** are a group of structurally related proteins, which are characterised by a bundle of shorter (approximately 15 aa) and longer (approximately 25 aa) α-helices, usually four α-helix bundle cytokines, of which many are interleukins:
   (a) Examples of shorter α-helices cytokines are the interleukins IL-2, IL-3, IL-4, IL-5, Il-7, IL-9, IL-13 and IL-15 and granulocyte-macrophage colony-stimulating factor.
   (b) Examples of longer α-helices cytokines are the interleukins IL-6, IL-10, IL-11 and IL-12 and granulocyte colony-stimulating factor.
   (c) The four α-helix bundle hormones erythropoietin and thrombopoietin (for details see Chapter 13).
2. ***Interferons*** (IFN) are a group of cytokines, which are all characterised by a helix bundle structure, among them IFN-α, IFN-β and IFN-γ.
3. ***Tumor necrosis factors*** (TNF) are a group of cytokines characterised by a β-sheet structure, among them TNF-α and TNF-β.
4. ***Chemokines*** are a group of cytokines usually characterised by a β-sheet structure and a single α-helix; among them are interleukin IL-8, platelet factor 4, platelet basic protein and cytokines A2 and A5.

As in hormones the cytokines can be divided into three different groups, depending on their place of action:

(a) ***Autocrine.*** The cytokine acts on the cell that secretes it.
(b) ***Paracrine.*** The action of the cytokine is restricted to the close vicinity of its secretion.
(c) ***Endocrine.*** The cytokine acts at a distance from the releasing place and is transported to the target cells and tissues via the vascular system.

*Human Blood Plasma Proteins: Structure and Function* Johann Schaller, Simon Gerber, Urs Kämpfer, Sofia Lejon and Christian Trachsel

**Table 14.1** Characteristics of a selection of cytokines and growth factors in human plasma.

| Family | Origin | Size (# aa) | Major effects |
|---|---|---|---|
| ***Interleukin families*** | | | |
| Interleukin-1 α/β (IL-1 α/β) | Most nucleated cells | 159/153 | Inflammatory response, T cell activation, activation macrophages |
| Interleukin-2 (IL-2)/T cell growth factor | T cells | 133 | T cell proliferation, antigen specific immune response |
| Interleukin-3 (IL-3)/hematopoietic growth factor | T cells | 133 | Proliferation and differentiation hematopoietic cells |
| Interleukin-4 (IL-4)/B cell stimulatory factor | T cells and mast cells | 129 | B cell activation and other cells |
| Interleukin-13 (IL-13) | T cells | 112 | Inhibition inflammatory cytokine production |
| Interleukin-5 (IL-5)/eosinophil differentiation factor | T cells and mast cells | 115 | Growth and differentiation of eosinophil cells |
| Interleukin-6 (IL-6)/B cell stimulatory factor 2 | Wide variety of cells | 183 | Multifunctional cytokine acting on a wide variety of cells |
| Interleukin-11(IL-11)/adipogenesis inhibitory factor | Most cells | 178 | Stimulation of proliferation hematopoietic stem cells and megakaryocyte progenitor cells |
| Interleukin-12 α-chain/β-chain (IL-12A/IL-12B) | B cells, macrophages | 197/306 | Activation natural killer cells |
| Granulocyte colony-stimulating factor (G-CSF) | Variety of cells | 177 | Production of neutrophilic granulocytes |
| Granulocyte–macrophage colony-stimulating factor (GM-CSF) | Variety of tissue cells | 127 | Stimulation of growth and differentiation of hematopoietic precursor cells |
| Interleukin-7 (IL-7) | Bone marrow | 152 | Stimulation of proliferation of lymphoid progenitors |
| Interleukin-9 (IL-9) | T cells | 126 | Support of growth of helper T cells |
| Interleukin-10 (IL-10)/cytokine synthesis inhibitory factor | Variety of cells | 160 | Inhibition of a number of cytokines |
| Interleukin-15 (IL-15) | T cells | 114 | Stimulation of the proliferation of T lymphocytes |
| ***Tumour necrosis factor family*** | | | |
| Tumour necrosis factor (TNF-α)/cachectin (soluble form) | Macrophages, natural killer cells | 157 | Local inflammation and potent pyrogen |
| Lymphotoxin-α (LT-α)/tumour necrosis factor-β | Lymphocytes | 171 | Cytotoxic for many types of tumour cells |
| Lymphotoxin-β (LT-β)/tumour necrosis factor C | T cells and B cells | 244 | Probably in immune response regulation |
| *Interferon Family* | | | |
| Interferon α-1/13 (IFN-α)/interferon α-D | Macrophages | 166 | Antiviral activity |
| Interferon β (IFN-β)/Fibroblast interferon | Fibroblasts | 166 | Antiviral, antibacterial and anticancer activity |
| Interferon γ (IFN-γ)/immune interferon | T lymphocytes | 146 | Antiviral and immunoregulatory activity |
| ***Chemokines*** | | | |
| Interleukin-8 (IL-8)/neutrophil-activating protein 1 | Macrophages, others | 77 | Chemotactic factor |
| Platelet factor 4 (PF-4)/oncostatin A | Megakaryocytes | 70 | Neutralisation of heparin anticoagulant activity |
| Platelet basic protein/small inducible cytokine B7 | Megakaryocytes | 94 | Various biological functions |
| Small inducible cytokine A2 | Macrophages, others | 76 | Chemotactic factor |
| Small inducible cytokine A5/RANTES (3–68) | T cells, blood platelets | 68 | Chemoattractant for different cells |
| ***Platelet-derived growth factor/vascular endothelial growth factor family*** | | | |
| Platelet-derived growth factor A chain/B chain | Platelet α-granules | 125/109 | Potent mitogen for mesenchymal cells |
| Vascular endothelial growth factor A (VEGF-A) | Many tissues | 206 | Involved in angiogenesis, vasculogenesis, endothelial cell growth |
| Vascular endothelial growth factor B (VEGF-B) | All tissues except liver | 186 | Growth factor for endothelial cells |
| Vascular endothelial growth factor C (VEGF-C) | Many tissues | 116 | Involved in angiogenesis and endothelial cell growth |

(*Continued*)

**Table 14.1** (*Continued*).

| Family | Origin | Size (# aa) | Major effects |
|---|---|---|---|
| Vascular endothelial growth factor D (VEGF-D) | Many tissues | 117 | Involved in angiogenesis and endothelial cell growth |
| ***Transforming growth factor β family*** | | | |
| Transforming growth factor β1, β2, β3 (TGF-β1, -β2, -β3) | Bone, monocytes | All 112 | Multifunctional cytokine: control proliferation and differentiation of many cell types |
| Transforming growth factor α | Macrophages, brain | 50 | Mitogenic cytokine, acts synergistically with TGF-β |
| ***Miscellaneous*** | | | |
| Nerve growth factor β (β-NGF) | Fibroblasts, neurons | 120 | Important for sympathetic and sensory nervous system |
| Pro-epidermal growth factor (EGF) | Submaxillary gland | 1185 | Important for cell growth, proliferation and differentiation |
| Insulin-like growth factor binding proteins 1–7 | Many tissues | 213–289 | Regulation insulin growth factors |

Cytokines act via specific receptors on the cell surface of the target cells. As a characteristic feature all these cytokine receptors contain at least one transmembrane segment. The cytokine receptors can be divided into different classes:

1. **Type I: hematopoietin receptors**. Hematopoietin receptors usually consist of one or two chains, which are responsible for the binding of the cytokine, and a further chain, which is required for the signal transduction into the target cell. They contain various types of domains, among them cytokine receptor domains, FN3 domains, Ig domains and sushi/CCP/SCR domains.
2. **Type II: interferon receptors**. Interferon receptors consist of two chains, similar to the hematopoietin receptors.
3. **Type III: tumour necrosis factor receptors** (TNF). The TNF receptors are single-chain receptors containing Cys-rich regions of the nerve growth factor receptor family.
4. **Type IV: Ig superfamily receptors**. The Ig receptors are normally monomers containing sequences with tyrosine kinase activity.
5. **Chemokine receptors**. Chemokine receptors are single-chain receptors containing seven transmembrane segments, typical for receptors linked to G proteins.
6. **Transforming growth factor receptors** (TGF). The TGF receptors are receptors exhibiting serine kinase and threonine kinase activity.

Cytokines are involved in many biological processes and play a key role in many of these processes, among them the development and functioning of the immune system. In both the innate and the adaptive immune response, they play a major role in many immunological, inflammatory and infectious diseases and are important in developmental processes, e.g. embryogenesis. The various cytokines and growth factors will be discussed according to the classification compiled in Table 14.1.

## 14.2 INTERLEUKIN FAMILIES

Interleukins were first detected as expression products of leukocytes, but it became clear later that interleukins are produced by a wide variety of cells. The efficient functioning of the immune system is largely dependent on interleukins and a series of immune diseases and immune deficiencies are related to interleukin deficiencies. The various interleukin families are compiled in Table 14.2 and the most important ones will be discussed in some detail.

### 14.2.1 Interleukin-1 family

The interleukin-1 family has four members: two forms of interleukin-1 (IL-1), IL-1α (hematopoietin) and IL-1β (catabolin), interleukin-18 (IL-18; 157 aa: Q14116) and interleukin-33 (IL-33: O95760) (see also Chapter 5). IL-1 is produced by most

**Table 14.2** Interleukin families.

| Family | Members. | Code | Signature |
|---|---|---|---|
| IL-1 | IL-1α/IL-1β/IL-18/IL-33 | P01583/P01584/Q14116/O95760 | PS00253 |
| IL-2 | IL-2 | P60568 | PS00424 |
| IL-3 | IL-3 | P08700 | |
| IL-4/IL-13 | IL-4/IL-13 | P05112/P35225 | PS00838 |
| IL-5 | IL-5 | P05113 | |
| IL-6 | IL-6/IL-11/IL-12A/IL-23A G-CSF | P05231/P20809/P29459/Q9NPF7 P09919 | PS00254 |
| IL-7/IL-9 | IL-7/IL-9 | P13232/P15248 | PS00255 |
| IL-10 | IL-10/IL-19/IL-20, IL-22/IL-24/IL-26 | P22301/Q9UHD0/Q9NYY1/Q9GZX6/Q13007/Q9NPH9 | PS00520 |
| IL-15 | IL-15 | P40933 | |
| IL-17 | IL-17/IL-17B/IL-17C/IL-17D/IL-17F/IL-25 | Q16552/Q9UHF5/Q9P0M4/Q8TAD2/Q96PD4/Q9H293 | |
| IL-21 | IL-21 | Q9HBE4 | |
| IL-28/IL-29 | IL-28A/IL-28B/IL-29 | Q8IZJ0/Q8IZI9/Q8IU54 | |

nucleated cells and is important in the inflammatory response of the body to infection. IL-1α and IL-1β are pleiotropic mediators of the host response to infections and injuries. Both are involved in various biological processes like T lymphocyte and B lymphocyte activation and proliferation, induction of prostaglandin synthesis and both act as pyrogens.

**Interleukin-1α** (IL-1α) (see Data Sheet) or hematopoietin and **interleukin-1β** (IL-1β) (see Data Sheet) or catabolin originate from two different genes and are mainly produced by macrophages. Both are synthesised as precursors lacking the classical hydrophobic signal sequences (IL-1α: 271 aa; IL-1β: 269 aa) and are processed to their mature forms by proteolytic cleavage of a single peptide bond (IL-1α: Arg 112–Ser 113; IL-1β: Asp 116–Ala 117 in the precursors) resulting in mature IL-1α (159 aa: Ser 1–Ala 159) and IL-1β (153 aa: Ala 1–Ser 153). Both are single-chain monomers that are usually nonglycosylated and contain no disulfide bridges. Both enhance locally many cell activities and promote hematopoiesis, amplify immunological reactions, induce fever, hypoglycemia and hypotension and stimulate hepatic acute-phase response.

The 3D structure of entire interleukin-1β (153 aa: Ala 1–Ser 153) was determined by X-ray diffraction (PDB: 9ILB) and is shown in Figure 14.1. The structure is characterised by 12 antiparallel β-strands folded into a six-stranded β-barrel similar in architecture to the Kunitz-type trypsin inhibitor. The 3D structure of IL-1α (PDB: 2ILA) is similar to IL-1β.

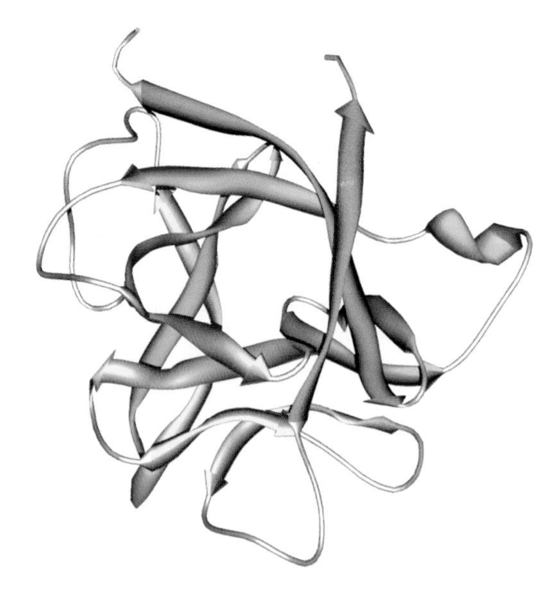

**Figure 14.1 3D structure of interleukin-1β (PDB: 9ILB)**
Interleukin-1β contains 12 antiparallel β-strands folded into a six-stranded β-barrel.

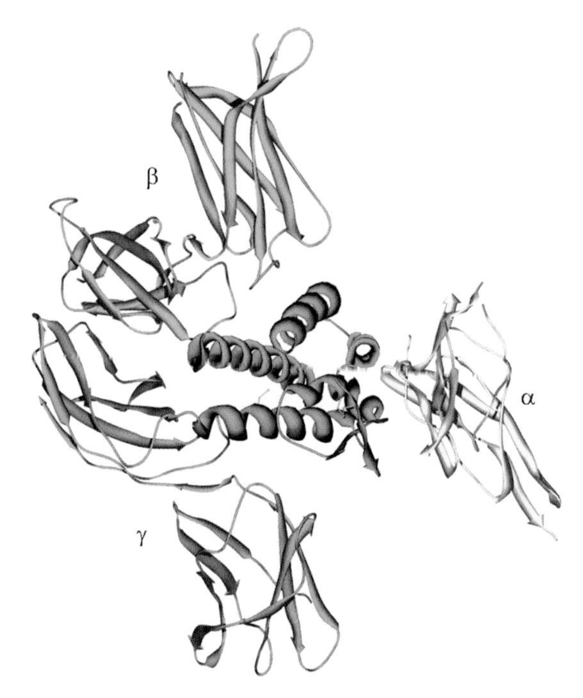


**Figure 14.2 3D structure of interleukin-2 in complex with the extracellular domain of the IL-2 receptor (PDB: 2B5I)**
IL-2 (in blue) is characterised by a α-helix bundle (in blue) containing a single disulfide bridge (in pink) and one free Cys residue (in light blue). The extracellular domain of the IL-2 receptor consists of three chains: IL-2Rα (in yellow), IL-2Rβ (in green) and IL-2Rγ (in red).

### 14.2.2 Interleukin-2 family

Interleukin-2 (IL-2) is the only so far known member of the IL-2 family (see also Chapter 5). IL-2 is a cytokine or hormone that is crucial for the natural response of the body to microbial infection and in the discrimination process between foreign and self.

**Interleukin-2** (IL-2) (see Data Sheet) or T cell growth factor synthesised by T cells is present as a single-chain glycoprotein containing a single disulfide bridge (Cys 58–Cys 105) and a free Cys residue (aa 125) of unknown function. IL-2 acts as a growth factor for activated T lymphocytes and some B lymphocytes and nonactivated natural killer cells.

The 3D structure of entire interleukin-2 (133 aa: Ala 1–Thr 133) in complex with the extracellular domain of the IL-2 receptor consisting of three chains: (a) the extracellular domain of the α-chain IL-2Rα (217 aa: Glu 1–Glu 217: P01589), (b) the extracellular domain of the β-chain IL-2Rβ (214 aa: Ala 1–Thr 214: P14784) and (c) part of the extracellular domain of the γ-chain IL-2Rγ (199 aa: Pro 34–Asn 232: P31785) was determined by X-ray diffraction (PDB: 2B5I) and is shown in Figure 14.2. The structure of IL-2 is characterised by an α-helix bundle (in blue) and contains a single intrachain disulfide bridge (in pink) and a free Cys residue (in light blue). The IL-2Rβ (in green) and IL-2Rγ (in red) are typical type I cytokine receptors each containing an FN3 domain characterised by a β-sandwich consisting of seven antiparallel strands. The bases of IL-2Rβ and IL-2Rγ converge and form a Y-shaped structure and IL-2 binds to the fork of the structure. IL-2Rα (in yellow) is not a typical cytokine receptor and consists of two sushi domains, each containing six β-strands, and IL-2Rα sits on top of IL-2 and seems not to make any contacts with IL-2Rβ and IL-2Rγ.

### 14.2.3 Interleukin-3 family

The only known member of the interleukin-3 family is interleukin- 3 (IL-3) itself. IL-3 is a cytokine and growth factor that is involved in hematopoiesis.

**Interleukin-3** (IL-3) (see Data Sheet) or hematopoietic growth factor primarily produced by activated T cells is present as a single-chain glycoprotein containing a single disulfide bridge (Cys 16–Cys 84). IL-3 is a cytokine and growth factor that stimulates the proliferation of hematopoietic pluripotent progenitor cells.

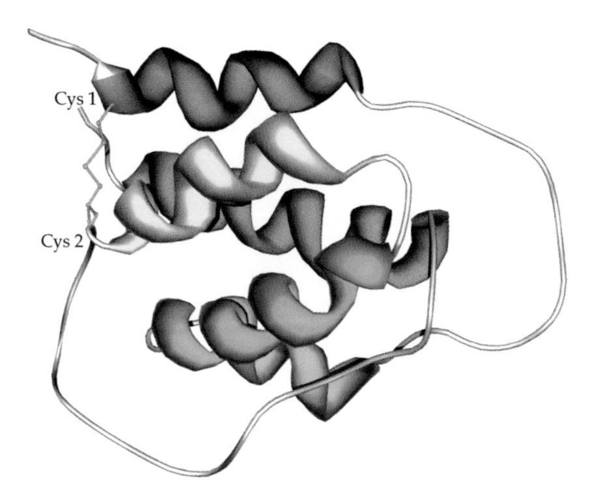


**Figure 14.3 3D structure of interleukin-3 (PDB: 1JLI)**
IL-3 is characterised by a left-handed four-helix bundle and a single disulfide bridge (in pink).

The 3D structure of a mutant and N- and C-terminally truncated interleukin-3 (112 aa: Val14Ala–Gln 125) was determined by NMR spectroscopy (PDB: 1JLI) and is shown in Figure 14.3. The structure is characterised by a left-handed four-helix bundle stabilised by a single disulfide bridge (in pink).

### 14.2.4 Interleukin-4/interleukin-13 family

The interleukin-4/interleukin-13 family has two members: interleukin-4 (IL-4) plays an important role in the control and regulation of the immune and inflammatory system and interleukin-13 (IL-13) is a lymphokine that regulates inflammatory and immune responses. IL-4 and IL-13 are distantly related (see also Chapter 5).

**Interleukin-4** (IL-4) (see Data Sheet) or B cell stimulatory factor 1 produced by activated T cells and mast cells is a single-chain glycoprotein containing three intrachain disulfide bridges (Cys 3–Cys 127, Cys 24–Cys 65, Cys 46–Cys 99). IL-4 stimulates the proliferation of activated B cells and T cells and is an important regulator in humoral-mediated and adaptive immunity. It induces the expression of MHC class II molecules on resting B cells and enhances the secretion and expression of IgE and IgG1.

The 3D structure of interleukin-4 (129 aa: His 1–Ser 129) was determined by X-ray diffraction (PDB: 2INT) and is shown in Figure 14.4. The compact globular structure is characterised by a left-handed antiparallel four-helix bundle connected by two double-stranded antiparallel β-sheets. IL-4 contains three intrachain disulfide bridges (in pink) arranged in the pattern **Cys 1–Cys 6, Cys 2–Cys 4, Cys 3–Cys 5**.

**Interleukin-13** (IL-13) (see Data Sheet) expressed by T lymphocytes is a single-chain glycoprotein containing two disulfide bridges (Cys 28–Cys 56, Cys 44–Cys 70). The function of IL-13 considerably overlaps with that of IL-4. IL-13 seems to play a critical role in the regulation of inflammatory and immune responses. Together with interleukin-2 it regulates the synthesis of interferon γ.

The 3D structure of interleukin-13 (113 aa: Met 0–Asn 112) was determined by NMR spectroscopy (PDB: 1IJZ) and is shown in Figure 14.5. IL-13 is characterised by a short-chain left-handed four-helix bundle stabilised by two intrachain disulfide bridges (in pink) arranged in the pattern **Cys 1–Cys 3, Cys 2–Cys 4** and exhibits a significant similarity with the folding topology of IL-4.

### 14.2.5 Interleukin-5 family

The only known member of the interleukin-5 family is interleukin-5 (IL-5) itself. IL-5 is a cytokine involved in eosinophil production.

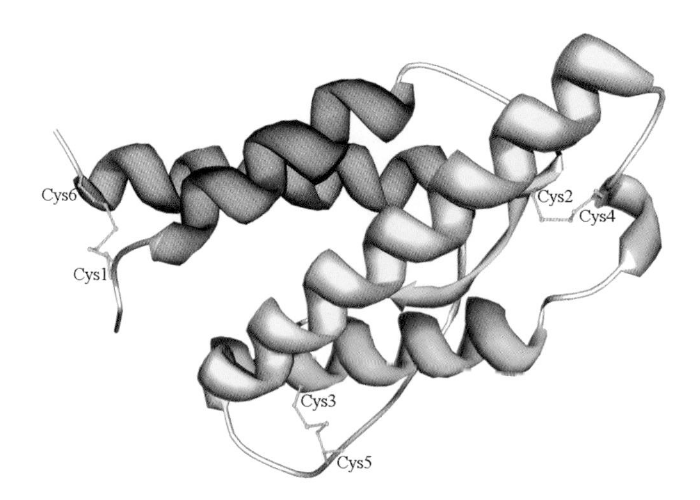


**Figure 14.4 3D structure of interleukin-4 (PDB: 2INT)**
IL-4 is characterised by a left-handed antiparallel four-helix bundle and contains three intrachain disulfide bridges (in pink).

**Interleukin-5** (IL-5) (see Data Sheet) or eosinophil differentiation factor produced by T cells and mast cells is a homodimeric, disulfide-linked (Cys 44–Cys 86) glycoprotein. IL-5 induces the production of eosinophils and acts as an eosinophil differentiation factor, a B cell growth factor and a T cell replacing factor.

The 3D structure of a shortened form of interleukin-5 (108 aa: Ile 5–Ile 112) was determined by X-ray diffraction (PDB: 1HUL) and is shown in Figure 14.6. The structure of homodimeric IL-5 (in yellow and blue) is characterised by two

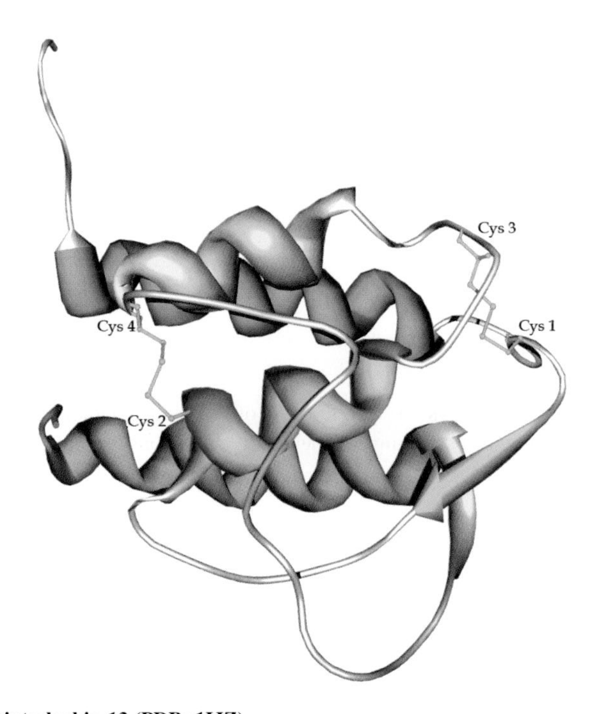


**Figure 14.5 3D structure of interleukin-13 (PDB: 1IJZ)**
IL-13 is characterised by a short chain left-handed four-helix bundle and contains two intrachain disulfide bridges (in pink).

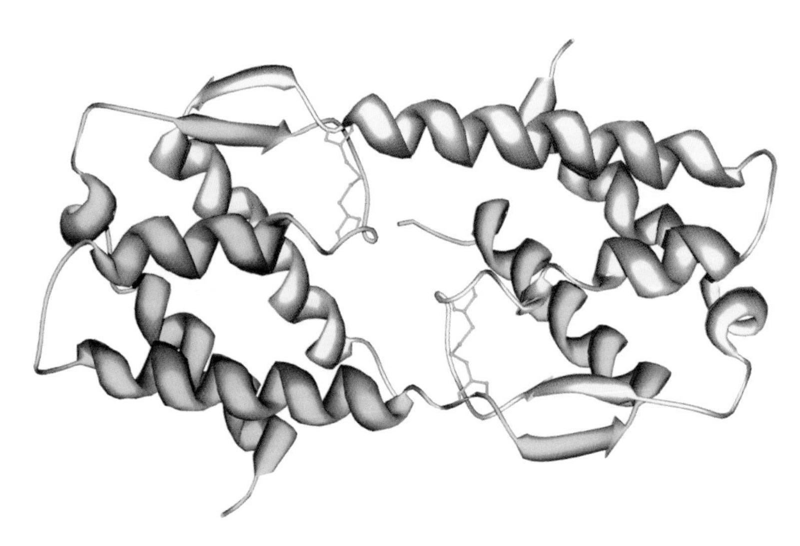

**Figure 14.6 3D structure of homodimeric interleukin-5 (PDB: 1HUL)**
IL-5 is characterised by two left-handed bundles of four $\alpha$-helices and two short $\beta$-sheets and the monomers are linked by two interchain disulfide bridges (in green).

left-handed bundles of four $\alpha$-helices in an end-to-end topology and two short $\beta$-sheets. The monomers are linked by two interchain disulfide bridges (in green) in a cross-wise manner.

### 14.2.6 Interleukin-6 family

The interleukin-6 family consists of several members, among them interleukin-6 (IL-6) itself, interleukin-11 (IL-11), interleukin-12 $\alpha$-chain (IL-12A), interleukin-23 (IL-23A; 170 aa: Q9NPF7) and granulocyte colony-stimulating factor (G-CSF). The 3D structure of the IL-6 family members is characterised by a long-chain four-helix bundle topology containing four conserved Cys residues involved in two intrachain disulfide bridges (see also Chapter 5).

**Interleukin-6** (IL-6) (see Data Sheet) or interferon $\beta_2$ produced by a wide variety of cells is a single-chain glycoprotein containing two intrachain disulfide bridges (Cys 43–Cys 49, Cys 72–Cys 82). IL-6 acts on many types of cells and participates in the immune response, in hematopoiesis and in the inflammatory response and induces antibody production, acute phase reactions and fever.

The 3D structure of interleukin-6 (186 aa: Val 1–Met 183 + N-terminal tripeptide) in complex with the extracellular domain of the IL-6 receptor consisting of the extracellular binding domains of the $\alpha$-chain (IL-6R$\alpha$, 201 aa: Glu 96–Trp 296, P08887) and the extracellular activation and binding domains of the $\beta$-chain (IL-6R$\beta$ or gp130, 299 aa: Glu 1–Thr 299, P40189) was determined by X-ray diffraction (PDB: 1P9M) and is shown in Figure 14.7(a). The ternary complex of IL-6 (in blue), IL-6R$\alpha$ (in red) and Il-6R$\beta$ (in green) forms a hexamer. The structure of IL-6 (PDB: 2IL6) is characterised by the expected four-helix bundle topology stabilised by two intrachain disulfide bridges (in pink) arranged in the pattern **Cys 1–Cys 2, Cys 3–Cys 4** and is shown in Figure 14.7(b).

**Interleukin-11** (IL-11) (see Data Sheet) or adipogenesis inhibitory factor produced by most cell types is a single-chain cytokine rich in Pro residues (12 %) but devoid of Cys residues, but nevertheless belongs to the IL-6 family with a four $\alpha$-helix bundle topology and a high helical content (57 %). IL-11 exhibits pleiotropic effects on multiple tissue types and directly stimulates the proliferation of hematopoietic stem cells and megakaryocytes progenitor cells and induces megakaryocyte maturation resulting in an increased platelet production.

**Interleukin-12** (IL-12) (see Data Sheet) produced by activated macrophages and B lymphocytes is a heterodimeric cytokine composed of the interleukin-12 $\alpha$-chain (IL-12$\alpha$) and $\beta$-chain (IL-12$\beta$). IL-12$\alpha$ containing three possible intrachain disulfide bridges (Cys 42–Cys 174, Cys 63–Cys 101 are assigned, Cys 15–Cys 88 is unassigned) is linked by a single interchain disulfide bridge (Cys 74 in $\alpha$-chain–Cys 177 in $\beta$-chain) to IL-$\beta$ containing four intrachain disulfide bridges (Cys 28–Cys 68, Cys 109–Cys 120, Cys 148–Cys 171, Cys 278–Cys 305) and a free Cys residue (aa 252) of

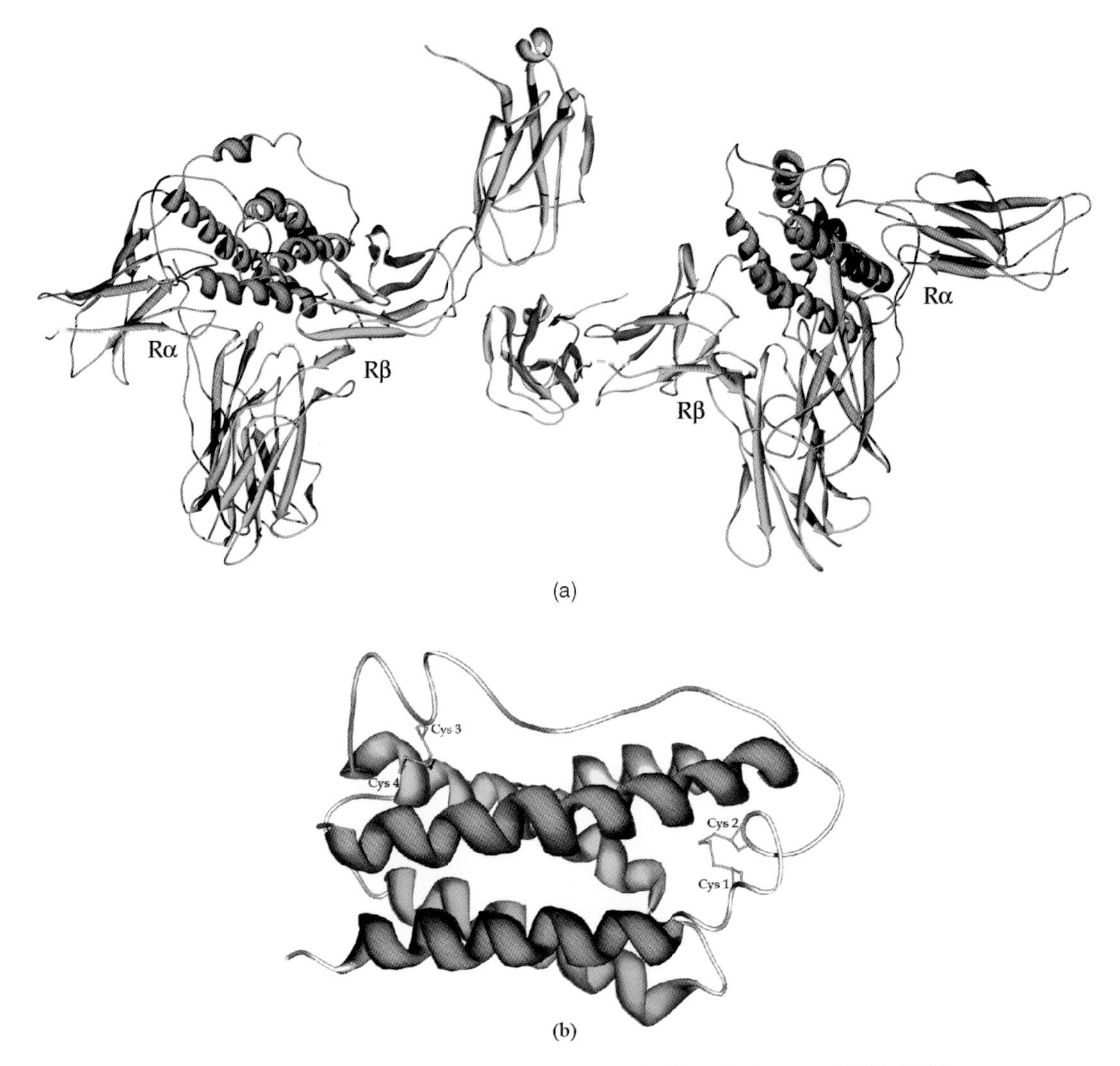


**Figure 14.7 3D structure of interleukin-6 with the extracellular domain of the IL-6 receptor (PDB: 1P9M)**
(a) The ternary complex of IL-6 (in blue) and the extracellular domains of IL-6Rα (in red) and Il-6Rβ (in green) forms a hexamer.
(b) The monomer of 1L-6 (PDB: 2IL6) is characterised by a four-helix bundle topology and contains two intrachain disulphide bridges (in pink).

unknown function. IL-12 exhibits numerous biological activities such as the increase of the cytolytic activity of natural killer cells, the production of interferon γ and the proliferation of activated T cells and natural killer cells.

The 3D structure of interleukin-12 consisting of the α-chain (197 aa: Arg 1–Ser 197) and the β-chain (306 aa: Ile 1–Ser 306) linked by a single disulfide bridge (in green) was determined by X-ray diffraction (PDB: 1F45) and is shown in Figure 14.8. IL-12α (in dark blue) exhibits the expected structure of the IL-6 family members with a four-helix bundle topology and IL-12β (in light blue) is similar in structure to the extracellular domain of the IL-6 receptor α-chain. IL-12β contains an N-terminal Ig-like C2 type domain and two FN3 domains characterised by β-sheet structures.

Defects in the interleukin-12B gene are the cause of a Mendelian susceptibility to mycobacteria disease, also known as familial disseminated atypical mycobacterial infection (for details see MIM: 209950).

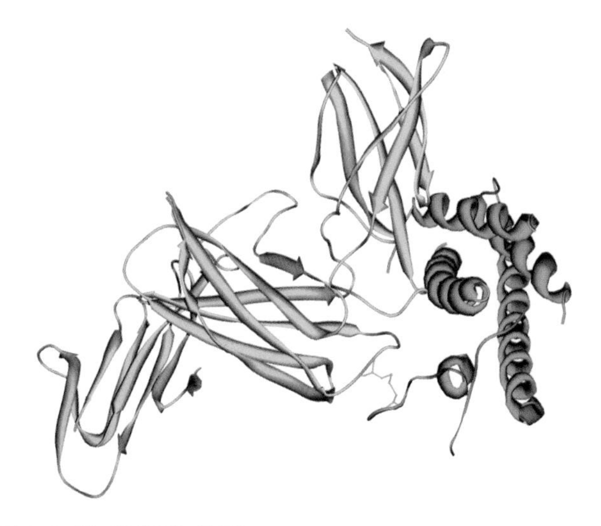

**Figure 14.8 3D structure of interleukin-12 (PDB: 1F45)**
IL-12 consists of two chains: IL-12α (in dark blue) exhibits a four-helix bundle topology and IL-12β (in light blue) contains a N-terminal Ig-like C2 type domain and two FN3 domains characterised by β-sheet structures.

**Granulocyte colony-stimulating factor** (G-CSF) (see Data Sheet) or pluripoietin produced by various types of cells (macrophages, endothelial cells, fibroblasts) is a single-chain glycoprotein containing two intrachain disulfide bridges (Cys 39–Cys 45, Cys 67–Cys 77) and a free Cys residue (aa 18) of unknown function. G-CSF is induced by bacterial infection and is involved in hematopoiesis by controlling the production, differentiation and function of neutrophil granulocytes and monocytes-macrophages.

The 3D structure of granulocyte colony-stimulating factor (174 aa: Thr 1–Pro 174) in complex with the N-terminal half of the extracellular domain of the G-CSF receptor (313 aa: Glu 1–Glu 308 + pentapeptide: Q99062) consisting of an Ig-like C2 type domain and two FN3 domains was determined by X-ray diffraction (PDB: 2D9Q) and is shown in Figure 14.9(a). The dimer of the complex between G-CSF (in blue) and G-CSF receptor (in red) looks like an unstable table with a tabletop and two diagonal legs and is similar in structure to the IL-6/IL-6 receptor complex (compare Figure 14.7). G-CSF exhibits the expected long-chain four-helix bundle topology and the structure is stabilised by two intrachain disulfide bridges (in pink) arranged in the pattern **Cys 1–Cys 2, Cys 3–Cys 4** and is shown in Figure 14.9(b).

### 14.2.7 Interleukin-7/interleukin-9 family

There are two members of the interleukin-7/interleukin-9 family, namely interleukin-7 (IL-7) and interleukin-9 (IL-9), which are evolutionary related and share the same disulfide bridge pattern (see also Chapter 5).

**Interleukin-7** (IL-7) (see Data Sheet) produced in the bone marrow is a single-chain glycoprotein containing three intrachain disulfide bridges (Cys 2–Cys 141, Cys 34–Cys 129, Cys 47–Cys 92). IL-7 is a hematopoietic growth factor involved in the proliferation of lymphoid progenitor cells, both B cells and T cells.

**Interleukin-9** (IL-9) (see Data Sheet) or T cell growth factor P40 produced by T cells is a single-chain glycoprotein containing ten Cys residues with an unassigned disulfide bridge pattern. IL-9 supports the IL-2 and IL-4 independent growth of helper T cells.

### 14.2.8 Interleukin-10 family

The interleukin-10 family consists of several members, among them interleukin-10 (IL-10) itself, interleukin-19 (IL-19; 153 aa: Q9UHD0), interleukin-20 (IL-20; 152 aa: Q9NYY1), interleukin-22 (IL-22; 146 aa: Q9GZX6), interleukin-24 (IL-24; 155 aa: Q13007) and interleukin-26 (IL-26; 150 aa: Q9NPH9). The IL-10 family shares two conserved disulfide bridges (see also Chapter 5).

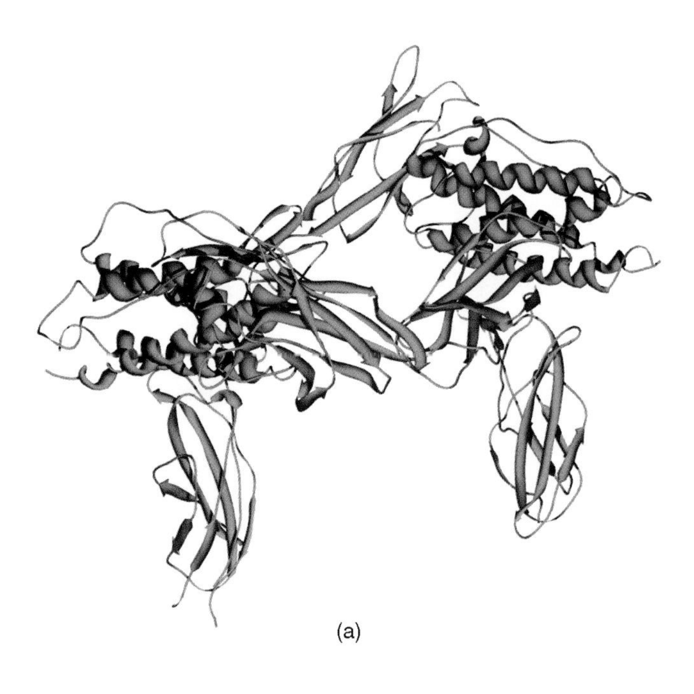

(a)

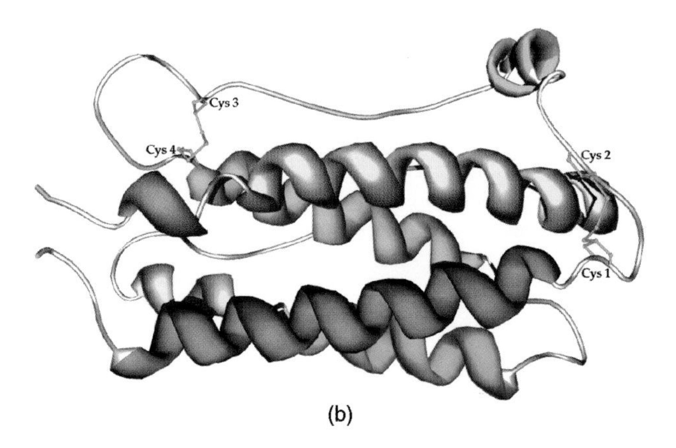


(b)

**Figure 14.9 3D structures of (a) the granulocyte colony-stimulating factor in complex with the N-terminal half of the extracellular domain of the G-CSF receptor (PDB: 2D9Q) and (b) the granulocyte colony-stimulating factor (PDB: 2D9Q)**

(a) The dimer of G-CSF (in blue) in complex with its receptor (in red) looks like an unstable table with a tabletop and two diagonal legs.

(b) G-CSF exhibits a long chain four-helix bundle topology and contains two intrachain disulfide bridges (in pink).

**Interleukin-10** (IL-10) (see Data Sheet) or cytokine synthesis inhibitory factor produced by a variety of cells (helper T cells, B cells, monocytes and macrophages) is a noncovalently associated homodimer with each chain containing two intrachain disulfide bridges (Cys 12–Cys 108, Cys 62–Cys 114). IL-10 is a cytokine inhibiting the synthesis of interferon γ, interleukin-2 and interleukin-3, tumor necrosis factor and granulocyte–macrophage colony-stimulating factor.

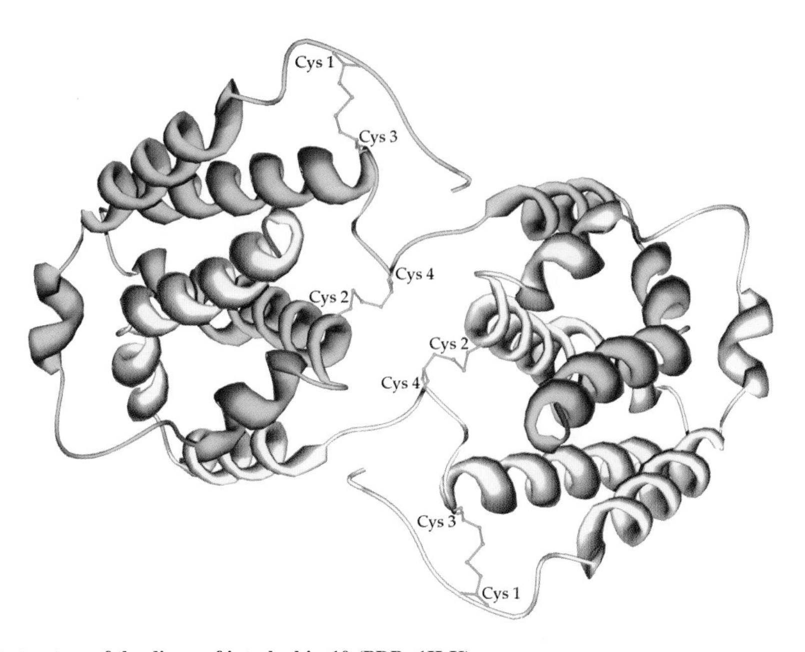


**Figure 14.10 3D structure of the dimer of interleukin-10 (PDB: 1ILK)**
IL-10 contains six α-helices, four forming a classical α-helix bundle and two intrachain disulfide bridges (in pink and green). The intertwined dimer (in yellow and blue) forms a V-shaped structure.

The 3D structure of entire interleukin-10 (160 aa: Ser 1–Asn 160) was determined by X-ray diffraction (PDB: 1ILK) and is shown in Figure 14.10. The monomer contains six α-helices, of which four form a classical α-helix bundle in the up–up down–down orientation stabilised by two intrachain disulfide bridges (in pink and green) arranged in the pattern **Cys 1–Cys 3, Cys 2–Cys 4**. The intertwined dimer (in yellow and blue) forms a V-shaped structure presumed to participate in receptor binding. IL-10 exhibits structural similarity with γ-interferon.

Defects in the interleukin-10 gene are the cause of susceptibility to Crohn's disease (for details see MIM: 266600), a form of remitting inflammatory bowel disease.

### 14.2.9 Interleukin-15 family

The only known member of the interleukin-15 family is interleukin-15 (IL-15) itself.

**Interleukin-15** (IL-15) (see Data Sheet), most abundant in placenta and skeletal muscle, is a single-chain glycoprotein containing two intrachain disulfide bridges (Cys 35–Cys 85, Cys 42–Cys 88) and a potential N-terminal propeptide (19 aa) is removed. IL-15 is involved in the stimulation and proliferation of T lymphocytes and shares many biological activities with interleukin-2.

### 14.2.10 Granulocyte–macrophage colony-stimulating factor family

Granulocyte–macrophage colony-stimulating factor (GM-CSF) shares no significant sequence similarity with other proteins, but because of its structural similarity with other interleukins exhibiting a four-helix bundle topology GM-CSF is included in this part of the chapter.

**Granulocyte–macrophage colony-stimulating factor** (GM-CSF) (see Data Sheet) or simply colony-stimulating factor produced by a variety of cells (endothelial cells, fibroblasts and T lymphocytes) is a single-chain monomeric glycoprotein containing two intrachain disulfide bridges (Cys 54–Cys 96, Cys 88–Cys 121).

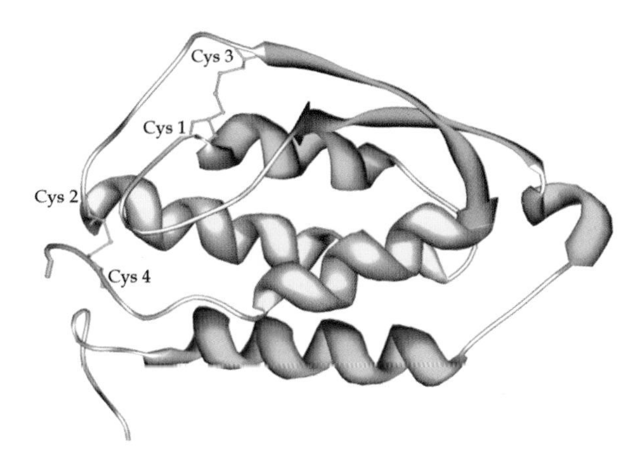


**Figure 14.11 3D structure of granulocyte–macrophage colony-stimulating factor (PDB: 2GMF)**
GM-CSF exhibits a short chain four-helix bundle topology with two connecting antiparallel β-strands and two intrachain disulfide bridges (in pink).

GM-CSF is characterised by a consensus sequence located around two conserved Cys residues involved in two disulfide bridges:

GM_CSF (PS00702): **C**–P–[LP]–T–x–E–[ST]–x–**C**

GM-CSF stimulates the growth and differentiation of hematopoietic stem and precursor cells such as granulocytes, macrophages, eosinophils and erythrocytes.

The 3D structure of entire granulocyte–macrophage colony-stimulating factor (127 aa: Ala 1–Glu 127) was determined by X-ray diffraction (PDB: 2GMF) and is shown in Figure 14.11. GM-CSF is characterised by a short-chain four-helix bundle structure with an up–up down–down topology and two overhand connecting loops, each with a short antiparallel β-sheet stabilised by two intrachain disulfide bridges (in pink) arranged in the pattern **Cys 1–Cys 3, Cys 2–Cys 4**.

## 14.3 TUMOUR NECROSIS FACTOR FAMILY

The tumour necrosis factor family has many members. Three important members are discussed here in some detail: tumour necrosis factor (TNF) itself, lymphotoxin α (LT-α) and lymphotoxin β (LT-β). The members of the family are structurally but also functionally related.

**Tumour necrosis factor** (TNF-α) (see Data Sheet) or cachectin produced by macrophages exists in two forms:

1. The membrane form (233 aa: Met 1–Leu 233) consists of a cytoplasmic region (35 aa), a potential transmembrane segment (21 aa) and an extracellular domain (177 aa).
2. The soluble form (157 aa: Val 1(77)–Leu 157(233)) is derived from the membrane form by proteolytic processing, removing an N-terminal peptide (76 aa) and thus consists of the extracellular domain only.

The soluble form of TNF-α is a single-chain cytokine containing a single intrachain disulfide bridge (Cys 69–Cys 101) existing as noncovalently linked homotrimer. TNF-α is involved in a wide variety of functions such as the cytolysis of certain tumour cell lines and the induction of cachexia and TNF-α is a potent pyrogen causing fever directly by stimulation of interleukin-1 secretion.

The 3D structure of the entire soluble form of tumour necrosis factor-α (157 aa: Val 1–Leu 157) was determined by X-ray diffraction (PDB: 1TNF) and is shown in Figure 14.12. The monomer forms an elongated, antiparallel β-pleated sheet sandwich with a 'jelly-roll' topology stabilised by a single intrachain disulfide bridge (in pink). The three monomers form a compact bell-shaped trimer.

**Figure 14.12 3D structure of the soluble form of trimeric tumour necrosis factor-α (PDB: 1TNF)**
TNF-α forms an elongated, antiparallel β-pleated sheet sandwich containing a single intrachain disulfide bridge (in pink) and the trimer exhibits a bell-shaped structure.

Cachexia is accompanied by several diseases such as cancer or infectious diseases and is characterised by weight loss, muscular atrophy, fatigue and weakness and anorexia. Inflammatory cytokines such as tumour necrosis factor α, interferon γ and interleukin-6 seem to play a role in cachexia.

**Lymphotoxin** (LT) (see Data Sheet) exists in two forms:

1. **Lymphotoxin α** (LT-α) or tumour necrosis factor-β (TNF-β) produced by lymphocytes is a secreted and soluble single-chain glycoprotein without disulfide bridges, usually present as noncovalently linked homotrimer. LT-α is cytotoxic for a wide variety of tumour cells (*in vitro* and *in vivo*).
2. **Lymphotoxin β** (LT-β) or tumour necrosis factor C produced in the spleen and thymus is a type II membrane protein with a potential transmembrane segment (30 aa) in the N-terminal region that provides a membrane anchor for the attachment to cell surfaces. LT-β forms heterotrimers of either two LT-β and one LT-α subunits or one LT-β and two LT-α subunits (less frequent). LT-β seems to play a specific role in the regulation of the immune response.

The 3D structure of an N-terminally truncated form of tumour necrosis factor-β (144 aa: Lys 28–Leu 171) in complex with an extracellular fragment of the tumour necrosis factor-β receptor (139 aa: Cys 23–Cys 161, P19438) was determined by X-ray diffraction (PDB: 1TNR) and is shown in Figure 14.13. The extracellular fragment of TNF-β receptor consisting of four similar folding domains (in red) arranged in a very elongated end-to-end assembly binds in the groove between two adjacent TNF-β subunits (in blue). TNF-β is characterised by a sandwich of two predominantly antiparallel β-sheets of a 'jelly-roll' Greek-key topology.

## 14.4 THE INTERFERON FAMILY

Interferons are cytokines that are produced by the immune system in response to challenges by viruses, bacteria, parasites and tumour cells. Interferons are glycoproteins that belong to two families, the interferon α/β family (type I) and the interferon γ family (type II) (see also Chapter 5).

### 14.4.1 The interferon α/β family

The interferon α/β family comprises interferon α (IFN-α) with the subtypes IFN-α1/13, IFN-α2, IFN-α4 to -α8, IFN-α10, IFN-α14, IFN-α16, IFN-α17, IFN-α21, interferon β (IFN-β), interferon δ (IFN-δ), interferon κ (IFN-κ) and interferon ω (IFN-ω), of which IFN-α1/13 and IFN-β will be discussed in detail.

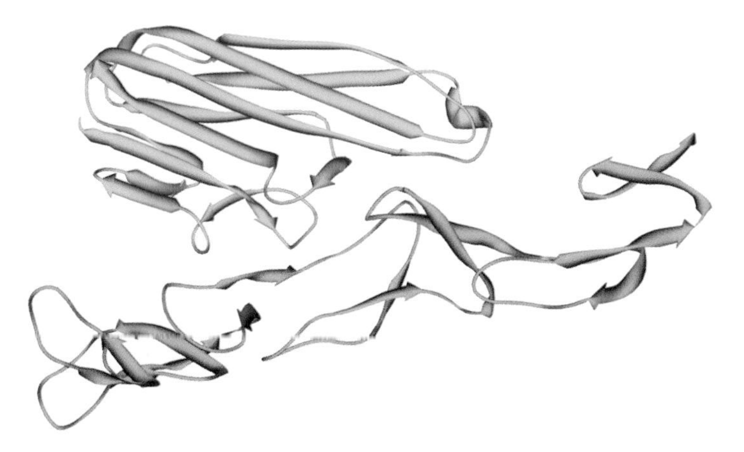

**Figure 14.13 3D structure of an N-terminally truncated form of tumour necrosis factor-β in complex with a fragment of the extracellular domain of the TNF-β receptor (PDB: 1TNR)**
A fragment of the extracellular domain of the TNF-β receptor (in red) binds in the groove between two adjacent TNF-β subunits (in blue).

**Interferon α-1/13** (IFN-α1/13) (see Data Sheet) or interferon α-D produced by macrophages is a single-chain nonglycosylated cytokine containing two intrachain disulfide bridges (Cys 1–Cys 99, Cys 29–Cys 139). IFN-α1/13 exhibits antiviral and antioncogenic activities and acts as a pyrogenic factor and stimulates the production of two enzymes, a protein kinase and a 2′-5′-oligoadenylate synthetase (400 aa: P00973).

The 3D structure of entire interferon α-2A (165 aa: Cys 1–Glu 165) was determined by NMR spectroscopy (PDB: 1ITF) and is shown in Figure 14.14. The structure is characterised by a cluster of five α-helices, of which four form a left-handed helix bundle in an up–up down–down topology and two overhand connections and the structure is stabilised by two intrachain disulfide bridges (in pink) arranged in the pattern **Cys 1–Cys 3, Cys 2–Cys 4**.

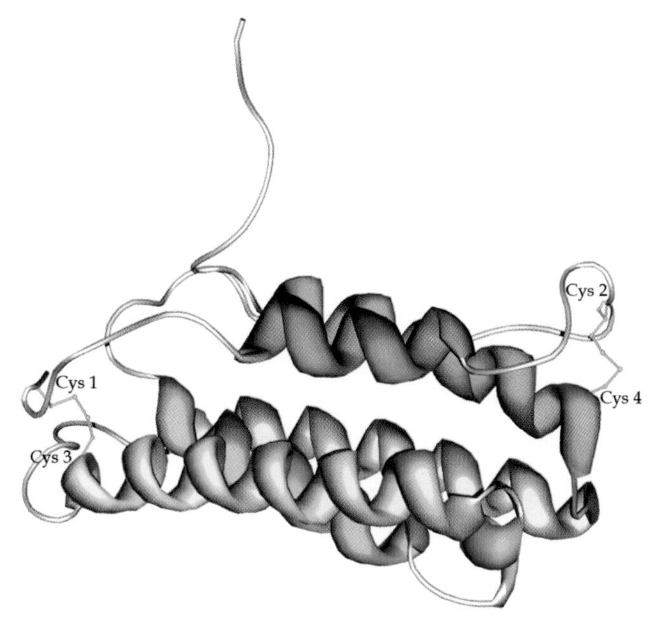


**Figure 14.14 3D structure of interferon α (PDB: 1ITF)**
Interferon-α forms a left-handed helix bundle and contains two intrachain disulfide bridges (in pink).

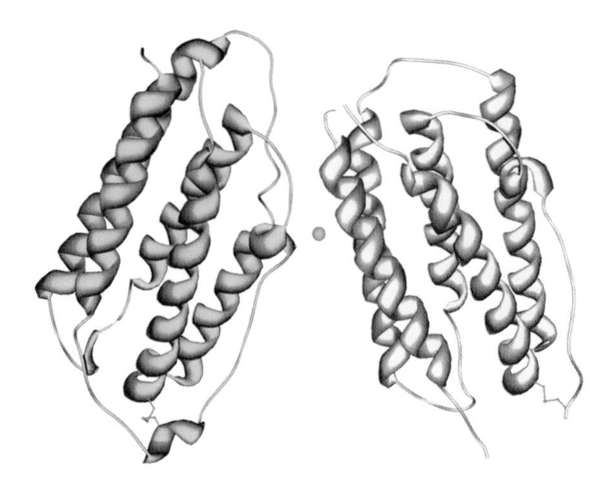

**Figure 14.15 3D structure of interferon β (PDB: 1AU1)**
Interferon β exhibits the classical helix bundle topology and contains a single disulfide bridge (in pink). The dimer binds a zinc ion (green ball) at the interface of two molecules.

**Interferon β** (IFN-β) (see Data Sheet) or fibroblast interferon produced by fibroblasts is a single-chain glycoprotein containing a single intrachain disulfide bridge (Cys 31–Cys 141). IFN-β exhibits antiviral, antibacterial and anticancer activities.

The 3D structure of entire interferon β (166 aa: Met 1–Asn 166) was determined by X-ray diffraction (PDB: 1AU1) and is shown in Figure 14.15. IFN-β adopts a fold similar to IFN-α2 characterised by the classical helix bundle topology and stabilised by a single disulfide bridge (in pink). The IFN-β dimer contains a zinc-binding site containing a zinc ion (green ball) at the interface of the two molecules in the asymmetric unit.

### 14.4.2 The interferon γ family

The interferon γ (type II interferon) family contains only interferon γ (IFN-γ).

**Interferon γ** (IFN-γ) (see Data Sheet) or immune interferon produced by T lymphocytes and natural killer cells is a single-chain glycoprotein containing no Cys residues and thus is devoid of disulfide bridges. Mature IFN-γ is generated from the proform by releasing a C-terminal pentapeptide and carries an N-terminal pyro-Glu residue. Proteolytic processing of mature IFN-γ (138 aa) results in two C-terminally truncated forms (127 aa: Gln 1–Gly 127; 134 aa: Gln 1–Met 134). IFN-γ is a pleiotropic cytokine with multiple effects and exhibits antiviral, antiproliferative and immunomodulatory activities. IFN-γ is an important modulator of the immune system by activation of macrophages, by induction of MHC class II antigens and is an important factor in the production of antibodies.

The 3D structure of a C-terminally truncated interferon γ (134 aa: Met 0–Gln 133) in complex with the extracellular domain of the interferon γ receptor α-chain (245 aa: Glu 1–Ser 245, P15260) was determined by X-ray diffraction (PDB: 1FG9) and is shown in Figure 14.16. The complex consists of two intertwined IFN-γ molecules (in yellow and blue) and two attached receptor molecules (in red and pink). The IFN-γ monomer consists of six α-helices connected by short loops. In the homodimer the first four helices of one monomer and the last two helices of the second monomer form a structural domain. The extracellular domain of the IFN-γ receptor α-chain consists of two FN3 domains each characterised by a β-sandwich structure consisting of a three-stranded and four-stranded β-sheet structure.

Genetic variation in the interferon γ gene in Caucasians is associated with the risk of aplastic anemia (for details see MIM: 609135) caused by a damage in stem cells in the bone marrow, resulting in the reduction of circulating blood cells.

## 14.5 CHEMOKINES

Chemokines are a family of small cytokines that are secreted by various types of cells and exhibit mitogenic, chemotactic or inflammatory activities. Chemokines are small cationic proteins of approximately 70–100 residues length characterised by

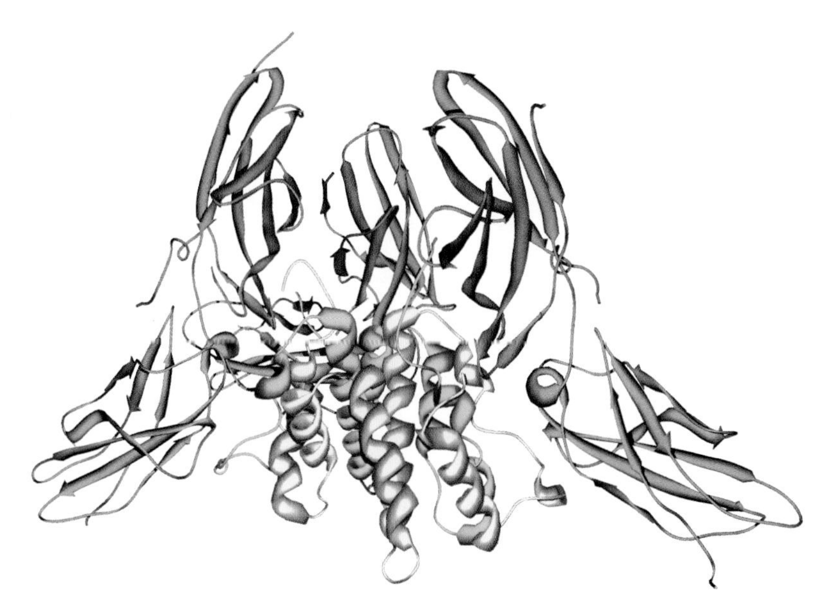

**Figure 14.16 3D structure of a C-terminally truncated interferon γ in complex with the extracellular domain of the interferon γ receptor α-chain (PDB: 1FG9)**
The complex consists of two intertwined IFN-γ molecules (in yellow and blue) and two attached receptor molecules (in red and pink).

two conserved intrachain disulfide bridges. Chemokines share a similar structural fold characterised by an extended N-terminal loop, three strands of antiparallel β-sheet connected to a C-terminal α-helix (see also Chapter 5). There are two main families:

1. The intercrine α/chemokine CxC family, where the Cys 1 and Cys 2 are separated by a single residue (**Cys**–Xaa–**Cys**).
2. The intercrine β/chemokine CC family, where the Cys 1 and Cys 2 are adjacent (**Cys–Cys**).

### 14.5.1 The intercrine α/chemokine CxC family

The intercrine α/chemokine CxC family comprises many members, among them interleukin-8 (IL-8), platelet factor 4 (PF-4) and platelet basic protein (PBP). Many of the CxC chemokines share a Glu–Leu–Arg sequence immediately preceding the CxC sequence, which is important for the induction of neutrophil migration.

**Interleukin-8** (IL-8) (see Data Sheet) or neutrophil-activating factor produced by many types of cells such as monocytes, macrophages and endothelium is a single-chain cytokine containing two intrachain disulfide bridges (Cys 12–Cys 39, Cys 14–Cys 55) usually present as noncovalently linked homodimer. Various N-terminally processed forms exist, some of which exhibit an increased biological activity compared to standard IL-8. IL-8 is a chemotactic factor attracting neutrophils, basophils and T cells and is involved in neutrophil activation.

The 3D structure of an N-terminally processed dimeric form of interleukin-8 (in yellow and blue) (72 aa: Ser 6–Ser 77) was determined by X-ray diffraction (PDB: 3IL8) and is shown in Figure 14.17. IL-8 exhibits the expected structural fold with an N-terminal loop region, an antiparallel β-sheet linked to a C-terminal α-helix. The structure is stabilised by two intrachain disulfide bridges (in pink and green) arranged in the pattern **Cys 1–Cys 3, Cys 2–Cys 4**.

**Platelet factor 4** (PF-4) (see Data Sheet) or oncostatin A produced by megakaryocytes is a single-chain cytokine containing two intrachain disulfide bridges (Cys 10–Cys 36, Cys 12–Cys 52) and is present as noncovalently linked homotetramer bound to a proteoglycan. PF-4 neutralises the anticoagulant effect of heparin because of its strong binding to heparin.

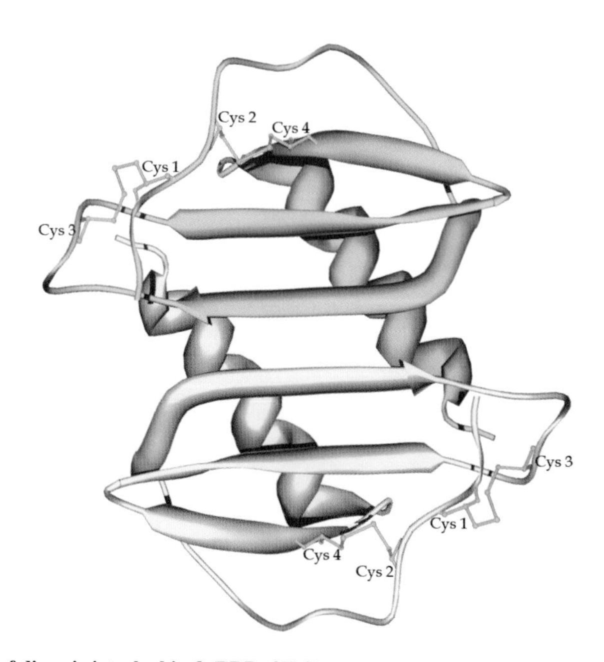


**Figure 14.17 3D structure of dimeric interleukin-8 (PDB: 3IL8)**
IL-8 contains a N-terminal loop region, an antiparallel β-sheet linked to a C-terminal α-helix and two intrachain disulphide bridges (in pink and green).

The 3D structure of entire platelet factor 4 (70 aa: Glu 1–Ser 70) was determined by X-ray diffraction (PDB: 1RHP) and is shown in Figure 14.18. In the tetramer the N-terminal regions form antiparallel β-sheet-like structures, which in turn form noncovalent interactions between the dimers. The structure is characterised by two extended six-stranded β-sheets.

**Platelet basic protein** (PBP) (see Data Sheet) or small inducible cytokine B7 produced by megakaryocytes is a single-chain cytokine containing two intrachain disulfide bridges (Cys 29–Cys 55, Cys 31–Cys 71). There are several N- and/or C-terminally processed forms:

1. **Connective tissue-activating peptides III** (CTAP-III: Asn 10–Asp 94, Asn 10–Asp 90) exhibit synovial cell stimulating activity.
2. **β-Thromboglobulin** (Gly 14–Asp 94) is a noncovalently associated homotetramer.
3. **Neutrophil-activating peptides 2** (NAP-2: Ser 21–Asp 94, Asp 22–Asp 94, Ala 25–Asp 94, Ala 25–Asp 90, Ala 25–Leu 87) activate neutrophils, which seems to be important for local inflammatory responses.

The 3D structure of neutrophil-activating peptide 2 (70 aa: Ala 25–Asp 94) was determined by X-ray diffraction (PDB: 1NAP) and is shown in Figure 14.19. The structure is similar to IL-8 and PF-4 and is characterised by an extended N-terminal loop, followed by three antiparallel β-strands arranged in a Greek-key fold linked to a single C-terminal α-helix. The structure is stabilised by two intrachain disulfide bridges (in pink) arranged in the pattern **Cys 1–Cys 3, Cys 2–Cys 4**.

### 14.5.2 The intercrine β/chemokine CC family

The intercrine-β/chemokine CC family comprises many members, among them the small inducible cytokines A2 (CCL2) and A5 (CCL5), which are described as examples. The CC chemokines induce the migration of monocytes and other cell types such as natural killer cells and dendritic cells.

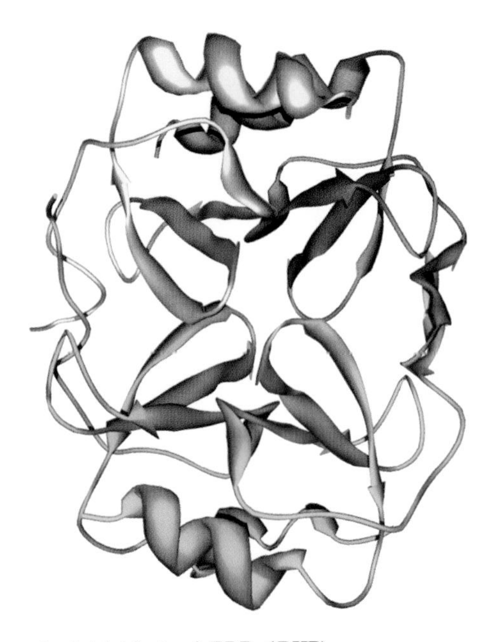

**Figure 14.18 3D structure of tetrameric platelet factor 4 (PDB: 1RHP)**
The structure is characterised by two extended six-stranded β-sheets. The N-terminal regions exhibit antiparallel β-sheet-like structures forming non-covalent interactions between the dimers.

**Small inducible cytokine A2** (CCL2) (see Data Sheet) or monocyte chemotactic protein 1 (MCP-1) produced by macrophages is a single-chain glycoprotein containing two intrachain disulfide bridges (Cys 11–Cys 36, Cys 12–Cys 52) and an N-terminal pyro-Glu residue and is present either as monomer or noncovalently associated homodimer. Several N-terminally processed forms exist that are able to regulate receptor and target cell selectivity. CCL2 is a chemotactic factor that attracts monocytes and basophils but not neutrophils and eosinophils. CCL2 induces monocytes to leave the bloodstream and enter the surrounding tissue. Deletion of the N-terminal pyro-Glu residue converts CCL2 from an activator of basophil mediator release to an eosinophil chemoattractant. Several basic amino acids are involved in the binding of glycosamino-glycan side chains of proteoglycans on endothelial cells.

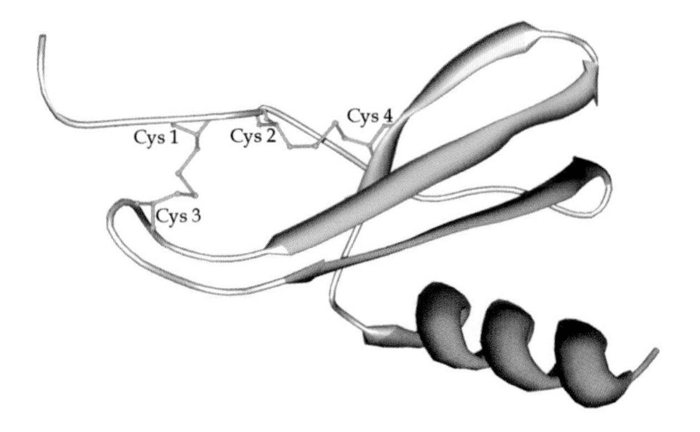


**Figure 14.19 3D structure of neutrophil-activating peptide 2 (PDB: 1NAP)**
NAP contains an extended N-terminal loop, three antiparallel β-strands linked to a single C-terminal α-helix and two intrachain disulfide bridges (in pink).

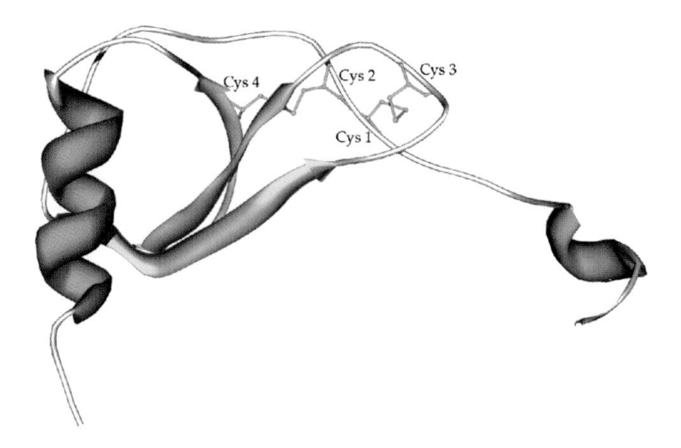


**Figure 14.20 3D structure of small inducible cytokine A2 (PDB: 1DOK)**
Small inducible cytokine A2 contains two short N-terminal β-strands linked to the central three-stranded antiparallel β-sheet packed to the C-terminal α-helix and two intrachain disulphide bridges (in pink).

The 3D structure of entire small inducible cytokine A2 (76 aa: Gln 1–Thr 76) was determined by X-ray diffraction (PDB: 1DOK) and is shown in Figure 14.20. The monomer is characterised by two short N-terminal β-strands linked to the central three-stranded antiparallel β-sheet, which is packed to the C-terminal α-helix. The structure is stabilised by two intrachain disulfide bridges (in pink) arranged in the pattern **Cys 1–Cys 3, Cys 2–Cys 4**.

**Small inducible cytokine A5** (CCL5) (see Data Sheet) or RANTES, produced primarily by T cells and blood platelets, is a single-chain glycoprotein containing two intrachain disulfide bridges (Cys 10–Cys 34, Cys 11–Cys 50). CCL5 is a chemoattractant for blood monocytes, memory helper T cells and eosinophils and causes the release of histamine.

**RANTES (3–68)** is generated by N-terminal proteolytic processing of CCL5 by dipeptidyl peptidase IV. RANTES is chemotactic for T cells, eosinophils and basophils.

The 3D structure of entire RANTES (68 aa: Ser 1–Ser 68) was determined by NMR spectroscopy (PDB: 1RTN) and is shown in Figure 14.21. The monomer is characterised by two short N-terminal β-strands connected to a three-stranded

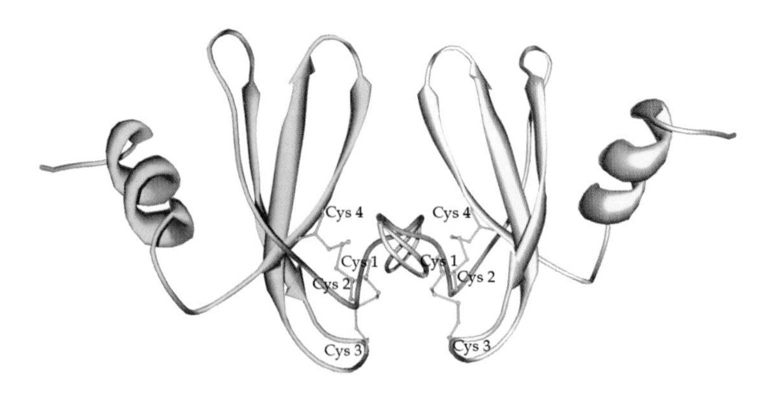


**Figure 14.21 3D structure of dimeric RANTES (PDB: 1RTN)**
RANTES contains two short N-terminal β-strands connected to a three-stranded antiparallel β-sheet packed against a C-terminal α-helix and two intrachain disulfide bridges (in pink and green).

antiparallel β-sheet packed against a C-terminal α-helix. The structure is stabilised by two intrachain disulfide bridges (in pink and green) arranged in the pattern **Cys 1–Cys 3, Cys 2–Cys 4**. Dimerisation occurs between the N-terminal regions of the two monomers (in yellow and blue).

## 14.6 THE PLATELET-DERIVED GROWTH FACTOR/VASCULAR ENDOTHELIAL GROWTH FACTOR FAMILY

The platelet-derived growth factor/vascular endothelial growth factor family contains two main groups of proteins, which are structurally related, namely:

1. The platelet-derived growth factor (PDGF) is composed of an A-chain and a B-chain, which are different but closely related.
2. The vascular endothelial growth factors (VEGF) have four members: VEGF-A, VEGF-B, VEGF-C and VEGF-D.

In addition, there is a placenta growth factor (203 aa: P49763) that also belongs to this family.

**Platelet-derived growth factor** (PDGF) (see Data Sheet) expressed in platelets is composed of an A-chain and a B-chain, which form disulfide-linked homodimers AA and BB and disulfide-linked heterodimers AB. The A-chain contains three intrachain disulfide bridges (Cys 10–Cys 54, Cys 43–Cys 91, Cys 47–Cys 93) and the AA dimer is linked by two interchain disulfide bridges arranged in a cross-wise manner (Cys 37–Cys 46). The B-chain also contains three intrachain disulfide bridges (Cys 16–Cys 60, Cys 49–Cys 97, Cys 53–Cys 99) and the BB dimer is also linked by two interchain disulfide bridges arranged in a cross-wise manner (Cys 43–Cys 52). In both chains the intrachain disulfide bridges are arranged in a tight knot-like cystine topology. The AB heterodimer is also arranged in an antiparallel manner implying a cross-wise linkage of the disulfide bridges (Cys 37 in A-chain–Cys 52 in B-chain, Cys 46 in A-chain–Cys 43 in B-chain). Chain A is generated from the precursor form by removal of a 66 residue N-terminal propeptide and chain B by removal of two propeptides, a 61 residue N-terminal and a 51 residue C-terminal propeptide. PDGF is a potent mitogen and stimulates the proliferation of connective tissue cells and glial cells. PDGF stimulates the growth of the placenta and the embryo during development and wound healing in the adult.

The 3D structure of the platelet-derived growth factor B-chain homodimer (109 aa: Ser 1–Thr 109) was determined by X-ray diffraction (PDB: 1PDG) and is shown in Figure 14.22. PDGF is folded into two highly twisted antiparallel pairs of β-strands with an unusual knot of three intrachain disulfide bonds (in pink) arranged in the pattern **Cys 1–Cys 4, Cys 2–Cys 5, Cys 3–Cys 6**. The homodimer (in yellow and blue) is linked by two interchain disulfide bridges (in green) linked in a cross-wise manner.

**Vascular endothelial growth factors** (VEGFs) (see Data Sheet) are a group of four growth factors: VEGF-A (206 aa), VEGF-B (186 aa), VEGF-C (116 aa) and VEGF-D (117 aa) involved in vasculogenesis, angiogenesis and endothelial cell growth. VEGFs induce endothelial cell proliferation, promote cell migration, inhibit apoptosis and induce the permeabilisation of blood vessels.

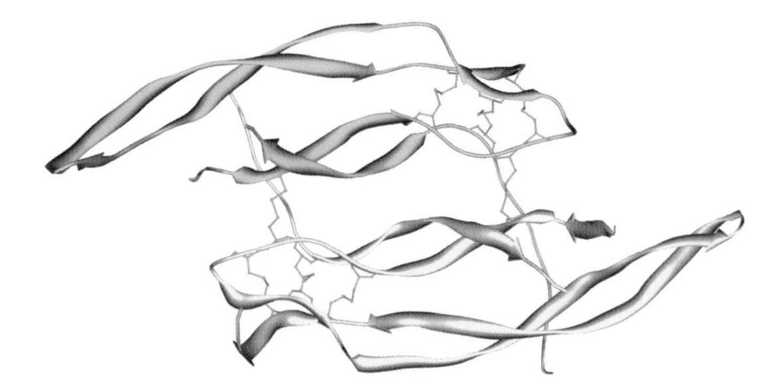

**Figure 14.22 3D structure of platelet-derived growth factor B-chain homodimer (PDB: 1PDG)**
PDGF is folded into two highly twisted antiparallel pairs of β-strands with an unusual knot of three intrachain disulfide bonds (in pink). The homodimer is linked by two interchain disulfide bridges (in green).

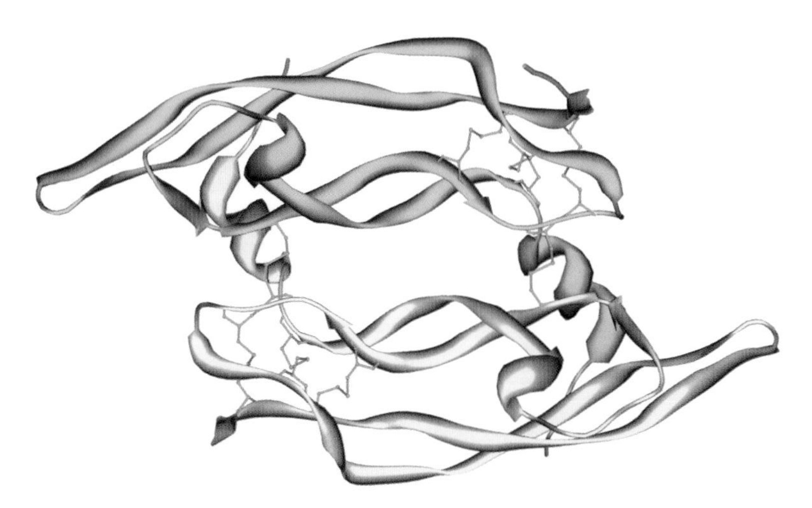

**Figure 14.23 The 3D structure of homodimeric vascular endothelial growth factor-A121 isoform (PDB: 2VPF)**
VEGF is folded into antiparallel pairs of β-strands containing a cystine-knot structure of three intrachain disulfide bridges (in pink) and two interchain disulfide bridges (in green).

**VEGF-A/VEGF-B** are single-chain growth factors consisting of an N-terminal region structurally related to PDGF containing three intrachain disulfide bridges (Cys 26–Cys 68, Cys 57–Cys 102/101, Cys 61–Cys 104/103) and two interchain disulfide bridges (Cys 51–Cys 60) forming the homodimer. There are at least nine splice variants of VEGF-A, of which VEGF-A189 (189 aa), VEGF-A165 (165 aa) and VEGF-A121 (121 aa) are the most widely expressed.

**VEGF-C/VEGF-D** are single-chain growth factors containing three intrachain disulfide bridges (Cys 20/23–Cys 64/65, Cys 51/54–Cys 98/101, Cys 55/58–Cys 100/103) generated from the proforms by releasing N-terminal (80/67 aa) and C-terminal (192/149 aa) propeptides. Fully processed VEGF-C seems to be mainly a disulfide-linked homodimer (Cys 45–Cys 54), whereas VEGF-D is predominantly a noncovalently associated homodimer, but a disulfide-linked homodimer (Cys 48–Cys 57) is possible.

The 3D structure of a truncated form of the VEGF-A121 isoform (102 aa: Gly 8–Asp 109) was determined by X-ray diffraction (PDB: 2VPF) and is shown in Figure 14.23. The homodimer (in yellow and blue) exhibits a very similar structure to PDGF characterised by antiparallel pairs of β-strands containing a cystine-knot structure of three intrachain disulfide bridges (in pink) arranged in the pattern **Cys 1–Cys 4, Cys 2–Cys 5, Cys 3–Cys 6** and two interchain disulfide bridges (in green) arranged in a cross-wise manner.

## 14.7 TRANSFORMING GROWTH FACTOR-β FAMILY

The transforming growth factor-β (TGF-β) family contains four TGF-β proteins: TGF-β1, TGF-β2 and TGF-β3 are very closely related, whereas TGF-β4 (280 aa: O00292) is more distantly related. The TGF-β family contains other members: the group of the bone morphogenetic proteins, the group of the growth/differentiation factors and the inhibins.

Although the transforming growth factor-α (TGF-α) is structurally not related to the transforming growth factor-β (TGF-β) family, TGF-α acts synergistically with TGF-β and is therefore discussed in this part of the chapter.

**Transforming growth factor-β** (TGF-β) (see Data Sheet) contains three main forms: **TGF-β1, TGF-β2 and TGF-β3**, which are all very closely related. They are all single-chain glycoproteins containing four intrachain disulfide bridges all located at the same positions (Cys 7–Cys 16, Cys 15–Cys 78, Cys 44–Cys 109, Cys 48–Cys 111) and are present as disulfide-linked homodimers (Cys 77–Cys 77). The TGF-βs are generated from approximately 400 residues long precursor proteins by proteolytic processing at Arg–Lys/His–Arg/Lys–Arg cleavage sites. The TGF-βs are involved in cell differentiation and proliferation, in tissue morphogenesis and embryonic development. Especially, TGF-β1 regulates many other peptide growth factors and plays an important role in bone remodelling, TGF-β2 suppresses the growth of interleukin-2 dependent T cells and TGF-β3 is involved in embryogenesis and cell proliferation.

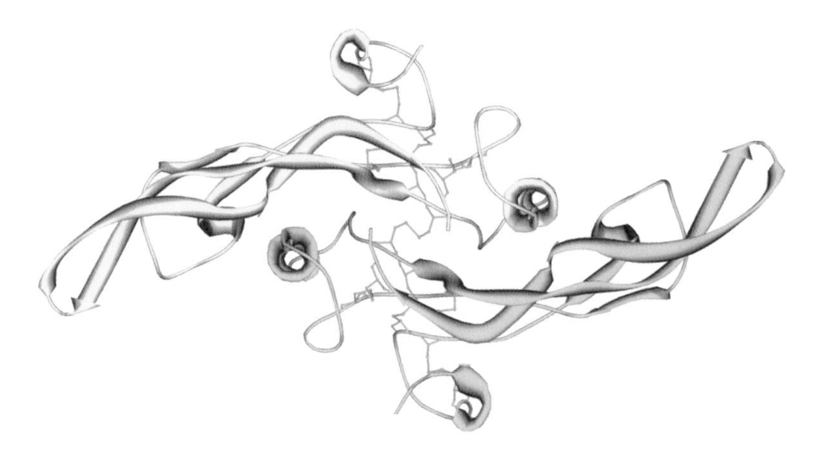

**Figure 14.24 3D structure of the transforming growth factor β1 (PDB: 1KLD)**
TGF-β1 consists of two antiparallel β-strands and a separate long α-helix and a disulfide-rich core with four intrachain disulfide bridges (in pink) and the homodimer is connected by a single interchain disulfide bridge (in green).

The 3D structure of the transforming growth factor-β1 (112 aa: Ala 1–Ser 112) was determined by NMR spectroscopy (PDB: 1KLD) and is shown in Figure 14.24. The monomer consists of two antiparallel β-strands forming a flat curved surface and a separate long α-helix and a disulfide-rich core with four intrachain disulfide bridges (in pink) arranged in the pattern **Cys 1–Cys 3, Cys 2–Cys 6, Cys 4–Cys 7, Cys 5–Cys 8**. The homodimer (in yellow and blue) is connected by a single interchain disulfide bridge (in green) and the helix of one monomer interacts with the concave β-sheet surface of the other monomer. The structure is similar to other disulfide-knot type proteins.

Defects in the transforming growth factor-β1 gene are the cause of the Camurati–Engelmann disease, also known as progressive diaphyseal dysplasia (for details see MIM: 131300). The disease is autosomal dominant, characterised by hyperosteosis and sclerosis in the diaphyses (central or mid section) of long bones.

**Transforming growth factor-α** (TGF-α) (see Data Sheet) expressed in macrophages and the brain is a type I membrane protein in its precursor form containing a extracellular domain, a potential transmembrane segment and a C-terminal cytoplasmic domain. Mature TGF-α consisting of the extracellular domain only contains three intrachain disulfide bridges (Cys 8–Cys 21, Cys 16–Cys 32, Cys 34–Cys 43) and is generated from the precursor form by removing N-terminal (16 aa) and C-terminal (71 aa) propeptides. TGF-α is a mitogenic polypeptide that is able to bind to the EGF receptor and acts synergistically with TGF-β.

The 3D structure of the transforming growth factor-α (50 aa: Val 1–Ala 50) in complex with the extracellular domain of the epidermal growth factor receptor (501 aa: Leu 1–Ser 501, P00533) was determined by X-ray diffraction (PDB: 1MOX) and is shown in Figure 14.25(a). The dimer contains two TGF-α molecules (in pink) and two extracellular domains of the EGF receptor (in yellow and blue) comprising three structural domains, namely L1, CR1 and L2, where TGF-α interacts with both L1 and L2. TGF-α is characterised by a central antiparallel β-sheet loosely attached to a third N-terminal strand and a short C-terminal antiparallel β-sheet and contains three intrachain disulfide bridges (in pink) arranged in the pattern **Cys 1–Cys 3, Cys 2–Cys 4, Cys 5–Cys 6** and is shown in Figure 14.25(b).

## 14.8 MISCELLANEOUS

### 14.8.1 The nerve growth factor-β family

The nerve growth factor-β (NGF-β) family contains as main component the NGF-β protein itself and the structurally related brain-derived neurotrophic factor (119 aa: P23560) and several neurotrophins. The members of the NGF-β family contain three conserved intrachain disulfide bridges arranged in the following pattern:

**Cys** 1–**Cys** 4, **Cys** 2–**Cys** 5, **Cys** 3–**Cys** 6

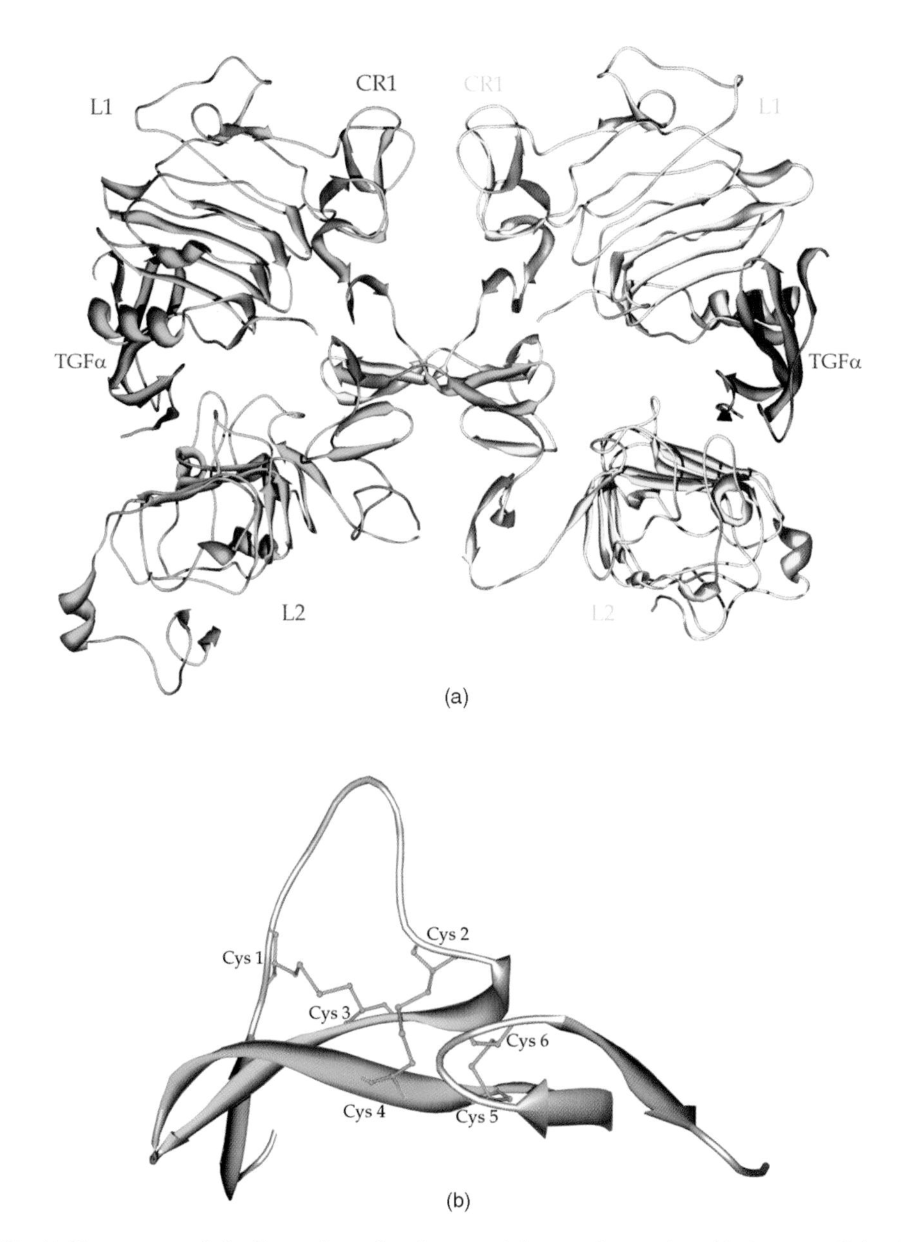


**Figure 14.25 (a) 3D structures of the dimer of transforming growth factor α in complex with the extracellular domain of the epidermal growth factor receptor (PDB: 1MOX) and (b) Monomer of transforming growth factor (PDB: 1MOX)**
(a) The dimer contains two TGF-α molecules (in pink) and two extracellular domains of the EGF receptor (in yellow and blue) comprising three structural domains: L1, CR1 and L2.
(b) TGF-α contains a central antiparallel β-sheet loosely attached to a third N-terminal strand and a short C-terminal antiparallel β-sheet and three intrachain disulfide bridges (in pink).

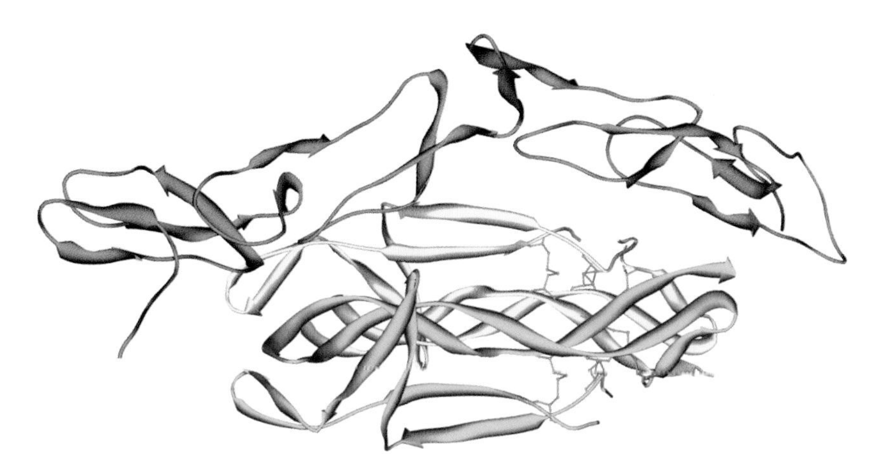

**Figure 14.26 3D structure of nerve growth factor-β in complex with the extracellular domain of the rat neurotrophin receptor p75 (PDB: 1SG1)**
The homodimer of NGF-β (in yellow and blue) comprising seven β-strands arranged in three antiparallel pairs of twisted β-strands stabilised by a knot of three intrachain disulfide bridges (in pink) is bound to a single molecule of the extracellular domain of the rat neurotrophin receptor p75 (in red).

In addition, the NGF-β family is characterised by a consensus sequence located in the central region of the protein containing two Cys residues involved in the intrachain disulfide bridge pattern:

NGF_1 (PS00248): [GSRE]–**C**–[KRL]–G–[LIVT]–[DE]–x(3)–[YW]–x–S–x–**C**

**Nerve growth factor-β** (NGF-β) (see Data Sheet), expressed primarily by fibroblasts and Schwann cells, is a single-chain glycoprotein containing three intrachain disulfide bridges (Cys 15–Cys 80, Cys 58–Cys 108, Cys 68–Cys 110) and is present as noncovalently linked homodimer. NGF-β is generated from the precursor by removing an N-terminal 103 residue long propeptide. NGF-β is important for the development and maintenance of the sympathetic and sensory nervous system and stimulates the division and differentiation of sympathetic and embryonic sensory neurons.

The 3D structure of the entire nerve growth factor-β (120 aa: Ser 1–Ala 120) in complex with the C-terminally truncated extracellular domain of the rat neurotrophin receptor p75 (161 aa: Lys 1–Glu 161, P07174) was determined by X-ray diffraction (PDB: 1SG1) and is shown in Figure 14.26. The topology of monomeric NGF-β comprises seven β-strands arranged in three antiparallel pairs of twisted β-strands stabilised by a knot of three intrachain disulfide bridges (in pink) arranged in the pattern **Cys 1–Cys 4, Cys 2–Cys 5, Cys 3–Cys 6**. The complex is composed of an NGF-β homodimer (in blue and yellow) asymmetrically bound to a single molecule of the extracellular domain of the rat neurotrophin receptor p75 (in red). The binding of the p75 domain to one side of the NGF-β induces a bending of the entire NGF-β molecule: The central β-sheets remain intact, but the head and tail loops are bent away from the p75 domain.

### 14.8.2 Epidermal growth factor

The epidermal growth factor-like (EGF-like) domain is found in many plasma proteins, primarily in blood coagulation and fibrinolysis and in the complement system, but the EGF-like domain was originally found in the epidermal growth factor (for details see Chapter 4).

**Epidermal growth factor** (EGF) (see Data Sheet) produced primarily by the submaxillary gland is a single-chain growth factor containing three intrachain disulfide bridges (Cys 6–Cys 20, Cys 14–Cys 31, Cys 33–Cys 42). EGF is derived by limited proteolysis from the single-chain type I membrane glycoprotein pro-EGF consisting of nine LDLRB repeats and nine EGF-like domains and a potential 21 residues long transmembrane segment (Val 1011–Gly 1031). EGF plays an important role in the regulatory process of cell growth, proliferation and differentiation.

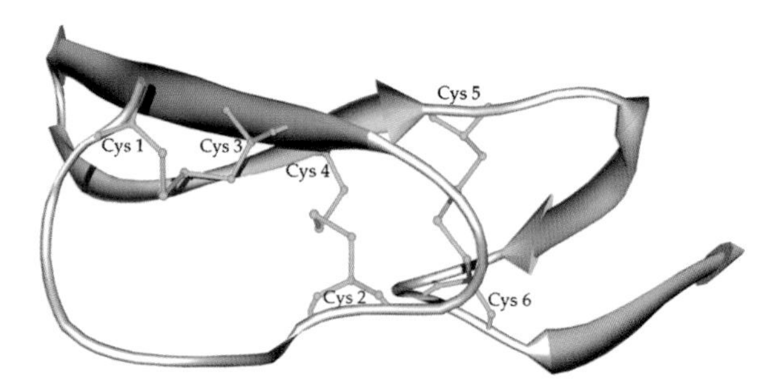


**Figure 14.27 3D structure of the epidermal growth factor (PDB: 1JL9)**
EGF contains an antiparallel β-sheet in the N-terminal region and a short antiparallel β-sheet in the C-terminal portion and three intrachain disulfide bridges (in pink).

The 3D structure of the epidermal growth factor (51 aa: Asn 1–Glu 51) was determined by X-ray diffraction (PDB: 1JL9) and is shown in Figure 14.27. The structure is characterised by an antiparallel β-sheet in the N-terminal region and a short antiparallel β-sheet in the C-terminal portion of the molecule and three intrachain disulfide bridges (in pink) arranged in the pattern **Cys 1–Cys 3, Cys 2–Cys 4, Cys 5–Cys 6** that stabilise the structure.

### 14.8.3 Insulin-like growth factor binding proteins

(Pro)-insulin and the structurally and functionally related insulin-like growth factors have already been discussed in the chapter on hormones (for details see Chapter 13). There are at least seven insulin-like growth factor binding proteins (IGFBPs): IGFBP1–IGFBP7. The structurally related IGFBP1–IGFBP6 proteins contain relatively conserved N-terminal IGFBP domains, C-terminal thyroglobulin type 1 domains and a structurally distinct central domain, whereas IGFBP7 is less closely related. The IGFBPs are characterised by a consensus sequence located in the conserved Cys-rich region in the N-terminal section of the proteins:

IGFBD_M_1 (PS00222): [GP]–C–[GSET]–[CE]–[CA]–x(2)–C–[ALP]–x(6)–C

IGFBPs are involved in the regulation of insulin growth factors.

**Insulin-like growth factor binding proteins** (IGFBP1–IGFBP6) (see Data Sheet) expressed in various cell types are single-chain proteins comprising a relatively conserved N-terminal IGFBP domain characterised by six intrachain disulfide bridges (only five in IGFBP6), a structurally distinct central domain and a quite conserved C-terminal thyroglobulin type 1 domain characterised by three intrachain disulfide bridges. IGFBPs are characterised by additional specific features: IGFBP1 and IGFBP2 are nonglycosylated proteins carrying an **RGD** sequence as cell attachment sites and in addition IGFBP1 has three phosphorylation sites (Ser 101, 119, 169). IGFBP3 is a glycoprotein with two Ser/Thr-rich regions in the central domain and two possible phosphorylation sites (Ser 111 and 113) and IGFBP4–IGFBP6 are glycoproteins. The structurally less related IGFBP7 contains an N-terminal IGFBP domain, a central Kazal-like domain and a C-terminal Ig-like C2-type domain. IGFBPs are involved in the regulation of the effects of insulin growth factors either by inhibiting or stimulating their effects on cells.

## REFERENCES

Bradshaw *et al.*, 1993, Nerve growth factor revisited, *Trends Biochem. Sci.*, **18**, 48–52.
Hwa *et al.*, 1999, The insulin-like growth factor-binding protein (IGFBP) superfamily, *Endocrine Rev.*, **20**, 761–787.
Walter *et al.*, 1992, Three-dimensional structure of recombinant human granulocyte–macrophage colony-stimulating factor, *J. Mol. Biol.*, **224**, 1075–1085.

DATA SHEETS

## *Epidermal growth factor*

### Fact Sheet

| | | | |
|---|---|---|---|
| ***Classification*** | EGF | ***Abbreviations*** | EGF |
| ***Structures/motifs*** | 9 LDLRB; 9 EGF-like | ***DB entries*** | EGF_HUMAN/P01133 |
| ***MW/length*** | Pro-EGF: 131 572 Da/1185 aa<br>EGF: 6222 Da/53 aa | ***PDB entries*** | 1JL9 |
| ***Concentration*** | | ***Half-life*** | |
| ***PTMs*** | 9 N-CHO (potential) | | |
| ***References*** | Bell *et al.*, 1986, *Nucleic Acids Res.*, **14**, 8427–8446.<br>Lu *et al.*, 2001, *J. Biol. Chem.*, **276**, 34913–34917. | | |

### Description

**Epidermal growth factor** (EGF) primarily produced in the submaxillary is involved in cell growth, proliferation and differentiation.

### Structure

EGF (53 aa) is a single-chain growth factor containing three intrachain disulfide bridges (Cys 6–Cys 20, Cys 14–Cys 31, Cys 33–Cys 42) and is derived from Pro-EGF by limited proteolysis. Pro-EGF is a large type I membrane glycoprotein containing 9 LDLRB repeats and 9 EGF-like domains and a potential 21 residues long transmembrane segment (Val 1011–Gly 1031). The 3D structure of EGF is characterised by an antiparallel β-sheet in the N-terminal region and a short antiparallel β-sheet in the C-terminal portion of the molecule (**1JL9**).

### Biological Functions

EGF plays an important role in the regulation of cell growth, proliferation and differentiation.

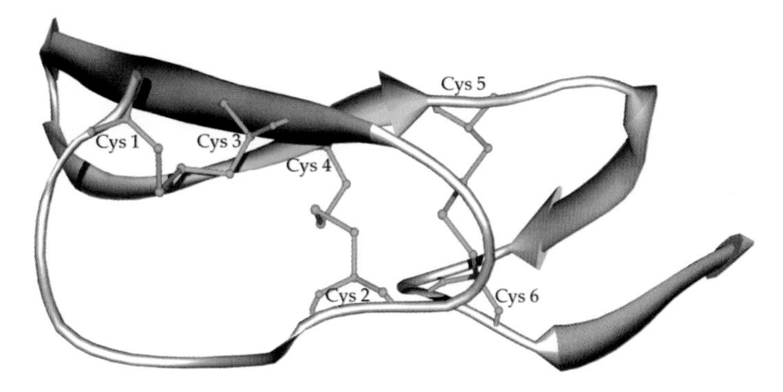


Epidermal growth factor

## *Granulocyte colony-stimulating factor (Pluripoietin)*

### Fact Sheet

| | | | |
|---|---|---|---|
| ***Classification*** | IL-6 | ***Abbreviations*** | G-CSF |
| ***Structures/motifs*** | | ***DB entries*** | CSF3_HUMAN/P09919 |
| ***MW/length*** | 19 058 Da/178 aa | ***PDB entries*** | 2D9Q |
| ***Concentration*** | approx. 20 ng/l | ***Half-life*** | several hrs |
| ***PTMs*** | 1 O-CHO | | |
| ***References*** | Nagata *et al.*, 1986, *Nature*, **319**, 415–418.<br>Tamada *et al.*, 2006, *Proc. Natl. Acad. Sci. USA*, **103,** 3135–3140. | | |

### Description

**Granulocyte colony-stimulating factor** (G-CSF) is synthesised in a variety of cells in response to bacterial infection. G-CSF exhibits significant similarity with interleukin-6.

### Structure

G-CSF is a single-chain monomeric glycoprotein with two intrachain disulfide bridges (Cys 40–Cys 46, Cys 68–Cys 78) and a single free Cys residue (aa 18). The 3D structure of G-CSF is characterised by a long-chain four-helical bundle topology stabilised by two intrachain disulfide bridges (**2D9Q**).

### Biological Functions

G-CSF is a cytokine that acts in hematopoiesis by controlling the production, differentiation and function of granulocytes and monocytes–macrophages.

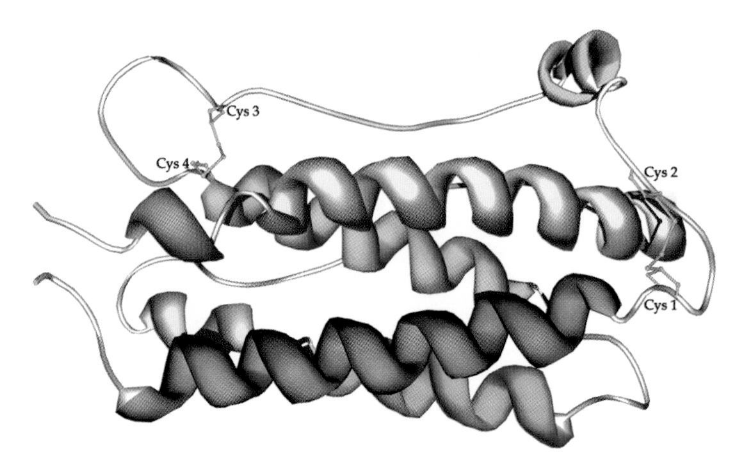


Granulocyte colony-stimulating factor

## *Granulocyte–macrophage colony-stimulating factor*

### Fact Sheet

| | | | |
|---|---|---|---|
| ***Classification*** | GM-CSF | ***Abbreviations*** | GM-CSF |
| ***Structures/motifs*** | | ***DB entries*** | CSF2_HUMAN/P04141 |
| ***MW/length*** | 14 478 Da/127 aa | ***PDB entries*** | 2GMF |
| ***Concentration*** | < 1 μg/l | ***Half-life*** | min–h |
| ***PTMs*** | 2 N-CHO; 4 O-CHO (1 partial) | | |
| ***References*** | Rozwarski *et al.*, 1996, *Proteins*, **26**, 304–313. | | |
| | Wong *et al.*, 1985, *Science*, **228**, 810–815. | | |

### Description

**Granulocyte–macrophage colony-stimulating factor** (GM-CSF) is synthesised in a variety of tissues stimulating the growth and differentiation of hematopoietic precursor cells.

### Structure

GM-CSF is a single-chain monomeric glycoprotein with two intrachain disulfide bridges (Cys 54–Cys 96, Cys 88–Cys 121). The 3D structure is characterised by a short-chain four-helical bundle structure with an up–up down–down topology (**2GMF**).

### Biological Functions

GM-CSF is a cytokine that is involved in the growth and proliferation of various types of cells such as granulocytes, macrophages, eosinophils and erythrocytes.

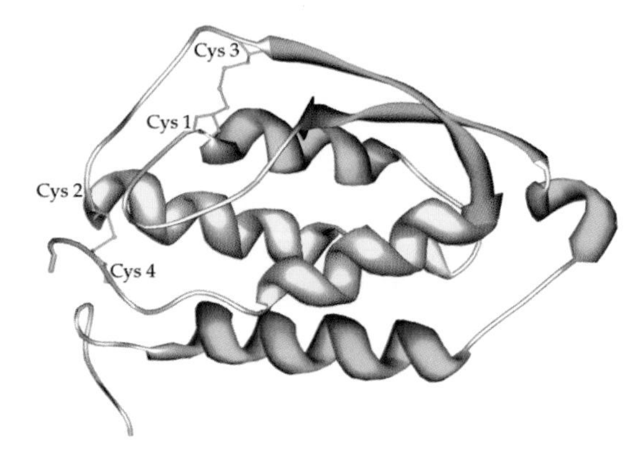


Granulocyte–macrophage colony-stimulating factor

## *Insulin-like growth factor binding proteins*

### Fact Sheet

| | | | |
|---|---|---|---|
| ***Classification*** | | ***Abbreviations*** | IGFBP1; IGFBP2; IGFBP3; IGFBP4; IGFBP5; IGFBP6; IGFBP7 |
| ***Structures/motifs*** | IGFBP1–IGFBP6: 1 IGFBP; 1 thyroglobulin type-I<br>IGFBP7: 1 Ig-like C2-type; 1 IGFBP; 1 Kazal-like; | ***DB entries*** | IBP1_HUMAN/P08833<br>IBP2_HUMAN/P18065<br>IBP3_HUMAN/P17936<br>IBP4_HUMAN/P22692<br>IBP5_HUMAN/P2459<br>3IBP6_HUMAN/P24592<br>IBP7_HUMAN/Q16270 |
| ***MW/length*** | IGFBP1: 25 270 Da/234 aa<br>IGFBP2: 31 376 Da/289 aa<br>IGFBP3: 28 736 Da/264 aa<br>IGFBP4: 25 975 Da/237 aa<br>IGFBP5: 28 573 Da/252 aa<br>IGFBP6: 22 607 Da/213 aa<br>IGFBP7: 26 493 Da/256 aa | ***PDB entries*** | |
| ***Concentration*** | IGFBP1: 10–100 μg/l<br>IGFBP2: 70–450 μg/l<br>IGFBP3: 2–5 mg/l<br>IGFBP4: unknown<br>IGFBP5: 500–600 μg/l<br>IGFBP6: 100–300 μg/l | ***Half-life*** | IGFBP1: approx. 12 min<br>IGFBP2: approx. 14 min<br>IGFBP3: < 30 min<br>IGFBP4: unknown<br>IGFBP5: unknown<br>IGFBP6: unknown |
| ***PTMs*** | IGFBP1: 3 P-Ser (aa 101, 119, 169);<br>IGFBP3: 3 N-CHO (1 potential);<br>IGFBP4: 1 N-CHO (potential);<br>IGFBP5: 1 O-CHO;<br>IGFBP6: 5 O-CHO (similarity);<br>IGFBP7: 1 N-CHO (probable) | | |
| ***References*** | Brinkmann *et al.*, 1988, *EMBO J.*, **7**, 2417–2423.<br>Cubbage *et al.*, 1990, *J. Biol. Chem.*, **265**, 12642–12649.<br>Kiefer *et al.*, 1991, *J. Biol. Chem.*, **266**, 9043–9049.<br>Kiefer *et al.*, 1991, *Biochem. Biophys. Res. Commun.*, **176**, 219–225.<br>Oh *et al.*, 1996, *J. Biol. Chem.*, **271**, 30322–30325.<br>Zapf *et al.*, 1990, *J. Biol. Chem.*, **265**, 14892–14898. | | |

### Description

**Insulin-like growth factor binding proteins** (IGFBPs) are expressed in many types of tissues and form a family of growth factors involved in the regulation of insulin growth factors.

### Structure

IGFBP1–IGFBP6 are single-chain proteins each containing one IGFBP domain and one thyroglobulin type I domain. The rather conserved N-terminal and C-terminal regions are characterised by five to six and three disulfide bridges, respectively. IGFBP1 and IGFBP2 each contain an **RGD** sequence as a cell attachment site. IGFBP7 contains three types of domains: one IGFBP, one Kazal-like and one Ig-like C2 type and is somehow different from the other IGFBPs.

### Biological Functions

IGFBPs are involved in the regulation of insulin growth factors by either inhibiting or stimulating the growth promoting effects of the insulin growth factors on cells.

## *Interferon α-1/13 (Interferon α-D)*

### Fact Sheet

| | | | |
|---|---|---|---|
| ***Classification*** | Interferon α/β | ***Abbreviations*** | IFN-α |
| ***Structures/motifs*** | | ***DB entries*** | IFNA1_HUMAN/P01562 |
| ***MW/length*** | 19 386 Da/166 aa | ***PDB entries*** | 1ITF |
| ***Concentration*** | | ***Half-life*** | approx. 5 h |
| ***PTMs*** | | | |
| ***References*** | Klaus *et al.*, 1997, *J. Mol. Biol.*, **274**, 661–675.<br>Taniguchi *et al.*, 1980, *Nature*, **285**, 547–549. | | |

### Description

**Interferon α** (IFNα) is produced by macrophages and has antiviral activity.

### Structure

IFN-α is a single-chain monomeric nonglycosylated protein with two intrachain disulfide bridges (Cys 1–Cys 99, Cys 29–139) and exists in many subtypes. Interferon α-1 and α-13 are identical. The 3D structure of interferon α-2A is characterised by a left-handed four-helix bundle in an up–up down–down topology and two over-hand connections (**1ITF**).

### Biological Functions

IFNα has the potent ability to confer a virus-resistant state on the target cells. It stimulates the production of two enzymes, a protein kinase and a 2′-5′-oligoadenylate synthetase.

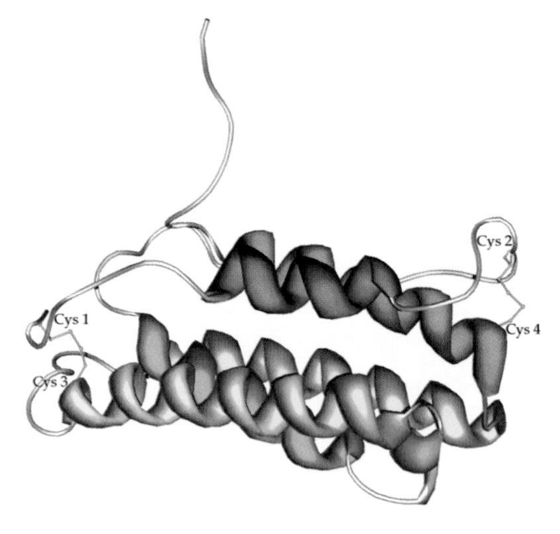


Interferon α-2A

# *Interferon β (Fibroblast interferon)*

## Fact Sheet

| | | | |
|---|---|---|---|
| ***Classification*** | Interferon α/β | ***Abbreviations*** | IFN-β |
| ***Structures/motifs*** | | ***DB entries*** | IFNB_HUMAN/P01574 |
| ***MW/length*** | 20 027 Da/166 aa | ***PDB entries*** | 1AU1 |
| ***Concentration*** | | ***Half-life*** | |
| ***PTMs*** | 1 N-CHO | | |
| ***References*** | Derynck *et al.*, 1980, *Nature*, **285**, 542–547.<br>Karpusas *et al.*, 1997, *Proc. Natl. Acad. Sci. USA*, **94**, 11813–11818. | | |

## Description

**Interferon β** (IFN-β) is produced by fibroblasts and exhibits antiviral and antibacterial activity. IFN-β belongs to the type I class of interferons.

## Structure

IFN-β is a single-chain monomeric glycoprotein with a single intrachain disulfide bridge (Cys 31–Cys 141). IFN-β adopts the classical helix bundle topology with a zinc-binding site at the interface of two monomers in the dimer (**1AU1**).

## Biological Functions

IFN-β has antiviral, antibacterial and anticancer activities.

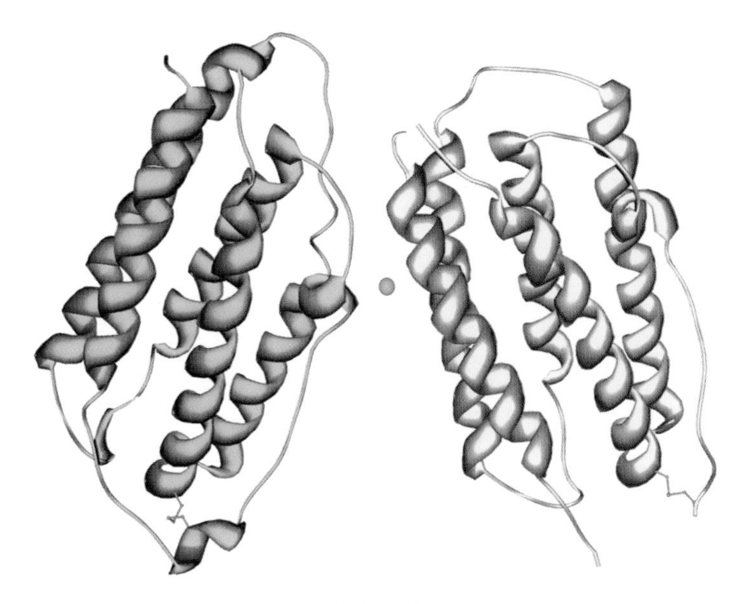

Interferon β

# *Interferon γ (Immune interferon)*

## Fact Sheet

| | | | |
|---|---|---|---|
| ***Classification*** | Type II (gamma) interferon | ***Abbreviations*** | IFN-γ |
| ***Structures/motifs*** | | ***DB entries*** | IFNG_HUMAN/P01579 |
| ***MW/length*** | 17 146 Da/146 aa | ***PDB entries*** | 1FG9, 1FYH |
| ***Concentration*** | Usually only traces in blood | ***Half-life*** | approx. 20 min |
| ***PTMs*** | 2 N-CHO; 1 Pyro-Glu | | |
| ***References*** | Gray *et al.*, 1982, *Nature*, **298**, 859–863.<br>Randal *et al.*, 2001, *Structure*, **9**, 155–163.<br>Rinderknecht *et al.*, 1984, *J. Biol. Chem.*, **259**, 6790–6797.<br>Thiel *et al.*, 2000, *Structure Fold Des.*, **8**, 927–936. | | |

## Description

**Interferon γ** (IFN-γ) is produced by T lymphocytes and exhibits immunoregulatory and antiviral activity.

### Structure

IFN-γ is a single-chain monomeric glycoprotein present as homodimer. Mature IFN-γ is generated from the proform by releasing a C-terminal pentapeptide and carries a pyro-Glu residue. Two C-terminally truncated forms of IFN-γ have been described. The IFN-γ monomer is characterised by six α-helices and in the homodimer (in yellow and blue) the helices are in an interdigitating fashion. The IFN-γ receptor α-chain consists of two fibronectin type III domains, each characterised by a β-sandwich consisting of a three-stranded and a four-stranded β-sheet structure (**1FG9**).

## Biological Functions

IFN-γ has antiviral, antiproliferative and immunomodulatory activity. IFN-γ is an important modulator of the immune system by activation of macrophages and by induction of MHC class II antigens and is an important factor in the production of antibodies.

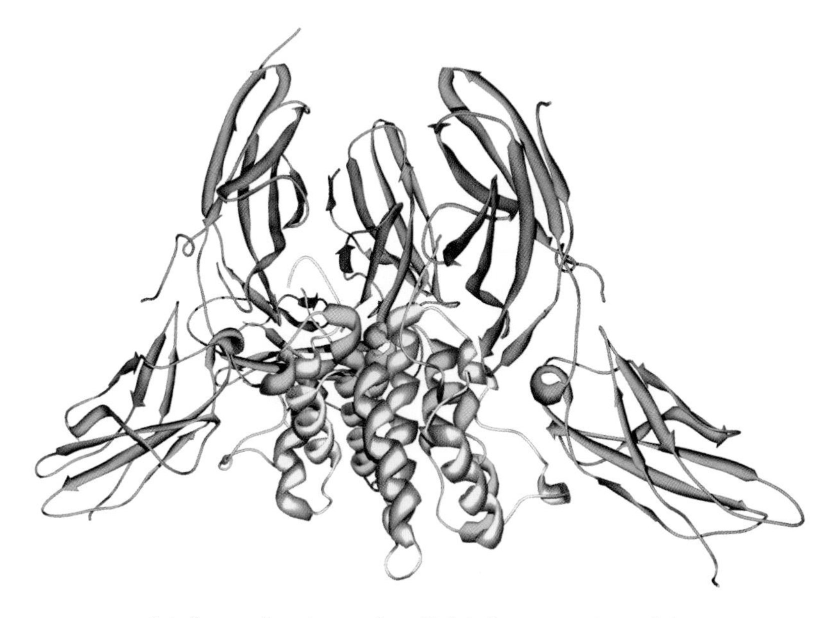

Interferon γ dimer in complex with Interferon γ receptor α-chain

# *Interleukin-1: IL-1α (Hematopoietin)/IL-1β (Catabolin)*

## Fact Sheet

| | | | |
|---|---|---|---|
| ***Classification*** | IL-1 | ***Abbreviations*** | IL-1α; IL-1β |
| ***Structures/motifs*** | | ***DB entries*** | IL1A_HUMAN/P01583<br>IL1B_HUMAN/P01584 |
| ***MW/length*** | IL-1α: 18 048 Da/159 aa<br>IL-1β: 17 377 Da/153 aa | ***PDB entries*** | 9ILB, 2ILA |
| ***Concentration*** | < 40 ng/l | ***Half-life*** | approx. 10 min |
| ***PTMs*** | IL-1α: 1 N-CHO (potential) | | |
| ***References*** | March *et al.*, 1985, *Nature*, **315**, 641–647.<br>Yu *et al.*, 1999, *Proc. Natl. Acad. Sci. USA*, **96**, 103–108. | | |

## Description

**Interleukin-1** (IL-1) exists in two different forms: IL-1α and IL-1β encoded by two different genes. IL-1 is produced by most nucleated cells, mainly by activated macrophages.

## Structure

IL-1α and IL-β are single-chain, monomeric proteins derived from their corresponding precursors but lacking the classical hydrophobic signal sequences by proteolytic processing via removal of two N-terminal propeptides (IL-1α: 112 aa; IL-β: 116 aa). The 3D structure of IL-1β is characterised by 12 antiparallel β-strands folded into a six-stranded β-barrel (**9ILB**). The 3D structure of IL-1α (**2ILA**) is similar to IL-1β.

## Biological Functions

IL-1α and IL-β are pleiotropic mediators of the host response to infections and injuries. IL-1 is a pyrogen, induces prostaglandin synthesis and is involved in T lymphocyte activation and proliferation and via interleukin-2 also in B lymphocytes.

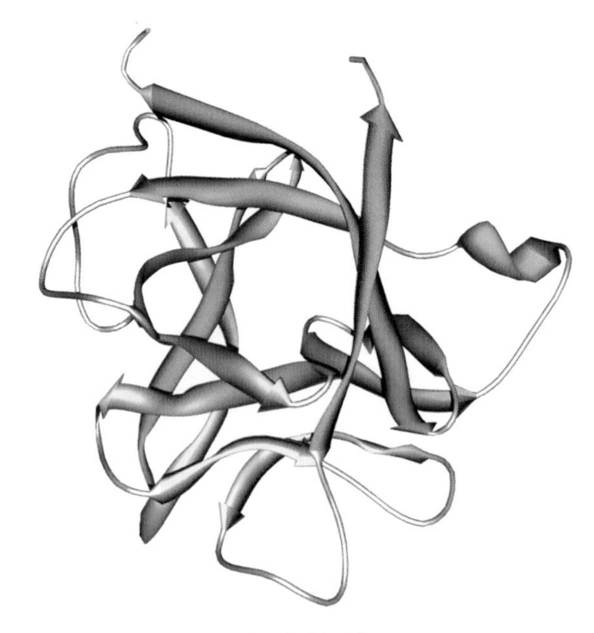

Interleukin-1β

## *Interleukin-2 (T-cell growth factor)*

### Fact Sheet

| | | | |
|---|---|---|---|
| ***Classification*** | IL-2 | ***Abbreviations*** | IL-2 |
| ***Structures/motifs*** | | ***DB entries*** | IL2_HUMAN/P60568 |
| ***MW/length*** | 15 418 Da/133 aa | ***PDB entries*** | 2B5I |
| ***Concentration*** | Not detectable | ***Half-life*** | |
| ***PTMs*** | 1 O-CHO | | |
| ***References*** | Taniguchi *et al.*, 1983, *Nature*, **302**, 305–310.<br>Wang *et al.*, 2005, *Science*, **310**, 1159–1163. | | |

### Description

**Interleukin-2** (IL-2) is produced by T cells and is involved in the antigen specific immune response.

### Structure

IL-2 is a single-chain monomeric glycoprotein with a single intrachain disulfide bridge (Cys 58–Cys 105) and a single free Cys residue (125 aa) of unknown biological function. The 3D structure of IL-2 is characterised by an α-helix bundle (in blue) and forms a complex with the extracellular domain of the IL-2 receptor consisting of three chains: IL-2Rα (in yellow), IL-2Rβ (in green) and IL-2Rγ (in red) (**2B5I**).

### Biological Functions

IL-2 is a multifunctional lymphocytotrophic hormone involved in the antigen specific immune response. IL-2 serves as a growth factor for activated T lymphocytes and some B lymphocytes.

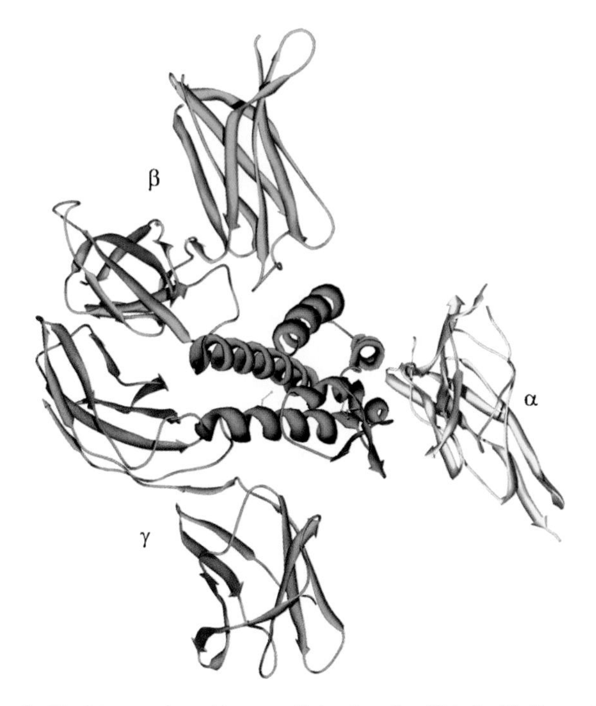


Interleukin-2 in complex with extracellular domain of Interleukin-2 receptor

## *Interleukin-3 (Hematopoietic growth factor)*

### Fact Sheet

| | | | |
|---|---|---|---|
| ***Classification*** | IL-3 | ***Abbreviations*** | IL-3 |
| ***Structures/motifs*** | | ***DB entries*** | IL3_HUMAN/P08700 |
| ***MW/length*** | 15 091 Da/133 aa | ***PDB entries*** | 1JLI |
| ***Concentration*** | Not detectable | ***Half-life*** | min |
| ***PTMs*** | 2 N-CHO (potential) | | |
| **References** | Feng *et al.*, 1996, *J. Mol. Biol.*, **259**, 524–541.<br>Yang *et al.*, 1986, *Cell*, **47**, 3–10. | | |

### Description

**Interleukin-3** (IL-3) is primarily produced by activated T cells and acts as a growth factor in hematopoiesis.

### Structure

IL-3 is a monomeric single-chain glycoprotein with one essential intrachain disulfide bridge (Cys 16–Cys 84). The 3D structure is characterised by a left-handed four-helix bundle stabilised by a single disulfide bridge (**1JLI**).

### Biological Functions

Il-3 is a growth factor involved in the proliferation and differentiation of early hematopoietic progenitor cells.

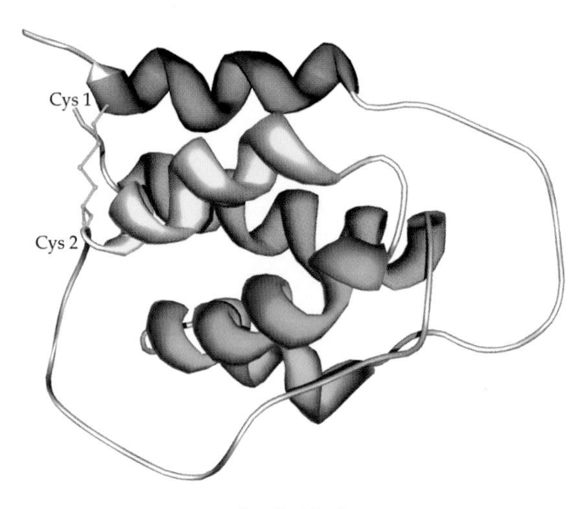


Interleukin-3

## *Interleukin-4 (B cell stimulatory factor 1)*

### Fact Sheet

| | | | |
|---|---|---|---|
| ***Classification*** | IL-4/IL-13 | ***Abbreviations*** | IL-4 |
| ***Structures/motifs*** | | ***DB entries*** | IL4_HUMAN/P05112 |
| ***MW/length*** | 14 963 Da/129 aa | ***PDB entries*** | 2INT |
| ***Concentration*** | < 10 ng/l | ***Half-life*** | |
| ***PTMs*** | 1 N-CHO | | |
| ***References*** | Walter *et al.*, 1992, *J. Biol. Chem.*, **267**, 20371–20376.<br>Yokota *et al.*, 1986, *Proc. Natl. Acad. Sci. USA*, **83**, 5894–5898. | | |

### Description

**Interleukin-4** (IL-4) is produced by activated T cells and mast cells with multiple functions like the immune response and activation of various types of cells.

### Structure

IL-4 is a monomeric single-chain glycoprotein with three intrachain disulfide bridges (Cys 3–Cys 127, Cys 24–Cys 65, Cys 46–Cys 99). The 3D structure is characterised by a left-handed, antiparallel four-helix bundle (**2INT**).

### Biological Functions

IL-4 participates in the activation of certain B cells and other cell types. It induces the expression of MHC class II molecules on resting B cells and enhances the secretion and expression of IgE and IgG1.

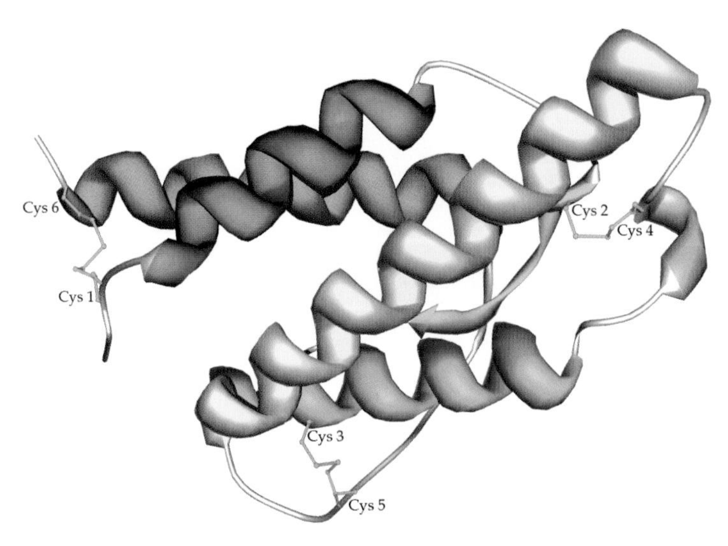


Interleukin-4

## *Interleukin-5 (Eosinophil differentiation factor)*

### Fact Sheet

| | | | |
|---|---|---|---|
| ***Classification*** | IL-5 | ***Abbreviations*** | IL-5 |
| ***Structures/motifs*** | | ***DB entries*** | IL5_HUMAN/P05113 |
| ***MW/length*** | 13 149 Da/115 aa | ***PDB entries*** | 1HUL |
| ***Concentration*** | | ***Half-life*** | |
| ***PTMs*** | 1 N-CHO; 1 O-CHO | | |
| ***References*** | Azuma *et al.*, 1986, *Nucleic Acids Res.*, **14**, 9149–9158.<br>Milburn *et al.*, 1993, *Nature*, **363**, 172–176. | | |

### Description

**Interleukin-5** (IL-5) is produced by T cells and mast cells and is involved in the induction of eosinophil production.

### Structure

IL-5 is a homodimeric glycoprotein with two interchain disulfide bridges (Cys 44–Cys 86). The 3D structure is characterised by two four-helix bundles and two short β-sheets (**1HUL**).

### Biological Functions

IL-5 acts as an eosinophil differentiation factor, a B cell growth factor and a T cell replacing factor.

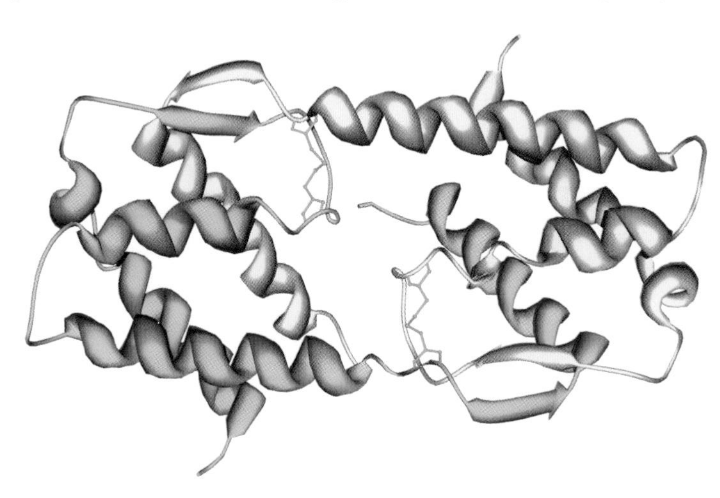

Interleukin-5

## *Interleukin-6 (Interferon $\beta_2$)*

### Fact Sheet

| | | | |
|---|---|---|---|
| ***Classification*** | IL-6 | ***Abbreviations*** | IL-6 |
| ***Structures/motifs*** | | ***DB entries*** | IL6_HUMAN/P05231 |
| ***MW/length*** | 20 813 Da/183 aa | ***PDB entries*** | 2IL6 |
| ***Concentration*** | < 100 ng/l | ***Half-life*** | |
| ***PTMs*** | 3 N-CHO; 3 O-CHO | | |
| ***References*** | Boulanger *et al.*, 2003, *Science*, **300**, 2101–2104.<br>Hirano *et al.*, 1986, *Nature*, **324**, 73–76. | | |

### Description

**Interleukin-6** (IL-6) is produced by a wide variety of cells and exhibits significant similarity with granulocyte colony-stimulating factor (G-CSF).

### Structure

IL-6 is a single-chain monomeric glycoprotein with two intrachain disulfide bridges (Cys 43–Cys 49, Cys 72–Cys 82). The 3D structure of IL-6 is characterised by a four-helix bundle topology stabilised by two intrachain disulfide bridges (**2IL6**).

### Biological Functions

IL-6 is a multifunctional cytokine acting on a wide variety of cells that participate in the induction of acute-phase reactions and antibody production.

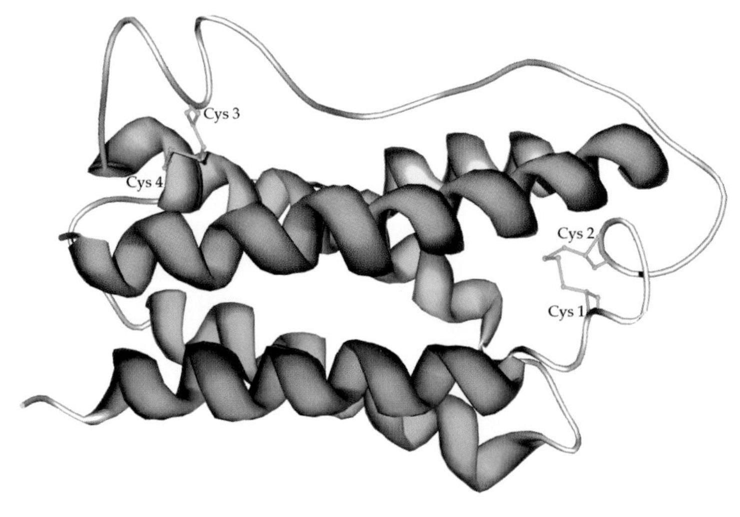


Interleukin-6

# *Interleukin-7*

## Fact Sheet

| | | | |
|---|---|---|---|
| ***Classification*** | IL-7/IL-9 | ***Abbreviations*** | IL-7 |
| ***Structures/motifs*** | | ***DB entries*** | IL7_HUMAN/P13232 |
| ***MW/length*** | 17 387 Da/152 aa | ***PDB entries*** | |
| ***Concentration*** | | ***Half-life*** | |
| ***PTMs*** | 3 N-CHO (potential) | | |
| ***References*** | Goodwin *et al.*, 1989, *Proc. Natl. Acad. Sci. USA*, **86**, 302–306. | | |

## Description

**Interleukin-7** (IL-7) is produced in the bone marrow and is important during certain stages of B cell maturation.

## Structure

IL-7 is a single-chain glycoprotein containing three intrachain disulfide bridges (Cys 2–Cys 141, Cys 34–Cys 129, Cys 47–Cys 92).

## Biological Functions

IL-7 is capable of stimulating the proliferation of lymphoid progenitor cells and is important for the proliferation of B cell maturation during certain stages.

## *Interleukin-8 (Neutrophil-activating factor)*

### Fact Sheet

| | | | |
|---|---|---|---|
| ***Classification*** | Intercrine alpha (chemokine CxC) | ***Abbreviations*** | IL-8; NAP |
| ***Structures/motifs*** | | ***DB entries*** | IL8_HUMAN/P10145 |
| ***MW/length*** | 8922 Da/77 aa | ***PDB entries*** | 3IL8 |
| ***Concentration*** | < 10 ng/l | ***Half-life*** | |
| ***PTMs*** | | | |
| ***References*** | Baggiolini *et al.*, 1992, *FEBS Lett.*, **307**, 97–101.<br>Baldwin *et al.*, 1991, *Proc. Natl. Acad. Sci. USA*, **88**, 502–506. | | |

### Description

**Interleukin-8** (IL-8) is produced by monocytes, macrophages and endothelial cells and is involved in the stimulation of neutrophils.

### Structure

IL-8 is a single-chain chemokine containing two intrachain disulfide bridges (Cys 12–Cys 39, Cys 14–Cys 55) present in solution as a dimer. There are several N-terminally processed forms that exhibit in several cases an increased biological activity. The 3D structure exhibits the expected fold characteristic for CxC chemokines (**3IL8**).

### Biological Functions

IL-8 is a chemotactic factor that exhibits interaction with neutrophils, basophils and T cells and is important in neutrophil activation.

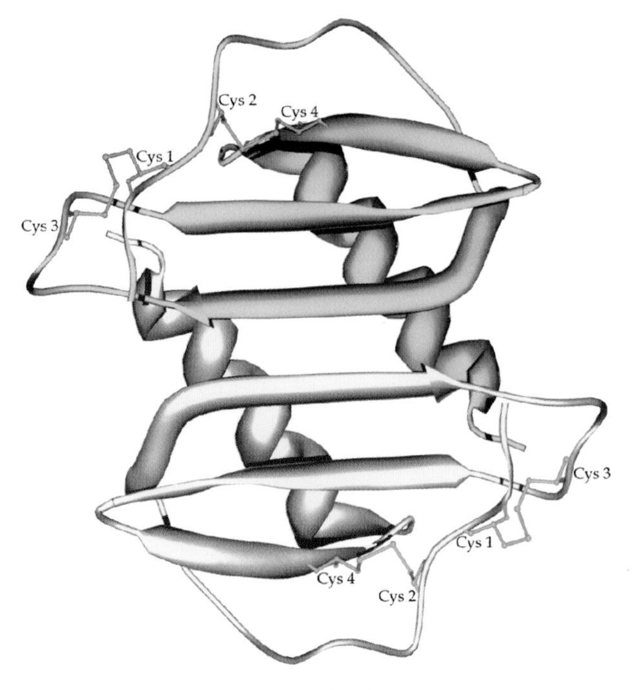


Interleukin-8

## *Interleukin-9 (T cell growth factor P40)*

### Fact Sheet

| | | | |
|---|---|---|---|
| ***Classification*** | IL-7/IL-9 | ***Abbreviations*** | IL-9 |
| ***Structures/motifs*** | | ***DB entries*** | IL9_HUMAN/P15248 |
| ***MW/length*** | 14 121 Da/126 aa | ***PDB entries*** | |
| ***Concentration*** | | ***Half-life*** | |
| ***PTMs*** | 4 N-CHO (potential) | | |
| ***References*** | Yang *et al.*, 1989, *Blood*, **74**, 1880–1884. | | |

### Description

**Interleukin-9** (IL-9) is produced by T cells and supports the growth of helper T cells.

### Structure

IL-9 is a single-chain glycoprotein containing 10 Cys residues with an unassigned disulfide bridge pattern.

### Biological Functions

IL-9 supports the IL-2 and IL-4 independent growth of helper T cells.

## *Interleukin-10 (Cytokine synthesis inhibitory factor)*

### Fact Sheet

| | | | |
|---|---|---|---|
| ***Classification*** | IL-10 | ***Abbreviations*** | IL-10 |
| ***Structures/motifs*** | | ***DB entries*** | IL10_HUMAN/P22301 |
| ***MW/length*** | 18 647 Da/160 aa | ***PDB entries*** | ILK |
| ***Concentration*** | < 5 ng/l | ***Half-life*** | |
| ***PTMs*** | 1 N-CHO (potential) | | |
| ***References*** | Vieira *et al.*, 1991, *Proc. Natl. Acad. Sci. USA*, **88**, 1172–1176.<br>Zdanov *et al.*, 1996, *Protein Sci.*, **5**, 1955–1962. | | |

### Description

**Interleukin-10** (IL-10) is produced by a variety of cells and is involved in the inhibition of the production of various cytokines.

### Structure

IL-10 is a non-covalently-linked homodimer with each chain containing two intrachain disulfide bridges (Cys 12–Cys 108, Cys 62–Cys 114). The 3D structure of the monomer exhibits the classical four-helix bundle topology and the intertwined dimer forms a V-shaped structure (**1ILK**).

### Biological Functions

IL-10 inhibits the synthesis of a number of cytokines such as interferon γ, interleukin-2 and interleukin-3, tumour necrosis factor and granulocyte–macrophage colony-stimulating factor.

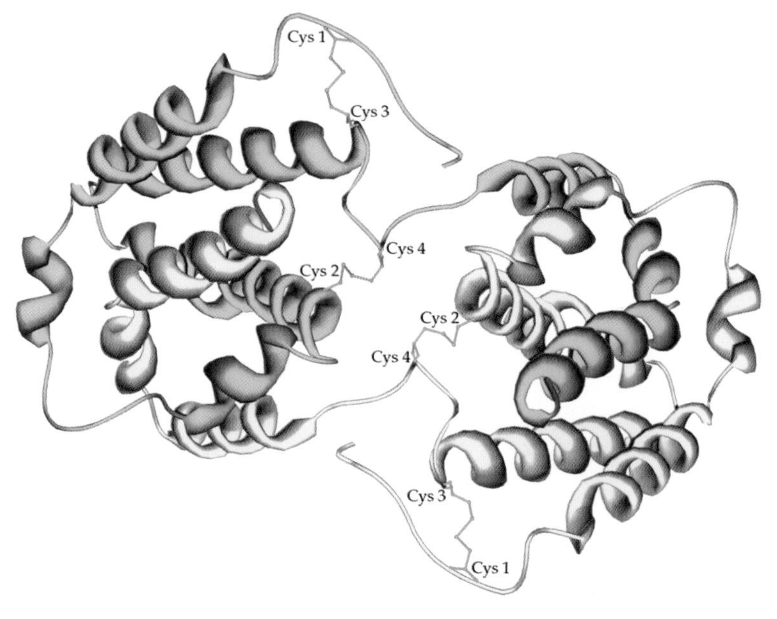


Interleukin-10

## *Interleukin-11 (Adipogenesis inhibitory factor)*

### Fact Sheet

| | | | |
|---|---|---|---|
| ***Classification*** | IL-6 | ***Abbreviations*** | IL-11 |
| ***Structures/motifs*** | | ***DB entries*** | IL11_HUMAN/P20809 |
| ***MW/length*** | 19 144 Da/178 aa | ***PDB entries*** | |
| ***Concentration*** | Not detectable | ***Half-life*** | approx. 12 h |
| ***PTMs*** | | | |
| ***References*** | Paul *et al.*, 1990, *Proc. Natl. Acad. Sci. USA*, **87**, 7512–7516. | | |

### Description

**Interleukin-11** (IL-11) is expressed in most cells with pleiotropic effects on many tissues.

### Structure

IL-11 is a single-chain monomeric cytokine rich in proline (12 %) but devoid of cysteine with a high helical content (57 %).

### Biological Functions

IL-11 directly stimulates the proliferation of hematopoietic stem cells and megakaryocyte progenitor cells and induces megakaryocyte maturation resulting in an increased platelet production.

## *Interleukin-12: IL-12 α-chain/IL-12 β-chain*

### Fact Sheet

| | | | |
|---|---|---|---|
| ***Classification*** | IL-12α: IL-6; IL-12β: Type I cytokine receptor | ***Abbreviations*** | IL-12α; IL-12β |
| ***Structures/motifs*** | IL-12β: 1 FN3; 1 Ig-like C2 type | ***DB entries*** | IL12A_HUMAN/P29459<br>IL12B_HUMAN/P29460 |
| ***MW/length*** | IL-12α: 22 542 Da/197 aa<br>IL-12β: 34 697 Da/306 aa | ***PDB entries*** | 1F45 |
| ***Concentration*** | | ***Half-life*** | hours |
| ***PTMs*** | IL-12α: 2 N-CHO (potential); IL-12β: 2 N-CHO (1 potential); 1 C-CHO (Trp) | | |
| ***References*** | Gubler *et al.*, 1991, *Proc. Natl. Acad. Sci. USA*, **88**, 4143–4147.<br>Yoon *et al.*, 2000, *EMBO J.*, **19**, 3530–3541. | | |

### Description

**Interleukin-12** (IL-12) is secreted from activated macrophages and B lymphocytes and exhibits numerous biological activities.

### Structure

IL-12 is heterodimeric cytokine consisting of two-chains: IL-12α containing three possible intrachain disulfide bridges (assigned: Cys 42–Cys 174, Cys 63–Cys 101; unassigned: Cys 15–Cys 88) and IL-12β containing four intrachain disulfide bridges (Cys 28–Cys 68, Cys 109–Cys 120, Cys 148–Cys 171, Cys 278–Cys 305) are derived from two distinct but unrelated genes. The two chains are linked by a single disulfide bridge: Cys 74 in IL-12α and Cys 177 in IL-12β. The 3D structure of IL-12 is similar to type I cytokine receptors: IL-12α (in blue) is characterised by a four-helix bundle topology and IL-12β (in light blue) by β-sheet structures (**1F45**).

### Biological Functions

Among the many biological functions IL-12 has been shown to increase the cytolytic activity of natural killer cells, to induce interferon γ production and to stimulate the proliferation of activated T and natural killer cells.

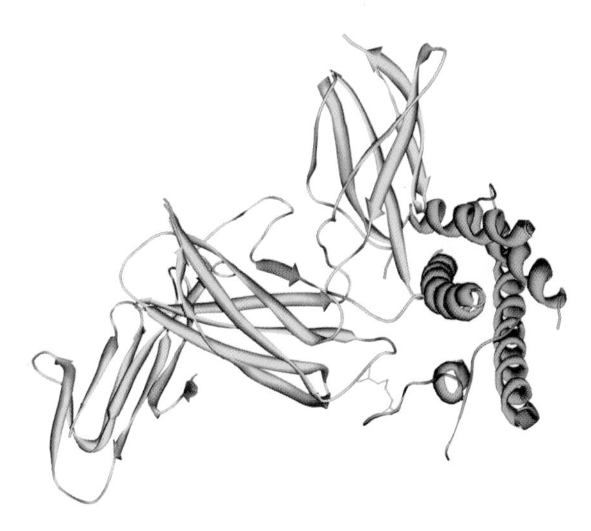

Interleukin-12

# *Interleukin-13*

## Fact Sheet

| | | | |
|---|---|---|---|
| ***Classification*** | IL-4/IL-13 | ***Abbreviations*** | IL-13 |
| ***Structures/motifs*** | | ***DB entries*** | IL13_HUMAN/P35225 |
| ***MW/length*** | 12 344 Da/112 aa | ***PDB entries*** | 1IJZ |
| ***Concentration*** | | ***Half-life*** | |
| ***PTMs*** | 4 N-CHO (potential) | | |
| ***References*** | Minty *et al.*, 1993, *Nature*, **362**, 248–250.<br>Moy *et al.*, 2001, *J. Mol. Biol.*, **310**, 219–230. | | |

## Description

**Interleukin-13** (IL-13) is expressed by T lymphocytes and regulates the function of monocytes and B cells.

## Structure

IL-13 is a single-chain monomeric glycoprotein with two intrachain disulfide bridges (Cys 28–Cys 56, Cys 44–Cys 70). The 3D structure is characterised by a short-chain left-handed four-helix bundle topology (**1IJZ**) similar to IL-4.

## Biological Functions

IL-13 seems to play a critical role in the regulation of inflammatory and immune responses. Together with interleukin-2 it regulates the synthesis of interferon $\gamma$.

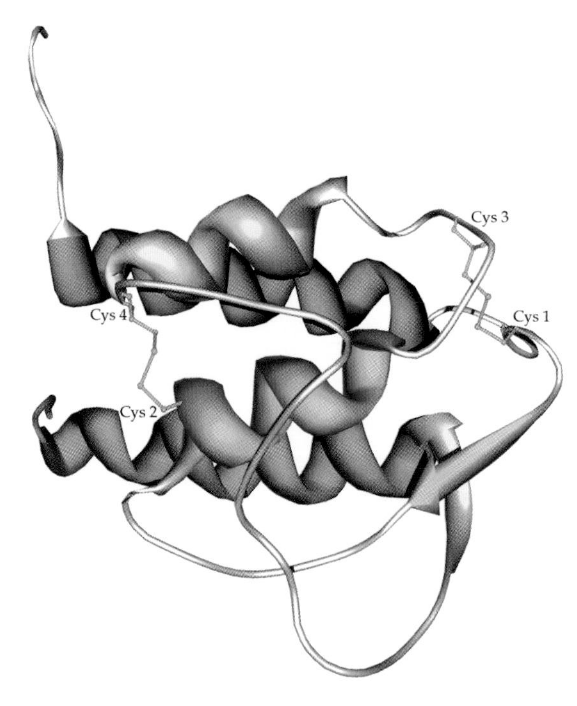


Interleukin-13

# *Interleukin-15*

## Fact Sheet

| | | | |
|---|---|---|---|
| ***Classification*** | IL-15 | ***Abbreviations*** | IL-15 |
| ***Structures/motifs*** | | ***DB entries*** | IL15_HUMAN/P40933 |
| ***MW/length*** | 12 744 Da/114 aa | ***PDB entries*** | |
| ***Concentration*** | | ***Half-life*** | |
| ***PTMs*** | 1 N-CHO (potential) | | |
| ***References*** | Grabstein *et al.*, 1994, *Science*, **264**, 965–968. | | |

## Description

**Interleukin-15** (IL-15) is most abundant in placenta and skeletal muscle and stimulates the proliferation of T lymphocytes.

## Structure

IL-15 is a single-chain monomeric glycoprotein with two intrachain disulfide bridges (Cys 35–Cys 85, Cys 42–Cys 88). A potential N-terminal 19 residue propeptide is removed.

## Biological Functions

IL-15 is a cytokine that is involved in the stimulation and proliferation of T lymphocytes and shares many biological activities with interleukin-2.

## *Lymphotoxin (LT): Lymphotoxin-α (LT-α) and Lymphotoxin-β (LT-β)*

### Fact Sheet

| | | | |
|---|---|---|---|
| ***Classification*** | Tumour necrosis factor | ***Abbreviations*** | LT-α; LT-β |
| ***Structures/motifs*** | | ***DB entries*** | TNFB_HUMAN/P01374<br>TNFC_HUMAN/Q06643 |
| ***MW/length*** | LT-α: 18 661 Da/171 aa<br>LT-β: 25 390 Da/244 aa | ***PDB entries*** | 1TNR |
| ***Concentration*** | | ***Half-life*** | |
| ***PTMs*** | LT-α: 1 N-CHO; 1 O-CHO (partial); LT-β: 1 N-CHO (potential) | | |
| ***References*** | Banner *et al.*, 1993, *Cell*, **73**, 431–445.<br>Browning *et al.*, 1993, *Cell*, **72**, 847–856.<br>Gray *et al.*, 1984, *Nature*, **312**, 721–724. | | |

### Description

There exist two forms of **lymphotoxin** (LT): Lymphotoxin α (LT-α) and lymphotoxin β (LT-β) are expressed by lymphocytes and are cytotoxic for a wide range of tumour cells.

### Structure

LT-α exists either as secreted homotrimer or as membrane-associated heterotrimers consisting of either one LT-α and two LT-β or two LT-α and one LT-β. In the N-terminal region LT-β contains a potential 30 residues transmembrane segment that provides a membrane anchor for the attachment to the cell surface. The extracellular fragment of the LT receptor (in red) binds in the groove between two adjacent LT-α subunits (in blue) (**1TNR**).

### Biological Functions

LT-α is a cytotoxin for many types of tumour cells. LT-β seems to play a specific role in the regulation of the immune response.

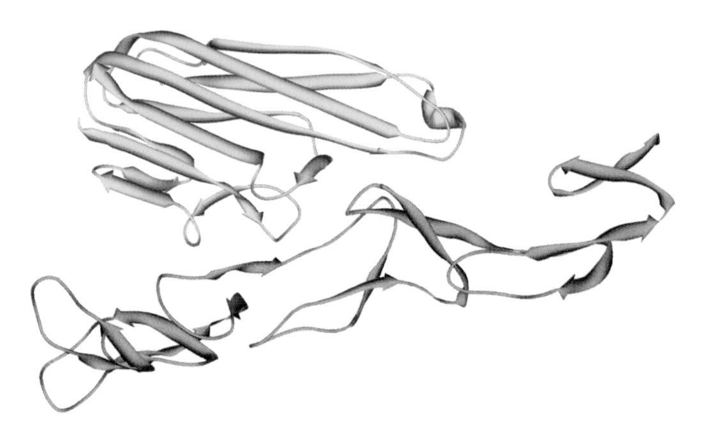

Tumor necrosis factor-β + Tumor necrosis factor-β receptor

# *Nerve growth factor-β*

## Fact Sheet

| | | | |
|---|---|---|---|
| ***Classification*** | NGF-beta | ***Abbreviations*** | NGF-β |
| ***Structures/motifs*** | | ***DB entries*** | NGF_HUMAN/P01138 |
| ***MW/length*** | 13 494 Da/120 aa | ***PDB entries*** | 1SG1 |
| ***Concentration*** | approx. 2 μg/l | ***Half-life*** | |
| ***PTMs*** | 2 N-CHO (potential) | | |
| ***References*** | He and Garcia, 2004, *Science*, **304**, 870–875.<br>Robinson *et al.*, 1995, *Biochemistry*, **34**, 4139–4146.<br>Ulrich *et al.*, 1983, *Nature*, **303**, 821–825. | | |

## Description

**Nerve growth factor-β (NGF-β) is secreted by fibroblasts and other cells and is important for the nervous system.**

## Structure

NGF-β is a single-chain glycoprotein containing three intrachain disulfide bridges (Cys 15–Cys 80, Cys 58–Cys 108, Cys 68–Cys 110) and is present as a non-covalently-linked homodimer. NGF-β is derived from its precursor by the removal of a 103 residue N-terminal propeptide. The 3D structure of monomeric NGF-β is characterised by seven twisted β-strands arranged in three antiparallel pairs. The homodimer (in blue and yellow) binds asymmetrically to the extracellular domain of the rat neurotrophin receptor p75 domain (in red) (**1SG1**).

## Biological Functions

NGF-β is important for the development and maintenance of the sympathetic and sensory nervous system. NGF-β stimulates the division and differentiation of sympathetic and embryonic sensory neurons.

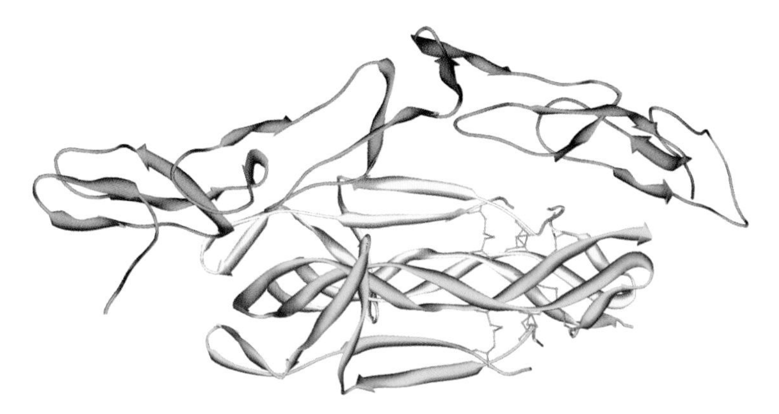

Nerve growth factor-β dimer in complex with neurotrophin receptor p75

# *Platelet basic protein (Small inducible cytokine B7)*

## Fact Sheet

| | | | |
|---|---|---|---|
| ***Classification*** | Intercrine alpha (chemokine CxC) | ***Abbreviations*** | PBP |
| ***Structures/motifs*** | | ***DB entries*** | SCYB7_HUMAN/P02775 |
| ***MW/length*** | 10 266 Da/94 aa | ***PDB entries*** | 1NAP |
| ***Concentration*** | 20–40 ng/l | ***Half-life*** | 10 min–1 h |
| ***PTMs*** | | | |
| ***References*** | Malkowski *et al.*, 1995, *J. Biol. Chem.*, **270**, 7077–7087.<br>Wenger *et al.*, 1989, *Blood*, **73**, 1498–1503. | | |

## Description

**Platelet basic protein** (PBP) is produced by megakaryocytes and stored in the α-granules of platelets.

## Structure

PBP is a single-chain chemokine containing two intrachain disulfide bridges (Cys 29–Cys 55, Cys 31–Cys 71). There are several N- and/or C-terminally processed forms: (1) connective tissue-activating peptides III (CTAP-III: Asn 10–Asp 94, Asn 10–Asp 90), (2) β-thromboglobulin (Gly 14–Asp 94) and (3) neutrophil-activating peptides 2 (NAP-2: Ser 21–Asp 94, Asp 22–Asp 94, Ala 25–Asp 94, Ala 25–Asp 90, Ala 25–Leu 87). The 3D structure of NAP-2 is characterised by a central antiparallel three-stranded β-sheet and a C-terminal α-helix (**1NAP**).

## Biological Functions

PBP and its proteolytic products CATP-III, β-thromboglobulin and NAP-2 are released from activated platelets and exhibit various biological functions.

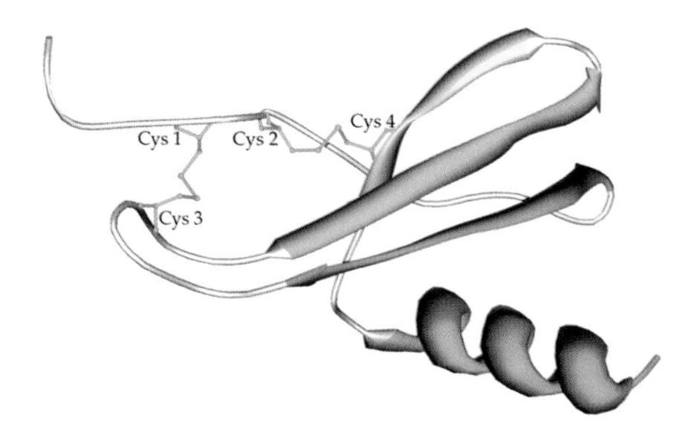


Neutrophil activating peptide 2

## *Platelet-derived growth factor: A-chain/B-chain*

### Fact Sheet

| | | | |
|---|---|---|---|
| ***Classification*** | PDGF/VEGF growth factor | ***Abbreviations*** | PDGFA; PDGFB |
| ***Structures/motifs*** | | ***DB entries*** | PDGFA_HUMAN/P04085<br>PDGFB_HUMAN/P01127 |
| ***MW/length*** | PDGF A-chain: 14 306 Da/125 aa<br>PDGF B-chain: 12 294 Da/109 aa | ***PDB entries*** | 1PDG |
| ***Concentration*** | | ***Half-life*** | mins |
| ***PTMs*** | PDGF A-chain: 1 N-CHO | | |
| ***References*** | Andersson *et al.*, 1992, *J. Biol. Chem.*, **267**, 11260–11266.<br>Betsholtz *et al.*, 1986, *Nature*, **320**, 695–699.<br>Collins *et al.*, 1985, *Nature*, **316**, 748–750.<br>Oefner *et al.*, 1992, *EMBO J.*, **11**, 3921–3926. | | |

### Description

**Platelet-derived growth factor** (PDGF) is released by platelets upon injuries and is a potent mitogen for cells of mesenchymal origin.

### Structure

The PDGF A-chain and the PDGF B-chain form homodimers and heterodimers: the A–A and the B–B homodimers are each linked by two interchain disulfide bridges in a cross-wise manner (Cys 37–Cys 46 in A–A and Cys 43–Cys 52 in B–B) and A–B heterodimers are arranged in an antiparallel manner resulting in a cross-wise linkage of the two interchain disulfide bridges. Each chain contains three intrachain disulfide bridges arranged in a tight knot-like structure (A-chain: Cys 10–Cys 54, Cys 43–Cys 91, Cys 47–Cys 93; B-chain: Cys 16–Cys 60, Cys 49–Cys 97, Cys 53–Cys 99). Chain A is generated from the precursor by removal of a 66 residue N-terminal propeptide and chain B by removal of two propeptides, a 61 residue N-terminal and a 51 residue C-terminal propeptide. PDGF B is folded into two highly twisted antiparallel pairs of β-strands (**1PDG**).

### Biological Functions

PDGF stimulates the proliferation of connective tissues and glial cells. PDGF stimulates the growth of the placenta and the embryo during development and wound healing in the adult.

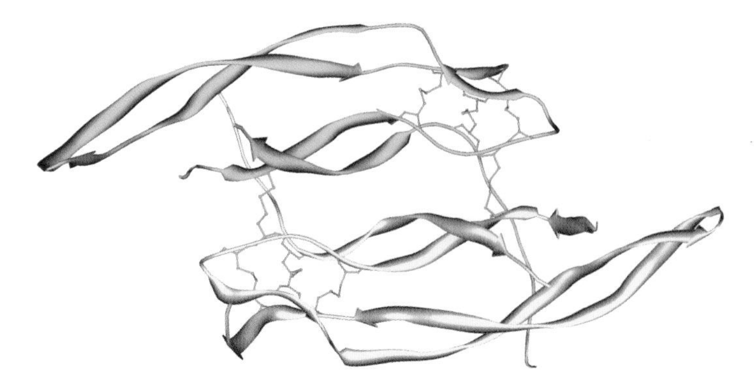

Platelet-derived growth factor

## *Platelet factor 4 (Oncostatin A)*

### Fact Sheet

| | | | |
|---|---|---|---|
| ***Classification*** | Intercrine alpha (chemokine CxC) | ***Abbreviations*** | PF-4 |
| ***Structures/motifs*** | | ***DB entries*** | PLF4_HUMAN/P02776 |
| ***MW/length*** | 7769 Da/70 aa | ***PDB entries*** | 1RHP |
| ***Concentration*** | approx. 10 µg/l | ***Half-life*** | 20–80 min |
| ***PTMs*** | | | |
| ***References*** | Poncz *et al.*, 1987, *Blood,* **69**, 219–223.<br>Zhang *et al.*, 1994, *Biochemistry*, **33**, 8361–8366. | | |

### Description

**Platelet factor 4** (PF-4) is produced by megakaryocytes and is stored in the α-granules of platelets. PF-4 is noncovalently bound to a proteoglycan molecule.

### Structure

PF-4 is a single-chain chemokine containing two intrachain disulfide bridges (Cys 10–Cys 36, Cys 12–Cys 52) present as homotetramer. The 3D structure is characterised by two extended six-stranded β-sheets (**1RHP**).

### Biological Functions

PF-4 neutralises the anticoagulant activity of heparin. It acts as a chemotactic molecule for neutrophils and monocytes.

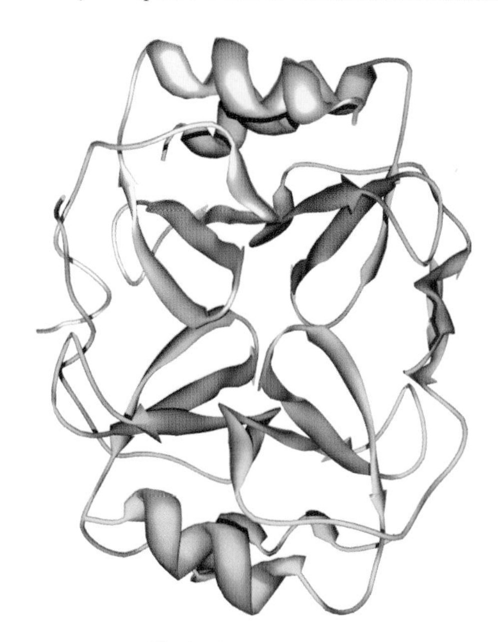

Platelet factor 4 tetramer

## *Small inducible cytokine A2 (Monocyte chemotactic protein 1)*

### Fact Sheet

| | | | |
|---|---|---|---|
| ***Classification*** | Intercrine beta (chemokine CC) | ***Abbreviations*** | CCL2 |
| ***Structures/motifs*** | | ***DB entries*** | CCL2_HUMAN/P13500 |
| ***MW/length*** | 8685 Da/76 aa | ***PDB entries*** | 1DOK |
| ***Concentration*** | | ***Half-life*** | |
| ***PTMs*** | 1 Pyro-Glu; 1 N-CHO (potential) | | |
| ***References*** | Lubkowski *et al.*, 1997, *Nat. Struct. Biol.*, **4**, 64–69.<br>Robinson *et al.*, 1989, *Proc. Natl. Acad. Sci. USA*, **86**, 1850–1854. | | |

### Description

**Small inducible cytokine A2** (CCL2) is produced by macrophages and other cells and acts as a chemotactic molecule on monocytes.

### Structure

CCL2 is a single-chain chemokine with two intrachain disulfide bridges (Cys 11–Cys 36, Cys 12–Cys 52) and an N-terminal pyro-Glu and is present either as a monomer or as a homodimer. Several N-terminally processed forms exist that are able to regulate receptor and target cell selectivity. The 3D structure is characterised by two short N-terminal β-strands, a central three-stranded antiparallel β-sheet and a C-terminal α-helix (**1DOK**).

### Biological Functions

CCL2 is a chemotactic factor that attracts monocytes and basophils but not neutrophils and eosinophils. Several amino acids are involved in the binding of glycosaminoglycans side chains of proteoglycans on endothelial cells.

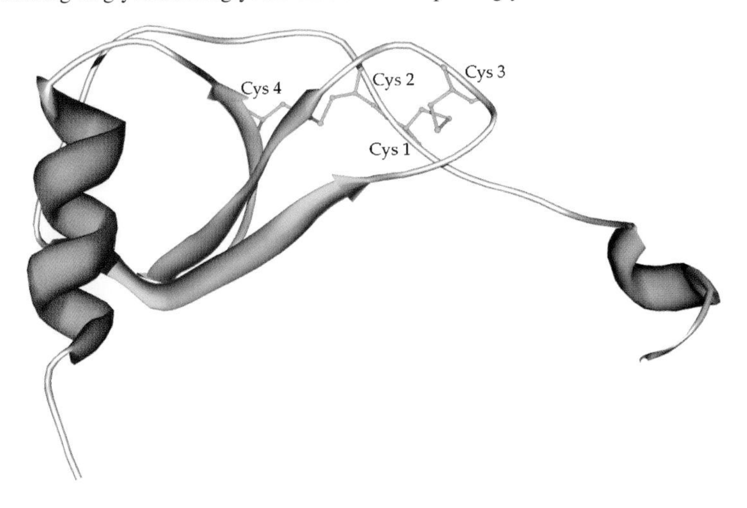


Small inducible cytokine A2

## *Small inducible cytokine A5 (RANTES)/RANTES (3–68)*

### Fact Sheet

| | | | |
|---|---|---|---|
| ***Classification*** | Intercrine beta (chemokine CC) | ***Abbreviations*** | CCL5 |
| ***Structures/motifs*** | | ***DB entries*** | CCL5_HUMAN/P13501 |
| ***MW/length*** | CCL5: 7851 Da/68 aa<br>RANTES (3–68): 7667 Da/66 aa | ***PDB entries*** | 1RTN |
| ***Concentration*** | | ***Half-life*** | |
| ***PTMs*** | 2 O-CHO | | |
| ***References*** | Schall *et al.*, 1988, *J. Immunol.*, **141**, 1018–1025.<br>Skelton *et al.*, 1995, *Biochemistry*, **34**, 5329–5342. | | |

### Description

**Small inducible cytokine A5** (CCL5) is primarily produced by T cells and blood platelets and acts as a chemoattractant for various cell types.

### Structure

CCL5 is a single-chain chemokine with two intrachain disulfide bridges (Cys 10–Cys 34, Cys 11–Cys 50). There is an N-terminal processed form RANTES (3–68) most likely produced via proteolytic cleavage by dipeptidyl peptidase 4. The monomeric structure is characterised by two short N-terminal β-strands linked to a three-stranded antiparallel β-sheet packed to a C-terminal α-helix (**1RTN**).

### Biological Functions

CCL5 is a chemoattractant for blood monocytes, memory helper T cells and eosinophils and causes the release of histamines from basophils and activated eosinophils.

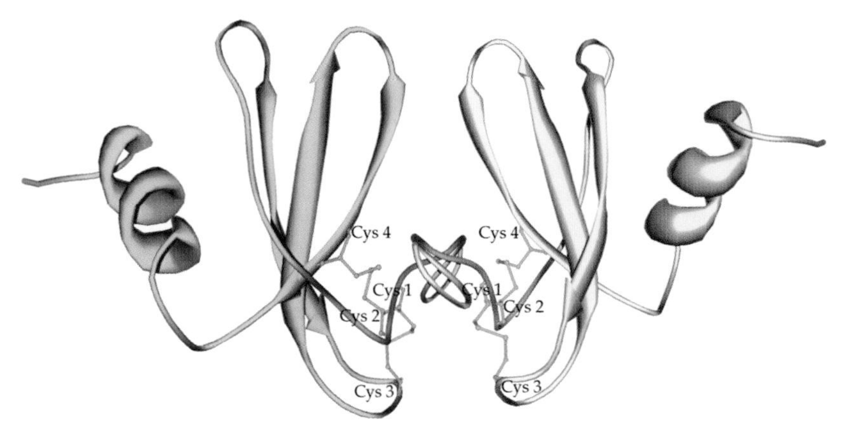


RANTES

## *Transforming growth factor α*

### Fact Sheet

| | | | |
|---|---|---|---|
| ***Classification*** | | ***Abbreviations*** | TGF-α |
| ***Structures/motifs*** | 1 EGF-like | ***DB entries*** | TGFA_HUMAN/P01135 |
| ***MW/length*** | 5552 Da/50 aa | ***PDB entries*** | 2TGF; 1MOX |
| ***Concentration*** | | ***Half-life*** | |
| ***PTMs*** | | | |
| ***References*** | Derynck *et al.*, 1984, *Cell*, **38**, 287–297.<br>Garrett *et al.*, 2002, *Cell*, **110**, 763–773.<br>Harvey *et al.*, 1991, *Eur. J. Biochem.*, **198**, 555–562. | | |

### Description

**Transforming growth factor α** (TGF-α) is a mitogenic polypeptide.

### Structure

TGF-α is a single-chain polypeptide containing one EGF-like domain. TGF-α is generated from a precursor type I membrane protein by removal of N-terminal (16 aa) and C-terminal (71aa) propeptides. The 3D structure of TGF-α is characterised by a central antiparallel β-sheet and a short C-terminal antiparallel β-sheet. TGF-α binds the extracellular domain of the EGF receptor in a 1:1 complex (**1MOX**).

### Biological Functions

TGF-α is a mitogen that is able to bind to the EGF receptor and to act synergistically with TGF-β.

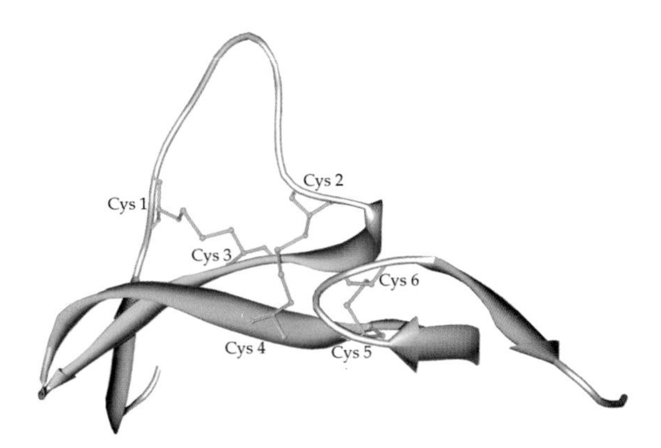


Transforming growth factor α

# *Transforming growth factors β*

## Fact Sheet

| | | | |
|---|---|---|---|
| ***Classification*** | TGF-beta | ***Abbreviations*** | TGF-β1; TGF-β2; TGF-β3 |
| ***Structures/motifs*** | | ***DB entries*** | TGFB1_HUMAN/P01137<br>TGFB2_HUMAN/P61812<br>TGFB3_HUMAN/P10600 |
| ***MW/length*** | TGFB1: 12 795 Da/112 aa<br>TGFB2: 12 720 Da/112 aa<br>TGFB3: 12 723 Da/112 aa | ***PDB entries*** | 1KLD |
| ***Concentration*** | | ***Half-life*** | 30–90 min |
| ***PTMs*** | | | |
| ***References*** | de Martin *et al.*, 1987, *EMBO J.*, **6**, 3673–3677.<br>Derynck *et al.*, 1985, *Nature*, **316**, 701–705.<br>Hinck *et al.*, 1996, *Biochemistry*, **35**, 8517–8534.<br>ten Dijke *et al.*, 1988, *Proc. Natl. Acad. Sci. USA*, **85**, 4715–4719. | | |

## Description

**Transforming growth factors β** (TGF-βs) are synthesised by many cells like T cells and monocytes. TGF-βs form a family of very similar growth factors involved in various biological functions like proliferation and differentiation in many cell types.

## Structure

TGF-βs are single-chain growth factors present as disulfide-linked homodimers in their active form and characterised by one interchain (Cys 77–Cys 77) and four intrachain disulfide bridges (Cys 7–Cys 16, Cys 15–Cys 78, Cys 44–Cys 109, Cys 48–Cys 111), all located at the same position in all TGF-βs. TGF-βs belong to the disulfide-knot proteins characterised by three α-helices and four distorted β-sheets. TGF-βs are derived from approx. 400 residues long precursors by removing large N-terminal propeptides at basic proteolytic cleavage sites (**R–[KH]–[RK]–R**). The 3D structure of the monomer is characterised by two antiparallel β-strands and a separate long α-helix containing four intrachain disulfide bridges (in pink) and the homodimer is linked by a single interchain disulfide bridge (in green) (**1KLD**).

## Biological Functions

TGF-βs are multifunctional cytokines: TGF-β1 regulates many other peptide growth factors and plays an important role in bone remodelling. TGF-β2 suppresses the growth of interleukin-2-dependent T cells. TGF-β3 is involved in embryogenesis and cell proliferation.

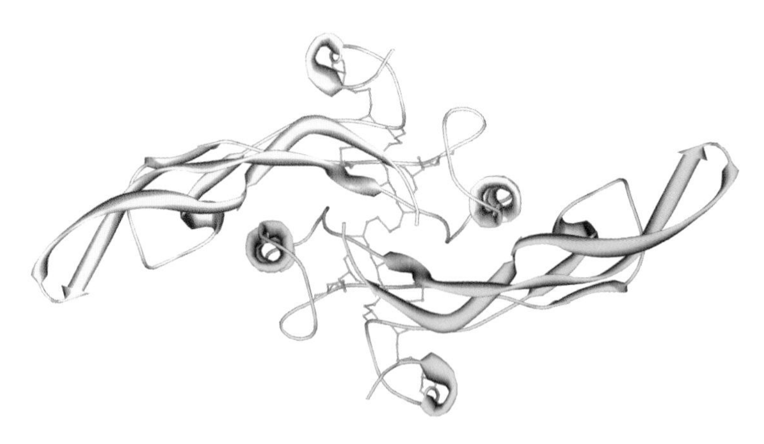

Transforming growth factor β

# *Tumour necrosis factor (Cachectin)*

## Fact Sheet

| | | | |
|---|---|---|---|
| ***Classification*** | Tumour necrosis factor | ***Abbreviations*** | TNF; TNF-α |
| ***Structures/motifs*** | | ***DB entries*** | TNFA_HUMAN/P01375 |
| ***MW/length*** | Membrane form: 25 644 Da/233 aa<br>Soluble form: 17 353/157 aa | ***PDB entries*** | 1TNF |
| ***Concentration*** | approx. 10 ng/l | ***Half-life*** | approx. 15 mins |
| ***PTMs*** | | | |
| ***References*** | Eck *et al.*, 1989, *J. Biol. Chem.*, **264**, 17595–17605.<br>Pennica *et al.*, 1984, *Nature*, **312**, 724–729. | | |

## Description

**Tumour necrosis factor** (TNF) is secreted by macrophages and is a potent pyrogen causing fever. It shows similarity with lymphotoxin (TNF-β).

## Structure

There are two forms of TNF: (1) a membrane form (233 aa) with a potential 21 residue transmembrane segment in the N-terminal region and (2) a soluble form (157 aa) that is derived from the membrane form by proteolytic processing of a 76 residue N-terminal peptide. TNF exhibits an elongated antiparallel β-pleated sheet sandwich structure with a 'jelly-roll' topology containing a single intrachain disulfide bridge (Cys 69–Cys 101) and the three monomers form a bell-shaped homotrimer (**1TNF**).

## Biological Functions

TNF is a major mediator of many immune and inflammatory responses such as fever and shock. It can induce cell death of certain tumour cell lines.

Tumour necrosis factor α

## *Vascular endothelial growth factors*

### Fact Sheet

| | | | |
|---|---|---|---|
| ***Classification*** | PDGF/VEGF growth factor | ***Abbreviations*** | VEGF-A; VEGF-B; VEGF-C; VEGF-D |
| ***Structures/motifs*** | | ***DB entries*** | VEGFA_HUMAN/P15962<br>VEGFB_HUMAN/P49765<br>VEGFC_HUMAN/P49767<br>VEGFD_HUMAN/O43915 |
| ***MW/length*** | VEGF-A: 23 895 Da/206 aa<br>VEGF-B: 19 360 Da/186 aa<br>VEGF-C: 13 104 Da/116 aa<br>VEGF-D: 13 109 Da/117 aa | ***PDB entries*** | 2VPF |
| ***Concentration*** | | ***Half-life*** | |
| ***PTMs*** | VEGF-A: 1 N-CHO; VEGF-B: O-CHO<br>VEGF-C: 3 N-CHO (potential)<br>VEGF-D: 3 N-CHO (potential) | | |

***References***


Joukov *et al.*, 1996, *EMBO J.*, **15**, 290–298.
Joukov *et al.*, 1997, *EMBO J.*, **16**, 3898–3911.
Leung *et al.*, 1989, *Science*, **246**, 1306–1309.
Muller *et al.*, 1997, *Structure*, **5**, 1325–1338.
Olofsson *et al.*, 1996, *Proc. Natl. Acad. Sci. USA*, **93**, 2576–2581.
Stacker *et al.*, 1999, *J. Biol. Chem.*, **274**, 32127–32136.
Yamada *et al.*, 1997, *Genomics*, **42**, 483–488.


### Description

**Vascular endothelial growth factors** (VEGFs) are a family of growth factors involved in angiogenesis, vasculogenesis and endothelial cell growth.

### Structure

VEGFs belong to a family of growth factors that form homodimers: VEGF-A and VEGF-B form an antiparallel homodimer linked by two interchain disulfide bridges (Cys 51–Cys 60), each chain containing three intrachain disulfide bonds. VEGF-C forms mainly disulfide-linked homodimers, whereas VEGF-D homodimers are predominantly non-covalently-linked. Mature VEGF-C and VEGF-D are generated from the precursor forms by proteolytic processing, resulting in the removal of N-terminal and C-terminal propeptides. The 3D structure of the disulfide-linked homodimer of VEGF-A121 isoform is similar to PDGF (**2VPF**).

### Biological Functions

VEGFs induce endothelial cell proliferation, promote cell migration, inhibit apoptosis and induce the permeabilisation of blood vessels.

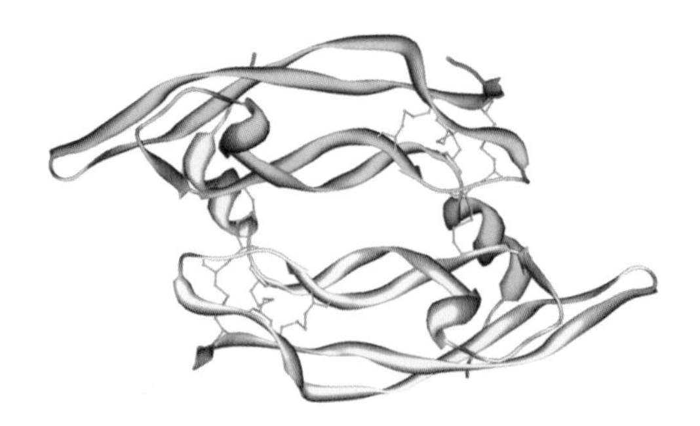

Vascular endothelial growth factors

# 15

# Transport and Storage

## 15.1 INTRODUCTION

The cardiovascular system, the closed circulatory system of the body, represents an excellent tool for the transport of any type of compound from the place of synthesis (organs, glands, tissues) to the place of use or to the place of storage. If the stored compounds are required again they are transported back to the places of need. In addition, metabolites and degradation products are also transported via the vascular system for recycling and reuse or for excretion to the corresponding organs. Almost any type of compound can be transported through the vascular system, large or small, polar or unpolar, charged or uncharged, showing very variable structures and exhibiting different functions:

1. High molecular compounds: proteins, polysaccharides and polynucleotides.
2. Intermediate molecular compounds: peptides, oligosaccharides and oligonucleotides.
3. Low molecular compounds: sugars, amino acids, nucleotides, fatty acids, lipids, vitamins, hormones and others.
4. Salts and ions: sodium, calcium, iron, bicarbonate, chloride and many others.
5. Gases: oxygen, carbon dioxide and monoxide, nitric oxide.
6. Water.
7. Others: trace elements, synthetic drugs.

Many compounds require a specific vehicle for their transport, very often specific proteins. A selection of plasma proteins involved in the transport of various types of compounds via the vascular system are discussed in this chapter. The various plasma proteins involved in transport and storage are given in the following sections.

## 15.2 THE SERUM ALBUMIN FAMILY

*Serum albumin*, the most abundant and one of the most important proteins in plasma, belongs to the serum albumin family, together with afamin, α-fetoprotein and vitamin D-binding protein (see also Chapter 5). They are evolutionary, structurally and functionally related, are major transport proteins in plasma and bind and transport various types of compounds through the vascular system. The members of the serum albumin family are structurally characterised by three homologous domains of approximately 190 residues length and each domain contains five to six disulfide bridges arranged in the following pattern:

**Cys 1–Cys 3, Cys 2–Cys 4, Cys 5–Cys 7, Cys 6–Cys 8, Cys 9–Cys 11, Cys 10–Cys 12**

**Serum albumin** (HSA) (see Data Sheet) or simply albumin, the major circulating plasma protein, is synthesised in the liver and is present in all body fluids. HSA is a single-chain protein containing three albumin domains characterised by six disulfide bridges, but is devoid of any posttranslational modifications (except in some variants). HSA has a good binding capacity for different types of compounds: water, ions ($Ca^{2+}$, $Na^{+}$, $K^{+}$) and heavy metal ions ($Cu^{2+}$, $Zn^{2+}$), fatty acids, hormones, bilirubin, nitric oxide, synthetic drugs and many other compounds. Because of its high concentration in plasma (35–50 g/l) and its excellent binding capacity, HSA is the major transport protein in plasma and regulates and maintains the

*Human Blood Plasma Proteins: Structure and Function* Johann Schaller, Simon Gerber, Urs Kämpfer, Sofia Lejon and Christian Trachsel

colloidal osmotic pressure in blood, which is essential for the distribution of body fluids between intravascular compartments and tissues, which in turn is important for the regulation of body temperature. In addition, HSA is by far the most important protein contributing to the acid/base balance in plasma. The concentration of HSA is a marker for good nutrition and longevity and a decline is a negative acute-phase marker of illness, trauma and infection.

The extremely high concentration of HSA interferes with the isolation protocols of middle- or low-abundance proteins and is often carried on as a contaminant and thus might nurse false hopes in the detection of new proteins with new properties.

The 3D structure of entire serum albumin (585 aa: Asp 1–Leu 585) in complex with fatty acids (monounsaturated oleic acid (C18:1) and polyunsaturated arachidonic acid (C20:4)) was determined by X-ray diffraction (PDB: 1GNI) and is shown in Figure 15.1(a). HSA exhibits a heart-shaped structure characterised by a high content of $\alpha$-helices (67 %) and no $\beta$-strands. The unsaturated fatty acids (in pink) occupy the seven binding sites 1 to 7 distributed across the protein, which are also bound by saturated fatty acids. Site 1 is in subdomain IB, site 2 is at the interface between subdomains IA and IIA, sites 3 and 4 are both within subdomain IIIA (known as drug site II), site 5 is within subdomain IIIB, site 6 is a shallow trench at the interface between subdomains IIA and IIB, and site 7 is entirely contained within the drug-binding pocket of domain IIA (known as drug site I). As an example, the surface representation of the binding of oleic acid within the binding pocket 4 located in subdomain IIIA is shown in Figure 15.1(b).

Defects in the albumin gene are the cause of various diseases:

1. Familial dysalbuminemic hyperthyroxinemia is due to an increased affinity of albumin for thyroxin ($T_4$) and is the most common cause of inherited euthyroid hyperthyroxinemia in the Caucasian population (for details see MIM: 103600).
2. Hyperzincemia is due to a variant structure of albumin, leading to an increased binding of zinc to albumin and resulting in an asymptotic increase of zinc in blood (for details see MIM: 194470).
3. Hypoalbuminemia (abnormally low level of albumin) is a specific form of hypoproteinemia (abnormally low level of protein) and may lead to an increased morbidity or mortality. A low level of albumin can be an indicator of chronic malnutrition.

**$\alpha$-Fetoprotein** (AFP) (see Data Sheet) or $\alpha$-fetoglobulin is synthesised in the fetal liver and yolk sack and is present after birth only in trace amounts (<50 μg/l). AFP is a single-chain glycoprotein containing three albumin domains with the characteristic disulfide bridge pattern and with His 4 involved in the binding of copper and nickel. AFP has a similar function in the fetus as serum albumin in adults and reaches the highest level during weeks 12–16 in pregnancy. AFP binds copper and nickel and fatty acids as well as serum albumin does. In addition, it binds estradiol to prevent the fetal brain from damage.

**Afamin** (see Data Sheet) or $\alpha$-albumin is involved in the transport of yet unknown ligands. Afamin is a single-chain plasma glycoprotein containing three albumin domains with the characteristic disulfide bridge pattern.

**Vitamin D-binding protein** (DBP) (see Data Sheet) is predominantly synthesised in the liver and is found in many body fluids and cell types. DBP is a single-chain plasma glycoprotein containing three albumin domains (the albumin 3 domain is only half size) with the characteristic disulfide bridge pattern and there are over 80 variants. DBP is a multifunctional protein and binds and transports vitamin D and its metabolites and fatty acids in plasma and prevents the polymerisation of actin by binding to its monomer.

The 3D structure of entire vitamin D-binding protein (458 aa: Leu 1–Leu 458) in complex with $\alpha$-actin-1 from skeletal muscle (375 aa: Asp 1–Phe 375, P68133) was determined by X-ray diffraction (PDB: 1MA9) and is shown in Figure 15.2. The structure of DBP with three albumin domains (in yellow, green and blue; this domain is only half size) is highly helical and binds to the actin subdomains 1 and 3 (in pink) and occludes the cleft at the interface between the subdomains.

## 15.3 THE GLOBIN FAMILY

Globins are heme-containing proteins that are involved in the binding and transport of oxygen. Globins exist in almost all organisms and can be divided into the following groups (see also Chapter 5):

1. ***Hemoglobins*** (Hb) from vertebrates are responsible for the transport of oxygen from the lungs to the tissues and are present as a tetramer of two nonidentical chains. Most vertebrates express specific embryonic or fetal forms with an increased affinity for oxygen.
2. ***Myoglobins*** (Mb) from vertebrates are responsible for the storage of oxygen in muscles and are present as monomer.
3. ***Invertebrate globins***. A wide variety of globins are found in invertebrates.
4. ***Leghemoglobins*** (Lg) originate from the root nodules of leguminous plants providing oxygen for bacteroids.
5. ***Flavohemoproteines*** are present in bacteria and fungi. They consist of two distinct domains, an N-terminal globin domain and a C-terminal FAD-containing reductase domain.

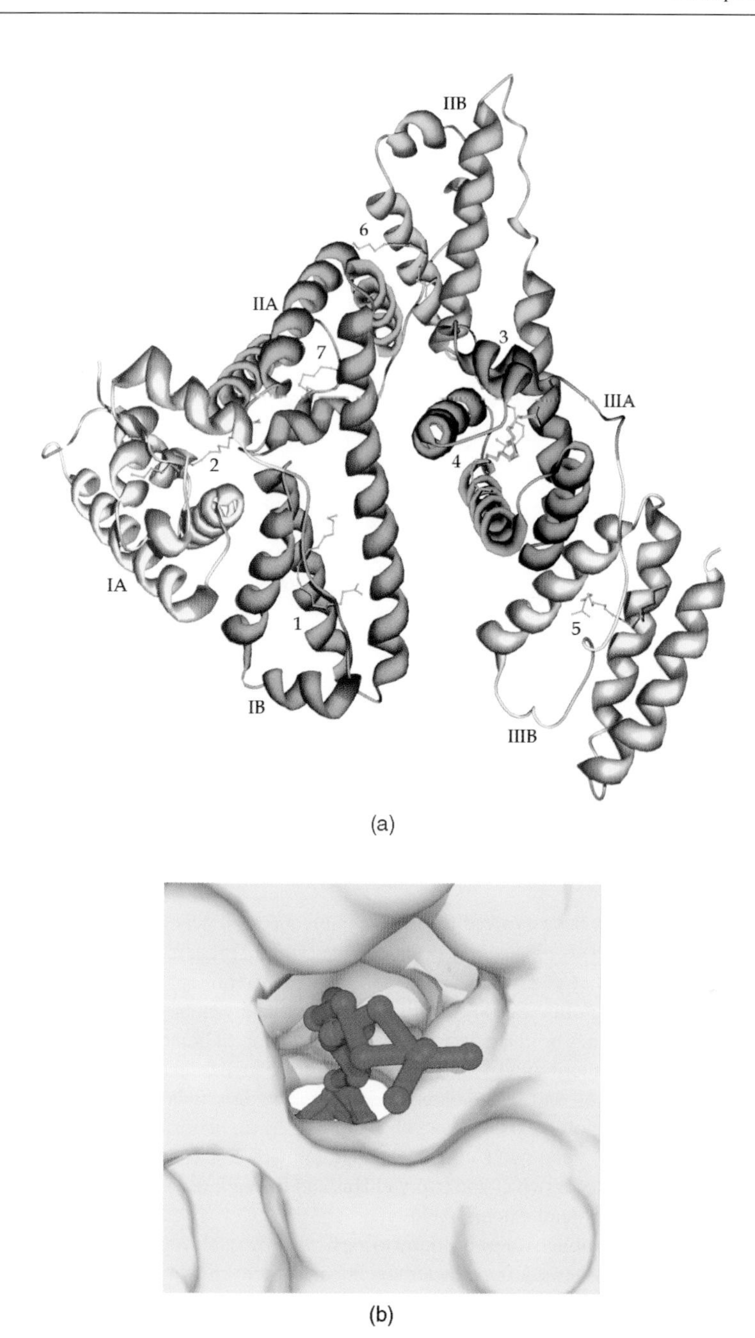


**Figure 15.1 3D structure of serum albumin in complex with unsaturated fatty acids (PDB: 1GNI)**
(a) Serum albumin exhibits a heart-shaped structure characterised by a high α-helix content. The unsaturated fatty acids (in pink) occupy the seven binding sites 1 – 7 distributed across the six subdomains IA + IB, IIA + IIB and IIIA + IIIB.
(b) Surface representation of the binding of oleic acid within binding pocket 4 located in subdomain IIIA.

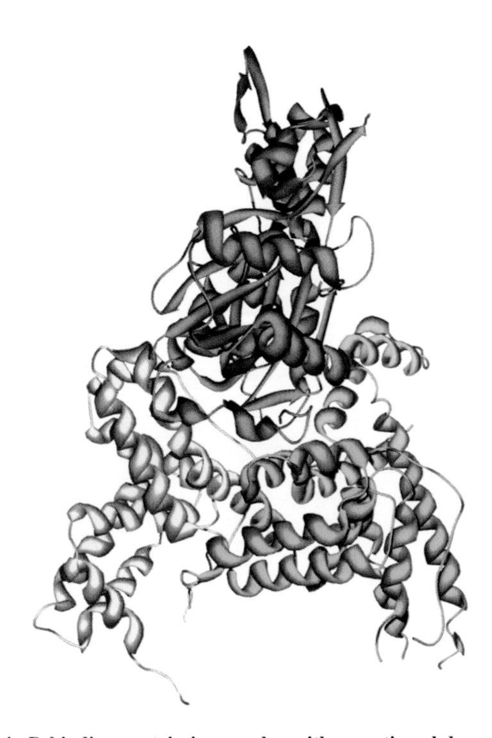

**Figure 15.2 3D structure of vitamin D-binding protein in complex with an actin subdomain (PDB: 1MA9)**
The three albumin domains (in yellow, green and blue) in the vitamin D-binding protein are highly helical and the actin subdomains 1 and 3 (in pink) occlude the cleft between the subdomains.

It is very likely that all these globins have evolved from a common ancestral gene.

**Hemoglobin** (Hb) (see Data Sheet) is exclusively produced in erythroid precursor cells and represents approximately 95 % of the total protein content of mature erythrocytes. Hb is a tetramer of noncovalently linked chains, two $\alpha$- and two $\beta$-chains. Besides the common hemoglobin several other types exist, especially embryonic and fetal types that are especially adapted to a higher affinity for oxygen. The characteristics of the major types of hemoglobin are summarised in Table 15.1.

The most common type of adult hemoglobin HbA ($\alpha_2\beta_2$) is composed of two $\alpha$- and two $\beta$-chains. A variant, hemoglobin HbA2 ($\alpha_2\delta_2$), is composed of two $\alpha$- and two $\delta$-chains and represents less than 3.5 % of adult hemoglobin. Several embryonic and fetal forms of hemoglobin exist that are characterised by an increased affinities for oxygen: hemoglobin Gower-1 ($\zeta_2\varepsilon_2$) is composed of two $\zeta$- and two $\varepsilon$-chains, hemoglobin Gower-2 ($\alpha_2\varepsilon_2$) contains two $\alpha$- and two $\varepsilon$-chains, hemoglobin Portland-1 ($\zeta_2\gamma_2$) consists of two $\zeta$- and two $\gamma$-chains and hemoglobin HbF ($\alpha_2\gamma_2$) is composed of two $\alpha$- and two $\gamma$-chains and is also present to a small extent in adults.

The four chains in hemoglobin are noncovalently bound to each other by salt bridges, hydrogen bonds and hydrophobic interactions, and each chain contains a heme group, a heterocyclic ring system known as porphyrin, each containing one iron atom. Each iron ion is coordinated to His 58 (distal) and His 87 (proximal) in the $\alpha$-chains and to His 63 (distal) and His 92 (proximal) in the $\beta$-chains.

The affinity of hemoglobin for oxygen is dependent on the pH, the $pCO_2$ and organic phosphates such as 2,3-bisphosphoglycerate. The pH dependency of oxygen binding is known as the Bohr effect. The optimal uptake of oxygen in the lungs is due to $CO_2$ expulsion, resulting in an increased pH and a subsequent optimal release of oxygen in the capillary circulation due to a $CO_2$ influx from respiring tissues, leading to a decreased pH. Hemoglobin is also an important blood buffer protein by transporting large amounts of $CO_2$ from the tissues to the lungs, with a moderate decrease of pH. The presence of two pairs of unlike chains leads to a cooperative effect known as the heme–heme interaction, permitting the unloading of increased amounts of oxygen over a narrow range of oxygen tension and thus facilitating the oxygen release to the respiring tissue. Carbon monoxide, nitric oxide and other heme ligands are the main inhibitors of hemoglobin.

**Table 15.1** Types of Hemoglobins: adult, fetal and embryonic forms.

| Type of hemoglobin | Chains (#aa) | Databank entry | Origin | Comments |
|---|---|---|---|---|
| HbA: $\alpha_2\beta_2$ | $\alpha$: 141 | P69905 | Adult | Common |
| | $\beta$: 146 | P68871 | | |
| HbF: $\alpha_2\gamma_2$ | $\alpha$: 141 | P69905 | Fetal | No Bohr effect |
| | $\gamma$1: 146 | P69891 | Adult | No heme–heme cooperativity |
| | $\gamma$2: 146 | P69892 | | |
| HbA2: $\alpha_2\delta_2$ | $\alpha$: 141 | P69905 | Adult | <3.5 % |
| | $\delta$: 146 | P02042 | | |
| HbGower-2: $\alpha_2\varepsilon_2$ | $\alpha$: 141 | P69905 | Embryonic | Early embryonic |
| | $\varepsilon$: 146 | P02100 | | |
| HbGower-1: $\zeta_2\varepsilon_2$ | $\zeta$. 141 | P02008 | Embryonic | Early embryonic |
| | $\varepsilon$: 146 | P02100 | | |
| HbPortland-1: $\zeta_2\gamma_2$ | $\xi$: 141 | P02008 | Embryonic | |
| | $\gamma$: 146 | P69891/P69892 | | |
| HbQ1: $\theta$ | $\theta$: 141 | P9105 | Fetal | Early embryonic |

The 3D structure of a mutant hemoglobin ($\alpha$-chain (141 aa): Val 1–Arg 141; $\beta$-chain (146 aa): Val1Met–His 146) was determined by X-ray diffraction (PDB: 1A00) and is shown in Figure 15.3. The two $\alpha$-chains (in yellow and red) and the two $\beta$-chains (in green and blue) each contain a heme group (in pink), where the iron ions are coordinated to His 58 (distal) and His 87 (proximal) in the $\alpha$-chain and to His 63 (distal) and His 92 (proximal) in the $\beta$-chain. Both chains are characterised by a high content of $\alpha$-helices but no $\beta$-strands.

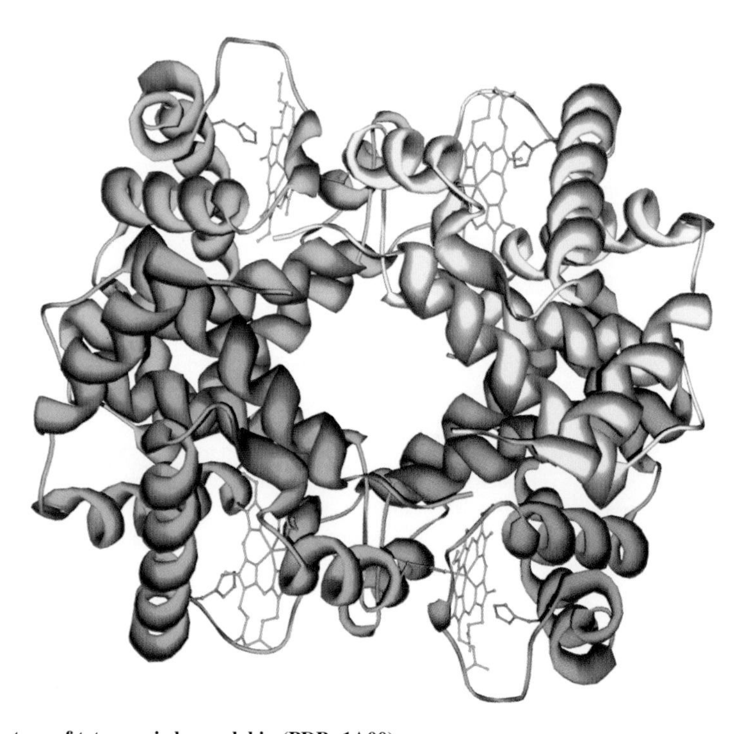

**Figure 15.3 3D structure of tetrameric hemoglobin (PDB: 1A00)**
The two $\alpha$-chains (in yellow and red) and the two $\beta$-chains (in green and blue) contain each a heme group (in pink) and the iron ions are coordinated to His residues.

The 3D structure of embryonic hemoglobin HbGower-2 ($\alpha_2\varepsilon_2$) ($\alpha$-chain (141 aa): Val 1–Arg 141; $\varepsilon$-chain (146 aa): Val 1–His 146) was determined by X-ray diffraction (PDB: 1A9W). The sequence identity between adult HbA and HbGower-2 amounts to 75 %. The structure of the embryonic $\varepsilon$-chain is very similar to that of the homologous adult $\beta$-chain and the fetal $\gamma$-chain. The embryonic hemoglobins have a very similar oxygen affinity and binding rates relative to the adult forms. However, the response to allosteric effectors such as chloride ions is reduced in prenatal hemoglobins, resulting in a higher oxygen affinity of embryonic hemoglobins.

Defects in the HBA1 and HBA2 genes (missing $\alpha$-chain genes in the gene cluster) are the cause of $\alpha$-thalassemia with two main forms:

1. In $\alpha(0)$-thalassemia no protein is synthesised.
2. In $\alpha(+)$-thalassemia some but less than normal amounts of protein are produced.

The phenotype is due to an unstable $\alpha$-chain which is rapidly degraded by proteolysis prior to the formation of $\alpha/\beta$ dimers. The disease is primarily found in the Mediterranean basin, Central Africa, India and the Far East (for more details see MIM: 141800 and 141850). Defects in the HBB gene are the cause of $\beta$-thalassemia due to an imbalance of the globin chain production in adult HbA. There are two main forms of $\beta$-thalassemia:

1. $\beta(0)$-thalassemia is characterised by the absence of the $\beta$-chain.
2. $\beta(+)$-thalassemia is characterised by a reduced amount of $\beta$-chain.

The severe forms of $\beta$-thalassemia are characterised by the accumulation of excess $\alpha$-chains in the erythroid precursor cells, which in turn leads to a large increase in erythroid apoptosis, causing ineffective erythropoiesis and severe microcytic hypochromic anemia (for details see MIM: 141900).

Defects in the hemoglobin $\beta$-chain gene are the cause of sickle cell anemia characterised by abnormally shaped erythrocytes (sickle shape) resulting in chronic anemia, periodic pain, serious infections and damage of vital organs. The abnormal sickle cell hemoglobin HbS carries a point mutation (Glu 6 $\rightarrow$ Val 6) resulting in a more hydrophobic surface, thus enabling a hydrophobic interaction of Val 6 with Phe 85 and Leu 88 in the hydrophobic cleft of a neighbouring molecule and leading to the aggregation of hemoglobin. The flow of the stiffer, sickle-shaped erythrocytes is restricted and leads to microvascular occlusions, resulting in the cut-off of the blood supply to tissues. The disease is widely spread among the black population (incidence: approximately 4 in 1000) and in certain African areas sickle cell anemia is linked to malaria (for details see MIM: 603903).

Some $\gamma$-chain variants are the cause of severe jaundice and cyanosis in premature and newborn babies. Elevated levels of hemoglobin HbA2 are characteristic for $\beta$-thalassemia.

**Myoglobin** (Mb) (see Data Sheet) is a single-chain globin containing a single Cys residue (aa 110) of unknown function and is present as monomer. The iron ion in heme is coordinated to His 64 (distal) and His 93 (proximal). Myoglobin is the oxygen storage protein in muscles and facilitates the mobility of oxygen in tissues. Myoglobin exhibits a higher affinity for oxygen than hemoglobin, resulting in a decreased free oxygen concentration in tissues and leading to a diffusion of oxygen from the blood capillaries to the surrounding tissues. Carbon monoxide and cyanide are the main inhibitors of myoglobin.

The 3D structure of a recombinant myoglobin mutant (153 aa: Gly 1–Gly 153) was determined by X-ray diffraction (PDB: 2MM1) and is shown in Figure 15.4(a). The $\alpha$-helical myoglobin contains eight $\alpha$-helices and a single heme group (in pink), which is coordinated to two His residues (in green). The surface representation in Figure 15.4(b) shows the binding of the heme group in the binding pocket. Human myoglobin exhibits a very similar structure to myoglobins from other species.

Muscle damage and destruction are the cause of myoglobinemia, characterised by an increased release of myoglobin into the bloodstream and as a consequence the presence of myoglobin in urine (called myoglobinuria).

### 15.3.1 Haptoglobin

Although haptoglobin is not a member of the globin family it is closely related to hemoglobin functionally. Haptoglobin forms a complex with free hemoglobin released from the erythrocytes, thus preventing the loss of iron through the kidneys and protecting the kidneys from the oxidative activity of hemoglobin. In addition, haptoglobin renders hemoglobin accessible for proteolytic degradation.

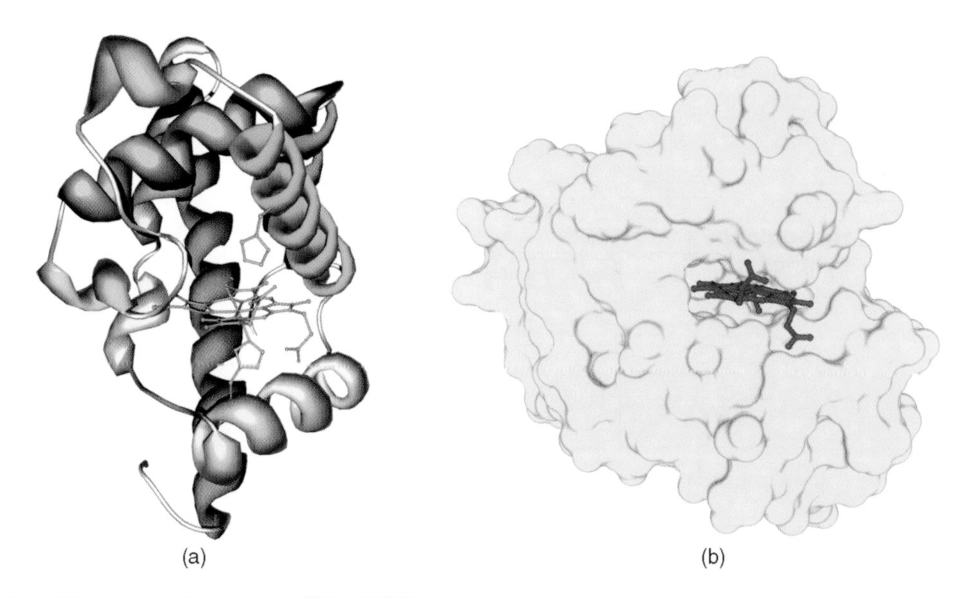


**Figure 15.4 3D structure of myoglobin (PDB: 2MM1)**
(a) Monomeric myoglobin contains eight β-helices and a single heme group (in pink) is coordinated to two His residues (in green).
(b) Surface representation of the heme binding pocket in myoglobin.

**Haptoglobin** (Hp) (see Data Sheet) is synthesised in the liver and is present in plasma as a tetramer of two α- and two β-chains. The two α-chains are linked by two interchain disulfide bridges (Cys 15–Cys 15, Cys 74–Cys 74) and the α- and the β-chains are linked by a single interchain disulfide bridge (α-chain: Cys 131, β-chain: Cys 105). The α-chain (142 aa: Val 1–Gln 142) contains two sushi/CCP/SCR domains which, in contrast to standard sushi domains with two disulfide bridges (for details see Chapter 4), have only a single disulfide bridge (sushi 1: Cys 34–Cys 68, sushi 2: Cys 93–Cys 137). The β-chain (245 aa: Ile 1–Asn 245) belongs to the peptidase S1 family but exhibits no proteolytic activity because in the catalytic triad His 41 is replaced by Lys and Ser 194 by Ala.

Two main allelic forms of haptoglobin exist, Hp1 ($\alpha_1$-chain: 83 aa) and Hp2 ($\alpha_2$-chain: 142 aa), where the latter arose from a partial gene duplication of the former.

## 15.4 IRON TRANSPORT AND STORAGE

Iron is involved in many important biological processes and therefore the transport, storage and metabolism of iron ions ($Fe^{2+}$/$Fe^{3+}$) are carefully controlled and regulated. The transport and storage of iron ions is carried out by specific iron transport and iron storage proteins. The transport of iron ions in the bloodstream and the storage in the cells is depicted schematically in Figure 15.5. Ferric iron ($Fe^{3+}$) from food intake is reduced in the intestine to the more soluble ferrous iron ($Fe^{2+}$) and is bound to highly glycosylated intestinal mucins (5159 aa: Q02817). Mucosal uptake of iron is facilitated by integrins and is transported in the cell by mobilferrin and is stored in ferritin in the ferric state ($Fe^{3+}$). The ferritin-bound iron is reduced by the ferritin reductase and released as ferrous iron ($Fe^{2+}$), which is used for hemoglobin synthesis. $Fe^{2+}$ is again oxidised by ceruloplasmin.

The main iron-binding and iron transport proteins are the transferrins. Transferrins can transport iron ions only in the ferric ($Fe^{3+}$) state. Ceruloplasmin catalyses the oxidation of ferrous iron ($Fe^{2+}$) to ferric iron ($Fe^{3+}$). The main iron storage protein in cells is ferritin.

### 15.4.1 Iron oxidation: ceruloplasmin

Iron ions can only be transported by transferrins in the ferric ($Fe^{3+}$) state. The enzyme ceruloplasmin is able to oxidise a great variety of substances but its main catalytic function is the oxidation of ferrous ($Fe^{2+}$) iron to ferric ($Fe^{3+}$) iron:

$$4Fe^{2+} + 4H^{+} + O_2 \Longleftrightarrow 4Fe^{3+} + 2H_2O$$

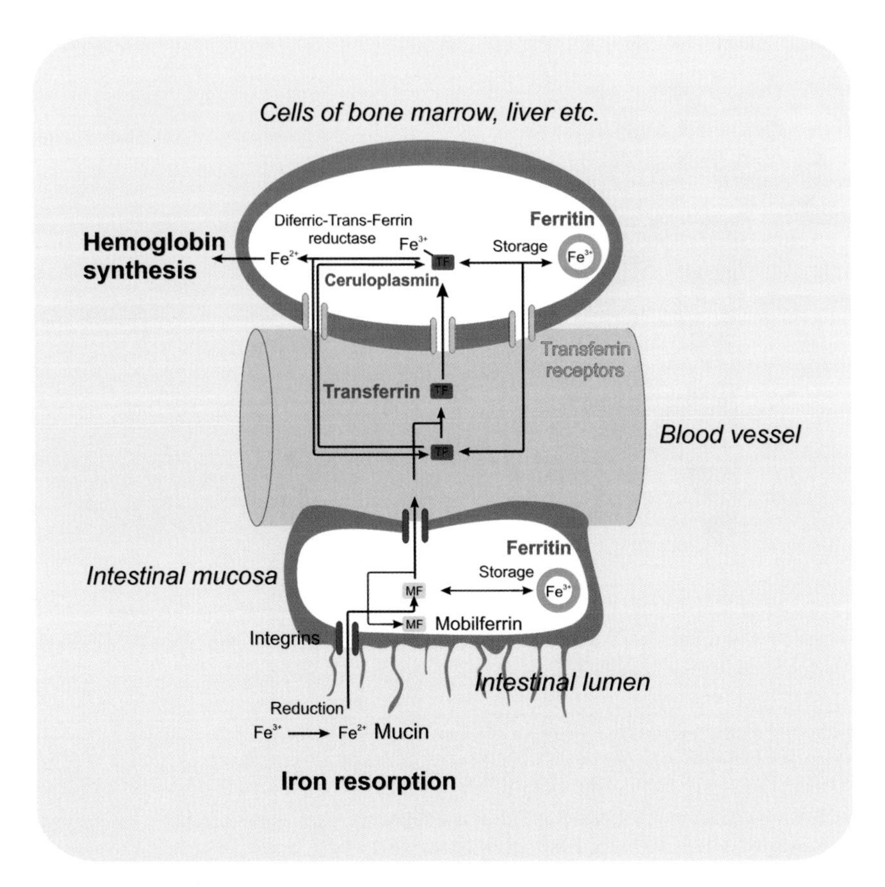


**Figure 15.5 Transport and storage of iron**
Ferric iron ($Fe^{3+}$) from food intake is reduced in the intestine to the more soluble ferrous iron ($Fe^{2+}$). Iron is stored in ferritin in the ferric state ($Fe^{3+}$). The ferritin-bound iron is reduced by the ferritin reductase and released as ferrous iron ($Fe^{2+}$), which is again oxidised by ceruloplasmin.

Besides ferroxidase activity, ceruloplasmin also exhibits amine oxidase and superoxide dismutase activity and is involved in copper transport and homeostasis and participates in the acute phase reaction to stress. Ceruloplasmin belongs to the multicopper oxidase family (see also Chapter 5). Although ceruloplasmin is a copper-binding glycoprotein with three different copper centres the main carrier of copper in plasma is albumin.

**Ceruloplasmin** (CP, EC 1.16.3.1) (see Data Sheet) or ferroxidase is a single-chain glycoprotein (approximately 10 % CHO) primarily synthesised in the liver. It contains three F5/8 type A domains, of which each domain consists of two plastocyanin-like domains. Ceruloplasmin contains five intrachain disulfide bridges and binds six copper ions in three distinct copper-binding centres called type 1 (blue), type 2 (normal) and type 3 (coupled binuclear). The residues involved in copper binding in the three centres are tabulated in Table 15.2. The copper atoms are primarily coordinated to His residues, but three Cys, two Met and a Leu residue are also involved.

**Table 15.2** Copper binding in ceruloplasmin.

| Copper (#) | Type | Residues involved | Comments |
|---|---|---|---|
| 1 | 2 (normal) | His 101, 103, 978, 980 | |
| 2 | 3 (coupled binuclear) | His 103, 161, 1022 | |
| 3 | 3 (coupled binuclear) | His 163, 980, 1020 | |
| 4 | 1 (blue) | His 276, 324; Cys 319 | |
| 5 | 1 (blue) | His 637, 685; Cys 680; Met 690 | |
| 6 | 1 (blue) | His 975, 1026; Leu 974; Cys 1021; Met 1031 | Leu: Carbonyl carbon |

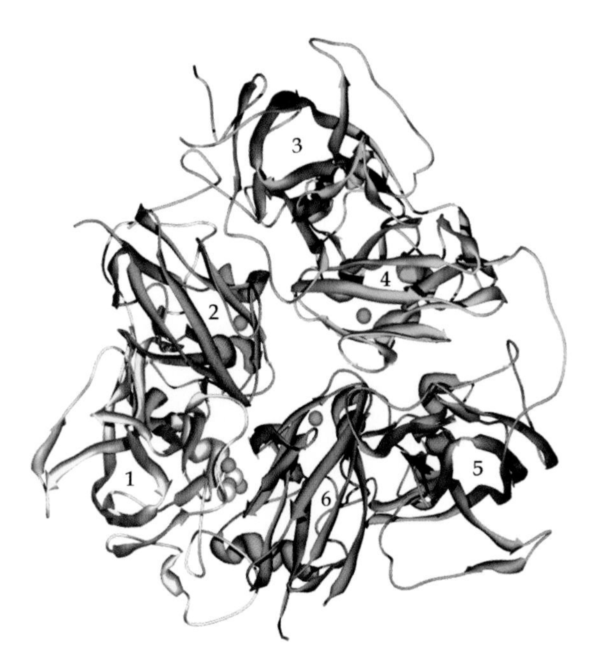


**Figure 15.6 3D structure of ceruloplasmin (PDB: 1KCW)**
Ceruloplasmin contains six plastocyanin-like domains (each domain is shown in a different colour) and the six copper atoms are indicated as pink balls and bicarbonate as grey balls.

The 3D structure of entire ceruloplasmin (1046 aa: Lys 1–Gly 1046) was determined by X-ray diffraction (PDB: 1KCW) and is shown in Figure 15.6. The structure comprises six plastocyanin-like domains arranged in a triangular array (each domain is shown in a different colour). Three of the six copper atoms (pink balls) with bicarbonate as counter ions (grey balls) form a trinuclear cluster at the interface of domains 1 and 6. The other three copper atoms form mononuclear sites in domains 2, 4 and 6 (type 1 coppers) (see also Chapter 5).

Defects in the ceruloplasmin gene are the cause of aceruloplasminemia, which is an autosomal recessive disorder of the iron metabolism characterised by iron accumulation in the brain and visceral organs, leading to the triad of retinal degeneration, diabetes mellitus and neurological disturbances (for details see MIM: 604290). In hereditary disorders such as Wilson's and Menkes disease the level of ceruloplasmin is decreased and copper cannot be incorporated into ceruloplasmin in the liver due to mutations in the intracellular copper-transporting ATPase 2.

### 15.4.2 Iron transport proteins: transferrins

Transferrins are a family of eukaryotic iron-binding glycoproteins controlling the level of free iron in biological fluids. There are several types of transferrins depending on their place of origin:

1. Blood serotransferrin or siderophilin.
2. Milk lactotransferrin or lactoferrin.
3. Egg white ovotransferrin or conalbumin (686 aa: P02789).
4. Membrane-associated melanotransferrin (690 aa: P08582).

Transferrins are approximately 700 residues in size and have evolved by duplication of a domain of approximately 340 aa. Each domain binds one iron atom and the iron atom is coordinated to four conserved residues: **Asp, two Tyr, His**. All Cys in both domains form intrachain disulfide bridges. Three consensus patterns are centered on the iron-binding residues (residues highlighted in red are iron ligands):

TRANSFERRIN _1 (PS00205): **Y**– x(0, 1) – [VAS] – V–[IVAC]–[IVA] –[IVA]–[RKH]–[RKS]–[GDENSA]

TRANSFERRIN _2 (PS 00206): [**YI**]–x – G – A–[FLI] – [KRHNQS]–**C** – L – x(3,4) – G – [DENQ] – V – [GAT] – [FYW]

TRANSFERRIN_3 (PS00207): [DENQ] – [ YF] – x –[LY] –L – **C** – x – [DN] – x (5, 8) – [LIV] – x (4, 5) – **C** – x (2) – A – x (4) – [**H**QR] – x – [LIVMFYW] – [LIVM]

**Serotransferrin** (TF) (see Data Sheet) or siderophilin is primarily synthesised in the liver and transports iron $Fe^{3+}$ in blood and in the interstitial fluid from the sites of absorption, storage or erythrocyte destruction to erythropoietic, proliferating and iron storage cells. TF is a single-chain glycoprotein (approximately 6 % CHO) containing two homologous domains, which probably have evolved by gene duplication. Domain 1 (336 aa: Val 1–Thr 336) contains eight and domain 2 (343 aa: Asp 337–Pro 679) 11 intrachain disulfide bridges and each domain contains one iron $Fe^{3+}$-binding site and one anion-binding site (carbonate/bicarbonate):

***Binding sites in domain 1***:

$Fe^{3+}$: **Asp 63, Tyr 95 and 188, His 249**

Bicarbonate: **Thr 120, Arg 124, Ala 126 and Gly 127** (both amide nitrogen)

***Binding sites in domain 2***:

$Fe^{3+}$: **Asp 392, Tyr 426 and 517, His 585**

Bicarbonate: **Thr 452, Arg 456, Ala 458 and Gly 459** (both amide nitrogen)

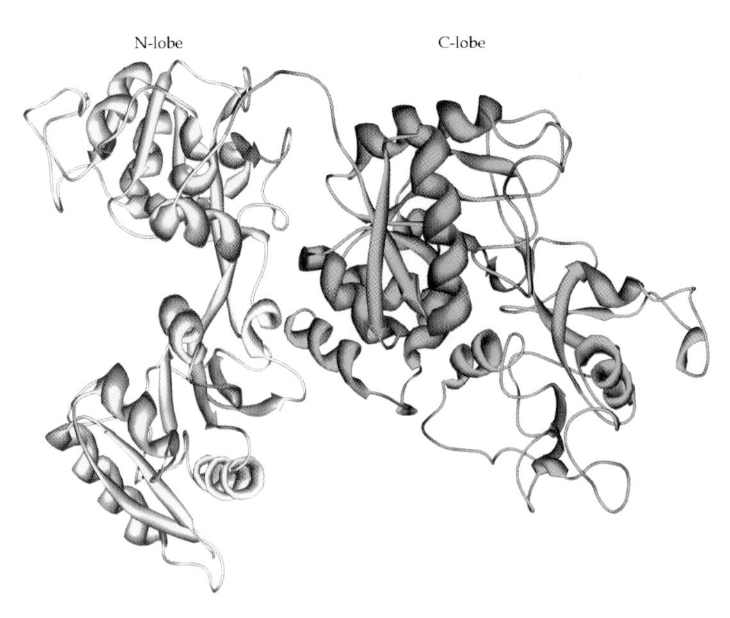


**Figure 15.7 3D structure of serotransferrin (PDB: 2HAV)**
Serotransferrin consists of two lobes, a N-terminal (in yellow) and a C-terminal (in green) connected by a linker region (in red) and the binding and release of iron results in a large conformational change, in which each lobe closes and opens with a rigid motion around a hinge region.

The 3D structure of glycosylated/nonglycosylated serotransferrin (676 aa: Lys 4–Pro 679) was determined by X-ray diffraction (PDB: 2HAV) and is shown in Figure 15.7. Serotransferrin consists of two lobes, an N-terminal (in yellow) and a C-terminal (in green) domain, connected by a linker region (in red). The binding and release of iron results in a large conformational change, in which each lobe closes and opens with a rigid motion around a hinge region. The structure suggests that differences in the hinge region of both lobes influence the iron-release/binding rate.

Defects in the serotransferrin gene are the cause of atransferrinemia, which is a rare autosomal recessive disorder characterised by iron overload and hypochromic anemia (for details see MIM: 209300).

**Lactotransferrin** (LF, EC 3.4.21.-) (see Data Sheet) or lactoferrin is primarily synthesised in secretory epithelial cells and is a multifunctional protein that binds and transports iron, exhibits antimicrobial activity and is part of the innate immune defence system; it is also involved in inflammatory events. LF is a single-chain glycoprotein (6 % CHO) and, like TF, contains two homologous domains. Domain 1 (345 aa: Gly 1–Arg 345) contains six and domain 2 (347 aa: Asp 346–Lys 692) 11 intrachain disulfide bridges and each domain contains one iron $Fe^{3+}$-binding site and one anion-binding site (carbonate/bicarbonate):

***Binding sites in domain 1***:

$Fe^{3+}$: **Asp 61, Tyr 93 and 193, His 254**

Bicarbonate: **Thr 118, Arg 122, Ala 124 and Gly 125** (both amide nitrogen)

***Binding sites in domain 2***:

$Fe^{3+}$: **Asp 396, Tyr 436 and 519, His 598**

Bicarbonate: **Thr 462, Arg 466, Ala 468 and Gly 469** (both amide nitrogen)

Lactotransferrin is the parent protein of **lactoferroxin A** (6 aa: Tyr 1–Tyr 6), **lactoferroxin B** (5 aa: Arg 1–Tyr 5) and **lactoferroxin C** (7 aa: Lys 1–Tyr 7), which exhibit opioid antagonist activity. In addition, there are two peptides, **kaliocin-1** (31 aa: Phe 152–Ala 182 in LF) and **lactoferricin** (48 aa: Gly 1–Ala 48 in LF), which, like the parent protein lactotransferrin, exhibit antimicrobial activity.

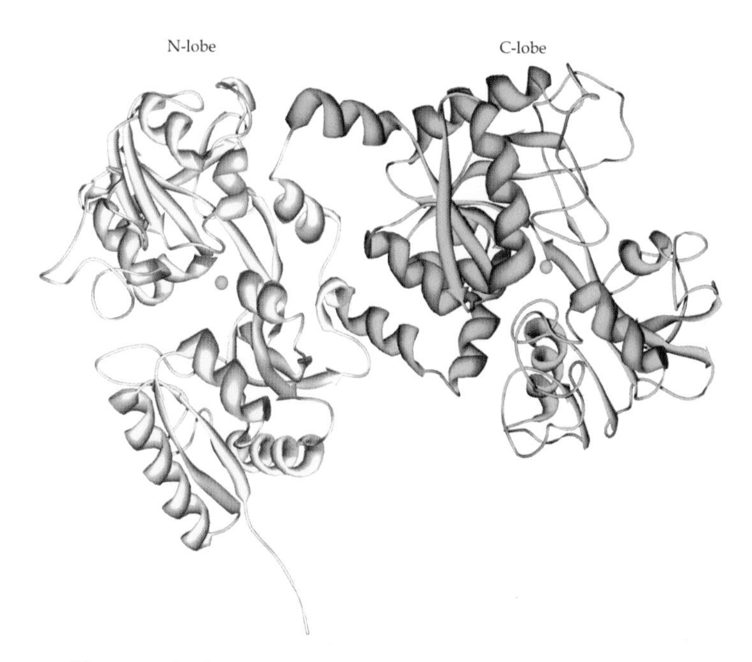


**Figure 15.8 3D structure of lactotransferrin (PDB: 2BJJ)**
The structure clearly shows the two domains, the N-lobe (in yellow) and the C-lobe (in green) connected by a single α-helix (in red) and each domain contains a single iron atom (pink balls).

The 3D structure of entire recombinant lactotransferrin (692 aa: Gly 1–Lys 692) was determined by X-ray diffraction (PDB: 2BJJ) and is shown in Figure 15.8. The structure clearly shows the two domains, the N-lobe (in yellow) and the C-lobe (in green) connected by a single α-helix (in red) and each domain contains a single iron atom (pink balls).

### 15.4.3 Iron storage

Iron is primarily stored in the liver in soluble form complexed to the main iron storage protein ferritin or as insoluble deposits of hemosiderin ($Fe^{3+}$-oxyhydroxyde) if the iron storage capacity of ferritin is exceeded. Free iron is toxic to cells and therefore the body has developed several protective mechanisms to bind iron in various tissues. An excess iron burden in the body may lead to excess hemosiderin formation, which is deposited in different organs such as the liver and the heart and may impair these organs. Apoferritin (iron-free ferritin) consists of 24 subunits of two different classes, the L (light chain) and H (heavy chain). The heteropolymeric ferritin forms a shell with a variable subunit content, principally in the range of $H_{24}L_0$ to $H_0L_{24}$, with L-rich polymers predominantly in the liver and spleen and H-rich polymers in the heart. The protein shell composed of 24 subunits has a roughly spherical structure of 12 nm diameter with a central nearly spherical cavity of 8 nm diameter with a binding capacity of up to 4500 iron atoms ($Fe^{3+}$).

There are several conserved regions in ferritin molecules. There are two consensus patterns: the first is located around three conserved Glu residues (shown in red) in the central part of the molecule, thought to be involved in iron binding, and the second is located in the C-terminal region forming a hydrophobic channel through which iron and small molecules have access to the central cavity of ferritin:

FERRITIN_1 (PS00540): **E**–x–[KR]–**E**–x(2)–**E**–[KR]–[LF]–[LIVMA]–x(2)–Q–N–x–R–x–G–R

FERRITIN_2 (PS00204): D–x(2)–[LIVMF]–[STACQV]–[DH]–[FYMI]–[LIV]–[EN]–x(2)–[FYC]–L–x(6) –[LIVMQ]–[KNER]

**Ferritin light chain** (L subunit) and **ferritin heavy chain** (H subunit, EC 1.16.3.1) (see Data Sheet) are present in almost all cells but are primarily synthesised in the liver. The light chain contains a ferritin-like diiron domain with five Glu residues (53, 56, 57, 60, 63) as potential iron-binding sites and an N-terminal *N*-acetyl-Ser residue. The heavy chain also contains a ferritin-like diiron domain with two iron-binding centres, iron 1 (Glu 27, Glu 61 and His 65) and iron 2 (Glu 62, 107 and 140), as iron-binding sites and the heavy chain is phosphorylated (P-Ser 178). The L- and the H-chains exhibit a 57 % sequence identity.

The heavy chain catalyses the first step of the iron storage, the oxidation of $Fe^{2+}$ to $Fe^{3+}$:

$$4Fe^{2+} + 4H^{+} + O_2 \Longleftrightarrow 4Fe^{3+} + 2H_2O$$

The light chain promotes the nucleation of the iron, thus enabling the storage of $Fe^{3+}$ in the cavity of the protein shell.

The 3D structure of the entire recombinant ferritin heavy chain (183 aa: Met 0–Ser 182) present as a 24-oligomer was determined by X-ray diffraction (PDB: 2FHA). The monomeric structure is characterised by a four-helix bundle, shown in Figure 15.9(a), and the oligomeric ferritin structure exhibits the expected spherical shape with the expected central nearly spherical cavity for iron binding, shown in Figure 15.9(b).

### 15.4.4 Heme transport and iron recovery: hemopexin

Of all heme-binding proteins hemopexin exhibits the highest affinity for heme and is the main intravascular heme transporter to the liver for breakdown and for iron recovery. Free heme is very toxic to the cells and hemopexin is a powerful antioxidant, preventing heme-catalysed tissue damage. Due to the high concentration of hemoglobin in the erythrocytes and in plasma the recovery of iron from hemoglobin is a very important process.

Hemopexin is structurally related to vitronectin (see Chapter 7) and to many members of the matrix metalloproteinase family (matrixins). All these proteins contain hemopexin-like domains, which are thought to facilitate binding to various molecules and proteins, and are characterised by the following consensus sequence:

HEMOPEXIN (PS00024): [LIFAT] – { IL } – x (2) – W – x (2,3) – [PE] – x – { VF } – [LIVMFY] – [DENQS] – [STA] – [AV] – [LIVMFY]

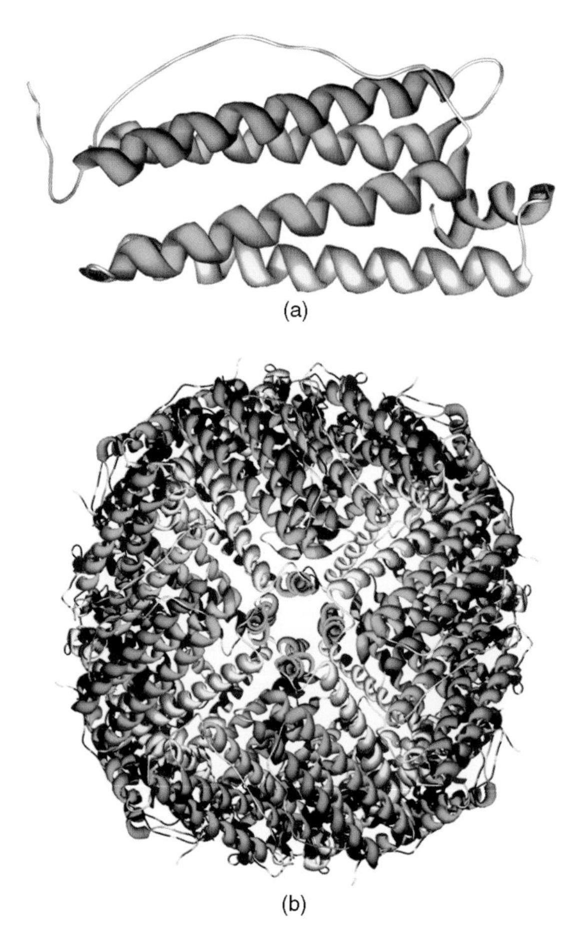


**Figure 15.9 3D structures of (a) the monomeric ferritin heavy chain and (b) the 24-mer ferritin heavy chain (PDB: 2FHA)**
(a) The ferritin heavy chain monomer exhibits a classical four-helix bundle structure.
(b) The oligomeric ferritin structure exhibits a spherical shape with a central nearly spherical cavity for iron binding.

**Hemopexin** (HPX) (see Data Sheet) or β-1B-glycoprotein is a single-chain glycoprotein (20 % CHO) mainly synthesised in the liver. HPX contains five hemopexin-like domains and six intrachain disulfide bridges and iron is coordinated to His 56 and His 127.

Instead of human hemopexin, the 3D structure of rabbit hemopexin (435 aa: Val 1–His 435 + 25 aa signal peptide, P20058) determined by X-ray diffraction (PDB: 1QHU) is given in Figure 15.10. Human and rabbit hemopexin share an 80 % sequence identity. The heme–hemopexin complex reveals a binding site formed between two similar four-bladed β-propeller domains (in yellow and green) connected by an interdomain linker peptide (in red, the dashed grey line is modelled). The heme group (in pink) is coordinated to two His residues (in blue) in a binding pocket dominated by aromatic and basic residues.

## 15.5 TRANSPORT OF HORMONES, STEROIDS AND VITAMINS

The transport of rather apolar/nonpolar (lipophilic) compounds in blood is usually carried out by transport proteins, the most prominent being the 'universal' transport protein albumin, which has a binding affinity for many different types of compounds, among them nonpolar compounds such as hormones, steroids and vitamins (see Section 15.3).

The nonpolar hormones triiodothyronine ($T_3$) and thyroxine ($T_4$) are synthesised in the thyroid gland in the parent protein thyroglobulin (for details see Chapter 13). The nonpolar $T_3$ and $T_4$ are transported in blood by the major thyroid

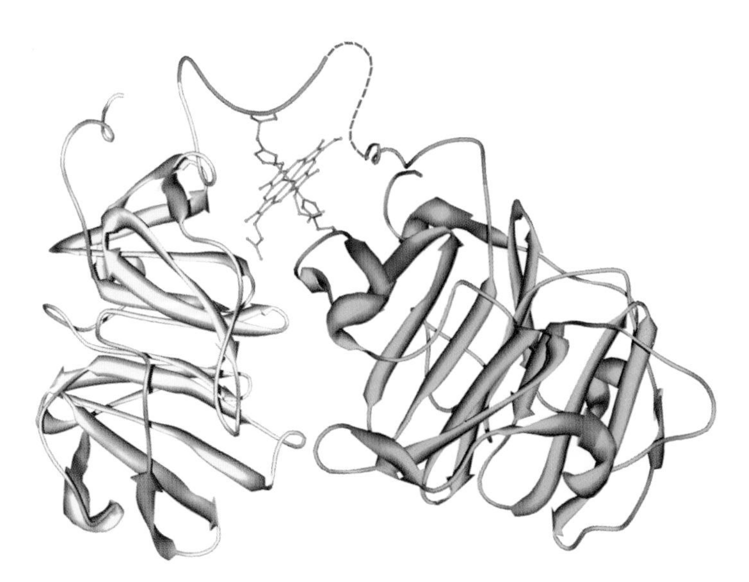

**Figure 15.10 3D structure of rabbit hemopexin (PDB: 1QHU)**
Human and rabbit hemopexin share an 80 % sequence identity. The heme group (in pink) is located between two similar four-bladed β-propeller domains (in yellow and green) connected by an interdomain linker peptide (in red, the dashed grey line is modelled) and is coordinated to two His residues (in blue).

hormone-binding protein thyroxine-binding globulin, and also by transthyretin and by albumin. Steroids such as glucocorticoids, androgens and estrogens are mainly transported by transcortin and albumin and gonadal steroids by the sex hormone-binding globulin. The lipophilic vitamin A is transported by the plasma retinol-binding globulin, primarily in complex with transthyretin, preventing its loss through the kidney glomeruli.

### 15.5.1 Hormone transport proteins

Thyroxine-binding globulin (TBG) is the major transport protein of the nonpeptide (nonpolar) thyroid hormones triiodothyronine ($T_3$) and thyroxine ($T_4$), which are derived from the parent protein hormone thyroglobulin (for details see Chapter 13). Transthyretin (TTR) and albumin are also involved in the transport of thyroid hormones, but TGB possesses the highest affinity and is the main carrier for $T_3$ and $T_4$ but has the lowest concentration in plasma. TTR has a lower affinity but a higher concentration than TBG and albumin has the lowest affinity of all but by far the highest concentration.

Transthyretin is highly conserved in vertebrates and there are two consensus sequences, one in the N-terminal region with a Lys residue involved in thyroxine binding and one in the C-terminal end:

TRANSTHYRETIN_1 (PS00768): [**K**H] – [IV] – L– [DN] – x (3) – G – x – P – [AG] – x(2) – [LIVM] – x – [IV]

TRANSTHYRETIN_2 (PS00769): [YWF]–[TH]–[IVT]–[AP]–x(2)–[LIVM]–[STA]–[PQ]–[FYWG]–[GS]–[FY]–[QST]

**Thyroxine-binding globulin** (TBG) (see Data Sheet) or T4-binding globulin is primarily synthesised in the liver and is a single-chain glycoprotein (18–20 % CHO) with five Cys residues but no documented disulfide bridges. TBG carries a single binding site for $T_3/T_4$ and structurally belongs to the serpin family, but exhibits no known inhibitory activity. There are three relatively common and well-document variants:

1. TBG-Poly (common polymorphism): Leu 283 → Phe. This variant is difficult to distinguish from the wild type TBG-C (common) and is found with variable frequency in different races.

2. TBG-S (slow): Asp 171 → Asn. This variant is functionally indistinguishable from TBG-C, but can be separated electrophoretically from TBG-C due to the loss of a negative charge. TBG-S is primarily found in the black population and in Pacific islanders.
3. TBG-A: Ala 191 → Thr. This variant has a 51 % allele frequency in Australian aborigines and exhibits an approximately 50 % reduced affinity for thyroid hormones; it has a reduced stability.

The 3D structure of an N-terminally truncated form of TBG (379 aa: Leu 19–Ala 395 + N-terminal dipeptide Gly–Ser from the thrombin cleavage site) in complex with thyroxine was determined by X-ray diffraction (PDB: 2CEO) and is shown in Figure 15.11. The structure of TBG is different from other serpins and contains a binding pocket for thyroxine (in pink) on the surface of the molecule, as can be seen in the surface representation in Figure 15.11(b). The upper half of its main β-sheet is fully open, allowing the reactive loop peptide to move in and out of the sheet readily and thus enabling an equilibrated binding and release of thyroxine (Figure 15.11(a)). In TBG the serpin inhibitory mechanism is adapted to a reversible flip-flop transition from a high affinity to a low affinity form.

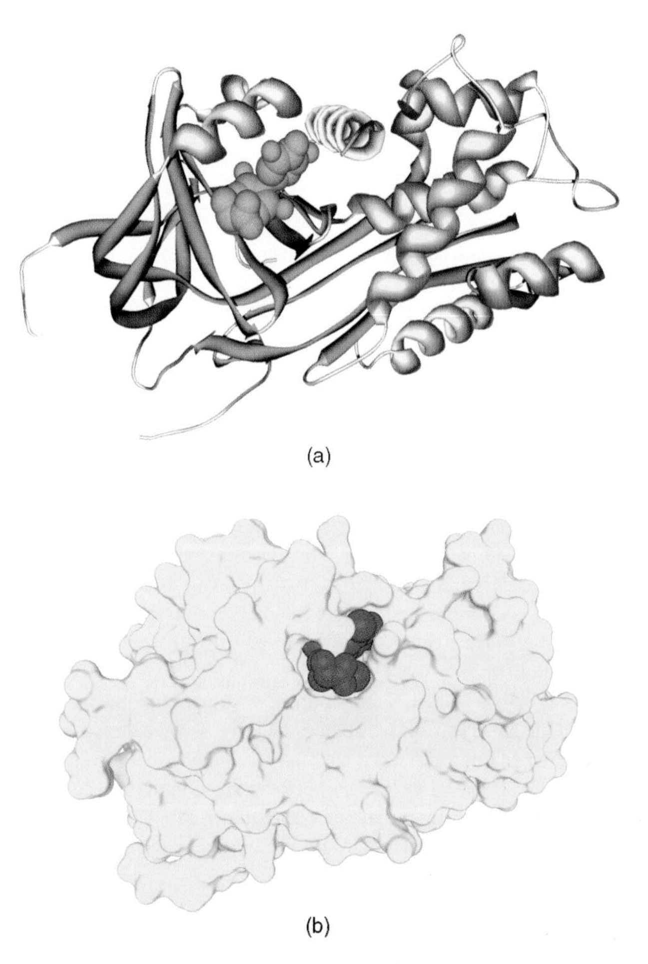


**Figure 15.11 3D structure of thyroxine-binding globulin in complex with thyroxine (PDB: 2CEO)**
(a) The structure of thyroxine-binding globulin is different from other serpins and contains a thyroxine-binding pocket at the surface of the molecule containing a thyroxine molecule (in pink).
(b) Surface representation of the thyroxine binding pocket.

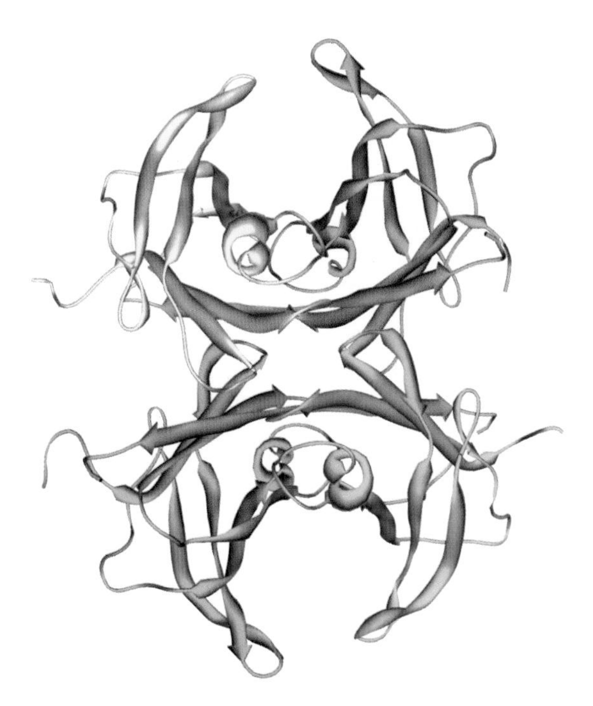

**Figure 15.12 3D structure of tetrameric transthyretin (PDB: 1RLB)**
Homotetrameric transthyretin is characterised by a central channel, in which two high-affinity binding sites for thyroid hormones are located. Each monomer consists of two four-stranded β-sheets with the shape of a prolate ellipsoid (each monomer is shown in a different colour).

Defects in the TBG gene are the cause of TBG deficiency, resulting in reduced or elevated protein-bound iodine, but has a normal thyroid function (for details see MIM: 314200 and 188600).

**Transthyretin** (TTR) (see Data Sheet) or prealbumin is synthesised in the liver, the brain and the eye as a single-chain protein with no disulfide bridges but one free Cys of unknown function and circulates in blood as a noncovalently linked homotetramer with a dimer of dimers configuration. There are two thyroxine-binding sites per tetramer located in a channel formed at the dimer interface and $T_4$ is coordinated to Lys 15 and Glu 54 in the monomer. Approximately 40 % of TTR circulates as a 1:1 complex with the plasma retinol-binding protein (RBP) protecting the relatively small RBP from glomerular filtration and renal catabolism.

The 3D structure of TTR (127 aa: Gly 1–Glu 127) in complex with mutated plasma retinol-binding protein lacking a C-terminal nonapeptide (174 aa: Glu 1 – Cys 174) was determined by X-ray diffraction (PDB: 1RLB). The structure of homotetrameric transthyretin is shown in Figure 15.12. Each TTR monomer consists of two four-stranded β-sheets with the shape of a prolate ellipsoid. Antiparallel β-sheet interactions between the monomers result in two stable dimers, which form the tetramer (each monomer is shown in a different colour). The quaternary TTR structure is characterised by a central channel that runs through the tetramer in which two high-affinity binding sites for thyroid hormones are located.

Defects in the TTR gene are the cause of different diseases:

1. Familial amyloidotic polyneuropathy is a late-onset disease with a dominant mode of inheritance with protein fibril formation in many tissues and amyloid deposition around nerves, which are the cause of familial amyloidotic cardiomyopathy and carpel tunnel syndrome (for details see MIM: 176300).
2. Amyloidosis VII or leptomeningeal/meningocerebrovacular amyloidosis is distinct from other forms of transthyretin amyloidosis because of the primary involvement of the central nervous system (for details see MIM: 105210).
3. Hyperthyroxinemia (for details see MIM: 176300).

### 15.5.2 Steroid transport proteins

Two transport proteins are mainly responsible for the transport of steroids:

1. Corticosteroid-binding globulin (CBG) or transcortin is the major transport protein in plasma for the transport of glucocorticoids and progestins. CBG binds large portions of cortisol and aldosterone in plasma (approximately 75 % and 60 %, respectively) and also binds progesterone. CBG regulates the access of glucocorticoid hormones to target cells.
2. Sex hormone-binding globulin (SHBG) binds and transports sex steroids, especially testosterone and estradiol, and regulates their access to tissues and target cells.

**Corticosteroid-binding globulin** (CBG) (see Data Sheet) or transcortin is synthesised in the liver as single-chain glycoprotein containing two Cys residues, of which conserved Cys 238 is involved in the single steroid-binding site. CBG structurally belongs to the serpin family but without any known inhibitory activity and is closely related to the other transport protein in plasma belonging to the serpin family, with thyroxine-binding globulin sharing a sequence identity of 43 %.

Defects in the transcortin gene are the cause of corticosteroid-binding globulin deficiency, characterised by a low affinity for cortisol (for details see MIM: 122500).

**Sex hormone-binding globulin** (SHBG) (see Data Sheet) or sex steroid-binding protein is primarily synthesised in hepatocytes as a single-chain glycoprotein (8 % CHO) and is present in plasma as a noncovalently linked homodimer that binds one steroid molecule. SHBG contains two laminin G-like domains, each with a single intrachain disulfide bridge (Cys 164–Cys 188 in domain 1 and Cys 333–Cys 361 in domain 2).

The 3D structure of the N-terminal laminin G-like domain of SHBG (189 aa: Leu 1–Asp 189) in complex with 2-methoxyestradiol, a biologically active estrogen metabolite, was determined by X-ray diffraction (PDB: 1LHW) and is shown in Figure 15.13. The structure is characterised by two seven-stranded β-sheets and the ligand 2-methoxyestradiol (in pink) intercalates between them.

### 15.5.3 Vitamin A transport protein

Vitamin A is transported in plasma in the form of retinol by plasma retinol-binding protein (RBP) from the site of its storage in the liver to the peripheral target cells such as epithelial cells. In plasma, RBP forms a noncovalently linked 1:1 complex with

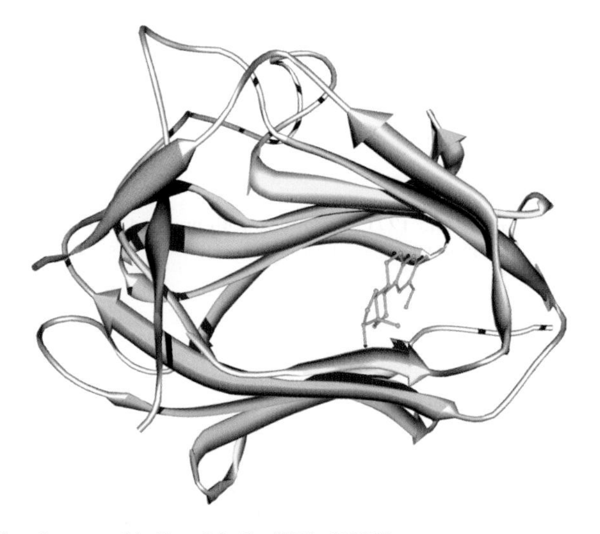

**Figure 15.13 3D structure of sex hormone-binding globulin (PDB: 1LHW)**
The structure consists of two seven-stranded β-sheets and 2-methoxyestradiol (in pink) is located in-between.

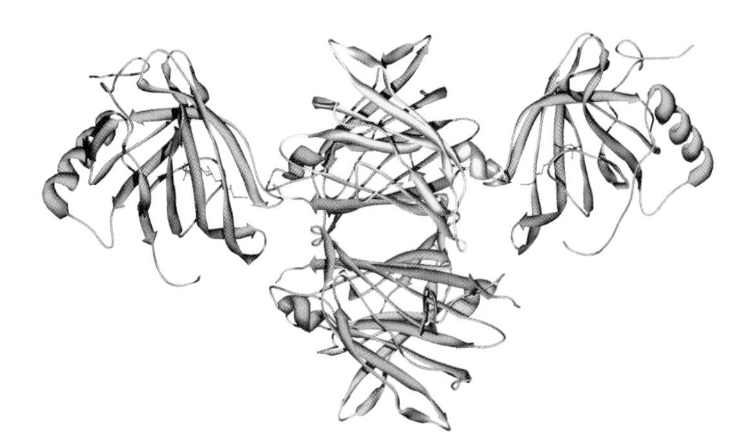

**Figure 15.14 3D structure of plasma retinol-binding protein in complex with transthyretin (PDB: 1RLB)**
Two molecules of plasma retinol-binding protein (in pink) each containing a retinol molecule (in blue) form a complex with homotetrameric transthyretin.

transthyretin, thus preventing a premature loss by filtration through the kidney glomeruli and subsequent renal catabolism due to its rather small size.

**Plasma retinol-binding protein** (RBP) (see Data Sheet) is primarily synthesised in the liver as a single-chain protein containing three intrachain disulfide bridges (Cys 4–Cys 160, Cys 70–Cys 174, Cys 120–Cys 129). RBP belongs to the lipocalin family and exists in several C-terminally truncated forms (C-1, C-2, C-4 and C-7 residues).

The 3D structure of RBP (182 aa: Glu 1–Leu 182) in complex with retinol was determined by X-ray diffraction (PDB: 1RBP). The structure is characterised by an antiparallel eight-stranded up-and-down β-barrel that encapsulates the retinol ligand (in blue). Two molecules of RBP (in pink) form a complex with homotetrameric transthyretin (PDB: 1RLB) shown in Figure 15.14.

Defects in the RBP gene are the cause of retinol-binding protein deficiency resulting in night vision problems due to a progressed atrophy of the retinal pigment epithelium (for details see (MIM: 180250). Vitamin A deficiency blocks the secretion of the binding protein, resulting in a defective delivery and supply of vitamin A to the epidermal cells.

## 15.6 OTHER TRANSPORT PROTEINS IN PLASMA

### 15.6.1 Selenoproteins

Selenium is vital to humans and animals and many Se-containing proteins have been characterised, where Se is present as Se-Cys, which is encoded by the opal codon UGA. In addition, Se is incorporated unspecifically as Se-Met instead of Met. Selenium is essential for the synthesis of selenoproteins, but excess selenium has toxic effects and leads to selenium poisoning. The threshold between essential and toxic concentrations of Se is rather low, with a factor in the range of 10–100.

**Selenoprotein P** (SeP) (see Data Sheet) is mainly synthesised in the liver and the heart as single-chain glycoprotein containing ten Se-Cys residues and two 14 residues long His-rich stretches (His 185–His 198 with 9 His, His 225–His 238 with 7 His). SeP seems to be involved in the transport of selenium and the Se supply to tissues such as brain and testis. In addition, SeP is probably involved in the extracellular antioxidant defence properties of selenium.

## REFERENCES

Crichton *et al.*, 1987, Iron transport and storage, *Eur. J. Biochem.*, **164**, 485–506.
Hellman and Gitlin, 2002, Ceruloplasmin metabolism and function, *Annu. Rev. Nutr.*, **22**, 439–458.
Kryukov *et al.*, 2003, Characterization of mammalian selenoproteomes, *Science*, **300**, 1439–1443.
Schreiber *et al.*, 1997, The evolution of gene expression structure and function of transthyretin, *Comp. Biochem. Physiol.*, **116B**, 137–160.

DATA SHEETS

# *Afamin (α-Albumin)*

## Fact Sheet

| | | | |
|---|---|---|---|
| ***Classification*** | ALB/AFP/VDB | ***Abbreviations*** | α-Alb |
| ***Structures/motifs*** | 3 Albumin | ***DB entries*** | AFAM_HUMAN/P43652 |
| ***MW/length*** | 66 577 Da/578 aa | ***PDB entries*** | |
| ***Concentration*** | 30 mg/l | ***Half-life*** | |
| ***PTMs*** | 4 N-CHO (3 potential) | | |
| ***References*** | Lichenstein *et al.*, 1994, *J. Biol. Chem.*, **269**, 18149–18154. | | |

## Description

**Afamin** is a blood plasma glycoprotein involved in ligand transport. It exhibits structural and functional similarity with serum albumin, α-fetoprotein and vitamin D-binding protein.

## Structure

Afamin is a single-chain plasma glycoprotein containing three albumin domains with the characteristic disulfide bridge pattern.

## Biological Functions

Afamin is involved in the transport of ligands of yet unknown origin.

# *Ceruloplasmin (Ferroxidase; EC 1.16.3.1)*

## Fact Sheet

| | | | |
|---|---|---|---|
| ***Classification*** | Multicopper oxidase | ***Abbreviations*** | CP |
| ***Structures/motifs*** | 3 F5/8 type A; 6 plastocyanin-like | ***DB entries*** | CERU_HUMAN/P00450 |
| ***MW/length*** | 120 085 Da/1046 aa | ***PDB entries*** | 1KCW |
| ***Concentration*** | 210–450 µg/l | ***Half-life*** | 22 h |
| ***PTMs*** | 4 N-CHO | | |
| ***References*** | Koschinsky *et al.*, 1986, *Proc. Natl. Acad. Sci. USA*, **83**, 5086–5090.<br>Zaitseva *et al.*, 1996, *J. Biol. Inorg. Chem.*, **1**, 15–23. | | |

## Description

**Ceruloplasmin** (CP) is expressed in the lung during development and in the liver in adults. It is an abundant plasma protein that participates in the acute-phase reaction to stress.

## Structure

CP is a large single-chain plasma glycoprotein (approx. 10 % CHO) containing three F5/8 type A, of which each domain consists of two plastocyanin-like domains. CP binds six $Cu^{2+}$ ions (in pink) via three distinct copper-binding centres known as type 1 (blue), type 2 (normal) and type 3 (coupled binuclear). The 3D structure is characterised by a triangular array of the six plastocyanin-like domains (shown in different colours) and three of the six copper atoms form a trinuclear cluster at the interface of domains 1 and 6 and the other three copper atoms form three mononuclear sites in domains 2, 4 and 6 (**1KCW**).

## Biological Functions

CP has four possible functions: ferroxidase activity, amine oxidase activity, copper transport and homeostasis, superoxide dismutase activity. CP catalyses the reaction: $4\ Fe^{2+} + 4\ H^+ + O_2 \Leftrightarrow 4\ Fe^{3+} + 2\ H_2O$.

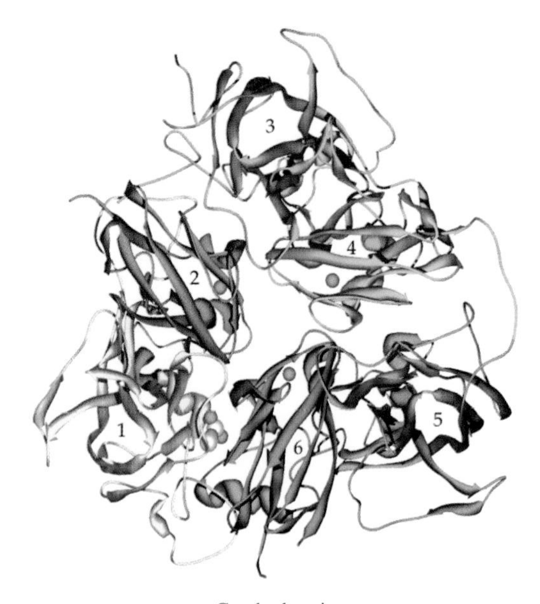


Ceruloplasmin

## *Corticosteroid-binding globulin (Transcortin)*

### Fact Sheet

| | | | |
|---|---|---|---|
| ***Classification*** | Serpin | ***Abbreviations*** | CBG |
| ***Structures/motifs*** | | ***DB entries*** | CBG_HUMAN/P08185 |
| ***MW/length*** | 42 639 Da/383 aa | ***PDB entries*** | |
| ***Concentration*** | approx. 70 mg/l | ***Half-life*** | |
| ***PTMs*** | 6 N-CHO (5 potential) | | |
| ***References*** | Hammond *et al.*, 1987, *Proc. Natl. Acad. Sci. USA*, **84**, 5153–5157. | | |

### Description

**Corticosteroid-binding globulin** (CBG) is synthesised in the liver and is involved in the transport of glucocorticoids and exhibits structural similarity with serpins.

### Structure

CBG is a single-chain plasma glycoprotein belonging to the serpin family but has no known inhibitory activity. Conserved Cys 228 is involved in steroid binding. CBG is very likely to be very closely related to the other transport proteins in plasma of the serpin family without any inhibitory activity, e.g. thyroxine-binding globulin.

### Biological Functions

CBG is a major transport protein for glucocorticoids and progestins and regulates their concentration in plasma.

## *Ferritin: light chain and heavy chain (EC 1.16.3.1)*

### Fact Sheet

| | | | |
|---|---|---|---|
| ***Classification*** | Ferritin | ***Abbreviations*** | |
| ***Structures/motifs*** | Each chain: 1 ferritin-like diiron | ***DB entries*** | FRIL_HUMAN/P02792<br>FRIH_HUMAN/P02794 |
| ***MW/length*** | Light chain: 19 888 Da/174 aa<br>Heavy chain: 21 094 Da/182 aa | ***PDB entries*** | 2FHA |
| ***Concentration*** | 10–250 μg/l | ***Half-life*** | 3–4 days in tissues |
| ***PTMs*** | Light chain: *N*-acetyl-Ser (aa 1); heavy chain: P-Ser (aa 178) | | |
| ***References*** | Boyd *et al.*, 1985, *J. Biol. Chem.*, **260**, 11755–11761.<br>Hempstead *et al.*, 1997, *J. Mol. Biol.*, **268**, 424–448. | | |

### Description

**Ferritin** is an iron storage protein synthesised in all human cells with the highest concentration in the liver. It exists as an oligomer of 24 subunits with two types of subunits, a light and a heavy chain.

### Structure

The ferritin light chain carries an N-terminal *N*-acetyl-Ser and contains five Glu residues (53, 56, 57, 60, 63) in close proximity as potential iron-binding sites. The ferritin heavy chain carries a P-Ser (aa 178) and contains two iron-binding centres: iron 1 (Glu 27, 62 and His 65) and iron 2 (Glu 62, 107, 140). Both chains contain one ferritin-like diiron domain. The 3D structure of the monomeric ferritin heavy chain is highly helical (four-helix bundle) and the 24-oligomer exhibits a spherical shape (12 nm) with a central cavity (8 nm) for iron binding (**2FHA**).

### Biological Functions

Ferritin stores iron in a soluble nontoxic and readily available form. Ferritin catalyses the reaction: $4\,Fe^{2+} + 4\,H^{+} + O_2 \Leftrightarrow 4\,Fe^{3+} + 2\,H_2O$. Ferritin is important for iron homeostasis.

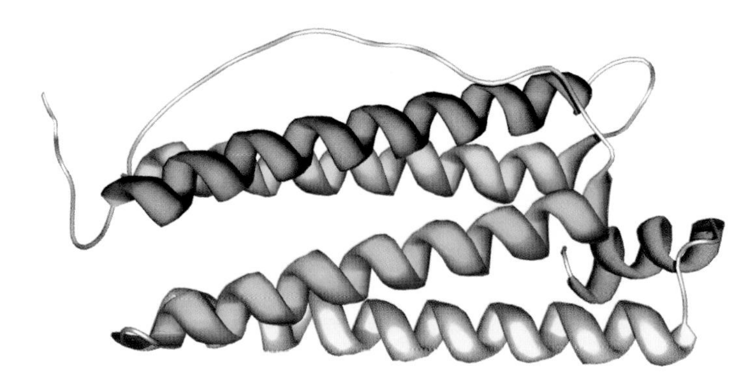

Ferritin

## *α-Fetoprotein (α-Fetoglobulin)*

### Fact Sheet

| | | | |
|---|---|---|---|
| ***Classification*** | ALB/AFP/VDB | ***Abbreviations*** | AFP |
| ***Structures/motifs*** | 3 Albumin | ***DB entries*** | FETA_HUMAN/P02771 |
| ***MW/length*** | 66 478 Da/591 aa | ***PDB entries*** | |
| ***Concentration*** | <50 μg/l | ***Half-life*** | |
| ***PTMs*** | 1 N-CHO | | |
| ***References*** | Morinaga *et al.*, 1983, *Proc. Natl. Acad. Sci. USA*, **80**, 4604–4608. | | |

### Description

**α-Fetoprotein** (AFP) is synthesised in the fetal liver and the yolk sac and is present in plasma after birth only in trace amounts. AFP exhibits structural and functional similarity with serum albumin, afamin and vitamin D-binding protein.

### Structure

AFP is a single-chain plasma glycoprotein containing three albumin domains with the characteristic disulfide bridge pattern. His 4 is involved in the binding of copper and nickel.

### Biological Functions

AFP has a similar function in the fetus as serum albumin in adults. AFP binds several types of ligands like copper and nickel, fatty acids and bilirubin. It specifically binds estradiol to prevent the damage of the fetal brain.

## *Haptoglobin*

### Fact Sheet

| | | | |
|---|---|---|---|
| ***Classification*** | Peptidase S1 | ***Abbreviations*** | |
| ***Structures/motifs*** | α-chain: 2 sushi/CCP/SCR | ***DB entries*** | HPT_HUMAN/P00738 |
| ***MW/length*** | α-chain: 15 946 Da/142 aa<br>β-chain: 27 265 Da/245 aa | ***PDB entries*** | |
| ***Concentration*** | approx. 1.4 g/l | ***Half-life*** | |
| ***PTMs*** | 4 N-CHO | | |
| ***References*** | Yang *et al.*, 1983, *Proc. Natl. Acad. Sci. USA*, **80**, 5875–5879. | | |

### Description

**Haptoglobin** is synthesised in the liver and is present in plasma as a tetramer of two α- and two β-chains.

### Structure

The α-chain of haptoglobin contains two sushi/CCP/SCR domains and the β-chain comprises the serine protease part. Haptoglobin exhibits no catalytic activity: His and Ser in the catalytic centre are replaced by Lys and Ala, respectively. The two α-chains in haptoglobin are linked by two interchain disulfide bridges (Cys 15–Cys 15, Cys 74–Cys 74) and the α- and the β-chain are linked by a single interchain disulfide bridge (α-chain: Cys 131; β-chain: Cys 105).

### Biological Functions

Haptoglobin binds free hemoglobin in plasma, thus preventing the loss of iron via the kidneys and protecting them from damage by hemoglobin.

# Hemoglobin

## Fact Sheet

| | | | |
|---|---|---|---|
| ***Classification*** | Globin | ***Abbreviations*** | Hb |
| ***Structures/motifs*** | | ***DB entries*** | HBA_HUMAN/P69905<br>HBB_HUMAN/P68871<br>HBG1_HUMAN/P69891<br>HBG2_HUMAN/P69892<br>HBD_HUMAN/P02042<br>HBE_HUMAN/P02100<br>HBAT_HUMAN/P09105<br>HBAZ_HUMAN/P02008 |
| ***MW/length*** | $\alpha$-chain: 15 126 Da/141 aa; $\beta$-chain: 15 867 Da/146 aa<br>$\gamma_1$-chain: 16 009 Da/146 aa; $\gamma_2$-chain: 15 995 Da/146 aa<br>$\delta$-chain: 15 924 Da/146 aa; $\varepsilon$-chain: 16 072 Da/146 aa<br>$\theta$-chain: 15 377 Da/141 aa; $\zeta$-chain: 15 506 Da/141 aa | ***PDB entries*** | 1A00 |
| ***Concentration*** | 150 g/l (in blood); 330 g/l (in erythrocytes) | ***Half-life*** | 120 days |
| ***PTMs*** | $\beta$-chain: 1 Nitroso-Cys (93 aa);<br>$\gamma$-chains: *N*-Acetyl-Gly (minor); $\zeta$-chain: *N*-Acetyl-Ser | | |
| ***References*** | Baralle *et al.*, 1980, *Cell*, **21**, 621–626.<br>Hill *et al.*, 1962, *J. Biol. Chem.*, **237**, 3151–3156.<br>Hsu *et al.*, 1988, *Nature*, **331**, 94–96.<br>Kavanaugh *et al.*, 1998, *Biochemistry*, **37**, 4358–4373.<br>Lawn *et al.*, 1980, *Cell*, **21**, 647–651.<br>Proudfoot et al., 1982, *Cell*, **31**, 553–563.<br>Slighton *et al.*, 1980, *Cell*, **21**, 627–638.<br>Spritz *et al.*, 1980, *Cell*, **21**, 639–646.<br>Sutherland-Smith *et al.*, 1998, *J. Mol. Biol.*, **280**, 475–484. | | |

## Description

**Hemoglobin** is a plasma protein that is exclusively produced in erythroid precursor cells. Hemoglobin comprises approx. 95 % of the total proteins in mature red blood cells. Hemoglobin is responsible for the oxygen transport in blood and its delivery to the tissues.

## Structure

Adult hemoglobin is a non-covalently-linked heterotetramer (designation: $\alpha_2\beta_2$) composed of two $\alpha$-chains (141 aa) and two $\beta$-chains (146 aa). His 58 and 87 in the $\alpha$-chain and His 63 and 92 in the $\beta$-chain are the iron-binding sites. A heterotetramer of two $\alpha$- and two $\delta$-chains represents less than 3.5 % of adult hemoglobin HbA2. There are several forms of embryonic hemoglobins that exhibit a higher affinity for oxygen: a heterotetramer of two $\alpha$- and either two $\gamma_1$- or two $\gamma_2$-chains in fetal hemoglobin HbF, a heterotetramer of two $\zeta$- and two $\varepsilon$-chains in early embryonic hemoglobin Gower-1 and a heterotetramer of two $\alpha$- and two $\varepsilon$-chains in early embryonic hemoglobin Gower-2 and a heterotetramer of two $\zeta$- and two $\gamma$-chains in hemoglobin Portland-1. The two $\alpha$-chains (in yellow and red) and the two $\beta$-chains (in green and blue) each contain a heme group (in pink), where the iron ions are coordinated to His 58 and 87 in the $\alpha$-chain and to His 63 and 92 in the $\beta$-chain (**1A00**).

## Biological Functions

Hemoglobin binds oxygen reversibly and transports it from the lung to the peripheral tissues. In reverse, hemoglobin transports the produced $CO_2$ from the tissues to the lung. The affinity of hemoglobin for oxygen is influenced by pH, $pCO_2$ and organic phosphates. Hemoglobin functions as an important blood buffer protein. Inhibitors are carbon monoxide, nitric oxide and other heme ligands. Elevated levels of HbA2 are characteristic for $\beta$-thalassemia traits.

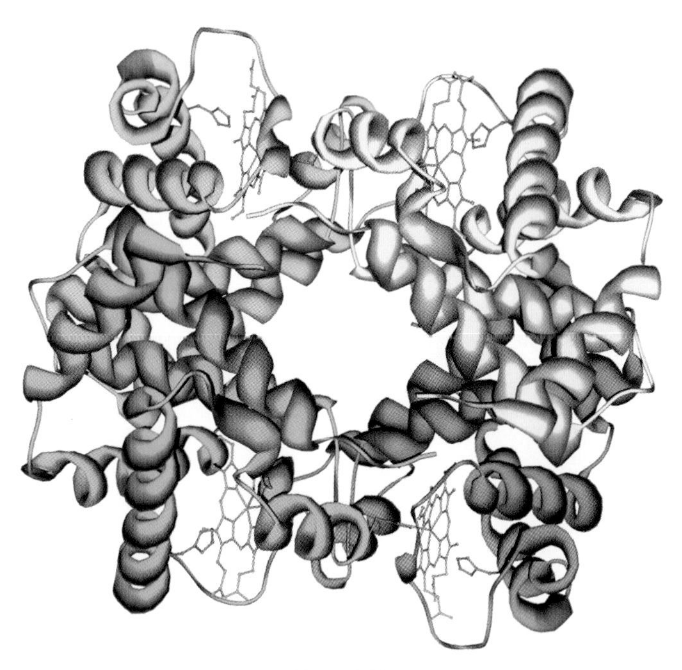

Hemoglobin

## *Hemopexin (β-1B-glycoprotein)*

### Fact Sheet

| | | | |
|---|---|---|---|
| ***Classification*** | | ***Abbreviations*** | HPX |
| ***Structures/motifs*** | 5 Hemopexin-like | ***DB entries*** | HEMO_HUMAN/P02790 |
| ***MW/length*** | 49 295 Da/439 aa | ***PDB entries*** | 1QHU |
| ***Concentration*** | 500–1000 mg/l | ***Half-life*** | 7 days |
| ***PTMs*** | 5 N-CHO; 1 O-CHO | | |
| ***References*** | Takahashi *et al.*, 1985, *Proc. Natl. Acad. Sci. USA*, **82**, 73–77. | | |
| | Paoli *et al.*, 1999, *Nat. Struct. Biol.*, **6**, 926–931. | | |

### Description

**Hemopexin** (HPX) is the major intravascular heme transporter that is mainly synthesised in the liver. HPX exhibits structural similarity with vitronectin.

### Structure

HPX is a single-chain plasma glycoprotein (20 % CHO) containing five hemopexin-like domains with His 56 and 127 involved in iron binding (heme as axial ligand). The 3D structure of the rabbit heme–hemopexin complex reveals a binding site formed between two four-bladed β-propeller domains (in yellow and green) connected by an interdomain linker peptide (in red, the dashed line is modelled) (**1QHU**).

### Biological Functions

HPX binds heme and transports it to the liver for breakdown and iron recovery. HPX has the highest affinity for heme of all heme-binding proteins and is a scavenger of circulating heme. HPX is a powerful antioxidant preventing heme-catalysed tissue damage.

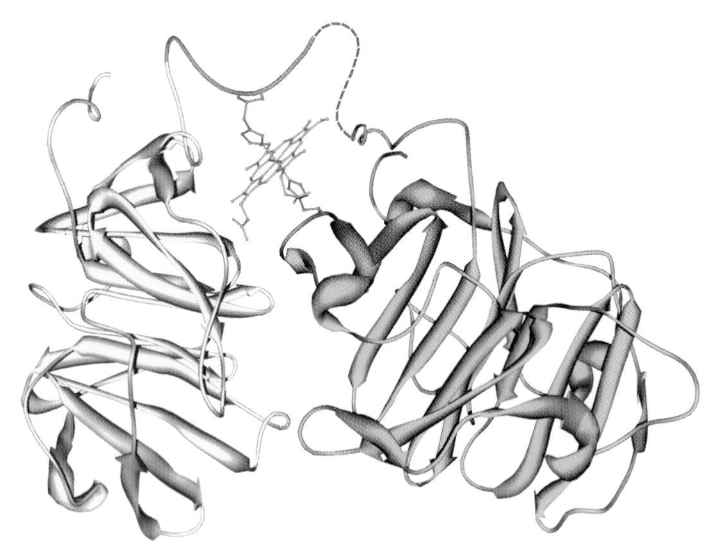

Hemopexin

## *Lactotransferrin (Lactoferrin; EC 3.4.21.-)*

### Fact Sheet

| | | | |
|---|---|---|---|
| *Classification* | Transferrin | *Abbreviations* | LF |
| *Structures/motifs* | | *DB entries* | TRFL_HUMAN/P02788 |
| *MW/length* | 76 321 Da/692 aa<br>Lactoferroxin A: 659 Da/5 aa<br>Lactoferroxin B: 721 Da/5 aa<br>Lactoferroxin C: 868 Da/7 aa<br>Kaliocin-1: 3193 Da/31 aa<br>Lactoferricin: 5588 Da/48 aa | *PDB entries* | 2BJJ |
| *Concentration* | 20 2500 μg/l | *Half-life* | very rapidly cleared |
| *PTMs* | 2 N-CHO | | |
| *References* | Metz-Boutigue *et al.*, 1984, *Eur. J. Biochem.*, **145**, 659–666.<br>Thomassen *et al.*, 2005, *Transgen. Res.*, **14**, 397–405. | | |

### Description

**Lactotransferrin** (LF) is an iron-binding and transport protein present in almost all biological fluids. It exhibits structural and functional similarity with serotransferrin, melanotransferrin and ovotransferrin.

### Structure

LF is a single-chain plasma glycoprotein (6 % CHO) consisting of two homologous domains each with one iron-binding site coordinated to one Asp and His and two Tyr residues and one carbonate-binding site coordinated to one Thr, Arg, Ala and Gly residue. LF is the parent protein of lactoferroxins A (6 aa), B (5 aa) and C (7 aa), which exhibit opioid antagonist activity. In addition, LF is also the parent protein of kaliocin-1 (31 aa) and lactoferricin (48 aa), which exhibit antimicrobial activity. The 3D structure clearly shows the N- (in yellow) and C-lobe (in green) connected by a single α-helix (in red) and each domain contains one iron ion (pink balls) (**2BJJ**).

### Biological Functions

LF binds two $Fe^{3+}$ ions in association with an anion, usually carbonate/bicarbonate. LF is involved in numerous inflammatory events and immune response functions and exhibits antimicrobial activity.

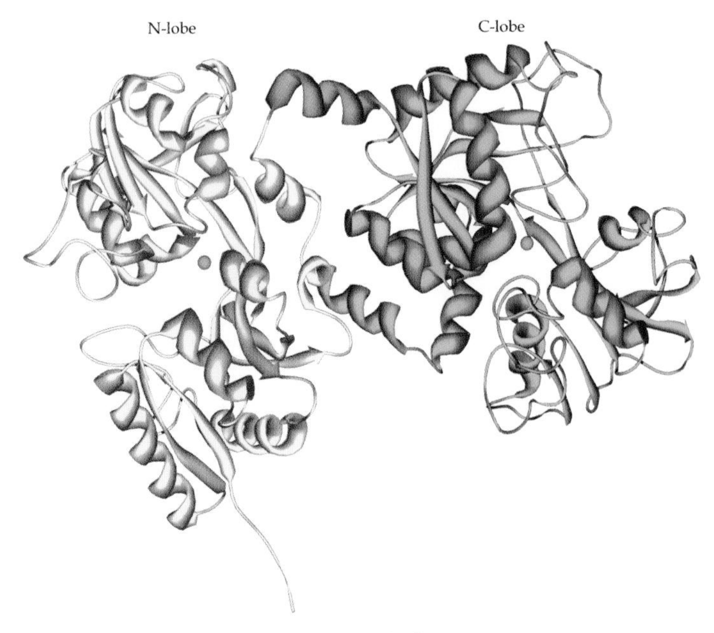


Lactotransferrin

# *Myoglobin*

## Fact Sheet

| | | | |
|---|---|---|---|
| ***Classification*** | Globin | ***Abbreviations*** | Mb |
| ***Structures/motifs*** | | ***DB entries*** | MYG_HUMAN/P02144 |
| ***MW/length*** | 17 053 Da/153 aa | ***PDB entries*** | 2MM1 |
| ***Concentration*** | approx. 6 mg/g tissue | ***Half-life*** | 80–90 days (muscle) |
| ***PTMs*** | | | |
| ***References*** | Hubbard *et al.*, 1990, *J. Mol. Biol.*, **213**, 215–218.<br>Romero-Herrera *et al.*, 1971, *Nature New Biol.*, **232**, 149–152. | | |

## Description

**Myoglobin** is an oxygen storage protein primarily found in muscle tissues and belongs to the globin superfamily. It serves as a reserve supply of oxygen.

## Structure

Myoglobin is a single-chain globin containing a single free Cys 110 residue of unknown function. His 64 and 93 are the iron binding sites. The 3D structure is characterised by eight α-helices and no β-sheets (**2MM1**).

## Biological Functions

Myoglobin binds oxygen with a higher affinity than hemoglobin. Myoglobin binds oxygen in tissues, thus reducing the local free oxygen concentration. This leads to the diffusive transport of oxygen from the blood capillaries to the surrounding tissue. Myoglobin is inhibited by carbon monoxide and cyanide.

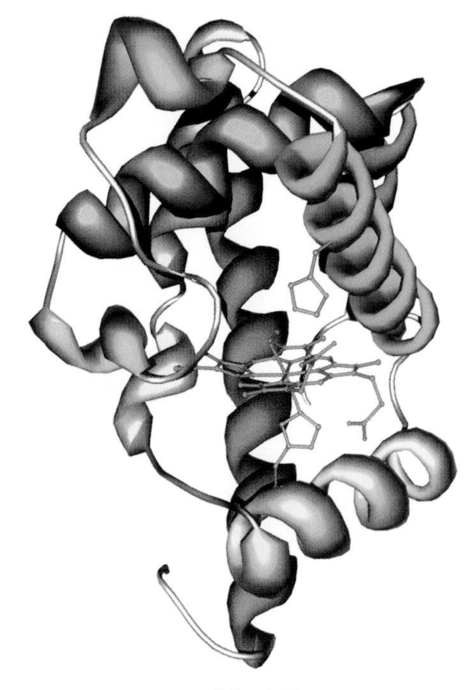

Myoglobin

## *Plasma retinol-binding protein*

### Fact Sheet

| | | | |
|---|---|---|---|
| ***Classification*** | Lipocalin | ***Abbreviations*** | RBP |
| ***Structures/motifs*** | | ***DB entries*** | RETBP_HUMAN/P02753 |
| ***MW/length*** | 21 072 Da/183 aa | ***PDB entries*** | 1RBP; 1RLB |
| ***Concentration*** | 35–60 mg/l | ***Half-life*** | 12 h |
| ***PTMs*** | | | |
| ***References*** | Colantuoni *et al.*, 1983, *Nucleic Acids Res.*, **11**, 7769–7776.<br>Cowan *et al.*, 1990, *Proteins*, **8**, 44–61. | | |

### Description

**Plasma retinol-binding protein** (RBP) is mainly synthesised in the liver and is present in plasma noncovalently bound in a 1:1 complex to transthyretin.

### Structure

RBP is a single-chain nonglycosylated plasma protein containing three intrachain disulfide bridges (Cys 4–Cys 160, Cys 70–Cys 174, Cys 120–Cys 129). The 3D structure is characterised by an antiparallel eight-stranded β-barrel that encapsulates the retinol ligand (in blue). Two molecules of RBP (in pink) form a complex with homotetrameric transthyretin (**1RLB**).

### Biological Functions

RBP transports vitamin A in the form of retinol from its storage sites in the liver to the peripheral tissues. The formation of the RBP–transthyretin complex prevents a premature loss by filtration through the kidney glomeruli.

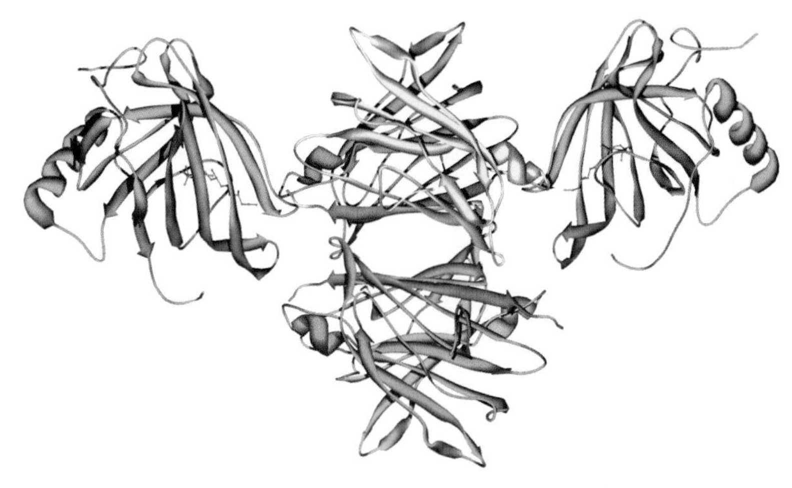

Plasma retional-binding protein in complex with transthyretin tetramer

## *Selenoprotein P*

### Fact Sheet

| | | | |
|---|---|---|---|
| ***Classification*** | | ***Abbreviations*** | SeP |
| ***Structures/motifs*** | | ***DB entries*** | SEPP1_HUMAN/P49908 |
| ***MW/length*** | 40 764 Da/362 aa | ***PDB entries*** | |
| ***Concentration*** | | ***Half-life*** | |
| ***PTMs*** | 6 N-CHO (4 potential); 10 Se-Cys | | |
| ***References*** | Mostert, 2000, *Arch. Biochem. Biophys.*, **376**, 433–438. | | |

### Description

**Selenoprotein P** (SeP) is synthesised in the liver and in the heart and seems to be involved in the transport and supply of selenium to tissues.

### Structure

SeP is a single-chain glycoprotein containing 10 Se-Cys residues that are all encoded by the opal codon UGA. It contains two 14 residue His-rich stretches.

### Biological Functions

SeP might be responsible for some of the extracellular antioxidant defence characteristics of selenium and seems to be involved in selenium transport and its supply to various tissues such as the brain and testis.

# Serotransferrin (Siderophilin)

## Fact Sheet

| | | | |
|---|---|---|---|
| *Classification* | Transferrin | *Abbreviations* | TF |
| *Structures/motifs* | | *DB entries* | TRFE_HUMAN/P02787 |
| *MW/length* | 75 184 Da/679 aa | *PDB entries* | 2HAV |
| *Concentration* | 1.8–2.7 g/l | *Half-life* | 8–10 days |
| *PTMs* | 2 N-CHO; 1 O-CHO | | |
| *References* | Yang *et al.*, 1984, *Proc. Natl. Acad. Sci. USA*, **81**, 2752–2756.<br>Wally *et al.*, 2006, *J. Biol. Chem.*, **281**, 24934–24944. | | |

## Description

**Serotransferrin** (TF) is an iron-binding and transport protein synthesised in the liver. TF exhibits structural and functional similarity with lactotransferrin, melanotransferrin and ovotransferrin.

## Structure

TF is a single-chain plasma glycoprotein (approx. 6 % CHO) consisting of two homologous domains. Each domain contains one iron-binding site coordinated to an Asp, His and two Tyr residues and one bicarbonate-binding site coordinated to one Thr, Arg, Ala and Gly residue. The 3D structure suggests that differences in the hinge region of the N- an C-terminal lobes influences the iron-release/binding rate (**2HAV**).

## Biological Functions

TF binds two $Fe^{3+}$ ions in association with an anion, usually carbonate/bicarbonate. TF is responsible for the transport of iron from the site of absorption and heme degradation to the sites of storage and utilisation.

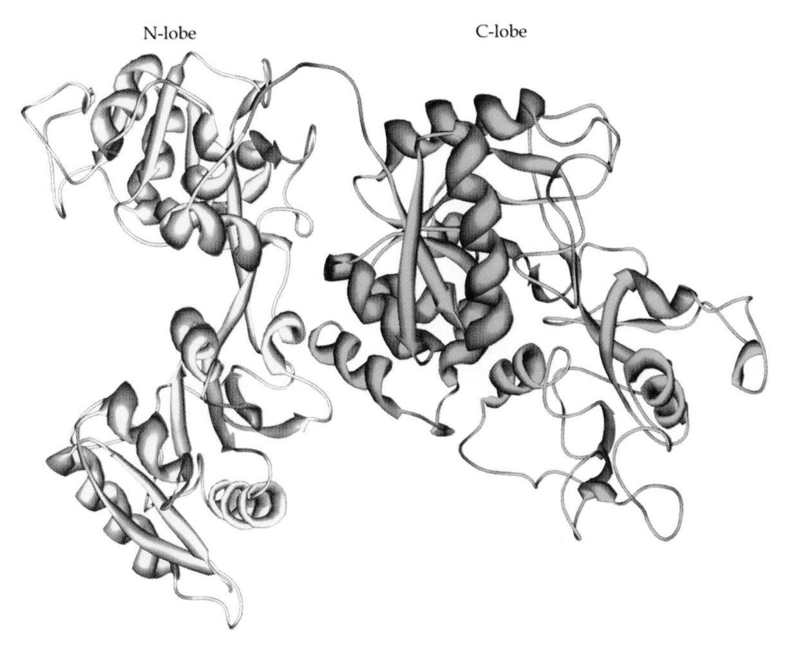


Serotransferrin

# *Serum albumin*

## Fact Sheet

| | | | |
|---|---|---|---|
| ***Classification*** | ALB/AFP/VDB | ***Abbreviations*** | HSA |
| ***Structures/motifs*** | 3 Albumin | ***DB entries*** | ALBU_HUMAN/P02768 |
| ***MW/length*** | 66 472 Da/585 aa | ***PDB entries*** | 1GNI |
| ***Concentration*** | 35–50 g/l | ***Half-life*** | 19 days |
| ***PTMs*** | 2 N-CHO (only in some variants); Acetylsalicylic acid (Lys 223) | | |
| ***References*** | Lawn *et al.*, 1981, *Nucleic Acids Res.*, **9**, 6103–6114.<br>Petitpas *et al.*, 2001, *J. Mol. Biol.*, **314**, 955–960.<br>Zunszain *et al.*, 2003, *BMC Structural Biology*, **3**, 6. | | |

## Description

**Serum albumin** (HSA) is the major circulating blood plasma protein. HSA is synthesised in the liver and is present in all body fluids. It exhibits structural and functional similarity with α-fetoprotein, afamin and vitamin D-binding protein.

## Structure

HSA is a single-chain plasma protein containing three albumin domains with a characteristic disulfide bridge pattern (5 or 6 disulfide bridges). HSA carries no posttranslational modifications except in some variants. The 3D structure is characterised by a high content of α-helices and no β-sheets. The seven binding sites 1–7 for fatty acids are distributed across the six subdomains IA + IB, IIA + IIB and IIIA + IIIB (**1GNI**).

## Biological Functions

HSA has a good binding capacity for many different types of ions (calcium, sodium, potassium, copper) and molecules like water, fatty acids, hormones, bilirubins and drugs and thus acts as a major transport protein in plasma. Its main function is the regulation of the colloidal osmotic pressure of blood and HSA is the main protein compound of the acid/base balance in the vascular system.

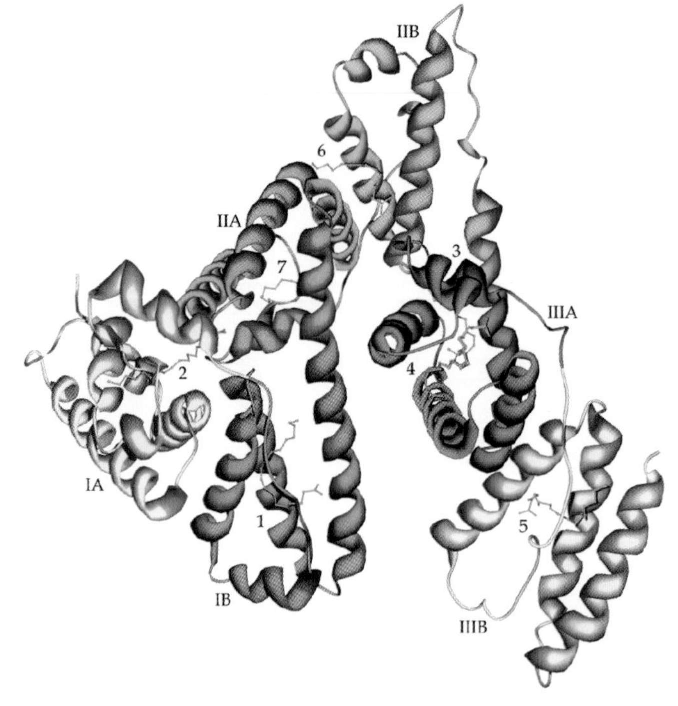


Serum albumin

# *Sex hormone-binding globulin (Sex steroid-binding protein)*

## Fact Sheet

| | | | |
|---|---|---|---|
| ***Classification*** | | ***Abbreviations*** | SHBG |
| ***Structures/motifs*** | 2 Laminin G-like | ***DB entries*** | SHBG_HUMAN/P04278 |
| ***MW/length*** | 40 468 Da/373 aa | ***PDB entries*** | 1LHW |
| ***Concentration*** | Variable: age-depending | ***Half-life*** | |
| ***PTMs*** | 3 N-CHO; 1 O-CHO | | |
| ***References*** | Avvakumov *et al.*, 2002, *J. Biol. Chem.*, **277**, 45219–45225. Walsh *et al.*, 1986, *Biochemistry*, **25**, 7584–7590. | | |

## Description

**Sex hormone-binding globulin** (SHBG) is primarily synthesised in hepatocytes and functions as a transport protein of sex steroids.

## Structure

SHBG is a single-chain glycoprotein (8 % CHO) containing two laminin G-like domains each containing one intrachain disulfide bridge (Cys 164–Cys 188, Cys 333–Cys 361) and is present as non-covalently-linked homodimer. Each dimer binds one molecule of steroid. The 3D structure of the N-terminal laminin G-like domain in complex with 2-methoxy-estradiol is characterised by two seven-stranded β-sheets and the ligand intercalates between them (**1LHW**).

## Biological Functions

SHBG binds and transports sex steroids, e.g. testosterone and estradiol, and regulates their access to tissues and target cells.

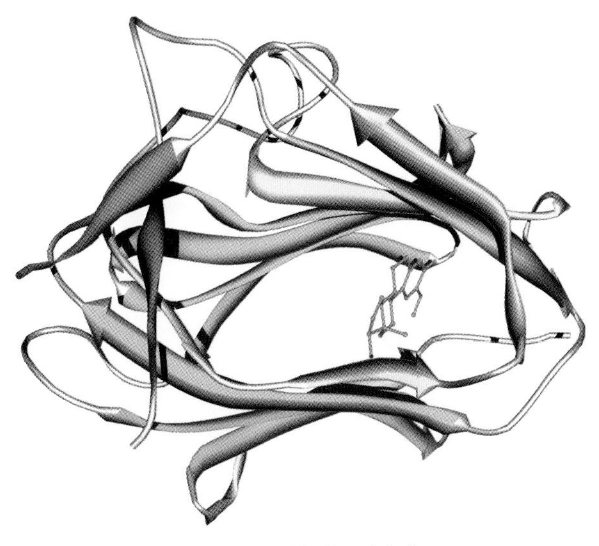

Sex hormone-binding globulin

## *Thyroxine-binding globulin (T4-binding globulin)*

### Fact Sheet

| | | | |
|---|---|---|---|
| ***Classification*** | Serpin | ***Abbreviations*** | TBG |
| ***Structures/motifs*** | | ***DB entries*** | THBG_HUMAN/P05543 |
| ***MW/length*** | 44 102 Da/395 aa | ***PDB entries*** | 2CEO |
| ***Concentration*** | 10–20 mg/l | ***Half-life*** | 5–6 days |
| ***PTMs*** | 5 N-CHO | | |
| ***References*** | Flink *et al.*, 1986, *Proc. Natl. Acad. Sci. USA*, **83**, 7708–7712.<br>Zhou *et al.*, 2006, *Proc. Natl. Acad. Sci. USA*, **103**, 13321–13326. | | |

### Description

**Thyroxine-binding globulin** (TBG) is synthesised in the liver and is involved in the thyroid hormone transport and exhibits structural similarity with serpins.

### Structure

TBG is a single-chain plasma glycoprotein (18–20 % CHO) with a single binding site for $T_3/T_4$ and exhibits structural similarity with serpins but has no known inhibitory activity. There are three well-documented variants: TBG-Poly (Leu 283 → Phe), TBG-S (Asp 171 → Asn) and TBG-A (Ala 191 → Thr). The 3D structure of TBG is different from other serpins, with the binding pocket for thyroxine (in pink) on the surface of the molecule (**2CEO**).

### Biological Functions

TBG is the major thyroid hormone ($T_3$ and $T_4$) transport protein.

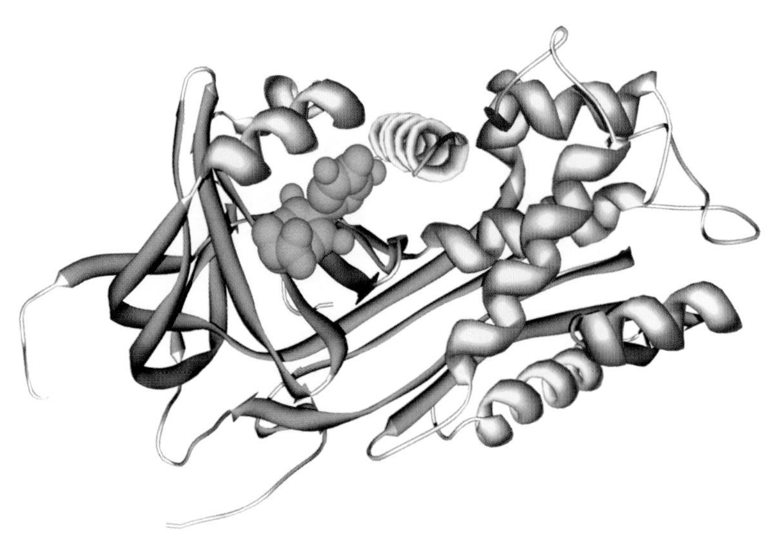

Thyroxine-binding globulin

# *Transthyretin (Prealbumin)*

## Fact Sheet

| | | | |
|---|---|---|---|
| ***Classification*** | Transthyretin | ***Abbreviations*** | TTR |
| ***Structures/motifs*** | | ***DB entries*** | TTHY_HUMAN/P02766 |
| ***MW/length*** | 13 761 Da/127 aa | ***PDB entries*** | 1RLB |
| ***Concentration*** | 250 mg/l | ***Half-life*** | 2 days |
| ***PTMs*** | | | |
| ***References*** | Monaco *et al.*, 1995, *Science*, **268**, 1039–1041.<br>Sasaki *et al.*, 1985, *Gene*, **37**, 191–197. | | |

## Description

**Transthyretin** (TTR) is synthesised in the liver, in the choroids plexus of the brain and the eye. TTR is a thyroid hormone-binding transport protein.

## Structure

TTR is a single-chain protein present as non-covalently-linked homotetramer that forms in plasma a 1:1 molar complex with retinol-binding protein (RBP). The 3D structure of monomeric TTR is characterised by two four-stranded β-sheets. Tetrameric TTR has a dimer of dimers configuration with a central channel containing two high-affinity binding sites for thyroid hormones (**1RLB**).

## Biological Functions

TTR is a thyroid hormone-binding protein that is involved in the transport of thyroxine from the bloodstream to the brain. TTR is an important component of the TTR–RBP complex transporting retinol.

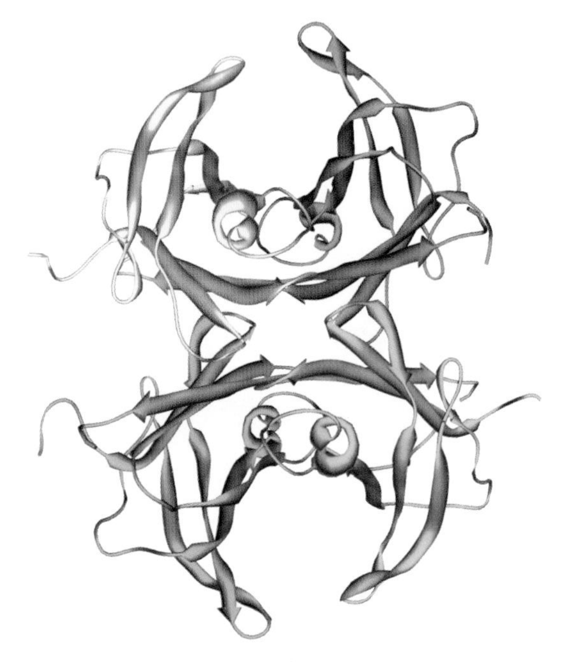

Transthyretin tetramer

# *Vitamin D-binding protein*

## Fact Sheet

| | | | |
|---|---|---|---|
| ***Classification*** | ALB/AFP/VDB | ***Abbreviations*** | DBP |
| ***Structures/motifs*** | 3 Albumin | ***DB entries*** | VTDB_HUMAN/P02774 |
| ***MW/length*** | 51 243 Da/458 aa | ***PDB entries*** | 1MA9 |
| ***Concentration*** | approx. 400 mg/l | ***Half-life*** | approx. 2.5 days |
| ***PTMs*** | 1 N-CHO (potential) | | |
| ***References*** | Cooke *et al.*, 1985, *J. Clin. Invest.*,**76**, 2420–2424.<br>Verboven *et al.*, 2003, *Acta Crystallogr. D*, **59**, 263–273. | | |

## Description

**Vitamin D-binding protein** (DBP) is a multifunctional blood plasma protein predominantly synthesised in the liver. It exhibits structural and functional similarity with serum albumin, afamin and α-fetoprotein.

## Structure

DBP is a single-chain plasma glycoprotein containing three albumin domains (domain 3 is only half size) with the characteristic disulfide bridge pattern. The 3D structure of DBP with three albumin domains (in yellow, green and blue; this domain is only half size) is highly helical and binds to subdomains 1 and 3 of α-actin shown in pink (**1MA9**).

## Biological Functions

DBP is the main carrier and reservoir of vitamin D sterols. It binds and transports fatty acids. It also binds monomeric actin and thus prevents its polymerisation.

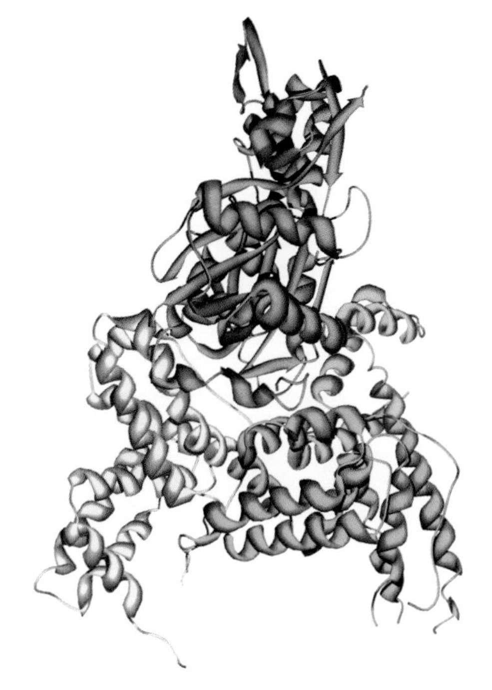

Vitamin D-binding protein in complex with α-actin sub domains

# 16

# Additional Proteins

## 16.1 INTRODUCTION

Many proteins exist in blood plasma that are difficult to assign unambiguously to one of the previous chapters. However, the proteins discussed in this chapter are nevertheless important blood plasma proteins. The assignment to this chapter is based on information obtained from literature and on personal considerations. Although 3D structural data are not available for all proteins presented in this chapter, a brief discussion of their structural features and function seems to be adequate.

## 16.2 ADDITIONAL BLOOD PLASMA PROTEINS

**$\alpha_2$-HS-glycoprotein** ($\alpha_2$-HS) (see Data Sheet) or fetuin-A synthesised in the liver is derived from the precursor by limited proteolysis removing an internal 40 residue connecting peptide, thus resulting in two-chain $\alpha_2$-HS linked by a single interchain disulfide bridge (Cys 14 in chain A, Cys 18 in chain B). Chain A (282 aa) contains two cystatin-like domains with five intrachain disulfide bridges (Cys 71–Cys 82, Cys 96–Cys 114, Cys 128–Cys 131, Cys 190–Cys 201, Cys 212–Cys 239) and an S-Ser residue (120 aa, partial). Fetuin-A is closely related to fetuin-B (367 aa: Q9UGM5).

Fetuins are characterised by two consensus sequences both located in the cystatin-like domains containing the conserved Cys residues, which are involved in disulfide bridges:

FETUIN-1(PS01254): **C**–[DN]–[DE]–x(54)–**C**–H–x(9)–**C**–x(12,14)–**C**–x(17,19)–**C**–x(13)–**C**–x(2)–**C**

FETUIN-2(PS01255): [ND]– x–L–E–T–x–**C**–H– x–L

$\alpha_2$-HS promotes endocytosis and possesses opsonic properties. It influences the mineral phase of bones and accumulates in bones and teeth. $\alpha_2$-HS has an especially high concentration during fetal development and is one of the few negative acute-phase marker proteins.

**Chromogranin A** (CgA) (see Data Sheet) present in endocrine cells and neurons is a single-chain highly acidic glycoprotein (approximately 5 % O-CHO) containing four P-Ser residues (aa 200, 252, 304, 315) and a single intrachain disulfide bridge (Cys 17–Cys 38). Chromogranin A is the parent protein of several peptides derived by limited proteolysis at dibasic dipeptides. Among them are **vasostatin-1** (76 aa: Leu 1–Gln 76), **vasostatin-2** (113 aa: Leu 1–Glu 113) and **pancreastatin** (48 aa: Ser 254–Gly-$NH_2$ 301). The amide group of Gly-$NH_2$ in pancreastatin is provided from the adjacent C-terminal Gly residue in chromogranin A. Chromogranin A is structurally and functionally closely related to secretogranin-1 (chromogranin B) and secretogranin-2 (587 aa: P13521).

CgA seems to be a helper protein involved in the intracellular packaging and processing of peptide hormones and neuropeptides. Pancreastatin inhibits the glucose-induced insulin release from pancreas. CgA is the most widespread marker of the matrix of neuroendocrine secretory granules.

**Secretogranin-1** (SgI) (see Data Sheet) or chromogranin B present in endocrine cells and neurons is a single-chain highly acidic glycoprotein (approximately 5 % O-CHO) containing two P-Ser residues (aa 129, 385), two S-Tyr residues (aa 153 (potential), aa 321) and a single intrachain disulfide bridge (Cys 16–Cys 37). SgI is the parent protein of biologically active

*Human Blood Plasma Proteins: Structure and Function* Johann Schaller, Simon Gerber, Urs Kämpfer, Sofia Lejon and Christian Trachsel

**Figure 16.1 3D structure of cytoplasmic horse gelsolin (PDB: 2FGH)**
3D structure of cytoplasmic horse gelsolin contains six globular gelsolin-like repeats (each in a different colour) and ATP (in pink) binds in a positively charged pocket.

peptides such as the **GAWK peptide** (74 aa: Phe 420–Glu 493 in SgI) and the **CCB peptide** (60 aa: Ser 597–Arg 656) generated by limited proteolysis at basic residues. SgI seems to be a helper protein involved in the intracellular packaging and processing of peptide hormones and neuropeptides. Secretogranin-1 is structurally and functionally closely related to secretogranin-2 and chromogranin A.

**Gelsolin** (see Data Sheet) or brevin exists in two isoforms derived from the same gene by alternative splicing: isoform 1 or secreted plasma gelsolin and isoform 2 or cytoplasmic gelsolin. Plasma gelsolin contains six gelsolin-like repeats, a single intrachain disulfide bridge (Cys 188–Cys 201) and five P-Tyr residues (*in vitro*). Gelsolin is a calcium-regulated protein that binds to actin and fibronectin and is involved in the regulation of actin polymerisation. Plasma gelsolin is involved in the clearing of actin filaments from the circulation, a process termed 'actin scavenging'.

The 3D structure of entire human gelsolin is not available. However, the 3D structure of entire cytoplasmic horse gelsolin (731 aa: Met 0–Ala 730, Q28372) containing a bound ATP molecule was determined by X-ray diffraction (PDB: 2FGH) and is shown in Figure 16.1. Horse gelsolin exhibits a sequence identity of 92 % with the human species. Gelsolin contains six globular gelsolin-like repeats G1–G6 (each in a different colour) characterised by a central β-sheet structure flanked by a large and a small α-helix. ATP (in pink) binds in a positively charged pocket on the surface of gelsolin in domains G3, G4 and G5.

Defects in the gelsolin gene are the cause of a variety of systemic amyloidosis. The Finnish type is characterised by cranial neuropathy and lattice corneal dystrophy manifested by facial paresis (local amyloid deposition, for details see MIM: 105120).

**Ficolin-3** (see Data Sheet) synthesised in the liver and the lung is a single-chain glycoprotein containing one collagen-like and one fibrinogen C-terminal domain, is N-terminally modified (pyro-Glu) and carries six OH-Pro residues. Ficolin-3 is present as disulfide-linked homopolymer and is involved in the serum lectin activity; it exhibits an affinity for various mono- or oligosaccharides.

**Leptin** (LEP) (see Data Sheet) or obesity factor present in adipose tissue is a single-chain protein containing a single intrachain disulfide bridge (Cys 96–Cys 146). LEP seems to be part of the signalling pathway regulating the amount of the body fat depot and seems to be involved in the regulation of the energy balance as part of the homeostatic mechanism to maintain a constant adipose mass.

The 3D structure of entire leptin (146 aa: Val 1–Cys 146) was determined by X-ray diffraction (PDB: 1AX8) and is shown in Figure 16.2. The structure is characterised by a classical four-helix bundle topology similar to that of long-chain helical cytokines and the structure is stabilised by a single intrachain disulfide bridge (in pink). (grey dashed line is modelled).

Defects in the leptin gene may be the cause of autosomal recessive obesity (for details see MIM: 601665).

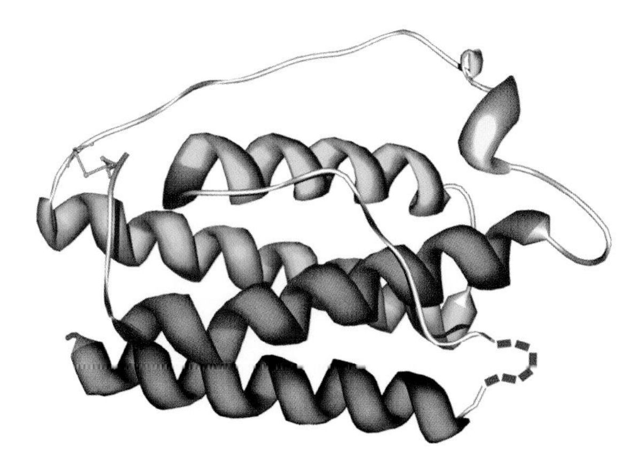

**Figure 16.2 3D structure of leptin (PDB: 1AX8)**
The 3D structure of leptin is characterised by a classical four-helix bundle topology similar to that of long-chain helical cytokines and the structure is stabilised by a single intrachain disulphide bridge (in pink). (grey dashed line is modelled).

**Osteonectin** (see Data Sheet) or SPARC or BM-40, most abundant in bone tissue, is a single-chain glycoprotein (approximately 5 % CHO) containing one follistatin-like domain and an EF-hand domain and seven intrachain disulfide bridges. The N-terminal Asp/Glu-rich region (53 aa, 18 acidic but no basic residues) binds five to eight calcium ions with low affinity and an EF-hand loop one $Ca^{2+}$ with high affinity. Osteonectin is involved in tissue remodelling and seems to regulate cell growth via interactions with the extracellular matrix. The binding to thrombospondin, plasminogen and tPA suggests an important role in formation and regulation of the extracellular matrix.

The 3D structure of osteonectin lacking the N-terminal Asp/Glu-rich region (233 aa: Pro 54–Ile 286) was determined by X-ray diffraction (PDB: 1BMO) and is shown in Figure 16.3. Osteonectin contains two main domains: the elongated

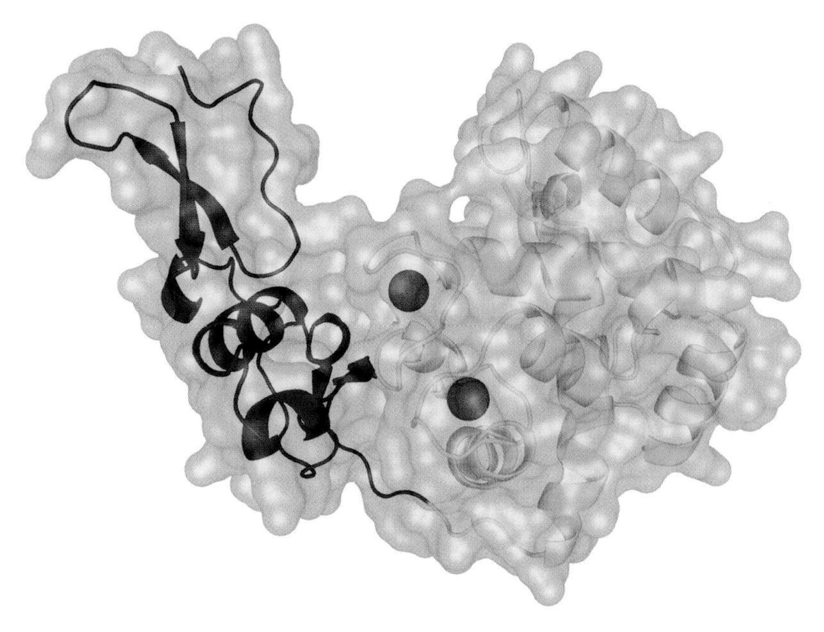

**Figure 16.3 3D structure of osteonectin lacking the N-terminal Asp/Glu-rich region (PDB: 1BMO)**
The 3D structure of osteonectin lacking the N-terminal Asp/Glu-rich region (233 aa: Pro 54 – Ile 286) contains an elongated follistatin-like domain (in blue) structurally related to the Kazal-type serine protease inhibitors and an EF-hand calcium-binding domain (in yellow) that binds to calcium ions (pink balls).

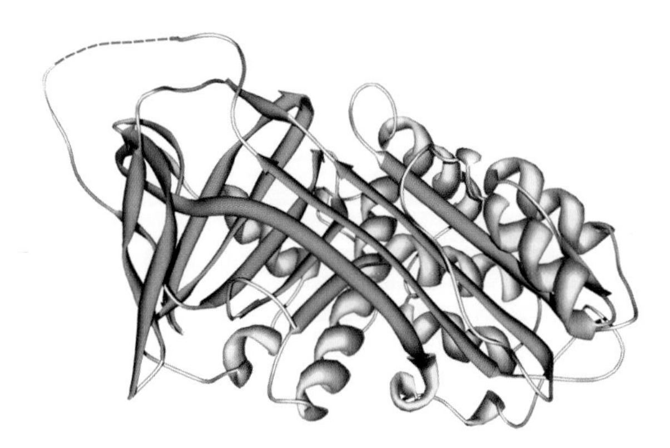

**Figure 16.4 3D structure of pigment epithelial-derived factor (PDB: 1IMV)**
The 3D structure of pigment epithelial-derived factor exhibits the overall topology of a serpin. (The dashed grey line is modelled).

follistatin-like domain (in blue) consisting of a β-hairpin and a small hydrophobic core of α/β structure structurally related to the Kazal-type serine protease inhibitors and the EF-hand calcium-binding domain (in yellow) characterised by helices containing a pair of EF-hand calcium-binding sites (calcium ions are shown as pink balls).

**Pigment epithelium-derived factor** (PEDF) (see Data Sheet) is a single-chain glycoprotein carrying an N-terminal pyro-Glu residue and belongs to the serpin family, but exhibits no inhibitory activity as it does not undergo the conformational change from the stressed to the relaxed state (S → R transition). PEDF is the most potent inhibitor of angiogenesis in the ocular compartment and exhibits neurotrophic activity.

The 3D structure of pigment epithelial-derived factor (398 aa: Asn 2–Pro 399) was determined by X-ray diffraction (PDB: 1IMV) and is shown in Figure 16.4. PEDF exhibits the overall topology of a serpin but unlike any other serpin it has a striking asymmetric charge distribution, which might be of functional importance.

**$\alpha_1$-B-glycoprotein** (see Data Sheet) is a glycoprotein (13 % CHO) consisting of five Ig-like V type domains with each domain containing a single intrachain disulfide bridge (Cys 28–Cys 72, Cys 118–Cys 161, Cys 211–Cys 258, Cys 304–Cys 353, Cys 402–Cys 449). Although structurally closely related to the Ig-like domains in immunoglobulins, which exhibit binding properties, the exact function of $\alpha_1$-B-glycoprotein remains unknown.

**Leucine-rich $\alpha_2$-glycoprotein** (LRG) (see Data Sheet), present in neutrophils, is a single-chain leucine-rich (21 %) glycoprotein (23 % CHO) containing 13 segments of 24 residues each, eight of which exhibit a periodic pattern of leucine distribution termed leucine-rich repeats (LRRs), with the consensus sequence

P–X(2)–**L**–**L**–X(5)–**L**–X(2)–**L**– X–**L**–X(2)–N–X–**L**–X(2)–**L**

and two intrachain disulfide bridges (Cys 8–Cys 21, Cys 268–Cys 294). LRG seems to be involved in early neutrophilic granulocyte differentiation and protein–protein interactions, but its physiological role is unknown.

**Pregnancy-specific $\beta_1$-glycoprotein 1** (PSBG1) (see Data Sheet) produced by the placenta in high quantity during pregnancy is a member of the PSBG family (PSBG1–PSGB11). PSBG1 is a single-chain glycoprotein containing one Ig-like V type and three Ig-like C2 type domains and three intrachain disulfide bridges (Cys 135–Cys 183, Cys 228–Cys 276, Cys 320–Cys 360). Seven members of the family contain an **RGD** sequence as a potential cell attachment site. PSBG has been used as a marker to diagnose pregnancy and to predict certain pregnancy-related complications.

**Lipopolysaccharide-binding protein** (LBP) (see Data Sheet) is a single-chain glycoprotein that binds to the lipopolysaccharides (LPS) in bacteria and participates in the LPS-mediated activation of cells. LBP is increased at least by a factor of 10 from 1–10 to 50–100 mg/l during the acute phase response.

**Clusterin** (see Data Sheet) or apolipoprotein J, synthesised in various types of tissues and platelets as a single-chain precursor glycoprotein (approximately 30 % CHO), is processed into mature two-chain clusterin by cleaving the Arg 205–Ser 206 peptide bond. The two chains are arranged in antiparallel orientation and are linked by five interchain disulfide bridges: Cys 80 (β)–Cys 86 (α), Cys 91 (β)–Cys 78 (α), Cys 94 (β)–Cys 75 (α), Cys 99 (β)–Cys 68 (α) and Cys 107 (β)–Cys 58 (α). Clusterin is associated with hydrophobic complexes such as the membrane attack complex C5b,6, 7 of the complement system, immunoglobulin complexes, high-density lipoproteins and neutral lipids.

**Insulin-like growth factor-binding protein complex acid labile chain** (ALS, see Data Sheet) is a single-chain leucine-rich (22 %) glycoprotein containing 20 leucine-rich repeats (LRRs) characterised by a distinct consensus sequence that is similar to but not identical with the corresponding consensus sequence in leucine-rich $\alpha_2$-glycoprotein:

P–P– x–A–F–x–G–**L**–G–x–**L**– x(2)–**L**–x–**L**–S–x–N–x–**L**–x(2)–**L**

ALS forms a heterotrimeric complex composed of either insulin-like growth factor I or II (IGF-I or IGF-II) and IGF-binding protein 3. The complex serves to bind IGF in order to increase its molecular weight and to inhibit the insulin-like function of IGF. ALS is involved in protein–protein interactions resulting in protein complex formation, receptor–ligand binding or cell adhesion.

DATA SHEETS

## *Chromogranin A*

### Fact Sheet

| | | | |
|---|---|---|---|
| ***Classification*** | Chromogranin/secretogranin | ***Abbreviations*** | CgA |
| ***Structures/ motifs*** | | ***DB entries*** | CMGA_HUMAN/ P10645 |
| ***MW/length*** | Chromogranin A: 48 960 Da/439 aa<br>Vasostatin-1: 8556 Da/76 aa<br>Vasostatin-2: 12702 Da/113 aa<br>Pancreastatin: 5080 Da/48 aa | ***PDB entries*** | |
| ***Concentration*** | approx. 40 μg/l | ***Half-life*** | approx.16 min |
| ***PTMs*** | Chromogranin: 3 O-CHO; 4 P-Ser; pancreastatin: Gly-$NH_2$ | | |
| ***References*** | Konecki *et al.*, 1987, *J. Biol. Chem.*, **262**, 17026–17030. | | |

### Description

**Chromogranin A** (CgA) is present in endocrine cells and neurons and seems to be involved in the processing of peptide hormones and neuropeptides. CgA exhibits structural and functional similarity with chromogranin B and secretogranin-2.

### Structure

CgA is a highly acidic single-chain glycoprotein (approx. 5 % O-CHO) containing four P-Ser residues and exhibits calcium-binding affinity. CgA is the parent protein of vasostatin-1 (76 aa), vasostatin-2 (113 aa) and pancreastatin (48 aa) and other peptides that are generated by limited proteolysis at dibasic dipeptides.

### Biological Functions

CgA seems to be a helper protein involved in the intracellular packaging and processing of peptide hormones and neuropeptides. Pancreastatin inhibits the glucose-induced insulin release from pancreas. CgA is the most widespread marker of the matrix of neuroendocrine secretory granules.

## *Clusterin (Apolipoprotein J)*

### Fact Sheet

| | | | |
|---|---|---|---|
| ***Classification*** | Clusterin | ***Abbreviations*** | CLI |
| ***Structures/motifs*** | | ***DB entries*** | CLUS_HUMAN/P10909 |
| ***MW/length*** | 50 063 Da/427 aa | ***PDB entries*** | |
| ***Concentration*** | 35–105 mg/l | ***Half-life*** | |
| ***PTMs*** | 6 N-CHO | | |
| ***References*** | Kirszbaum *et al.*, 1989, *EMBO J.*, **8**, 711–718. | | |

### Description

**Clusterin** is widely expressed in various tissues and is capable of binding hydrophobic complexes.

### Structure

Precursor single-chain clusterin is proteolytically processed at the Arg 205–Ser 206 peptide bond, resulting in mature two-chain glycoprotein (approx. 30 % CHO) clusterin consisting of the β-chain (205 aa) and the α-chain (222 aa). The two chains are linked by five interchain disulfide bridges in an antiparallel orientation.

### Biological Functions

Clusterin binds to hydrophobic complexes such as the C5b,6,7 complex of the complement system, to neutral lipids and high-density lipoproteins.

## *Ficolin-3*

### Fact Sheet

| | | | |
|---|---|---|---|
| ***Classification*** | Ficolin lectin | ***Abbreviations*** | |
| ***Structures/motifs*** | 1 collagen-like; 1 fibrinogen C-terminal | ***DB entries*** | FCN3_HUMAN/O75636 |
| ***MW/length*** | 30 340 Da/276 aa | ***PDB entries*** | |
| ***Concentration*** | | ***Half-life*** | |
| ***PTMs*** | 1 N-CHO (potential); 6 OH-Pro; 1 Pyro-Glu | | |
| ***References*** | Sugimoto *et al.*, 1998, *J. Biol. Chem.*, **273**, 20721–20727 | | |

### Description

**Ficolin-3** is synthesised in the liver and the lung and is present as disulfide-linked homopolymer.

### Structure

Ficolin-3 is a single-chain N-terminally modified (pyro-Glu) glycoprotein containing one collagen-like and one fibrinogen C-terminal domain and carries six OH-Pro residues.

### Biological Functions

Ficolin-3 is involved in serum lectin activity and has affinity for mono- and oligosaccharides.

# *Gelsolin (Brevin)*

## Fact Sheet

| | | | |
|---|---|---|---|
| ***Classification*** | Villin/gelsolin | ***Abbreviations*** | GSN |
| ***Structures/motifs*** | 6 Gelsolin-like | ***DB entries*** | GELS_HUMAN/P06396 |
| ***MW/length*** | 82 959 Da/755 aa | ***PDB entries*** | 2FGH |
| ***Concentration*** | 100–400 mg/l | ***Half-life*** | approx. 2.3 days |
| ***PTMs*** | 5 P-Tyr (*in vitro*) | | |
| ***References*** | Kwiatkowski *et al.*, 1986, *Nature*, **323**, 455–458.<br>Urosev *et al.*, 2006, *J. Mol. Biol.*, **357**, 765–772. | | |

## Description

**Gelsolin** exists in two isoforms derived from the same gene by alternative splicing: isoform 1, the secreted, plasma form and isoform 2, the cytoplasmic form. Gelsolin is a calcium-regulated actin-modulating protein.

## Structure

Plasma gelsolin (isoform 1) is a single-chain protein containing six gelsolin-like repeats and a single intrachain disulfide bridge (Cys 188–Cys 201). Gelsolin contains six globular gelsolin-like repeats characterised by a central β-sheet structure flanked by a large and a small α-helix (2FGH). ATP (in pink) is bound into a positively charged pocket on the surface of the molecule.

## Biological Functions

Gesolin binds to actin and fibronectin. Plasma gelsolin is involved in the clearing of actin filaments from circulation, a process termed 'actin scavenging'.

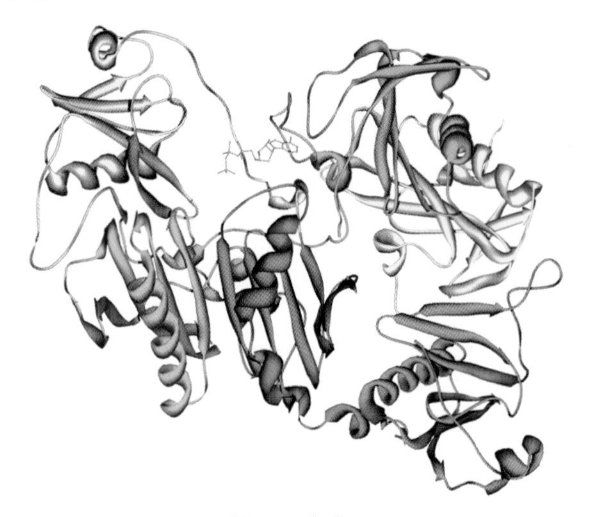

Horse gelsolin

# *$\alpha_1$-B-glycoprotein*

## Fact Sheet

| | | | |
|---|---|---|---|
| ***Classification*** | | ***Abbreviations*** | |
| ***Structures/motifs*** | 5 Ig-like V type | ***DB entries*** | A1BG_HUMAN/P04217 |
| ***MW/length*** | 51 941 Da/474 aa | ***PDB entries*** | |
| ***Concentration*** | 220 mg/l | ***Half-life*** | |
| ***PTMs*** | 4 N-CHO | | |
| ***References*** | Ishioka *et al.*, 1986, *Proc. Natl. Acad. Sci. USA*, **83**, 2363–2367. | | |

## Description

**$\alpha_1$-B-glycoprotein** is a plasma glycoprotein.

## Structure

$\alpha_1$-B-glycoprotein is a single-chain glycoprotein (13 % CHO) containing five Ig-like V type domains and each domain contains a single intrachain disulfide bridge.

## Biological Functions

Although structurally related to the Ig-like domain immunoglobulins, the function of $\alpha_1$-B-glycoprotein is unknown.

# *$\alpha_2$-HS-glycoprotein (Fetuin-A)*

## Fact Sheet

| | | | |
|---|---|---|---|
| ***Classification*** | Fetuin | ***Abbreviations*** | $\alpha_2$-HS |
| ***Structures/motifs*** | 2 Cystatin-like | ***DB entries*** | FETUA_HUMAN/P02765 |
| ***MW/length*** | Chain A: 30 222 Da/282 aa<br>Chain B: 27 40 Da/27 aa | ***PDB entries*** | |
| ***Concentration*** | 400–850 mg/l | ***Half-life*** | approximately 1.5 days |
| ***PTMs*** | Chain A: 2 N-CHO; 2 O-CHO; 1 P-Ser (partial, aa 120); chain B: 1 O-CHO | | |
| ***References*** | Lee *et al.*, 1987, *Proc. Natl. Acad. Sci. USA*, **84**, 4403–4407. | | |

## Description

$\alpha_2$-HS-glycoprotein ($\alpha_2$-HS) is synthesised in the liver and plays a role in bone formation and resorption. It is closely related to fetuin-B in structure and function.

## Structure

$\alpha_2$-HS is a two-chain glycoprotein derived from the precursor by removing a 40 residue connecting propeptide. Chain A contains two cystatin-like domains with five intrachain disulfide bridges and is partially phosphorylated at Ser 120. The two chains are linked by a single disulfide bridge (Cys 14 in chain A and Cys 18 in chain B).

## Biological Functions

$\alpha_2$-HS promotes endocytosis, possesses opsonic properties and influences the mineral phase of bones. $\alpha_2$-HS is especially high in fetal development and is one of the few negative acute-phase marker proteins.

## *Insulin-like growth factor-binding protein complex acid labile chain (ALS)*

### Fact Sheet

| | | | |
|---|---|---|---|
| ***Classification*** | | ***Abbreviations*** | ALS |
| ***Structures/motifs*** | 20 LRR (leucine-rich) | ***DB entries*** | ALS_HUMAN/P35858 |
| ***MW/length*** | 63 247 Da/578 aa | ***PDB entries*** | |
| ***Concentration*** | approx. 25 mg/l | ***Half-life*** | |
| ***PTMs*** | 6 N-CHO (potential) | | |
| ***References*** | Leong *et al.*, 1992, *Mol. Endocrinol.*, **6**, 870–876. | | |

### Description

The **insulin-like growth factor-binding protein complex acid labile chain** (ALS) is involved in protein–protein interactions.

### Structure

ALS is a single-chain leucine-rich (22 %) glycoprotein containing 20 LRR repeats. ALS forms a heterotrimeric complex with insulin-like growth factors I or II and insulin-like growth factor-binding protein 3.

### Biological Functions

ALS is involved in protein–protein interactions resulting in protein complexes, receptor–ligand binding or cell adhesion.

## *Leptin (obesity factor)*

### Fact Sheet

| | | | |
|---|---|---|---|
| ***Classification*** | Leptin | ***Abbreviations*** | LEP |
| ***Structures/motifs*** | | ***DB entries*** | LEP_HUMAN/P41159 |
| ***MW/length*** | 16 026 Da/146 aa | ***PDB entries*** | 1AX8 |
| ***Concentration*** | approx. 10 μg/l | ***Half-life*** | |
| ***PTMs*** | | | |
| ***References*** | Zhang *et al.*, 1994, *Nature*, **372**, 425–432.<br>Zhang *et al.*, 1997, *Nature*, **387**, 206–209. | | |

### Description

**Leptin** (LEP) is an obesity factor present in adipose tissues.

### Structure

LEP is a single-chain plasma protein containing a single intrachain disulfide bridge (Cys 96–Cys 146). The 3D structure is characterised by a classical four-helix bundle topology similar to the long-chain helical cytokines (**1AX8**).

### Biological Functions

LEP seems to be part of the pathway regulating the size of the body fat depot. It seems to be involved in the regulation of the energy balance as part of the homeostatic mechanism to maintain a constant adipose mass.

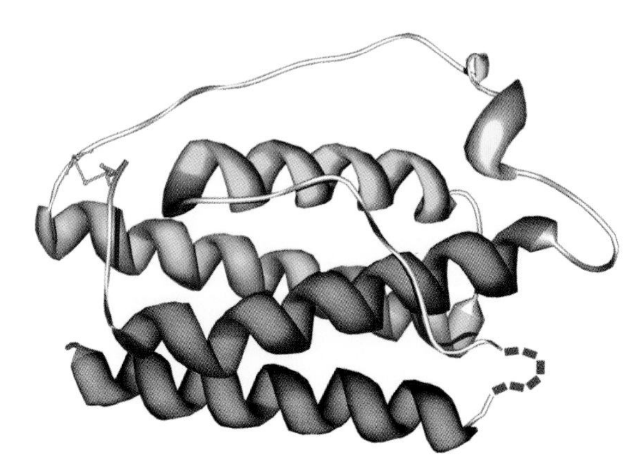

Leptin

## *Leucine-rich $\alpha_2$-glycoprotein*

### Fact Sheet

| | | | |
|---|---|---|---|
| ***Classification*** | | ***Abbreviations*** | LRG |
| ***Structures/motifs*** | 8 LRR (leucine-rich) | ***DB entries*** | A2GL_HUMAN/P02750 |
| ***MW/length*** | 34 346 Da/312 aa | ***PDB entries*** | |
| ***Concentration*** | approx. 20 mg/l | ***Half-life*** | |
| ***PTMs*** | 5 N-CHO (1 potential); 1 O-CHO | | |
| ***References*** | Takahashi *et al.*, 1985, *Proc. Natl. Acad. Sci. USA*, **82**, 1906–1910. | | |

### Description

**Leucine-rich $\alpha_2$-glycoprotein** (LRG) is present in neutrophils but its function is still unknown.

### Structure

LRG is a single-chain glycoprotein (23 % CHO) containing eight leucine-rich repeats (LRR) and two intrachain disulfide bridges (Cys 8–Cys 21, Cys 268–Cys 294).

### Biological Functions

LRG seems to be involved in early neutrophilic granulocyte differentiation but its physiological role is unknown.

## *Lipopolysaccharide-binding protein*

### Fact Sheet

| | | | |
|---|---|---|---|
| ***Classification*** | BPI/LBP | ***Abbreviations*** | LBP |
| ***Structures/motifs*** | | ***DB entries*** | LBP_HUMAN/P18428 |
| ***MW/length*** | 50 875 Da/456 aa | ***PDB entries*** | |
| ***Concentration*** | 1–10 mg/l | ***Half-life*** | |
| ***PTMs*** | 4 N-CHO (4 potential) | | |
| ***References*** | Schumann *et al.*, 1990, *Science*, **249**, 1429–1431. | | |

### Description

**Lipopolysaccharide-binding protein** (LBP) binds lipopolysaccharides (LPSs) in bacteria.

### Structure

LBP is a single-chain plasma glycoprotein.

### Biological Functions

LBP binds to lipopolysaccharides in the outer membrane of all gram-negative bacteria. During an acute-phase response LBP increases to 50–100 mg/l.

# *Osteonectin (SPARC)*

## Fact Sheet

| | | | |
|---|---|---|---|
| ***Classification*** | SPARC | ***Abbreviations*** | SPARC |
| ***Structures/motifs*** | 1 EF-hand; 1 follistatin-like | ***DB entries*** | SPRC_HUMAN/P09486 |
| ***MW/length*** | 32 698 Da/286 aa | ***PDB entries*** | 1BMO |
| ***Concentration*** | 2 μg/2 × $10^8$ platelets | ***Half-life*** | |
| ***PTMs*** | 1 N-CHO (probable) | | |
| ***References*** | Lankat-Buttgereit *et al.*, 1988, *FEBS Lett.*, **236**, 352–356.<br>Hohenester *et al.*, 1997, *EMBO J.*, **16**, 3778–3786. | | |

## Description

**Osteonectin** (SPARC) is most abundant in bone tissue and is involved in tissue remodelling.

## Structure

SPARC is a single-chain glycoprotein (approx. 5 % CHO) containing one follistatin-like and one EF-hand domain and an Asp/Glu-rich N-terminal region that binds calcium. The 3D structure of osteonectin lacking the Asp/Glu-rich N-terminal region contains an elongated follistatin-like domain (in blue) resembling a Kazal-type structure and an EF-hand calcium-binding domain (in yellow) interacting with two calcium ions (pink balls) (**1BMO**).

## Biological Functions

The binding to thrombospondin, plasminogen and tissue-type plasminogen activator suggests an important role in extracellular matrix formation and regulation.

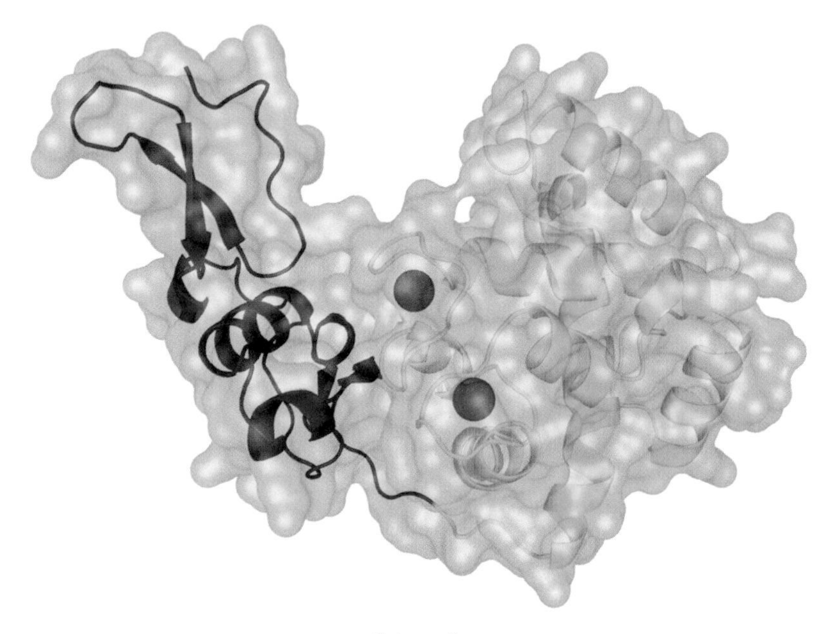

Osteonectin

# *Pigment epithelium-derived factor*

## Fact Sheet

| | | | |
|---|---|---|---|
| ***Classification*** | Serpin | ***Abbreviations*** | PEDF |
| ***Structures/motifs*** | | ***DB entries*** | PEDF_HUMAN/P36955 |
| ***MW/length*** | 44 418 Da/399 aa | ***PDB entries*** | 1IMV |
| ***Concentration*** | approx. 5 mg/l | ***Half-life*** | |
| ***PTMs*** | 1 N-CHO; 1 Pyro-Glu | | |
| ***References*** | Steele *et al.*, 1993, *Proc. Natl. Acad. Sci. USA*, **90**, 1526–1530.<br>Simonovic *et al.*, 2001, *Proc. Natl. Acad. Sci. USA*, **98**, 11131–11135. | | |

## Description

**Pigment epithelium-derived factor** (PEDF) present in retinal pigment epithelial cells and in blood plasma is a serpin with no inhibitory activity.

## Structure

PEDF is a single-chain plasma glycoprotein carrying an N-terminal pyro-Glu residue that shows structural similarity with serpins but exhibits no known inhibitory activity because it does not undergo the conformational transition from the stressed to the relaxed conformation. The 3D structure resembles that of other serpins but the charge distribution is asymmetrical (**1IMV**).

## Biological Functions

PEDF is the most potent inhibitor of angiogenesis in the ocular compartment and exhibits neurotrophic activity.

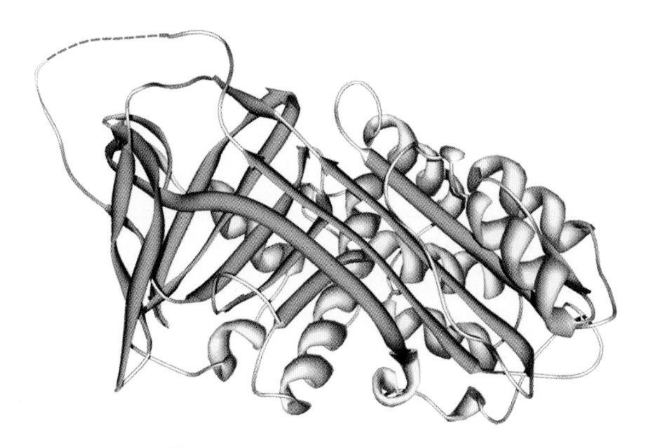

Pigment epithelium-derived factor

## *Pregnancy-specific $\beta_1$-glycoprotein 1*

### Fact Sheet

| | | | |
|---|---|---|---|
| ***Classification*** | Immunoglobulin | ***Abbreviations*** | PSBG1 |
| ***Structures/motifs*** | 3 Ig-like C2 type; 1Ig-like V type | ***DB entries*** | PSG1_HUMAN/P11464 |
| ***MW/length*** | 43496 Da/385 aa | ***PDB entries*** | |
| ***Concentration*** | 200–400 mg/l | ***Half-life*** | |
| ***PTMs*** | 7 N-CHO (potential) | | |
| ***References*** | Streydio *et al.*, 1988, *Biochem. Biophys. Res. Commun.*, **154**, 130–137. | | |

### Description

The family of **pregnancy-specific $\beta_1$-glycoproteins** (PSBGs) comprises eleven glycoproteins (PSBG1–PSBG11) that are produced in high quantity during pregnancy.

### Structure

PSBG1 is a single-chain glycoprotein containing one Ig-like V type and three Ig-like C2 type domains. PSBG1 has three probable interchain disulfide bridges (Cys 135–Cys 183, Cys 228–Cys 276, Cys 320–Cys 360). All eleven PSBGs contain one Ig-like V type and two (PSBG2, 5 and 11) or three (all others) Ig-like C2 type domains and seven PSBGs have a potential cell attachment site (**RGD** sequence).

### Biological Functions

PSBGs are major secretory products of the human placenta and are used clinically to diagnose pregnancy and to predict pregnancy-related complications.

## *Secretogranin-1(Chromogranin B)*

### Fact Sheet

| | | | |
|---|---|---|---|
| ***Classification*** | Chromogranin/secretogranin | ***Abbreviations*** | SgI |
| ***Structures/motifs*** | | ***DB entries*** | SCG1_HUMAN/P05060 |
| ***MW/length*** | Secretogranin-1: 76296 Da/657 aa<br>GAWK peptide: 8793 Da/74 aa<br>CCB peptide: 6975 Da/60 aa | ***PDB entries*** | |
| ***Concentration*** | | ***Half-life*** | |
| ***PTMs*** | 2 P-Ser (129, aa 385); 2 S-Tyr (153, aa 321) | | |
| ***References*** | Benedum *et al.*, 1987, *EMBO J.*, **6**, 1203–1211.<br>Benjannet *et al.*, 1987, *FEBS Lett.*, **224**, 142–148. | | |

### Description

**Secretogranin-1** (SgI) is present in endocrine cells and neurons and seems to be involved in the processing of peptide hormones and neuropeptides. SgI exhibits structural and functional similarity with chromogranin A and secretogranin-2.

### Structure

SgI is a highly acidic single-chain glycoprotein (approx. 5 % O-CHO) containing two P-Ser and two S-Tyr residues. CgB is the parent protein of the GAWK peptide and the CCB, which are generated by limited proteolysis at basic dipeptides.

### Biological Functions

SgI seems to be a helper protein involved in the intracellular packaging and processing of peptide hormones and neuropeptides.

# Appendix

*Human Blood Plasma Proteins: Structure and Function* Johann Schaller, Simon Gerber, Urs Kämpfer, Sofia Lejon and Christian Trachsel

**Table A.1** Additional human proteins.

| Name | Function | SwissProt/PDB | Chapter (number) |
|---|---|---|---|
| Blood group Rh(CE) polypeptide | Rhesus blood group system | P18577 | Blood Components (2) |
| Blood group Rh(D) polypeptide | Rhesus blood group system | Q02161 | Blood Components (2) |
| Fibulin-1 | Haemostasis and thrombosis | P23142 | Domains, Motifs and Repeats (4) |
| Hepatocyte growth factor | Growth factor various types tissues | P14210 | Domains, Motifs and Repeats (4) |
| Hepatocyte growth factor-like | Hyaluronan-binding | P26927 | Domains, Motifs and Repeats (4) |
| Hepatocyte growth factor activator | Activator hepatocyte growth factor | Q04756 | Domains, Motifs and Repeats (4) |
| Hepatocyte growth factor activator-like | Probably activator growth factors | Q14520 | Domains, Motifs and Repeats (4) |
| Tenascin | Extracellular matrix protein | P24821 | Domains, Motifs and Repeats (4) |
| (Pro)-relaxin H1/H2/H3 | Hormones | P04808/P04090/Q8WXF3 | Protein Families (5) |
| Insulin-like peptides INSL5/L6 | Development | Q9Y5Q6/Q9Y581 | Protein Families (5) |
| Calcitonin gene-related peptide I/II | Vasodilatation | P06881; 1LS7/P10092 | Protein Families (5) |
| Islet amyloid polypeptide | Inhibition glucose | P10997 | Protein Families (5) |
| Interferon delta | Cytokine | P37290 | Protein Families (5) |
| Serine/threonine protein phosphatases (EC 3.1.3.16) | Dephosphorylation | P62136 | Posttranslational Modifications (6) |
| Dual specificity protein phosphatases 1 (EC 3.1.3.48) | Dephosphorylation | P28562 | Posttranslational Modifications (6) |
| Lysyl hydroxylases 1 (EC 1.14.11.4) | Hydroxylation Lys | Q02809 | Posttranslational Modifications (6) |
| Protein-tyrosine sulfotransferase 1 (EC 2.8.2.20) | O-sulfation Tyr | O60507 | Posttranslational Modifications (6) |
| Pyroglutamyl-peptidase 1 (EC 3.4.19.3) | Release Pyro-Glu | Q9NXJ5 | Posttranslational Modifications (6) |
| Vitamin K-dependent $\gamma$-carboxylase (EC 6.4.-.-) | Carboxylation Glu | P38435 | Posttranslational Modifications (6) |
| Spectrin: $\alpha$-chain/$\beta$-chain | Platelet cytoskeleton | P02549/P11277 | Blood Coagulation and Fibrinolysis (7) |
| $\beta$-Actin | Platelet cytoskeleton | P60709 | Blood Coagulation and Fibrinolysis (7) |
| Band 4.1 protein | Platelet cytoskeleton | P11171 | Blood Coagulation and Fibrinolysis (7) |
| Thrombospondin-2 | Adhesive glycoprotein | P35442 | Blood Coagulation and Fibrinolysis (7) |
| Laminin: $\alpha$-chain/$\beta$-chain/$\gamma$-chain | Cell-binding | P25391/P07942/P11047 | Blood Coagulation and Fibrinolysis (7) |
| Vitronectin receptor | Receptor | P06756 | Blood Coagulation and Fibrinolysis (7) |
| Urokinase plasminogen activator surface receptor | Receptor | Q03405 | Blood Coagulation and Fibrinolysis (7) |
| C1-binding protein | Complement-binding | Q07021 | Blood Coagulation and Fibrinolysis (7) |
| Thrombin-activatable fibrinolysis inhibitor (pancreatic) | Carboxypeptidase | P15086 | Blood Coagulation and Fibrinolysis (7) |
| Annexin A8 | Anticoagulant protein | P13928 | Blood Coagulation and Fibrinolysis (7) |
| Angiotensin-converting enzyme (testis) | Carboxypeptidase | P22966 | Blood Coagulation and Fibrinolysis (7) |
| Polymeric-immunoglobulin receptor | Receptor immunoglobulins | P01833 | The Immune System (9) |
| Lymphocyte antigen 96 | Innate immune response | Q9Y6Y9 | The Immune System (9) |
| HLA class II histocompatibility antigen $\gamma$-chain | HLA class II antigen stabilisation | P04233/1A6A | The Immune System (9) |
| Toll-like receptor 4 | Innate immune response | O00206 | The Immune System (9) |
| Glutathione reductase (EC 1.8.1.7) | Reduction oxidised glutathione | P00390 | Enzymes (10) |
| Epididymal secretory glutathione peroxidase (EC 1.11.1.9) | Protection oxidative damage | O75715 | Enzymes (10) |
| Alanine aminotransferase 1 (EC 2.6.1.2) | Nitrogen metabolism | P24298 | Enzymes (10) |
| Carboxypeptidase A1 (EC 3.4.17.1) | Release C-terminal amino acids | P15085 | Enzymes (10) |
| Carboxypeptidase A2 (EC 3.4.17.15) | Release C-terminal amino acids | P48052 | Enzymes (10) |
| Tryptase $\alpha_1$ (EC 3.4.21.59) | Mast cell serine protease | P15157 | Enzymes (10) |
| Tryptase $\beta_1$ (EC 3.4.21.59) | Mast cell serine protease | Q15661 | Enzymes (10) |
| Tryptase $\delta$ (EC 3.4.21.59) | Mast cell serine protease | Q9BZJ3 | Enzymes (10) |
| Cathepsin S (EC 3.4.22.27) | Cysteine protease | P25774 | Enzymes (10) |
| Gastricsin (EC 3.4.23.3) | Aspartyl protease | P20142 | Enzymes (10) |

| | | | |
|---|---|---|---|
| Poly [ADP-ribose] polymerase 4 (EC 2.4.2.30) | Polymerase | Q9UKK3 | Enzymes (10) |
| Ribonuclease inhibitor | Inhibition RNase and angiogenin | P13489 | Enzymes (10) |
| Clathrin light chain A/B | Major protein coated pits | P09496/P09497 | Lipoproteins (12) |
| Clathrin heavy chain 1/2 | Major protein coated pits | Q00610/P53675 | Lipoproteins (12) |
| Pancreatic triacylglycerol lipase (EC 3.1.1.3) | Lipoprotein metabolism | P16233 | Lipoproteins (12) |
| Serum amyloid A-4 protein | Apolipoprotein HDL | P35542 | Lipoproteins (12) |
| Pancreatic polypeptide | Antagonist cholecystokinin | P01298 | Hormones (13) |
| Gastrin-releasing peptide | Release gastrin | P07492 | Hormones (13) |
| Motilin | Gastrointestinal motility | P12872 | Hormones (13) |
| Parathyroid hormone-related protein | Hormone | P12272 | Hormones (13) |
| Prolactin-releasing peptide | Hormone | P81277 | Hormones (13) |
| Islet amyloid polypeptide | Hormone | P10997 | Hormones (13) |
| Thyroid peroxidase (EC 1.11.1.8) | Thyroxine formation | P07202 | Hormones (13) |
| Follicle-stimulating hormone receptor | Hormone receptor | P23945/1XWD | Hormones (13) |
| Growth hormone receptor | Hormone receptor | P10912/1A22 | Hormones (13) |
| Atrial natriuretic peptide clearance receptor | Hormone receptor | P17342/1YK0 | Hormones (13) |
| Interleukin-18 | Cytokine: interferon γ inducer | Q14116 | Cytokines and Growth Factors (14) |
| Interleukin-33 | Cytokine: interleukin-1 like | O95760 | Cytokines and Growth Factors (14) |
| Interleukin-2 receptor α-chain | Cytokine receptor | P01589/2B5I | Cytokines and Growth Factors (14) |
| Interleukin-2 receptor β-chain | Cytokine receptor | P14784/2B5I | Cytokines and Growth Factors (14) |
| Interleukin-2 receptor γ-chain | Cytokine receptor | P31785/2B5I | Cytokines and Growth Factors (14) |
| Interleukin-23 subunit α | Cytokine | Q9NPF7 | Cytokines and Growth Factors (14) |
| Interleukin-6 receptor α-chain | Cytokine receptor | P08887/1P9M | Cytokines and Growth Factors (14) |
| Interleukin-6 receptor subunit β | Cytokine receptor | P40189/1P9M | Cytokines and Growth Factors (14) |
| Granulocyte colony-stimulating factor receptor | Cytokine receptor | Q99062/2D9Q | Cytokines and Growth Factors (14) |
| Interleukin-19 | Cytokine | Q9UHD0 | Cytokines and Growth Factors (14) |
| Interleukin-20 | Cytokine | Q9NYY1 | Cytokines and Growth Factors (14) |
| Interleukin-22 | Cytokine | Q9GZX6 | Cytokines and Growth Factors (14) |
| Interleukin-24 | Cytokine | Q13007 | Cytokines and Growth Factors (14) |
| Interleukin-26 | Cytokine | Q9NPH9 | Cytokines and Growth Factors (14) |
| Tumor necrosis factor receptor | Cytokine receptor | P19348/1TNR | Cytokines and Growth Factors (14) |
| Interferon alpha-2 | Cytokine | P01563 | Cytokines and Growth Factors (14) |
| Interferon gamma receptor α-chain | Cytokine receptor | P15260/1FG9 | Cytokines and Growth Factors (14) |
| Interferon delta-1 | Cytokine | P37290 | Cytokines and Growth Factors (14) |
| 2′-5′-Oligoadenylate synthetase 1 (EC 2.7.7.-) | Mediate resistance virus infection | P00973 | Cytokines and Growth Factors (14) |
| Placenta growth factor | Growth factor | P49763 | Cytokines and Growth Factors (14) |
| Platelet-derived growth factor D | Growth factor | Q9GZP0 | Cytokines and Growth Factors (14) |
| Transforming growth factor beta-4 | Growth factor | O00292 | Cytokines and Growth Factors (14) |
| Epidermal growth factor receptor | Growth factor receptor | P00533/1MOX | Cytokines and Growth Factors (14) |
| Brain-derived neurotrophic factor | Growth factor | P23560 | Cytokines and Growth Factors (14) |
| α-Actin-1 | Skeletal muscle | P68135/1MA9 | Transport and Storage (15) |
| Ovotransferrin | Iron-binding | P02789 | Transport and Storage (15) |
| Melanotransferrin | Iron-binding | P08582 | Transport and Storage (15) |
| Mucin-2 | Protection of epithelia | Q02817 | Transport and Storage (15) |
| Secretogranin-2 | Precursor biologically active peptides | P13521 | Additional Proteins (16) |
| Fetuin-B | Negative acute-phase marker | Q9UGM5 | Additional Proteins (16) |

**Table A.2** Additional nonhuman proteins.

| Name | Species | Function | SwissProt/PDB | Chapter (number) |
|---|---|---|---|---|
| Pancreatic trypsin inhibitor | Bovine | Serine protease inhibitor | P00974/1BZX | Protein Families (5) |
| Anionic trypsin | Salmon | Serine protease | P35031/1BZX | Protein Families (5) |
| Pancreatic lipase related protein 1 | Dog | Lipase (no activity) | P06857/1RP1 | Protein Families (5) |
| Myelomonocytic growth factor | Avian | Hematopoietic growth factor | P13854 | Protein Families (5) |
| Protein disulfide isomerase (EC5.3.4.1) | Yeast | Disulfide rearrangement | P32474 | Posttranslational Modifications (6) |
| Endoplasmic oxidoreductin-1 (EC1.8.4.-) | Yeast | Disulfide bond formation | Q03103 | Posttranslational Modifications (6) |
| Disulfide bond formation protein B | *E. coli* | Disulfide bond formation | P0A6M2 | Posttranslational Modifications (6) |
| Streptokinase A | *S. pyogenes* | Activation plasminogen | P10520 | Blood Coagulation and Fibrinolysis (7) |
| Streptokinase C | *S. equisimilis* | Activation plasminogen | P00779 | Blood Coagulation and Fibrinolysis (7) |
| Ecotin | *E. coli* | Protease inhibitor | P23827/1XX9 | Blood Coagulation and fibrinolysis (7) |
| Antistasin | Leech | Inhibitor FXa | P15358 | Blood Coagulation and Fibrinolysis (7) |
| Hirudin | Leech | Inhibitor thrombin | P01050 | Blood Coagulation and Fibrinolysis (7) |
| Trypsin inhibitor | Soybean | Inhibitor trypsin | P01070 | Blood Coagulation and Fibrinolysis (7) |
| Monocyte differentiation antigen CD14 | Mouse | Innate immune response | P10810 | The Immune System (9) |
| Aspartate aminotransferase, cytoplasmic | Porcine | Transfer nitrogenous groups | P00503 | Enzymes (10) |
| Papain (EC 3.4.22.2) | Papaya | Cysteine protease | P00784 | Enzymes (10) |
| Chymotrypsinogen A+B | Bovine | Serine protease | P00766/P00767 | Enzymes (10) |
| Ovalbumin | Chicken | Serpin (no inhibitory function) | P01012 | Inhibitors (11) |
| Calcitonin | Eel | Hormone (calcium metabolism) | P01262 | Hormones (13) |
| Oxytocin | Bovine | Contraction smooth muscle | P01175 | Hormones (13) |
| Vasopressin | Bovine | Regulation osmolarity | P01180 | Hormones (13) |
| Tumour necrosis factor receptor member 16 | Rat | Nerve growth factor receptor | P07174/1SG1 | Cytokines and Growth Factors (14) |
| Ovotransferrin | Chicken | Iron transport | P02789 | Transport and Storage (15) |
| Hemopexin | Rabbit | Heme transport/iron recovery | P20058/1QHU | Transport and Storage (15) |
| Gelsolin | Horse | Regulation actin polymerisation | Q28372/2FGH | Additional Proteins (16) |

**Table A.3** Blood coagulation and fibrinolysis (Chapter 7).

| Name | Function | Swiss-Prot |
|---|---|---|
| von Willebrand factor | Primary haemostasis/molecular glue | P04275 |
| Fibronectin | Primary haemostasis/adhesive protein | P02751 |
| Thrombospondin-1 | Primary haemostasis/adhesive protein | P07996 |
| Vitronectin | Primary haemostasis/adhesive protein | P04004 |
| Fibrinogen (factor I) | Common pathway/precursor | P02671/P02675/P02679 |
| Prothrombin (factor II, EC 3.4.21.5) | Common pathway/zymogen | P00374 |
| Tissue factor (factor III) | Extrinsic pathway/cofactor | P13726 |
| Coagulation factor V | Common pathway/cofactor | P12259 |
| Coagulation factor VII (EC 3.4.21.21) | Extrinsic pathway/zymogen | P08709 |
| Coagulation factor VIII | Common pathway/cofactor | P00451 |
| Coagulation factor IX (EC 3.4.21.22) | Intrinsic pathway/zymogen | P00740 |
| Coagulation factor X (EC 3.4.21.6) | Common pathway/zymogen | P00742 |
| Coagulation factor XI (EC 3.4.21.27) | Intrinsic pathway/zymogen | P03951 |
| Coagulation factor XII (EC 3.4.21.38) | Intrinsic pathway/zymogen | P00748 |
| Coagulation factor XIII (EC 2.3.2.13) | Common pathway/transglutaminase | P00488/P05160 |
| Plasma kallikrein (EC 3.4.21.34) | Intrinsic pathway/zymogen | P03952 |
| Vitamin K-dependent protein C (EC 3.4.21.69) | Regulation/zymogen | P04070 |
| Vitamin K-dependent protein S | Regulation/cofactor | P07225 |
| Vitamin K-dependent protein Z | Inhibition/cofactor | P22891 |
| Thrombomodulin | Anticoagulant/membrane glycoprotein | P07204 |
| Plasminogen (EC 3.4.21.7) | Fibrinolysis/zymogen | P00747 |
| Tissue-type plasminogen activator (EC 3.4.21.68) | Activation plasminogen/zymogen | P00750 |
| Urokinase-type plasminogen activator (EC 3.4.21.73) | Activation plasminogen/zymogen | P00749 |
| Thrombin-activatable fibrinolysis inhibitor (EC 3.4.17.20) | Regulation kinins/carboxypeptidase | Q96IY4 |
| Histidine-rich glycoprotein | Intrinsic activation | P04196 |
| Tetranectin | Kringle-binding protein | P05452 |
| Annexin A5 | Anticoagulant | P08758 |
| $\beta_2$-Glycoprotein | Regulation intrinsic pathway | P02749 |
| Kininogen | Precursor bradykinin/thiol protease | P01042 |
| Angiotensinogen | Precursor angiotensins | P01019 |
| Renin (EC 3.4.23.15) | Generation angiotensin I/peptidase | P00797 |
| Angiotensin-converting enzyme (EC 3.4.15.1) | Regulation/generation angiotensin II | P12821 |

**Table A.4** The complement system (Chapter 8).

| Name | Function | Swiss-Prot |
|---|---|---|
| Complement C1q subcomponent | Classical pathway/binding IgM and IgG | P02745/P02746/P02747 |
| Complement C1r subcomponent (EC 3.4.21.41) | Classical pathway/zymogen | P00736 |
| Complement C1s subcomponent (EC 3.4.21.42) | Classical pathway/zymogen | P09871 |
| Complement C2 (EC 3.4.21.43) | Classical pathway/zymogen | P06681 |
| Complement C3 | Classical and alternative pathway/precursor | P01024 |
| Complement C4 | Classical pathway/precursor | P01028 |
| Complement factor B (EC 3.4.21.47) | Alternative pathway/zymogen | P00751 |
| Complement factor D (EC 3.4.21.46) | Alternative pathway/zymogen | P00746 |
| Complement C5 | Membrane attack complex/precursor | P01031 |
| Complement component C6 | Membrane attack complex/binding C5b | P13671 |
| Complement component C7 | Membrane attack complex/binding C5b,6 | P10643 |
| Complement component C8 | Membrane attack complex/binding C5b,6,7 | P07357/P07358/P07360 |
| Complement component C9 | Membrane attack complex/polymerisation | P02748 |
| Perforin | Cytolytic protein | P14222 |
| Mannose-binding protein C | Complement activation | P11226 |
| Complement-activating component (EC 3.4.21.-) | Complement activation/zymogen | P48740 |
| Mannan-binding lectin serine protease 2 | Complement activation/zymogen | O00187 |
| C-reactive protein | Complement activation/acute phase | P02741 |
| Complement decay-accelerating factor | Complement regulation | P08174 |
| C4b-binding protein | Regulation classical pathway | P04003/P20851 |
| Membrane cofactor protein | Complement regulation/membrane protein | P15529 |
| Complement receptor type 1+2 | Complement regulation/membrane protein | P17927/P20023 |
| Complement factor H | Complement regulation | P08603 |
| Complement factor I (EC 3.4.21.45) | Complement regulation/serine protease | P05156 |
| CD59 glycoprotein | Complement regulation/GPI-anchor | P13987 |
| Properdin | Regulation alternative pathway | P27918 |

**Table A.5** The immune system (Chapter 9).

| Name | Function | Swiss-Prot |
|---|---|---|
| Immunoglobulin IgG1/IgG2/IgG3/IgG4 | Most important antibody | P01857/P01859/P01860/P01861 |
| Immunoglobulin IgA1/IgA2 | Antibody | P01876/P01877 |
| Immunoglobulin IgM | Pentameric antibody | P01871 |
| Immunoglobulin IgD | Antibody | P01880 |
| Immunoglobulin J chain | Component polymeric immunoglobulins | P01591 |
| Immunoglobulin IgE | Antibody | P01854 |
| HLA class I histocompatibility antigen | Presentation of pathogen-derived peptides | P30443 |
| HLA class II histocompatibility antigen | Presentation of pathogen-derived peptides | P01903/P01912 |
| $\beta_2$-Microglobulin | β-Chain of HLA class I | P61769 |
| Zn-$\alpha_2$-glycoprotein | Immune response | P25311 |
| Serum amyloid P-component | Innate immune system/amyloid deposits | P02743 |
| Monocyte differentiation antigen CD14 | Innate immune system | P08571 |
| Neutrophil defensin 1 | Innate immune system | P59665 |
| β-Defensin 1 | Innate immune system | P60022 |
| $\alpha_1$-Acid glycoprotein 1+2 | Modulation immune system | P02763/P19652 |
| $\alpha_1$-Microglobulin | Immunoregulation | P02760 |

**Table A.6** Enzymes (Chapter 10).

| Name | Function | Swiss-Prot |
|---|---|---|
| Catalase (EC 1.11.1.6) | Protection oxidative damage/peroxidase | P04040 |
| Myeloperoxidase (EC 1.11.1.7) | Host defence system/peroxidase | P05164 |
| Glutathione peroxidase 3 (EC 1.11.1.9) | Protection peroxide damage/peroxidase | P22352 |
| γ-Glutamlytransferase 1 (EC 2.3.2.2) | Glutathione breakdown | P19440 |
| Aspartate aminotransferase (EC 2.6.1.1) | Amino acid metabolism | P17174 |
| Creatine kinase M-type/B-type (EC 2.7.3.2) | Phosphotransferase | P06732/P12277 |
| Cholinesterase (EC 3.1.1.8) | Protection against ingested poisons | P06276 |
| Serum paraoxonase/arylesterase 1 (EC 3.1.1.2/EC 3.1.8.1) | Resistance to poisoning | P27169 |
| Angiogenin (EC 3.1.27.-) | Induction vascularisation | P03950 |
| Lysozyme C (EC 3.2.1.17) | Bacteriolytic function/glycosidase | P61626 |
| Carboxypeptidase B (EC 3.4.17.2) | Release C-terminal basic residues | P15086 |
| Carboxypeptidase N (EC 3.4.17.3) | Protection inflammatory peptides | P15169/P22792 |
| Chymotrypsinogen B (EC 3.4.21.1) | Digestive tract/zymogen | P17538 |
| Trypsin I+II (EC 3.4.21.4) | Activation pancreatic zymogens | P07477/P07478 |
| Leukocyte elastase (EC 3.4.21.37) | Degradation elastin, collagen | P08246 |
| Cathepsin G (EC 3.4.21.20) | Degradation connective tissue | P08311 |
| Chymase (EC 3.4.21.39) | Chymotrypsin-like | P23946 |
| Tryptase $\beta_2$ (EC 3.4.21.59) | Trypsin-like | P20231 |
| Prostate-specific antigen (EC 3.4.21.77) | Kallikrein-like/marker prostate cancer | P07288 |
| Cathepsin L (EC 3.4.22.15) | Protein degradation/cysteine protease | P07711 |
| Caspase-1 (EC 3.4.22.36) | Inflammatory response/cysteine protease | P29466 |
| Caspase-3 (EC 3.4.22.-) | Apoptosis/cysteine protease | P42574 |
| Pepsin A (EC 3.4.23.1) | Digestive enzyme/aspartic protease | P00790 |
| Cathepsin D (EC 3.4.23.5) | Protein breakdown/aspartic protease | P07339 |
| Protein disulfide isomerase (EC 5.3.4.1) | Rearrangement disulfide bonds | P07237 |

**Table A.7** Inhibitors (Chapter 11).

| Name | Function | Swiss-Prot |
|---|---|---|
| Antithrombin III | Inhibition thrombin/serpin | P01008 |
| Heparin cofactor II | Inhibition thrombin/serpin | P05546 |
| $\alpha_1$-Antitrypsin | Inhibition elastase/serpin | P01009 |
| Kallistatin | Inhibition kallikrein/serpin | P29622 |
| Plasma serine protease inhibitor | Inhibition protein C | P05154 |
| Tissue factor pathway inhibitor | Inhibition coagulation cascade/Kunitz type | P10646 |
| Protein Z-dependent protease inhibitor | Inhibition coagulation factor Xa/serpin | Q9UK55 |
| $\alpha_2$-Antiplasmin | Inhibition plasmin/serpin | P08697 |
| Plasminogen activator inhibitor-1 | Inhibition tPA/uPA/serpin | P05121 |
| Plasminogen activator inhibitor-2 | Inhibition uPA/serpin | P05120 |
| Neuroserpin | Inhibition plasminogen activators/serpin | Q99574 |
| Plasma protease C1 inhibitor | Regulation classical pathway/serpin | P05155 |
| $\alpha_2$-Macroglobulin | Inhibition of all types of proteases | P01023 |
| Pregnancy zone protein | Inhibition of all types of proteases | P20742 |
| $\alpha_1$-Antichymotrypsin | Inhibition chymotrypsin/cathepsin G/serpin | P01011 |
| Leukocyte elastase inhibitor | Inhibition neutrophil elastase/serpin | P30740 |
| Pancreatic secretory trypsin inhibitor | Inhibition trypsin/Kazal type | P00995 |
| Inter-$\alpha$-trypsin inhibitor light chain (bikunin) | Inhibition various serine proteases/Kunitz type | P02760 |
| Inter-$\alpha$-trypsin inhibitor heavy chain: H1/H2/H3/H4 | Part of various plasma protease inhibitors | P19827/P19823/Q06033/Q14624 |
| Cystatin A/B | Inhibition thiol proteases | P01040/P04080 |
| Cystatin C | Inhibition cysteine proteases | P01034 |

**Table A.8** Lipoproteins (Chapter 12).

| Name | Function | Swiss-Prot |
|---|---|---|
| Lipoprotein lipase (EC 3.1.1.34) | Hydrolysis triacylglycerols/lipase | P06858 |
| Hepatic triacylglycerol lipase (EC 3.1.1.3) | Metabolism high-density lipoproteins/lipase | P11150 |
| Phosphatidylcholine-sterol acyltransferase (EC 2.3.1.43) | Plasma lipoprotein metabolism/transferase | P04180 |
| Apolipoprotein A-I | Major protein in HDL particles | P02647 |
| Apolipoprotein A-II | Major protein in HDL particles | P02652 |
| Apolipoprotein A-IV | Major protein in HDL particles/chylomicrons | P06727 |
| Apolipoprotein B-100/apolipoprotein B-48 | Major protein in LDL particles | P04114 |
| Apolipoprotein C-I | Major protein in VLDL particles | P02654 |
| Apolipoprotein C-II | Major protein in VLDL particles | P02655 |
| Apolipoprotein C-III | Main protein in VLDL particles | P02656 |
| Apolipoprotein C-IV | Lipoprotein metabolism | P55056 |
| Apolipoprotein D | Component of HDL particles | P05090 |
| Apolipoprotein E | Component of VLDL particles/chylomicrons | P02649 |
| Apolipoprotein F | Associated with LDL particles | Q13790 |
| Apolipoprotein L1 | Associated with HDL particles | O14791 |
| Apolipoprotein(a) (EC 3.4.21.-) | Regulation fibrinolysis | P08519 |
| Low-density lipoprotein receptor | Major cholesterol-carrying lipoprotein | P01130 |
| Very low-density lipoprotein receptor | Binds to VLDL particles | P98155 |
| Serum amyloid A protein | Major acute phase protein in inflammation | P02735 |

**Table A.9** Hormones (Chapter 13).

| Name | Function | Swiss-Prot |
|---|---|---|
| Glucagon | Regulation blood glucose level/pancreatic | P01275 |
| Insulin | Promotion CHO metabolism/pancreatic | P01308 |
| Somatostatin | Inhibition somatotropin/pancreatic | P61278 |
| Gastrins | Secretion HCl/digestive enzymes/gastrointestinal | P01350 |
| Secretin | Secretion bicarbonate/gastrointestinal | P09683 |
| Cholecystokinins | Secretion digestive enzymes/bicarbonate/gastrointestinal | P06307 |
| Gastric inhibitory polypeptide | Stimulation insulin release | P09681 |
| Parathyroid hormone | Hypercalcaemic hormone/calcium metabolism | P01270 |
| Calcitonin | Hypocalcaemic hormone/calcium metabolism | P01258 |
| Corticoliberin | Release corticotropin/releasing factor | P06850 |
| Corticotropin-lipotropin | Release cortisol | P01189 |
| Thyroliberin | Release thyrotropin/releasing factor | P20396 |
| Thyrotropin | Release triiodothyronine/thyroxine | P01215/P01222 |
| Thyroglobulin | Matrix thyroid hormone formation | P01266 |
| Gonadoliberin 1 and 2 | Release gonadotropins/releasing factor | P01148/O43555 |
| Follitropin | Development follicles and sperma | P01215/P01225 |
| Lutropin | Promotion spermatogenesis and ovulation | P01215/P01229 |
| Somatoliberin | Release growth hormone/releasing factor | P01286 |
| Somatotropin | Growth control | P01241 |
| Insulin-like growth factors (somatomedins) | Insulin-like activity | P01343/P05019/P01344 |
| Somatostatin | Release somatotropin | P61278 |
| Choriogonadotropin | Maintenance maternal progesterone level | P01215/P01233 |
| Prolactin | Promotion lactation | P01236 |
| Vasopressin-neurophysin 2-copeptin | Stimulation water resorption | P01185 |
| Oxytocin-neurophysin 1 | Contraction smooth muscle uterus and mammary gland | P01178 |
| Proenkephalins A and B | Precursor endogenous opiates | P01210/P01213 |
| Atrial natriuretic factor | Control water and sodium concentration in body | P01160 |
| Vasoactive intestinal peptide | Vasodilatation/lowering blood pressure | P01282 |
| Calcitonin gene-related peptides I+II | Induction vasodilatation | P06881/P10092 |
| Substance P | Vasodilatator | P20366 |
| Erythropoietin | Erythrocyte differentiation | P01588 |
| Thrombopoietin | Regulation megakaryocyte and platelet production | P40225 |
| Adiponectin | Regulation hematopoiesis immune system | Q15848 |

**Table A.10** Cytokines and Growth Factors (Chapter 14).

| Name | Function | Swiss-Prot |
|---|---|---|
| Interleukin-1: IL-1α/IL-1β | Pleiotropic mediators in infections | P01583/P01584 |
| Interleukin-2 | Antigen specific immune response | P60568 |
| Interleukin-3 | Growth factor in hematopoiesis | P08700 |
| Interleukin-4 | Immune response/activation of various types of cells | P05112 |
| Interleukin-13 | Regulation inflammatory and immune response | P35225 |
| Interleukin-5 | Induction eosinophil production | P05113 |
| Interleukin-6 | Induction acute phase reactions/antibody production | P05231 |
| Interleukin-11 | Pleiotropic effects on many tissues | P20809 |
| Interleukin-12: IL-12α/IL-12β | Numerous biological activities | P29459/P29460 |
| Granulocyte colony-stimulating factor | Control granulocytes and monocytes/macrophages | P09919 |
| Interleukin-7 | B cell maturation | P13232 |
| Interleukin-9 | Growth helper T cells | P15248 |
| Interleukin-10 | Inhibition production certain cytokines | P22301 |
| Interleukin-15 | Stimulation proliferation T lymphocytes | P40933 |
| Granulocyte-macrophage colony-stimulating factor | Growth/differentiation hematopoietic precursor cells | P04141 |
| Tumour necrosis factor | Mediator immune and inflammatory response | P01375 |
| Lymphotoxin: LT-α/LT-β | Cytotoxin certain tumour cells | P01374/Q06643 |
| Interferon α-1/13 | Antiviral activity | P01562 |
| Interferon β | Antiviral and antibacterial activity | P01574 |
| Interferon γ | Immunoregulatory and antiviral activity | P01579 |
| Interleukin-8 | Stimulation neutrophils | P10145 |
| Platelet factor 4 | Chemotactic for neutrophils and monocytes | P02776 |
| Platelet basic protein | Various biological functions | P02775 |
| Small inducible cytokine A2 | Chemotactic on monocytes | P13500 |
| Small inducible cytokine A5 | Chemoattractant for various cell types | P13501 |
| Platelet-derived growth factor: A/B | Potent mitogen | P04085/P01127 |
| Vascular endothelial growth factors: A/B/C/D | Angiogenesis/vasculogenesis/endothelial cell growth | P15962/P49765/P49767/O43915 |
| Transforming growth factor β: 1/2/3 | Multifunctional cytokines | P01137/P61812/P10600 |
| Transforming growth factor α | Mitogenic polypeptide | P01135 |
| Nerve growth factor β | Development nervous system | P01138 |
| Epidermal growth factor | Regulation cell growth, proliferation, differentiation | P01133 |
| Insulin-like growth factor-binding proteins | Regulation insulin-like growth factors | P08833/P18065/P17936/P22692/P24593/P24592/Q16270 |

**Table A.11** Transport and storage (Chapter 15).

| Name | Function | Swiss-Prot |
|---|---|---|
| Serum albumin | Major transport protein | P02768 |
| α-Fetoprotein | Functions as fetal serum albumin | P02771 |
| Afamin | Transport of ligands | P43652 |
| Vitamin D-binding protein | Carrier vitamin D sterols | P02774 |
| Hemoglobin: variants | Oxygen transport | P69905/P68871/P69891/P69892/ P02042/P02100/P09105/P02008 |
| Myoglobin | Reserve supply of oxygen | P02144 |
| Haptoglobin | Binds free hemoglobin | P00738 |
| Ceruloplasmin (EC 1.16.3.1) | Oxidation ferrous iron | P00450 |
| Serotransferrin | Iron transport | P02787 |
| Lactotransferrin (EC 3.4.21.-) | Iron transport | P02788 |
| Ferritin: light/heavy chain (EC 1.16.3.1) | Iron storage | P02792/P02794 |
| Hemopexin | Heme transport | P02790 |
| Thyroxine-binding globulin | Thyroid hormone transport | P05543 |
| Transthyretin | Thyroid hormone transport | P02766 |
| Transcortin/corticosteroid-binding globulin | Transport glucocorticoids/serpin | P08185 |
| Sex hormone-binding globulin | Transport sex steroids | P04278 |
| Plasma retinol-binding protein | Transport vitamin A | P02753 |
| Selenoprotein P | Transport and supply selenium | P49908 |

**Table A.12** Additional proteins (Chapter 16).

| Name | Function | Swiss-Prot |
|---|---|---|
| $\alpha_2$-HS-glycoprotein | Bone formation and resorption | P02765 |
| Chromogranin A | Processing peptide hormones and neuropeptides | P10645 |
| Secretogranin-1 | Processing peptide hormones and neuropeptides | P05060 |
| Gelsolin | Actin-modulating protein | P06396 |
| Ficolin-3 | Serum lectin activity | O75636 |
| Leptin | Regulation body fat depot | P41159 |
| Osteonectin | Tissue remodelling | P09486 |
| Pigment epithelium-derived factor | Potent inhibitor of angiogenesis/serpin | P36955 |
| $\alpha_1$-B-glycoprotein | | P04217 |
| Leucine-rich $\alpha_2$-glycoprotein | | P02750 |
| Pregnancy-specific $\beta_1$-glycoprotein 1 | Diagnosis pregnancy | P11464 |
| Lipopolysaccharide-binding protein | Binding lipopolysaccharides | P18428 |
| Clusterin | Binding hydrophobic complexes | P10909 |
| Insulin-like growth factor-binding protein complex acid labile chain | Protein–protein interactions | P35858 |

**Table A.13** Other useful databases.

| Database | Description | URL |
|---|---|---|
| Plasma DB | Plasma protein database | www-lmmb.ncifcrf.gov/plasmaDB |
| Human Protein Atlas | Localisation of proteins in human normal tissues and cancer cells | www.proteinatlas.org |
| DIP™ Database | Database of interacting proteins | http://dip.doe-mbi.ucla.edu |
| DSDBASE | Disulfide database | http://caps.ncbs.res.in/dsdbase |
| REACTOME | A curated knowledge base of biological pathways | www.reactome.org |
| RESID Database | Protein modifications | www.ebi.ac.uk/RESID |

# Glossary

**Ab initio** Latin: from the beginning. A calculation based on established laws of nature without additional assumptions.

**Acute-phase protein** A class of proteins whose plasma concentration increase / decrease in response to inflammation.

**Adaptive immune system** The development of specific immunity to a particular pathogen by producing specific antibodies and memory cells.

**Adhesive proteins** A protein capable of sticking to surfaces of substances linking them together.

**Aggregation** The interaction between molecules or particles forming larger structures. In proteins: often insoluble aggregates.

**Agonist** Hormone, neurotransmitter or drug responsible for triggering a response in a cell by stimulating a receptor (opposite: antagonist).

**Amphipathic** Greek: *amphis*: both; *philia*: love, friendship. A compound possessing both hydrophilic and hydrophobic properties.

**Amyloidosis** The disproportionate accumulation of the amyloid protein leading to amyloid plaques.

**Anaphylatoxin** The fragments C3a, C4a, C5a of the complement components C3, C4, C5. Anaphylatoxins induce local inflammatory reactions.

**Anemia** Greek: without blood. The deficiency of red blood cells and/or hemoglobin.

**Angiogenesis** The formation of new blood vessels.

**Angiotensin / Rennin system** The system that regulates the vascular tonus, the volume and the mineral balance of body fluids.

**Antagonist** Any substance that interferes with or prevents the action of a hormone, neurotransmitter or drug (opposite: agonist).

**Antibody** Immunoglobulin with a large Y-shaped structure capable of recognising antigens.

**Antigen** From: **anti**body **gen**eration. A substances that can be recognised by the adaptive immune system.

**Aspartyl (aspartic) protease** A protease carrying two Asp residues in the active site.

**Atherosclerosis** A disease affecting the arterial blood vessel, caused by a chronic inflammatory response.

**Autocatalytic / Autoactivation** The catalysis of a chemical reaction by one of its own products.

*Human Blood Plasma Proteins: Structure and Function* Johann Schaller, Simon Gerber, Urs Kämpfer, Sofia Lejon and Christian Trachsel

**Autocrine** Acting on the signal releasing cell (opposite: endocrine).

**β-Barrel** Several β-strands associate into extensively curved sheets connected by helices.

**β-Hairpin** Two antiparallel β-strands linked by a short loop.

**β-Sandwich** An assembly of interacting β-strands at an angle connected by short loops.

**β-Sheet** An assembly of β-strands interacting via hydrogen bonds. In an **antiparallel** arrangement the successive β-strands alternate directions and in a **parallel** arrangement the strands are in the same direction.

**Blood plasma** The yellow coloured, liquid part of blood (approx. 55 % of blood, by vol.).

**Cardiovascular system** The heart and blood vessels considered as a whole.

**Catalytic chain / domain** The chain / domain of an enzyme containing the active site region.

**Catalytic triad** The active site residues (Ser, Asp, His) in serine proteases.

**Cell proliferation** The rapid reproduction of living cells. Cell growth in a cell population.

**Class switching** The mechanism of changing the constant heavy chain regions in immunoglobulins.

**Coagulation cascade** An enzymatic pathway leading to the formation of a blood clot to prevent blood loss from a ruptured blood vessel. Consists of:

- the extrinsic pathway
- the intrinsic pathway
- the common pathway

**Cofactor** A substance required for the catalytic activity of an enzyme.

**Coiled coil** A structural motif of at least two α-helices arranged like the strands of a rope.

**Combinatorial joining** The recombination of the D, J and V segments in the variable domains of immunoglobulins.

**Common pathway** The final pathway in the coagulation cascade leading to clot formation.

**Complement system** A system of blood proteins, which become activated on infection and aid in clearing bacteria and other pathogens from the body. Consists of:

- the classical pathway
- the alternative pathway
- the lectin pathway

**Complementarity determining regions (CDRs)** The hypervariable regions in the variable domains of immunoglobulins.

**Configuration** The spatial arrangement of bonds in a molecule that can not be altered without breaking bonds.

**Conformation** The overall shape of a molecule. The rotation around single bonds results in different, closely related states.

**Consensus pattern** A conserved amino acid sequence pattern that is characteristic for related proteins.

**Cysteine (thiol) protease** A protease carrying a Cys residue in the active site.

**Cytokine** A protein/peptide that is used as signalling compound in cellular communication. Cytokines are comparable to hormones and growth factors.

**De novo** Latin: afresh, anew. Refers to the synthesis of complex compounds from basic molecules.

**Denaturation** The loss of tertiary and secondary structure resulting in a less ordered state. Often combined with the loss of biological function.

**Differentiation** The process by which a cell acquires a certain morphology and turn into a specific cell type.

**Dimer** An assembly of two subunits, usually without covalent bonds. Exceptions: Linkage with disulfide bridges. A **homodimer** is built of identical, a **heterodimer** of different subunits.

**Distal** Most far away from the point of contact/ connection.

**Disulfide bridge** Covalent bond between the sulphur atoms of two Cys side chains.
- **Interchian:** bridges between two polypeptide chains.
- **Intrachain:** bridges within one polypeptide chain.

**Diuresis** The excessive discharge of urine.

**Domain** In proteins: an autonomously folding, functional unit (module).

**Embolus / emboli** An air bubble, blood clot or foreign body that obstructs or occluds blood vessels.

**Embryogenesis** The development and growth of an embryo.

**Endocrine** Acting on cells in a distance of the signal releasing cells (opposite: autocrine).

**Endocytosis** The process by which eukaryotic cells take up extracellular material by invagination of the plasma membrane to form vesicles enclosing the external material (opposite: exocytosis).

**Endothelial cells / endothelium** A layer of specialised cells lining the interior surface of blood vessels.

**Enzyme** A protein with a catalytic activity containing an active site.

**Erythrocyte** A biconcave discoid shaped red blood cell without a nucleus.

**Extracellular matrix (ECM)** A structure of glycoproteins and proteoglycans which are attached to cell surfaces.

**Extrinsic pathway** The activation of the coagulation cascade by contact of blood with free endothelial cells (collagen).

**Fab / Fc / (Fab')$_2$ fragments** The fragments generated by limited proteolysis in the hinge regions of immunoglobulins: With papain: two **Fab** (**f**ragment **a**ntigen **b**inding) and one **Fc** (**f**ragment **c**rystallisable). With pepsin: one **(Fab')$_2$**.

**Feedback**
 **Positive feedback:** a reaction that results in amplification or growth of the output signal.
 **Negative feedback:** a reaction in opposite phase with (decreasing) the input signal.

**Fibrin clot** A fibrin network forming a haemostatic plug containing activated platelets.

**Fibrinolysis** The proteolytic degradation of fibrin networks by plasmin.

**Globular protein** A protein with a globe-like compact structure distinguished from extended and filamentous structures.

**Glycoprotein** A protein containing N- and/or O-linked CHO moieties.

**Glycosylphosphatidylinositol (GPI) anchor** Membrane anchor usually attached to the C-terminus of a protein.

**Granulocyte** A granules containing leukocyte: neutrophils, eosinophils, basophils.

**Greek-key** Four adjacent antiparallel β-strands.

**Growth factor** A protein / peptide that stimulates cellular proliferation and cellular differentiation. Often difficult to distinguish from cytokines.

**H(a)emostasis** The balanced process of blood clot formation, dissolution of blood clots and repair of injured tissues.

**Hematopoiesis** The process of blood production by hematopietic cells.

**Hemorrhage** Excessive bleeding.

**Heparin** A glycosaminoglycan preventing blood clot formation.

**Hepatic** Related to the liver.

**Homologous** In biology: sharing structural similarity due to common ancestry.

**Hormone (Peptide-)** Greek: to set in motion. A messenger molecule in intracellular communication.

**Human Leukocyte Antigen (HLA)** Human major histocompatibility complex (MHC).

**Hydrolase** An enzyme that catalyses the hydrolysis of a chemical bond.

**Hypertension** High blood pressure.

**Hypotension** Low blood pressure.

**In silico** Performed on a computer or via a computer simulation.

**In vitro** Latin: (with)in the glass. Performing an experiment outside a living organism under controlled conditions.

**In vivo** Latin: (with)in the living. Performing an experiment in a living organism (tissue) or cell system.

**Inflammation** A protective attempt by the organism to remove a harmful stimuli as well as initiate the healing process.

**Inhibitor** A substance capable of stopping / retarding a chemical reaction. In enzymes: substance inhibiting the catalytic activity of an enzyme.

**Innate immune system** The natural, non-adaptive immediate immune defence.

**Intrinsic pathway** Contact phase activation of the coagulation cascade with negative charged surfaces (artificial or natural).

**Ischemic stroke / stroke** Medical term for ''brain attack'': blood cut-off leading to loss of brain function.

**Isomerase** An enzyme that catalyses the interconversion of isomers.

**Isopeptide bond** Similar to peptide bond. In proteins: crosslinking of two polypeptide chains via side chains of Lys and Gln.

**Jelly roll** Several $\alpha$-helices arranged in a curved pattern resembling a jelly roll fold.

**Junction diversity** Diversity generated at the joining sites of the D, J and V segments in immunoglobulins.

**Kinin / (Kallikrein) system** A system involved in inflammation, blood coagulation and blood pressure control. Contains the vasodilatator bradykinin.

**Leukocytes** The white blood cells consisting of three main types: Granulocytes, monocytes / macrophages and lymphocytes.

**Ligand** A molecule capable of binding to a binding site via intermolecular interactions.

**Ligase** An enzyme that catalyses the linkage of two molecules by forming new chemical bonds (requires ATP).

**Lipolysis** The hydrolysis of lipids.

**Lyase** An enzyme that catalyses the non-hydrolytic addition / removal of functional groups.

**Lymphocytes** White blood cells consisting of three main types: B cells, T cells and natural killer (NK) cells.

**Major histocompatibility complex (MHC)** A protein complex that presents antigen- derived peptides.

**Mediator** A substance that acts at low concentrations as a specific transmitter or signal.

**Membrane Attack Complex (MAC)** A protein complex consisting of complement components C5b, C6, C7 and C8 linked to up to 18 C9 molecules.

**Metabolism / Metabolites Metabolism:** the integrated network of biochemical reactions that supports life in living organism. **Metabolites:** the intermediates and products of biochemical pathways.

**Metalloprotease** An enzyme that contains a metal ion (often zinc or calcium) in their active site.

**Michaelis-like complex** An enzyme-substrate-like complex.

**Module** Sequence motif occurring in unrelated proteins or in repetitive form within the same protein.

**Monocyte / Macrophage** A large devouring leukocyte.

**Myocardial infarction** Medical term for ''heart attack'': Blood cut-off leading to necrosis of heart tissue.

**Natriuresis** The excessive excretion of sodium into urine.

**Natural killer cells** Cells that are able to kill cells which do not carry a recognition signal (foreign tissue).

**Necrosis** The death (cell disruption) of tissue by injury, disease or pathologic state.

**Opioid (peptides)** A substance with morphine-like effects in the body.

**Ortholog / orthologous** Similar genes in different species originating from a common ancestor.

**Oxidoreductase** An enzyme that catalyses oxidation / reduction reactions.

**Oxyanion hole** An additional binding pocket in serine proteases. E.g. in chymotrypsin: formed by the backbone hydrogens of Gly 193 and Ser 195.

**Paracrine** Acting on cells in close proximity of the signal releasing cell.

**Paralog / paralogous** A homologous gene sequence.

**Pathogen** An infectious agent such as bacteria, viruses, fungi or their metabolites, causing diseases in the host.

**Phagocytosis** The cellular process of engulfing solid particles by the cell membrane.

**Phosphorylation** The esterification with phosphoric acid.

**Physicochemical parameters:** Properties of molecules such as mass, charge, isoelectric point, binding constants.

**Plasma (B) cell** A cell capable of secreting large amounts of antibodies.

**Platelet (thrombocyte)** A discoid shaped blood cell without nucleus. An activated platelet exhibits a spherical shape exposing spines (pseudopodia).

**Point mutation** A mutation involving a change at a single base-pair in DNA.

**Polymorphism** The discontinuous genetic variation in a species resulting in more than one form or type of an individual.

**Posttranslational modification (PTM)** The chemical modification of a protein after its translation.

**Pre-(pro-) protein** The precursor form of a protein, usually inactive / non-functional. Contains a **prosequence / propeptide** sequence which is usually removed by proteolytic processing in order to generate the mature (active) protein.

**Proteinuria** Excessive amounts of protein in the urine.

**Proteoglycan** Polysaccharide containing protein (opposite: glycoprotein).

**Proximal** Closest to the point of contact / connection.

**Pyrogen** A fever inducing substance.

**Receptor** A protein capable of binding ligands, thus inducing a cellular response to the ligand.

**RGD sequence** Recognition sequence Arg – Gly – Asp as binding motif.

**Rhesus factor (Rh-factor)** Refers to the five main Rhesus antigens C, c, D, E and e.

**Rossmann fold** An open, twisted β-sheet surrounded by α-helices on both sides.

**Salt bridges** In proteins: ionic interaction between positively and negatively charged amino acids.

**Serine protease** A protease carrying a Ser residue in the catalytic triad.

**Serpin** High molecular weight **ser**ine **p**rotease **in**hibitor.

**Signal sequence / Signal peptide** A sequence that directs the transport of a protein after it's synthesis. Is usually removed by signal peptidases.

**Signal transduction** The conversion of signals / stimuli by cells into other signals, often involving biochemical reactions inside the cells.

**Somatic hypermutation** A somatic point mutation in the variable regions of antibodies.

**Stem cell** A cell that has the ability to renew itself and differentiate into a wide range of specialised cell types.

**Structural motif** A 3D fold occurring in unrelated proteins or in repetitive form within the same protein.

**Subendothelium** The cells in the blood vessel wall (intima) located below the endothelium and the extracellular matrix.

**Suicide inhibition** The irreversible enzyme inhibition by serpins.

**Synergistic** Greek: *synergos*: working together. Agents acting together creating an effect greater than predicted.

**T cell** T lymphocytes exhibiting distinct functions: cytotoxic, helper, memory and regulatory T cells.

**Thrombosis** Blood clot (thrombus) formation.

**Topology** In proteins: different shapes.

**Transferase** An enzyme that catalyses the transfer of functional groups.

**Vascular system** A circulatory system such as blood vessels transporting substances.

**Vascularisation** The process whereby tissue becomes rich in blood vessels and capillaries.

**Vasoactive** A substance that exhibitis vasoconstrictive or vasodilative activity.

**Vasoconstriction / Vasodilatation** The constriction / relaxation of blood vessels by reduction / enlargement of its lumen.

**Vertebrate / Invertebrate** An animal with / without a spinal column.

**White / red clot** Blood clot containing primarily platelets / erythrocytes.

**Zymogen** Inactive precursor of an enzyme. Requires an activation step to become an active enzyme.

# Index



*Human Blood Plasma Proteins: Structure and Function* Johann Schaller, Simon Gerber, Urs Kämpfer, Sofia Lejon and Christian Trachsel